AF247476

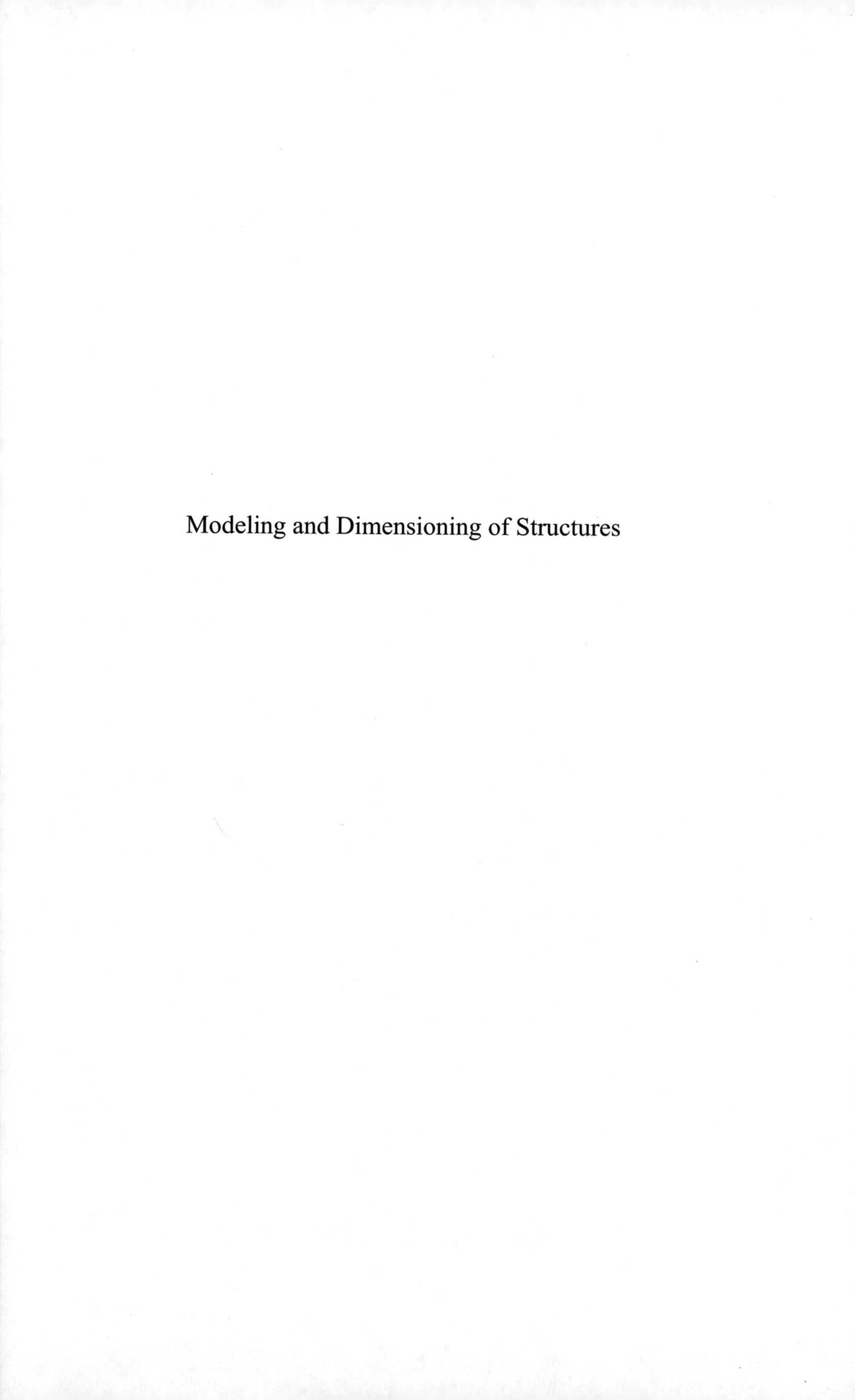

Modeling and Dimensioning of Structures

Modeling and Dimensioning of Structures

A Practical Approach

Daniel Gay

Jacques Gambelin

Part of this book adapted from "Dimensionnement des structures: une introduction" published in France by Hermes Science Publications in 1999

First published in Great Britain and the United States in 2008 by ISTE Ltd and John Wiley & Sons, Inc.

ISTE Ltd
6 Fitzroy Square
London W1T 5DX
UK

www.iste.co.uk

John Wiley & Sons, Inc.
111 River Street
Hoboken, NJ 07030
USA

www.wiley.com

Library of Congress Cataloging-in-Publication Data

Modeling and dimensioning of structures: a practical approach/Daniel Gay, Jacques Gambelin.
    p. cm.
"Part of this book adapted from "Dimensionnement des structures: une introduction" published in France by Hermes Science Publications in 1999."
Includes bibliographical references and index.
ISBN 978-1-84821-040-0
1. Structural engineering--Data processing. 2. Structural engineering--Mathematics. 3. Structural analysis (Engineering) 4. Structural frames--Mathematical models. I. Gay, Daniel, 1942- II. Gambelin, Jacques.
TA640.S77 2007
624.1'7--dc22

2007009432

British Library Cataloguing-in-Publication Data
A CIP record for this book is available from the British Library
ISBN: 978-1-84821-040-0

Printed and bound in Great Britain by Antony Rowe Ltd, Chippenham, Wiltshire.

Table of Contents

Preface

This book is aimed at teachers, students, engineers and technicians working in the field of mechanical design.

The modeling and sizing of structures and mechanical assemblies nowadays and universally resort to specific computation software based on finite element analysis. The contents and the scheduling of this work are thus meant to enable the reader to understand and prepare the use of computer code under the best possible conditions.

This work proposes the basics of the dimensioning of structures to readers who will not inevitably continue studies in mechanics of continuous media. This book acts as a support for courses, exercises and experimental work, and as preliminary practical work before calculating (practical modeling). We will find there as straightforward as possible an approach of principles and methods used nowadays to size mechanical structures and assemblies, at the various stages of their design (to learn how to dimension, with the aim of a better design).

With this posted goal, this book differs from a traditional text on the strength of materials. The study plan is not the same. Everything is done to lead rapidly to the concepts of flexibility and stiffness of a structure, in order to approach them in their matrix form.

This work has a "utility" purpose: the main practical results obtained at each stage are synthesized systematically in the form of summarized tables. Formalism and notations are reduced to a minimum. Basic elements for beginners are marked in the margin by a shaded band.

This book is composed of three main parts. The technical level increases from Part 1 to Part 2, and each part can be used independently. Part 1 corresponds to a 2-year undergraduate degree or Licence (L2). Part 2 is meant for an educational program of higher level in industrial design, i.e. corresponding to a 3-year undergraduate degree or Licence (L3) or bachelor's degree, if necessary finished by a one-year Master's (M1) diploma (EU studies).

In Part 3 supplements are given, particularly regarding the modeling and dimensioning of structural joints.

Thus, the book is useful for students, as well as for practical engineers who want to learn, on the job, the guidelines for the use of finite element software.

This work is intended to adapt and replace, correctly, the traditional subject which the strength of materials constitutes. It addresses the modeling of the behavior of structures by avoiding the often too abstract formalism of the mechanics of continuous media. It provides comprehensive coverage of both the analysis and design used in industry today. The content and structure of the book are intended to help the reader understand the use of finite element techniques and software, which are now essential for such a discipline, and supply the material necessary to make models and to interpret results.

PART 1

Level 1

In this first part, the shaded vertical borders mark the essential, basic elements, for beginners. This first level is meant for students having, in the European system (licence, Masters, PhD), a "first year of licence". In Chapters 3 and 4 we deal with the simplest finite elements, along with some applications showing the fundamental steps of calculation.

Chapter 1

The Basics of Linear Elastic Behavior

Mechanical structures consist of accurately organized groups of parts formed of "solid" media. These solids are obtained by the processing of proven construction materials: cements and steels for architects, metal alloys (steel, light alloys, titanium, etc.) for mechanical designers, more advanced materials (special alloys, composites) for certain kinds of precise activities (terrestrial transport, aeronautics, shipbuilding, etc.).

We arrange these parts of structures to "optimize" their performance and the cost returns of the complete structure:

– durability for a predefined time frame;

– the lowest possible cost returns and maintenance.

These are fundamental considerations, for they form the goals that designers try to achieve. It is therefore important to be well advised right from the beginning. We shall have the opportunity to review this[1].

At the beginning of this chapter we shall accept without demonstration, certain standard properties of a "loaded" elastic structure, that is to say, under mechanical loads. These properties are useful for a better understanding of what follows[2].

---

1 See Chapter 7.
2 For the proof of these properties, see Chapter 10.

1.1. Cohesion forces

In a real structure, it is necessary to accept the existence of a system of internal cohesion forces which originate from intermolecular actions and which allow, among other things, the preservation of the initial form of the structure.

In a structure made of a material we shall assume to be elastic[3], let us isolate a particle of matter specified by a *very small sphere* around a point M (Figure 1.1a).

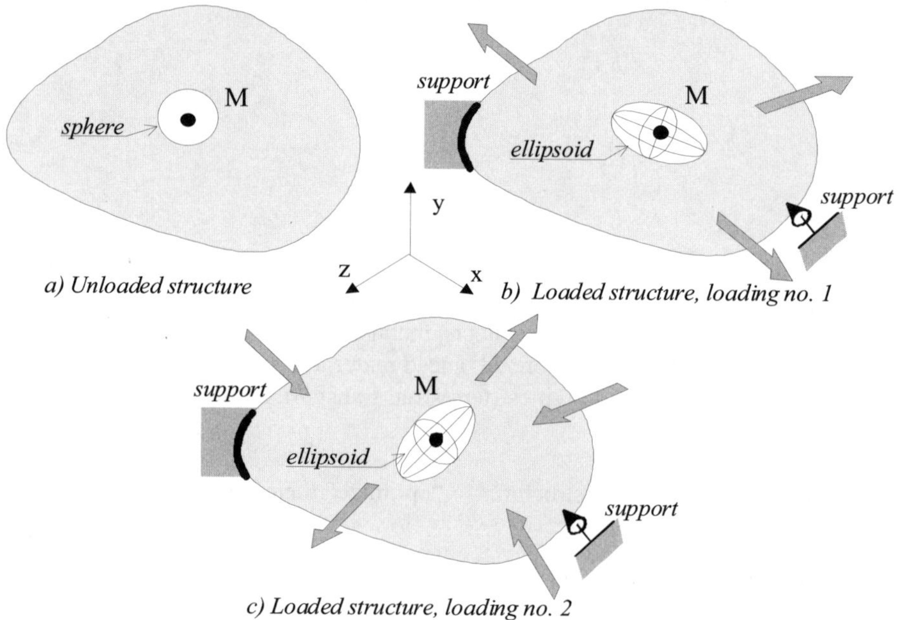

Figure 1.1. *A spherical domain's deformation around point M*

When this structure is loaded, point M undergoes a displacement (Figures 1.1b and 1.1c) which we shall assume to be very small when compared to the dimensions of the structure, so that the latter's shape does not vary perceptibly. It is shown for all materials made up of standard structures, that the small spherical domain around the point M first deforms weakly[4] becoming an ellipsoid. The shape and the orientation of that ellipsoid change not only with the position of point M in

---

3 We shall return later on (section 1.3) to this notion of elastic material.
4 The deforming steps (ellipsoid) of Figures 1.1 and 1.3 are greatly exaggerated for standard metal alloys; in reality the variation in the shape is imperceptible as the displacement of all the points such as M is of very small amplitude compared to the dimensions of the structure.

the structure but also with the nature of the loads as illustrated in Figure 1.1c. Such an isolated ellipsoid is shown in Figure 1.2.

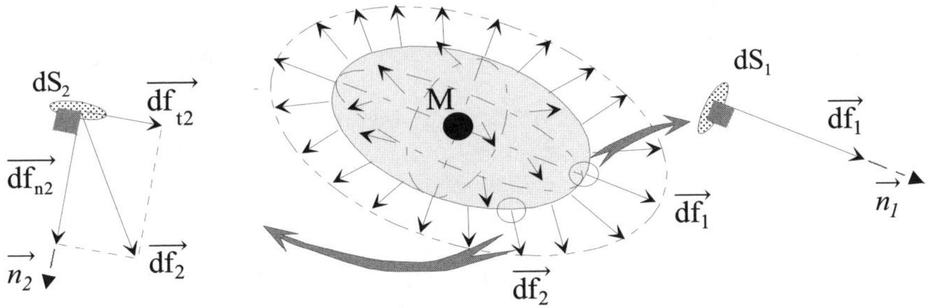

Figure 1.2. *Cohesion forces around the small domain*

Thus isolated, we must calculate the forces $\overrightarrow{df_i}$ exerted by the suppressed part of the structure on all the elementary surfaces dS_i with the outgoing normal unit $\overrightarrow{n_i}$ constituting the surface of the ellipsoid. These elementary forces, which assure the equilibrium of a particle, are referred to as *cohesion forces* or internal forces of the structure.

NOTE

❑ These cohesion forces lead the initially spherical domain to become ellipsoidal. The distribution of these forces on the surface therefore has symmetry properties.

❑ These cohesion forces are made up of:

– on the one hand the initial cohesion forces before the structure was loaded. This initial unloaded state is also called "neutral state";

– on the other hand the complementary cohesion forces created by the loading. We shall be exclusively interested in these complementary cohesion forces from now on.

❑ These complementary cohesion forces shall henceforth be referred to as "cohesion forces". Their distribution around any point in the structure is therefore very narrowly linked to the applied forces, as illustrated in Figure 1.3.

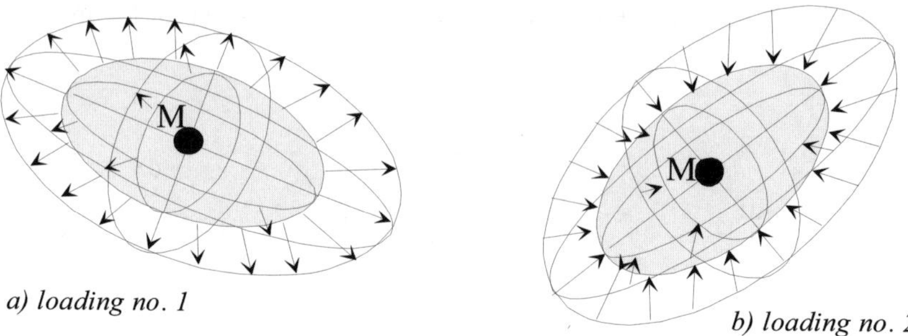

Figure 1.3. *Cohesion forces for two distinct loadings*

1.2. The notion of stress

1.2.1. *Definition*

For the small domain with center M, and for each direction $\vec{n_i}$, we shall define a vector called "stress vector" noted as:

$$\overrightarrow{C_{i\,(M,\vec{n_i})}} = \frac{\vec{df_i}}{dS_i} \qquad [1.1]$$

where:

➤ $\vec{df_i}$ is the cohesion force exerted by the extracted part, on one surface with the value dS_i of the ellipsoid with center M (Figure 1.2);

➤ $\vec{n_i}$ is the outgoing normal[5] on the surface dS_i of the ellipsoid on which $\vec{df_i}$ is exerted.

NOTE

❑   Dimensional   homogenity   of   the   stress   vector: $\dfrac{unit\ of\ force}{unit\ of\ surface} = \dfrac{1\,N}{1\,m^2} = 1\,Pa$ (Pascal) or, in a manner more adapted for design engineers: $\dfrac{1\,N}{1\,mm^2} = 1\,MPa$ (MegaPascal).

---

5 The choice of the output normal makes it possible, in the case of a partition within a structure, to characterize the part that has been isolated.

1.2.2. *Graphical representation*

In Figure 1.2, we observe that the surface dS_1 undergoes a cohesion force df_1 normal to it. The corresponding stress vector, denoted $\overrightarrow{C_{1(M,\vec{n}_1)}}$, is drawn in Figure 1.4. It is colinear to $\vec{n_1}$. We obtain $\overrightarrow{C_1}.\vec{n_1} = \sigma_1$. The algebraic value of σ_1 is referred to as normal stress.

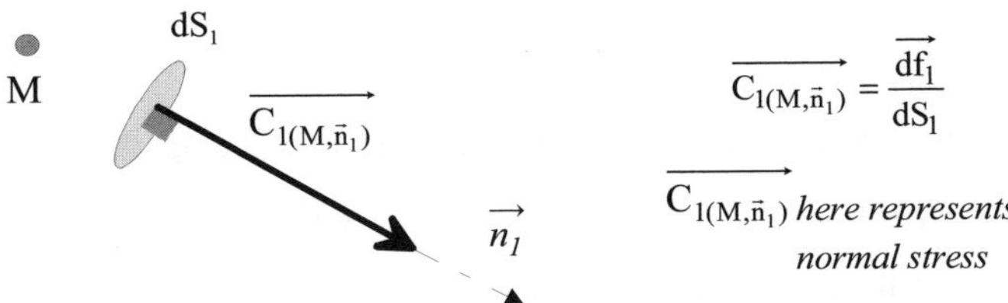

Figure 1.4. *A single normal component of the stress*

Remaining on Figure 1.2, we have represented surface dS_2, undergoing a cohesion force $\overrightarrow{df_2}$, which appears to have two components: one normal component $\overrightarrow{df_{n2}}$, and one tangential component $\overrightarrow{df_{t2}}$ [6]. The corresponding stress vector, denoted $\overrightarrow{C_{2(M,\vec{n}_2)}}$, is drawn in Figure 1.5. By defining a tangential unit vector $\vec{t_2}$ to the intersection of the plane of surface dS_2 and the plane formed by the vectors $\overrightarrow{C_2}$ and $\vec{n_2}$, we obtain $\overrightarrow{C_2}.\vec{n_2} = \sigma_2$ and $\overrightarrow{C_2}.\vec{t_2} = \tau_2$. The two components σ_2 and τ_2 are defined as:

➤ for σ_2 : *normal* stress;

➤ for τ_2 : *tangential* or *shear* stress.

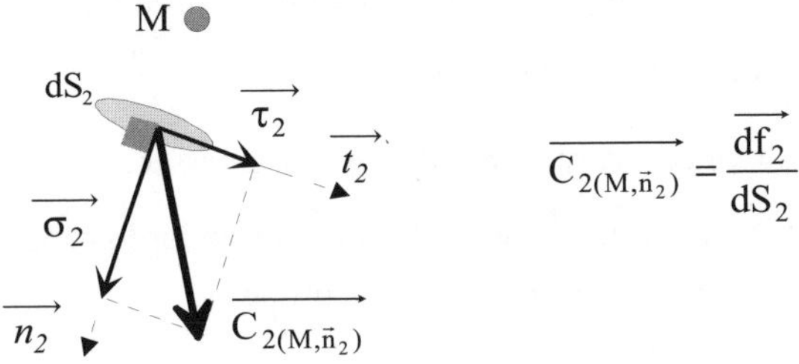

Figure 1.5. *Normal and tangential components of a stress*

---

6 $\overrightarrow{df_{t2}}$ is found in the plane of surface dS_2, tangential to the ellipsoid.

1.2.3. *Normal and shear stresses*

As we observe in Figure 1.5 we may note the vector $\overrightarrow{C_{2(M,\vec{n}_2)}}$ as the sum of the normal stress vector $\overrightarrow{\sigma_2}$ and the tangential or shear stress vector $\overrightarrow{\tau_2}$, i.e.:

$$\overrightarrow{C_{2(M,\vec{n}_2)}} = \overrightarrow{\sigma_2} + \overrightarrow{\tau_2}$$

NOTE

❏ We demonstrate[7] and we shall admit for the time being that if we isolate a surface dS_2' perpendicular to $\overrightarrow{\tau_2}$ (plane of Figure 1.5 re-examined in Figure 1.6), this surface will automatically undergo a shear stress $\overrightarrow{\tau_2}'$ such that $\left\|\overrightarrow{\tau_2}\right\| = \left\|\overrightarrow{\tau_2}'\right\|$ oriented as shown in Figure 1.6.

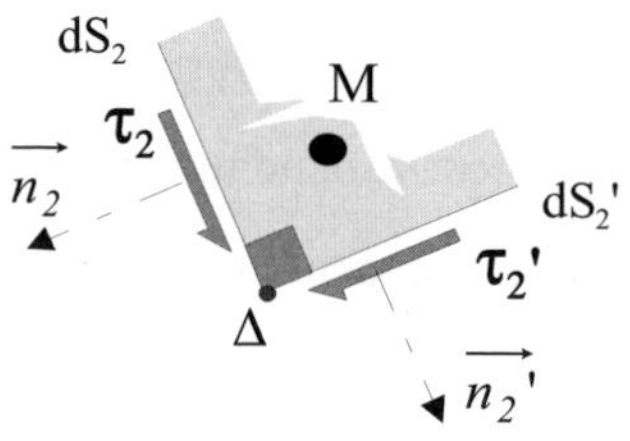

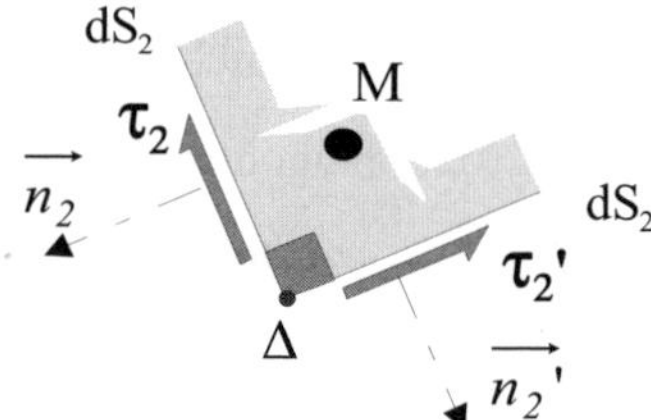

a) stresses converge
on the edge Δ of the dihedral

b) stresses diverge
from the edge Δ of the dihedral

Figure 1.6

We say there is reciprocity in the shear stresses:

➢ if $\overrightarrow{\tau_2}$ is directed towards edge Δ, the intersection between surfaces dS_2 and dS_2', then $\overrightarrow{\tau_2}'$ is directed towards edge Δ;

➢ if $\overrightarrow{\tau_2}$ is directed away from edge Δ, the intersection between surfaces dS_2 and dS_2', then $\overrightarrow{\tau_2}'$ is directed away from (this standard property is referred to as the reciprocity of shear stresses or Cauchy's property).

---

7 See Chapter 10, section 10.2.2 and [10.19].

❑ We may note in Figure 1.2 that we represented several points on the ellipsoid's surface specifically as shown in Figure 1.4 where $\overrightarrow{C_{1(M,\vec{n}_1)}}$ reaches an entirely normal stress. These points have been reworked and illustrated with their associated surfaces in Figure 1.7.

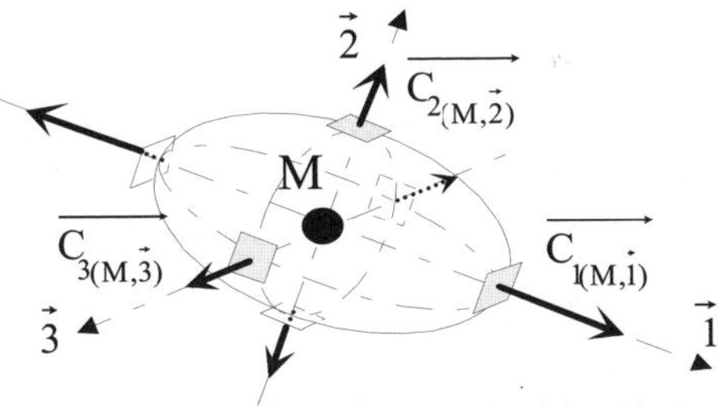

Figure 1.7. *Principal stresses and associated directions*

In this way we can note three specific normal directions denoted $\vec{1}$, $\vec{2}$, $\vec{3}$. From now on we consider that these directions are orthogonal to each other. We call them principal directions for the stresses. The stresses $\overrightarrow{C_{1(M,\vec{1})}}$, $\overrightarrow{C_{2(M,\vec{2})}}$ $\overrightarrow{C_{3(M,\vec{3})}}$ following these directions are called principal[8] stresses.

1.3. Hooke's law derived from a uniaxially applied force

1.3.1. *The stretch test*

A cylindrical slender sample is made from a standard material, a metal alloy for example. At a macroscopic level the structure of this piece remains the same throughout. It is therefore said to be *homogenous*. Moreover this material should respond to a mechanical load no matter how the test sample was manufactured from its original raw material. We call such a material *isotropic*.

Throughout this book we shall only refer to this type of material, i.e.:

Material: • *homogenous*

  • *isotropic*

---

8 These results are simply quoted here without demonstration: for justification, see Chapter 10, section 10.1.2, [10.2], Figure 10.4.

The sample represented in Figure 1.8 is composed of two heads with a larger diameter by which it is attached to the testing machine. We first measure a preliminary "initial length" ℓ between two reference points marked on one generator along the cylindrical part. A tension force $\vec{F}$ colinear to the $\vec{x}$ axis is then applied on the ends of the sample, i.e. $\vec{F} = F\vec{x}$. The initial length ℓ increases. Each cross-section of the slender cylindrical part moves, albeit very weakly, along the direction of the applied force (along $\vec{x}$) as represented in the figure. The displacement of the cross-section with abscissa x is denoted as u(x). We then state that with the increase in length denoted by $\Delta\ell$ there is a reduction in the diameter denoted by Δd, as a result of which we refer to Poisson's effect.

The force-displacement diagram registered during the drawing test may be sketched as in Figure 1.9, with:

* in abscissas, the values of the ratio $\dfrac{\Delta\ell}{\ell}$ noted ε_x (dimensionless), referred to as *dilatation* or *axial strain*;

* in ordinates, the values of quotient $\dfrac{F}{S}$ where S is the initial area[9]. It is homogenous to a stress, which we shall define by the ratio:

$$\frac{\vec{F}}{S} = \frac{F}{S}\vec{x}$$

A normal stress is therefore applied to the cross-section. The force F and the associated surface S here having finite values, this stress is said to be "average-stress". We may pose:

$$\frac{F}{S} = \sigma_x$$

---

9 We will note that during the test, the real area of the straight section slightly decreases. It is therefore different from S.

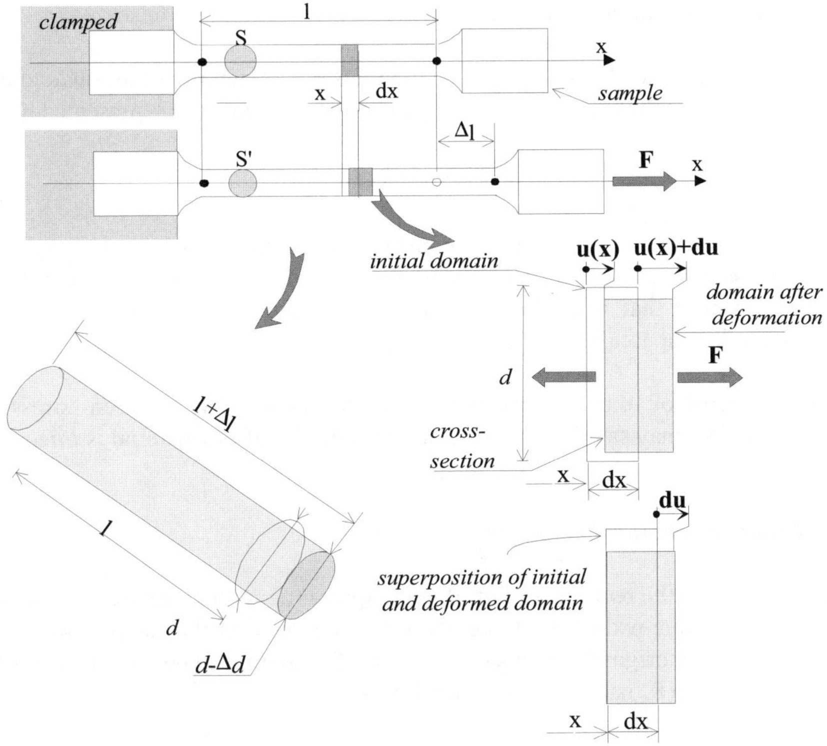

Figure 1.8. *A sample under traction*

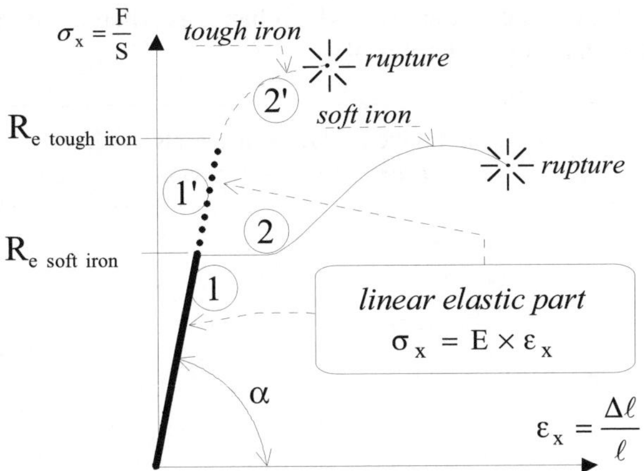

Figure 1.9. *Stretch test diagram*

1.3.2. *Linear mechanical behavior*

The rectilinear part (zone 1 or 1') of the registered diagram corresponds to a line with an equation of the type "y = ax" (linear relationship) which is represented here as:

$$\sigma_x = E \times \varepsilon_x$$

The value of this line's slope is a constant characteristic of each material and is referred to as *Young's modulus* or elastic modulus and denoted by E, (E = tan α). We may observe that E is dimensionally homogenous to a stress, for example numbered in Megapascals (MPa).

In this part of the test we note the proportionality between stress and deformation. We may say that the mechanical behavior of the material is *linear*.

1.3.3. *Elastic mechanical behavior*

Remaining on the rectilinear part of the diagram (up to the limit denoted by R_e in Figure 1.9), the removal of the force reverses the stretch of the sample, which now moves back to its original dimensions. We say that the mechanical behavior of the material is *elastic* (R_e is called the elastic limit).

The *linear* and *elastic* properties of the material components of a structure are the basis for what follows. Other mechanical behaviors may influence the sizing of structures, particularly the behavior that can be seen in zones 2 or 2' in Figure 1.9. We say that the behavior is *elastoplastic* in these zones; it is no longer linear. It continues until the sample fractures, which happens when the normal stress limit R_r is reached (resistance to the fracture)[10].

We shall limit ourselves to materials that have
elastic and ***linear*** behaviors

---

10 This non-linear behavior (elastoplastic behavior) is not studied further in this book (except for Chapter 7, section 7.2.3.2).

1.3.4. *Interpretation of the test at a macroscopic level*

Having assumed that the behavior is elastic and linear, we obtain (Figures 1.8 and 1.9):

$$\frac{F}{S} = E \times \frac{\Delta \ell}{\ell} \quad \Leftrightarrow \quad \sigma_x = E \times \varepsilon_x$$

This is what we refer to as Hooke's law, which may be re-written as:

$$\frac{\Delta \ell}{\ell} = \frac{1}{E} \times \frac{F}{S} \quad \Leftrightarrow \quad \varepsilon_x = \frac{1}{E} \times \sigma_x \qquad [1.2]$$

NOTE

❏ Poisson's effect: we have already pointed out the reduction of diameter Δd. We define the relative reduction by the ratio $\frac{\Delta d}{d}$. The test shows that this is proportional to the axial strain. We may therefore note:

$$\frac{\Delta d}{d} = -\nu \frac{\Delta \ell}{\ell} \qquad [1.3]$$

where ν is a dimensionless so-called "Poisson's" coefficient. For the current construction materials the usual values of Poisson's coefficient fall within 0.25 to 0.3[11].

❏ At the beginning of section 1.3.1 we specified that the material should be taken as isotropic, i.e., its behavior was independent of the manner in which the sample was manufactured from the raw material. A unique value for Young's modulus represents this here, as well as a unique value for Poisson's coefficient ν.

1.3.5. *Interpretation of the test at a mesoscopic[12] level*

We may note that all elementary slices of length dx of the sample in Figure 1.8 are subjected to the same load and as a consequence are subjected to the same elongation "du".

---

11 See Appendix B.
12 Or the "intermediate" level between the extreme "macroscopic" and "microscopic" levels.

Then we can write[13]:

$$\frac{du}{dx} = \frac{\int_{\ell} du}{\int_{\ell} dx} \qquad [1.4]$$

where $\int_{\ell} du(x) = \Delta\ell$, and $\int_{\ell} dx = \ell$, thereby: $\dfrac{du}{dx} = \dfrac{\Delta\ell}{\ell} = \varepsilon_x = C^{te}$, $\forall\, x$.

We see at the scale of the elementary slice of volume $S_x{\times}dx$ (or mesoscopic scale) that equation [1.2] can be re-written as:

$$\frac{du}{dx} = \frac{1}{E}\,\sigma_x \quad \Leftrightarrow \quad \varepsilon_x = \frac{1}{E}{\times}\sigma_x \qquad [1.5]$$

whereby we recognize the axial strain ε_x and the normal stress σ_x.

NOTE

❑ The strain is proportional to the stress in the elastic zone.

❑ During this stretch test, on the section with abscissa $x + dx$, the stress σ_x is in the same direction as the normal outgoing from the isolated element. Its algebraic value is therefore positive on this normal (tension state). If we reverse the direction of the force $\vec{F}$, the sample now being compressed, stress σ_x shall be represented in the direction opposite to the outgoing normal (Figure 1.10).

---

13 Remembering that $\dfrac{a}{b} = \dfrac{n \times a}{n \times b}$.

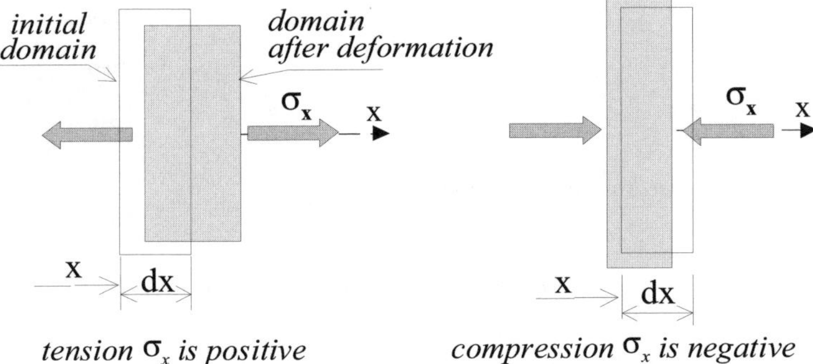

Figure 1.10. *The algebraic sign of the normal stress*

❏ For all sections S, the relative variation of diameter $\dfrac{\Delta d}{d}$ or the transverse strain may be written , according to [1.3], as:

$$\frac{\Delta d}{d} = -\nu \times \varepsilon_x = -\frac{\nu}{E} \times \sigma_x$$

❏ We may note that in Figure 1.8, after deformation, the length dx moves on to have the value: $(dx)' = dx + du = dx\left(1 + \dfrac{du}{dx}\right)$ or rather: $dx \to dx(1 + \varepsilon_x)$.

❏ As we record the diagram in Figure 1.9, we obtain the elongation $\Delta\ell$ thanks to an extensometer which measures the variation of length between the two references that define ℓ on the sample. We may also use an electric "extensometer gauge" as shown in Figure 1.11.

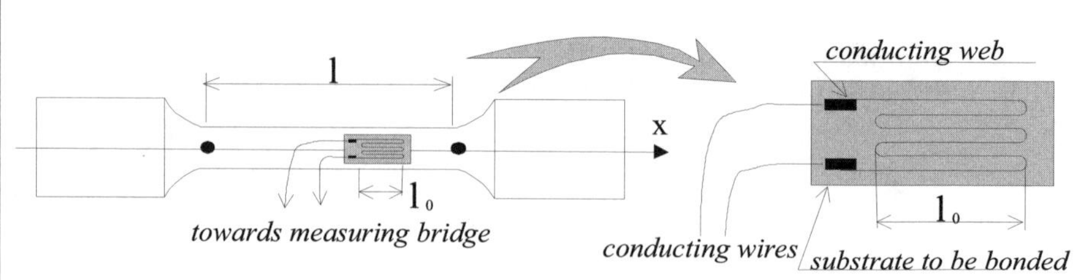

Figure 1.11. *Extensometer gauge*

A conducting grid of length ℓ_0 is attached to a soft substrate, which is then bonded to the zone that is to be measured on the sample. It can be said, without getting into the behavior of the gauge, that the variation of length $\Delta\ell_0$ of the grid as the sample elongates results in a proportional variation of electrical resistance ΔR of this grid, say: $\left(\dfrac{\Delta\ell_0}{\ell_0}\right)_{\text{wire}} = \dfrac{\Delta\ell}{\ell} = \lambda\dfrac{\Delta R}{R}$. By measuring $\dfrac{\Delta R}{R}$ thanks to a Wheatstone bridge, we obtain after calibrating the value $\left(\dfrac{\Delta\ell_0}{\ell_0}\right)_{\text{wire}} = \varepsilon_x$, when the gauge is perfectly bonded to the sample. The electrical extensometer gauges measure the local axial strains (compression or elongation) along the direction of the grid.

1.3.6. *Interpretation of the test at a microscopic level*

Returning to the elementary slice in the sample in Figure 1.8 with a volume of $S \times dx$, we may consider that it is made up of an infinity of small microscopic bits as in Figure 1.12a. This small elementary parallelepiped bit, with the orthonormal axes $\vec{x}, \vec{y}, \vec{z}$, becomes distorted as represented. Taking into account our previous considerations, it stretches along the direction $\vec{x}$, and shortens along the directions $\vec{y}$ and $\vec{z}$. The axial deformation along $\vec{x}$ being denoted by ε_x, we may denote the axial deformation along $\vec{y}$ and $\vec{z}$ by $-\nu\varepsilon_x$, as in Figure 1.12a. We obtain the final dimensions of the element as:

$$- \text{along } \vec{x}: \ dx(1+\varepsilon_x)$$

$$- \text{along } \vec{y}: \ dy(1-\nu\varepsilon_x) = dy(1+\varepsilon_y)$$

$$- \text{along } \vec{z}: \ dz(1-\nu\varepsilon_x) = dz(1+\varepsilon_z)$$

If we prefer to represent this initial small volume as a sphere (Figure 1.12b), it then takes the final shape of the already familiar ellipsoid. With the previous considerations, this ellipsoid becomes "stretched" along the $\vec{x}$ direction. In Figure 1.7, we had the principal stresses and associated directions for the ellipsoid. In our case, the ellipsoid in Figure 1.7 becomes that of 1.12b, where we may note that the axis $\vec{x}$ coincides with one of the primary directions.

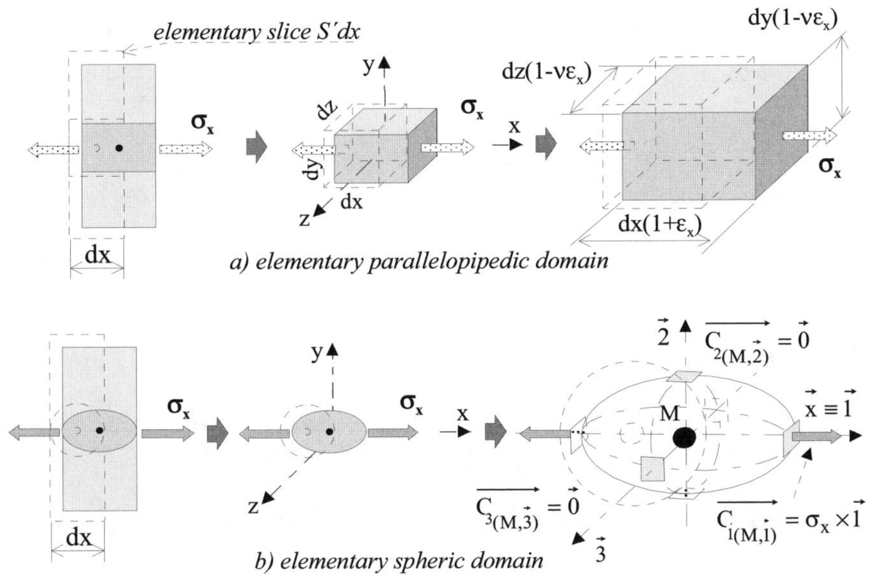

Figure 1.12. *Microscopic scale*

NOTE

❑ In Figure 1.12, all parallelepipedic domains included in the elementary slice with a volume of $S \times dx$ get distorted in the same way. This is the same for the spherical domains. The stress vector shown manifests identically on each of the elementary domains.

We say here that the state of stress is:

– uniaxial: as in cross-section S, the stresses are colinear to the $\vec{x}$ axis, which is normal to S;

– constant: as in cross-section S and $\forall$ x, the stresses have the same constant value. The distribution of the normal stresses on the cross-section S form what we may call a *uniform field*.

❑ For this simple stretch test case, the distribution of the normal stresses on a cross-section being uniform (see Figure 1.13), it coincides with the average value: $\sigma_x = \dfrac{F}{S}$.

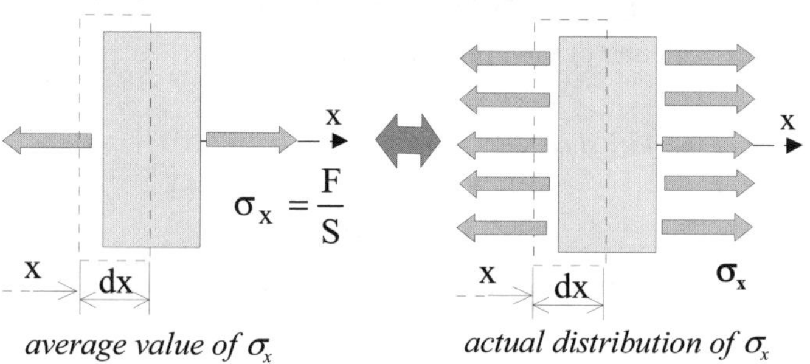

Figure 1.13. *Distribution of the normal stress*

1.3.7. *Summary*

materials:	homogenous isotropic elastic linear

Hooke's law for uniaxial traction or compression (along the $\vec{x}$ axis)	
Macroscopic scale	
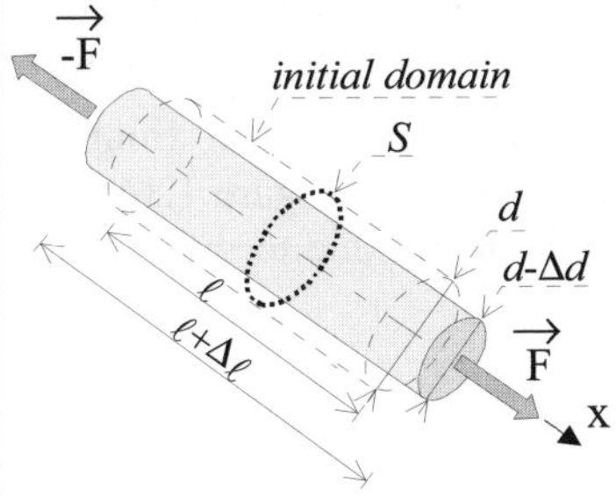 	$\dfrac{\Delta\ell}{\ell} = \dfrac{1}{E} \times \dfrac{F}{S}$ $\updownarrow \qquad \updownarrow$ $\varepsilon_x = \dfrac{1}{E} \times \sigma_x \Leftrightarrow \sigma_x = E \times \varepsilon_x$ $\dfrac{\Delta d}{d} = -\nu\,\dfrac{\Delta\ell}{\ell}$ $\updownarrow \qquad \updownarrow$ $\varepsilon_d = -\nu \times \varepsilon_x$ *E: Young's modulus or modulus of elasticity* *ν: Poisson's coefficient*
Mesoscopic scale	
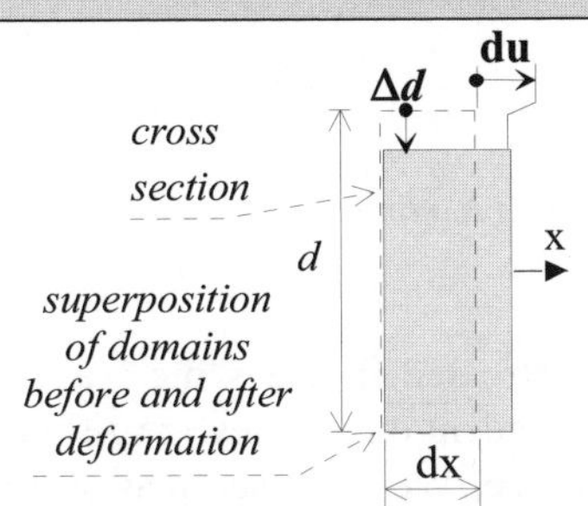 	$\dfrac{du(x)}{dx} = \dfrac{1}{E}\,\sigma_x$ $\updownarrow \qquad \updownarrow$ $\varepsilon_x = \dfrac{1}{E}\,\sigma_x \Leftrightarrow \sigma_x = E \times \varepsilon_x$ $\dfrac{\Delta d}{d} = -\nu \times \varepsilon_x$ The dimensions become: $dx \to dx(1 + \varepsilon_x)$ and $d \to d(1 - \nu\varepsilon_x)$
Microscopic scale	
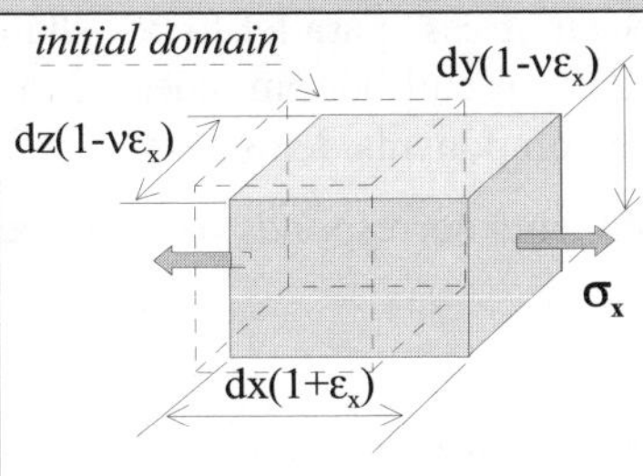 	$\varepsilon_x = \dfrac{1}{E}\,\sigma_x \Leftrightarrow \sigma_x = E \times \varepsilon_x$ $\varepsilon_y = -\nu\,\varepsilon_x\,;\ \varepsilon_z = -\nu\,\varepsilon_x$ The dimensions become: $dx \to dx(1 + \varepsilon_x)$ $dy \to dy(1 + \varepsilon_y)$ $dz \to dz(1 + \varepsilon_z)$

$$[1.6]$$

1.4. Plane state of stresses

1.4.1. *Definition*

We take into account (Figure 1.14a) a structure with the plane (xy) on the top surface and a constant width "e" along $\vec{z}$. It is loaded by some forces on its lateral sides, that are parallel to the plane (xy), and which we shall suppose uniformly distributed along the direction $\vec{z}$. We remove a small parallelpiped from this structure (dx × dy × e) having M as its center, as represented in Figure 1.14b, and shall illustrate the resulting stresses on the faces of this prism, due to the cohesion forces as defined in section 1.2.

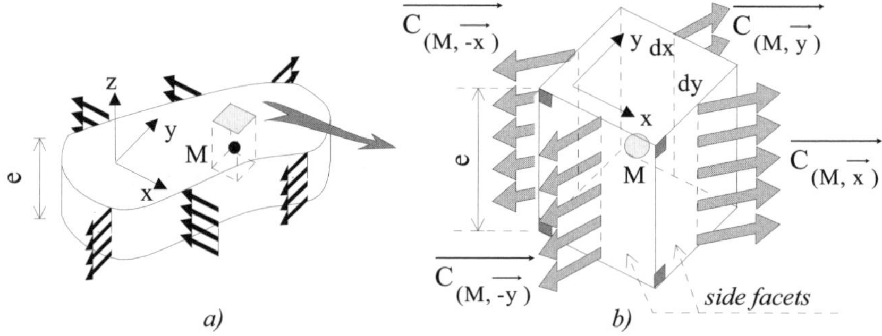

Figure 1.14. *Bidimensional structure and loading*

Due to the loading form, the structure shall be stressed in the same way no matter what the ordinate z is. The stresses are not dependent on z. They are zero on the top and bottom sides of the structure which are free. We shall see that on the lateral faces of the small prism, stresses are uniform along $\vec{z}$ as shown in the figure. Each stress vector is composed of a normal stress and a tangential or shear stress. We shall also note that the latter must necessarily be parallel to the plane (xy). Effectively, a non-zero vertical component of the shear stress (parallel to $\vec{z}$) shall imply the presence of reciprocal shear stresses, on the top and bottom sides of the prism, which is not the case here. We thereby get the components denoted σ_x, σ_y, τ_{xy} (or τ_{yx})[14] on Figure 1.15. Each of these components corresponds to what we shall call a simple state of stresses.

---

14 Expression τ_{xy} represents shear stress on the face with normal $\vec{x}$ acting in direction parallel to $\vec{y}$. Remember that $\tau_{xy} = \tau_{yx}$ (see section 1.2.3).

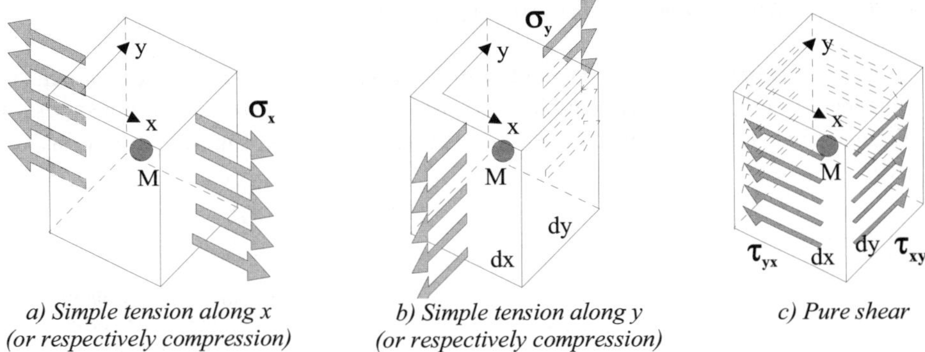

a) Simple tension along x
(or respectively compression)

b) Simple tension along y
(or respectively compression)

c) Pure shear

Figure 1.15. *Components of a plane state of stresses*

The complete plane state of stresses is represented in Figure 1.16 by superposing the three simple states.

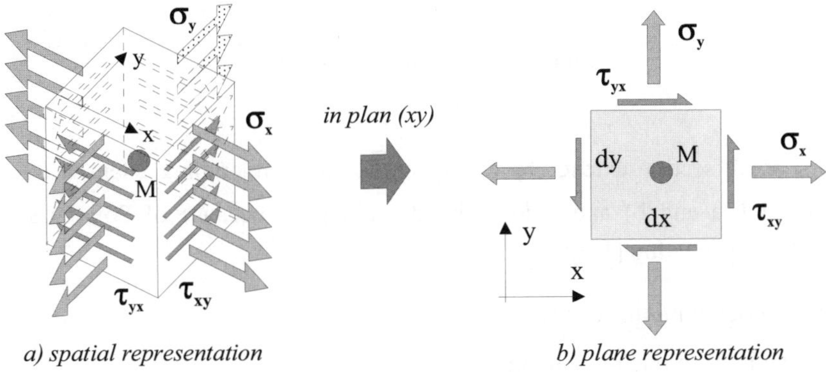

a) spatial representation

b) plane representation

Figure 1.16. *Complete plane state of stresses*

We may regroup the characteristic components in a column matrix[15]:

$$\left\{ \begin{array}{c} \sigma_x \\ \sigma_y \\ \tau_{xy} \end{array} \right\}.$$

---

15 Note: the terms of this column matrix are not the three components of the same stress vector. Remember that, in fact, $\overrightarrow{C_{(M,\vec{x})}} = \sigma_x \vec{x} + \tau_{xy} \vec{y}$ and $\overrightarrow{C_{(M,\vec{y})}} = \tau_{yx} \vec{x} + \sigma_y \vec{y}$.

1.4.2. *Behavior relationships for state of plane stresses*

Let us isolate a small element with stresses as defined earlier, which is represented in the plane (xy), and let us consider successive states of simple stresses (Figure 1.15), first σ_x, then σ_y, then τ_{xy} represented here in Figures 1.17, 1.18 and 1.19.

1.4.2.1. *Case 1: simple tension along $\vec{x}$* [16]

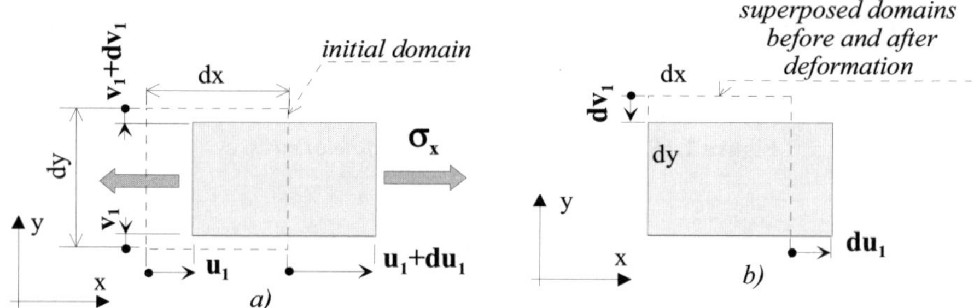

Figure 1.17. *Simple tension along $\vec{x}$*

The small section extends by du_1 along $\vec{x}$, upon undergoing a tensile stress σ_x (Figures 1.17a and b) and reduces by dv_1 along $\vec{y}$ because of Poisson's effect (according to equation [1.3]).

By analogy with the tensile test (see [1.6]) we note:

– for relative elongation along $\vec{x}$:

$$\frac{du_1}{dx} = \varepsilon_{x1} = \frac{1}{E}\sigma_x \qquad [1.7]$$

– for relative reduction along the transverse direction $\vec{y}$:

$$\frac{dv_1}{dy} = \varepsilon_{y1} = -\nu \times \varepsilon_{x1} = -\frac{\nu}{E}\sigma_x \qquad [1.8]$$

---

16 To establish behavior relationships the basic domain is represented as stressed through tension. Compression stresses would lead to formally identical results.

where ν is Poisson's coefficient marking the proportionality between the two deformations (see equation [1.3]).

NOTE

❐   We examine here an elementary domain with dimensions $dx \times dy \times e$ (intermediary or "mesoscopic" scale)[17].

To mark the deformations we have put forward an "elastic displacement field" of components u_1 and v_1 in the plane (xy) (as in Figure 1.17a).

These displacements are generally functions of x and y, i.e.:

$$u_1(x, y) \; ; \; v_1(x, y)$$

The most general increment of each of these components may be written[18] as:

$$du_1 = \frac{\partial u_1}{\partial x} dx + \frac{\partial u_1}{\partial y} dy \; ; \; dv_1 = \frac{\partial v_1}{\partial x} dx + \frac{\partial v_1}{\partial y} dy$$

In Figure 1.17a, we note that the increment denoted by du_1 is the same no matter what the height y is on the small domain. It does not depend on y. We therefore obtain here:

$$du_1 = \frac{\partial u_1}{\partial x} \times dx + 0 \times dy = \frac{\partial u_1}{\partial x} dx \; \left(\text{or } \frac{\partial u_1}{\partial y} = 0 \right)$$

Replacing in equation [1.7]:

$$\varepsilon_{x1} = \frac{du_1}{dx} = \frac{\partial u_1}{\partial x}$$

Similarly on Figure 1.17a the increment dv_1 is the same no matter what the abscissa x may be on the domain. It does not depend on x. We therefore have:

---

17 We note that, in fact, the isolated domain (see Figure 1.14b) has a finite thickness e.

18 If a function of several variables $U = f(x, y, z)$ admit partial derivatives f_x', f_y', f_z', the total differential of U, noted as dU, is:

$$dU = f_x' dx + f_y' dy + f_z' dz = \frac{\partial U}{\partial x} dx + \frac{\partial U}{\partial y} dy + \frac{\partial U}{\partial z} dz$$

$$dv_1 = 0 \times dx + \frac{\partial v_1}{\partial y} \times dy = \frac{\partial v_1}{\partial y} \, dy \ \left(\text{or} \ \frac{\partial v_1}{\partial x} = 0 \right)$$

Replacing in equation [1.8]:

$$\varepsilon_{y1} = \frac{dv_1}{dy} = \frac{\partial v_1}{\partial y}$$

1.4.2.2. *Case 2: simple tension along* $\vec{y}$

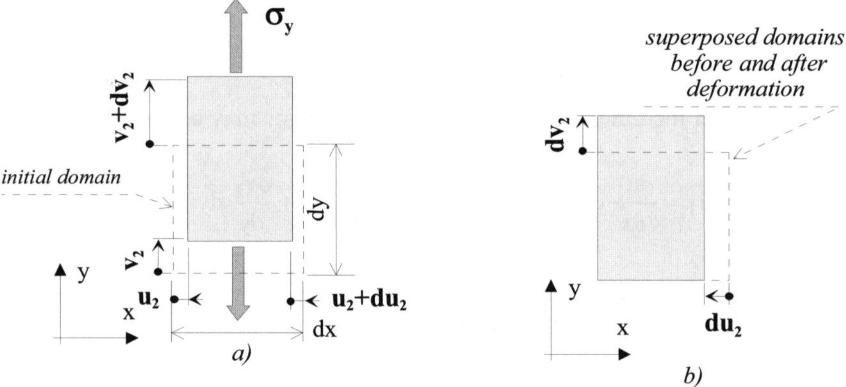

Figure 1.18. *Simple tension along* $\vec{y}$

The small section being subjected to a tensile stress σ_y we may write analogically:

– the relative elongation along $\vec{y}$:

$$\frac{dv_2}{dy} = \varepsilon_{y2} = \frac{1}{E} \sigma_y \tag{1.9}$$

– the relative reduction along $\vec{x}$:

$$\frac{du_2}{dx} = \varepsilon_{x2} = -\nu \times \varepsilon_{y2} = -\frac{\nu}{E} \sigma_y \tag{1.10}$$

NOTE

❏ As for the preceding case 1, we may make a comment here concerning the displacement field:

$$u_2(x,y) \quad ; \quad v_2(x,y)$$

We also have:

$$du_2 = \frac{\partial u_2}{\partial x}\,dx \quad \left(\frac{\partial u_2}{\partial y} = 0\right) ; \quad dv_2 = \frac{\partial v_2}{\partial y}\,dy \quad \left(\frac{\partial v_2}{\partial x} = 0\right)$$

Replacing in equations [1.9] and [1.10]

$$\varepsilon_{x2} = \frac{du_2}{dx} = \frac{\partial u_2}{\partial x} ; \quad \varepsilon_{y2} = \frac{dv_2}{dy} = \frac{\partial v_2}{\partial y}$$

1.4.2.3. *Case 3: pure shear*

Here the domain $dx \times dy \times e$ in axes $\vec{x}$, $\vec{y}$ is considered to be under pure shear stress. The figure also shows the reciprocal shear stresses already discussed in section 1.2.3[19]:

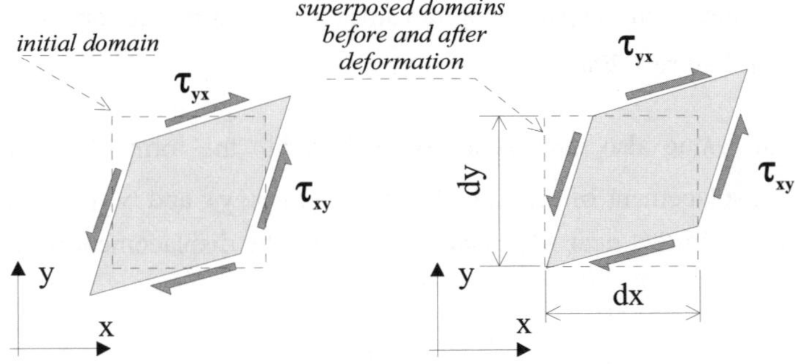

Figure 1.19. *Pure shear*

---

19 We note again that the deformation of the domain is strongly magnified in the figure.

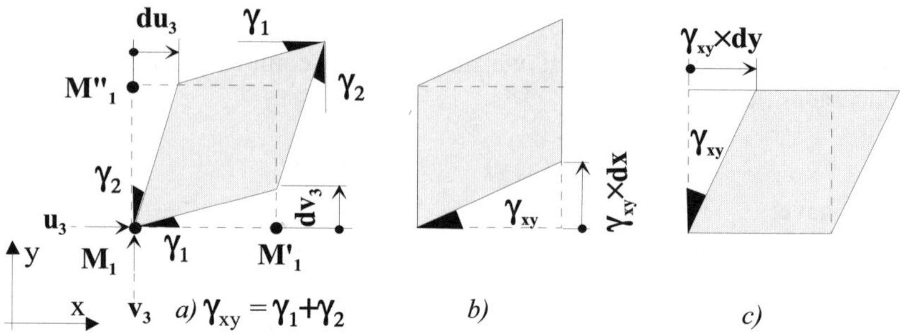

Figure 1.20. *Representation of angular distortion γ under pure shear*

The rectangular section becomes a parallelogram with the length of its sides, for example dx and dy, still conserved. Only the angles have changed as the deformation took place. We say that the section has undergone an angular distortion γ_{xy}. This is the amount the initial right angles have varied. This distortion γ_{xy} has been represented in different ways in Figure 1.20. It is very small, and that allows:

– the graphical representation in Figure 1.20;

– the assimilation of the parallelogram sides with their projections on axes $\vec{x}$ and $\vec{y}$ as shown in Figure 1.20a.

This small value also enables us to easily write the form of *distortion* γ_{xy}. Writing the displacement of a point $M_1(x,y)$ as $u_3(x,y)$ and $v_3(x,y)$ means that for any points infinitely near M_1 in the (x,y) plane, the displacement may be written as:

$$u_3 + du_3 \text{ with } du_3 = \frac{\partial u_3}{\partial x}\,dx + \frac{\partial u_3}{\partial y}\,dy$$

$$v_3 + dv_3 \text{ with } dv_3 = \frac{\partial v_3}{\partial x}\,dx + \frac{\partial v_3}{\partial y}\,dy$$

As a consequence we may note that in Figure 1.20 we have:

$$\text{at } M'_1: \ dv_3 = \frac{\partial v_3}{\partial x}\,dx + \frac{\partial v_3}{\partial y} \times 0 = \frac{\partial v_3}{\partial x}\,dx \quad (\text{as } dy = 0)$$

$$\text{at } M''_1: du_3 = \frac{\partial u_3}{\partial x} \times 0 + \frac{\partial u_3}{\partial y} dy = \frac{\partial u_3}{\partial y} dy \quad (\text{as } dx = 0)$$

On noting $\gamma_1 \times dx = dv_3$ and $\gamma_2 \times dy = du_3$ in Figure 1.20a, (the smallness of the distortion allowing us to note $\tan \gamma \cong \gamma$), the distortion form may be written as:

$$\gamma_{xy} = \gamma_1 + \gamma_2 = \frac{dv_3}{dx} + \frac{du_3}{dy} = \frac{\partial v_3}{\partial x} + \frac{\partial u_3}{\partial y} \qquad [1.11]$$

Thanks to the linear elasticity of the material, the angular distortion is proportional to the shear stress. We write:

$$\gamma_{xy} = \frac{1}{G} \times \tau_{xy} \quad \Leftrightarrow \quad \tau_{xy} = G \times \gamma_{xy} \qquad [1.12]$$

This constant G linking the shear stress and the angular distortion is called *shear modulus*[20]. This behavior relation is analogous in its form to Hooke's law.

NOTE

❑ As in the case of Hooke's law (Figure 1.9) this proportional relationship between the stress and the deformation has a limit. It is the shear elastic limit determined by the test and that we conventionally denote as R_{eg} (MPa). As we exceed this value, there is a shear plastification and then a rupture occurs for a maximum shear stress R_{rg} (MPa), called resistance to the shear rupture.

❑ As already noted, we shall suppose here that the lengths dx and dy of the sides of the initial section in Figure 1.20 do not change as the distortion takes place. This is linked to the absence of normal stresses. As a result we therefore have, for example:

$$\text{at } M'_1: du_3 = 0 = \frac{\partial u_3}{\partial x} dx \quad \Rightarrow \quad \frac{\partial u_3}{\partial x} = 0$$

$$\text{at } M''_1: dv_3 = 0 = \frac{\partial v_3}{\partial y} dy \quad \Rightarrow \quad \frac{\partial v_3}{\partial y} = 0$$

This note will be useful for what follows.

---

20 Or also "shear modulus" or "Coulomb's modulus".

1.4.2.4. *Complete state of stress (superposition)*

As the three simple stresses σ_x, σ_y and τ_{xy} are simultaneously applied to the small section we obtain what appears in Figure 1.21.

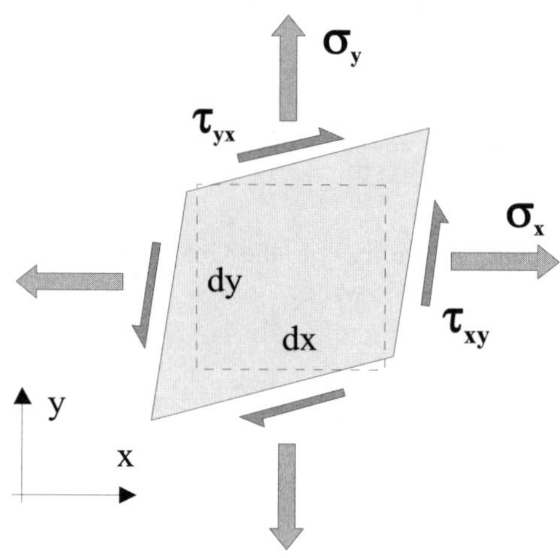

Figure 1.21. *The complete state of stresses*

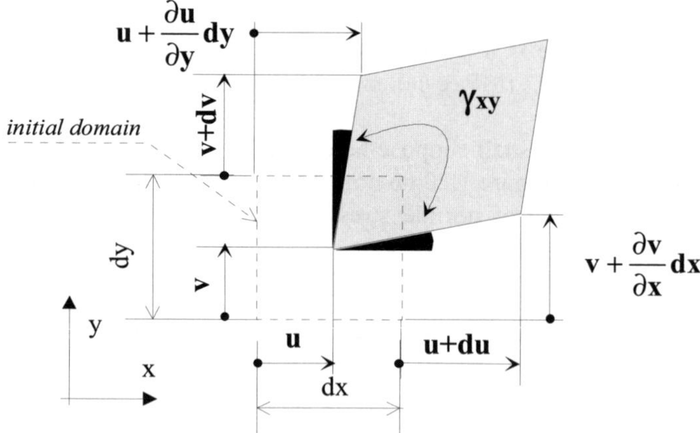

Figure 1.22. *Displacements and deformations*

The total deformation appears as the sum of the already discussed deformations for each simple state of stresses. It is represented in Figure 1.22. With equations [1.7] and [1.10], we have the relative elongation along $\vec{x}$ [21] as:

$$\frac{\partial u}{\partial x} = \frac{\partial u_1}{\partial x} + \frac{\partial u_2}{\partial x} + \frac{\partial u_3}{\partial x} = \frac{\partial u_1}{\partial x} + \frac{\partial u_2}{\partial x} + 0$$

For example:

$$\varepsilon_x = \varepsilon_{x1} + \varepsilon_{x2} = \frac{1}{E}\sigma_x - \frac{\nu}{E}\sigma_y \qquad [1.13]$$

With equations [1.8] and [1.9], we have along $\vec{y}$:

$$\frac{\partial v}{\partial y} = \frac{\partial v_1}{\partial y} + \frac{\partial v_2}{\partial y} + \frac{\partial v_3}{\partial y} = \frac{\partial v_1}{\partial y} + \frac{\partial v_2}{\partial y} + 0$$

For example:

$$\varepsilon_y = \varepsilon_{y1} + \varepsilon_{y2} = -\frac{\nu}{E}\sigma_x + \frac{1}{E}\sigma_y \qquad [1.14]$$

We also note that angular distortion equation [1.11] takes a more generalized[21] form here:

$$\gamma_{xy} = \frac{\partial v_3}{\partial x} + \frac{\partial u_3}{\partial y} = \left(\frac{\partial v_1}{\partial x} + \frac{\partial v_2}{\partial x} + \frac{\partial v_3}{\partial x}\right) + \left(\frac{\partial u_1}{\partial y} + \frac{\partial u_2}{\partial y} + \frac{\partial u_3}{\partial y}\right) = \frac{\partial v}{\partial x} + \frac{\partial u}{\partial y}$$
$$\quad\quad\Downarrow\quad\;\Downarrow\qquad\qquad\qquad\Downarrow\quad\;\Downarrow$$
$$\quad\quad 0\quad\;\; 0\qquad\qquad\qquad 0\quad\;\; 0$$

For example, with [1.12]:

$$\gamma_{xy} = \frac{1}{G}\tau_{xy} \qquad [1.15]$$

NOTE

❑ Poisson's effect (a contraction here) also appears along the axis $\vec{z}$ perpendicular to the plane of the stresses (Figure 1.23). We thereby observe the

---

21 See [1.7], [1.10] and notes in section 1.4.2.1.

deformation along $\vec{z}$ generated by $\sigma_x \Rightarrow -\nu\varepsilon_{x1}$ and by $\sigma_y \Rightarrow -\nu\varepsilon_{y2}$. The deformation along $\vec{z}$ is therefore equal to:

$$\varepsilon_z = -\nu\times\varepsilon_{x1} - \nu\times\varepsilon_{y2} = -\frac{\nu}{E}\left(\sigma_x + \sigma_y\right)$$

With equations [1.13] and [1.14] it can be rewritten as: $\varepsilon_z = -\dfrac{\nu}{1-\nu}\left(\varepsilon_x + \varepsilon_y\right)$

❑ As a result the small square prism as in Figure 1.14 remains a square prism after deformation. However, width "e'" is different from the original width "e" as represented on Figure 1.23.

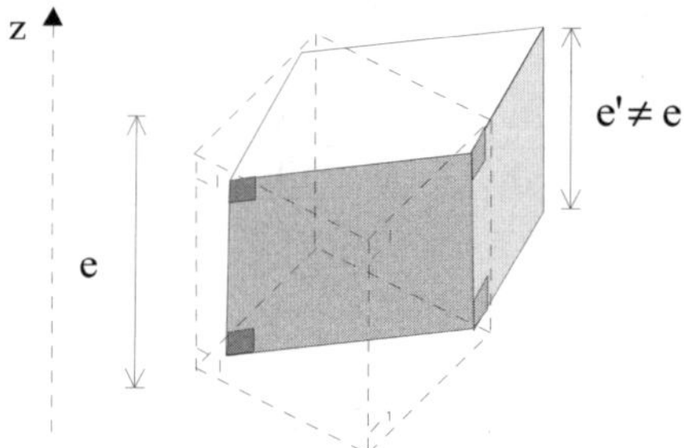

Figure 1.23. *Spatial representation of the deformed elementary prism*

❑ Equations [1.13] to [1.15] may be regrouped as a single matrix equation:

$$\left\{\begin{array}{c}\varepsilon_x \\ \varepsilon_y \\ \gamma_{xy}\end{array}\right\} = \left[\begin{array}{ccc} \dfrac{1}{E} & -\dfrac{\nu}{E} & 0 \\ -\dfrac{\nu}{E} & \dfrac{1}{E} & 0 \\ 0 & 0 & \dfrac{1}{G} \end{array}\right] \bullet \left\{\begin{array}{c}\sigma_x \\ \sigma_y \\ \tau_{xy}\end{array}\right\} \qquad [1.16]$$

A square symmetric matrix (3×3) appears, with the characteristic elastic coefficients E, G, v. Knowing that elsewhere we show the relation[22]:

$$G = \frac{E}{2(1+v)}$$

Just two distinct elastic coefficients, for example E, v, remain to be represented in equation [1.16], the linear elastic behavior.

❏ We also note that we may write equation [1.16] in the reverse way by inverting the elastic coefficients matrix:

After calculations we obtain[23]:

$$\left\{\begin{array}{c}\sigma_x \\ \sigma_y \\ \tau_{xy}\end{array}\right\} = \left[\begin{array}{ccc}\dfrac{E}{1-v^2} & \dfrac{vE}{1-v^2} & 0 \\ \dfrac{vE}{1-v^2} & \dfrac{E}{1-v^2} & 0 \\ 0 & 0 & G\end{array}\right] \bullet \left\{\begin{array}{c}\varepsilon_x \\ \varepsilon_y \\ \gamma_{xy}\end{array}\right\} \qquad [1.17]$$

❏ We reconsider the previous deformed parallepipedic element in Figure 1.24a. In Figure 1.24b we marked an initial and deformed small spherical element. We indicated earlier that the latter is an ellipsoid. By comparing the present one with that of Figure 1.7, we note that here the plane of stresses (xy) is parallel to the plane ($\vec{1}$, $\vec{2}$). This is represented in Figure 1.24c.

---

22 This propriety is shown in Chapter 10, section 10.1.5.2.
23 To invert a square matrix, see Chapter 12, section 12.1.2.4.

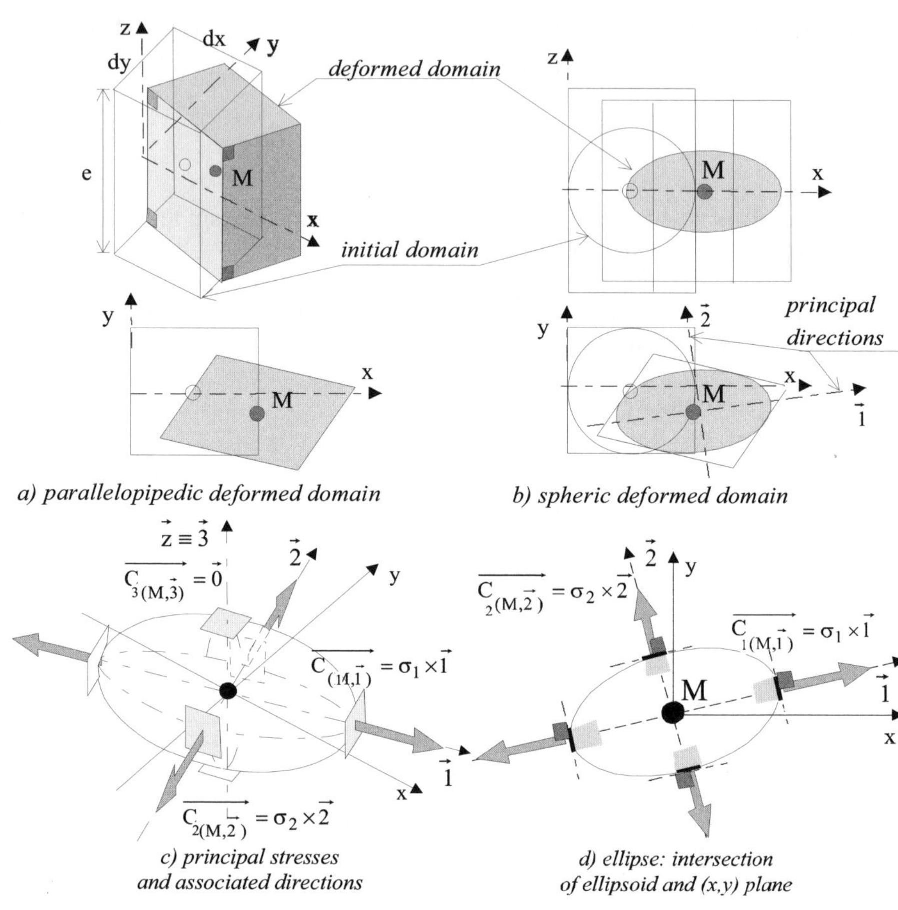

c) principal stresses
and associated directions

d) ellipse: intersection
of ellipsoid and (x,y) plane

Figure 1.24. *Principal stresses and directions for a plane state of stresses*

At the same time we note that the direction $\vec{z}$, perpendicular to the plane, becomes the third primary direction $\vec{z} = \vec{3}$. As a result the principal stress on any surface of normal $\vec{z}$ is zero.

❏ With the help of Figure 1.24d[24] we may imagine that there exists a rectangular prism with axis $\vec{z}$, made up of surfaces normal to $\vec{1}$ and $\vec{2}$, on which the tangential stress is zero. Therefore this rectangular prism does not undergo an

---

24 See also Chapter 10, section 10.1.2.

angular distortion γ. $\vec{1}$ and $\vec{2}$ are the principal directions for the stresses. They are orthogonal. We have already mentioned their existence in section 1.2.3 (Figure 1.7). We therefore note in principal axis ($\vec{1}, \vec{2}$):

$$\underset{\textit{no distortion}}{\textit{principal deformations}\quad \begin{Bmatrix} \varepsilon_1 \\ \varepsilon_2 \end{Bmatrix}} \quad \Leftrightarrow \quad \underset{\textit{principal directions}}{\begin{matrix}(\vec{1}, \vec{2})\end{matrix}} \quad \Leftrightarrow \quad \underset{\textit{no shear}}{\textit{principal stresses}\quad \begin{Bmatrix} \sigma_1 \\ \sigma_2 \end{Bmatrix}}$$

❐ We have seen in section 1.3.5 the use of electrical extensometer gauges (Figure 1.11) to directly measure the deformation in the same direction as the gauge's wire. We fix two small extensometric gauges at point M onto the structure of Figure 1.25a, as shown:

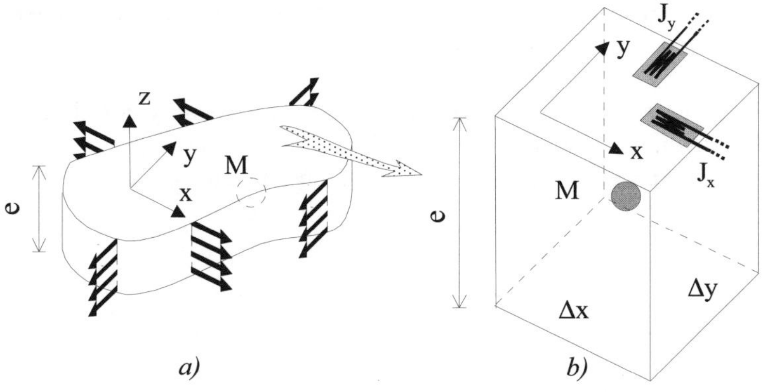

Figure 1.25. *Measuring the deformations* ε_x *and* ε_y

These are extremely small gauges (from 1 mm to a few mms in length). The element around point M is small as well, with a volume $\Delta x \times \Delta y \times e$. We get closer to the basic element in Figure 1.14b: $dx \times dy \times e$. The gauges will give us the local deformations (expansion or contraction):

$$\varepsilon_x \text{ with the gauge } J_x, \text{ and } \varepsilon_y \text{ with the gauge } J_y$$

❑ We now calculate the values of stresses σ_x and σ_y, starting from ε_x and ε_y, thanks to equation [1.17]:

$$\begin{Bmatrix} \sigma_x \\ \sigma_y \end{Bmatrix} = \frac{E}{1-v^2} \begin{bmatrix} 1 & v \\ v & 1 \end{bmatrix} \bullet \begin{Bmatrix} \varepsilon_x \\ \varepsilon_y \end{Bmatrix}$$

However, we cannot determine τ_{xy}. In fact, the gauges cannot measure angular distortions such as γ_{xy}, but only deformations (expansion or contraction) along the length of their wires[25].

❑ Principle of superpositioning of the effects of applied forces.

A loading applied to a structure as shown in Figure 1.14 leads to a plane state of stresses and small displacements at all points M described earlier. If several loadings act on this structure simultaneously, it will result in a superposition of states of plane stresses caused by each of these loadings.

We have just seen how Hooke's law of proportionality correlates the stresses and deformations. Therefore, deformations at all points of the structure appear as the sum of deformations caused by each of these loadings.

---

25 The addition of a third gauge allows us to calculate the value of the distortion γ_{xy} (see Chapter 10, section 10.1.6).

1.4.3. *Summary*

For a plane state of stresses (plane (xy)), the behavior of a homogenous, isotropic, elastic and linear material is summed up in the following table.

<table>
<tr><td colspan="2" align="center">The mechanical loading on a structure leads to</td></tr>
<tr><td colspan="2">

small elastic displacements of any point M(x,y,z) denoted:

$$u(x,y) \text{ along } \vec{x}$$
$$v(x,y) \text{ along } \vec{y}$$
$$w(z) \text{ along } \vec{z}$$

these components constitute "a small elastic displacement field".
</td></tr>
<tr><td>

a stress state at any point M, in any plane parallel to the plane (xy)

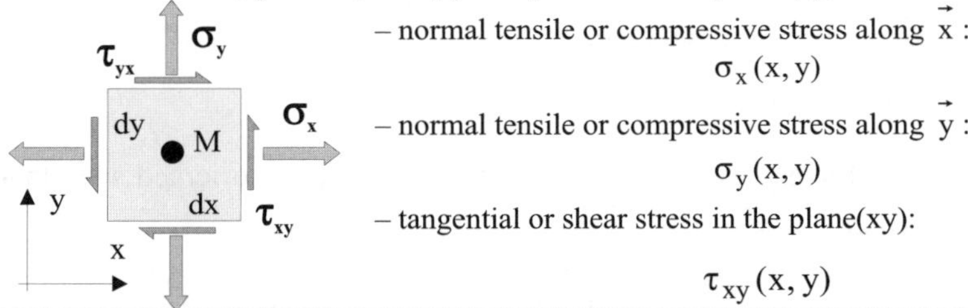

</td><td>

– normal tensile or compressive stress along $\vec{x}$:
$$\sigma_x(x, y)$$
– normal tensile or compressive stress along $\vec{y}$:
$$\sigma_y(x, y)$$
– tangential or shear stress in the plane(xy):
$$\tau_{xy}(x, y)$$
</td></tr>
<tr><td colspan="2">

these stresses create:
</td></tr>
<tr><td colspan="2">

a state of strain or deformation at any point M:

– dilatation or contraction along $\vec{x}$: $\varepsilon_x = \dfrac{\partial u}{\partial x}$

– dilatation or contraction along $\vec{y}$: $\varepsilon_y = \dfrac{\partial v}{\partial y}$

– angular distortion in the plane (xy): $\gamma_{xy} = \dfrac{\partial u}{\partial y} + \dfrac{\partial v}{\partial x}$
</td></tr>
<tr><td colspan="2">

the deformations and stresses are linked by a "behavior equation":
</td></tr>
<tr><td align="center">

deformations-stresses
</td><td align="center">

stresses-deformations
</td></tr>
</table>

$$
\begin{Bmatrix} \varepsilon_x \\ \varepsilon_y \\ \gamma_{xy} \end{Bmatrix}
=
\begin{bmatrix}
\dfrac{1}{E} & -\dfrac{\nu}{E} & 0 \\
-\dfrac{\nu}{E} & \dfrac{1}{E} & 0 \\
0 & 0 & \dfrac{1}{G}
\end{bmatrix}
\bullet
\begin{Bmatrix} \sigma_x \\ \sigma_y \\ \tau_{xy} \end{Bmatrix}
\qquad
\begin{Bmatrix} \sigma_x \\ \sigma_y \\ \tau_{xy} \end{Bmatrix}
=
\begin{bmatrix}
\dfrac{E}{1-\nu^2} & \dfrac{\nu E}{1-\nu^2} & 0 \\
\dfrac{\nu E}{1-\nu^2} & \dfrac{E}{1-\nu^2} & 0 \\
0 & 0 & G
\end{bmatrix}
\bullet
\begin{Bmatrix} \varepsilon_x \\ \varepsilon_y \\ \gamma_{xy} \end{Bmatrix}
$$

$$[1.18]$$

1.5. Particular case of straight beams

1.5.1. *Preliminary observations*

1.5.1.1. *Geometric characteristics*

We give here the main definitions relating to straight beams[26]:

♦ The beam is a simple slender cylindrical structure, made up of a single material.

♦ We call cross-section S of a beam the intersection of the cylindrical structure by any plane orthogonal to the rectilinear generators. Figure 1.26 shows a beam with a rectangular cross-section. The notion of cross-section plays an important role in the study of the beam's behavior.

♦ With any cross-section is associated:

– a center of gravity: G;

– two orthogonal axes having their origin at G: $\vec{y}$ and $\vec{z}$, oriented such that the following is verified:

$$\int_S y\,dS = \int_S z\,dS = \int_S yz\,dS = 0$$

♦ We call quadratic principal moments the quantities:

$$I_y = \int_S z^2\,dS \; ; I_z = \int_S y^2\,dS$$

♦ The line constituted by the geometric centers is called the mean line of the beam. It allows the building of a third axis $\vec{x}$ completing the orthonormal trihedral $\vec{x}, \vec{y}$ and $\vec{z}$, so-called *local coordinates* of the beam (Figure 1.27).

1.5.1.2. *Resultant force and moment for cohesion forces*

Taking into account the special geometry of the beams, specific hypothesis allows us to gain the knowledge of stresses under loading, as well as the resulting deformations.

The cross-section of a beam with its linked local coordinates are fundamental geometric elements for the study of the behavior. As the beam undergoes external

---

26 A detailed study of straight beams appears in Chapter 9.

forces, we consider the elementary cohesion forces on a cross-section, and we combine them in the form of resultant force and moment of "cohesion".

The steps are as follows[27]. We represent a beam (Figure 1.26a) linked to certain surroundings (structures S_1 and S_2) at its two ends 1 and 2 and undergoing a mechanical loading. The beam is isolated (Figure 1.26b) and the surroundings are replaced by applied resultant force and moment, taken at the geometric centers G_1 and G_2 of the corresponding end sections. In Figure 1.26c we have isolated a segment I of the beam and shown the cross-section of this segment. At the geometric center G of this cross-section, the resultant force and moment of cohesion is equal to that created by all the mechanical forces acting on the beam segment II. This is noted:

$$\{Coh_{II/I}\}_G = \{\mathcal{F}_{ext/II}\}_G \qquad\qquad [1.19]$$

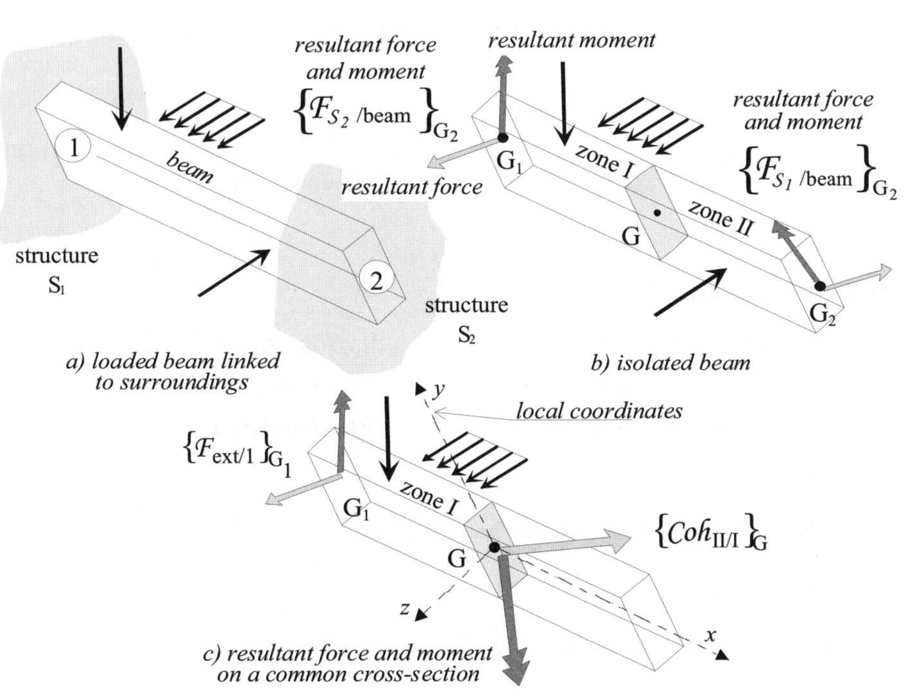

Figure 1.26. *Resultant force and moment on local axis*

27 For a detailed and more complete approach, see section 9.1.4, and for an application, see section 9.4.

As we project the resultant force and moment on local axes, we obtain the usual characteristic resultants for beams (Figure 1.27).

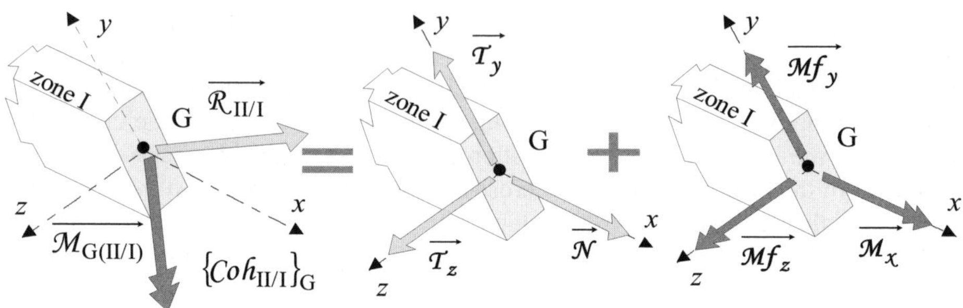

Figure 1.27. *Projection of the resultant force and moment on local axis*

which gives in the local coordinates ($\vec{x}$, $\vec{y}$, $\vec{z}$) denoted "*r*" or (xyz):

$$\left\{Coh_{\mathrm{II/I}}\right\}_G = \left\{ \begin{array}{l} \overrightarrow{\mathcal{R}_{\mathrm{II/I}}} = \mathcal{N}\,\vec{x} + \mathcal{T}_y\,\vec{y} + \mathcal{T}_z\,\vec{z} \\ \overrightarrow{\mathcal{M}_{G(\mathrm{II/I})}} = \mathcal{M}_x\,\vec{x} + \mathcal{M}f_y\,\vec{y} + \mathcal{M}f_z\,\vec{z} \end{array} \right\}_{G,\,r} \qquad [1.20]$$

with the terminology:

$\mathcal{N}$: *normal resultant force* $\mathcal{M}_x$: *longitudinal moment*

$\mathcal{T}_y$: *shear resultant force* $\mathcal{M}f_y$: *bending moment*

$\mathcal{T}_z$: *shear resultant force* $\mathcal{M}f_z$: *bending moment*

1.5.2. *Effects linked to the resultant forces and moments*

○ Displacements

We will not study in detail here the effects of resultant forces and moments[28].

---

28 This study appears in Chapter 9.

Put simply, under these loads a beam deforms, and small displacements of a cross-section beam of abscissa x are characterized by:

• the displacement of the geometric center G of the section: this has the following components in local axes, considered before applying the loading (local coordinate $\vec{x}, \vec{y}, \vec{z}$)[29]:

$$u(x); \qquad v(x); \qquad w(x)$$

• the orientation of the considered section; this is located by the (small) rotation of the section, which has the following components on local axes, considered before applying the loading (local coordinate $\vec{x}, \vec{y}, \vec{z}$):

$$\theta_x(x); \qquad \theta_y(x); \qquad \theta_z(x)$$

These displacements are illustrated in Figure 1.28.

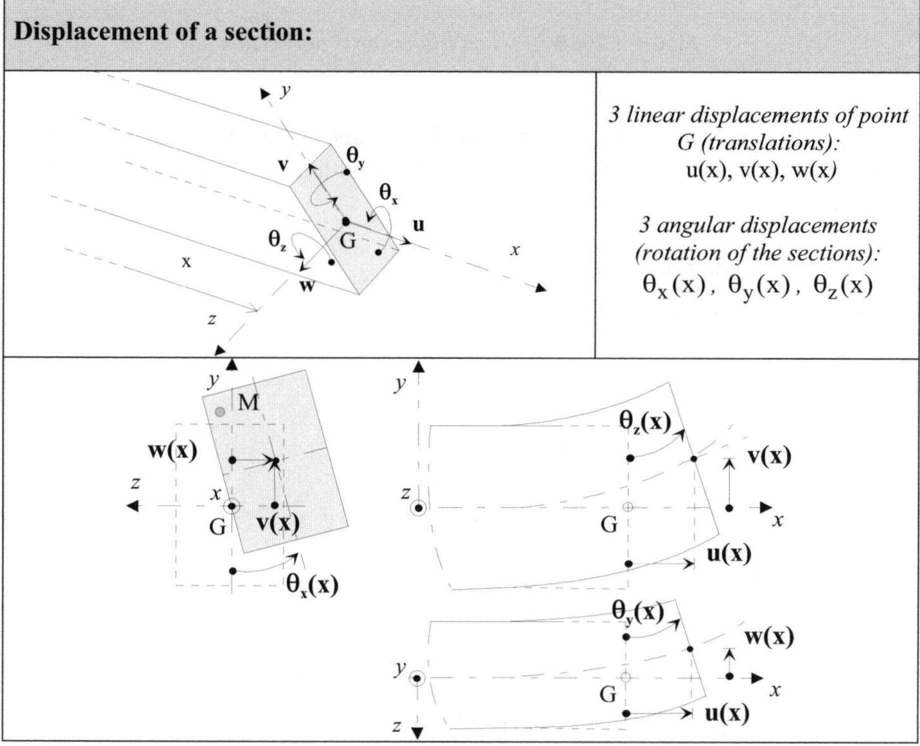

Figure 1.28. *Small displacements of a common cross-section*

---

29 These displacements and rotations are considered to be very small here when compared to the size of the beam. These are small enough that the variation of the shape of the beam remains imperceptible.

○ Stresses

As shown in Figure 1.29, the stress at any point of the cross-section has the components σ_x, τ_{xy}, τ_{xz}.

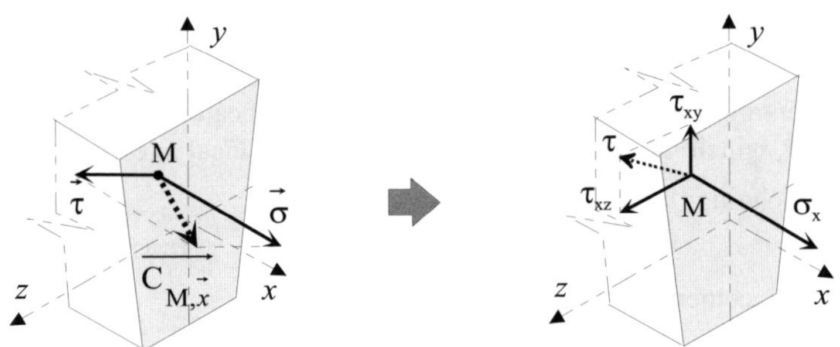

Figure 1.29. *Stresses on a common cross-section*

The following figures sum up in detail for each resultant force and moment [1.20] all the essential results related to the deformations of beams and stresses on cross-sections.

1.5.2.1. *Normal resultant*[30]

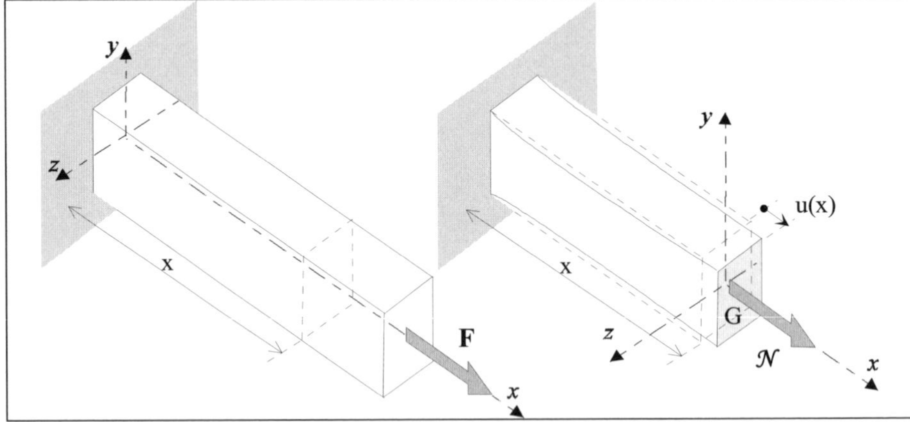

___

30 To develop these results, see Chapter 9, section 9.3.1 and [9.5].

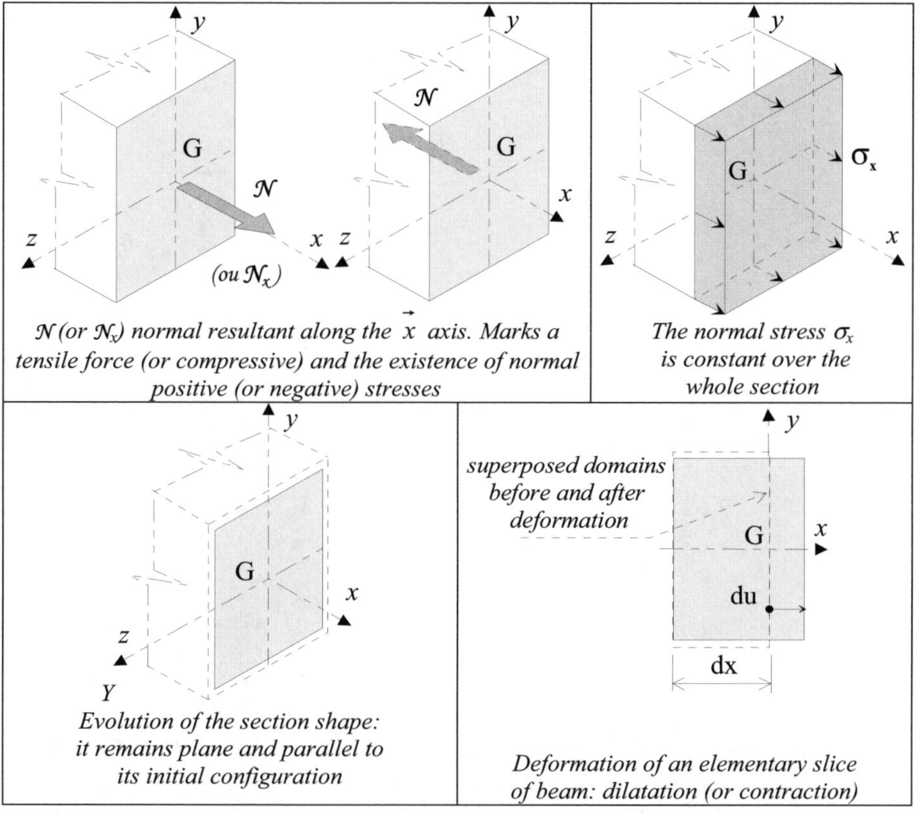

$\mathcal{N}$ (or $\mathcal{N}_x$) normal resultant along the $\vec{x}$ axis. Marks a
tensile force (or compressive) and the existence of normal
positive (or negative) stresses

The normal stress σ_x
is constant over the
whole section

Evolution of the section shape:
it remains plane and parallel to
its initial configuration

superposed domains
before and after
deformation

Deformation of an elementary slice
of beam: dilatation (or contraction)

⇨ *deformation:*

$$\varepsilon_x = \frac{du}{dx}$$

⇨ *behavior relation for normal resultant:*

$$\mathcal{N} = ES\frac{du}{dx}$$

E: elastic modulus; S: area of the cross-section

⇨ *normal stress:*

$$\sigma_x = \frac{\mathcal{N}}{S}$$

Figure 1.30. *Normal resultant $\mathcal{N}$ and its consequences*

1.5.2.2. Shear resultant $\mathcal{T}_y$ [31]

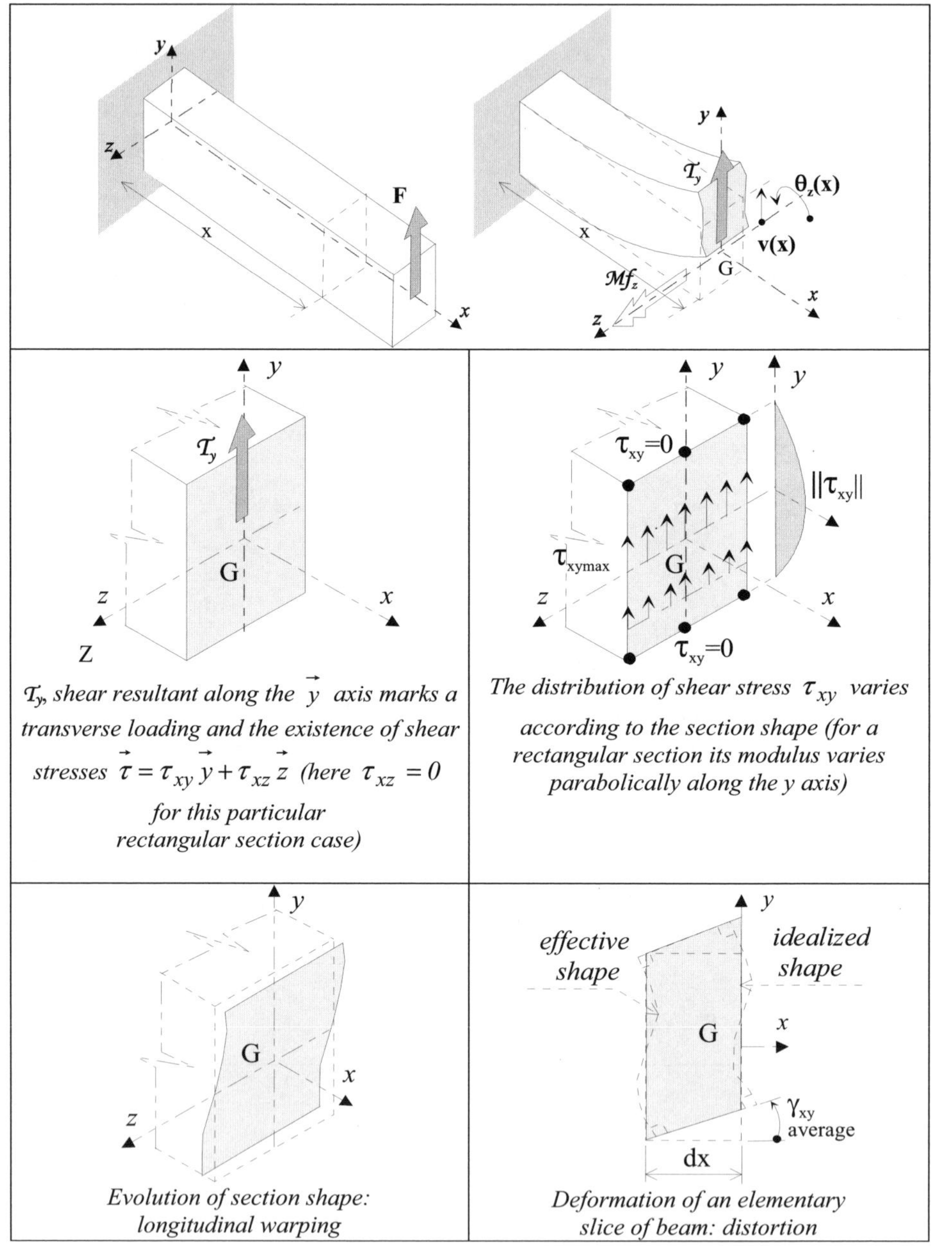

$\mathcal{T}_y$, shear resultant along the $\vec{y}$ axis marks a transverse loading and the existence of shear stresses $\vec{\tau} = \tau_{xy}\,\vec{y} + \tau_{xz}\,\vec{z}$ (here $\tau_{xz} = 0$ for this particular rectangular section case)

The distribution of shear stress τ_{xy} varies according to the section shape (for a rectangular section its modulus varies parabolically along the y axis)

Evolution of section shape: longitudinal warping

Deformation of an elementary slice of beam: distortion

31 For these results, see Chapter 9, section 9.3.4 and [9.59].

⇨ *deformation:*

$$\gamma_{xy_{average}} = \left(\frac{dv}{dx} - \theta_z\right) \; average \; distortion$$

⇨ *behavior relation for shear resultant* $\mathcal{T}_y$:

$$\mathcal{T}_y = G \times k_y \times S\left(\frac{dv}{dx} - \theta_z\right)$$

G: shear modulus; S: area of the cross-section
k_y: *shear coefficient of the section or "coefficient of reduced section" <1*

⇨ *shear stress:*

$$\tau_{xy} = \mathcal{T}_y \times \frac{\partial f}{\partial y} \; ; \qquad \tau_{xz} = \mathcal{T}_y \times \frac{\partial f}{\partial z}$$

where f(y,z) is a function depending on the shape of the section and the direction $\vec{y}$ *of the shear resultant.*

⇨ *Maximum shear stress: located on the* $\vec{z}$ *axis and its value depends on the section shape. Here, for a rectangular section:* $\tau_{xy\max} = \dfrac{3}{2}\dfrac{\mathcal{T}_y}{S}$. [32]

Figure 1.31. *Resultant shear* $\mathcal{T}_y$ *and its consequences*

1.5.2.3. *Shear resultant* $\mathcal{T}_z$ [33]

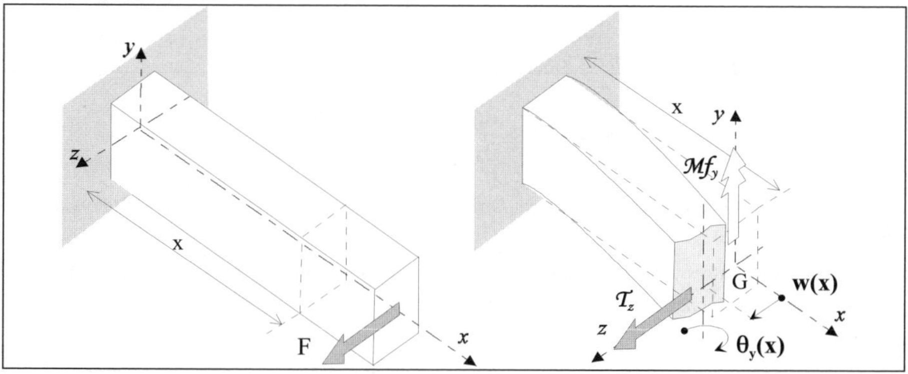

---

32 See also [9.57].
33 For these results, see Chapter 9, section 9.3.4.7 and [9.59].

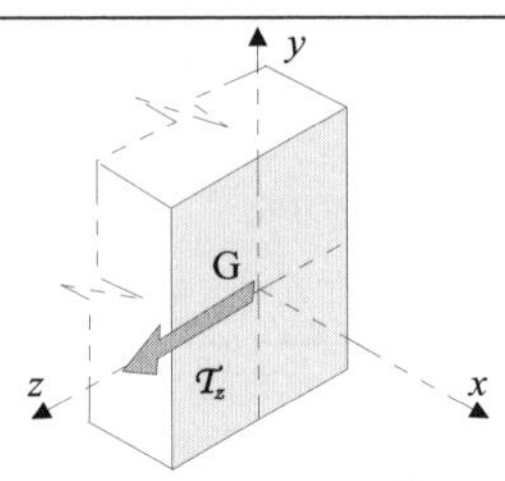

$\mathcal{T}_z$, resultant shear along the $\vec{z}$ axis, marks transverse loading and the existence of shear stress $\vec{\tau} = \tau_{xy}\,\vec{y} + \tau_{xz}\,\vec{z}$ (here $\tau_{xy} = 0$ for this particular rectangular section)

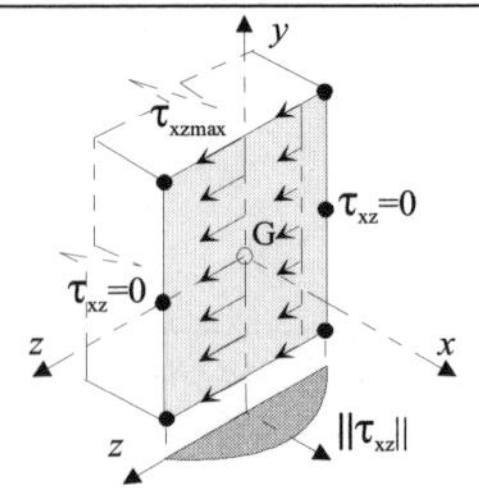

Distribution of shear stress τ_{xz} varies according to the section shape (for a rectangular section, its modulus varies parabolically along the z axis)

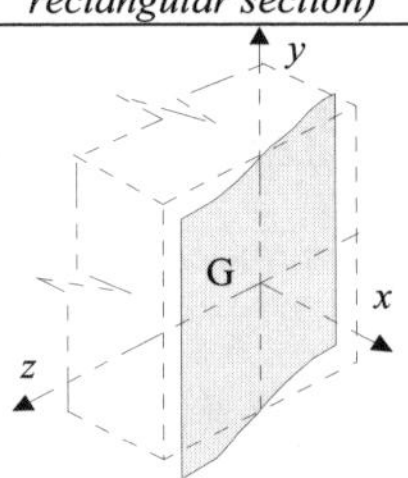

Evolution of section shape: longitudinal warping

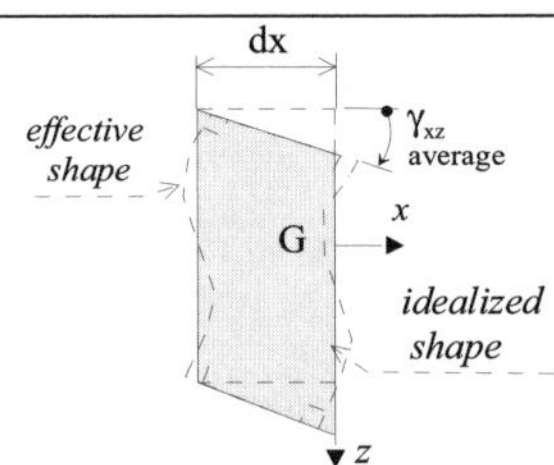

Deformation of an elementary slice of beam: distortion

⇨ *deformation:*

$$\gamma_{xz_{average}} = \left(\frac{dw}{dx} + \theta_y \right) \text{ average distortion}$$

⇨ *behavior relation for the resultant shear* $\mathcal{T}_z$:

$$\mathcal{T}_z = G \times k_z \times S\left(\frac{dw}{dx} + \theta_y \right)$$

G: shear modulus; S: area of the cross-section
k_y: *shear coefficient of the section or "coefficient of reduced section" <1*

⇨ *shear stress:*

$$\tau_{xz} = \mathcal{T}_z \times \frac{\partial g}{\partial z} \; ; \; \tau_{xy} = \mathcal{T}_z \times \frac{\partial g}{\partial y} \,,$$

where g(y,z) is a function depending on the form of the section and the direction $\vec{z}$ *of the resultant shear.*

⇨ *Maximum shear stress: is located on the* $\vec{y}$ *axis and its value depends on the section shape. Here for a rectangular section:*

$$\tau_{xz\,max} = \frac{3}{2}\frac{\mathcal{T}_z}{S}.\,{}^{34}$$

Figure 1.32. *Resultant shear* $\mathcal{T}_z$ *and its consequences*

---

34 See also section 9.3.4.5/*NOTE*.

1.5.2.4. *Torsion moment Mt*

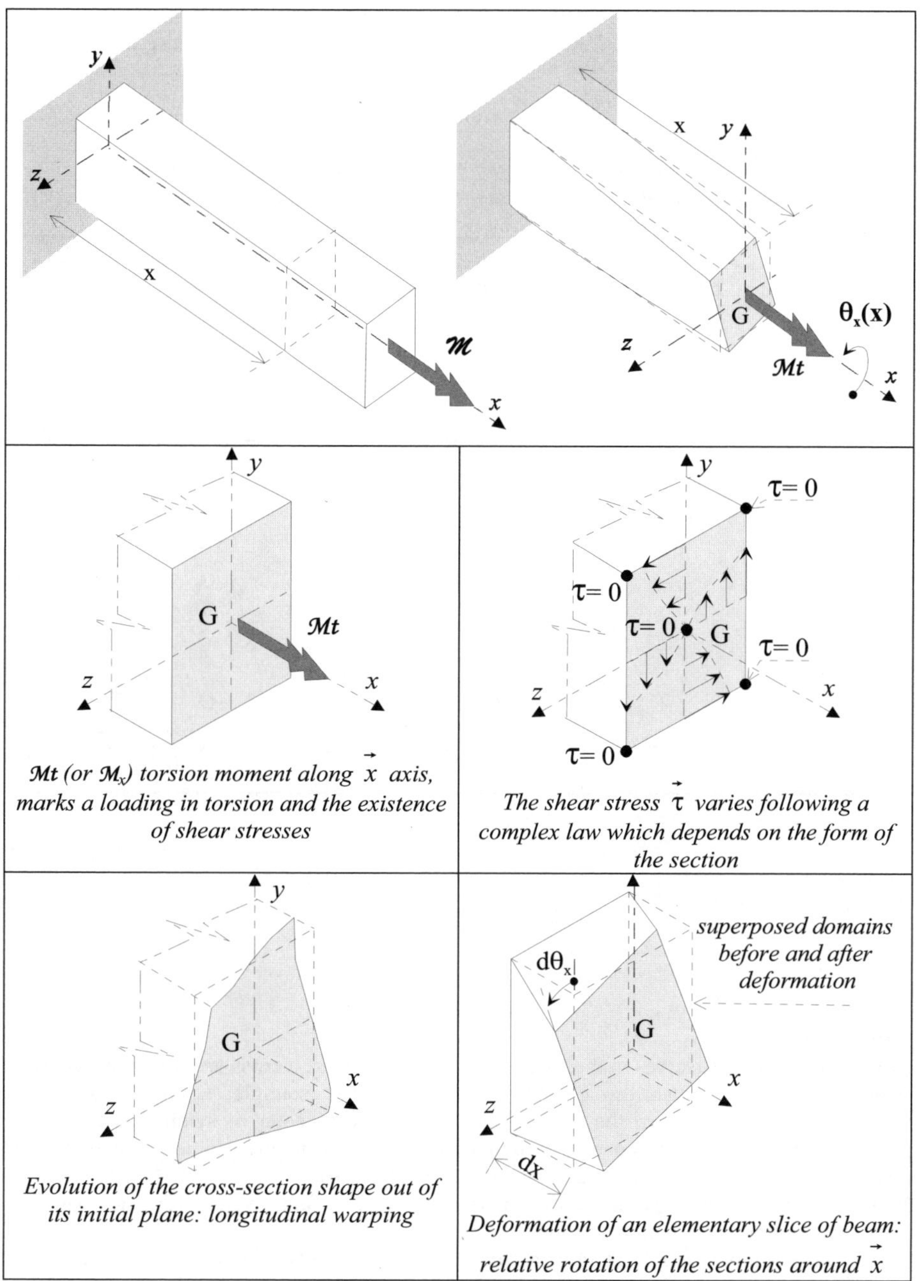

Mt (or M_x) torsion moment along $\vec{x}$ axis, marks a loading in torsion and the existence of shear stresses

The shear stress $\vec{\tau}$ varies following a complex law which depends on the form of the section

Evolution of the cross-section shape out of its initial plane: longitudinal warping

Deformation of an elementary slice of beam: relative rotation of the sections around $\vec{x}$

⇨ deformation:

$$\frac{d\theta_x}{dx} \quad \textit{"rate of rotation" of cross-sections}$$

⇨ behavior relation for the torsion moment[35]:

$$\mathcal{M}t = G \times J \frac{d\theta_x}{dx}$$

G: shear modulus; J: "torsion constant" depending on the shape of the section[36]

⇨ shear stress: distribution of shear stresses is generally calculated numerically, except for some simple geometric sections[37]

Figure 1.33. *Torsion moment* $\mathcal{M}t$ *and its consequences*

1.5.2.5. *Bending moment* $\mathcal{M}f_y$ [38]

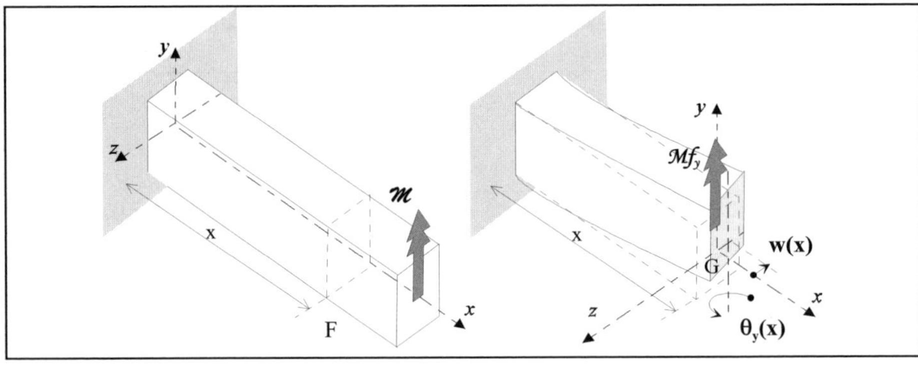

---

35 We are limiting ourselves in the case where the beams have a cross-section with a center of symmetry (which is also the geometric center G), for example: a circular section, a rectangular tube, etc. When the section does not enable a center of symmetry, the torsion rotation of the cross-sections occurs around a point distinct from the geometric center G. This other point is called the center of torsion (see section 9.3.2.4).

36 The calculation principle of J is summarized in [9.27]. Some values for J for particular geometric sections are given in section 9.3.2.5.

37 The calculation principle is summarized [9.27]. For results relating to several particular sections, see section 9.3.2.5.

38 To develop these results, see Chapter 9, [9.33] and [9.55].

<table>
<tr>
<td>

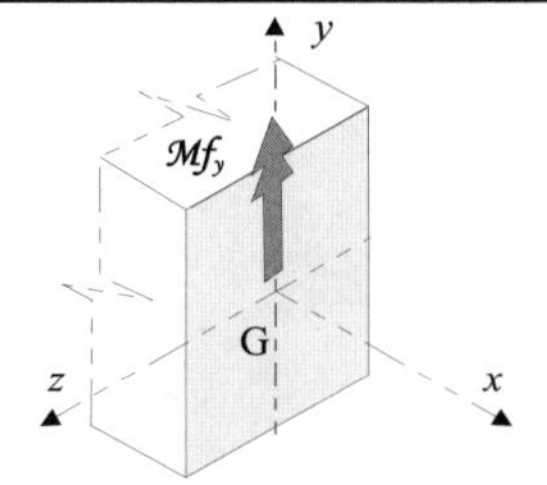

Mf$_y$, bending moment along the $\vec{y}$ axis marks a bending load and the existence of normal stresses

</td>
<td>

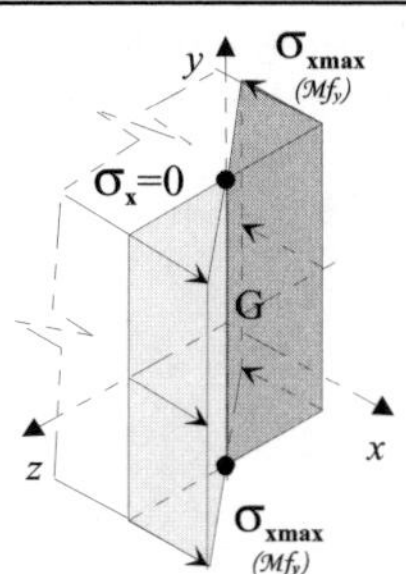

Normal stress σ_x follows a linear law from neutral axis $\vec{y}$

</td>
</tr>
<tr>
<td>

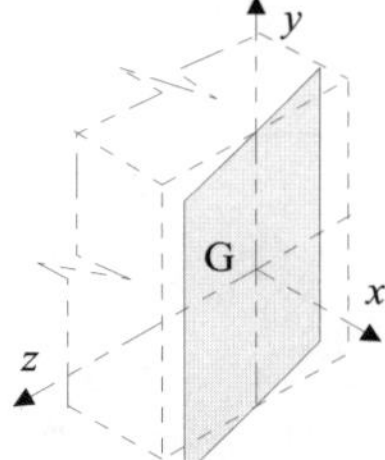

Evolution of the cross-section: it remains plane and pivots around $\vec{y}$ with reference to its initial position

</td>
<td>

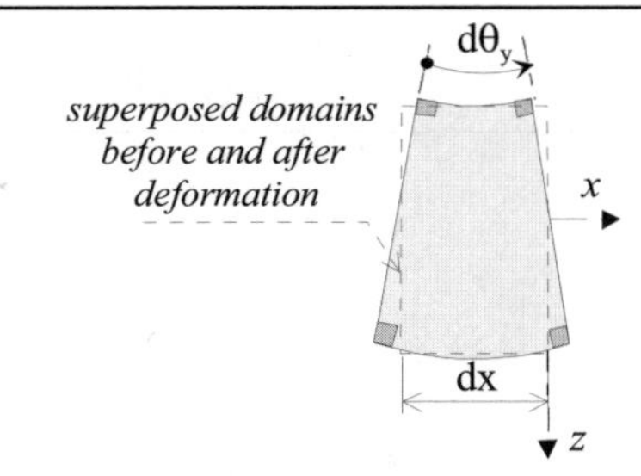

Deformation of an elementary slice of beam

</td>
</tr>
</table>

⇨ *deformation:*

$$\varepsilon_x = z \times \frac{d\theta_y}{dx}$$

⇨ *behavior relation for the bending moment Mf_y:*

$$Mf_y = E \times I_y \frac{d\theta_y}{dx}$$

and within Bernoulli's hypothesis $(\theta_y \cong -\frac{dw}{dx})$:

$$Mf_y = -E \times I_y \frac{d^2w}{dx^2}$$

E: elastic modulus; $I_y = \int_S z^2\, dS$ *principal quadratic moment*

⇨ *normal stress:*

$$\sigma_x = \frac{Mf_y \times z}{I_y}$$

Figure 1.34. *Bending moment Mf_y and its consequences*

1.5.2.6. *Bending moment* Mf_z [39]

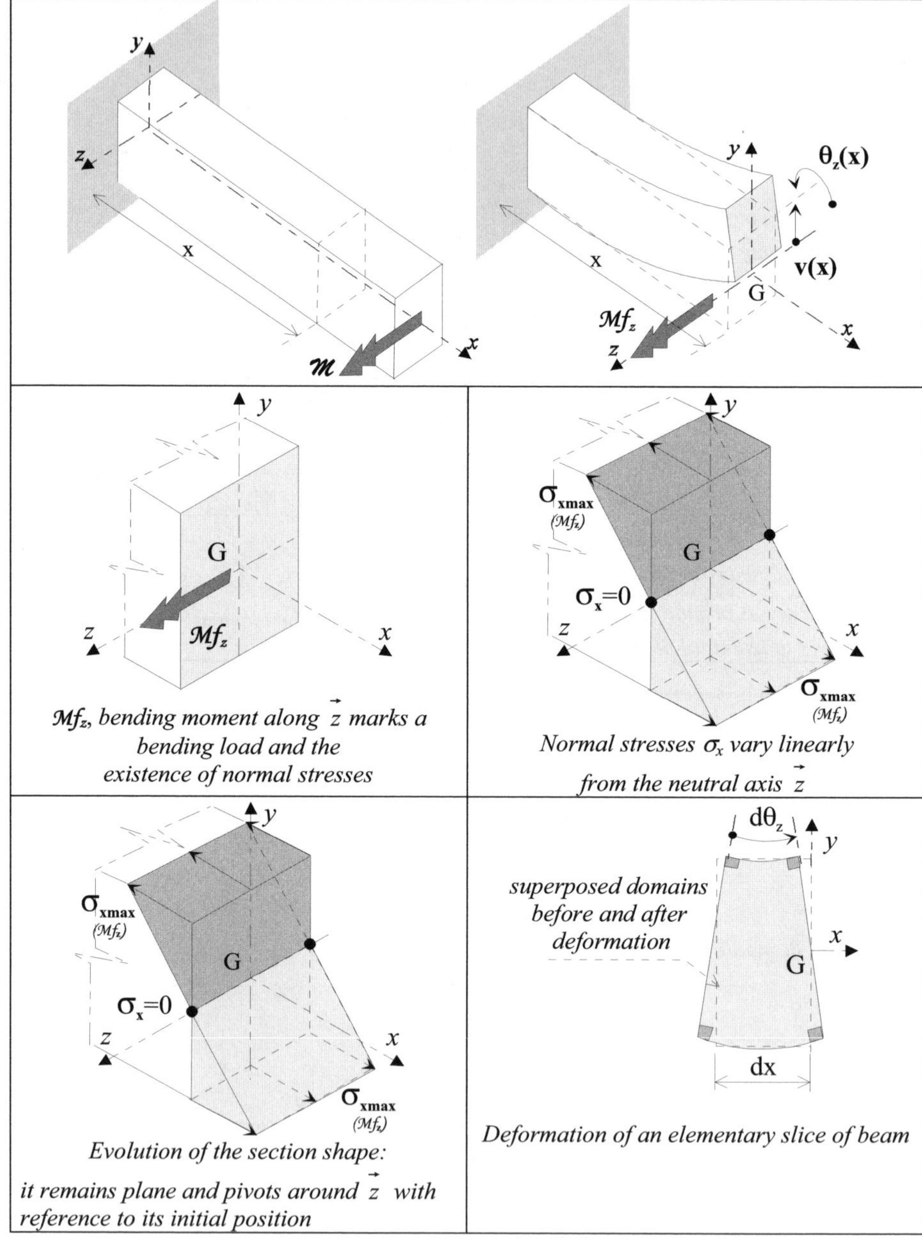

Mf_z, bending moment along $\vec{z}$ marks a bending load and the existence of normal stresses

Normal stresses σ_x vary linearly from the neutral axis $\vec{z}$

Evolution of the section shape: it remains plane and pivots around $\vec{z}$ with reference to its initial position

superposed domains before and after deformation

Deformation of an elementary slice of beam

---

39 To develop these results, see Chapter 9, [9.33] and [9.55].

$\Rightarrow$ *deformation:*

$$\varepsilon_x = -y\frac{d\theta_z}{dx}$$

$\Rightarrow$ *behavior relation for the bending moment $\mathcal{M}f_z$:*

$$\mathcal{M}f_z = E \times I_z \frac{d\theta_z}{dx}$$

or within the Bernoulli's hypothesis ($\theta_z \cong \dfrac{dv}{dx}$):

$$\mathcal{M}f_z = E \times I_z \frac{d^2v}{dx^2}$$

E: elastic modulus; $I_z = \displaystyle\int_S y^2\,dS$ principal quadratic moment

$\Rightarrow$ *normal stress:*

$$\sigma_x = -\frac{\mathcal{M}f_z \times y}{I_z}$$

Figure 1.35. *Bending moment $\mathcal{M}f_z$ and its consequences*

Chapter 2

Mechanical Behavior of Structures: An Energy Approach

We will now use the notions of elasticity from the previous chapter in order to find expressions for the deformation energy stored in an elastic structure under loading. This is a fundamental step which will help us to characterize the behavior under loads of an elastic structure by means of a stiffness matrix and later to explain the notion of discretization of a structure into finite elements with the aim of studying its elastic behavior when loaded.

2.1. Work and energy

2.1.1. *Elementary work developed by a force*

A force $\vec{F}$ produces work when its point of application A becomes displaced (see Figure 2.1). If $\vec{d\ell}$ represents the corresponding displacement vector, assumed to be very small, the produced elementary algebraic work is expressed as:

$$\vec{F} \cdot \vec{d\ell} \tag{2.1}$$

units: F in Newtons (N), $d\ell$ in meters (m) and $\vec{F} \cdot \vec{d\ell}$ in Joules (J).

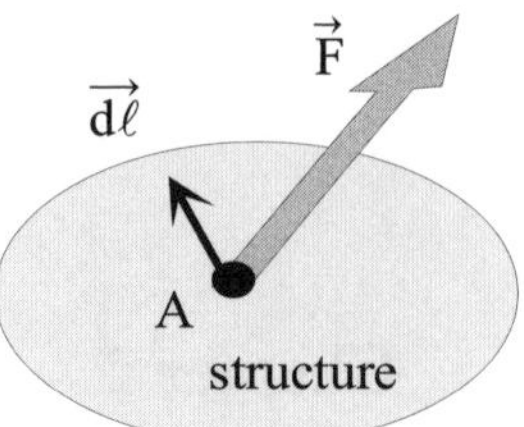

Figure 2.1. *Elementary work of a force:* $\vec{F} \cdot \vec{d\ell}$

2.1.2. *Elementary work developed by a moment*

Let $\overrightarrow{\mathcal{M}}_{/A}$ be a moment defined at a point A of an undeformable surface domain ΔS (see Figure 2.2). During an elementary rotation $\overrightarrow{d\Theta}$ of the domain, the moment produces an elementary algebraic moment:

$$\overrightarrow{\mathcal{M}}_{/A} \cdot \overrightarrow{d\Theta}^{\,1}$$

[2.2]

units: $\mathcal{M}_{/A}$ in Newton $\times$ meter (N.m), $d\Theta$ in radians (rd) and $\overrightarrow{\mathcal{M}}_{/A} \cdot \overrightarrow{d\Theta}$ in Joules (J).

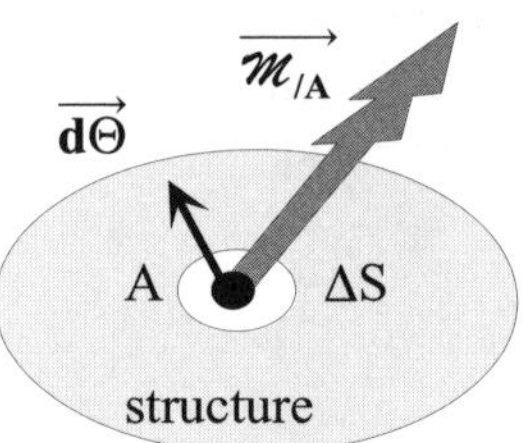

Figure 2.2. *Elementary work of a moment:* $\overrightarrow{\mathcal{M}}_{/A} \cdot \overrightarrow{d\Theta}$

---

1 To define the work of a force, we only need the displacement of the point of application. On the other hand, to define the work of a moment, we need to take into account a material domain around the point where the moment is considered; otherwise the rotation $\overrightarrow{d\Theta}$ of a point would have no physical sense.

2.2. Conversion of work into energy

2.2.1. *Potential energy of deformation*

Let us consider in Figure 2.3a a bar assumed to be homogenous and linearly elastic (see section 1.3). It is fixed at B and subjected to a traction $\overrightarrow{F_{ext}}$ at A. This force passes gradually from an initial value $\overrightarrow{F_{1ext}}$ (the initial state being denoted as 1) to a final value $\overrightarrow{F_{2ext}}$ (the final state being denoted as 2 in Figure 2.3d). The bar becomes longer, and the work of the force $\overrightarrow{F_{ext}}$ appears in the form of an accumulation of basic works[2] of type [2.1].

Let[3]:

$$W_{ext1-2} = \int_{state\ 1}^{state\ 2} \left(\overrightarrow{F_{ext}} \cdot \overrightarrow{d\ell} \right)$$

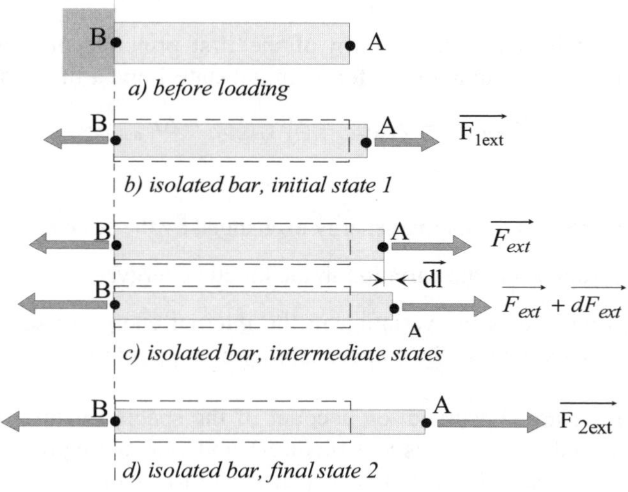

Figure 2.1. *Elastic bar*

---

2 Please note that the elementary work can be expressed indiscriminately:

$$\overrightarrow{F_{ext}} \cdot \overrightarrow{d\ell} \cong \left(\overrightarrow{F_{ext}} + \overrightarrow{dF_{ext}} \right) \cdot \overrightarrow{d\ell}$$

since the second order term $\overrightarrow{dF_{ext}} \cdot \overrightarrow{d\ell}$ can be neglected.

3 In Figure 2.5 where the bar is isolated, please note that the point B is fixed. Thus, the force at B, required to ensure the equilibrium of the bar, does not produce any work.

Taking the elastic nature of the material into account (see section 1.3), the corresponding work denoted as W_{ext1-2} is fully converted into energy stored within the bar, which we will call a variation of potential elastic (or deformation) energy ΔE_{pot1-2}. It is referred to as potential energy since it represents a recoverable energy reserve.

The principle of energy conservation is here expressed as:

$$W_{ext1-2} = \Delta E_{pot.\ of\ deformation\ 1-2} = E_{pot.2} - E_{pot.1} \qquad [2.3]$$

NOTE

❑ The above can be applied to any complex elastic structure with a loading that involves an initial state and a final state. We then pass from the initial to the final state by a sequence of successive states in equilibrium.

❑ In practice, when sizing a structure, state 1 is the state before loading. The structure is at that time free of any loading, and we can consider its initial potential elastic energy as zero. The previous expression can then be re-written as:

$$W_{ext1-2} = E_{pot.2} \quad \text{or more simply} \quad W_{ext1-2} = E_{pot} \qquad [2.4]$$

❑ We thus obtain a simplified form of the first principle of thermodynamics applied to a material system, that is for an initial state 1 and a final state 2:

$$W_{ext.1-2} + Q_{ext1-2} = \Delta E_{kine.1-2} + \Delta E_{init.1-2}$$

where:

– $Q_{ext.1-2}$ represents the thermal energy exchanged by the system with the outside;

– $\Delta E_{kine.1-2}$ represents the variation in the kinetic energy;

– $\Delta E_{init.1-2}$ represents the variation in the internal energy (associated with the elastic nature of a structure, the compressibility of a gas, etc.).

As far as we are concerned, on account of the specific nature of our area of study (the special case of elastic structures) and our assumptions (*progressive* application of the load, absence of energy dissipation) we obtain:

$$\Delta E_{kine.1-2} = 0 \qquad \text{and} \qquad Q_{ext1-2} = 0$$

We then see relation [2.3] with $\Delta E_{init.1-2} = E_{pot.2} - E_{pot.1}$.

2.2.2. *Potential energy for a spring*

Figure 2.4a represents a helicoidal spring, unloaded or in the neutral state, with its left end connected to a support. In Figure 2.6b we show an equilibrium position of the spring deformed under a tensile force $\vec{F}$.

To any value of stress $\vec{F} = F\vec{x}$ corresponds an elongation x, and for two values of the force F_i and F_j, the linear elastic nature of the constituent material leads us to the property:

$$\frac{F_i}{x_i} = \frac{F_j}{x_j} = C^{te} = k$$

We can then define what we refer to as the "behavior relationship" of the spring as:

$$x = \frac{1}{k} \times F \quad \Leftrightarrow \quad F = k \times x \qquad [2.5]$$

where k is the rigidity or stiffness of the spring and x its elongation measured from the neutral position[4].

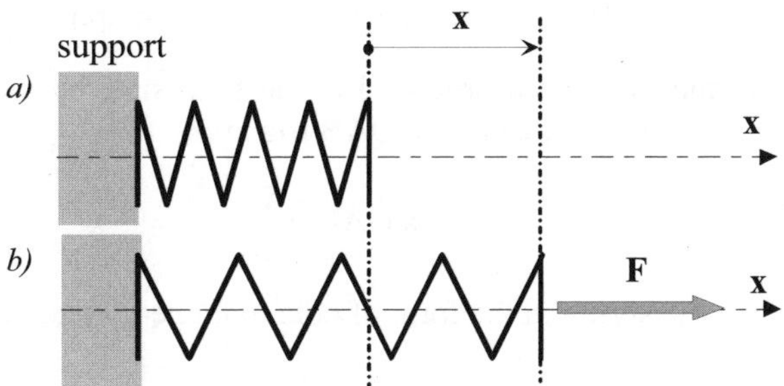

Figure 2.4. *Behavior of a helicoidal spring*

○ Analytical expression of work W_{ext} produced by the force $\vec{F}$

For an elementary displacement *dx* and taking into account expression [2.1]:

---

4 For a helicoidal spring, we can show that: $k = \dfrac{G \times d^4}{8 \times n \times D^3}$, a formula in which we note: the nature of the material (shear modulus G), the geometry of the spring (diameter of the wire d, winding diameter D, number of useful coils n).

$$\vec{F} \cdot dx\vec{x} = F \times dx = k \times x \times dx = dW_{ext}$$

where F and x (in *italics*) represent the intermediary values of the force and the elongation of the spring.

Since state 1, as mentioned earlier, refers to the state before loading and state 2 corresponds to the elongation "x" we will obtain:

$$W_{ext.1-2} = \int_0^x k \times x \times dx = \frac{k}{2} x^2$$

and taking into account [2.4], the potential deformation energy (energy accumulated in the spring) is also expressed as:

$$E_{pot.} = \frac{1}{2} kx^2 \qquad\qquad [2.6]$$

○ Graphical interpretation of the work produced by the force

In the "force–elongation" diagram which reflects the behavior relation [2.5] of the spring (see Figure 2.7), we can see that for an increase in dx of the elongation of the spring, the elementary work produced $\vec{F}$ is equal to:

$$dW_{ext} = \vec{F} \cdot dx\vec{x} = F \times dx \cong area\, "mnpq"$$

and we thus obtain the total work produced at the final state by the summation of these elementary areas, i.e. the area of the OAB triangle:

$$W_{ext.1-2} = area\, OAB = \frac{1}{2} F \times x = \frac{k}{2} x^2$$

fully accumulated in the spring in the form of potential energy (see [2.6]).

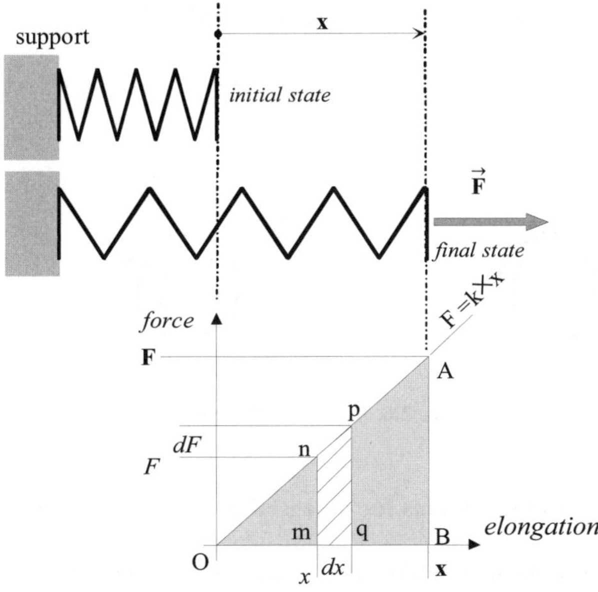

Figure 2.5. *Force-elongation diagram of a helicoidal spring*

NOTE

❑ The potential deformation energy stored in a spiral spring loaded by an axial moment $\overrightarrow{\mathcal{M}} = \mathcal{M}\vec{z}$ can be expressed in a similar manner (see Figure 2.6).

The behavior relationship of such a spring can in fact be written as:

$$\Theta = \frac{1}{k}\times\mathcal{M} \quad \Leftrightarrow \quad \mathcal{M} = k\times\Theta \tag{2.7}$$

which we will compare with expression [2.5].

With expression [2.2], we obtain the basic work:

$$dW_{ext} = M\times d\Theta$$

where M and $d\Theta$ (in *italics*) are the intermediate values of the moment and of the elementary rotation of the spiral spring.

Thus, the total work produced between the states 1 and 2 is:

$$W_{ext.1-2} = \frac{1}{2}\mathcal{M}\times\Theta = \frac{k}{2}\times\Theta^2$$

the potential deformation energy is therefore expressed as:

$$E_{\text{pot.}} = \frac{1}{2} k \times \Theta^2$$

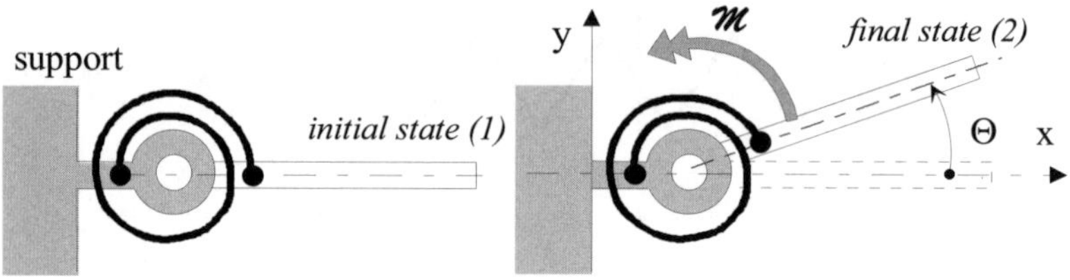

Figure 2.6. *Behavior of a spiral spring*

2.3. Some standard expressions for potential deformation energy

This notion of potential deformation energy is essential for understanding what follows. We will see in fact that it is closely linked to the principle of the method of finite elements which we will develop later. It is for this reason that we will now focus on finding expressions for potential deformation energy for some typical cases of loaded structures.

2.3.1. *Deformation energies in a straight beam*

2.3.1.1. Traction (or compression)

○ Case of the whole beam (macroscopic scale)

In Figure 2.7, we have a beam of length ℓ, with cross-section S. It is fixed at one end and subjected at the other end to an external force $\vec{F} = F\vec{x}$ applied *progressively* from the zero value to the value F. The corresponding elongation of the beam is denoted as u_ℓ.

An intermediate increase dF in the force provokes an increase in du_ℓ of the elongation u_ℓ (see Figure 2.8). During the elongation du_ℓ, the work of the intermediate force F is approximately equal to $F \times du_\ell$ (area of the dotted trapezium "mnpq" in Figure 2.8). The total work produced by the external force is represented as the sum of these elementary areas, i.e. the area of the OAB triangle. We have:

$$W_{ext1-2} = \frac{1}{2}F \times u_\ell$$

from which (see [2.4]):

$$E_{pot.} = \frac{1}{2}F \times u_\ell \qquad [2.8]$$

We can note here that using Hooke's law we obtain (see [1.2]):

$$u_\ell = \frac{1}{E} \times \frac{F}{S} \times \ell$$

then [2.8] can be re-written as:

$$E_{pot.} = \frac{1}{2}\frac{F^2 \times \ell}{E \times S} \qquad [2.9]$$

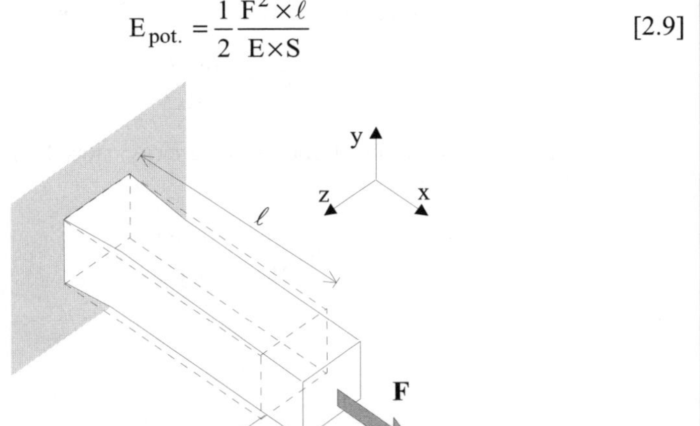

Figure 2.7. *Beam under traction*

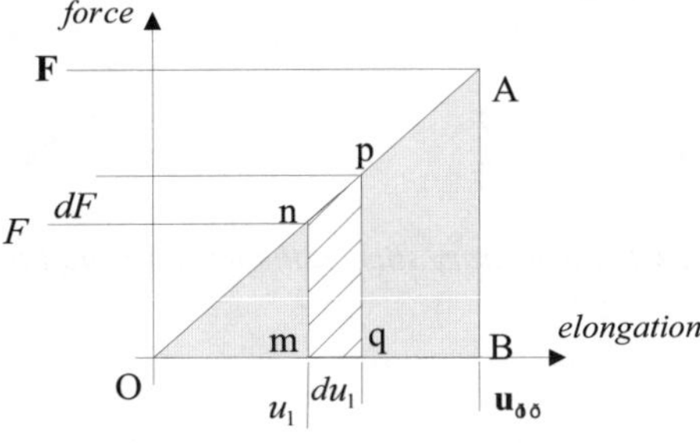

Figure 2.8. *Load-displacement diagram*

○ Case of an elementary slice of beam (mesoscopic scale)

Let us take an element of the previous beam, bound by two cross-sections spaced apart by a length dx (see Figure 2.2). Then as per [1.19] and [1.20] the resultant force and moment of cohesion on the section with outgoing normal $\vec{x}$ limiting the considered beam slice is reduced to the component:

$$\mathcal{N} = F$$

The elongation of this slice is denoted "du". Relation [2.8] becomes on the scale of the beam slice:

$$dE_{pot.\mathcal{N}} = \frac{1}{2}\mathcal{N} \times du(x) \qquad [2.10]$$

We have mentioned (see section 1.5.2.1) the behavior relation for normal force:

$$\mathcal{N} = E \times S \times \frac{du(x)}{dx}$$

Thus, the expression for deformation energy stored in the beam element dx becomes:

$$dE_{pot.\mathcal{N}} = \frac{1}{2}\frac{\mathcal{N}^2}{E \times S} \times dx \qquad [2.11]$$

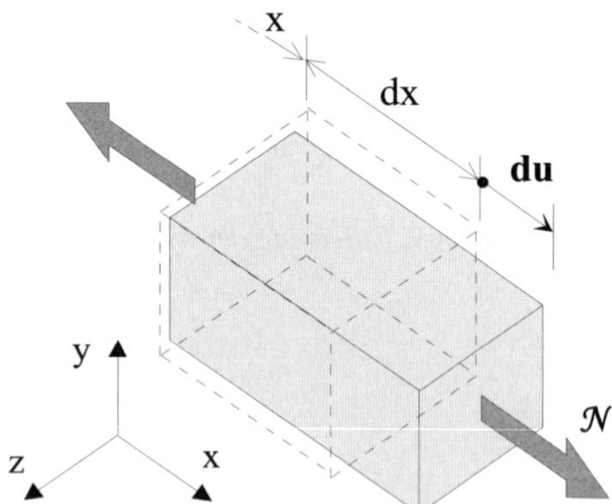

Figure 2.2. *Elementary slice of the beam in equilibrium*

NOTE

❏ Relation [2.11] is valid locally, and thus extensible to beams where, unlike the previous beam of length ℓ uniformly stretched, the normal force is likely to vary with x, i.e. $\mathcal{N} = \mathcal{N}(x)$. Thus, the deformation energy stored in any beam can be obtained from [2.11] under the more general form:

$$E_{\text{pot.}\mathcal{N}} = \frac{1}{2} \int_0^{\ell} \frac{\mathcal{N}^2}{E \times S} \times dx \qquad [2.12]$$

❏ In view of the subsequent use of relation [2.12] for the finite element method, we can express this potential according to the deformation $\varepsilon_x = \dfrac{du(x)}{dx}$ [5]. Then

relation $\mathcal{N} = E \times S \times \dfrac{du(x)}{dx}$ immediately gives us:

$$E_{\text{pot.}\mathcal{N}} = \frac{1}{2} \int_0^{\ell} E \times S \times \left(\frac{du(x)}{dx}\right)^2 \times dx \qquad [2.13]$$

❏ Special case: if $\mathcal{N}(x) = C^{te} = F$ with constant characteristics E and S along the beam, relation [2.12] gives us:

$$E_{\text{pot.}\mathcal{N}} = \frac{1}{2} \frac{F^2}{E \times S} \int_0^{\ell} dx$$

i.e.:

$$E_{\text{pot.}\mathcal{N}} = \frac{1}{2} \frac{F^2 \times \ell}{E \times S} \qquad [2.14]$$

we thus come back to expression [2.9] established for a beam of finite length.

2.3.1.2. *Torsion*

Here, we will restrict ourselves to cases of beams whose cross-section accepts a symmetric center (which is also the geometric center G), for example a circular section, a rectangular tube, etc.[6]

---

5 See section 1.5.2.1.

6 When the section does not accept a symmetric center, the rotation of torsion of the cross-sections takes place around a point distinct from the geometric center G. This other specific point is the center of torsion (see section 9.3.2.4).

○ Case of the whole beam (macroscopic scale)

Figure 2.12 shows a beam of length ℓ with one end clamped and with the other end subjected to an external moment $\overrightarrow{\mathcal{M}} = \mathcal{M}\vec{x}$ applied *progressively* from the zero value to the value $\mathcal{M}$. We denote as θ_x the angle of rotation of the end section around the axis $\vec{x}$.

An intermediate increase dM of the moment causes an increase $d\theta_x$ of the angle of rotation $\theta_x(x)$. During the rotation $d\theta_x$ the work of the intermediate moment is approximately equal to $M \times d\theta_x$ (dotted area "mnpq" in Figure 2.11). The total work produced by the external moment is represented by the sum of these elementary areas, i.e. the area of the OAB triangle:

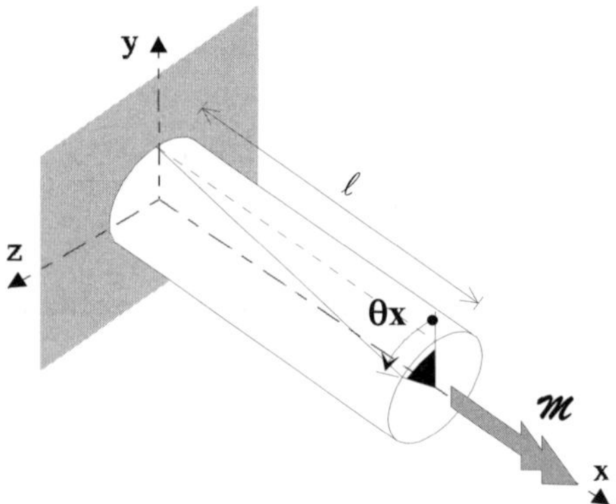

Figure 2.10. *Beam under torsion*

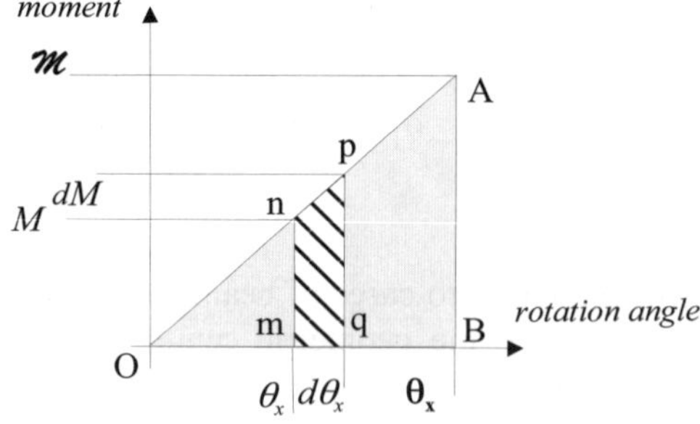

Figure 2.11. *Load-displacement diagram*

$$W_{\text{ext.1-2}} = \frac{1}{2}\mathcal{M} \times \theta_x$$

or (see [2.4]):

$$E_{\text{pot.}} = \frac{1}{2}\mathcal{M} \times \theta_x \qquad [2.15]$$

○ Case of an elementary slice of beam (mesoscopic scale)

Let us take a slice of the previous beam of length dx. Thus, according to [1.19] and [1.20] the resultant force and moment of cohesion on the section with outgoing normal $\vec{x}$ limiting the considered beam slice is reduced to the unique component:

$$\mathcal{M}_x = \mathcal{M}$$

which we will call "torsional moment" [7], i.e. $\mathcal{M}t = \mathcal{M}_x$ (see Figure 2.12), for the circular section of the beam considered here. As shown in Figure 2.12, the relative rotation of the two sections limiting the beam slice is denoted "$d\theta_x$".

Relation [2.15] considered on the scale of the beam slice becomes:

$$dE_{\text{pot.}\mathcal{M}t} = \frac{1}{2}\mathcal{M}t \times d\theta_x \qquad [2.16]$$

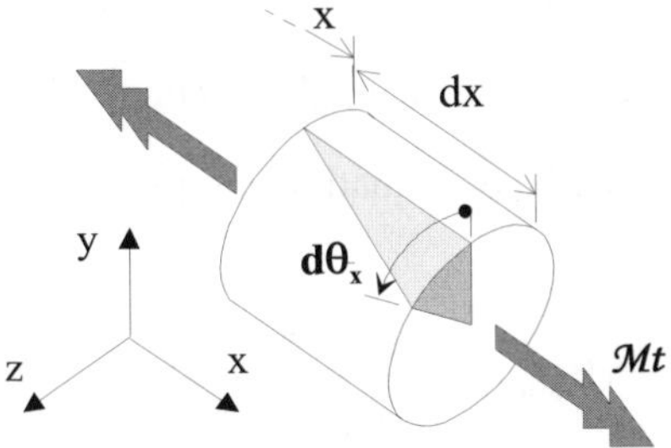

Figure 2.12. *Elementary beam slice in equilibrium*

We have mentioned earlier (see section 1.5.2.4) the behavior relationship associated with the torsional moment:

---

7 The same applies to all shapes of sections having a center of symmetry. See section 9.3.2.4, [9.21]: case of cross-sections where the longitudinal moment $\mathcal{M}_x$ coincides with the torsional moment $\mathcal{M}t$.

$$\mathcal{M}t = G \times J \frac{d\theta_x(x)}{dx}$$

where J is the torsional constant characteristic of the shape of the section.

Hence the expression for the deformation energy stored in the beam element dx becomes:

$$dE_{pot.\,\mathcal{M}t} = \frac{1}{2}\frac{\mathcal{M}t^2}{G \times J} \times dx \qquad [2.17]$$

NOTES

❑ Relation [2.17] is valid locally, and is therefore applicable to beams where the torsional moment is likely to vary with x, i.e. $\mathcal{M}t = \mathcal{M}t(x)$[8], unlike the previous simplified example concerning the whole beam.

In this case, the deformation energy stored in the whole beam during torsion can be obtained from [2.17] as:

$$E_{pot.\,\mathcal{M}t} = \frac{1}{2}\int_0^{\ell}\frac{\mathcal{M}t^2}{G \times J} \times dx \qquad [2.18]$$

❑ Special case: when the beam has a circular section, the torsional constant J is equal to I_0, polar quadratic moment of the section[9] ($I_0 = \frac{\pi d^4}{32}$ for a section of diameter d). Then the potential deformation energy by torsion is expressed as:

$$E_{pot.\,\mathcal{M}t} = \frac{1}{2}\int_0^{\ell}\frac{\mathcal{M}t^2}{G \times I_0} \times dx \qquad [2.19]$$

---

8 We must bear in mind that the standard results of the torsion of beams are established by assuming a constant torsional moment i.e. independent of the abscissa x (see section 9.3.2). For non-circular sections, when the torsional moment is likely to vary along the mean line of the beam, phenomena other than standard torsion phenomena are produced, especially the appearance of normal balanced stresses on cross-sections. These phenomena are neglected for traditional design of beam structures. The only case where these "complications" disappear is when the section of the beam is circular (it does not warp). We can then apply without reservation a torsional moment variable with x.

9 We have mentioned in Chapter 9, section 9.3.2.5 the values of the torsional constant J for a few section shapes.

❒ In view of the subsequent use of relation [2.18] for the finite element method, the torsional potential can be expressed as a function of $\dfrac{d\theta_x(x)}{dx}$. Then relation $Mt = G \times J \dfrac{d\theta_x(x)}{dx}$ leads us immediately with [2.17] to:

$$E_{pot.\,Mt} = \frac{1}{2} \int_0^\ell G \times J \times \left(\frac{d\theta_x(x)}{dx} \right)^2 \times dx \qquad [2.20]$$

❒ Special case: if $M_t(x) = C^{te} = \mathcal{M}$ with constant G and J characteristics along the beam, relation [2.19] leads us to:

$$E_{pot.\,Mt} = \frac{1}{2} \frac{Mt^2}{G \times J} \int_0^\ell dx$$

i.e.:

$$E_{pot.\,Mt} = \frac{1}{2} \frac{\mathcal{M}^2 \times \ell}{G \times J}$$

2.3.1.3. *Pure bending (xy plane)*

○ Case of the whole beam (macroscopic scale)

Let us consider in Figure 2.13 a beam of length ℓ, with a symmetry plane (xy), with one end clamped and the other subjected to an external moment $\overrightarrow{\mathcal{M}} = \mathcal{M}\,\vec{z}$, applied *progressively* from the zero value till the value $\mathcal{M}$. We denote as θ_z the angle through which this end section turns around the $\vec{z}$ axis. An intermediate increase of the moment dM causes an increase $d\theta_z$ of the angle of rotation (see Figure 2.16). During the rotation $d\theta_z$ the work of the intermediate moment M is equal to $M \times d\theta_z$ (which corresponds approximately to the dotted area "mnpq" in Figure 2.14).

The total work done by the external moment is represented by the sum of these elementary areas, i.e. of the OAB triangle, i.e.:

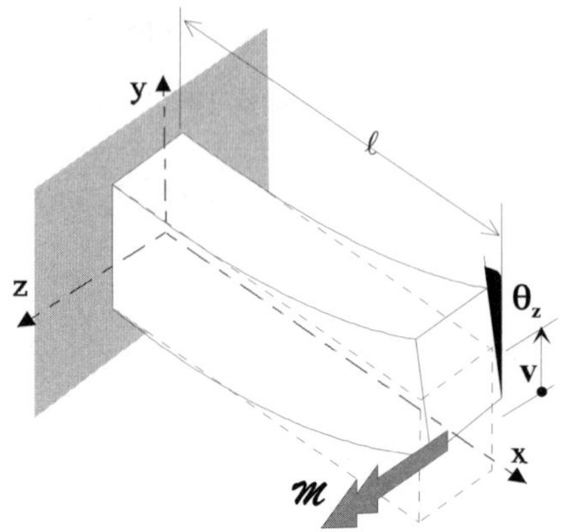

Figure 2.13. *Beam subjected to pure bending*

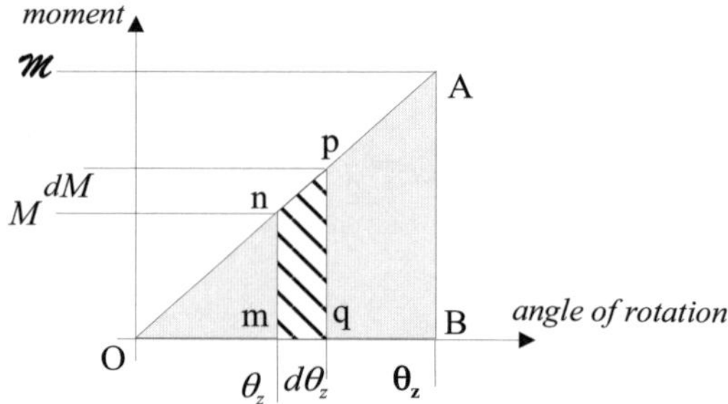

Figure 2.14. *Load-displacement diagram*

$$W_{\text{ext.}1-2} = \frac{1}{2}\mathcal{M} \times \theta_z$$

or even (see [2.4]):

$$E_{\text{pot.}} = \frac{1}{2}\mathcal{M} \times \theta_z \qquad [2.21]$$

○ Case of an elementary slice of beam (mesoscopic scale)

Let us take an element of length dx of the previous beam, (see Figure 2.15).

Thus, according to [1.19] and [1.20], the resultant force and moment of cohesion on section $\vec{x}$ limiting the considered beam slice is reduced to a unique component:

$$\mathcal{M}f_z = \mathcal{M}$$

which is a bending moment.

As indicated in Figure 2.15, the relative rotation of the two sections limiting the elementary beam slice is denoted as "$d\theta_z$" (see Figure 1.35).

Relation [2.21], considered at the scale of the beam slice, becomes:

$$dE_{pot.bending} = \frac{1}{2} \mathcal{M}f_z \times d\theta_z$$

We have mentioned earlier (see section 1.5.2.6) the behavior relationship with respect to the bending moment:

$$\mathcal{M}f_z = E \times I_z \frac{d\theta_z}{dx} \qquad [2.22]$$

where I_z is the quadratic moment of the section with respect to the main axis $\vec{z}$.

Then the deformation energy stored in the beam element dx becomes:

$$dE_{pot.bending} = \frac{1}{2} \frac{\mathcal{M}f_z{}^2}{E \times I_z} \times dx \qquad [2.23]$$

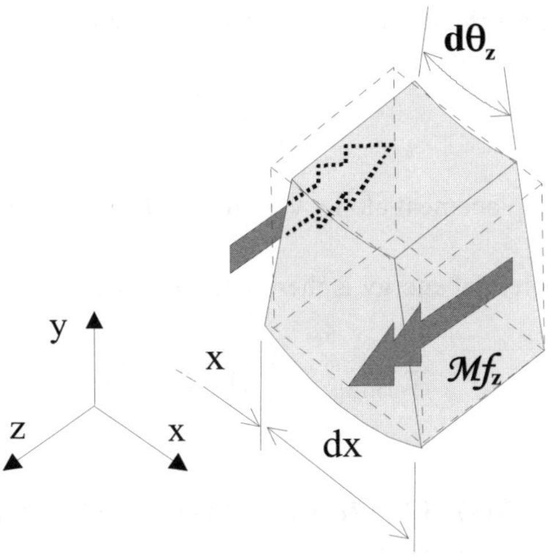

Figure 2.15. *Elementary beam slice in equilibrium*

NOTE

❒ Relation [2.23] is valid locally, and is therefore applicable to beams where the bending moment is likely to vary with x, i.e. $Mf_z = Mf_z(x)$, unlike the previous simplified example concerning the whole beam.

In this case the potential energy of deformation, or "strain potential energy", stored in the whole beam during pure bending can be obtained from [2.23] as:

$$E_{pot.\,Mf_z} = \frac{1}{2}\int_0^\ell \frac{Mf_z^{\,2}}{E\times I_z}\times dx \qquad [2.24]$$

❒ In view of the subsequent use of relation [2.24] for the finite element method, we can express pure bending potential as according to $\dfrac{d\theta_z}{dx}$. Then the behavior relationship $Mf_z = E\times I_z\dfrac{d\theta_z}{dx}$ leads us immediately to:

$$E_{pot.\,Mf_z} = \frac{1}{2}\int_0^\ell E\times I_z\times\left(\frac{d\theta_z}{dx}\right)^2\times dx$$

❒ In pure bending, the cross-sections remain normal to the deformed mean line. Since the displacements are very small on the beam scale (section 1.5.2.6), we obtain:

$$\theta_z \cong \frac{dv}{dx}$$

where v(x) is the displacement along $\vec{y}$ of the center of the section in Figure 2.13.

The previous potential energy is therefore written as:

$$E_{pot.\,Mf_z} = \frac{1}{2}\int_0^\ell E\times I_z\times\left(\frac{d^2v}{dx^2}\right)^2\times dx \qquad [2.25]$$

❒ Special case: if $Mf_z(x)= C^{te} = M$ with constant E and I_z characteristics along the beam, relation [2.24] leads us to:

$$E_{pot.\,Mf_z} = \frac{1}{2}\frac{Mf_z^{\,2}}{E\times I_z}\int_0^\ell dx$$

i.e.:

$$E_{\text{pot.}\,\mathcal{M}f_z} = \frac{1}{2}\frac{\mathcal{M}^2 \times \ell}{E \times I_z}$$

which we can compare with expression [2.21]. We can therefore see that the rotation of the end section of the beam equals: $\theta_z = \dfrac{\mathcal{M} \times \ell}{EI_z}$

❑ By studying pure bending in the plane (x,z) with the relationship given in section 1.5.2.5, we would obtain similar results, such as:

$$E_{\text{pot.}\,\mathcal{M}f_y} = \frac{1}{2}\int_0^\ell \frac{\mathcal{M}f_y^{\;2}}{E \times I_z}\,dx = \frac{1}{2}\int_0^\ell E \times I_y \times \left(\frac{d\theta_y}{dx}\right)^2 \times dx = \frac{1}{2}\int_0^\ell E \times I_y \times \left(\frac{d^2 w}{dx^2}\right)^2 \times dx$$

2.3.1.4. *Plane bending (xy plane)*

In the previous section, we assumed that the only load on the beam in Figure 2.13 would be reduced to a pure moment $\overrightarrow{\mathcal{M}} = \mathcal{M}\,\vec{z}$. In reality, beams are most often subjected to transverse forces. Let us then assume a load on the previous beam, (as per Figure 2.16), made up of a force $\vec{F} = F\vec{y}$ and a moment $\overrightarrow{\mathcal{M}} = \mathcal{M}\,\vec{z}$. Let us take at the abscissa x a beam element of length dx. Then according to [1.20] the resultant force and moment of cohesion on the section with outgoing normal $\vec{x}$ of the considered beam element now has the following components;

♦ a shearing force: $\mathcal{T}_y = F$

♦ a bending moment: $\mathcal{M}f_z = \mathcal{M} + F(\ell - x)$

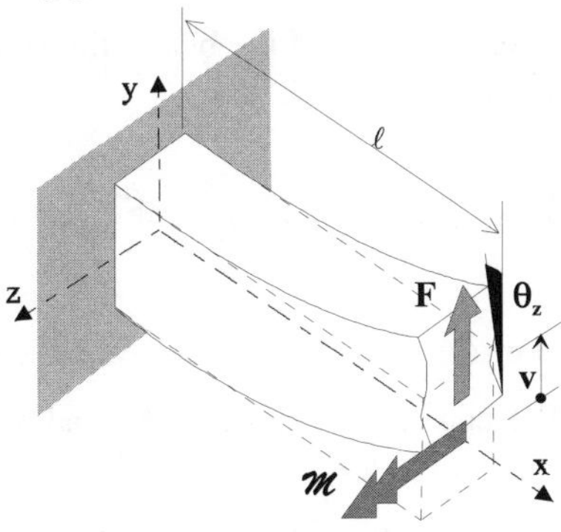

a) beam subjected to plane bending

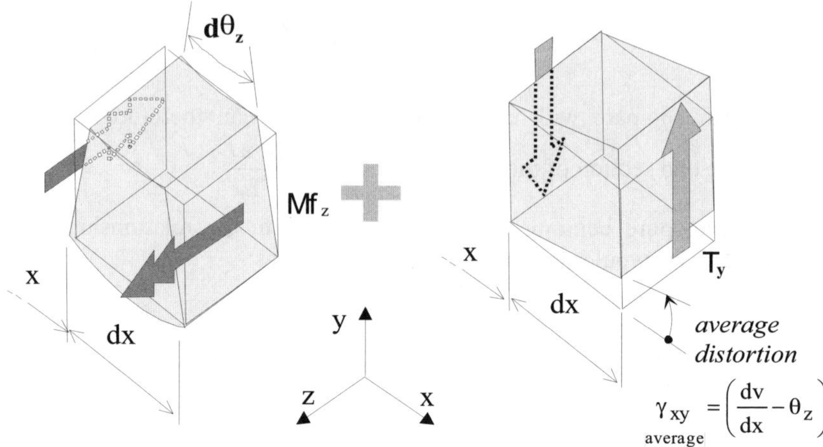

b) stresses on an elementary beam block slice

Figure 2.16. *Plane bending*

The beam bends along a symmetry plane (xy) (plane bending) and the deformation of the element dx (see Figure 2.18b) appears as a result of:

– an elementary rotation $d\theta_z(x)$;

– an average distortion mainly due to the shearing strain (see Figure 1.31). This mean distortion equals $\left(\dfrac{dv}{dx} - \theta_z\right)$, where v(x) is the displacement along $\vec{y}$ of the mean line of the beam (see Figure 1.28).

On the scale of the elementary slice of the beam, the strain potential energy must be completed with respect to expression [2.23]:

$$dE_{pot.\,bending} = dE_{pot.\mathcal{M}f_z} + dE_{pot.\mathcal{T}_y} \qquad [2.26]$$

The first contribution, i.e. $dE_{pot.\mathcal{M}f_z}$, is known (equation [2.23]). Let us evaluate the second, i.e. $dE_{pot.\mathcal{T}_y}$. We can see in Figure 2.16b that due to distortion, the shear force $\mathcal{T}_y$ is displaced by $\left(\gamma_{xy\,average} \times dx\right)$. The linear elastic nature of the material also leads us to an expression of the form:

$$dE_{pot.\mathcal{T}_y} = \frac{1}{2}\mathcal{T}_y \times \left(\gamma_{xy} \times dx \right)_{average}$$ [2.27]

Let us consider the behavior relation for the shear force given in section 1.5.2.2, i.e.:

$$\mathcal{T}_y = G \times k_y \times S\left(\frac{dv}{dx} - \theta_z \right)$$ [2.28]

where S is the area of the cross-section and k_y is the shape coefficient called "shear coefficient"[10]. The strain potential energy can therefore be written as:

$$dE_{pot.\mathcal{T}_y} = \frac{1}{2}\mathcal{T}_y \times \frac{\mathcal{T}_y}{G \times k_y \times S} \times dx$$

i.e.:

$$dE_{pot.\mathcal{T}_y} = \frac{1}{2}\frac{\mathcal{T}_y^{\,2}}{G \times k_y \times S} \times dx$$ [2.29]

In total, for plane bending, the elementary strain potential energy [2.26] can be written with [2.23] and [2.29]:

$$dE_{pot.bending} = \frac{1}{2}\frac{\mathcal{M}f_z^{\,2}}{E \times I_z} \times dx + \frac{1}{2}\frac{\mathcal{T}_y^{\,2}}{G \times k_y \times S} \times dx$$ [2.30]

NOTE

❏ Relation [2.30] is valid locally and is thus applicable to beams where the bending moment and the shear force are likely to vary with the x abscissa. In this case the strain potential energy stored in a beam of length ℓ during plane bending can be obtained from [2.30] in the form:

$$dE_{pot.bending} = \frac{1}{2}\frac{\mathcal{M}f_z^{\,2}}{E \times I_z} \times dx + \frac{1}{2}\frac{\mathcal{T}_y^{\,2}}{G \times k_y \times S} \times dx$$

❏ Keeping in mind its subsequent use, we can express this potential as a function of the displacements $\theta_z(x)$ and $v(x)$. Then performance relationship [2.22] and [2.28] leads us immediately to:

____

10 The principle for calculating the coefficient of reduced section – or shear coefficient – is given in Chapter 9.

$$E_{\text{pot.bending}} = \frac{1}{2}\int_0^\ell E \times I_z \times \left(\frac{d\theta_z}{dx}\right)^2 \times dx + \frac{1}{2}\int_0^\ell G \times k_y \times S \times \left(\frac{dv}{dx} - \theta_z\right)^2 \times dx \qquad [2.31]$$

❏ Bernoulli hypothesis for beams: for homogenous beams and studies concerning the behavior of structures in the static field, the contributions in energy due to shear distortion (second term of expression [2.31]) is small and most often negligible[11]. We can see here that by neglecting the distortion $\left(\dfrac{dv}{dx} - \theta_z\right)$ we obtain:

$$\theta_z \cong \frac{dv(x)}{dx} \quad \text{or also} \quad \frac{d\theta_z}{dx} \cong \frac{d^2 v}{dx^2} \qquad [2.32]$$

This is the standard Bernoulli's[12] hypothesis for beams. The strain potential energy is then reduced to:

$$E_{\text{pot.bending}} \cong \frac{1}{2}\int_0^\ell E \times I_z \times \left(\frac{d^2 v}{dx^2}\right)^2 \times dx \qquad [2.33]$$

We can see that with this approximation we are restricting the energy to expression [2.25] corresponding to only the bending moment being taken into account.

❏ The study of plane bending in the plane (xz) would lead us to the given performance relationship or similar expressions, such as:

$$E_{\text{pot.bending}} = \frac{1}{2}\int_0^\ell \frac{\mathcal{M}f_y^{\,2}}{E \times I_y} \times dx + \frac{1}{2}\int_0^\ell \frac{\mathcal{T}_z^{\,2}}{G \times k_z \times S} \times dx$$

$$\text{or } E_{\text{pot.bending}} = \frac{1}{2}\int_0^\ell E \times I_y \times \left(\frac{d\theta_y}{dx}\right)^2 \times dx + \frac{1}{2}\int_0^\ell G \times k_z \times S \times \left(\frac{dw}{dx} + \theta_y\right)^2 \times dx$$

and with Bernoulli's approximation $\left(\theta_y \cong -\dfrac{dw(x)}{dx}\right)$:

---

11 Except for certain categories of problems mentioned (see [9.53]).
12 See Chapter 9, section 9.1.5.2.

$$E_{pot.bending} \cong \frac{1}{2} \int_0^\ell E \times I_y \times \left(\frac{d^2 w}{dx^2}\right)^2 \times dx$$

Summary

The elementary potential energies stored in a slice dx of a beam subjected to traditional loads are summarized in the following table.

<table>
<tr><td colspan="2" align="center">Elementary potential energies in the domain (S×dx) of a straight beam</td></tr>
<tr><td colspan="2">

⇨ *traction-compression*:

$$dE_{pot.\,\mathcal{N}} = \frac{1}{2}\frac{\mathcal{N}^2}{ES}\,dx = \frac{1}{2}ES\left(\frac{du}{dx}\right)^2 dx$$

</td></tr>
<tr><td colspan="2">

⇨ *torsion*:

$$dE_{pot.\,\mathcal{M}t} = \frac{1}{2}\frac{\mathcal{M}t^2}{GJ}\,dx = \frac{1}{2}GJ\left(\frac{d\theta_x}{dx}\right)^2 dx$$

</td></tr>
<tr><td>

⇨ *pure bending (xy plane)*:

$$dE_{pot.bending\atop(xy)} = \frac{1}{2}\frac{\mathcal{M}f_z^2}{EI_z}\,dx = \frac{1}{2}EI_z\left(\frac{d^2 v}{dx^2}\right)^2 dx$$

</td><td>

⇨ *pure bending (xz plane)*:

$$dE_{pot.bending\atop(xz)} = \frac{1}{2}\frac{\mathcal{M}f_y^2}{EI_y}\,dx = \frac{1}{2}EI_y\left(\frac{d^2 w}{dx^2}\right)^2 dx$$

</td></tr>
<tr><td>

⇨ *plane bending (xy plane)*:

$$dE_{pot.bending\,(xy)} = \frac{1}{2}\frac{\mathcal{M}f_z^2}{EI_z}\,dx + \frac{1}{2}\frac{\mathcal{T}_y^2}{Gk_y S}\,dx$$

or:

$$dE_{pot.bending\,(xy)} = \frac{1}{2}EI_z\left(\frac{d\theta_z}{dx}\right)^2 dx + ...$$

$$..... \frac{1}{2}Gk_y S\left(\frac{dv}{dx} - \theta_z\right)^2 dx$$

current simplified form:

$$dE_{pot.bending\,(xy)} \cong \frac{1}{2}EI_z\left(\frac{d^2 v}{dx^2}\right)^2 dx$$

</td><td>

⇨ *plane bending (xz plane)*:

$$dE_{pot.bending\,(xz)} = \frac{1}{2}\frac{\mathcal{M}f_y^2}{EI_y}\,dx + \frac{1}{2}\frac{\mathcal{T}_z^2}{Gk_z S}\,dx$$

or:

$$dE_{pot.bending\,(xz)} = \frac{1}{2}EI_y\left(\frac{d\theta_y}{dx}\right)^2 dx + ...$$

$$..... \frac{1}{2}Gk_z S\left(\frac{dw}{dx} + \theta_y\right)^2 dx$$

current simplified form:

$$dE_{pot.bending\,(xz)} \cong \frac{1}{2}EI_y\left(\frac{d^2 w}{dx^2}\right)^2 dx$$

</td></tr>
</table>

$$[2.34]$$

2.3.2. *Deformation energy under plane stresses*

We have seen in section 1.4.2 how a small prismatic element, of volume $dV = dx \times dy \times e$, deforms under the action of plane stresses σ_x, σ_y, τ_{xy}. Let us consider once again the small element in Figure 1.14 and subject it to a combination of the following cohesive forces:

– traction along axis $\vec{x}$: $dF_x = \sigma_x \times e \times dy$ (Figure 2.17);

– traction along axis $\vec{y}$: $dF_y = \sigma_y \times e \times dx$ (Figure 2.18);

– shearing along plane (xy): $dF_{xy} = \tau_{xy} \times e \times dy$ (Figure 2.19).

These three forces are applied successively: first dF_x then dF_y followed by dF_{xy}. We then obtain the following three cases of loading:

– case 1: dF_x;

– case 2: dF_x then dF_y;

– case 3: dF_x then dF_y followed by dF_{xy}.

We will express the strain potential energy for this small isolated element by using the notions mentioned in section 1.4.2.

2.3.2.1. *Case 1: dF_x (Figure 2.17)*

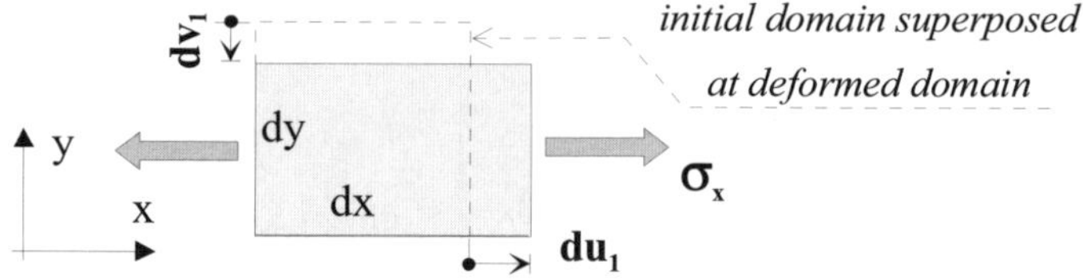

Figure 2.17. *Displacement when dF_x is applied*

Based on the approach mentioned in section 2.3.1.1, when a stress σ_x is progressively applied from an initial zero value, the deformation energy can be written as (see [2.10]):

$$dE_{pot.1} = \frac{1}{2} dF_x \times du_1$$

Taking into account the definition [1.7] of axial strain ε_x:

$$dE_{pot.1} = \frac{1}{2} dF_x \times \varepsilon_{x1} \times dx = \frac{1}{2}\left(\sigma_x \times e \times dy\right) \times \varepsilon_{x1} \times dx$$

where $dx \times dy \times e = dV$, volume of the isolated particle:

$$dE_{pot.1} = \frac{1}{2}\sigma_x \times \varepsilon_{x1} \times dV$$

or according to [1.18]:

$$dE_{pot.1} = \frac{1}{2}\frac{\sigma_x^{\,2}}{E} \times dV \qquad\qquad [2.35]$$

2.3.2.2. Case 2: dF_x then dF_y (Figure 2.18)

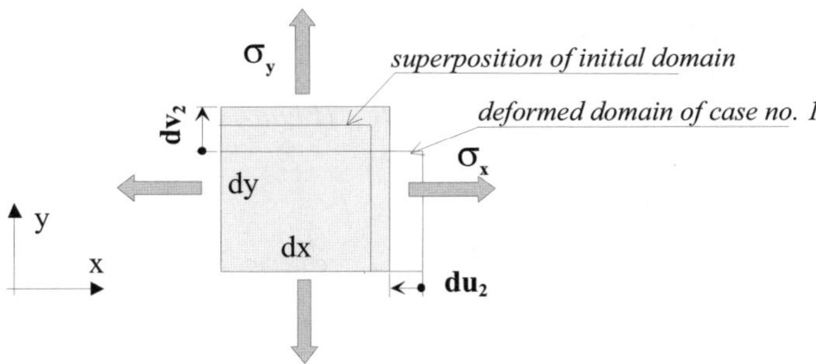

Figure 2.18. *Displacement when dF_x is applied followed by dF_y*

In addition to energy [2.35] expressed previously and resulting from the application of dF_x, we can see the appearance of an additional energy $dE_{pot\cdot2}$ resulting from dF_y which can be written as (see Figure 2.18):

$$dE_{pot.2} = \frac{1}{2} dF_y \times dv_2 + dF_x \times du_2 \quad^{13}$$

where du_2 represents the displacement along $\vec{x}$ due to Poisson's effect when a stress is applied along $\vec{y}$, or even:

---

13 Please note the absence of the $\dfrac{1}{2}$ coefficient, in front of the product $dF_x \times du_2$: in fact, dF_x is already present when the displacement du_2 takes place due to the progressive application of dF_y.

$$dE_{pot.2} = \frac{1}{2}\left(\sigma_y \times e \times dx\right) \times dv_2 + (\sigma_x \times e \times dy) \times du_2$$

$$dE_{pot.2} = \left(\frac{1}{2} \times \sigma_y \times \frac{dv_2}{dy} + \sigma_x \times \frac{du_2}{dx}\right) \times dV$$

Taking into account definitions [1.9] and [1.10] of longitudinal deformation and relative contraction associated with Poisson's effect, we obtain:

$$dE_{pot.2} = \left(\frac{1}{2} \times \sigma_y \times \varepsilon_{y2} + \sigma_x \times (-\nu \times \varepsilon_{y2})\right) \times dV$$

or according to [1.18]:

$$dE_{pot.2} = \left(\frac{1}{2} \times \frac{\sigma_y^{\ 2}}{E} - \frac{\nu}{E} \times \sigma_x \times \sigma_y\right) \times dV \qquad [2.36]$$

which can be added to [2.35].

2.3.2.3. *Case 3: dF_x then dF_y followed by dF_{xy} (Figure 2.19)*

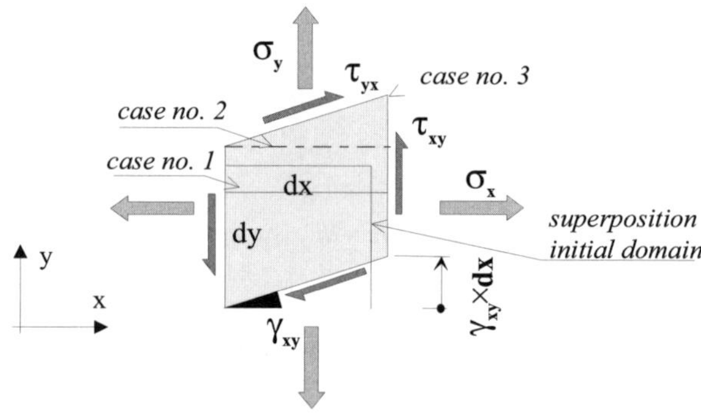

Figure 2.19. *Displacements when dF_x is applied followed by dF_y, and dF_{xy}*

In addition to the energies previously mentioned, [2.35] and [2.36], an additional energy denoted as $dE_{pot.3}$ appears. The length of the sides of the element (see Figure 2.19) remains invariable when it passes from state 2 to state 3 (pure shearing[14]). dF_x and dF_y, already applied, thus no longer produce work. Thus with the notations in Figure 2.19 we obtain:

---

14 See section 1.4.2.3.

$$dE_{pot.3} = \frac{1}{2} dF_{xy} \times (\gamma_{xy} \times dx)$$

$$dE_{pot.3} = \frac{1}{2} \left(\tau_{xy} \times e \times dy \right) \times (\gamma_{xy} \times dx)$$

$$dE_{pot.3} = \left(\frac{1}{2} \times \tau_{xy} \times \gamma_{xy} \right) \times dV$$

or, taking into account behavior relationship [1.18]:

$$dE_{pot.3} = \left(\frac{1}{2} \times \frac{\tau_{xy}^{\ 2}}{G} \right) \times dV \qquad [2.37]$$

Finally, when the three forces dF_x, dF_y, dF_{xy} are applied simultaneously:

$$dE_{pot.} = dE_{pot.1} + dE_{pot.2} + dE_{pot.3}$$

or with expressions [2.35], [2.36], [2.37]:

$$dE_{pot.} = \frac{1}{2} \left(\frac{\sigma_x^{\ 2}}{E} + \frac{\sigma_y^{\ 2}}{E} - 2 \frac{\nu}{E} \times \sigma_x \times \sigma_y + \frac{\tau_{xy}^{\ 2}}{G} \right) \times dV \qquad [2.38]$$

2.3.2.4. *Different expressions for potential energy: quadratic forms*

We can see that expression [2.38] is a second degree expression for stress. This is why it is referred to as the *quadratic form of stress*. We can also express it as[15]:

$$dE_{pot.} = \frac{1}{2} \underbrace{\begin{bmatrix} \sigma_x & \sigma_y & \tau_{xy} \end{bmatrix}}_{line\ matrix\ [1\times3]} \bullet \underbrace{\left\{ \begin{array}{c} \dfrac{1}{E}\sigma_x - \dfrac{\nu}{E}\sigma_y \\[2mm] -\dfrac{\nu}{E}\sigma_x + \dfrac{1}{E}\sigma_y \\[2mm] \dfrac{1}{G}\tau_{xy} \end{array} \right\}}_{column\ matrix\ [3\times1]} \times dV \qquad [2.39]$$

or in the matrix form:

$$dE_{pot.} = \frac{1}{2} \begin{bmatrix} \sigma_x & \sigma_y & \tau_{xy} \end{bmatrix} \bullet \begin{bmatrix} \dfrac{1}{E} & -\dfrac{\nu}{E} & 0 \\[2mm] -\dfrac{\nu}{E} & \dfrac{1}{E} & 0 \\[2mm] 0 & 0 & \dfrac{1}{G} \end{bmatrix} \bullet \left\{ \begin{array}{c} \sigma_x \\ \sigma_y \\ \tau_{xy} \end{array} \right\} \times dV$$

---

15 See Chapter 12, section 12.1.3 and [12.2].

or by using the transposed notation[16]:

$$dE_{pot.} = \frac{1}{2}\begin{Bmatrix}\sigma_x \\ \sigma_y \\ \tau_{xy}\end{Bmatrix}^T \bullet \begin{bmatrix}\dfrac{1}{E} & -\dfrac{\nu}{E} & 0 \\[2mm] -\dfrac{\nu}{E} & \dfrac{1}{E} & 0 \\[2mm] 0 & 0 & \dfrac{1}{G}\end{bmatrix} \bullet \begin{Bmatrix}\sigma_x \\ \sigma_y \\ \tau_{xy}\end{Bmatrix} \times dV \qquad [2.40]$$

Thus, the *quadratic form of stress* [2.38] appears "built" on the square and symmetric matrix consisting of the coefficients E, ν and G, which we have already seen in Chapter 1 (see [1.18]). This matrix appeared in the behavior relationship under plane stresses, which is recalled again here:

$$\begin{Bmatrix}\varepsilon_x \\ \varepsilon_y \\ \gamma_{xy}\end{Bmatrix} = \begin{bmatrix}\dfrac{1}{E} & -\dfrac{\nu}{E} & 0 \\[2mm] -\dfrac{\nu}{E} & \dfrac{1}{E} & 0 \\[2mm] 0 & 0 & \dfrac{1}{G}\end{bmatrix} \bullet \begin{Bmatrix}\sigma_x \\ \sigma_y \\ \tau_{xy}\end{Bmatrix} = \begin{Bmatrix}\dfrac{1}{E}\sigma_x - \dfrac{\nu}{E}\sigma_y \\[2mm] -\dfrac{\nu}{E}\sigma_x + \dfrac{1}{E}\sigma_y \\[2mm] \dfrac{1}{G}\tau_{xy}\end{Bmatrix}$$

where the total strains ε_x, ε_y and γ_{xy} of the elementary domain appear under the combined stresses σ_x, σ_y and τ_{xy}.

By comparing the previous expression with [2.39] we can see that the potential deformation energy can also be written as:

$$dE_{pot.} = \frac{1}{2}\begin{bmatrix}\sigma_x & \sigma_y & \tau_{xy}\end{bmatrix} \bullet \begin{Bmatrix}\varepsilon_x \\ \varepsilon_y \\ \gamma_{xy}\end{Bmatrix} \times dV = \frac{1}{2}\begin{Bmatrix}\sigma_x \\ \sigma_y \\ \tau_{xy}\end{Bmatrix}^T \bullet \begin{Bmatrix}\varepsilon_x \\ \varepsilon_y \\ \gamma_{xy}\end{Bmatrix} \times dV \qquad [2.41]$$

or, by expanding this expression, we obtain:

$$dE_{pot.} = \frac{1}{2}\left(\sigma_x \times \varepsilon_x + \sigma_y \times \varepsilon_y + \tau_{xy} \times \gamma_{xy}\right) \times dV$$

Still referring to [1.18], we obtain:

---

16 See footnote 15.

$$\begin{Bmatrix} \sigma_x \\ \sigma_y \\ \tau_{xy} \end{Bmatrix} = \begin{bmatrix} \dfrac{E}{1-v^2} & \dfrac{vE}{1-v^2} & 0 \\ \dfrac{vE}{1-v^2} & \dfrac{E}{1-v^2} & 0 \\ 0 & 0 & G \end{bmatrix} \bullet \begin{Bmatrix} \varepsilon_x \\ \varepsilon_y \\ \gamma_{xy} \end{Bmatrix}$$

transposing this expression[17]:

$$\begin{Bmatrix} \sigma_x \\ \sigma_y \\ \tau_{xy} \end{Bmatrix}^T = \begin{Bmatrix} \varepsilon_x \\ \varepsilon_y \\ \gamma_{xy} \end{Bmatrix}^T \bullet \begin{bmatrix} \dfrac{E}{1-v^2} & \dfrac{vE}{1-v^2} & 0 \\ \dfrac{vE}{1-v^2} & \dfrac{E}{1-v^2} & 0 \\ 0 & 0 & G \end{bmatrix}$$

then the deformation energy [2.41] takes the following form:

$$dE_{pot.} = \frac{1}{2} \begin{Bmatrix} \varepsilon_x \\ \varepsilon_y \\ \gamma_{xy} \end{Bmatrix}^T \bullet \begin{bmatrix} \dfrac{E}{1-v^2} & \dfrac{vE}{1-v^2} & 0 \\ \dfrac{vE}{1-v^2} & \dfrac{E}{1-v^2} & 0 \\ 0 & 0 & G \end{bmatrix} \bullet \begin{Bmatrix} \varepsilon_x \\ \varepsilon_y \\ \gamma_{xy} \end{Bmatrix} \times dV \qquad [2.42]$$

We obtain here a *quadratic form of the strains* associated with the square and symmetric matrix consisting of coefficients E, v and G of the behavior relationship [1.18].

By expanding [2.42] we obtain:

$$dE_{pot.} = \frac{1}{2} \left[\frac{E}{1-v^2} \left(\varepsilon_x^2 + \varepsilon_y^2 + 2v \times \varepsilon_x \times \varepsilon_y \right) + G \times \gamma_{xy}^2 \right] \times dV \qquad [2.43]$$

---

17 See Chapter 12.

Summary

The deformation energy of an elementary domain of volume $dV = dx \times dy \times e$ under plane stress (xy plane) can indiscriminately take the following forms.

<table>
<tr><td align="center">Different expressions for potential energy under plane stress</td></tr>
</table>

$$dE_{pot.} = \frac{1}{2}\left(\frac{\sigma_x^{\,2}}{E} + \frac{\sigma_y^{\,2}}{E} - 2\frac{\nu}{E} \times \sigma_x \times \sigma_y + \frac{\tau_{xy}^{\,2}}{G} \right) \times dV$$

$$\Updownarrow$$

$$dE_{pot.} = \frac{1}{2} \begin{Bmatrix} \sigma_x \\ \sigma_y \\ \tau_{xy} \end{Bmatrix}^T \bullet \begin{bmatrix} \dfrac{1}{E} & -\dfrac{\nu}{E} & 0 \\ -\dfrac{\nu}{E} & \dfrac{1}{E} & 0 \\ 0 & 0 & \dfrac{1}{G} \end{bmatrix} \bullet \begin{Bmatrix} \sigma_x \\ \sigma_y \\ \tau_{xy} \end{Bmatrix} \times dV$$

$$dE_{pot.} = \frac{1}{2}\left(\sigma_x \times \varepsilon_x + \sigma_y \times \varepsilon_y + \tau_{xy} \times \gamma_{xy} \right) \times dV$$

$$dE_{pot.} = \frac{1}{2}\left[\frac{E}{1-\nu^2}\left(\varepsilon_x^2 + \varepsilon_y^2 + 2\nu \times \varepsilon_x \times \varepsilon_y \right) + G \times \gamma_{xy}^2 \right] \times dV$$

$$\Updownarrow$$

$$dE_{pot.} = \frac{1}{2} \begin{Bmatrix} \varepsilon_x \\ \varepsilon_y \\ \gamma_{xy} \end{Bmatrix}^T \bullet \begin{bmatrix} \dfrac{E}{1-\nu^2} & \dfrac{\nu E}{1-\nu^2} & 0 \\ \dfrac{\nu E}{1-\nu^2} & \dfrac{E}{1-\nu^2} & 0 \\ 0 & 0 & G \end{bmatrix} \bullet \begin{Bmatrix} \varepsilon_x \\ \varepsilon_y \\ \gamma_{xy} \end{Bmatrix} \times dV$$

$$[2.44]$$

2.4. Work produced by external forces on a structure

NOTES:

❐ We still remain within the framework of the assumptions in section 1.3.3, i.e. we are considering structures having an elastic and linear domain behavior, globally and in all their constituent parts. Consequently, the work produced by these external forces also corresponds to a potential elastic energy which will be stored in the considered structure (see [2.4]).

❐ A structure constituted in this manner will be deformed under external forces, and consequently almost all the parts will undergo "small displacements". Figure 2.20 shows that the potential deformation energy can be stored globally or locally. We can see that the corresponding deformations, even localized, lead to the displacement of all the points of the structure

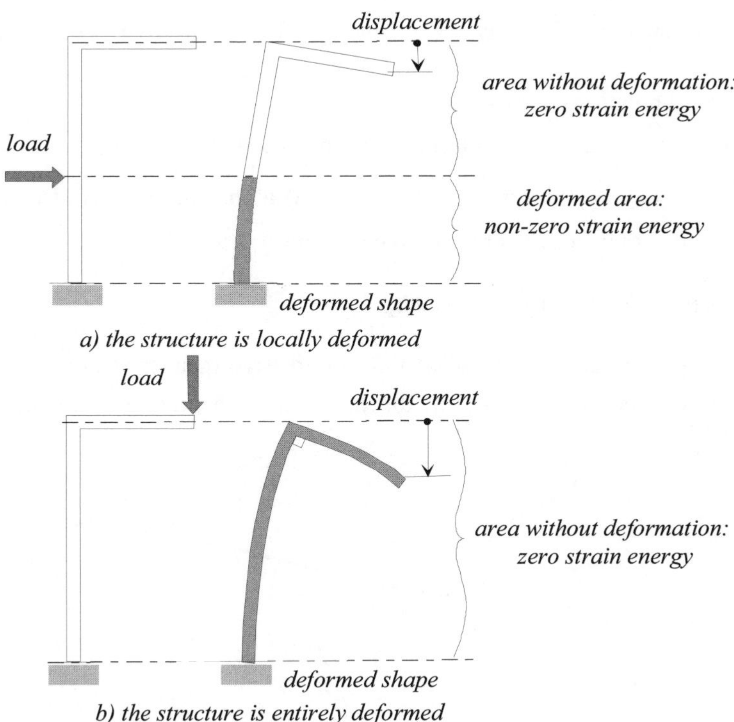

Figure 2.20. *Displacements and deformations*

❏ We will now examine simple structures in order to show the methodology and expression of results in the *matrix form*, which enables us to derive a *generalized expression* of the strain potential energy.

2.4.1. *Beam under plane bending subjected to two forces*

The following examples concern straight beams with a symmetry plane (xy), loaded and bending in that plane. The left end of these beams is clamped while the right is free. We will call them "clamped-free beams" or "cantilever". In all these examples, we will not consider the field of gravity (so that the specific weight of the beam does not come into play).

2.4.1.1. *Example 1*

Let us choose two points (1) and (2) on the mean line of such a beam, each bearing a transversal force $\vec{F_1} = F_1\vec{y}$ and $\vec{F_2} = F_2\vec{y}$ respectively (see Figure 2.21).

Indexed notations used are:

– $\vec{F_i} = F_i\vec{y}$ (with i = 1, 2) force along the $\vec{y}$ axis at point (i);

– $\vec{v_{ij}} = v_{ij}\vec{y}$: displacement (or deflection) along the $\vec{y}$ axis at point (i), (with i = 1, 2), resulting from a force applied at point (j), (with j = 1, 2).

◯ Sequential load of $\vec{F_1}$ followed by $\vec{F_2}$

➢ *load 1:* force $\vec{F_1}$ is applied in a progressive manner at point (1). When this force reaches its final value F_1, the displacement of its point of application is v_{11} (see Figure 2.21).

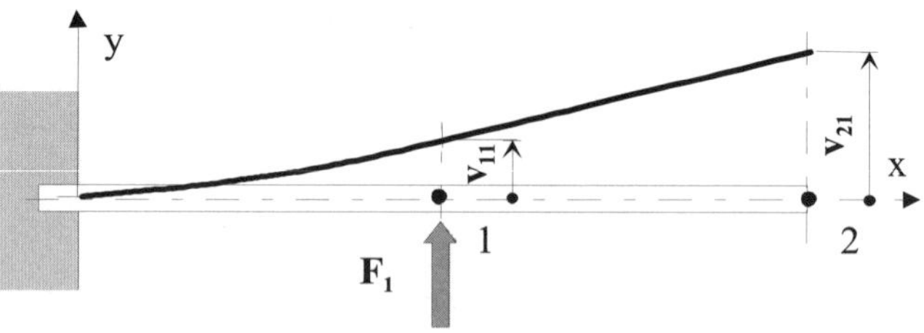

Figure 2.21. *Application of load 1*

The work produced by this force is fully transformed into potential deformation energy, which is expressed as:

$$W_1 = \frac{1}{2} F_1 \times v_{11} = E_{pot.1}$$

This corresponds to the OAB area in Figure 2.22.

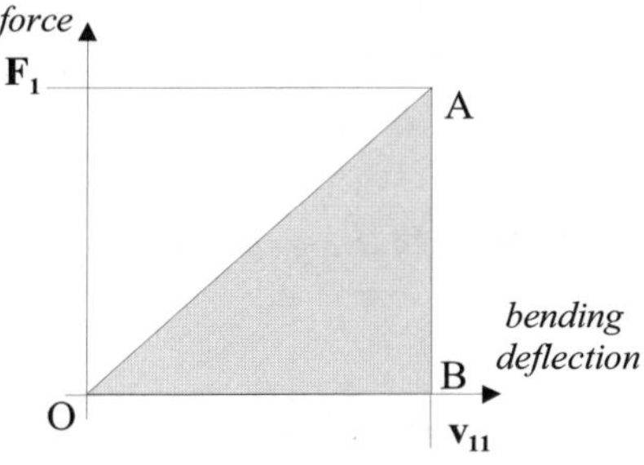

Figure 2.22. *Diagram for load 1*

➢ *load 2:* force $\vec{F_2}$ applied progressively at point (2) is added to load 1. The work produced by this force is fully transformed into potential deformation energy. All the points on the beam acquire additional deflections. The point of application of $\vec{F_1}$ gets displaced again and we therefore have to add the work produced by $\vec{F_1}$. Here $\vec{F_1}$ is a constant since it has already been applied to the beam (see Figure 2.23).

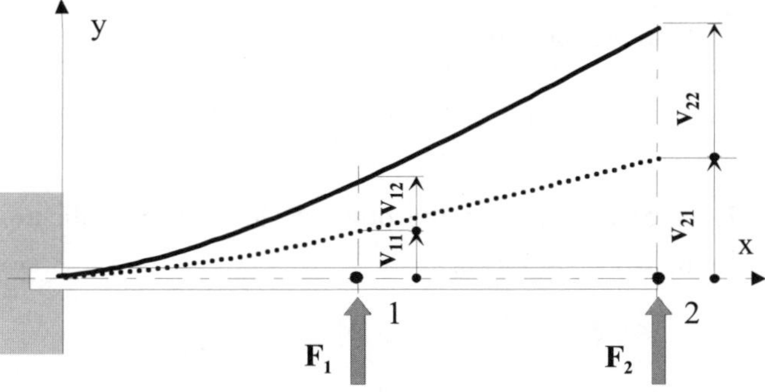

Figure 2.23. *Application of load 2*

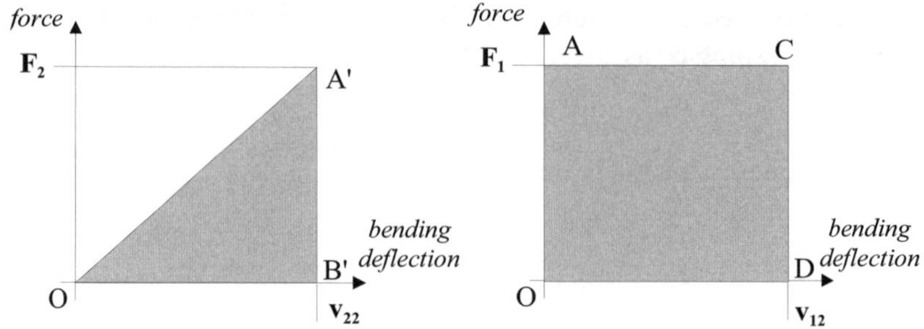

Figure 2.24. *Diagram for load 2*

We thus have:

$$W_2 = \frac{1}{2} F_2 \times v_{22} + F_1 \times v_{12} = E_{pot.2}$$

which corresponds to the OA'B' and OACD areas in Figure 2.24.

The work produced by $\vec{F_1}$ and $\vec{F_2}$ on the beam is fully transformed into potential deformation energy. We thus have in total:

$$W = W_1 + W_2 = \frac{1}{2} F_1 \times v_{11} + \frac{1}{2} F_2 \times v_{22} + F_1 \times v_{12} = E_{pot.} \qquad [2.45]$$

○ Reversed loading procedure

If we reverse the order of loading by applying first $\vec{F_2}$ at point 2

➢ *load 1:* $\vec{F_2}$ which produces work

$$W'_1 = \frac{1}{2} F_2 \times v_{22} = E'_{pot.1}$$

➢ *load 2:* $\vec{F_1}$ superimposed on the previous load. The application of this force corresponds to the work produced:

$$W'_2 = \frac{1}{2} F_1 \times v_{11} + F_2 \times v_{21} = E'_{pot.2}$$

The work produced by $\vec{F_2}$ and $\vec{F_1}$ on the beam is fully transformed into strain potential energy. This is expressed as:

$$W' = W'_1 + W'_2 = \frac{1}{2} F_2 \times v_{22} + \frac{1}{2} F_1 \times v_{11} + F_2 \times v_{21.} = E'_{pot.} \qquad [2.46]$$

Since the final state is the same (same deformation for the beam irrespective of the order of loading), the two final works W and W' are identical. The comparison of expressions [2.45] and [2.46] thus leads us to[18]:

$$F_1 \times v_{12} = F_2 \times v_{21} \qquad [2.47]$$

○ Total displacements under complete load sequence

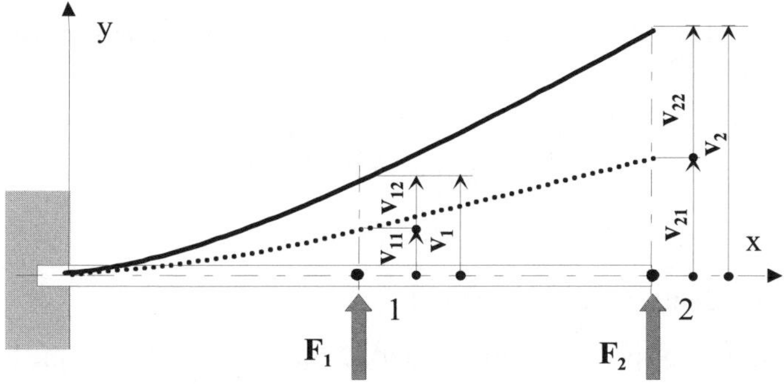

Figure 2.25. *Total displacements*

From the previously described loadings, we can show the total displacements when the loading sequence of the beam is complete (see Figure 2.25). We can in fact see in the figure that the final displacements can be written as:

$$v_1 = v_{11} + v_{12} \quad \text{and} \quad v_2 = v_{21} + v_{22} \qquad [2.48]$$

The expressions for work [2.45] or [2.46] can be re-written, taking into account [2.47], for example:

$$W = \frac{1}{2}F_1 \times v_{11} + \frac{1}{2}F_2 \times v_{22} + \left(\frac{1}{2}F_1 \times v_{12} + \frac{1}{2}F_2 \times v_{21}\right) = E_{pot.}$$

$$W = \frac{1}{2}F_1(v_{11} + v_{12}) + \frac{1}{2}F_2(v_{21} + v_{22}) = E_{pot.}$$

or with [2.48]:

$$W = \frac{1}{2}(F_1 \times v_1 + F_2 \times v_2) = E_{pot.} \qquad [2.49]$$

This scalar expression can also appear in the matrix form:

---

18 By using load in this manner, we return to the traditional "Maxwell-Betti" theorem (also known as the reciprocal theorem).

$$W = \frac{1}{2}\begin{bmatrix} F_1 & F_2 \end{bmatrix} \bullet \begin{Bmatrix} v_1 \\ v_2 \end{Bmatrix} \quad \text{or even} \quad W = \frac{1}{2}\begin{Bmatrix} F_1 \\ F_2 \end{Bmatrix}^T \bullet \begin{Bmatrix} v_1 \\ v_2 \end{Bmatrix} \qquad [2.50]$$

or by condensing the matrix:

$$W = \frac{1}{2}\{F\}^T \bullet \{d\} \qquad [2.51]$$

an expression where the loads and the displacements are summarized by column matrices or vectors[19]:

$$\{F\} = \begin{Bmatrix} F_1 \\ F_2 \end{Bmatrix} \quad , \quad \{d\} = \begin{Bmatrix} v_1 \\ v_2 \end{Bmatrix} \qquad [2.52]$$

NOTE

❏ To summarize the previous approach:

– we chose two specific points (1) and (2) on the mean line of the beam, which we can refer to as "nodes";

– we chose for each of these points a direction, to be specific the $\bar{y}$ axis direction;

– we applied a load at these points or nodes and along that direction;

– we noted displacements v_1 and v_2 along the direction of each force applied.

We will refer to v_1 and v_2 as the *degrees of freedom* (dof) associated with points (1) and (2) and forces $\vec{F_1}$ and $\vec{F_2}$. This is summarized in the table below.

Points on the structure	*Forces on these points*	*Displacements along the direction of these forces*
(1) (2) *(nodes)*	$\vec{F_1}$ $\vec{F_2}$ *(loading)*	$\vec{v_1}$ $\vec{v_2}$ *(dof)*

---

19 See Chapter 12, section 12.1.1.1.

○ Flexibility coefficients

The previous forces $\vec{F_1}$ and $\vec{F_2}$ can take on an infinite number of values in the elastic behavior domain of the beam. Let us consider the special case of "unit" forces, i.e. where $F_1 = F_2 = 1N$. In this case we will denote the displacements as α_{ij} instead of v_{ij} in order to differentiate them from the case of general loading.

This special case can be thus summarized:

$$v_{11} \rightarrow \alpha_{11}$$

$$F_1 \rightarrow 1N \qquad \qquad v_{12} \rightarrow \alpha_{12}$$
$$\Rightarrow$$
$$F_2 \rightarrow 1N \qquad \qquad v_{22} \rightarrow \alpha_{22}$$

$$v_{21} \rightarrow \alpha_{21}$$

The coefficients α_{ij} are displacements per unit force. Consequently, when $\vec{F_1}$ and $\vec{F_2}$ are any two forces, the linear elastic nature of the studied structure gives us:

$$v_{11} = \alpha_{11} \times F_1$$
$$v_{12} = \alpha_{12} \times F_2$$
$$v_{22} = \alpha_{22} \times F_2 \qquad\qquad [2.53]$$
$$v_{21} = \alpha_{21} \times F_1$$

In addition, relation [2.47] can be written as: $\dfrac{v_{21}}{F_1} = \dfrac{v_{12}}{F_2}$, i.e.:

$$\alpha_{21} = \alpha_{12} \qquad\qquad [2.54]$$

Let us now express the total displacements v_1 and v_2 (see [2.48]):

$$v_1 = v_{11} + v_{12}$$

$$v_2 = v_{21} + v_{22}$$

with [2.53] we will immediately obtain:

$$v_1 = \alpha_{11} \times F_1 + \alpha_{12} \times F_2 \qquad\qquad [2.55]$$

$$v_2 = \alpha_{21} \times F_1 + \alpha_{22} \times F_2$$

or in the matrix form:

$$\begin{Bmatrix} v_1 \\ v_2 \end{Bmatrix} = \begin{bmatrix} \alpha_{11} & \alpha_{12} \\ \alpha_{21} & \alpha_{22} \end{bmatrix} \bullet \begin{Bmatrix} F_1 \\ F_2 \end{Bmatrix} \qquad\qquad [2.56]$$

NOTE

❏ For the same forces F_1 and F_2, the displacements observed v_1 and v_2 increase with higher coefficients α_{ij}. The latter therefore characterizes what we may refer to as the "*flexibility*" of this structure. For this reason we will call them "flexibility coefficients". The notion of flexibility seems to depend on the characteristics of the structure (geometry and material) and not on the loading intensity.

❏ According to [2.54], we can note the symmetry of the coefficients α_{ij} of the square matrix which appears in [2.56]: we will refer to this *square* and *symmetric* matrix as a "*flexibility matrix*" and will summarize it in its condensed form $[\alpha]$. Keeping in mind the writing convention [2.52], relation [2.56] can be expressed in the condensed matrix form:

$$\{d\} = [\alpha] \bullet \{F\} \tag{2.57}$$

❏ We may note on each line of relation [2.56] the force F_i and its associated degree of freedom v_i. Let us recall that F_i and v_i have the same support, which is here is the direction $\vec{y}$. The flexibility matrix $[\alpha]$ appears linked to the two points chosen on this beam and to the two associated directions, here identical (direction $\vec{y}$).

❏ In this flexibility matrix, the non-diagonal coefficients are called *coupling terms*. We can of course see in expression [2.56] that these terms help to quantify the contribution of a force F_i with respect to the displacement v_j at a point j ($i \neq j$).

❏ As already mentioned, in relation [2.56] we have on each row the load and its associated degree of freedom, i.e.:

– force F_1 and its degree of freedom v_1;

– force F_2 and its degree of freedom v_2.

It is important to note that this association is double:

➤ *geometric* association: same direction for the load and its associated degree of freedom;

➤ *energy* association: the work results from the product of the loads and their associated degree of freedom (see [2.49]).

○ Relationship between work and flexibility

Expression [2.49] of the work produced by $\vec{F_1}$ and $\vec{F_2}$ can be written as a function of the flexibility coefficients (see [2.55]):

$$W = \frac{1}{2} F_1 \times \alpha_{11} \times F_1 + \frac{1}{2} F_2 \times \alpha_{22} \times F_2 + F_1 \times \alpha_{12} \times F_2 = E_{pot.} \qquad [2.58]$$

We obtain a quadratic form[20] in F_1 and F_2 which can be re-written as:

$$W = \frac{1}{2} \begin{bmatrix} F_1 & F_2 \end{bmatrix} \bullet \begin{Bmatrix} \alpha_{11} \times F_1 + \alpha_{12} \times F_2 \\ \alpha_{12} \times F_1 + \alpha_{22} \times F_2 \end{Bmatrix} = E_{pot.} \qquad [2.59]$$

or by using the flexibility matrix (with $\alpha_{12} = \alpha_{21}$) (see [2.54]):

$$W = \frac{1}{2} \begin{Bmatrix} F_1 \\ F_2 \end{Bmatrix}^T \bullet \begin{bmatrix} \alpha_{11} & \alpha_{12} \\ \alpha_{21} & \alpha_{22} \end{bmatrix} \bullet \begin{Bmatrix} F_1 \\ F_2 \end{Bmatrix} = E_{pot.}$$

Thus, the quadratic form in F_1 and F_2 which expresses work is built on the flexibility matrix $[\alpha]$, or its condensed form:

$$W = \frac{1}{2} \{F\}^T \bullet [\alpha] \bullet \{F\} = E_{pot.} \qquad [2.60]$$

NOTE
❏ We can associate relation [2.57] with the previous expression, which can then be written as:

$$W = \frac{1}{2} \{F\}^T \bullet \{d\} = E_{pot.}$$

We thus return to relation [2.51].

○ Application: explicit calculation of the flexibility matrix

Let us elaborate more explicitly the case of this beam loaded by $\vec{F_1}$ and $\vec{F_2}$. Let us fix the geometric and material characteristics of the beam: quadratic moment around the $\vec{z}$ axis: I_z, denoted here as I and Young's modulus: E.

➤ Let us first consider this beam loaded by a force $\vec{F_1} = F_1 \vec{y}$ at the abscissa x_1 (see Figure 2.26).

---

20 See Chapter 12, section 12.1.3.

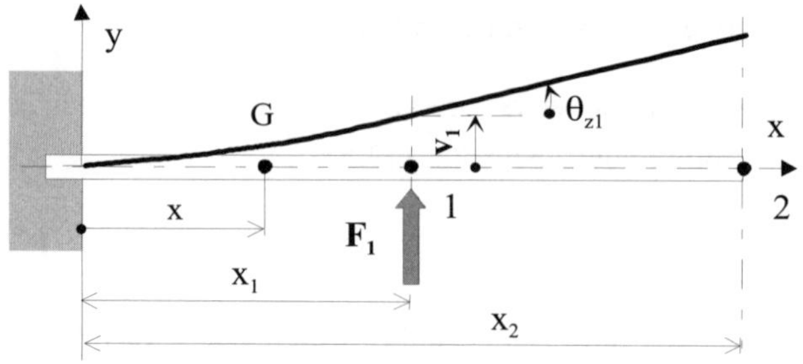

Figure 2.26. *Application of load.1*

The resultant force and moment at the geometric center G of a current section of abscissa $0 \le x \le x_1$ can be written as:

$$\{Coh\}_G = \begin{cases} \overrightarrow{T_y} = F_1\,\vec{y} \\ \\ \overrightarrow{Mf_z} = F_1(x_1 - x)\vec{z} \end{cases}_G$$

The behavior relation for the bending moment is written as (see section 1.5.2.6)[21]:

$$\frac{d^2 v}{dx^2} = \frac{Mf_z}{EI}$$

where $v(x)$ is the displacement along $\vec{y}$ of any point on the mean line of the beam.

In the zone $0 \le x \le x_1$ displacement $v(x)$ thus proves the differential equation:

$$EI\frac{d^2 v}{dx^2} = Mf_z = F_1(x_1 - x)$$

We thus successively obtain (the integration constants disappear since clamped at $x = 0$):

$$EI\frac{dv}{dx} = -F_1\frac{x^2}{2} + F_1 x_1 x$$

$$EIv(x) = -F_1\frac{x^3}{6} + F_1 x_1 \frac{x^2}{2} = F_1\frac{x^2}{2}\left(x_1 - \frac{x}{3}\right) \qquad [2.61]$$

---

21 We have used here Bernoulli's hypothesis where the deformations due to shear force are neglected (see section 9.1.5.2).

We can particularly see at abscissa x_1 a deflection:

$$v(x_1) = F_1 \frac{x_1^{\,3}}{3EI}$$

and a section rotation of: $\theta_z(x_1) \cong \dfrac{dv(x_1)}{dx} = F_1 \dfrac{x_1^{\,2}}{2EI}$.

The displacement at abscissa $x_2 > x_1$, with the portion of the beam $x_1 \leq x \leq x_2$ remaining rectilinear is therefore written as:

$$v(x_2) = v(x_1) + \frac{dv(x_1)}{dx}(x_2 - x_1)$$

$$v(x_2) = F_1 \frac{x_1^{\,3}}{3EI} + F_1 \frac{x_1^{\,2}}{2EI}(x_2 - x_1)$$

i.e.:

$$v(x_2) = F_1 \frac{x_1^{\,2}}{2EI}\left(x_2 - \frac{x_1}{3}\right)$$

➢ Let us now consider the same beam loaded with two forces $\vec{F_1}$ and $\vec{F_2}$ (see Figure 2.27), with $\vec{F_1} = F_1 \vec{y}$ at abscissa x_1 and $\vec{F_2} = F_2 \vec{y}$ at abscissa x_2:

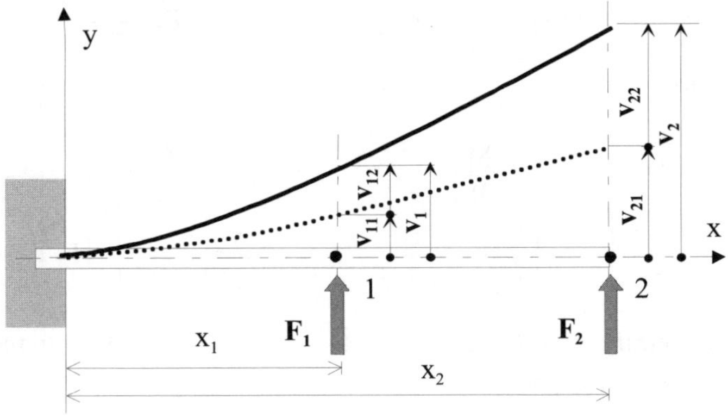

Figure 2.27. *Application of load 2*

Using the notations in Figure 2.29 the results that we have just obtained for $\vec{F_1}$ become:

– at point (1): $v_{11} = F_1 \dfrac{x_1^{\,3}}{3EI}$

– at point (2): $v_{21} = F_1 \dfrac{x_1^{\,2}}{2EI}\left(x_2 - \dfrac{x_1}{3}\right)$

Let us now consider force $\vec{F_2}$ at abscissa x_2. By considering the fact that $x_2 > x_1$, we can use expression [2.61] in which we replace F_1 by F_2 and x_1 by x_2. We obtain:

– at point (1): $v_{12} = F_2 \dfrac{x_1^{\,2}}{2EI}\left(x_2 - \dfrac{x_1}{3}\right)$

– at point (2): $v_{22} = F_2 \dfrac{x_2^{\,3}}{3EI}$

The total displacements can therefore be written as:

$$v_1 = v_{11} + v_{12} = F_1 \frac{x_1^{\,3}}{3EI} + F_2 \frac{x_1^{\,2}}{2EI}\left(x_2 - \frac{x_1}{3}\right)$$

$$v_2 = v_{21} + v_{22} = F_1 \frac{x_1^{\,2}}{2EI}\left(x_2 - \frac{x_1}{3}\right) + F_2 \frac{x_2^{\,3}}{3EI}$$

or in matrix form:

$$\begin{Bmatrix} v_1 \\ v_2 \end{Bmatrix} = \begin{bmatrix} \dfrac{x_1^{\,3}}{3EI} & \dfrac{x_1^{\,2}}{2EI}\left(x_2 - \dfrac{x_1}{3}\right) \\ \dfrac{x_1^{\,2}}{2EI}\left(x_2 - \dfrac{x_1}{3}\right) & \dfrac{x_2^{\,3}}{3EI} \end{bmatrix} \bullet \begin{Bmatrix} F_1 \\ F_2 \end{Bmatrix} \qquad [2.62]$$

where we can identify the form [2.57], i.e. $\{d\} = [\alpha] \bullet \{F\}$, notably with the symmetry of coefficients $\alpha_{12} = \alpha_{21}$.

2.4.1.2. *Example 2*

Let us consider once again the previous console beam, with two sections whose centers are (1) and (2). It is subjected to a force $\overrightarrow{F_1} = F_1\overrightarrow{y}$ at (1) and a moment $\overrightarrow{\mathcal{M}_2} = \mathcal{M}_2\overrightarrow{z}$ [22] at (2), the $\overrightarrow{z}$ axis completing the direct trihedral (xyz).

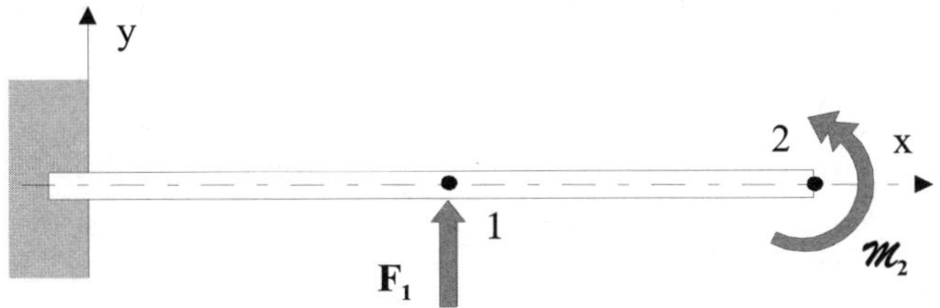

Figure 2.28. *Full loading*

We can follow an approach that is similar to the previous example (section 2.4.1.1).

○ Sequential loading of $\overrightarrow{F_1}$ followed by $\overrightarrow{\mathcal{M}_2}$

After progressive application of only $\overrightarrow{F_1}$ (see Figure 2.29) we obtain the work produced: $W_1 = \dfrac{1}{2}F_1 v_{11} = E_{pot.1}$.

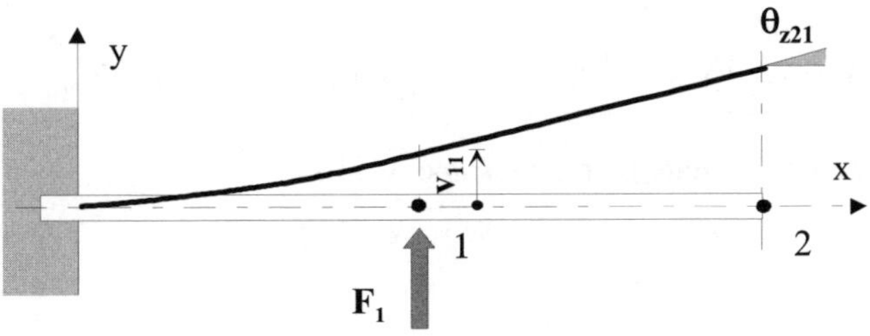

Figure 2.29. *Displacements under a force* $\overrightarrow{F_1}$

---

22 $\overrightarrow{\mathcal{M}_2} = \mathcal{M}_2\overrightarrow{z}$ is applied on the cross-section of abscissa x_2, i.e. on a material domain surrounding point (2) (see section 2.1.2).

After a progressive application of $\overrightarrow{\mathcal{M}_2}$, $\overrightarrow{F_1}$ remains in position, we obtain the work produced (see Figure 2.30):

$$W_2 = \frac{1}{2}\mathcal{M}_2\theta_{z22} + F_1 v_{12} = E_{pot.2}{}^{23}$$

where θ_{z22} is the rotation of section (2) around axis $\vec{z}$ due to the moment $\overrightarrow{\mathcal{M}_2}$.

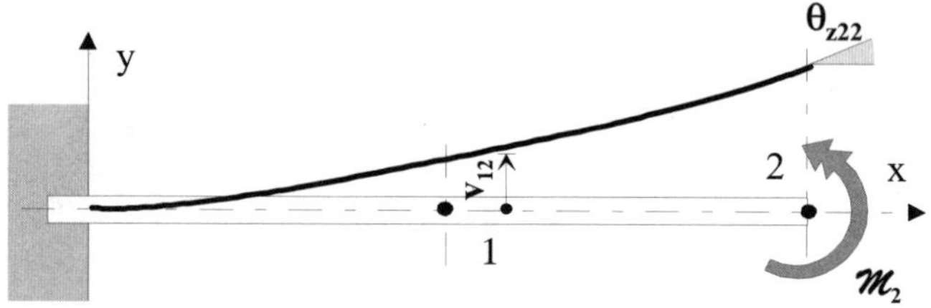

Figure 2.30. *Displacements under a moment* $\overrightarrow{\mathcal{M}_2}$

or in total:

$$W = W_1 + W_2 = \frac{1}{2}F_1 v_{11} + \frac{1}{2}\mathcal{M}_2\theta_{z22} + F_1 v_{12} = E_{pot.} \qquad [2.63]$$

○ Reverse loading procedure: $\overrightarrow{\mathcal{M}_2}$ followed by $\overrightarrow{F_1}$

Following an approach similar to section 2.4.1.1 by taking into account Figures 2.29 and 2.30, we can see that:

$$W' = W_1' + W_2' = \frac{1}{2}\mathcal{M}_2\theta_{z22} + \frac{1}{2}F_1 v_{11} + \mathcal{M}_2\theta_{z21} = E_{pot.} \qquad [2.64]$$

The final deformed state being the same, we have:

$$W = W'$$

---

23 On account of the linear elastic nature of the material, the progressive application of $\mathcal{M}_2$ leads us to the following diagram (see also [2.21]):

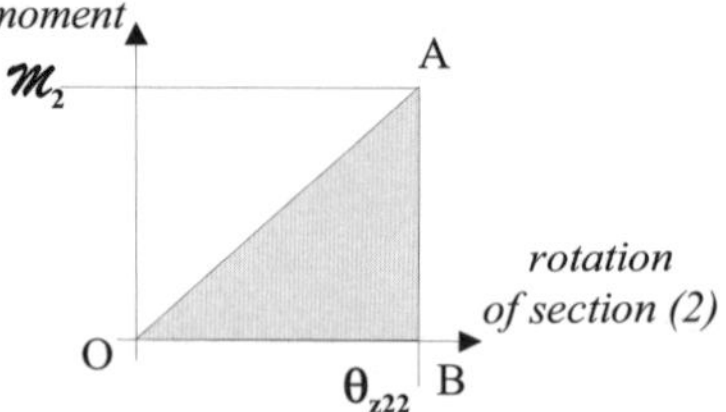

which leads us to:

$$F_1 v_{12} = \mathcal{M}_2 \theta_{z21} \qquad [2.65]$$

○ Total displacements under full load

These can be written as (see Figure 2.31):

$$v_1 = v_{11} + v_{12}$$
$$\theta_{z2} = \theta_{z21} + \theta_{z22} \qquad [2.66]$$

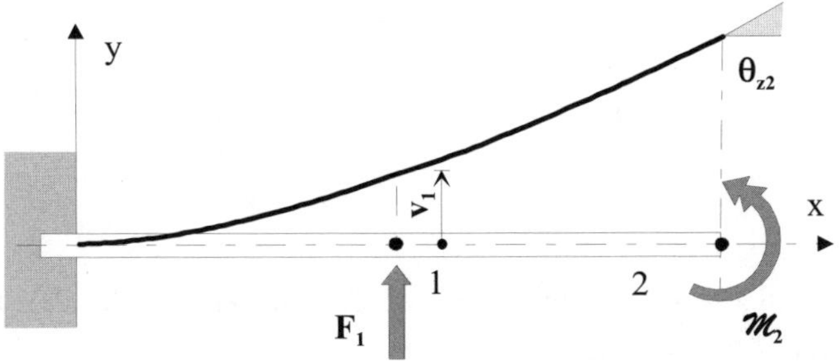

Figure 2.31. *Total displacements*

Considering [2.65], the expressions of work [2.63] and [2.64] can be re-written as:

$$W = \frac{1}{2}F_1 v_{11} + \frac{1}{2}\mathcal{M}_2\theta_{z22} + \frac{1}{2}\left(F_1 v_{12} + \mathcal{M}_2\theta_{z21}\right) = E_{pot.}$$

$$W = \frac{1}{2}F_1\left(v_{11} + v_{12}\right) + \frac{1}{2}\mathcal{M}_2\left(\theta_{z21} + \theta_{z22}\right) = E_{pot.}$$

or with [2.66]:

$$W = \frac{1}{2}F_1 v_1 + \frac{1}{2}\mathcal{M}_2\theta_{z2} = E_{pot.}$$

or in matrix form:

$$W = \frac{1}{2}\begin{bmatrix}F_1 & \mathcal{M}_2\end{bmatrix} \bullet \begin{Bmatrix}v_1 \\ \theta_{z2}\end{Bmatrix} = \frac{1}{2}\begin{Bmatrix}F_1 \\ \mathcal{M}_2\end{Bmatrix}^T \bullet \begin{Bmatrix}v_1 \\ \theta_{z2}\end{Bmatrix} \qquad [2.67]$$

which we can summarize by the general expression already used in the previous example (see [2.51]):

$$W = \frac{1}{2}\{F\}^T \bullet \{d\} \qquad [2.68]$$

where $\{F\} = \left\{ \begin{array}{c} F_1 \\ \mathcal{M}_2 \end{array} \right\}$ is the load vector and $\{d\} = \left\{ \begin{array}{c} v_1 \\ \theta_{z2} \end{array} \right\}$ the degree of freedom vector.

○ Flexibility coefficients.

If the loading is reduced to unit forces, we can find the connections:

$$\begin{array}{ccc} F_1 \rightarrow 1\ N & & v_{11} \rightarrow \alpha_{11} \\ \mathcal{M}_2 \rightarrow 1\ N.m & \Rightarrow & \begin{array}{c} v_{12} \rightarrow \alpha_{12} \\ \theta_{z22} \rightarrow \alpha_{22} \\ \theta_{z21} \rightarrow \alpha_{21} \end{array} \end{array}$$

Thus, we will obtain for any values of F_1 and $\mathcal{M}_2$ in the domain of linear elastic behavior of the structure:

$$\begin{aligned} v_{11} &= \alpha_{11} \times F_1 \\ v_{12} &= \alpha_{12} \times \mathcal{M}_2 \\ \theta_{z22} &= \alpha_{22} \times \mathcal{M}_2 \\ \theta_{z21} &= \alpha_{21} \times F_1 \end{aligned} \qquad [2.69]$$

and [2.65] can be written as:

$$\frac{F_1}{\theta_{z21}} = \frac{\mathcal{M}_2}{v_{12}}$$

i.e.:

$$\alpha_{21} = \alpha_{12} \qquad [2.70]$$

then [2.66] and [2.69] lead to:

$$\begin{aligned} v_1 &= \alpha_{11} \times F_1 + \alpha_{12} \times \mathcal{M}_2 \\ \theta_{z2} &= \alpha_{21} \times F_1 + \alpha_{22} \times \mathcal{M}_2 \end{aligned}$$

or even:

$$\left\{ \begin{array}{c} v_1 \\ \theta_{z2} \end{array} \right\} = \left[\begin{array}{cc} \alpha_{11} & \alpha_{12} \\ \alpha_{21} & \alpha_{22} \end{array} \right] \bullet \left\{ \begin{array}{c} F_1 \\ \mathcal{M}_2 \end{array} \right\} \qquad [2.71]$$

Let us introduce this relation in expression [2.68] of the work produced by $\vec{F_1}$ and $\overrightarrow{\mathcal{M}_2}$:

$$W = \frac{1}{2}\begin{Bmatrix} F_1 \\ \mathcal{M}_2 \end{Bmatrix}^T \bullet \begin{bmatrix} \alpha_{11} & \alpha_{12} \\ \alpha_{21} & \alpha_{22} \end{bmatrix} \bullet \begin{Bmatrix} F_1 \\ \mathcal{M}_2 \end{Bmatrix} = E_{pot.} \qquad [2.72]$$

We can also find here the possibility of expressing this work in the condensed form already seen earlier (see [2.60]):

$$W = \frac{1}{2}\{F\}^T \bullet [\alpha] \bullet \{F\} = E_{pot.} \qquad [2.73]$$

NOTES

❑ As in the previous example, we can see in relation [2.71] or in the expression for potential deformation energy [2.72] a flexibility matrix which is square and symmetric since $\alpha_{12} = \alpha_{21}$ (see [2.70]).

❑ In relation [2.71] we have on each row the load and its associated degree of freedom, i.e. here:

– force F_1 and its degree of freedom v_1;

– moment $\mathcal{M}_2$ and its degree of freedom θ_{z2}.

It is important to note that this association is double:

➢ *geometric* association: same direction for the load and its associated degree of freedom;

➢ *energy* association: the work results from the product of the loads and their associated degrees of freedom (see [2.67]).

❑ From this example, we can see that the notion of load and associated degree of freedom extends to the case of moments and section rotations of beams.

❑ Dimensional homogenity of flexibility coefficients: relation [2.71] shows that:

– α_{11} linear displacement per unit force is expressed in $m \times N^{-1}$;

– α_{22} angular displacement per unit moment is expressed in $rad \times (N.m)^{-1}$, i.e. in $(N \times m)^{-1}$;

– α_{12} (coupling term) is expressed in $rad \times N^{-1}$ and α_{21} in $m \times (N.m)^{-1}$, or for these two coefficients: N^{-1}.

○ Application: explicit calculation of the flexibility matrix

Let us consider characteristics of beams identical to the example in the previous section (Young's modulus E, quadratic moment around $\vec{z}$ axis: I_z denoted I).

Figure 2.26 is still representative of force F_1, and we obtain for $x \leq x_1$ (see [2.61]):

$$EIv(x) = F_1 \frac{x^2}{2}(x_1 - \frac{x}{3})$$

from which:

– at point (1): $v_{11} = F_1 \dfrac{x_1^3}{3EI}$

– at point (2) (rectilinear zone 1-2): $\theta_{z21} = \theta_z(x_1) \cong \dfrac{dv(x_1)}{dx} = F_1 \dfrac{x_1^2}{2EI}$

Let us now consider only the moment $\mathcal{M}_2$ at abscissa x_2. In zone $0 \leq x \leq x_2$, deformation $v(x)$ proves the differential equation (see section 1.5.2.6):

$$EI \frac{d^2v}{dx^2} = \mathcal{M}f_z = \mathcal{M}_2$$

We therefore successively obtain (the integration constants disappear due to clamping at $x = 0$):

$$EI \frac{dv(x)}{dx} = \mathcal{M}_2 x$$

$$EIv(x) = \mathcal{M}_2 \frac{x^2}{2}$$

We can particularly note:

– at point (1): $v_{12} = \mathcal{M}_2 \dfrac{x_1^2}{2EI}$

– at point (2): $\theta_{z22} = \theta_z(x_2) \cong \dfrac{dv(x_2)}{dx} = \mathcal{M}_2 \dfrac{x_2}{EI}$

The total displacements are then written as:

$$v_1 = v_{11} + v_{12} = F_1 \frac{x_1^3}{3EI} + \mathcal{M}_2 \frac{x_1^2}{2EI}$$

$$\theta_{z2} = \theta_{z21} + \theta_{z22} = F_1 \frac{x_1^2}{2EI} + \mathcal{M}_2 \frac{x_2}{EI}$$

or in the matrix form:

$$\begin{Bmatrix} v_1 \\ \theta_{z2} \end{Bmatrix} = \begin{bmatrix} \dfrac{x_1^{\,3}}{3EI} & \dfrac{x_1^{\,2}}{2EI} \\ \dfrac{x_1^{\,2}}{2EI} & \dfrac{x_2}{EI} \end{bmatrix} \bullet \begin{Bmatrix} F_1 \\ \mathcal{M}_2 \end{Bmatrix}$$

[2.74]

we can identify once again form [2.71], or $\{d\} = [\alpha] \bullet \{F\}$, notably with the symmetry of coefficients $\alpha_{12} = \alpha_{21}$.

2.4.2. *Beam in plane bending subject to "n" forces*

Let us consider in Figure 2.32 a beam with symmetry plane (xy) again embedded on the left end and free on the right, where "n" cross-sections with centers (1), (2),…,(n) have been marked. At these points, forces $\vec{F_1}, \vec{F_2}, \ldots, \vec{F_n}$ which are all colinear to the $\vec{y}$ axis have been applied.

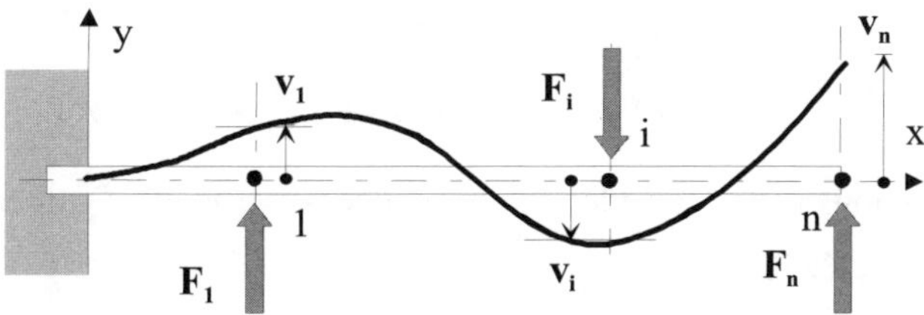

Figure 2.32. *Beam subject to "n" forces*

We can consider this global loading $\vec{F_1}, \ldots \vec{F_n}$, as a resultant of the partial successive loads:

– $\vec{F_1}$, then $\vec{F_2}$, followed by $\vec{F_3}, \ldots \vec{F_n}$;

– or $\vec{F_2}$, then $\vec{F_1}$, followed by $\vec{F_3}, \ldots \vec{F_n}$;

– or all sequences taken from the possible combinations of $\vec{F}_i$.

The final state will always be the same (giving the same deformation for the beam).

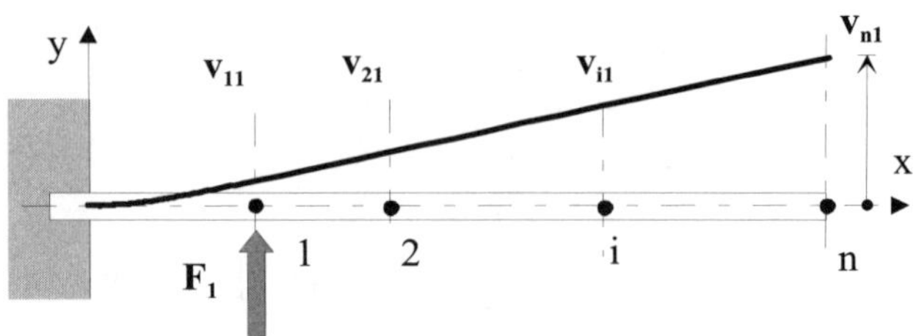

Figure 2.33. *Application of load 1*

If we consider for example the two successive loads $\overrightarrow{F_1}$ and $\overrightarrow{F_2}$ we return to the example of section 2.4.1.1 with the same notations (see Figure 2.33) and results, particularly relations [2.53] which we recall here:

$$v_{11} = \alpha_{11} \times F_1 ; \quad v_{12} = \alpha_{12} \times F_2 ; \quad v_{22} = \alpha_{22} \times F_2 ; \quad v_{21} = \alpha_{21} \times F_1 ; \quad \alpha_{12} = \alpha_{21}{}^{24}$$

If we consider for example only the successive loads $\overrightarrow{F_1}$ and $\overrightarrow{F_3}$, we will get similar results where index 3 replaces index 2, i.e.:

$$v_{11} = \alpha_{11} \times F_1 ; \quad v_{13} = \alpha_{13} \times F_3 ; \quad v_{33} = \alpha_{33} \times F_3 ; \quad v_{31} = \alpha_{31} \times F_1 ; \quad \alpha_{13} = \alpha_{31}$$

By generalizing, if we consider the two successive loads $\overrightarrow{F_i}$ then $\overrightarrow{F_j}$ alone (i and j $\in [1,....n]$), we will obtain:

$$v_{ii} = \alpha_{ii} \times F_i; \ v_{ij} = \alpha_{ij} \times F_j; \ v_{jj} = \alpha_{jj} \times F_j; \ v_{ji} = \alpha_{ji} \times F_i ; \alpha_{ij} = \alpha_{ji} \qquad [2.75]$$

The results of the example from section 2.4.1.1 can then be easily generalized to the present case.

○ Total displacement under complete load sequence

Thus, the total displacements [2.48] here become:

$$v_1 = v_{11} + v_{12} + v_{13} +......+ v_{1n} = \alpha_{11} \times F_1 + \alpha_{12} \times F_2 +.....+ \alpha_{1n} \times F_n$$

$$v_2 = v_{21} + v_{22} + v_{23} +......+ v_{2n} = \alpha_{21} \times F_1 + \alpha_{22} \times F_2 +.....+ \alpha_{2n} \times F_n$$

$$...$$

$$v_n = v_{n1} + v_{n2} + v_{n3} +......+ v_{nn} = \alpha_{n1} \times F_n + \alpha_{n2} \times F_2 +.....+ \alpha_{nn} \times F_n$$

---

24 Let us remember that the coefficients α_{ij} are displacements per unit force, i.e. resulting from the application of forces $F_i = F_j = 1$ N.

This can be more conveniently grouped together in the matrix form:

$$\begin{Bmatrix} v_1 \\ v_2 \\ .... \\ v_n \end{Bmatrix} = \begin{bmatrix} \alpha_{11} & \alpha_{12} & ... & \alpha_{1n} \\ \alpha_{21} & \alpha_{22} & ... & \alpha_{2n} \\ ... & ... & ... & ... \\ \alpha_{n1} & \alpha_{n2} & ... & \alpha_{nn} \end{bmatrix} \bullet \begin{Bmatrix} F_1 \\ F_2 \\ .... \\ F_n \end{Bmatrix} \qquad [2.76]$$

an expression where we find once again a "flexibility matrix" $[\alpha]$ square and symmetric (as per [2.75]), with "n" rows and "n" columns.

○ Relationship between work and flexibility

By following the stages of calculation which had led for example to expression [2.45], the work developed (or the deformation energy stored) becomes for the "n" forces $\overrightarrow{F_1}$ .... $\overrightarrow{F_n}$ [25]:

$$W = \frac{1}{2}(F_1 \times v_{11} + F_1 \times v_{12} + F_1 \times v_{13} + ... + F_1 \times v_{1n} .....$$
$$...... + F_2 \times v_{22} + F_2 \times v_{21} + F_2 \times v_{23} + ... + F_2 \times v_{2n} ......$$
$$...... + ...$$
$$...... + F_n \times v_{nn} + F_n \times v_{n1} + F_n \times v_{n2} + ... + F_n \times v_{n(n-1)}) = E_{pot.}$$

or even with the flexibility coefficients appearing in [2.75]:

$$W = \frac{1}{2}(F_1 \times \alpha_{11} \times F_1 + F_1 \times \alpha_{12} \times F_2 + F_1 \times \alpha_{13} \times F_3 + ... + F_1 \times \alpha_{1n} \times F_n .....$$
$$... + ...$$
$$... + F_n \times \alpha_{nn} \times F_n + F_n \times \alpha_{n1} \times F_1 + F_n \times \alpha_{n2} \times F_2 + ... + F_n \times \alpha_{n(n-1)} \times F_{(n-1)}) = E_{pot.}$$

which can be re-written as:

$$W = \frac{1}{2}\underset{\substack{line \ matrix \\ (1\times n)}}{\{F_1 \quad F_2 \quad .. \quad F_n\}} \bullet \underset{coloumn \ matrix \ (n\times 1)}{\begin{Bmatrix} \alpha_{11} \times F_1 + \alpha_{12} \times F_2 + \alpha_{13} \times F_3 + ... + \alpha_{1n} \times F_n \\ .... \\ .... \\ \alpha_{n1} \times F_1 + \alpha_{n2} \times F_2 + \alpha_{n3} \times F_3 + ... + \alpha_{nn} \times F_n \end{Bmatrix}} = E_{pot.}$$

---

25 We may recall that we can write for example $F_1 \times v_{12} = \frac{1}{2}(F_1 \times v_{12} + F_2 \times v_{21})$ on account of the equality [2.47]: $F_1 \times v_{12} = F_2 \times v_{21}$, or in a general manner: $F_i \times v_{ij} = \frac{1}{2}(F_i \times v_{ij} + F_j \times v_{ji})$.

or even:

$$W = \frac{1}{2}\begin{Bmatrix} F_1 \\ F_2 \\ .... \\ F_n \end{Bmatrix}^T \bullet \begin{bmatrix} \alpha_{11} & \alpha_{12} & ... & \alpha_{1n} \\ \alpha_{21} & \alpha_{22} & ... & \alpha_{2n} \\ ... & ... & ... & ... \\ \alpha_{n1} & \alpha_{n2} & ... & \alpha_{nn} \end{bmatrix} \bullet \begin{Bmatrix} F_1 \\ F_2 \\ .... \\ F_n \end{Bmatrix} = E_{pot.} \qquad [2.77]$$

$$\quad\;\; (1{\times}n) \qquad\qquad\qquad (n{\times}n) \qquad\qquad\qquad (n{\times}1)$$

We thus get a quadratic form[26] in F_i and F_j built on flexibility matrix $[\alpha]$, which is square and symmetric and of dimension $(n \times n)$.

NOTE

❏ By introducing total displacements v_1, v_2,..., v_n given by [2.76], [2.77] can be written as:

$$W = \frac{1}{2}\begin{Bmatrix} F_1 \\ F_2 \\ ... \\ F_n \end{Bmatrix}^T \bullet \begin{Bmatrix} v_1 \\ v_2 \\ .. \\ v_n \end{Bmatrix} = E_{pot.} \qquad [2.78]$$

or in expanded form:

$$W = \frac{1}{2}(F_1 \times v_1 + F_2 \times v_2 + ... + F_n \times v_n) = E_{pot.}$$

❏ As in the previous examples, we can see that each row of relation [2.76] exhibits a force F_i and its associated degree of freedom v_i. Let us note that even in this more general case, we still find this double association:

➤ *geometric* association: same direction for the load and its associated degree of freedom;

➤ *energy* association: the work results from the product of the loads and their associated degree of freedom (see [2.67]).

We can condense expressions [2.76], [2.77], and [2.78] by noting as in the previous examples:

---

26 See Chapter 12, section 12.1.3.

$$\{F\} = \begin{Bmatrix} F_1 \\ F_2 \\ ... \\ F_n \end{Bmatrix} \quad ; \quad \{d\} = \begin{Bmatrix} v_1 \\ v_2 \\ .. \\ v_n \end{Bmatrix} \quad ; \quad [\alpha] = \begin{bmatrix} \alpha_{11} & \alpha_{12} & ... & \alpha_{1n} \\ \alpha_{21} & \alpha_{22} & ... & \alpha_{2n} \\ ... & ... & ... & ... \\ \alpha_{n1} & \alpha_{n2} & ... & \alpha_{nn} \end{bmatrix}$$

load vector   degrees of freedom
(dof) vectors   flexibility matrix (n×n)

we then get an expression which we have already seen, summarized below:

$$\{d\} = [\alpha] \bullet \{F\}$$

$$W = \frac{1}{2}\{F\}^T \bullet [\alpha] \bullet \{F\} = E_{pot.} ; \quad W = \frac{1}{2}\{F\}^T \bullet \{d\} = E_{pot.} \qquad [2.79]$$

2.4.3. Generalization to any structure

2.4.3.1. Structure loaded by two forces $\vec{F_1}$ and $\vec{F_2}$

Let us consider any structure (see Figure 2.34), two points (1) and (2) belonging to this structure and two associated directions, Δ_1 in (1) and Δ_2 in (2), along which forces $\vec{F_1}$ and $\vec{F_2}$ are applied respectively. The total displacements at points (1) and (2) are denoted here $\vec{d_1}$ and $\vec{d_2}$. Their directions are in general *distinct* from the directions Δ_1 and Δ_2.

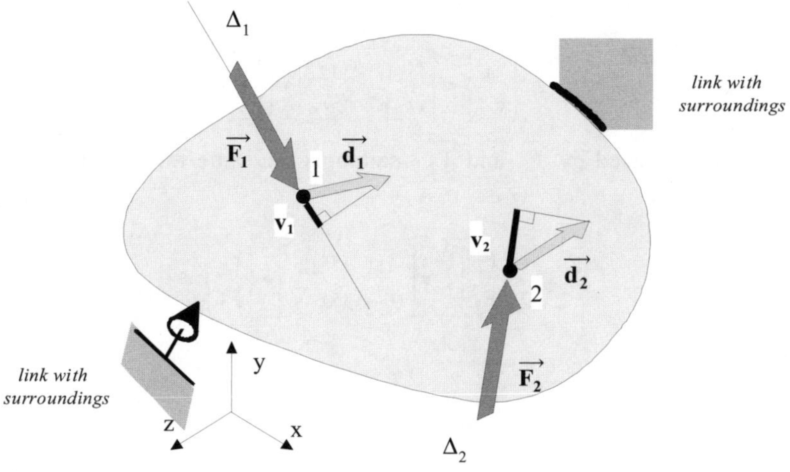

Figure 2.34. *A structure loaded by forces $\vec{F_1}$ and $\vec{F_2}$*

We can follow the now familiar approach used in the previous examples:

- let us first apply force $\vec{F_1}$ progressively. The structure is deformed and points (1) and (2) are displaced[27]. These displacements are *projected* along directions Δ_1 and Δ_2. We denote by:

   $- v_{11}$ the projection on Δ_1 of the displacement at (1),

   $- v_{21}$ the projection on Δ_2 of the displacement at (2);

- let us then apply force $\vec{F_2}$ progressively. The displacements of points (1) and (2) are *projected* along the directions Δ_1 and Δ_2. We denote by:

   $- v_{22}$ the projection on Δ_2 of the displacement at (2);

   $- v_{12}$ the projection on Δ_1 of the displacement at (1).

All the results mentioned in section 2.4.1.1 for the cantilever beam loaded by two forces are formally maintained for this case. We can particularly note:

$$v_{11} = \alpha_{11} \times F_1 ; \quad v_{12} = \alpha_{12} \times F_2 ; \quad v_{22} = \alpha_{22} \times F_2 ; \quad v_{21} = \alpha_{21} \times F_1 ; \quad \alpha_{21} = \alpha_{12}$$

The total displacements projected respectively on Δ_1 and Δ_2 (see Figure 2.34) can be written as:

$$v_1 = v_{11} + v_{12} = \alpha_{11} \times F_1 + \alpha_{12} \times F_2$$

$$v_2 = v_{21} + v_{22} = \alpha_{21} \times F_1 + \alpha_{22} \times F_2$$

$$\left\{ \begin{array}{c} v_1 \\ v_2 \end{array} \right\} = \left[\begin{array}{cc} \alpha_{11} & \alpha_{12} \\ \alpha_{21} & \alpha_{22} \end{array} \right] \bullet \left\{ \begin{array}{c} F_1 \\ F_2 \end{array} \right\} \tag{2.80}$$

The work produced by $\vec{F_1}$ and $\vec{F_2}$ maintains the same form expressed earlier, i.e.:

$$W = \frac{1}{2} \left\{ \begin{array}{c} F_1 \\ F_2 \end{array} \right\}^T \bullet \left[\begin{array}{cc} \alpha_{11} & \alpha_{12} \\ \alpha_{21} & \alpha_{22} \end{array} \right] \bullet \left\{ \begin{array}{c} F_1 \\ F_2 \end{array} \right\} = E_{pot.}$$

or with [2.80]:

---

27 We may recall that these elastic displacements are very small and do not modify in a perceptible manner the geometry of the studied structure.

$$W = \frac{1}{2}\begin{Bmatrix} F_1 \\ F_2 \end{Bmatrix}^T \bullet \begin{Bmatrix} v_1 \\ v_2 \end{Bmatrix} = E_{pot.}$$

2.4.3.2. Structure loaded by "n" forces $\overrightarrow{F_1},...\overrightarrow{F_n}$

Let us consider the previous structure (Figure 2.34) where we now mark not just two but "n" points (i) = 1, 2,…, n. Let us associate a direction Δ_i, with each of these points i.e. Δ_1 at (1), Δ_2 at (2),…, Δ_n at (n) . This structure is loaded by "n" forces $\overrightarrow{F_1},...\overrightarrow{F_n}$. A force $\overrightarrow{F_i}$ is applied at point (i) along the direction Δ_i.

If we consider for example the two successive loads $\overrightarrow{F_1}$ and $\overrightarrow{F_2}$, we obtain for the displacements along directions Δ_1 and Δ_2 the results of the previous example, in particular:

$$v_{11} = \alpha_{11} \times F_1; \quad v_{12} = \alpha_{12} \times F_2; \quad v_{22} = \alpha_{22} \times F_2; \quad v_{21} = \alpha_{21} \times F_1; \quad \alpha_{12} = \alpha_{21}$$

By generalizing, if we consider the two successive loads $\overrightarrow{F_i}$ then $\overrightarrow{F_j}$ acting alone (i and $j \in [1,..,n]$) we obtain:

$$v_{ii} = \alpha_{ii} \times F_i \; ; \quad v_{ij} = \alpha_{ij} \times F_j \; ; \quad v_{jj} = \alpha_{jj} \times F_j \; ; \quad v_{ji} = \alpha_{ji} \times F_i \; ; \quad \alpha_{ij} = \alpha_{ji} \qquad [2.81]$$

where v_{ij} is the projection on Δ_i of the displacement in (i) due to the application of the force $\overrightarrow{F_j}$ at point (j).

◯ Total displacements under complete load sequence

With [2.81], the total displacements along directions Δ_i can be written in a manner which is similar to what was seen in section 2.4.2:

$$v_1 = v_{11} + v_{12} + v_{13} +......+ v_{1n} = \alpha_{11} \times F_1 + \alpha_{12} \times F_2 +.....+ \alpha_{1n} \times F_n$$

$$v_2 = v_{21} + v_{22} + v_{23} +......+ v_{2n} = \alpha_{21} \times F_1 + \alpha_{22} \times F_2 +.....+ \alpha_{2n} \times F_n$$

$$\cdots\cdots\cdots\cdots\cdots\cdots\cdots\cdots\cdots\cdots\cdots\cdots\cdots\cdots\cdots\cdots\cdots\cdots$$

$$v_n = v_{n1} + v_{n2} + v_{n3} +......+ v_{nn} = \alpha_{n1} \times F_n + \alpha_{n2} \times F_2 +.....+ \alpha_{nn} \times F_n$$

i.e.:

$$\begin{Bmatrix} v_1 \\ v_2 \\ .... \\ v_n \end{Bmatrix} = \begin{bmatrix} \alpha_{11} & \alpha_{12} & ... & \alpha_{1n} \\ \alpha_{21} & \alpha_{22} & ... & \alpha_{2n} \\ ... & ... & ... & ... \\ \alpha_{n1} & \alpha_{n2} & ... & \alpha_{nn} \end{bmatrix} \bullet \begin{Bmatrix} F_1 \\ F_2 \\ .... \\ F_n \end{Bmatrix} \qquad [2.82]$$

where we get the square and symmetric flexibility matrix $[\alpha]\,(n \times n)$.

○ Relationship between work and flexibility

We will see that the work produced (the potential energy stored) by the "n" forces $\vec{F_1}, ..., \vec{F_n}$ can be formally written as in the application of section 2.4.2, i.e.:

$$W = \frac{1}{2} \begin{Bmatrix} F_1 \\ F_2 \\ .... \\ F_n \end{Bmatrix}^T \bullet \begin{bmatrix} \alpha_{11} & \alpha_{12} & ... & \alpha_{1n} \\ \alpha_{21} & \alpha_{22} & ... & \alpha_{2n} \\ ... & ... & ... & ... \\ \alpha_{n1} & \alpha_{n2} & ... & \alpha_{nn} \end{bmatrix} \bullet \begin{Bmatrix} F_1 \\ F_2 \\ .... \\ F_n \end{Bmatrix} = E_{pot.} \qquad [2.83]$$

NOTE

❑ By introducing the total displacements [2.82], [2.83] can be written as:

$$W = \frac{1}{2} \begin{Bmatrix} F_1 \\ F_2 \\ ... \\ F_n \end{Bmatrix}^T \bullet \begin{Bmatrix} v_1 \\ v_2 \\ .. \\ v_n \end{Bmatrix} = E_{pot.} \qquad [2.84]$$

❑ Each row of relation [2.82] exhibits a force F_i and its associated degree of freedom v_i. We must note that the latter is not generally the *real* displacement of point (i) where force F_i is applied. The degree of freedom v_i is only *the projection of this displacement* along direction Δ_i of force F_i.

❑ The double association already observed in the previous examples is still present:

➤ *geometric* association: same direction Δ_i for load F_i and its associated degree of freedom v_i.

➤ *energy* association: [2.84] results from the product of the loads and their associated degree of freedom.

❑ The "flexibility" of this structure, characterized by the flexibility matrix $[\alpha]$ seems to be associated with a particular loading at particular points. It probably depends closely on the geometric and material properties of the structure, but does not constitute an intrinsic property of this structure. In fact, if we modify the location and the number of points (i) as well as the direction of the forces, we will obtain a new associated flexibility matrix.

❏ We can condense expressions [2.82], [2.83], [2.84] in the forms mentioned earlier (see [2.78]).

$$\{d\} = [\alpha] \bullet \{F\}$$

$$W = \frac{1}{2}\{F\}^T \bullet [\alpha] \bullet \{F\} = E_{pot.} ; \quad W = \frac{1}{2}\{F\}^T \bullet \{d\} = E_{pot.} \qquad [2.85]$$

2.4.3.3. *A search for real displacements on a loaded structure*[28]

In the previous two examples the displacements considered were projections of the real displacements along the directions of the forces. These therefore provide only partial information on the real space displacements of the points of application of the forces.

Let us consider a case where we would like to know, not the displacements projected along fixed directions $\Delta_1,...,\Delta_n$, but the real displacements $\vec{d_1}$, $\vec{d_2},...,\vec{d_n}$ (see Figure 2.35). We can then associate with each point i three orthogonal directions $\Delta_{ix}, \Delta_{iy}, \Delta_{iz}$ parallel to each of the axis of reference coordinates $\vec{x}, \vec{y}, \vec{z}$), or global coordinates.

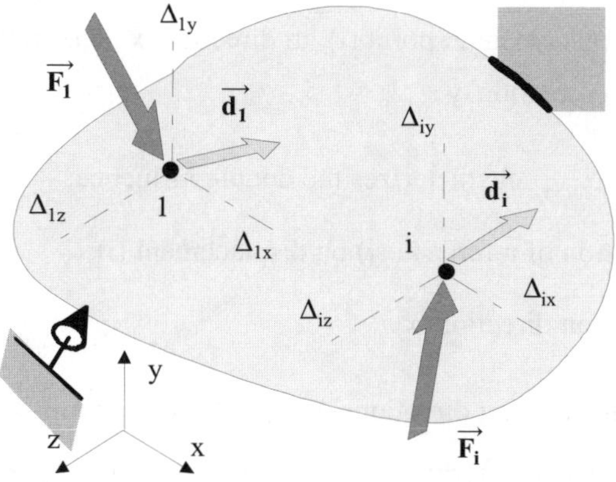

Figure 2.35. *Displacements and global coordinates*

---

28 Note: this case implies heavier notations than the previous cases.

At point (i), force $\vec{F_i} = X_i\,\vec{x} + Y_i\,\vec{y} + Z_i\,\vec{z}$ is applied and the corresponding total displacement is $\vec{d_i} = u_i\,\vec{x} + v_i\,\vec{y} + w_i\,\vec{z}$.

The results of the previous example can then be adapted in the following manner.

○ Total displacements under complete load sequence

♦ The total displacement along direction Δ_{1x} can be written as:

$$u_1 = \underbrace{u_{11x}}_{\substack{\text{contribution} \\ \text{of } X_1}} + \underbrace{u_{11y}}_{\substack{\text{contribution} \\ \text{of } Y_1}} + \underbrace{u_{11z}}_{\substack{\text{contribution} \\ \text{of } Z_1}} + \underbrace{u_{12x}}_{\substack{\text{contribution} \\ \text{of } X_2}} + \underbrace{u_{12y}}_{\substack{\text{contribution} \\ \text{of } Y_2}} + \underbrace{u_{12z}}_{\substack{\text{contribution} \\ \text{of } Z_2}}$$

$$+ \,\dots\, + \underbrace{u_{1nx}}_{\substack{\text{contribution} \\ \text{of } X_n}} + \underbrace{u_{1ny}}_{\substack{\text{contribution} \\ \text{of } Y_n}} + \underbrace{u_{1nz}}_{\substack{\text{contribution} \\ \text{of } Z_n}}$$

i.e.:

$$u_1 = \alpha_{11xx} \times X_1 + \alpha_{11xy} \times Y_1 + \alpha_{11xz} \times Z_1 + \alpha_{12xx} \times X_2 + \alpha_{12xy} \times Y_2 + \alpha_{12xz} \times Z_2$$
$$+ \dots + \alpha_{1nxz} \times Z_n$$

where as in the earlier case the coefficients $\alpha_{ijk\ell}$ are displacements per unit force, i.e. resulting from the action of forces $X_i = Y_i = Z_i = 1N$. We will have for example:

$$u_{ijy} = \alpha_{ijxy} \times Y_j \qquad\qquad [2.86]$$

where u_{ijy} is the displacement at point (i), in direction $\vec{x}$, due to the action of a force at point (j) acting in direction $\vec{y}$.

The coefficient α_{ijxy} characterizes the double influence:

– of the application of a force at (j) on displacement (i);

– of direction $\vec{y}$ on direction $\vec{x}$.

In a similar manner along directions Δ_{1y} and Δ_{1z} we have:

$$v_1 = \alpha_{11yx} \times X_1 + \alpha_{11yy} \times Y_1 + \alpha_{11yz} \times Z_1 + \alpha_{12yx} \times X_2 + \alpha_{12yy} \times Y_2 + \alpha_{12yz} \times Z_2.$$
$$+ \dots + \alpha_{1nyz} \times Z_n$$

$$w_1 = \alpha_{11zx} \times X_1 + \alpha_{11zy} \times Y_1 + \alpha_{11zz} \times Z_1 + \alpha_{12zx} \times X_2 + \alpha_{12zy} \times Y_2 + \alpha_{12zz} \times Z_2$$
$$+ \ldots\ldots\ldots + \alpha_{1nzz} \times Z_n$$

Thus, in a matrix form we obtain the equation:

$$\begin{Bmatrix} u_1 \\ v_1 \\ w_1 \\ u_2 \\ \cdots \\ v_n \\ w_n \end{Bmatrix} = \begin{bmatrix} \alpha_{11xx} & \alpha_{11xy} & \alpha_{11xz} & \alpha_{12xx} & \alpha_{12xy} & \cdots & \alpha_{1nxz} \\ \alpha_{11yx} & \alpha_{11yy} & \alpha_{11yz} & \alpha_{12yx} & \alpha_{12yy} & \cdots & \alpha_{1nyx} \\ \alpha_{11zx} & \alpha_{11zy} & \alpha_{11zz} & \alpha_{12zx} & \alpha_{12zy} & \cdots & \alpha_{1nzx} \\ \alpha_{21xx} & \cdots & & & & & \cdots \\ \cdots & & & & & & \\ \cdots & & & & & & \cdots \\ \alpha_{n1zx} & & & & & & \alpha_{nnzz} \end{bmatrix} \bullet \begin{Bmatrix} X_1 \\ Y_1 \\ Z_1 \\ X_2 \\ \cdots \\ Y_n \\ Z_n \end{Bmatrix} \quad [2.87]$$

in which a square and symmetric[29] flexibility matrix $[\alpha](3n \times 3n)$ appear.

◯ Relationship between work and flexibility

The work developed by the forces "n", $\vec{F_1}, \vec{F_2}, \ldots, \vec{F_n}$, or by the "$3 \times n$" components X_i, Y_i, Z_i, with i = 1, 2,…n, can be written in a manner similar to the one in all the previous examples:

$$W = \frac{1}{2} \underset{1 \times 3n}{\begin{Bmatrix} X_1 \\ Y_1 \\ \cdots \\ Z_n \end{Bmatrix}}^T \bullet \underset{3n \times 3n}{[\alpha]} \bullet \underset{3n \times 1}{\begin{Bmatrix} X_1 \\ Y_1 \\ \cdots \\ Z_n \end{Bmatrix}} = E_{pot.} \quad [2.88]$$

If [2.87] is combined with equation [2.88] we obtain:

---

29 The symmetry of the coefficients α_{ijxy} appears as in the previous examples: let us suppose that the load is reduced to:

    − a force X_i at point (i)

    − a force Y_j at point (j)

the transcription of the work developed by these two forces applied successively (first X_i then Y_j) will lead, as in the example in section 2.4.1.1 to:

$$W = \frac{1}{2} X_i u_{iix} + \frac{1}{2} Y_j v_{jjy} + X_i u_{ijy}$$

If the order of application is reversed, we will have:

$$W' = \frac{1}{2} Y_j v_{jjy} + \frac{1}{2} X_i u_{iix} + Y_j v_{jix}$$

from where using identification W = W', the equality $X_i u_{ijy} = Y_j v_{jix}$. That is, with [2.86] $X_i \times \alpha_{ijxy} \times Y_j = Y_j \times \alpha_{jiyx} \times X_i$ leading to the symmetry property $\alpha_{ijxy} = \alpha_{jiyx}$.

$$W = \frac{1}{2}\begin{Bmatrix} X_1 \\ Y_1 \\ Z_1 \\ X_2 \\ Y_2 \\ Z_2 \\ ... \\ Z_n \end{Bmatrix}^{T} \bullet \begin{Bmatrix} u_1 \\ v_1 \\ w_1 \\ u_2 \\ v_2 \\ w_2 \\ ... \\ w_n \end{Bmatrix} = E_{pot.}$$

[2.89]

that is:

$$W = \frac{1}{2}\left(X_1 \times u_1 + Y_1 \times v_1 + Z_1 \times w_1 + X_2 \times u_2 + ... + Z_n \times w_n\right) = E_{pot.}$$

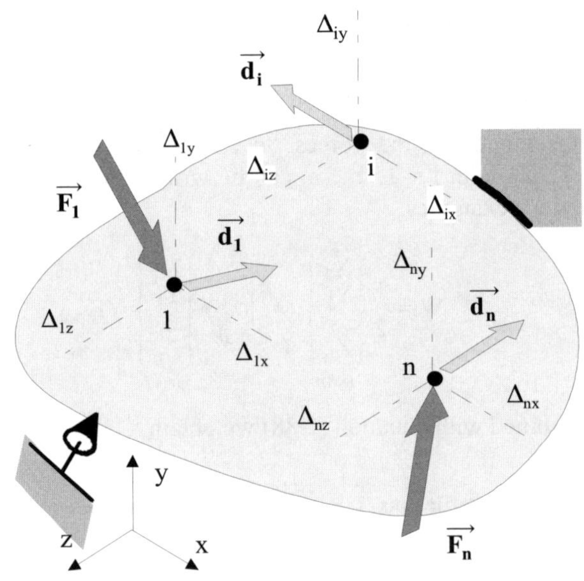

Figure 2.36. *Displacement $\vec{d_i}$ of a point (i) that is not loaded*

NOTES

❏ The spatial displacements can be reconstituted from the three components of displacement $\vec{d_i}$ at every point (i).

❏ Every line of equation [2.87] introduces a load X_i (or Y_i, or Z_i) and "its" associated degree of freedom u_i (or v_i, or w_i).

The double association always shows up:

➤ *geometric* association: same direction $\vec{x}$ (or $\vec{y}$, or $\vec{z}$) for the load and its degree of freedom;

➤ *energy* association: the work developed [2.89] results from the product of the loads and their associated displacements.

❏ In equation [2.87], the components X_i, Y_i, Z_i of the load can assume any algebraic[30] values, and in particular zero at certain points (i) (see Figure 2.36). Thus, points with incomplete or no loading at all can be considered: equation [2.87] will provide the displacements at these points.

❏ The "flexibility" of this structure, characterized by the flexibility matrix $[\alpha](3n \times 3n)$ of equation [2.87] appears linked to a set of special points. While it depends closely on the geometric properties and the material of the structure, it does not form an inherent property of it. In fact, if the location and the number of points (i) are modified, a new associated flexibility matrix is obtained.

❏ Equations [2.87], [2.88] and [2.89] can be condensed to the forms already mentioned (see [2.79] and [2.85]):

$$\{d\} = [\alpha] \bullet \{F\}$$

$$W = \frac{1}{2}\{F\}^T \bullet [\alpha] \bullet \{F\} = E_{pot.}; \quad W = \frac{1}{2}\{F\}^T \bullet \{d\} = E_{pot.} \qquad [2.90]$$

---

30 Provided the structure remains in the linear elastic behavior.

2.4.4. *Summary*

All the examples of the loaded structures in sections 2.4.1 to 2.4.3 have general properties that can be summarized in the following table.

structure	Loading and degrees of freedom of a structure
	❐ any geometric structure; ❐ composed of a linear elastic material; ❐ having links with its surroundings; ❐ associated with structural or global coordinates ($\overrightarrow{x},\overrightarrow{y},\overrightarrow{z}$); ❐ having "n" points (i) = [1,…n] within and on its external surface.
loads	
	❐ each point can be loaded by several force components, for example X_i, Y_i, Z_i at point (i); ❐ the load vector is represented as: $\quad \{F\} = \begin{Bmatrix} X_1 \\ Y_1 \\ Z_1 \\ ... \\ Z_n \end{Bmatrix};$ ❐ the loads can include moments acting on the small zones surrounding points (i).
degrees of freedom	
	❐ the displacement of every point (i) has as components u_i, v_i, w_i. These are the degrees of freedom (dof); ❐ the dof vector is represented as: $\quad \{d\} = \begin{Bmatrix} u_1 \\ v_1 \\ w_1 \\ ... \\ w_n \end{Bmatrix}$ ❐ the dof can include rotations of small areas surrounding points (i).
characteristic equations	

behavior equation:
 	
$\{d\} = [\alpha] \bullet \{F\}$
 	
work/potential energy:
 	
$W = \dfrac{1}{2}\{F\}^T \bullet [\alpha] \bullet \{F\} = E_{pot.}$
 	
$W = \dfrac{1}{2}\{F\}^T \bullet \{d\} = E_{pot.}$
	❐ $[\alpha]$ is the flexibility matrix. It is square and symmetric;
	❐ it has the same number of lines and columns as the degrees of freedom;
	❐ there is a dual association between a force component and its dof:
	   – *geometric* association: same geometric direction for the force and its dof;
	   – *energy* association: the work developed results from the product of the forces and their associated dof.

$$[2.91]$$

2.5. Links of a structure with its surroundings

All the cases dealt with previously represent structures linked to their surroundings (clamped-free beam for example). We will now carry out a closer examination of the influence of these links on the behavior[31].

2.5.1. *Example*

Let us take the case of the beam of section 2.4.1.1 with the same points and the same directions of loading.

In Figure 2.37, the previous clamping is now replaced by a perfect turning pair (pivot) with $\vec{z}$ axis connected to a spiral spring[32] of rigidity k placed between the wall and the beam.

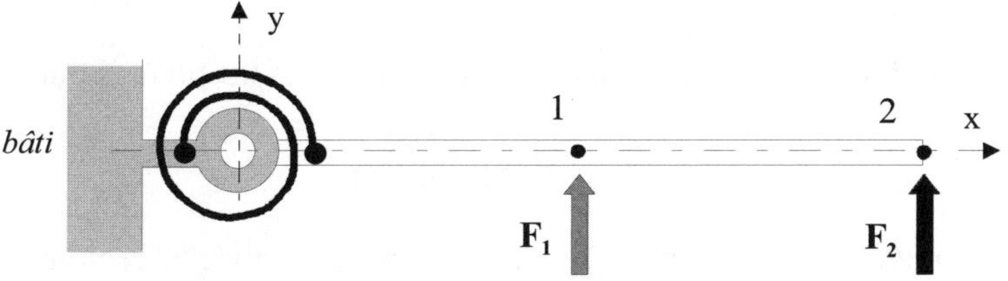

Figure 2.37. *Beam with a link of controlled stiffness*

Thus, a link resulting in a finite rigidity in rotation is obtained. Referring to the relation of proportionality between the moment applied on the axis of the pivot and the resultant rotation (here $\Theta_z \vec{z}$), we obtain

$$\mathcal{M}_{z\,beam\,/\,spring} = k \times \Theta_z$$

We now intend to determine the flexibility matrix of this structure thus linked to its surroundings. Let us isolate the beam. The equation of equilibrium in terms of moments around the $\vec{z}$ axis gives (Figure 2.38):

$$\mathcal{M}_{z\,beam\,/\,spring} + F_1 x_1 + F_2 x_2 = 0$$

i.e.:

---

31 We are still outside the gravitational field (non-intervention of weight).
32 See [2.7].

$$-k \times \Theta_z + F_1 x_1 + F_2 x_2 = 0$$

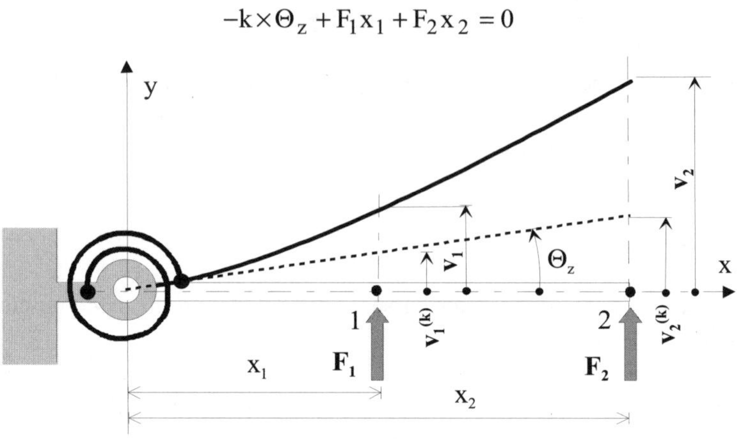

Figure 2.38. *Total displacements*

The value of the relative rotation of the constituents of the pivot is thus:

$$\Theta_z = \frac{F_1}{k} x_1 + \frac{F_2}{k} x_2$$

If the beam had been extremely rigid, it would have remained rectilinear and the respective vertical displacements (assumed to be small) would be observed perpendicular to points (1) and (2) as:

$$v_1^{(k)} \cong \Theta_z \times x_1 = \frac{F_1}{k} x_1^2 + \frac{F_2}{k} x_1 x_2$$

$$v_2^{(k)} \cong \Theta_z \times x_2 = \frac{F_1}{k} x_1 x_2 + \frac{F_2}{k} x_2^2$$

Since the beam also deforms due to plane bending, the total deflections perpendicular to points (1) and (2) are increased by the values obtained in section 2.4.1.1. Thus (see Figure 2.38):

$$v_1 = v_1^{(k)} + F_1 \frac{x_1^3}{3EI} + F_2 \frac{x_1^2}{2EI}\left(x_2 - \frac{x_1}{3}\right)$$

$$v_2 = v_2^{(k)} + F_1 \frac{x_1^3}{2EI}\left(x_2 - \frac{x_1}{3}\right) + F_2 \frac{x_2^2}{3EI}$$

i.e., by replacing:

$$v_1 = F_1\left(\frac{x_1^2}{k} + \frac{x_1^3}{3EI}\right) + F_2\left[\frac{x_1 x_2}{k} + \frac{x_1^2}{2EI}\left(x_2 - \frac{x_1}{3}\right)\right]$$

$$v_2 = F_1\left[\frac{x_1 x_2}{k} + \frac{x_1^2}{2EI}\left(x_2 - \frac{x_1}{3}\right)\right] + F_2\left(\frac{x_2^2}{k} + \frac{x_2^3}{3EI}\right)$$

or even in the matrix form $\{d\} = [\alpha] \bullet \{F\}$[33]:

$$\begin{Bmatrix} v_1 \\ v_2 \end{Bmatrix} = \begin{bmatrix} \dfrac{x_1^2}{k} + \dfrac{x_1^3}{3EI} & \dfrac{x_1 x_2}{k} + \dfrac{x_1^2}{2EI}\left(x_2 - \dfrac{x_1}{3}\right) \\ \dfrac{x_1 x_2}{k} + \dfrac{x_1^2}{2EI}\left(x_2 - \dfrac{x_1}{3}\right) & \dfrac{x_2^2}{k} + \dfrac{x_2^3}{3EI} \end{bmatrix} \bullet \begin{Bmatrix} F_1 \\ F_2 \end{Bmatrix} \qquad [2.92]$$

We note the flexibility matrix $[\alpha]$ associated with this beam and with this loading.

NOTES

❑ Thus, it can be noted that the same structure with the same nodes (1) and (2) and the same loading as section 2.4.1.1 is characterized here by a different flexibility matrix. This is due to the link characteristics of the structure with its surroundings, through the rigidity k of the spiral spring.

❑ Influence of the spring rigidity k on the flexibility matrix can be observed. If we try to greatly vary this stiffness, the following extremes can be seen:

➢ $k \to \infty$: then the flexibility matrix of equation [2.92] becomes identical to equation [2.62].

By increasing the rotational rigidity of the turning pair linkage (pivot), an infinitely rigid link or "perfect clamping" is established.

➢ *weak k* value: the coefficients of flexibility α_{ij} (see [2.92]) show that terms $\dfrac{x_1^2}{k}, \dfrac{x_1 x_2}{k}, \dfrac{x_2^2}{k}$, can become as large as necessary. In this case the flexibility terms $\dfrac{x_1^3}{3EI}, \dfrac{x_1^2}{2EI}\left(x_2 - \dfrac{x_1}{3}\right), \dfrac{x_2^3}{3EI}$, which are unchanged, become negligible compared to the contributions due to the rotation of the spiral spring.

---

33 See [2.91].

Thus, under the action of very weak forces $\vec{F_1}$ and $\vec{F_1}$, the displacements due to elastic deformations are "drowned" in a much more pronounced displacement where the rigid-body rotation is dominant compared to the deformation due to bending. The movement is closer to the rotational movement of an *almost rigid-body* around the axis of the pivot-link. This behavior no longer corresponds to the hypothesis of small displacements to which we wish to restrict ourselves.

➤ *k → 0:* all the terms α_{ij} of the flexibility matrices of equation [2.92] thus become infinite along with displacements v_1 and v_2.

This case corresponds to the simple omission of the spiral spring. The beam is pivoted in its surroundings and can no longer be in equilibrium under the action of finite forces $\vec{F_1}$ and $\vec{F_2}$: *the flexibility matrix is no longer defined.*

In short, equation [2.92] no longer allows the characterization of finite displacements v_1 and v_2 when finite forces $\vec{F_1}$ and $\vec{F_2}$ are applied to the beam.

❒ Thus, if equation [2.91] has to be established, there arises a necessity for a *complete positioning* of the structure under study before subjecting it to a loading. This positioning could include one (or more) links having finite rigidities. However, the small size of the displacements (translations or rotations), called degrees of freedom, implies links of high rigidity.

❒ From this example it was obvious that it was no longer possible to define the flexibility matrix when the beam structure's link with its environment enabled a *near rigid body movement*. It is easily understood that this particularity can be generalized to any type of structure. This will happen every time the links of the structure with its surroundings are insufficient to ensure its complete positioning in the absence of loading.

❒ Of course, a complete positioning of the beam structure studied in our example can be carried out in various ways. The following table summarizes what has been described.

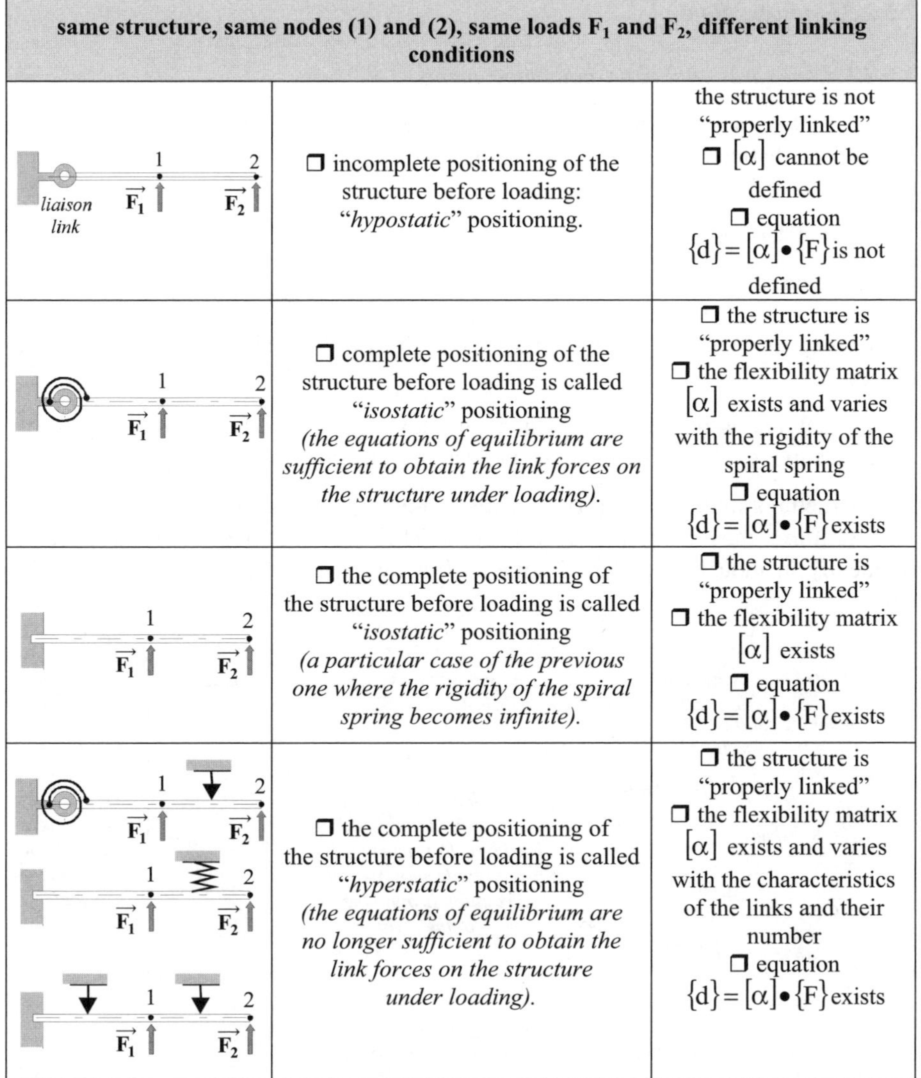

same structure, same nodes (1) and (2), same loads F_1 and F_2, different linking conditions		
	❐ incomplete positioning of the structure before loading: *"hypostatic"* positioning.	the structure is not "properly linked" ❐ $[\alpha]$ cannot be defined ❐ equation $\{d\} = [\alpha] \bullet \{F\}$ is not defined
	❐ complete positioning of the structure before loading is called *"isostatic"* positioning *(the equations of equilibrium are sufficient to obtain the link forces on the structure under loading).*	❐ the structure is "properly linked" ❐ the flexibility matrix $[\alpha]$ exists and varies with the rigidity of the spiral spring ❐ equation $\{d\} = [\alpha] \bullet \{F\}$ exists
	❐ the complete positioning of the structure before loading is called *"isostatic"* positioning *(a particular case of the previous one where the rigidity of the spiral spring becomes infinite).*	❐ the structure is "properly linked" ❐ the flexibility matrix $[\alpha]$ exists ❐ equation $\{d\} = [\alpha] \bullet \{F\}$ exists
	❐ the complete positioning of the structure before loading is called *"hyperstatic"* positioning *(the equations of equilibrium are no longer sufficient to obtain the link forces on the structure under loading).*	❐ the structure is "properly linked" ❐ the flexibility matrix $[\alpha]$ exists and varies with the characteristics of the links and their number ❐ equation $\{d\} = [\alpha] \bullet \{F\}$ exists

Figure 2.39. *Different connections to the surroundings for the same structure and loading*

2.5.2. *Generalization*

2.5.2.1. *Structures with rigid-body movements*

A structure without any link with its surroundings, i.e., completely free in space can have 6 distinctive movements analogous to rigid-body movements[34] defined in reference coordinates to be chosen:

- 3 translations,

- 3 rotations.

Two types of structures with the possibility of rigid-body movement are thus observed:

– structures without loading which can move as rigid-bodies. In such a case the structures move without being deformed and the strain potential energy is zero;

– structures under loading (thus with non-zero potential energy of deformation) with uniform movement (in equilibrium). This type of loaded structure occurs frequently in industrial reality. Figure 2.40 shows two typical examples of such a case.

types of structures	rigid–body movements	loading	displacements
machine shaft in continuous service 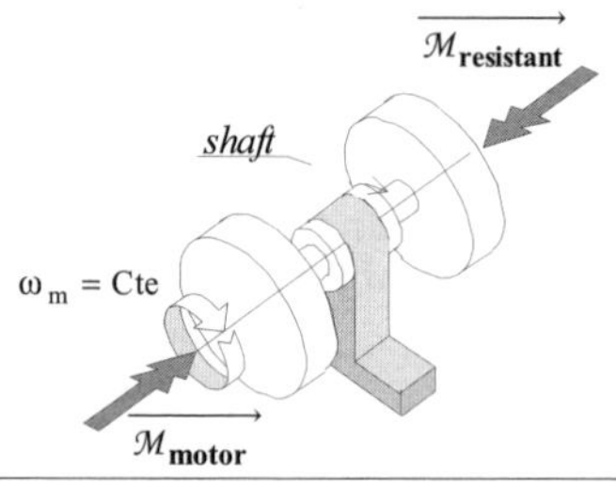	*the shaft is elastically deformed (torsional loading). The structure has a rigid-body movement that is a rotation around the axis of the pivot*	*motor torque and resisting couple*	*rotations of selected cross-sections in the shaft*
airplane at cruising speed 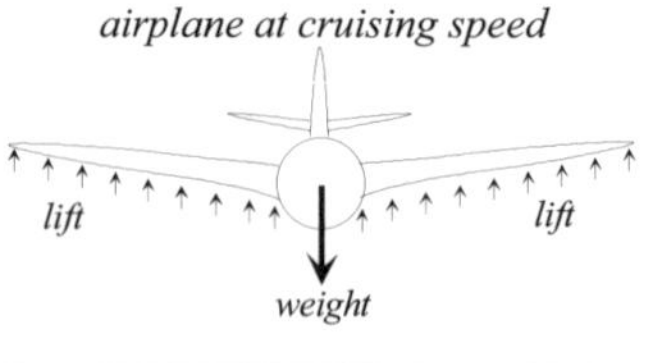	*the airplane is elastically deformed (aerodynamic forces and weights). The structure has 6 overall movements: 3 translations and 3 rotations with respect to a reference set of coordinates on the ground*	*aerodynamic forces, gravity*	*displacement vectors of points selected on the structure*

Figure 2.40. *Loaded structures in equilibrium capable of rigid-body movements*

---

34 This is valid for "monolithic" structures, i.e., structures consisting of elements or parts kinetically linked without internal mobility that could lead to internal hypostatism.

2.5.2.2. *"Properly linked" structure*

We have seen that if a structure can have one or more rigid-body movements, these movements can be very large under finite loads. As the flexibility matrix cannot thus be defined we shall call these structures not properly linked.

In short, this important feature should always be kept in mind:

> *If the linking conditions of a structure do not result in a complete positioning with respect to its surroundings (minimum isostatic linkage) there is a possibility of rigid-body movement(s) of this structure.*
>
> *Thus the flexibility matrix cannot be defined.*
>
> *The structure is not* **"properly linked"**[35] *for calculation.*

2.6. Stiffness of a structure

2.6.1. *Preliminary note*

In section 2.2.2 the behavior equation of a classical helical spring was defined, i.e. (see [2.5]):

$$F = k \times x$$

where F is the applied force and x is the corresponding elongation along the direction $\vec{x}$ (see Figure 2.4). In line with the previous sections, this spring can be considered as a structure:

– with a complete positioning before loading (clamped on the left);

– on which a point (extreme right) and a load $\vec{F}$ (along direction $\vec{x}$) are defined. This is shown again in Figure 2.41 below.

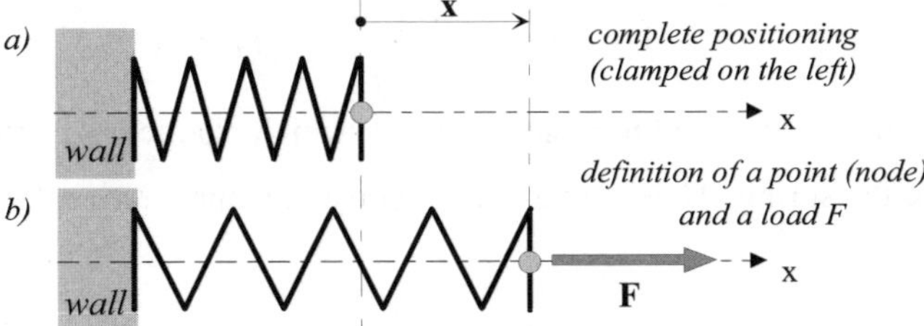

Figure 2.41. *The spring considered as a simple structure*

---

35 This notion shall be reconsidered and clarified in the following during subsequent discretization of the structure into finite elements.

The behavior of this structure is described in [2.91]. Particularly, equation $\{d\} = [\alpha] \bullet \{F\}$ should be interpreted here as:

$$\{d\} = [\alpha] \bullet \{F\} \Rightarrow x = \frac{1}{k} \times F$$

the only load F is associated with the only degree of freedom x, i.e.:

$$\{d\} = \underset{1 \times 1}{\{x\}} = x$$

and the flexibility matrix $[\alpha]$ is reduced to the single term:

$$[\alpha] = \underset{1 \times 1}{[\alpha]} = \alpha$$

from where we arrive at:

$$\frac{1}{k} = \alpha \qquad\qquad [2.93]$$

- $\dfrac{1}{k}$ which is actually the inverse of the rigidity of the spring, appears as a flexibility coefficient;

- as per [2.93], the notions of flexibility and of stiffness (or of rigidity) appear reciprocal to one another:

$$\alpha \times k = 1$$

It is also noted that the behavior equation of a spring is usually written in the form:

$$F = k \times x$$

this corresponds to the inverse of the general form $\{d\} = [\alpha] \bullet \{F\}$, i.e.:

$$F = k \times x \Leftrightarrow \{F\} = [\alpha]^{-1} \bullet \{d\} \qquad\qquad [2.94]$$

where matrix $[\alpha]^{-1}$ is reduced to a single term k which characterizes the rigidity or the stiffness of the spring. Thus matrix $[\alpha]^{-1}$ appears to characterize the rigidity.

2.6.2. *Stiffness matrix*

The behavior equation [2.94] of the spring can be extended to any structure by inverting [2.91], i.e.:

$$\{d\} = [\alpha] \bullet \{F\} \Leftrightarrow \{F\} = [\alpha]^{-1} \bullet \{d\} \tag{2.95}$$

Matrix $[\alpha]^{-1}$ will thus characterize the rigidity or the stiffness of the structure. From now on the notation:

$$[\alpha]^{-1} = [k] \tag{2.96}$$

where $[k]$ shall be the "stiffness matrix" of the structure.

Keeping in mind the properties of matrix $[\alpha]$ (square and symmetric matrix) stiffness matrix $[k]$ is also *square* and *symmetric*[36]:

Taking into account [2.96], formula [2.95] is written as:

$$\{F\} = [k] \bullet \{d\} \tag{2.97}$$

2.6.3. *Examples*

In the examples that follow, we will determine the stiffness matrix of simple structures (bar and beam) associated with various connections and loads.

2.6.3.1. *Example: beam under plane bending loaded by two forces*

This structure has already been discussed in section 2.4.1.1. Given below is the flexibility matrix [2.62] which was established at the end of the said section:

$$\{d\} = [\alpha] \bullet \{F\} \Leftrightarrow \begin{Bmatrix} v_1 \\ v_2 \end{Bmatrix} = \begin{bmatrix} \dfrac{x_1^{\,3}}{3EI} & \dfrac{x_1^{\,2}}{2EI}\left(x_2 - \dfrac{x_1}{3}\right) \\ \dfrac{x_1^{\,2}}{2EI}\left(x_2 - \dfrac{x_1}{3}\right) & \dfrac{x_2^{\,3}}{3EI} \end{bmatrix} \bullet \begin{Bmatrix} F_1 \\ F_2 \end{Bmatrix}$$

To simplify, let us take the special case where $x_2 = 2\ell$ (total length of the beam) and $x_1 = \ell$ (Figure 2.42).

---

36 See Chapter 12, section 12.1.2.4.

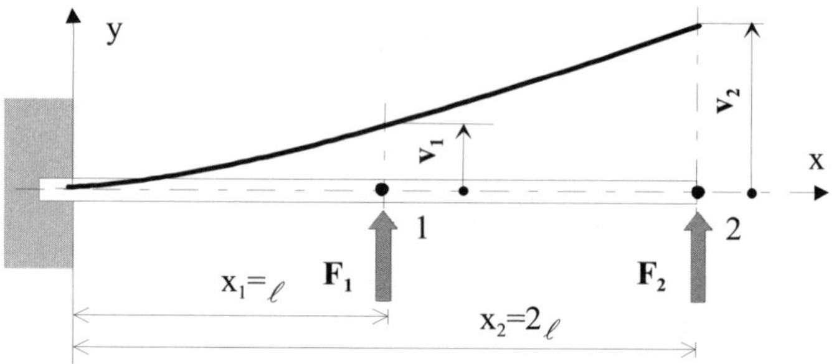

Figure 2.42. *Cantilever beam loaded by two forces*

The above formula can be re-written as:

$$\{d\} = [\alpha] \bullet \{F\} \Leftrightarrow \begin{Bmatrix} v_1 \\ \\ v_2 \end{Bmatrix} = \begin{bmatrix} \dfrac{\ell^3}{3EI} & \dfrac{5\ell^3}{6EI} \\ \dfrac{5\ell^3}{6EI} & \dfrac{8\ell^3}{3EI} \end{bmatrix} \bullet \begin{Bmatrix} F_1 \\ \\ F_2 \end{Bmatrix} \qquad [2.98]$$

which can be inverted to obtain[37]:

$$\{F\} = [\alpha]^{-1} \bullet \{d\} \Leftrightarrow \begin{Bmatrix} F_1 \\ \\ F_2 \end{Bmatrix} = \begin{bmatrix} \dfrac{96}{7}\dfrac{EI}{\ell^3} & -\dfrac{30}{7}\dfrac{EI}{\ell^3} \\ -\dfrac{30}{7}\dfrac{EI}{\ell^3} & \dfrac{12}{7}\dfrac{EI}{\ell^3} \end{bmatrix} \bullet \begin{Bmatrix} v_1 \\ \\ v_2 \end{Bmatrix} \qquad [2.99]$$

where the stiffness matrix $[k] = [\alpha]^{-1}$ of the structure under consideration appears:

$$[k] = \begin{bmatrix} \dfrac{96}{7}\dfrac{EI}{\ell^3} & -\dfrac{30}{7}\dfrac{EI}{\ell^3} \\ -\dfrac{30}{7}\dfrac{EI}{\ell^3} & \dfrac{12}{7}\dfrac{EI}{\ell^3} \end{bmatrix}$$

NOTE

❏ This stiffness matrix characterizes a beam structure associated with:

   – a link with the environment (clamped);

   – two cross-sections (1) and (2);

   – specified forces at (1) and at (2).

---

37 See section 12.1.2.4.

2.6.3.2. *Example: beam under plane bending loaded by a force and a moment*

Let us take the example studied in section 2.4.1.2 of a cantilever beam subjected to a force $\overrightarrow{F_1}$ at (1) and a moment $\overrightarrow{\mathcal{M}_2}$ at (2). The flexibility matrix which was established in formula [2.74] is recalled below:

$$\{d\}=[\alpha]\bullet\{F\} \Leftrightarrow \left\{\begin{array}{c} v_1 \\ \theta_{z2} \end{array}\right\} = \begin{bmatrix} \dfrac{x_1^{\,3}}{3EI} & \dfrac{x_1^{\,2}}{2EI} \\ \dfrac{x_1^{\,2}}{2EI} & \dfrac{x_2}{EI} \end{bmatrix} \bullet \left\{\begin{array}{c} F_1 \\ \mathcal{M}_2 \end{array}\right\}$$

To simplify, let us take the special case where $x_2 = 2\ell$ (total length of the beam) and $x_1 = \ell$ (Figure 2.43).

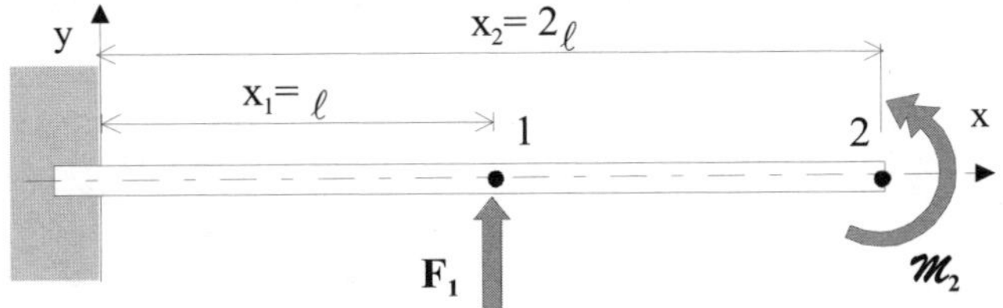

Figure 2.43. *Cantilever beam loaded by a force and a moment*

The formula above is re-written as:

$$\{d\}=[\alpha]\bullet\{F\} \Leftrightarrow \left\{\begin{array}{c} v_1 \\ \theta_{z2} \end{array}\right\} = \begin{bmatrix} \dfrac{\ell^3}{3EI} & \dfrac{\ell^2}{2EI} \\ \dfrac{\ell^2}{2EI} & \dfrac{2\ell}{EI} \end{bmatrix} \bullet \left\{\begin{array}{c} F_1 \\ \mathcal{M}_2 \end{array}\right\} \qquad [2.100]$$

which can be inverted to obtain:

$$\{F\}=[\alpha]^{-1}\bullet\{d\} \Leftrightarrow \left\{\begin{array}{c} F_1 \\ \mathcal{M}_2 \end{array}\right\} = \begin{bmatrix} \dfrac{24}{5}\dfrac{EI}{\ell^3} & -\dfrac{6}{5}\dfrac{EI}{\ell^2} \\ -\dfrac{6}{5}\dfrac{EI}{\ell^2} & \dfrac{4}{5}\dfrac{EI}{\ell} \end{bmatrix} \bullet \left\{\begin{array}{c} v_1 \\ \theta_{z2} \end{array}\right\} \qquad [2.101]$$

where the stiffness matrix $[k]=[\alpha]^{-1}$ appears.

$$[k] = \begin{bmatrix} \dfrac{24}{5} \dfrac{EI}{\ell^3} & -\dfrac{6}{5} \dfrac{EI}{\ell^2} \\ -\dfrac{6}{5} \dfrac{EI}{\ell^2} & \dfrac{4}{5} \dfrac{EI}{\ell} \end{bmatrix}$$

NOTES

❏ As in the previous example this stiffness matrix characterizes a certain beam structure associated with:
   – a link with the environment (clamped);
   – two cross-sections (1) and (2);
   – a specified load at (1) and (2).

❏ In view of the following, another interesting case is obtained in making in equation [2.74], $x_1 = x_2 = \ell$ for a beam of length ℓ. Point (1) merges with point (2) as represented in Figure 2.44:

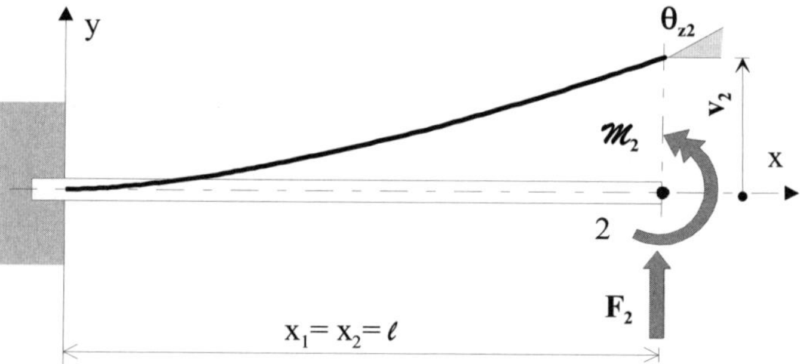

Figure 2.44. *Special case where point 2 merges with point 1*

The behavior equation becomes:

$$\{d\} = [\alpha] \bullet \{F\} \iff \begin{Bmatrix} v_2 \\ \theta_{z2} \end{Bmatrix} = \begin{bmatrix} \dfrac{\ell^3}{3EI} & \dfrac{\ell^2}{2EI} \\ \dfrac{\ell^2}{2EI} & \dfrac{\ell}{EI} \end{bmatrix} \bullet \begin{Bmatrix} F_2 \\ \mathcal{M}_2 \end{Bmatrix} \qquad [2.102]$$

which can be inverted to introduce the stiffness matrix $[k]$:

$$\begin{Bmatrix} F_2 \\ \mathcal{M}_2 \end{Bmatrix} = \begin{bmatrix} \dfrac{12EI}{\ell^3} & -\dfrac{6EI}{\ell^2} \\[2mm] -\dfrac{6EI}{\ell^2} & \dfrac{4EI}{\ell} \end{bmatrix} \bullet \begin{Bmatrix} v_2 \\ \theta_{z2} \end{Bmatrix}$$

2.6.3.3. *Generalization*

The most common case dealt with in section 2.4.2 leads to the same conclusions summarized in [2.91]. Based on the knowledge of the flexibility matrix $[\alpha]$, it is possible for each of them to obtain after inversion:

$$\{d\} = [\alpha] \bullet \{F\} \quad \Leftrightarrow \quad \{F\} = [k] \bullet \{d\}$$

and thus to define the stiffness matrix of the structure associated with the entire set of points or "nodes" (i) = 1,…, n chosen on this structure, and with boundary conditions (links and loading).

2.6.4. *Influence of the positioning*

NOTE

❏ In the equilibrium configurations in the following figures, it has been selected, as is conventional, to represent the forces and moments on the positive side of the reference axis.

2.6.4.1. *Example: bar working under traction-compression*

Figure 2.45 refers to a bar working under traction. The material and the geometric characteristics are: length ℓ, area of cross-section S and Young's modulus E. The centers of the end sections (or "nodes") are noted (1) and (2).

The bar is linked on the left (1) to its environment (wall) through an "elastic support" with finite rigidity, composed of a stiffness spring k_u. It is loaded by the forces $\overrightarrow{F_1} = F_1 \overrightarrow{x}$ at (1) and $\overrightarrow{F_2} = F_2 \overrightarrow{x}$ at (2).

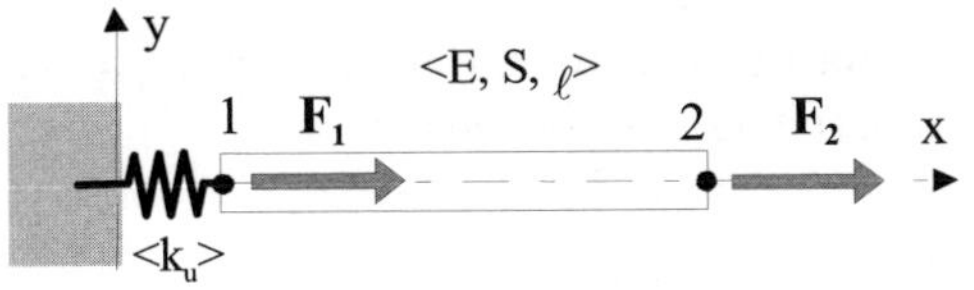

Figure 2.45. *Elastic support*

Figure 2.46b shows the isolated structure *composed of this single bar*. In accordance with their associative properties with loads (see [2.91]), the "degrees of freedom" are the displacements along $\vec{x}$ marked u_1 and u_2 of nodes (1) and (2).

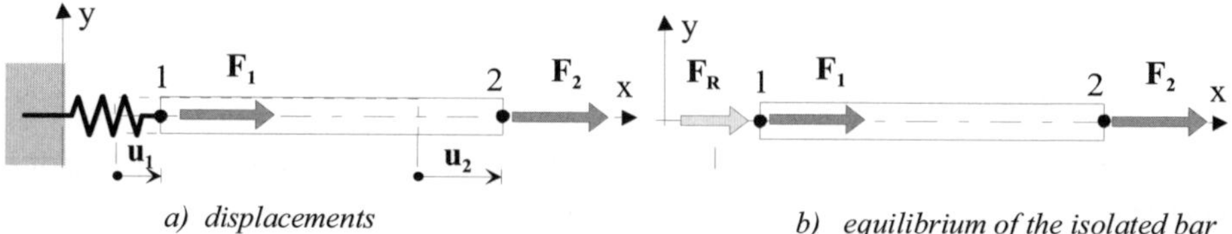

a) displacements   b) equilibrium of the isolated bar

Figure 2.46. *Nodal displacements and forces*

The force exerted by the spring on the bar at point (1) is written as (see [2.5]):

$$F_R = -k_u u_1 \qquad [2.103]$$

On the other hand the equilibrium of the bar along the $\vec{x}$ axis can be noted as:

$$(F_R + F_1) + F_2 = 0 \qquad [2.104]$$

Moreover, according to Hooke's law (see [1.2])

$$F_2 = \frac{ES}{\ell}(u_2 - u_1) \qquad [2.105]$$

Equations [2.104] and [2.105] can be re-written as[38]:

$$\left\{ \begin{array}{c} (F_R + F_1) \\ \\ F_2 \end{array} \right\} = \begin{bmatrix} \dfrac{ES}{\ell} & -\dfrac{ES}{\ell} \\ -\dfrac{ES}{\ell} & \dfrac{ES}{\ell} \end{bmatrix} \bullet \left\{ \begin{array}{c} u_1 \\ \\ u_2 \end{array} \right\} \qquad [2.106]$$

where $(F_R + F_1)$ is the resultant action exerted by the external medium on the bar at node (1)[39].

○ Extreme cases showing the influence of stiffness k_u of the elastic support

➤ If we assume a spring of zero stiffness, then $k_u = 0$. Thus, according to [2.103] $F_R = 0$, and equation [2.106] becomes:

---

38 In fact, we obtain $F_R + F_1 = -F_2 = -\dfrac{ES}{\ell}(u_2 - u_1)$.

39 It can be noted that in formula [2.106], only load F_1 and F_2 are arbitrary. In fact, when the values of these forces are fixed, we obtain based on [2.104]: $F_R = -(F_1 + F_2)$.

$$\{F\}=[K]\bullet\{d\} \iff \begin{Bmatrix} F_1 \\ F_2 \end{Bmatrix} = \begin{bmatrix} \dfrac{ES}{\ell} & -\dfrac{ES}{\ell} \\ -\dfrac{ES}{\ell} & \dfrac{ES}{\ell} \end{bmatrix} \bullet \begin{Bmatrix} u_1 \\ u_2 \end{Bmatrix} \qquad [2.107]$$

We obtain:

– a "floating" structure which is not linked to its environment, i.e., not properly linked, as in section 2.5.2.2. Particularly, it can have a rigid-body movement along the direction $\vec{x}$ of the load;

– a behavior equation [2.107] introducing the stiffness matrix $[k]$ of the "floating" structure.

As expected in section 2.5.2.2 it is observed that the inverse of this matrix, i.e. the flexibility matrix $[\alpha]$, cannot be calculated. In fact stiffness matrix $[k]$ of formula [2.107] is singular[40] because its main determinant is:

$$\left(\frac{ES}{\ell}\right)^2 - \left(\frac{ES}{\ell}\right)^2 = 0$$

➢ Let us take the case of an infinitely rigid spring $k_u \to \infty$. Equation [2.103] thus indicates:

$$u_1 = -\frac{F_R}{k_u} = 0$$

Thus the spring no longer deforms, and we obtain a rigid support as outlined in Figure 2.47:

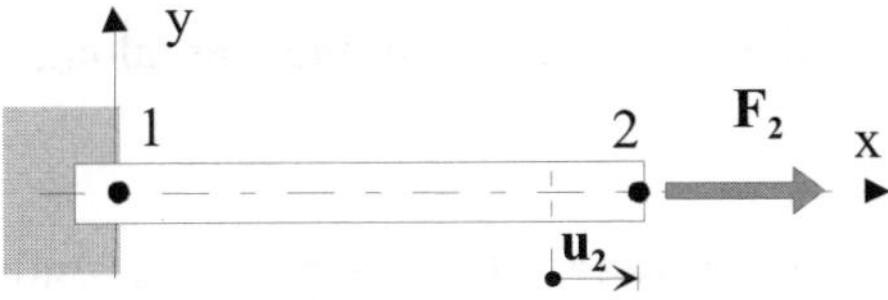

Figure 2.47. *Displacement of the bar with a rigid support on the left*

From equation [2.106][41] we understand:

---

40 See Chapter 12, section 12.1.2.4.

41 It is noted that $F_R = -k_u \times u_1 = \infty \times 0$ is not determined.

$$(F_R + F_1) = -\frac{ES}{\ell} u_2 \qquad [2.108]$$

$$F_2 = \frac{ES}{\ell} u_2 \qquad [2.109]$$

The same truss structure thus sees its stiffness matrix change when it is forced by a rigid link. In particular, in this last case the only degree of freedom u_2 remains and equation [2.109] should be interpreted as:

$$F_2 = \frac{ES}{\ell} \times u_2 \quad \Leftrightarrow \quad \{F^*\} = [k^*] \bullet \{d^*\}$$

where the new stiffness matrix $[k^*]$ is reduced to the single term:

$$k^* = \frac{ES}{\ell} \qquad [2.110]$$

NOTES

❏ Inversion of equation [2.109] is trivial. From this the value of the only degree of freedom u_2 is obtained:

$$u_2 = \frac{\ell}{ES} F_2$$

this value can then be introduced into [2.108], i.e.:

$$(F_R + F_1) = -\frac{ES}{\ell} \times \frac{\ell}{ES} F_2 = -F_2$$

$(F_R + F_1)$ appears as the link action on the bar along $\vec{x}$ at point (1), i.e.: $F_{1\ell} = (F_R + F_1)$.

❏ In a practical way we can observe that the previous results [2.109] and [2.108] are obtained if, in the matrix equation [2.106], the line and the column of the same row corresponding to $u_1 = 0$ in the stiffness matrix are deleted. Result [2.109] then remains:

$$F_2 = \frac{ES}{\ell} u_2$$

This enables the calculation of u_2.

The line corresponding to u_1 (equation [2.106]) which was first deleted enables the linking action $F_{1\ell} = (F_R + F_1)$, (equation [2.108]) to be calculated.

❑ It is seen that the bar in Figure 2.46b, with its two nodes and degrees of freedom, is characterized by a singular stiffness matrix as long as it is not properly linked.

Thus, equation [2.106] becomes usable only after the complete positioning is specified, for example here a fixed support which nullifies a degree of freedom ($u_1 = 0$), or a finite stiffness spring ($F_R = -k_u u_1$).

❑ The stiffness matrix $\left[k^*\right]$ of the bar structure fixed at 1 ($u_1 = 0$) (properly linked) (see [2.110]) appears as a sub-matrix of the stiffness matrix of the free bar structure [2.107].

❑ As noted above, when the element is "floating", its stiffness matrix $\left[k\right]$ is singular. Consequently, it cannot be inverted to give flexibility matrix $\left[\alpha\right]$. On the other hand, when the element is clamped at its left, it is characterized by the behavior relation $\{F^*\} = \left[k^*\right] \bullet \{d^*\}$ where $\left[k^*\right]$ is reduced to the only term $\dfrac{ES}{\ell}$. The inverse form of this relation is readily obtained, since inversion of equation [2.109] is trivial, and can be noted in the form $\{d^*\} = \left[\alpha^*\right] \bullet \{F^*\}$, where $\left[\alpha^*\right]$ appears as a sub-matrix of flexibility. $\left[\alpha^*\right]$ is now well-defined, reduced to the only term $\dfrac{\ell}{ES}$.

2.6.4.2. *Example: stiffness matrix of a beam structure under plane bending*

Figure 2.48 of a beam under plane bending in the (xy) plane is studied. The material and geometric characteristics are: length ℓ, quadratic moment around the main $\vec{z}$ axis I_z, (here denoted as I), Young's modulus E.

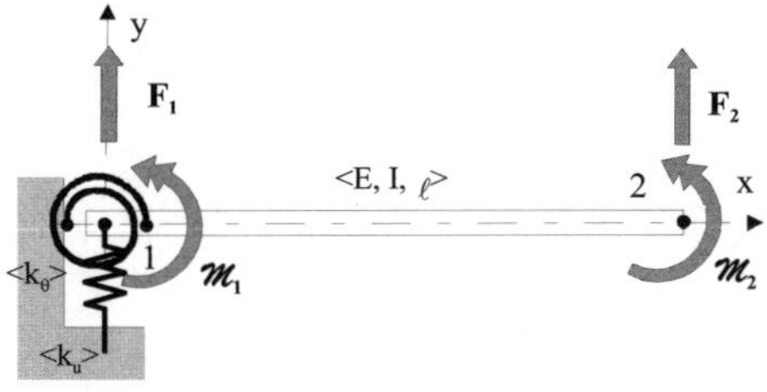

Figure 2.48. *Beam elastically clamped*

This beam is connected on its left (1) to a spring with a linear stiffness k_v along $\vec{y}$ and an angular stiffness k_θ along the $\vec{z}$ axis. It is loaded:

– at (1) by a force $\vec{F_1} = F_1\,\vec{y}$ and a moment $\vec{\mathcal{M}_1} = \mathcal{M}_1\,\vec{z}$;

– at (2) by a force $\vec{F_2} = F_2\,\vec{y}$ and a moment $\vec{\mathcal{M}_2} = \mathcal{M}_2\,\vec{z}$.

Let us isolate the structure *limited to the only beam* of Figure 2.49. Taking into account the load acting at (1) and (2), the associated[42] degrees of freedom can be defined. We have:

– at (1): v_1 and θ_{z1};

– at (2): v_2 and θ_{z2}.

The behavior equation of the two springs leads to:

– an action of the helical spring (traction-compression) $\vec{F_R} = F_R\,\vec{y}$ on the beam:

$$F_R = -k_v v_1$$

– an action of the spiral torsion spring $\vec{\mathcal{M}_R} = \mathcal{M}_R\,\vec{z}$ on the beam:

$$\mathcal{M}_R = -k_\theta \theta_{z1} \tag{2.111}$$

On the other hand the equations expressing the equilibrium of the beam can be represented along axes $\vec{y}$ and $\vec{z}$ as:

$$(F_R + F_1) + F_2 = 0$$

$$(\mathcal{M}_R + \mathcal{M}_1) + \mathcal{M}_2 + F_2 \ell = 0 \tag{2.112}$$

---

42 See [2.91].

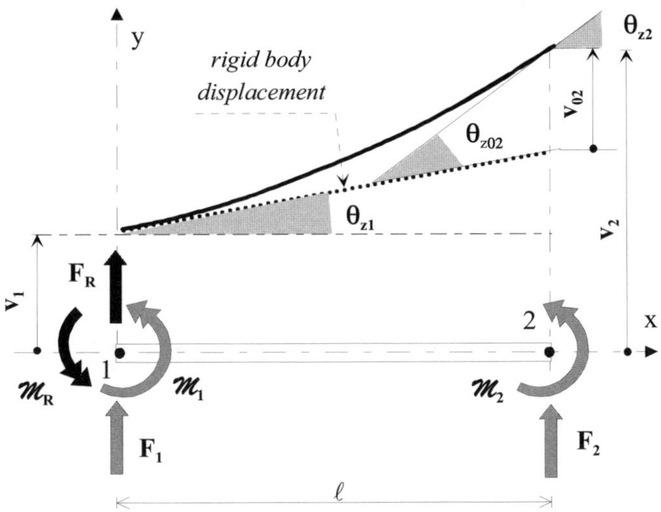

Figure 2.49. *Displacement in the case of an "elastic linkage"*

Figure 2.49 represents the final deformation of the beam as a consequence of:

– an intermediate displacement without deformation or *rigid-body* (represented by dotted lines);

– an elastic deformation similar to the one produced for the case in Figure 2.44, characterized by the linear and the angular displacements v_{02} and θ_{z02}. These are already known from equation [2.102]:

$$\left\{ \begin{array}{c} v_{02} \\ \\ \theta_{z02} \end{array} \right\} = \left[\begin{array}{cc} \dfrac{\ell^3}{3EI} & \dfrac{\ell^2}{2EI} \\ \\ \dfrac{\ell^2}{2EI} & \dfrac{\ell}{EI} \end{array} \right] \bullet \left\{ \begin{array}{c} F_2 \\ \\ \mathcal{M}_2 \end{array} \right\}$$

Thus, by following the notations in Figure 2.49 the equation can be written:

$$v_2 = v_1 + \ell\theta_{z1} + v_{02} = v_1 + \ell\theta_{z1} + \frac{\ell^3}{3EI}F_2 + \frac{\ell^2}{2EI}\mathcal{M}_2$$

$$\theta_{z2} = \theta_{z1} + \theta_{02} = \theta_{z1} + \frac{\ell^2}{2EI}F_2 + \frac{\ell}{EI}\mathcal{M}_2 \quad {}^{43}$$

---

43 As already mentioned several times, it is necessary to emphasize the small size of the elastic displacements, which are grossly enlarged in the figures. This small size enables displacements to be considered as their projections (for example $\ell\sin\theta_{z1} \cong \ell\theta_{z1}$, etc.).

i.e.:

$$v_2 - v_1 - \ell\theta_{z1} = \frac{\ell^3}{3EI}F_2 + \frac{\ell^2}{2EI}\mathcal{M}_2$$

$$\theta_{z2} - \theta_{z1} = \frac{\ell^2}{2EI}F_2 + \frac{\ell}{EI}\mathcal{M}_2 \qquad\qquad [2.113]$$

but by writing in a matrix form:

$$\begin{Bmatrix} v_2 - v_1 - \ell\theta_{z1} \\ \\ \theta_{z2} - \theta_{z1} \end{Bmatrix} = \begin{bmatrix} -1 & -\ell & 1 & 0 \\ \\ 0 & -1 & 0 & 1 \end{bmatrix} \bullet \begin{Bmatrix} v_1 \\ \theta_{z1} \\ v_2 \\ \theta_{z2} \end{Bmatrix}$$

the equations in [2.113] become:

$$\begin{bmatrix} \dfrac{\ell^3}{3EI} & \dfrac{\ell^2}{2EI} \\ \dfrac{\ell^2}{2EI} & \dfrac{\ell}{EI} \end{bmatrix} \bullet \begin{Bmatrix} F_2 \\ \\ \mathcal{M}_2 \end{Bmatrix} = \begin{bmatrix} -1 & -\ell & 1 & 0 \\ \\ 0 & -1 & 0 & 1 \end{bmatrix} \bullet \begin{Bmatrix} v_1 \\ \theta_{z1} \\ v_2 \\ \theta_{z2} \end{Bmatrix}$$

i.e.:
$$\begin{Bmatrix} F_2 \\ \\ \mathcal{M}_2 \end{Bmatrix} = \begin{bmatrix} \dfrac{\ell^3}{3EI} & \dfrac{\ell^2}{2EI} \\ \dfrac{\ell^2}{2EI} & \dfrac{\ell}{EI} \end{bmatrix}^{-1} \bullet \begin{bmatrix} -1 & -\ell & 1 & 0 \\ \\ 0 & -1 & 0 & 1 \end{bmatrix} \bullet \begin{Bmatrix} v_1 \\ \theta_{z1} \\ v_2 \\ \theta_{z2} \end{Bmatrix}$$

The inverse matrix given below has already appeared in section 2.6.3.2. We have:

$$\begin{Bmatrix} F_2 \\ \\ \mathcal{M}_2 \end{Bmatrix} = \underset{(2\times2)}{\begin{bmatrix} \dfrac{12EI}{\ell^3} & -\dfrac{6EI}{\ell^2} \\ -\dfrac{6EI}{\ell^2} & \dfrac{4EI}{\ell} \end{bmatrix}} \bullet \underset{(2\times4)}{\begin{bmatrix} -1 & -\ell & 1 & 0 \\ \\ 0 & -1 & 0 & 1 \end{bmatrix}} \bullet \begin{Bmatrix} v_1 \\ \theta_{z1} \\ v_2 \\ \theta_{z2} \end{Bmatrix}$$

after completing the product of the 2 matrices (2×2) and (2×4):

$$\begin{Bmatrix} F_2 \\ \\ \mathcal{M}_2 \end{Bmatrix} = \begin{bmatrix} -\dfrac{12EI}{\ell^3} & -\dfrac{6EI}{\ell^2} & \dfrac{12EI}{\ell^3} & -\dfrac{6EI}{\ell^2} \\ \dfrac{6EI}{\ell^2} & \dfrac{2EI}{\ell} & -\dfrac{6EI}{\ell^2} & \dfrac{4EI}{\ell} \end{bmatrix} \bullet \begin{Bmatrix} v_1 \\ \theta_{z1} \\ v_2 \\ \theta_{z2} \end{Bmatrix} \qquad [2.114]$$

The forms of F_2 and $\mathcal{M}_2$ thus obtained can be introduced into the two equations of equilibrium [2.112]. We then obtain:

$$(F_R + F_1) = -F_2 = \frac{12EI}{\ell^3} v_1 + \frac{6EI}{\ell^2} \theta_{z1} - \frac{12EI}{\ell^3} v_2 + \frac{6EI}{\ell^2} \theta_{z2}$$

$$(\mathcal{M}_R + \mathcal{M}_1) = -\mathcal{M}_2 - F_2\ell = \frac{6EI}{\ell^2} v_1 + \frac{4EI}{\ell} \theta_{z1} - \frac{6EI}{\ell^2} v_2 + \frac{2EI}{\ell} \theta_{z2} \qquad [2.115]$$

Finally, results [2.114] and [2.115] can be summarized in the matrix form:

$$\begin{Bmatrix} (F_1 + F_R) \\ (\mathcal{M}_1 + \mathcal{M}_R) \\ F_2 \\ \mathcal{M}_2 \end{Bmatrix} = \frac{EI}{\ell^3} \begin{bmatrix} 12 & 6\ell & -12 & 6\ell \\ 6\ell & 4\ell^2 & -6\ell & 2\ell^2 \\ -12 & -6\ell & 12 & -6\ell \\ 6\ell & 2\ell^2 & -6\ell & 4\ell^2 \end{bmatrix} \bullet \begin{Bmatrix} v_1 \\ \theta_{z1} \\ v_2 \\ \theta_{Z2} \end{Bmatrix} \qquad [2.116]$$

where the stiffness matrix of the structure appears.

○ Extreme cases which help in observing the influence of the elastic support

➢ Assuming that the stiffness springs are zero, $k_v = k_\theta = 0$. Thus, according to [2.111], $F_R = \mathcal{M}_R = 0$ and equation [2.116] becomes:

$$\{F\} = [k] \bullet \{d\} \Leftrightarrow \begin{Bmatrix} F_1 \\ \mathcal{M}_1 \\ F_2 \\ \mathcal{M}_2 \end{Bmatrix} = \frac{EI}{\ell^3} \begin{bmatrix} 12 & 6\ell & -12 & 6\ell \\ 6\ell & 4\ell^2 & -6\ell & 2\ell^2 \\ -12 & -6\ell & 12 & -6\ell \\ 6\ell & 2\ell^2 & -6\ell & 4\ell^2 \end{bmatrix} \bullet \begin{Bmatrix} v_1 \\ \theta_{z1} \\ v_2 \\ \theta_{Z2} \end{Bmatrix} \qquad [2.117]$$

Here we obtain:

– a "floating" structure, i.e. not linked properly as per section 2.5.2.2. In particular it can have a rigid-body movement in every direction of loading, that is a translation along $\vec{y}$ and a rotation around $\vec{z}$;

– a behavior equation [2.117] where $[k]$ is the stiffness matrix of the "floating" structure.

It is known (see section 2.5.2.2) that the flexibility matrix of this "not properly linked" structure cannot be defined. As it is obtained by inverting the stiffness matrix, it can be assumed that this last matrix is singular, i.e., its inverse cannot be calculated. In fact, it can be verified that the determinant of $[k]$ above is zero. It is not possible to invert equation [2.117] to find the displacements v_1, v_2 and θ_{z1}, θ_{z2} which as a result appear undefined.

➢ Let us take the case of a clamping with infinitely rigid springs, $k_v = k_\theta \to \infty$. Equations [2.111] then indicate: $v_1 = -\dfrac{F_R}{k_v} = 0$ and $\theta_{z1} = -\dfrac{\mathcal{M}_R}{k_\theta} = 0$. The springs no longer deform, and an infinitely rigid or "perfect" clamping is obtained, as represented in Figure 2.50.

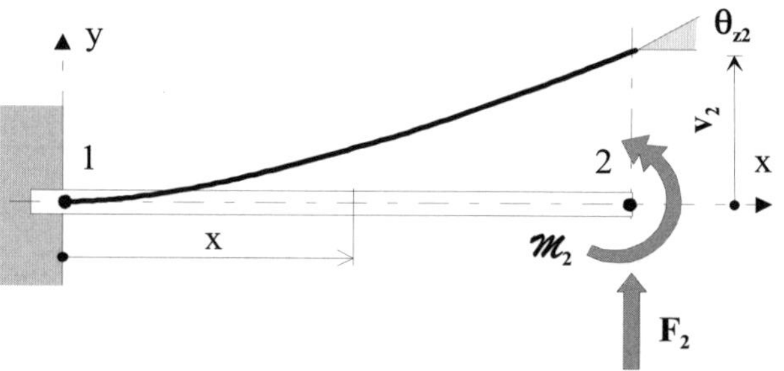

Figure 2.50. *Displacements in the case of a "rigid clamping"*

From equation [2.116][44] we obtain:

$$(F_1 + F_R) = \frac{12EI}{\ell^2} v_2 - \frac{6EI}{\ell^2} \theta_{z2}$$

$$(\mathcal{M}_1 + \mathcal{M}_R) = -\frac{6EI}{\ell^2} v_2 + \frac{2EI}{\ell} \theta_{z2} \qquad\qquad [2.118]$$

---

44 It is noted that the values $F_R = -k_v \times v_1 = \infty \times 0$ and $\mathcal{M}_R = -k_\theta \times \theta_{z1} = \infty \times 0$ are indeterminate.

$$\begin{Bmatrix} F_2 \\ \\ \mathcal{M}_2 \end{Bmatrix} = \begin{bmatrix} \dfrac{12EI}{\ell^3} & -\dfrac{6EI}{\ell^2} \\ -\dfrac{6EI}{\ell^2} & \dfrac{4EI}{\ell} \end{bmatrix} \bullet \begin{Bmatrix} v_2 \\ \\ \theta_{z2} \end{Bmatrix}$$

[2.119]

NOTE

❐ The inversion of equation [2.119] is immediate giving a flexibility matrix already seen in section 2.6.3.2 (see [2.102]). From this the value of the two degrees of freedom are obtained:

$$\begin{cases} v_2 = \dfrac{\ell^3}{3EI} F_2 + \dfrac{\ell^2}{2EI} \mathcal{M}_2 \\ \\ \theta_{z2} = \dfrac{\ell^2}{2EI} F_2 + \dfrac{\ell}{EI} \mathcal{M}_2 \end{cases}$$

These values can then be introduced into [2.118]. This enables calculating $(F_1 + F_R)$ and $(\mathcal{M}_1 + \mathcal{M}_R)$ that appear as linking actions exerted by the surroundings on the beam (clamping). These linking actions can be denoted from now on as: $(F_1 + F_R)$ and $(\mathcal{M}_1 + \mathcal{M}_R)$

❐ In a practical way it is observed that the previous results are obtained if in matrix equation [2.116], two lines corresponding to $v_1 = \theta_{z1} = 0$ and two columns of the same row are deleted. We then have formula [2.119] that enables the calculation of v_2 and θ_{z2} by inversion. By reverting to the two lines corresponding to v_1 and θ_{z1} which were deleted earlier, the linking actions $F_{1\ell} = (F_1 + F_R)$ and $\mathcal{M}_{1\ell} = (\mathcal{M}_1 + \mathcal{M}_R)$ are calculated (equation [2.118]).

❐ It is seen that the beam of Figure 2.49, with its two nodes and four degrees of freedom is characterized by a unique stiffness matrix as long as it is not properly linked.

Thus, equations [2.116] and [2.117] become usable only after the proper support conditions are specified, for example here a fixed support which cancels out all possibilities of rigid-body movements.

2.6.4.3. *Isostatism and hyperstatism*

Figure 2.51 shows the same beam fixed as before at point (1), and in addition simply supported and loaded by a moment $\overrightarrow{\mathcal{M}_2} = \mathcal{M}_2 \vec{z}$ at point (2).

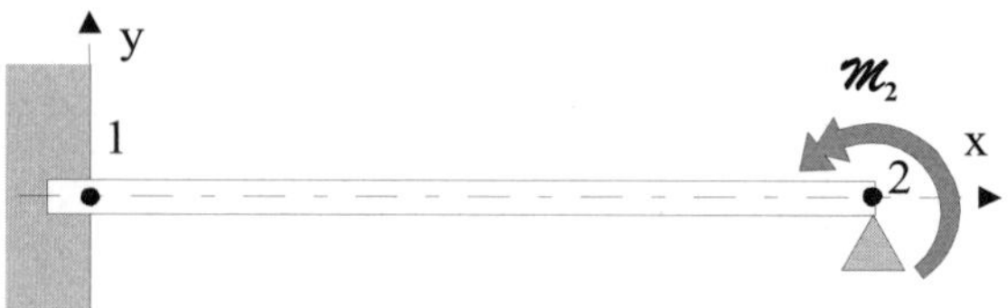

Figure 2.51. *Beam in a hyperstatic positioning, loaded by a moment*

In addition to $v_1 = \theta_{z1} = 0$, here $v_2 = 0$ at the support (2), which appears as an infinitely rigid spring.

The mechanical actions on the isolated beam are represented as before (Figure 2.52), and considering the notes from the previous section, we can apply:

– a linking force at node (1): $F_{1\ell}$;

– a linking moment at node (1): $\mathcal{M}_{1\ell}$;

– a linking force at node (2): $F_{2\ell}$.

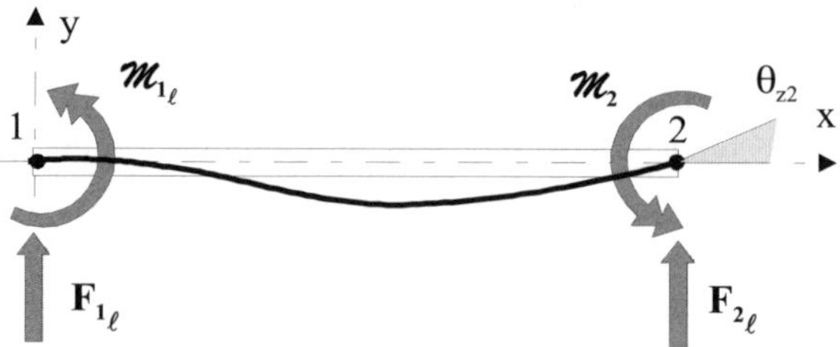

Figure 2.52. *Isolated beam*

Now three linking actions can be seen, whereas only two equations of equilibrium can be written (a resultant equation along $\vec{y}$ and a moment equation around the $\vec{z}$ axis). The linking conditions are called hyperstatic with degree $3 - 2 = 1$. Starting from [2.116] and [2.117], we can directly write:

$$\begin{Bmatrix} F_{1\ell} \\ \mathcal{M}_{1\ell} \\ F_{2\ell} \\ \mathcal{M}_2 \end{Bmatrix} = \frac{EI}{\ell^3} \begin{bmatrix} 12 & 6\ell & -12 & 6\ell \\ 6\ell & 4\ell^2 & -6\ell & 2\ell^2 \\ -12 & -6\ell & 12 & -6\ell \\ 6\ell & 2\ell^2 & -6\ell & 4\ell^2 \end{bmatrix} \cdot \begin{Bmatrix} v_1 = 0 \\ \theta_{z1} = 0 \\ v_2 = 0 \\ \theta_{z2} \end{Bmatrix} \qquad [2.120]$$

Let us follow the practical method described above:

a) let us delete the three lines and the three columns of the same row[45] corresponding to the zero dof in [2.120]:

– there remains:
$$\mathcal{M}_2 = \frac{4EI}{\ell}\,\theta_{z2},$$

– which leads to:
$$\theta_{z2} = \frac{\ell}{4EI}\,\mathcal{M}_2;$$

b) let us reconsider the three lines and the three columns previously deleted. They provide the linking actions owing to the value of θ_{z2} found below:

$$F_{1\ell} = \frac{6EI}{\ell^2} \times \frac{\ell}{4EI}\,\mathcal{M}_2 = \frac{3}{2\ell}\,\mathcal{M}_2$$

$$\mathcal{M}_{1\ell} = \frac{2EI}{\ell^2} \times \frac{\ell}{4EI}\,\mathcal{M}_2 = \frac{1}{2}\,\mathcal{M}_2$$

$$F_{2\ell} = -\frac{6EI}{\ell^2} \times \frac{\ell}{4EI}\,\mathcal{M}_2 = -\frac{3}{2\ell}\,\mathcal{M}_2$$

Conclusions

➤ On comparing the two recently studied types of linkage conditions summarized in Figure 2.53, it is observed that the method of calculation of the displacements under loading and of the linking actions is not influenced by the hyperstatic nature of the supported structure. The method remains the same. As already indicated in section 2.5.2 only the "hypostatic" positioning (structure not linked properly) poses a problem because it proves to be unusable (see Figure 2.39).

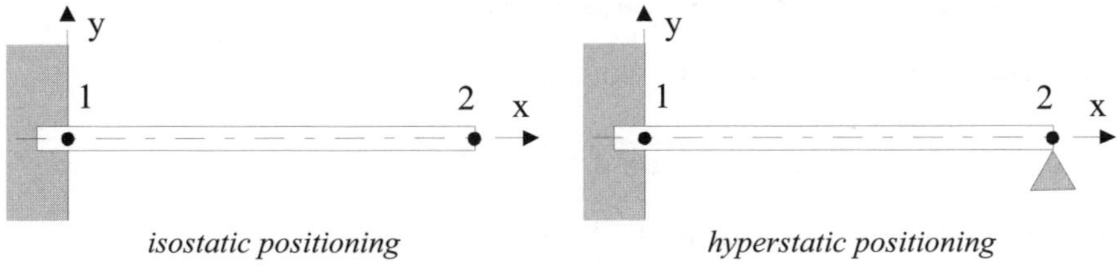

Figure 2.53. *The two types of positioning*

➤ The stiffness matrix of the beam structure properly linked always appears as a sub-matrix of the stiffness matrix of the free beam structure.

---

45 For a line no. i deleted, a column no. i is also deleted.

➤ If the number of blocked dof on a structure is increased, its stiffness also increases with the degree of hyperstatism. The structure becomes less and less "deformable". If all the dof are blocked, the stiffness of the structure apparently becomes infinite.

NOTE

❑ Since the initial stiffness matrix of equations [2.107] or [2.117] cannot be inverted, these relations become unusable. A usable equation could be deduced from the initial equation only through either isostatic or hyperstatic positioning on the structure. Let us summarize the corresponding procedure:

• the initial equation being $\{F\} = [k] \bullet \{d\}$, the structure is "floating", the stiffness matrix $[k]$ is singular;

• an isostatic or a hyperstatic positioning is carried out. Thereby certain dof are zero. These are denoted $\{d_0\} = \{0\}$;

• the lines and columns of the same row corresponding to the zero dof $\{d_0\} = \{0\}$ are then deleted in $[k]$. This leads to a "partitioning" of the stiffness matrix $[k]$ such that:

$$\{F\} = [k] \bullet \{d\} \quad \Rightarrow \quad \left\{\frac{F_\ell}{F^*}\right\} = \left[\begin{array}{c|c} [k_\ell] & [k_\ell^*] \\ \hline [k_\ell^*]^T & [k^*] \end{array}\right] \bullet \left\{\frac{\overset{zero\ dof}{d_o = 0}}{\underset{non\text{-}zero\ dof}{d^*}}\right\}$$

In this equation $\{d^*\}$ represents the column of the dof that remain free, and $\{F^*\}$ represents the external associated forces;

• a usable equation is derived from this:

$$\{F^*\} = [k^*] \bullet \{d^*\}$$

where the stiffness "sub-matrix" $[k^*]$ can be inverted (it is no longer singular), thus enabling the calculation of the non-zero degrees of freedom $\{d^*\}$;

• from the initial "partitioned" equation, we can thus deduce:

$$\{F_\ell\} = [k_\ell^*] \bullet \{d^*\}$$

which gives values of the linking actions $\{F_\ell\}$.

2.6.5. *Deformation energy and stiffness matrix*

2.6.5.1. *Example: beam from section 2.6.3.1*

Figure 2.54 shows the beam structure which has already been studied with its links, nodes, loading and its associated degrees of freedom.

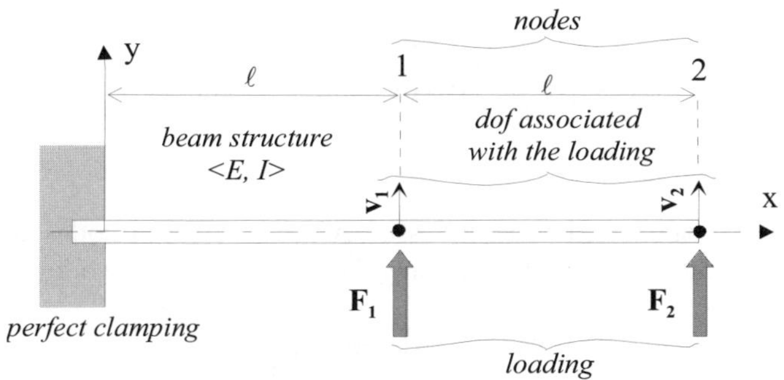

Figure 2.54. *Cantilever beam*

Equations [2.91] summarize the following general results for the potential energy of transformation:

$$\{d\} = [\alpha] \bullet \{F\} \; ; \quad W = \frac{1}{2}\{F\}^{T} \bullet \{d\} = \frac{1}{2}\{F\}^{T} \bullet [\alpha] \bullet \{F\} = E_{pot.}$$

Let us introduce the stiffness matrix defined in form [2.97], i.e.:

$$\{F\} = [k] \bullet \{d\}$$

The transpose of this equation is written as[46]:

$$\{F\}^{T} = \{d\}^{T} \bullet [k]^{47}$$

The strain potential energy becomes:

$$W = \frac{1}{2}\{F\}^{T} \bullet \{d\} = \frac{1}{2}\{d\}^{T} \bullet [k] \bullet \{d\} = E_{pot.} \qquad [2.121]$$

Thus, with terms [2.99] of the stiffness matrix, the strain potential energy for this example can be written as:

---

46 See Chapter 12, section 12.1.2.3.

47 It should be remembered that $[k]$ being square and symmetric it is identical to its transpose: $[k] = [k]^{T}$.

$$E_{pot.} = \frac{1}{2} \begin{Bmatrix} v_1 \\ v_2 \end{Bmatrix}^T \bullet \begin{bmatrix} \dfrac{96}{7}\dfrac{EI}{\ell^3} & -\dfrac{30}{7}\dfrac{EI}{\ell^3} \\ -\dfrac{30}{7}\dfrac{EI}{\ell^3} & \dfrac{12}{7}\dfrac{EI}{\ell^3} \end{bmatrix} \bullet \begin{Bmatrix} v_1 \\ v_2 \end{Bmatrix} = W$$

This is a quadratic form of the degrees of freedom which, when developed, can be written as:

$$E_{pot.} = \frac{1}{2} \times \frac{6EI}{7\ell^3} \left[16v_1{}^2 + 2v_2{}^2 - 10v_1 v_2 \right] = W$$

2.6.5.2. *Generalization*

For a properly supported structure associated with a set of nodes and a loading, results [2.91] introduce the following general expression of the work done:

$$W = \frac{1}{2}\{F\}^T \bullet \{d\} = E_{pot.}$$

where $\{d\}$ represents the vector of the degrees of freedom associated with the load. The behavior equation being written in form [2.97], i.e.:

$$\{F\} = [k] \bullet \{d\} \Leftrightarrow \{F\}^T = \{d\}^T \bullet [k]$$

we have: $W = \dfrac{1}{2}\{d\}^T \bullet [k] \bullet \{d\} = E_{pot.}$

NOTES

❏ It will be seen in Chapter 3 that this relation has the same form even if the structure is not properly linked.

❏ We saw (see [2.91]) the form of the same potential energy obtained from the flexibility matrix of the structure, i.e. $W = \dfrac{1}{2}\{F\}^T \bullet [\alpha] \bullet \{F\} = E_{pot.}$. Now based on what has previously been discussed, we note that this form is possible only if $[\alpha]$ exists, i.e. if the structure is properly linked.

Summary

In a given set of axis of reference (also known as structural axis of reference), a structure is subjected to a set of mechanical actions $\{F\}$ applied on "n" points of the structure. Along the directions associated with these loads, a set of displacements $\{d\}$ is observed. The corresponding characteristic relations are summarized in the following tables.

potential energy
$$W = \frac{1}{2}\{F\}^T \bullet \{d\} = E_{pot.}$$

$$W = \frac{1}{2}\{F\}^T \bullet \underset{\substack{\text{\textit{flexibity}}\\\text{\textit{matrix}}}}{[\alpha]} \bullet \{F\} = E_{pot.}$$ *equation can be written if and only if the structure is properly linked*	$$W = \frac{1}{2}\{d\}^T \bullet \underset{\substack{\text{\textit{stiffness}}\\\text{\textit{matrix}}}}{[k]} \bullet \{d\} = E_{pot.}$$ *equation can be written irrespective of the linking conditions of the structure*

"flexibility" approach of a structure	**"stiffness"** approach of a structure
"n" points or nodes selected on the structure	
$\{F\}$ loading (forces, moments) acting on "n" nodes	
$\{d\}$ degrees of freedom (linear, angular displacements) "associated" with the loading (see [2.91])	
structure not properly linked (not sufficiently linked to its environment, i.e. in an hypostatic manner)	
behavior equation:	
$$\{d\} = [\alpha] \bullet \{F\}$$ *the flexibility matrix $[\alpha]$, inverse of the stiffness matrix $[k]$, does not exist (it cannot be defined), this equation cannot be used*	$$\{F\} = [k] \bullet \{d\}$$ *the stiffness matrix $[k]$ exists, it is a singular matrix (it cannot be inverted)*
same structure properly linked (the previous structure is linked in an isostatic or hyperstatic manner to its surroundings	
behavior equation:	
$$\{d\} = [\alpha] \bullet \{F\}$$ equation not usable (see above) $$\{d^*\} = [\alpha^*] \bullet \{F^*\}$$ *the flexibility matrix $\left[\alpha^*\right]$, inverse of the stiffness sub-matrix $\left[k^*\right]$, exist*	to start with, same equation as above $$\{F\} = [k] \bullet \{d\}$$ $\Downarrow$ application of the linking conditions $\Downarrow$ $$\begin{Bmatrix} F_\ell \\ \hline F^* \end{Bmatrix} = \begin{bmatrix} [k_\ell] & \vdots & [k_\ell^*] \\ \hline [k_\ell^*]^T & \vdots & [k^*] \end{bmatrix} \bullet \begin{Bmatrix} d_0 = 0 \\ \hline d^* \end{Bmatrix}$$ *linking actions (unknown)* — *displacements prevented at the links*; *forces applied (know)* — *displacements unknown* $$\{F^*\} = [k^*] \bullet \{d^*\}$$ *the stiffness sub-matrix $\left[k^*\right]$ exists, and can be inverted*

In the following, in order to simplify the notations, and except in special cases, the transcriptions $\{F\}$, $\{d\}$, $\{k\}$ will also be used to describe the behavior of a properly linked structure

[2.122]

Chapter 3

Discretization of a Structure into Finite Elements

3.1. Preliminary observations

3.1.1. *Problem faced*

In the previous chapter we have seen some cases of elementary structures with simple boundary conditions (loads and links) for which the flexibility matrix corresponding to few dof could be easily written. The objective was to establish a behavior equation in the form (see [2.91]):

$$\{d\} = [\alpha] \bullet \{F\}$$

that makes it possible to obtain displacements or dof $\{d\}$ under a given loading $\{F\}$. Some examples (see section 2.6.4.3) also showed that these displacements enabled determination of the linking forces with indifferently isostatic or hyperstatic positioning of the structure. It will be seen later that some other characteristics of a loaded linear elastic structure can be obtained from these displacements.

Until recently only the "flexibility" approach was available for dimensioning the structures. Then, to manually treat more complex cases: hyperstatic positioning, planar or spatial assembly of beams, classic methods (Castigliano's method, Ménabréa's method, Bresse's formulae, theorem of the three moments, etc.) were used.

However, the calculations necessary for the definition of the flexibility matrix were rapidly becoming heavy and excessively voluminous once the structure became geometrically complicated. Moreover, it was practically impossible to associate sub-structures of different nature: beams, plates, massive parts.

Today these methods are no longer in use in industry and they have been replaced with the "stiffness" approach upon which we are going to expand. This approach along with modern calculation methods will help in dealing with more complex structures that can no longer be designed or dimensioned with the methods used earlier.

3.1.2. *Practical obtaining of the deformation energy for a complex structure*

When a structure is geometrically complicated, the following fundamental relation can neither be established nor *a fortiori* used in a direct manner (even if the structure is properly supported):

$$\{d\}_{\text{structure}} = [\alpha]_{\text{structure}} \bullet \{F\}_{\text{structure}}$$

Thus, in section 2.4.3, it is convincing that this equation exists and that it characterizes the behavior of the structure in Figure 2.34. However, it is difficult to find out how the coefficients α_{ij} can be clearly calculated. In a similar manner, for the inverse relationship, i.e.:

$$\{F\}_{\text{structure}} = [k]_{\text{structure}} \bullet \{d\}_{\text{structure}}$$

a direct method to calculate the terms of stiffness matrix [k] is not obvious either. The same thing applies to the calculation of the deformation energy at the complete structure level:

$$E_{\text{pot.} \atop \text{structure}} = \frac{1}{2} \{d\}^{T}_{\text{structure}} \bullet [k]_{\text{structure}} \bullet \{d\}_{\text{structure}}$$

It is to overcome this difficulty that a partition (or discretization) of the complex structure is looked at, by cutting it into sub-structures with simple geometric characteristics that will be specified later. These sub-structures will be of smaller dimensions compared to the entire structure, and of a finite number. As a consequence these will be called "finite elements".

The main principle is to evaluate separately, for each of these sub-structures, the potential energy of deformation known as elementary energy (potential energy of deformation stored in a finite element) i.e.:

$$E_{\substack{\text{pot.} \\ \text{element}}} = \frac{1}{2} \{d\}_{\text{element}}^{T} \bullet [k]_{\text{element}} \bullet \{d\}_{\text{element}}$$

equation in which $\{d\}_{\text{element}}$ represents the dof of the finite element and $[k]_{\text{element}}$ the corresponding stiffness matrix. The geometric simplicity of the element should enable explicit calculation (or at least numerical) of this stiffness matrix. The following step involves a summation of the elementary energies of elements in order to reconstitute the potential energy of deformation of the complete structure that one wishes to study. This principle is illustrated in Figure 3.1.

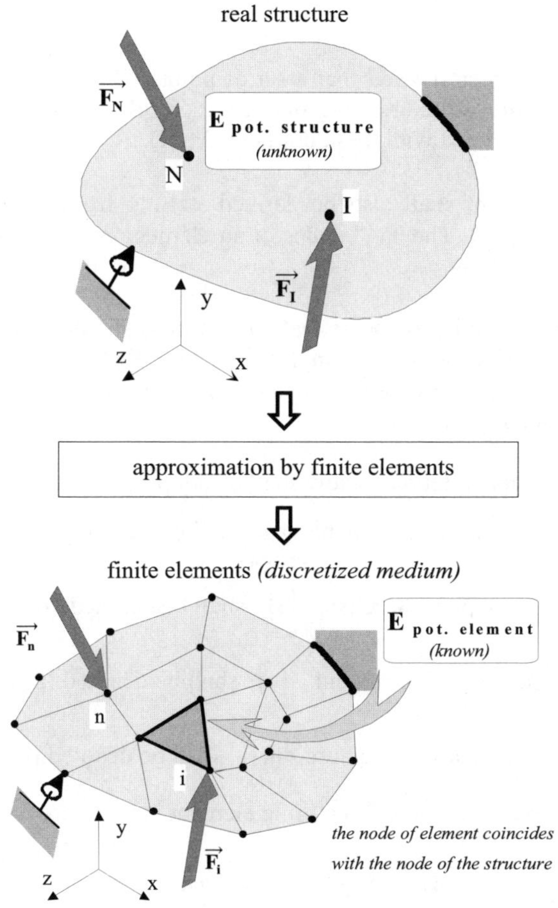

Figure 3.1. *Continuous initial structure and discretized structure*

NOTES

❏ To put this principle into practice:

  ◆ the geometry of the elements should be sufficiently simple or suitable so as to enable carrying out an analytical or a numerical calculation of the stiffness matrix when calculating the potential energy of deformation of every element;

  ◆ the degrees of freedom of the element $\{d\}_{\text{element}}$ should coincide with some of the degrees of freedom of the complete structure $\{d\}_{\text{structure}}$. In other words, $\{d\}_{\text{element}}$ should appear as a part of the complete list of the dof $\{d\}_{\text{structure}}$.

❏ The previous chapter showed that a set of points "i" or "nodes" (i = 1, ..., n) were chosen on the structure and one or more directions of loads (forces, moments) were associated with these dof (see [2.91]).

  Each finite element shall also be defined starting from a certain number of nodes "i" (i = 1,..., n_{el}). The "n_{el}" nodes of an element are a part of the "n" nodes of the complete structure.

❏ The partition into finite elements (called *meshing*) should enable, as exactly as possible, the reconstitution of the complete structure. Based on the geometry of the elements used, the approximation of the boundaries of the structure shall be more or less reliable (see Figure 3.1).

  In the following, the notations below shall be adopted:

  ◆ the dofs related to a finite element $\{d\}_{\text{element}}$ shall be denoted $\{d\}_{el}$;

  ◆ the dofs of the complete structure $\{d\}_{\text{structure}}$ shall be denoted $\{d\}_{str}$;

  ◆ the stiffness matrix of the element $[k]_{\text{element}}$ shall be denoted $[k]_{el}$;

  ◆ the stiffness matrix of the structure $[k]_{\text{structure}}$ shall be denoted $[K]_{str}$.

The deformation energy stored in a finite element is therefore written as:

$$E_{\substack{pot. \\ element}} = \frac{1}{2}\{d\}_{el}^{T} \bullet [k]_{el} \bullet \{d\}_{el}^{T}$$

  The total deformation energy of the structure consisting of "M" finite elements is then established as:

$$E_{\substack{pot. \\ structure}} = \sum_{m=1}^{M} E_{\substack{pot. \\ element\ m}}$$

i.e.:

$$\frac{1}{2}\{d\}_{str}^{T} \bullet [K]_{str} \bullet \{d\}_{str} = \sum_{m=1}^{M} \left\{ \frac{1}{2}\{d\}_{el}^{T} \bullet [k]_{el} \bullet \{d\}_{el} \right\}_{element\ m} \qquad [3.1]$$

♦ It will be seen later that this equation enables the "construction" of the global stiffness matrix of the complete structure $[K]_{str}$.

♦ With the knowledge of $[K]_{str}$, the behavior of a structure under loading as illustrated in the examples of section 2.6.3 can be studied only after giving a complete positioning to this structure[1]. Thus, it will be possible to access indirectly the flexibility matrix corresponding to the desired loads and links.

♦ To be put into practice this process requires an "assortment" or a catalog of finite elements, each of which is characterized by an elementary stiffness matrix $[k]_{el}$ that should have already been established. Section 3.2 shows the calculation method of such matrices for some simple elements[2].

3.1.3. *Local and global coordinates*

3.1.3.1. *Definition*

The diagram in Figure 3.2 shows a structure which can be modeled locally as a beam assembly. The mean lines of these beams[3] can have varied orientations in space. This is the same for the two main quadratic axes of a section which complete what is known as the "local coordinates" as specified in Figure 3.2b.

---

1 See section 2.5.2.2.
2 The description of other common finite elements can be found in Chapter 5.
3 See section 1.5.1.

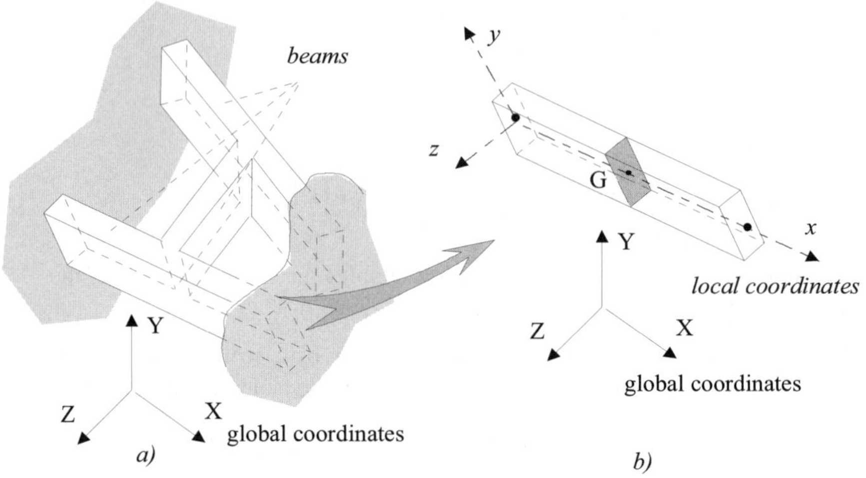

Figure 3.2. *Local and global coordinates*

It is well-known that a given vector (which can be a displacement vector, a translation or rotation, a force vector, a moment vector) can be expressed in the local coordinates (*xyz*) of a beam or in global coordinates (XYZ) connected to the structure. Its components in a coordinate system are derived from the components in the other system through a linear relation where the cosines of the angles appear, made by the axes of the coordinates systems.

➢ *In this chapter we shall limit ourselves to coplanar local and global coordinate systems, i.e., having a common axis* $\vec{z} = \vec{Z}$.

Figure 3.3 summarizes the correspondence that can be rapidly[4] written for a vector $\vec{V}$ in the (*xy*) plane ≡ plane (XY).

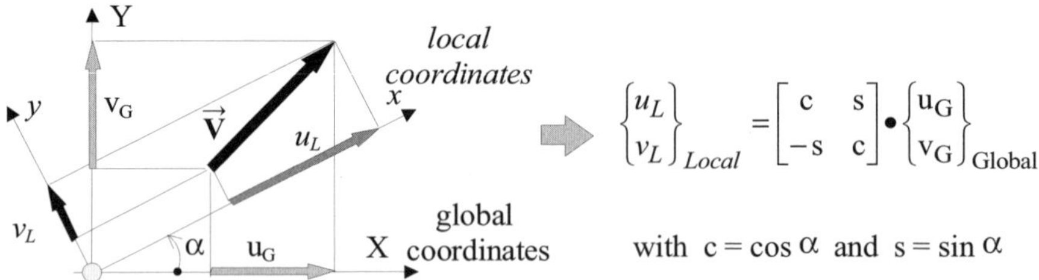

$$\left\{\begin{matrix} u_L \\ v_L \end{matrix}\right\}_{Local} = \begin{bmatrix} c & s \\ -s & c \end{bmatrix} \bullet \left\{\begin{matrix} u_G \\ v_G \end{matrix}\right\}_{Global}$$

with $c = \cos\alpha$ and $s = \sin\alpha$

Figure 3.3. *Components of a same vector in local and global coordinates*

---

4 See Chapter 12, section 12.2.

If vector $\vec{V}$ is considered in space, the notations in Figure 3.3 will give:

$$\vec{V} \rightarrow \begin{Bmatrix} u_L \\ v_L \\ w_L \end{Bmatrix}_{Local} = \begin{bmatrix} c & s & 0 \\ -s & c & 0 \\ 0 & 0 & 1 \end{bmatrix} \bullet \begin{Bmatrix} u_G \\ v_G \\ w_G \end{Bmatrix}_{Global}$$

An equation that can hereafter be noted as:

$$\begin{Bmatrix} u \\ v \\ w \end{Bmatrix}_{Local} = \begin{bmatrix} c & s & 0 \\ -s & c & 0 \\ 0 & 0 & 1 \end{bmatrix} \bullet \begin{Bmatrix} u \\ v \\ w \end{Bmatrix}_{Global} \qquad [3.2]$$

3.1.3.2. *Application to the elements of the structure*

For example, let us examine the beam in section 2.6.3.2 which is loaded by a force and a moment in conformity with Figure 2.44 that is reproduced below (Figure 3.4).

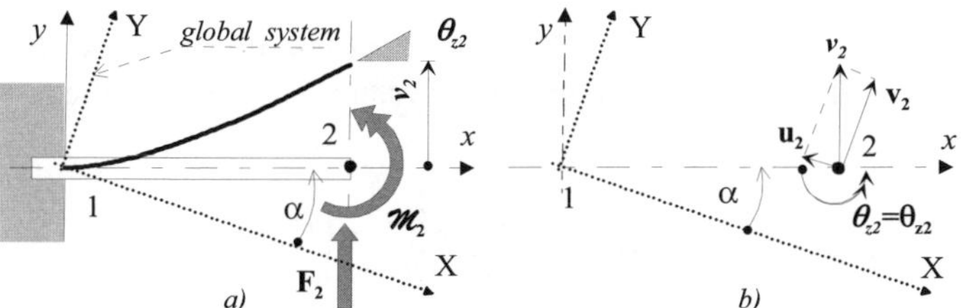

Figure 3.4. *Local and global coordinate system for a beam element*

Axes $\left(\vec{x}, \vec{y}, \vec{z}\right)$ form the local coordinate system, and we wish to analyze the behavior of the beam in the distinct global coordinates system (XYZ) with $\vec{Z} = \vec{z}$. Following [3.2], Figure 3.4 shows that it is possible to write at node (2) by replacing w_2 by θ_{z2}:

$$\begin{Bmatrix} u_2 = 0 \\ v_2 \\ \theta_{z2} \end{Bmatrix}_{Local} = \begin{bmatrix} c & s & 0 \\ -s & c & 0 \\ 0 & 0 & 1 \end{bmatrix} \bullet \begin{Bmatrix} u_2 \\ v_2 \\ \theta_{z2} \end{Bmatrix}_{Global} = [p] \bullet \begin{Bmatrix} u_2 \\ v_2 \\ \theta_{z2} \end{Bmatrix}_{Global} \qquad [3.3]$$

By repeating this exercise with each of the nodes of an element, it is possible to match the totality of the dof of an element in the local coordinate system, that is $\{d\}_{el\,Local}$ and then in the global one, that is $\{d\}_{el\,Global}$ by an equation of the type[5]:

---

5 The advantage of writing this equation from global to local coordinates shall be seen later.

$$\{d\}_{el} = [P] \bullet \{d\}_{el} \qquad\qquad [3.4]$$
$$\underset{Local}{\phantom{\{d\}_{el}}} \qquad\qquad \underset{Global}{\phantom{\{d\}_{el}}}$$

in which [P] is a "transfer matrix" obtained from $[p]$ in the form of:

$$[P] = \begin{bmatrix} [p] & 0 & 0 \\ 0 & [p] & 0 \\ 0 & 0 & \ddots \end{bmatrix} \begin{matrix} node\ 1 \\ node\ 2 \\ \dots \end{matrix}$$

It can be verified in equation [3.3] that the special feature[6] of this transfer matrix is:

$$[P]^{-1} = [P]^{T} \text{ (orthogonal matrix)}$$

In these conditions, let us consider the potential energy equation of an element:

$$\underset{element}{E_{pot.}} = \frac{1}{2}\{d\}_{el}^{T} \bullet [k]_{el} \bullet \{d\}_{el}$$

Now it is appropriate to specify the system of coordinates in which the dof are referred to. By putting ourselves in the local system which is specific to the element, we shall have:

$$\underset{element}{E_{pot.}} = \frac{1}{2}\{d\}_{el}^{T} \bullet [k]_{el} \bullet \{d\}_{el} \qquad\qquad [3.5]$$
$$ \underset{Local}{} \quad \underset{Local}{} \quad \underset{Local}{}$$

Except in special cases, the local coordinate systems differ from one element to the other. It is therefore believed that the method of summing up the elementary potential energies of equation [3.1] requires a unique definition of the dof of the elements in the global coordinate system specific to the whole structure.

By putting ourselves in the global system, the potential energy of an element should be written as:

$$\underset{element}{E_{pot.}} = \frac{1}{2}\{d\}_{el}^{T} \bullet [k]_{el} \bullet \{d\}_{el} \qquad\qquad [3.6]$$
$$ \underset{Global}{} \quad \underset{Global}{} \quad \underset{Global}{}$$

By introducing relation [3.4] in equation [3.5] we have:

$$\underset{element}{E_{pot.}} = \frac{1}{2}\left\{[P] \bullet \{d\}_{el}\right\}^{T} \bullet [k]_{el} \bullet [P] \bullet \{d\}_{el}$$
$$ \underset{Global}{} \qquad \underset{Local}{} \qquad \underset{Global}{}$$

____

6 See Chapter 12, section 12.2.

by noting that[7]:

$$\left\{ \underset{Global}{[P] \bullet \{d\}_{el}} \right\}^{T} = \underset{Global}{\{d\}_{el}^{T} \bullet [P]^{T}}$$

we obtain:

$$\underset{element}{E_{pot.}} = \frac{1}{2} \underset{Global}{\{d\}_{el}^{T}} \bullet [P]^{T} \bullet \underset{Local}{[k]_{el}} \bullet [P] \bullet \underset{Global}{\{d\}_{el}}$$

By comparing with [3.6], the stiffness matrix of the element defined in the global[8] coordinate system appears as:

$$\underset{Global}{[k]_{el}} = [P]^{T} \bullet \underset{Local}{[k]_{el}} \bullet [P] \qquad [3.7]$$

NOTES

❏ Let us verify that the transfer property from the local to the global coordinate system is also valid for the behavior equation of the element written in the form [2.122], i.e.:

$$\{F\}_{el} = [k]_{el} \bullet \{d\}_{el}$$

If this equation is written in the local system, it can be specified as:

$$\underset{Local}{\{F\}_{el}} = \underset{Local}{[k]_{el}} \bullet \underset{Local}{\{d\}_{el}} \qquad [3.8]$$

The loads are considered as vectors. The nodal forces therefore follow the same law of transformation as the associated dof, i.e. with transfer matrix [P] (see [3.4]):

$$\underset{Local}{\{F\}_{el}} = [P] \bullet \underset{Global}{\{F\}_{el}}$$

Therefore, [3.8] can be written as:

$$\underset{Global}{[P] \bullet \{F\}_{el}} = \underset{Local}{[k]_{el}} \bullet [P] \bullet \underset{Global}{\{d\}_{el}}$$

by recalling $[P]^{-1} = [P]^{T}$ (see above):

$$\underset{Global}{\{F\}_{el}} = [P]^{T} \bullet \underset{Local}{[k]_{el}} \bullet [P] \bullet \underset{Global}{\{d\}_{el}}$$

---

7 See section 12.1.2.3.

8 It is easily verified that the matrix obtained is symmetric. In fact we have (see Chapter 12, section 12.1.2.3):

$$\underset{Global}{[k]_{el}^{T}} = \left[[P]^{T} \bullet \underset{Local}{[k]_{el}} \bullet [P] \right]^{T} = [P]^{T} \bullet \underset{Local}{[k]_{el}^{T}} \bullet [P] = [P]^{T} \bullet \underset{Local}{[k]_{el}} \bullet [P] = \underset{Global}{[k]_{el}}$$

where we recognize the stiffness matrix [3.7] of the element defined in the global coordinate system. Thus we obtain the equivalent of [3.8] valid in the global or "structural" coordinate system:

$$\{F\}_{el}_{Global} = [k]_{el}_{Global} \bullet \{d\}_{el}_{Global}$$

3.1.3.3. *Summary*

<table>
<tr><td colspan="2" align="center">Case of coplanar local and global systems of coordinates</td></tr>
<tr><td>

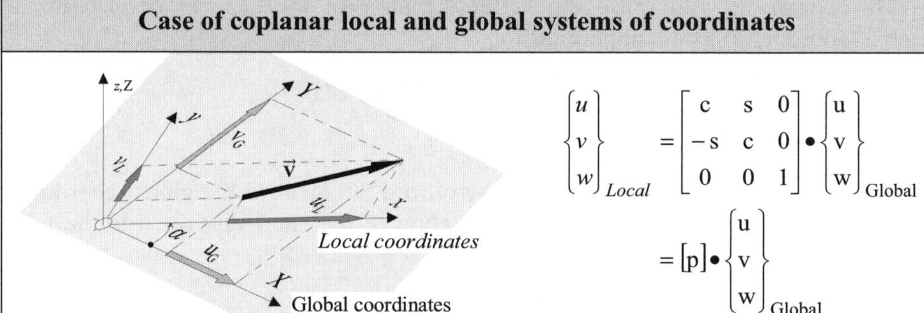

</td><td>

$$\begin{Bmatrix} u \\ v \\ w \end{Bmatrix}_{Local} = \begin{bmatrix} c & s & 0 \\ -s & c & 0 \\ 0 & 0 & 1 \end{bmatrix} \bullet \begin{Bmatrix} u \\ v \\ w \end{Bmatrix}_{Global}$$

$$= [p] \bullet \begin{Bmatrix} u \\ v \\ w \end{Bmatrix}_{Global}$$

</td></tr>
<tr><td colspan="2">

degrees of freedom of an element:

$$\{d\}_{el\ Local} = [P] \bullet \{d\}_{el\ Global} \quad \text{with } [P] = \begin{bmatrix} [p] & 0 & 0 \\ 0 & [p] & 0 \\ 0 & 0 & \ddots \end{bmatrix} \begin{matrix} node\ 1 \\ node\ 2 \\ \dots \end{matrix}$$

</td></tr>
<tr><td colspan="2">

nodal forces on an element:

$$\{F\}_{el}_{Local} = [P] \bullet \{F\}_{el}_{Global}$$

</td></tr>
<tr><td colspan="2" align="center">behavior equation of the element</td></tr>
<tr><td>

local system:

$$\{F\}_{el}_{Local} = [k]_{el}_{Local} \bullet \{d\}_{el}_{Local}$$

</td><td>

global system:

$$\{F\}_{el}_{Global} = [k]_{el}_{Global} \bullet \{d\}_{el}_{Global}$$

</td></tr>
<tr><td colspan="2" align="center">

$$[k]_{el}_{Global} = [P]^T \bullet [k]_{el}_{Local} \bullet [P]$$

</td></tr>
<tr><td colspan="2">

potential energy of deformation of the element:

$$E_{pot.\ element} = \frac{1}{2}\{d\}^T_{el}_{Local} \bullet [k]_{el}_{Local} \bullet \{d\}_{el}_{Local} = \frac{1}{2}\{d\}^T_{el}_{Global} \bullet [k]_{el}_{Global} \bullet \{d\}_{el}_{Global}$$

</td></tr>
</table>

[3.9]

3.2. Stiffness matrix of some simple finite elements

PRELIMINARY NOTES

❏ In this section the characteristics of some simple elements in their local coordinates will be established. Then, following the procedure above, we will reset these characteristics in distinct global coordinates, specific to the structure.

❏ To simplify and homogenize the figures, we represent conventionally the forces and the moments along the positive direction of the local axis of finite elements.

3.2.1. *Truss element loaded under traction (or compression)*

3.2.1.1. *Summary: pure traction (or compression) exerted on a beam*

Recalling[9] that the resultant force and moment at the center of gravity G of the cross-section of a beam with its mean line along the $\vec{x}$ axis, can be summarized as:

$$\{Coh\}_G = \left\{ \begin{array}{c} \mathcal{N}\,\vec{x} \\ \vec{0} \end{array} \right\}_G \text{, with } \mathcal{N}\text{: normal force}$$

corresponding displacements are elongations or compressions along the $\vec{x}$ axis denoted as u(x).

The behavior equation is written:

$$\mathcal{N} = ES\frac{du(x)}{dx}$$

3.2.1.2. *Stiffness matrix*

○ Definition of the element

The corresponding finite element in Figure 3.5 is a truss section of length ℓ. In terms of material property it has a longitudinal modulus of elasticity E and the area of its cross-section is S. This element shall be represented and modeled according to its mean line. The centers of end sections (1) and (2) are the nodes of the element. The coordinates (*xy*) linked to the element form the "local system".

---

9 See Chapter 1, section 1.5.2.1.

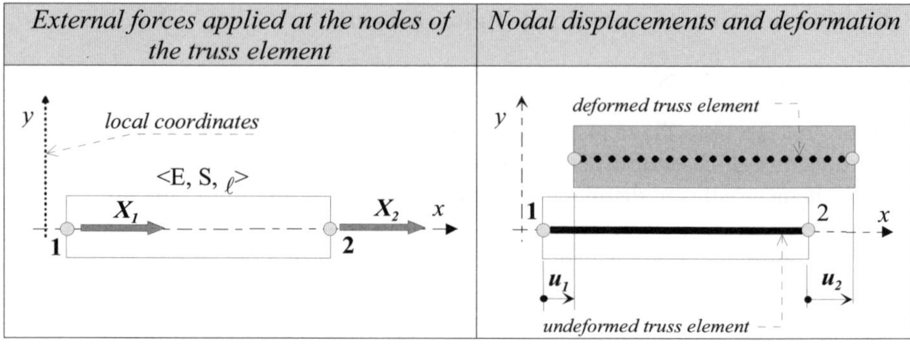

Figure 3.5. *Truss element*

The loading of the element consists of the forces $X_1\vec{x}$ at node (1) and $X_2\vec{x}$ at node (2). The associated degrees of freedom are *thus*[10] the displacements $u_1\vec{x}$ at node (1) and $u_2\vec{x}$ at node (2), that is a total of two dof for the element.

For any intermediate cross-section of the x axis ($0 \leq x \leq \ell$), the longitudinal displacement is indicated as u(x) with $u(0) = u_1$ and $u(\ell) = u_2$.

○ Displacement u(x)

Reference[11] is made to the form of the deformation energy $E_{pot.}$ based on the displacement u(x) of every cross-section along the x axis of the truss of length ℓ:

$$E_{pot.} = \frac{1}{2} \int_{\ell} ES\left(\frac{du(x)}{dx}\right)^2 dx$$

Calculation of E_{pot} requires the knowledge of the law of evolution u(x). It is a fundamental aspect of the procedure. In this case, the element being loaded only at its extremities, normal force $\vec{\mathcal{N}}$ is constant, i.e.:

$$\mathcal{N} = ES\frac{du(x)}{dx} = C^{te}$$

which leads to a very simple form of the displacement:

$$u(x) = ax + b$$

Let us calculate constants a and b in terms of displacements u_1 and u_2 of the nodes (or dof of the element) by introducing the boundary conditions at the ends:

– for x = 0:

$$u(0) = u_1, \text{ that is } u_1 = 0 + b, \text{ from where } b = u_1$$

– for x = ℓ:

$$u(\ell) = u_2, \text{ that is } u_2 = a\ell + u_1, \quad \text{from where } a = \frac{u_2 - u_1}{\ell}$$

The displacement function u(x) is written as:

$$u(x) = ax + b = \left(\frac{u_2 - u_1}{\ell}\right)x + u_1$$

which can be presented in matrix form:

$$u(x) = \underbrace{\left[\left(1 - \frac{x}{\ell}\right) \quad \frac{x}{\ell}\right]}_{\substack{row\ matrix \\ (1\times 2)}} \bullet \underbrace{\begin{Bmatrix} u_1 \\ u_2 \end{Bmatrix}}_{\substack{column\ matrix \\ (2\times 1)}}$$

The (1 × 2) row "space function" matrix (that is a function of the variable x) which appears here is called an *interpolation matrix*. This makes it possible to express the displacement of any cross-section in terms of displacements of the nodes.

○ Determination of the stiffness matrix

The differentiation of the displacement function is written as:

$$\frac{du(x)}{dx} = \frac{u_2 - u_1}{\ell} \tag{3.10}$$

then the energy of deformation referred to above becomes:

$$E_{pot.} = \frac{1}{2}\int_\ell ES\left(\frac{du(x)}{dx}\right)^2 dx = \frac{1}{2}\int_\ell ES\left(\frac{u_2 - u_1}{\ell}\right)^2 dx$$

$$E_{pot.} = \frac{1}{2}\frac{ES}{\ell}(u_2 - u_1)^2 = \frac{1}{2}\frac{ES}{\ell}\left(u_1^2 + u_2^2 - 2u_2 u_1\right) \tag{3.11}$$

A quadratic form of degrees of freedom u_1 and u_2 is obtained. It can be written as:

$$E_{\substack{pot. \\ element}} = \frac{1}{2}\{d\}_{el}^T \bullet [k]_{el} \bullet \{d\}_{el}$$

based on [3.11]:

$$E_{pot.} = \frac{1}{2}\underbrace{[u_1 \quad u_2]}_{\substack{row\ matrix\\(1\times 2)}} \bullet \underbrace{\left\{ \begin{array}{c} \dfrac{ES}{\ell}u_1 - \dfrac{ES}{\ell}u_2 \\[2mm] -\dfrac{ES}{\ell}u_1 + \dfrac{ES}{\ell}u_2 \end{array} \right\}}_{\substack{column\ matrix\\(2\times 1)}}$$

i.e.:

$$E_{pot.} = \frac{1}{2}\left\{ \begin{array}{c} u_1 \\ u_2 \end{array} \right\}^{T} \bullet \left[\begin{array}{cc} \dfrac{ES}{\ell} & -\dfrac{ES}{\ell} \\[2mm] -\dfrac{ES}{\ell} & \dfrac{ES}{\ell} \end{array} \right] \bullet \left\{ \begin{array}{c} u_1 \\ u_2 \end{array} \right\}$$

an equation which shows the stiffness matrix of the truss element under traction-compression:

$$[k]_{element} = \left[\begin{array}{cc} \dfrac{ES}{\ell} & -\dfrac{ES}{\ell} \\[2mm] -\dfrac{ES}{\ell} & \dfrac{ES}{\ell} \end{array} \right] \tag{3.12}$$

According to [2.122] the behavior of this element is written as:

$$\{F\}_{el} = [k]_{el} \bullet \{d\}_{el} \Leftrightarrow \left\{ \begin{array}{c} X_1 \\ X_2 \end{array} \right\} = \left[\begin{array}{cc} \dfrac{ES}{\ell} & -\dfrac{ES}{\ell} \\[2mm] -\dfrac{ES}{\ell} & \dfrac{ES}{\ell} \end{array} \right] \bullet \left\{ \begin{array}{c} u_1 \\ u_2 \end{array} \right\} \tag{3.13}$$

NOTES

❏ The developed equation [3.11] of the deformation energy can be condensed as below, by noting the terms of the stiffness matrix as k_{ij} [3.12] and then degrees of freedom u_i, as d_i:

$$E_{pot.} = \frac{1}{2}(k_{11}u_1^2 + k_{22}u_2^2 + 2k_{12}u_1u_2) = \frac{1}{2}\sum_{i=1}^{2}\sum_{j=1}^{2}k_{ij}d_id_j$$

❏ The truss element as represented in Figure 3.5 appears to be "floating" along the $\vec{x}$ axis. This corresponds to the notion of a structure "not properly linked", which is dealt with in section 2.5.2.2. From Figure 3.5 it is obvious that the absence of a link of the element with the environment gives rise to the possibility of a rigid-body movement in the form of a translation along the $\vec{x}$ direction of loading. Then the inverse of the stiffness matrix, i.e. the flexibility matrix does not exist. This can be verified here by calculating the main determinant of the stiffness matrix [3.12]. The result is:

$$\Delta_{\text{principal}} = \left(\frac{ES}{\ell}\right)^2 - \left(\frac{ES}{\ell}\right)^2 = 0$$

Thus, $[k]_{\text{element}}$ is singular (see [2.122]).

❏ Behavior of the truss element in a global coordinate system different from the local system. As specified in the beginning of section 3.2, equations [3.12] and [3.13] characterize the behavior of the truss element in its local coordinate system.

Let us observe this element with a certain orientation in the plane of a global coordinate system (XY) in Figure 3.6.

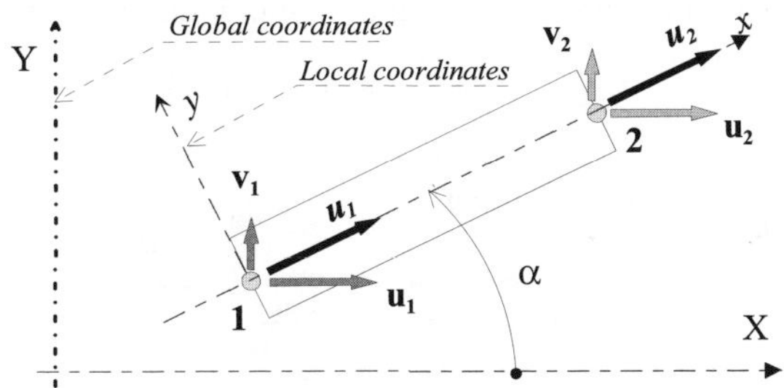

Figure 3.6. *Displacements in local and global coordinates*

Let us refer to the results in [3.9] by adapting them to the dof specific to this element. At nodes (1) and (2) we have:

$$\begin{Bmatrix} u_1 \\ v_1 = 0 \\ w_1 = 0 \end{Bmatrix}_L = \begin{bmatrix} c & s & 0 \\ -s & c & 0 \\ 0 & 0 & 1 \end{bmatrix} \bullet \begin{Bmatrix} u_1 \\ v_1 \\ w_1 = 0 \end{Bmatrix}_G \quad \text{and} \quad \begin{Bmatrix} u_2 \\ v_2 = 0 \\ w_2 = 0 \end{Bmatrix}_L = \begin{bmatrix} c & s & 0 \\ -s & c & 0 \\ 0 & 0 & 1 \end{bmatrix} \bullet \begin{Bmatrix} u_2 \\ v_2 \\ w_2 = 0 \end{Bmatrix}_G$$

i.e., by simplifying and regrouping so as to introduce the column of the dof of the element:

$$\begin{Bmatrix} u_1 \\ 0 \\ u_2 \\ 0 \end{Bmatrix}_L = \begin{bmatrix} c & s & 0 & 0 \\ -s & c & 0 & 0 \\ 0 & 0 & c & s \\ 0 & 0 & -s & c \end{bmatrix} \bullet \begin{Bmatrix} u_1 \\ v_1 \\ u_2 \\ v_2 \end{Bmatrix}_G \qquad [3.14]$$

where transfer matrix [P] from the global to the local system is observed:

$$[P] = \begin{bmatrix} c & s & 0 & 0 \\ -s & c & 0 & 0 \\ 0 & 0 & c & s \\ 0 & 0 & -s & c \end{bmatrix}$$

This transfer matrix shall be used according to the procedure described in [3.9]. Thus, when a plane structure has truss elements, these are characterized in the global coordinate system by 2 dof per node which is 4 dof for the element (see Figure 3.8).

❏ The only nodal loading that can be supported by a truss element is, by definition, a colinear force along its axis. For example, unlike a beam, it can neither support a transverse force nor a moment. That is why its links with its environment, wall or adjacent elements, should be modeled accordingly as in Figure 3.7.

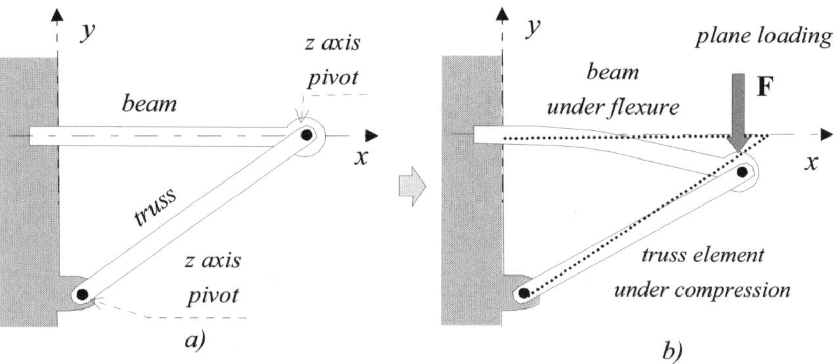

Figure 3.7. *Plane modeling of the links of a truss element*

❏ It can be noted that knowing the dof of a truss element helps in obtaining the normal stress in this element. The following results can be recalled[12]:

$$\sigma_x = \frac{\mathcal{N}}{S} = E\frac{du}{dx}$$

However, it has been noted above in [3.10]:

$$\frac{du(x)}{dx} = \frac{u_2 - u_1}{\ell}$$

---

12 See section 15.2.1.

Thus, the normal stress in the element is written as (in the local coordinate system):

$$\sigma_x = E\left(\frac{u_2 - u_1}{\ell}\right)$$

This stress is thus calculated in the postprocessors of the finite element codes.

❏ Modeling a structure into truss elements is above all of educational interest. For designing with calculation codes, the beam element defined later is preferred.

○ Conclusion

The behavior of the truss element under traction-compression is summarized in the following table.

Behavior in the local coordinate system

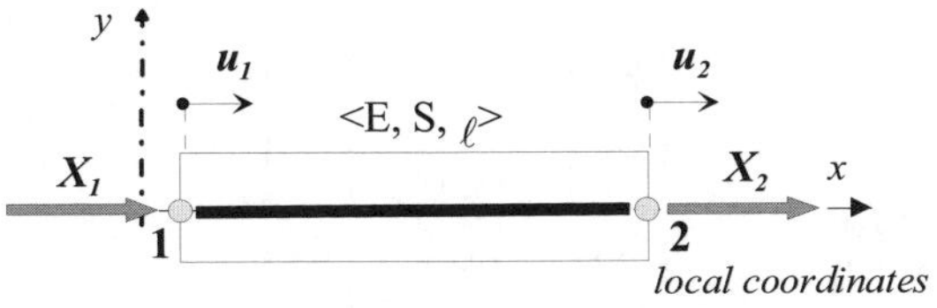

behavior equation:

$$\{F\}_{el} = [k]_{el} \bullet \{d\}_{el} \Leftrightarrow \begin{Bmatrix} X_1 \\ X_2 \end{Bmatrix}_{Local} = \begin{bmatrix} \dfrac{ES}{\ell} & -\dfrac{ES}{\ell} \\ -\dfrac{ES}{\ell} & \dfrac{ES}{\ell} \end{bmatrix} \bullet \begin{Bmatrix} u_1 \\ u_2 \end{Bmatrix}_{Local}$$

potential energy of deformation:

$$E_{pot.\ element} = \frac{1}{2}\{d\}_{el}^{T} \bullet [k]_{el} \bullet \{d\}_{l} \Leftrightarrow E_{pot.\ element} = \frac{1}{2}\begin{Bmatrix} u_1 \\ u_2 \end{Bmatrix}^{T}_{Local} \bullet \begin{bmatrix} \dfrac{ES}{\ell} & -\dfrac{ES}{\ell} \\ -\dfrac{ES}{\ell} & \dfrac{ES}{\ell} \end{bmatrix} \bullet \begin{Bmatrix} u_1 \\ u_2 \end{Bmatrix}_{Local}$$

normal stress on any cross-section of the element:

$$\sigma_x = E\left(\frac{u_2 - u_1}{\ell}\right)$$

Where the stiffness matrix is:

$$[k]_{el\ Local} = \begin{bmatrix} \dfrac{ES}{\ell} & -\dfrac{ES}{\ell} \\ -\dfrac{ES}{\ell} & \dfrac{ES}{\ell} \end{bmatrix}$$

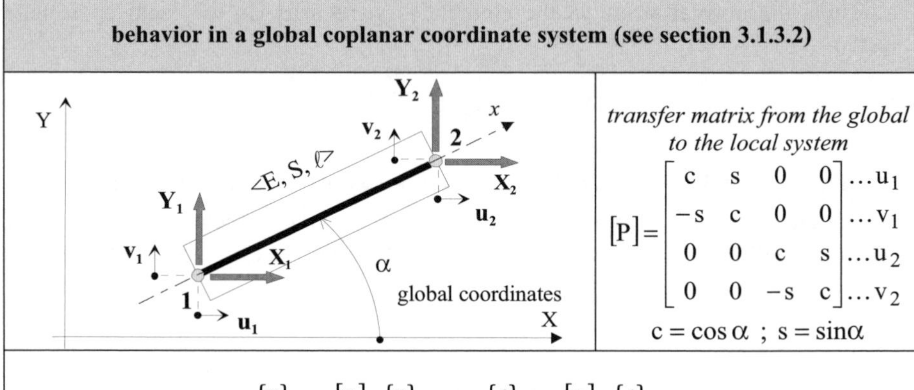

$$\{F\}_{el}^{Local} = [P] \bullet \{F\}_{el}^{Global} \; ; \quad \{d\}_{el}^{Local} = [P] \bullet \{d\}_{el}^{Global}$$

$$[k]_{el}^{Global} = [P]^T \bullet [k]_{el}^{Local} \bullet [P]$$

behavior equation:

$$\{F\}_{el}^{Global} = [k]_{el}^{Global} \bullet \{d\}_{el}^{Global} \quad \Leftrightarrow \quad \begin{Bmatrix} X_1 \\ Y_1 \\ X_2 \\ Y_2 \end{Bmatrix}_{Global} = [k]_{el}^{Global}_{(4\times4)} \bullet \begin{Bmatrix} u_1 \\ v_1 \\ u_2 \\ v_2 \end{Bmatrix}_{Global}$$

potential energy of deformation:

$$E_{pot.}^{element} = \frac{1}{2} \{d\}_{el}^{T\,Global} \bullet [k]_{el}^{Global} \bullet \{d\}_{el}^{Global}$$

Figure 3.8. *Behavior of a truss-element*

3.2.1.3. *Beam element under torsion*

Hereafter we shall limit ourselves to beams whose cross-section accepts a center of symmetry (which is also the geometric center G). For example a circular section (as in Figure 3.9), a rectangular section, an I-section, etc.[13]

---

13 When the section does not accept any center of symmetry, there is a torsional rotation of the sections around a point distinct from the geometric center G. This other point is called the torsion center (see section 9.3.2.4).

3.2.1.4. *Torsion loading on a beam element*[14]

We refer to the resultant force and moment at the center of gravity G of the cross-section of a beam with its mean line along $\vec{x}$, i.e.:

$$\{Coh\}_G = \left\{ \begin{array}{c} \vec{0} \\ \mathcal{M}t\vec{x} \end{array} \right\}_G \text{, with } \mathcal{M}t\text{: torsion moment}$$

The corresponding displacement is a rotation of the section around the $\vec{x}$ axis denoted $\theta_x(x)$, (angular displacement). The behavior equation is given as:

$$\mathcal{M}t = GJ \frac{d\theta_x(x)}{dx}$$

3.2.1.5. *Stiffness matrix*

⭕ Definition of the element

The finite element in Figure 3.9 is a beam of length ℓ. In terms of material property it has a shear modulus G and the cross-section is characterized by the torsion constant J. This element shall be represented and modeled according to its mean line. The centers of end sections (1) and (2) are the nodes of the element. The coordinates (*xyz*) linked to the element make up the local coordinate system.

The loading on the element consists of the moments $L_1\vec{x}$ at node (1) and $L_2\vec{x}$ at node (2). In line with the previous chapter (see [2.122]) the associated degrees of freedom are thus the section rotations $\theta_{x1}\vec{x}$ at node (1) and $\theta_{x2}\vec{x}$ at node (2), which makes a total of two dof for the element.

For any intermediate cross-section on the x axis ($0 \le x \le \ell$), the section rotates by $\theta_x(x)$ with $\theta_x(0) = \theta_{x1}$ and $\theta_x(\ell) = \theta_{x2}$.

---

14 See section 1.5.2.4.

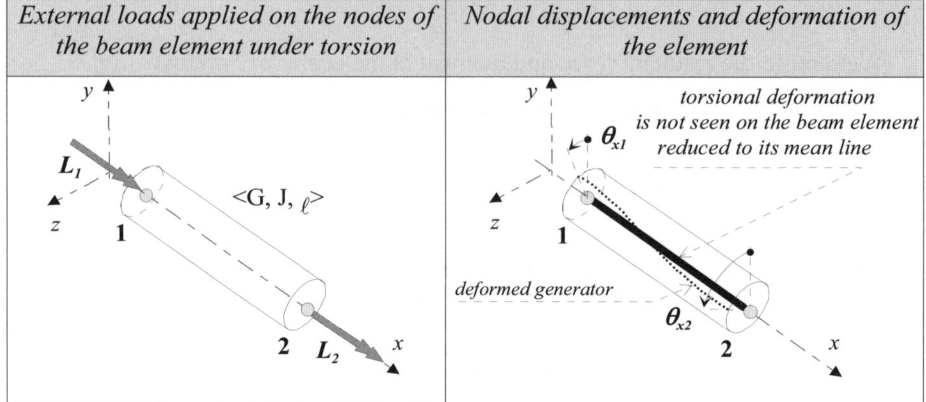

Figure 3.9. *Beam element under torsion*

○ Angular displacement $\theta_x(x)$

Reference (see [2.20]) is made to the deformation energy $E_{pot.}$ based on the rotation of cross-sections $\theta_x(x)$ along the x axis of the beam of length ℓ:

$$E_{pot.} = \frac{1}{2} \int_\ell GJ\left(\frac{d\theta_x(x)}{dx}\right)^2 dx$$

Calculation of this energy requires the knowledge of the law of evolution $\theta_x(x)$. As indicated for the previous truss element, this is a fundamental step towards characterization of the element. In this case, the element being loaded only at its extremities, the torsional moment $\overrightarrow{Mt}$ remains constant:

$$Mt = GJ\frac{d\theta_x(x)}{dx} = C^{te}$$

which leads to a simple expression of the angular displacement function:

$$\theta_x(x) = ax + b$$

Let us calculate constants a and b by introducing the boundary conditions at the ends, that is the displacements of the nodes (or dof) of the element denoted as θ_{x1} and θ_{x2}:

– for x = 0:

$$\theta_x(0) = \theta_{x1}, \text{ that is } \theta_{x1} = 0 + b; \text{ from where } b = \theta_{x1}$$

– for x = ℓ:

$$\theta_x(\ell) = \theta_{x2}, \text{ that is } \theta_{x2} = a\ell + \theta_{x1}, \text{ from where } a = \frac{\theta_{x2} - \theta_{x1}}{\ell}$$

The displacement function $\theta_x(x)$ is written as:

$$\theta_x(x) = ax + b = \left(\frac{\theta_{x2} - \theta_{x1}}{\ell}\right)x + \theta_{x1}$$

which can be presented in matrix form:

$$\theta_x(x) = \underbrace{\left[\left(1 - \frac{x}{\ell}\right) \quad \frac{x}{\ell}\right]}_{\substack{row\ matrix \\ (1\times2)}} \bullet \underbrace{\begin{Bmatrix} \theta_{x1} \\ \theta_{x2} \end{Bmatrix}}_{\substack{column\ matrix \\ (2\times1)}}$$

The "space function" matrix (1×2) (that is a function of variable x) which appears here is called interpolation matrix. It helps in expressing the angular displacement of any cross-section in terms of angular displacements of the nodes.

O Determination of the stiffness matrix

The differentiation of the displacement function is written as:

$$\frac{d\theta_x(x)}{dx} = \frac{\theta_{x2} - \theta_{x1}}{\ell}$$

then the potential energy of deformation [2.20] becomes:

$$E_{pot.} = \frac{1}{2}\int_\ell GJ\left(\frac{\theta_{x2} - \theta_{x1}}{\ell}\right)^2 dx$$

$$E_{pot.} = \frac{1}{2}\frac{GJ}{\ell}\left(\theta_{x2} - \theta_{x1}\right)^2 = \frac{1}{2}\frac{GJ}{\ell}\left(\theta_{x1}^2 + \theta_{x2}^2 - 2\theta_{x2}\theta_{x1}\right) \qquad [3.15]$$

A quadratic form of the degrees of freedom is obtained. It can be written as:

$$\underset{element}{E_{pot.}} = \frac{1}{2}\{d\}_{el}^T \bullet [k]_{el} \bullet \{d\}_{el}$$

This gives:

$$E_{pot.} = \frac{1}{2}\underbrace{[\theta_{x1} \quad \theta_{x2}]}_{\substack{row\ matrix \\ (1\times 2)}} \bullet \underbrace{\left\{ \begin{array}{c} \dfrac{GJ}{\ell}\theta_{x1} - \dfrac{GJ}{\ell}\theta_{x2} \\[2ex] -\dfrac{GJ}{\ell}\theta_{x1} + \dfrac{GJ}{\ell}\theta_{x2} \end{array} \right\}}_{\substack{column\ matrix \\ (2\times 1)}}$$

that is: $\displaystyle E_{pot.} = \frac{1}{2}\left\{\begin{array}{c}\theta_{x1}\\[1ex]\theta_{x2}\end{array}\right\}^{T} \bullet \left[\begin{array}{cc}\dfrac{GJ}{\ell} & -\dfrac{GJ}{\ell} \\[2ex] -\dfrac{GJ}{\ell} & \dfrac{GJ}{\ell}\end{array}\right] \bullet \left\{\begin{array}{c}\theta_{x1}\\[1ex]\theta_{x2}\end{array}\right\}$

an equation which shows the stiffness matrix of the beam element under torsion:

$$[k]_{element} = \left[\begin{array}{cc}\dfrac{GJ}{\ell} & -\dfrac{GJ}{\ell} \\[2ex] -\dfrac{GJ}{\ell} & \dfrac{GJ}{\ell}\end{array}\right] \tag{3.16}$$

As per [2.122], the behavior of the beam element under torsion is therefore written as:

$$\{F\}_{el} = [k]_{el} \bullet \{d\}_{el} \iff \left\{\begin{array}{c}L_1\\[1ex]L_2\end{array}\right\} = \left[\begin{array}{cc}\dfrac{GJ}{\ell} & -\dfrac{GJ}{\ell} \\[2ex] -\dfrac{GJ}{\ell} & \dfrac{GJ}{\ell}\end{array}\right] \bullet \left\{\begin{array}{c}\theta_{x1}\\[1ex]\theta_{x2}\end{array}\right\} \tag{3.17}$$

NOTES

❐ The developed form [3.15] of the deformation energy can be condensed as below, by noting the terms of the stiffness matrix as k_{ij} [3.16] and then the degrees of freedom θ_{xi} as d_i:

$$E_{pot.} = \frac{1}{2}\left(k_{11}\theta_{x1}{}^2 + k_{22}\theta_{x2}{}^2 + 2k_{12}\theta_{x1}\theta_{x2}\right) = \frac{1}{2}\sum_{i=1}^{2}\sum_{j=1}^{2}k_{ij}d_i d_j$$

❐ The beam element under torsion as represented in Figure 3.9 appears to be "floating" about the $\vec{x}$ axis. This corresponds to the notion of a structure "not properly linked" and which is dealt with in section 2.5.2. From Figure 3.9 it is obvious that absence of a link with the environment gives rise to the possibility of a rigid-body rotation around the $\vec{x}$ direction of loading. Thus the inverse of the stiffness matrix, the flexibility matrix, does not exist. This can be verified by calculating the main determinant of stiffness matrix [3.16]. The result is:

$$\Delta_{\text{principal}} = \left(\frac{GJ}{\ell}\right)^2 - \left(\frac{GJ}{\ell}\right)^2 = 0$$

Thus, $[k]_{\text{element}}$ is singular (see [2-122]).

☐ Local and global coordinate systems: As in the case of the previous truss element, when the center line of the beam element under torsion has any orientation in the coplanar global coordinates (XY), for every nodal rotation θ_{xL} in the local coordinates there are two components θ_{XG}, θ_{YG} in the global system.

Then an equation similar to the one mentioned for the truss element is observed, i.e. (see [3.14]):

$$\begin{Bmatrix} \theta_{x1} \\ 0 \\ \theta_{x2} \\ 0 \end{Bmatrix}_{L} = \begin{bmatrix} c & s & 0 & 0 \\ -s & c & 0 & 0 \\ 0 & 0 & c & s \\ 0 & 0 & -s & c \end{bmatrix} \bullet \begin{Bmatrix} \theta_{X1} \\ \theta_{Y1} \\ \theta_{X2} \\ \theta_{Y2} \end{Bmatrix}_{G}$$
$$\underset{(4\times1)}{} \qquad \underset{4\times4}{} \qquad \underset{4\times1)}{}$$

where the same transfer matrix [P] from the global to the local coordinate system appears and it is used according to the procedure described in [3.9]. It is observed that when a planar structure has beam elements under torsion, they are characterized in the global system by 2 dof per node, i.e. 4 dof for the element (see Figure 3.10).

☐ Knowledge of the dof of a beam element under torsion does not lead simply to the distribution of torsional shear stresses in the cross-sections for any cross-sectional shapes[15]. That is why the postprocessors of the finite element programs do not provide this kind of information.

☐ The only nodal loading that the beam element can handle under torsion is as per definition, a moment colinear to its axis (for example, application for the shafts of machines).

---

15 The calculation method of these stresses, detailed in Chapter 9, is summarized in [9.27]. In some special cases of section shapes it is possible to have a simple expression for these stresses (see section 9.3.2.5).

○ Conclusion

The behavior of the beam element under torsion is summarized in the following table.

<table>
<tr><td colspan="2" align="center">Behavior in local coordinates</td></tr>
<tr><td>

</td><td>

$$[k]_{\acute{e}l}^{Local} = \begin{bmatrix} \dfrac{GJ}{\ell} & -\dfrac{GJ}{\ell} \\[2mm] -\dfrac{GJ}{\ell} & \dfrac{GJ}{\ell} \end{bmatrix}$$

</td></tr>
<tr><td colspan="2">

behavior equation:

$$\{F\}_{el}^{Local} = [k]_{el}^{Local} \bullet \{d\}_{el}^{Local} \Leftrightarrow \begin{Bmatrix} L_1 \\ L_2 \end{Bmatrix}_{Local} = \begin{bmatrix} \dfrac{GJ}{\ell} & -\dfrac{GJ}{\ell} \\[2mm] -\dfrac{GJ}{\ell} & \dfrac{GJ}{\ell} \end{bmatrix} \bullet \begin{Bmatrix} \theta_{x1} \\ \theta_{x2} \end{Bmatrix}_{Local}$$

potential energy of deformation:

$$E_{pot.}^{element} = \frac{1}{2}\{d\}_{el}^{T} \bullet [k]_{el} \bullet \{d\}_{el} \Leftrightarrow E_{pot.}^{element} = \frac{1}{2}\begin{Bmatrix} \theta_{x1} \\ \theta_{x2} \end{Bmatrix}_{Local}^{T} \bullet \begin{bmatrix} \dfrac{GJ}{\ell} & -\dfrac{GJ}{\ell} \\[2mm] -\dfrac{GJ}{\ell} & \dfrac{GJ}{\ell} \end{bmatrix} \bullet \begin{Bmatrix} \theta_{x1} \\ \theta_{x2} \end{Bmatrix}_{Local}$$

</td></tr>
<tr><td colspan="2">

Shear stresses in a cross-section of the element:

τ_{xy} *and* τ_{xz} *cannot be calculated in the general case of any cross-sectional shapes*[16].

</td></tr>
</table>

---

16 The calculation method of these stresses, detailed in Chapter 9, is summarized in [9.27]. In certain special cases, of section forms it is possible to have a simple equation of these stresses; see section 9.3.2.5.

Behavior in global coplanar coordinates (see section 3.1.3.2)

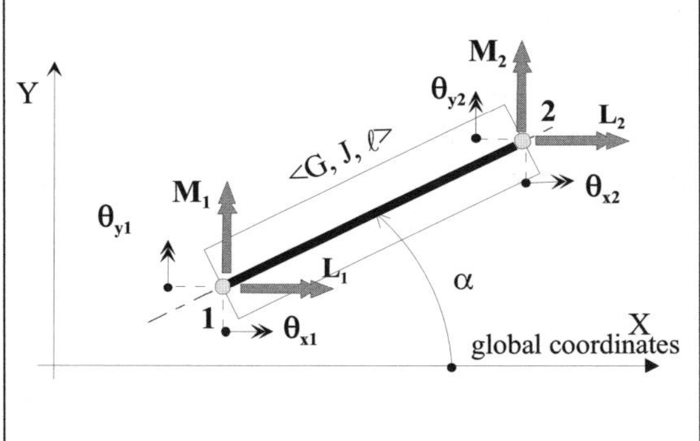

transfer matrix from the global to the local coordinate system

$$[P] = \begin{bmatrix} c & s & 0 & 0 \\ -s & c & 0 & 0 \\ 0 & 0 & c & s \\ 0 & 0 & -s & c \end{bmatrix} \begin{matrix} \cdots \theta_{x1} \\ \cdots \theta_{y1} \\ \cdots \theta_{x2} \\ \cdots \theta_{y2} \end{matrix}$$

$$c = \cos \alpha \; ; \; s = \sin \alpha$$

$$\{F\}_{el} = [P] \bullet \{F\}_{el} \; ; \quad \{d\}_{el} = [P] \bullet \{d\}_{el}$$
$$\underset{Local}{} \quad \underset{Global}{} \quad \underset{Local}{} \quad \underset{Global}{}$$

$$[k]_{el} = [P]^T \bullet [k]_{el} \bullet [P]$$
$$\underset{Global}{} \quad \underset{Local}{}$$

behavior equation:

$$\{F\}_{el} = [k]_{el} \bullet \{d\}_{el} \quad \hat{U} \quad \begin{Bmatrix} L_1 \\ M_1 \\ L_2 \\ M_2 \end{Bmatrix} = [k]_{el} \bullet \begin{Bmatrix} \theta_{x1} \\ \theta_{y1} \\ \theta_{x2} \\ \theta_{y2} \end{Bmatrix}$$

$$\underset{Global}{} \; \underset{Global}{} \; \underset{Global}{} \qquad\qquad \underset{Global}{} \qquad \underset{Global}{\underset{(4\times4)}{}} \qquad\qquad \underset{Global}{}$$

potential energy of deformation:

$$E_{pot.} = \frac{1}{2} \{d\}_{el}^T \bullet [k]_{el} \bullet \{d\}_{el}$$
$$\underset{element}{} \qquad \underset{Global}{} \; \underset{Global}{} \; \underset{Global}{}$$

Figure 3.10. *Behavior of a torsion element*

3.2.3. *Beam element under plane bending*

3.2.3.1. *Summary: plane bending of a beam*[17]

Consider the case of a beam bending in a main plane, for example (xy), whose mean line is along the $\vec{x}$ axis. We remember then the fact that the resultant force and moment at the geometric center G of the cross-section is reduced to:

$$\{Coh\}_G = \begin{Bmatrix} \mathcal{T}_y \vec{y} \\ \mathcal{M}f_z \vec{z} \end{Bmatrix}_G$$

where $\mathcal{T}_y$ is the shear force along the $\vec{y}$ axis and $\mathcal{M}f_z$ is the bending moment about the $\vec{z}$ axis.

Thus, the displacements of a section[18] are:

♦ v(x): linear displacement of the section center, or deflection along the $\vec{y}$ axis;

♦ θ_z(x): angular displacement or rotation of the section around the $\vec{z}$ axis.

This shows that if the lateral surface of the beam is free of force, equilibrium of an elementary beam slice leads to the equations[19]:

$$\frac{d\mathcal{M}f_z}{dx} + \mathcal{T}_y = 0 \text{ and } \frac{d\mathcal{T}_y}{dx} = 0 \qquad [3.18]$$

In accordance with the Bernoulli hypothesis, the behavior relation is written as[20]:

$$\theta_z(x) \cong \frac{dv}{dx} \Rightarrow \frac{d^2 v}{dx^2} = \frac{\mathcal{M}f_z}{EI_z} \qquad [3.19]$$

3.2.3.2. *Stiffness matrix*

◯ Definition of the finite element

The element is defined in Figure 3.11 as a beam of length ℓ. Its longitudinal elasticity module is E (material property) and the moment of inertia of any cross-

---

17 See section 2.3.1.4.
18 See Figures 1.26 and 2.16.
19 The local equilibrium extended to a cross-section is studied in Chapter 9, section 9.2 (see [9.12]).
20 See section 15.2.6 and [2.32].

section with respect to the $\vec{z}$ axis is I_z (cross-sectional property). This finite element shall be represented and modeled according to its mean line for calculation.

The centers of end sections (1) and (2) are the nodes of the element. Axis (xyz), linked to the element forms the local coordinate system.

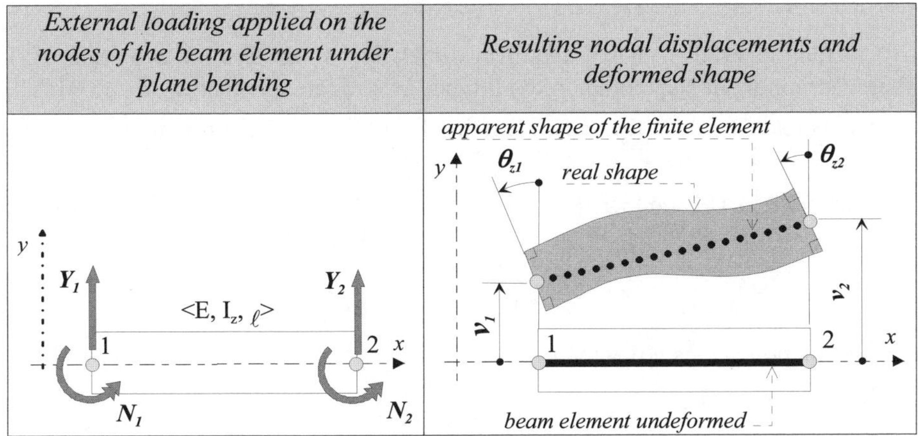

Figure 3.11. *Beam element under plane bending*

The loads on the element consist of transverse forces and moments respectively $Y_1\vec{y}$, $N_1\vec{z}$ at node (1) and $Y_2\vec{y}$, $N_2\vec{z}$ at node (2). As per the previous chapter (see [2.122]) the associated degrees of freedom are:

– translations $v_1\vec{y}$ at node (1) and $v_2\vec{y}$ at node (2);

– sectional rotations $\theta_{z1}\vec{z}$ at node (1) and $\theta_{z2}\vec{z}$ at node (2),

a total of $2\times2=4$ dof for the element.

○ Form of displacements

Every intermediate cross-section along x axis ($0\leq x\leq\ell$) has its displacement marked as:

♦ v(x) with $v(0)=v_1$ and $v(\ell)=v_2$;

♦ $\theta_z(x)$ with $\theta_z(0)=\theta_{z1}$ and $\theta_z(\ell)=\theta_{z2}$.

Reference (see [2.33]) is made to the energy of deformation $E_{pot.}$ based on the transverse movement v(x) of every cross-section along x axis for the beam of length ℓ:

$$E_{pot.} = \frac{1}{2} \int_{\ell} EI_z \left(\frac{d^2 v}{dx^2} \right)^2 dx \qquad [3.20]$$

Calculation of this energy requires knowledge of the law of evolution v(x). There, an important step common to all types of standard elements is observed. The element is loaded only at the ends. Then the shear force $\vec{T}_y$ is constant.

Thus, with [3.18] and [3.19]:

$$\frac{d\mathcal{T}_y}{dx} = -\frac{d^2 \mathcal{M}f_z}{dx^2} = -EI_z \frac{d^4 v}{dx^4} = 0$$

the deformation v(x) is governed by the equation:

$$\frac{d^4 v(x)}{dx^4} = 0$$

which after successive integrations leads to: $v(x) = ax^3 + bx^2 + cx + d$.

Following Bernoulli approximations mentioned in [3.19], i.e. $\theta_z(x) \cong \dfrac{dv}{dx}$, we obtain:

$$v(x) = ax^3 + bx^2 + cx + d$$
$$\theta_z(x) = 3ax^2 + 2bx + c$$

The constants a, b, c, d are calculated with the help of boundary conditions, that is the nodal displacements (or dof) of element v_1, v_2 and θ_{z1}, θ_{z2}:

– for x = 0:

$$v(0) = v_1, \quad \rightarrow v_1 = d$$
$$\theta_z(0) = \theta_{z1}, \quad \rightarrow \theta_{z1} = c$$

– for x = ℓ:

$$v(\ell) = v_2, \quad \rightarrow v_2 = a\ell^3 + b\ell^2 + c\ell + d$$

$$\theta_z(\ell) = \theta_{z2}, \quad \rightarrow \theta_{z2} = 3a\ell^2 + 2b\ell + c$$

A system of four equations with four unknowns a, b, c, d is obtained, the solution of which leads to the following regrouped equation with: $\bar{x} = \dfrac{x}{\ell}$

$$\left\{\begin{matrix} v(x) \\ \theta_z(x) \end{matrix}\right\} = \begin{bmatrix} 1-3\bar{x}^2+2\bar{x}^3 & \ell(\bar{x}-2\bar{x}^2+\bar{x}^3) & (3\bar{x}^2-2\bar{x}^3) & \ell(\bar{x}^3-\bar{x}^2) \\ \dfrac{6}{\ell}(\bar{x}^2-\bar{x}) & (3\bar{x}^2-4\bar{x}+1) & -\dfrac{6}{\ell}(\bar{x}^2-\bar{x}) & (3\bar{x}^2-2\bar{x}) \end{bmatrix} \bullet \left\{\begin{matrix} v_1 \\ \theta_{z1} \\ v_2 \\ \theta_{z2} \end{matrix}\right\}$$

$$\textit{interpolation matrix (2×4)}$$

$$[3.21]$$

This equation provides the displacements of any intermediate cross-section through an interpolation matrix (2×4).

◯ Determination of the stiffness matrix

By differentiating twice v(x) provided by the previous equation, the following form of $\dfrac{d^2 v}{dx^2}$ is obtained:

$$\frac{d^2 v}{dx^2} = \frac{6}{\ell^2}(2\bar{x}-1)v_1 + \frac{4}{\ell}\left(\frac{3\bar{x}}{2}-1\right)\theta_{z1} - \frac{6}{\ell^2}(2\bar{x}-1)v_2 + \frac{2}{\ell}(3\bar{x}-1)\theta_{z2} \qquad [3.22]$$

By substituting in the deformation energy [3.20], the following equation is reached:

$$E_{pot.} = \frac{1}{2} EI_z \left(\frac{12}{\ell^3} v_1^2 + \frac{4}{\ell}\theta_{z1}^2 + \frac{12}{\ell^3} v_2^2 + \frac{4}{\ell}\theta_{z2}^2 + \frac{12}{\ell^2} v_1\theta_{z1} \ldots\ldots \right.$$

$$\left. \ldots\ldots\ldots - \frac{24}{\ell^3} v_1 v_2 + \frac{12}{\ell^2} v_1\theta_{z2} - \frac{12}{\ell^2} v_2\theta_{z1} + \frac{4}{\ell}\theta_{z1}\theta_{z2} - \frac{12}{\ell^2} v_2\theta_{z2} \right) \qquad [3.23]$$

which is re-written as:

$$E_{pot.} = \frac{1}{2}\left\{\begin{matrix} v_1 \\ \theta_{z1} \\ v_2 \\ \theta_{z2} \end{matrix}\right\}^T \bullet \frac{EI_z}{\ell^3}\begin{bmatrix} 12 & 6\ell & -12 & 6\ell \\ 6\ell & 4\ell^2 & -6\ell & 2\ell^2 \\ -12 & -6\ell & 12 & -6\ell \\ 6\ell & 2\ell^2 & -6\ell & 4\ell^2 \end{bmatrix} \bullet \left\{\begin{matrix} v_1 \\ \theta_{z1} \\ v_2 \\ \theta_{z2} \end{matrix}\right\}$$

We obtain a quadratic form of degrees of freedom v_1, v_2 and θ_{z1}, θ_{z2} which should be identified with the equation:

$$E_{\substack{\text{pot.} \\ \text{element}}} = \frac{1}{2}\{d\}_{el}^{T} \bullet [k]_{el} \bullet \{d\}_{el}$$

Thus, the stiffness matrix is obtained:

$$[k]_{el} = \frac{EI_z}{\ell^3}\begin{bmatrix} 12 & 6\ell & -12 & 6\ell \\ 6\ell & 4\ell^2 & -6\ell & 2\ell^2 \\ -12 & -6\ell & 12 & -6\ell \\ 6\ell & 2\ell^2 & -6\ell & 4\ell^2 \end{bmatrix} \qquad [3.24]$$

According to [2.122] the behavior of the beam element under plane bending, written in its local coordinate system, is:

$$\{F\}_{el} = [k]_{el} \bullet \{d\}_{el} \quad \Leftrightarrow \quad \begin{Bmatrix} Y_1 \\ N_1 \\ Y_2 \\ N_2 \end{Bmatrix} = \frac{EI_z}{\ell^3}\begin{bmatrix} 12 & 6\ell & -12 & 6\ell \\ 6\ell & 4\ell^2 & -6\ell & 2\ell^2 \\ -12 & -6\ell & 12 & -6\ell \\ 6\ell & 2\ell^2 & -6\ell & 4\ell^2 \end{bmatrix} \bullet \begin{Bmatrix} v_1 \\ \theta_{z1} \\ v_2 \\ \theta_{z2} \end{Bmatrix} \quad [3.25]$$

NOTES

❑ We note on every line in equation [3.25] a load and its associated degree of freedom[21]. We have already seen that this constitutes a general property of form $\{F\} = [k] \bullet \{d\}$.

❑ The developed form [3.23] of the deformation energy can also be written by noting the stiffness matrix terms as k_{ij} and the degrees of freedom as "d_i":

$$E_{\text{pot.}} = \frac{1}{2}EI_z\Big(k_{11}v_1^2 + k_{22}\theta_{z1}^2 + k_{33}v_2^2 + k_{44}\theta_{z2}^2 + 2k_{12}v_1\theta_{z1}\text{.......}$$

$$\text{....} + 2k_{13}v_1v_2 + 2k_{14}v_1\theta_{z2} + 2k_{23}v_2\theta_{z1} + 2k_{24}\theta_{z1}\theta_{z2} + 2k_{34}v_2\theta_{z2}\Big)$$

$$= \frac{1}{2}\sum_{i=1}^{4}\sum_{j=1}^{4}k_{ij}d_id_j$$

❑ The beam element represented in Figure 3.11 appears to be "floating" in the plane of bending (*xy*). It is not properly linked as per section 2.5.2. The absence of a link with its surroundings gives rise to the possibility of rigid-body movements in the directions of loads and their associated degrees of freedom. In the present case, it means a translation along $\vec{y}$ and a rotation around an axis parallel to $\vec{z}$. Thus the flexibility matrix which is inverse to the stiffness matrix cannot be

---

21 See [2.91] and [2.122].

defined (see section 2.5.2.2). It can be verified that equation [3.25] cannot be inverted (the main determinant of [k] being zero).

❒ The method adopted to obtain equations [3.23] and [3.25] is not unique. For example, in section 2.6.4.2 we have established the same stiffness matrix $[k]_{el}$ through a (more direct) different approach.

❒ Local and global coplanar coordinate system:

In Figure 3.12 the local coordinate system (*xyz*) of the element has any orientation with respect to the global system (XYZ) with $\vec{Z} \equiv \vec{z}$.

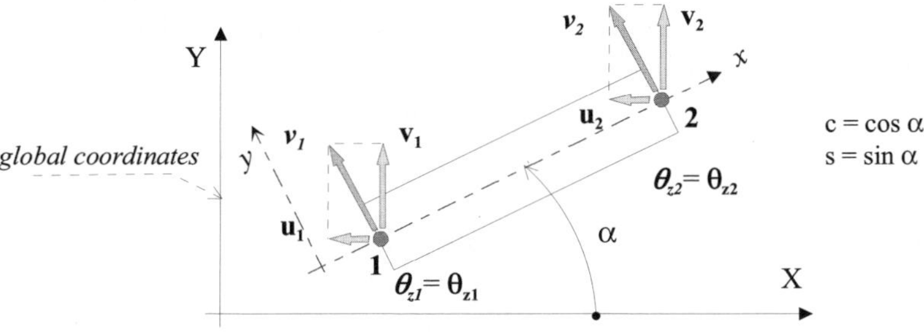

Figure 3.12. *Local and global systems*

By applying the results of [3.9] to dof specific to this element it is possible to write successively for the nodes (1) and (2) (see Figure 3.12):

$$\begin{Bmatrix} u_1 = 0 \\ v_1 \\ \theta_{1z} \end{Bmatrix} = \begin{bmatrix} c & s & 0 \\ -s & c & 0 \\ 0 & 0 & 1 \end{bmatrix} \bullet \begin{Bmatrix} u_1 \\ v_1 \\ \theta_{1z} \end{Bmatrix} \; ; \quad \begin{Bmatrix} u_2 = 0 \\ v_2 \\ \theta_{2z} \end{Bmatrix} = \begin{bmatrix} c & s & 0 \\ -s & c & 0 \\ 0 & 0 & 1 \end{bmatrix} \bullet \begin{Bmatrix} u_2 \\ v_2 \\ \theta_{2z} \end{Bmatrix}$$

Local   Global   Local   Global

By regrouping in order to show the dof column of the element:

$$\begin{Bmatrix} 0 \\ v_1 \\ \theta_{z1} \\ 0 \\ v_2 \\ \theta_{z2} \end{Bmatrix} = \begin{bmatrix} c & s & 0 & 0 & 0 & 0 \\ -s & c & 0 & 0 & 0 & 0 \\ 0 & 0 & 1 & 0 & 0 & 0 \\ 0 & 0 & 0 & c & s & 0 \\ 0 & 0 & 0 & -s & c & 0 \\ 0 & 0 & 0 & 0 & 0 & 1 \end{bmatrix} \bullet \begin{Bmatrix} u_1 \\ v_1 \\ \theta_{z1} \\ u_2 \\ v_2 \\ \theta_{z2} \end{Bmatrix}$$

(6×6)

Local   Global
(6×1)   (6×1)

Transfer matrix [P] is found in equation [3.9] as:

$$[P] = \begin{bmatrix} c & s & 0 & 0 & 0 & 0 \\ -s & c & 0 & 0 & 0 & 0 \\ 0 & 0 & 1 & 0 & 0 & 0 \\ 0 & 0 & 0 & c & s & 0 \\ 0 & 0 & 0 & -s & c & 0 \\ 0 & 0 & 0 & 0 & 0 & 1 \end{bmatrix}$$

Thus, when a plane structure has a beam element working in plane bending[22], the latter is generally characterized in the global coordinate system by 3 dof per node which is 6 dof for the element.

❏ Knowledge of the values of dof of the element in the local coordinate system leads to the normal stresses in every cross-section of this element. This can be written as (see section 1.5.2.6):

$$\sigma_x = -\frac{\mathcal{M}f_z \times y}{I_z}$$

with [3.19]:

$$\mathcal{M}f_z = E \times I_z \frac{d^2 v}{dx^2}$$

from where

$$\sigma_x = -y \times E \frac{d^2 v}{dx^2}$$

$\dfrac{d^2 v}{dx^2}$ is known from [3.22] with reference to dof v_1, v_2 and θ_{z1}, θ_{z2} in the local coordinate system. These dof being known, the stress σ_x for $0 \leq \overline{x} \leq 1$, i.e. $0 \leq x \leq \ell$, is obtained as:

$$\sigma_x = -y \times E \left\{ \frac{6}{\ell^2}(2\overline{x} - 1)v_1 + \frac{4}{\ell}\left(\frac{3\overline{x}}{2} - 1\right)\theta_{z1} - \frac{6}{\ell^2}(2\overline{x} - 1)v_2 + \frac{2}{\ell}(3\overline{x} - 1)\theta_{z2} \right\}$$

Nevertheless, knowing the dof values of the beam element is not generally sufficient in the case of any cross-sectional shape to reach the distribution of shear stresses due to the shear-force in the cross-section[23]. That is why the postprocessors of the finite element programs do not directly give this type of information.

---

22 We shall see later (see Chapter 5, section 5.2, [5.9]) the complete beam element. It has more degrees of freedom per node.
23 The form of these stresses is given in section 1.5.2.2 and the calculations detailed in Chapter 9, section 9.3.4.7 and [9.59].

❑ Study of the bending of this beam element in its second main plane (*xz*) shall be carried out in a similar manner. The local dof v and θ_z are changed to w and θ_y. Keeping in mind the sign conventions, the same deformation observed successively in plane (*xy*) and plane (*xz*) brings about a difference in sign, as indicated in Figure 3.13.

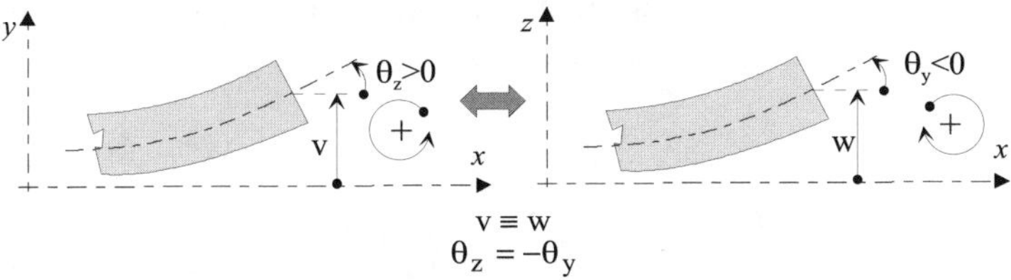

Figure 3.13. *Same deformation in every main plane*

With this observation, the potential energy of deformation [3.23] is re-written in the plane (*xz*):

$$E_{pot.} = \frac{1}{2} EI_y \left(\frac{12}{\ell^3} w_1{}^2 + \frac{4}{\ell} \theta_{y1}{}^2 + \frac{12}{\ell^3} w_2{}^2 + \frac{4}{\ell} \theta_{y2}{}^2 - \frac{12}{\ell^2} w_1 \theta_{y1} \dots\dots \right.$$

$$\left. \dots\dots\dots - \frac{24}{\ell^3} w_1 w_2 - \frac{12}{\ell^2} w_1 \theta_{y2} + \frac{12}{\ell^2} w_2 \theta_{y1} + \frac{4}{\ell} \theta_{y1} \theta_{y2} + \frac{12}{\ell^2} w_2 \theta_{y2} \right)$$

i.e.:

$$E_{pot.} = \frac{1}{2} \begin{Bmatrix} w_1 \\ \theta_{y1} \\ w_2 \\ \theta_{y2} \end{Bmatrix}^T \bullet \frac{EI_y}{\ell^3} \begin{bmatrix} 12 & -6\ell & -12 & -6\ell \\ -6\ell & 4\ell^2 & 6\ell & 2\ell^2 \\ -12 & 6\ell & 12 & 6\ell \\ -6\ell & 2\ell^2 & 6\ell & 4\ell^2 \end{bmatrix} \bullet \begin{Bmatrix} w_1 \\ \theta_{y1} \\ w_2 \\ \theta_{y2} \end{Bmatrix}$$

The stiffness matrix characterizing the bending in the (*xz*) plane is:

$$[k]_{\text{él}} = \frac{EI_y}{\ell^3} \begin{bmatrix} 12 & -6\ell & -12 & -6\ell \\ -6\ell & 4\ell^2 & 6\ell & 2\ell^2 \\ -12 & 6\ell & 12 & 6\ell \\ -6\ell & 2\ell^2 & 6\ell & 4\ell^2 \end{bmatrix} \qquad [3.26]$$

○ Conclusion

The behavior of the beam element bending in the plane (*xy*) is summarized in the following table.

<table>
<tr><td colspan="2" align="center">Behavior in the local coordinate system</td></tr>
<tr><td>

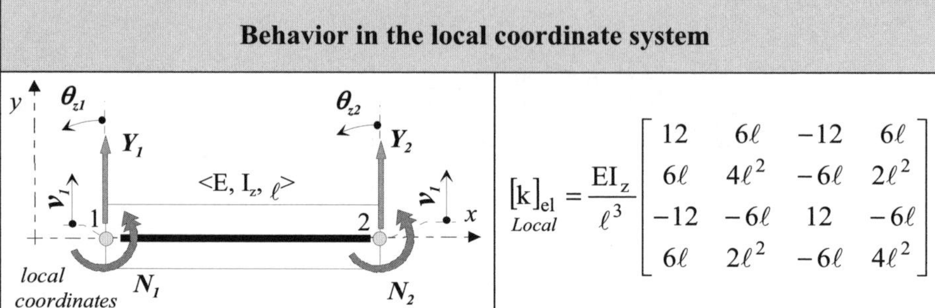

</td><td>

$$[k]_{el} \atop Local = \frac{EI_z}{\ell^3}\begin{bmatrix} 12 & 6\ell & -12 & 6\ell \\ 6\ell & 4\ell^2 & -6\ell & 2\ell^2 \\ -12 & -6\ell & 12 & -6\ell \\ 6\ell & 2\ell^2 & -6\ell & 4\ell^2 \end{bmatrix}$$

</td></tr>
</table>

behavior equation:

$$\{F\}_{el} \atop Local = [k]_{el} \atop Local \bullet \{d\}_{el} \atop Local \quad\Leftrightarrow\quad \begin{Bmatrix} Y_1 \\ N_1 \\ Y_2 \\ N_2 \end{Bmatrix}_{Local} = \frac{EI_z}{\ell^3}\begin{bmatrix} 12 & 6\ell & -12 & 6\ell \\ 6\ell & 4\ell^2 & -6\ell & 2\ell^2 \\ -12 & -6\ell & 12 & -6\ell \\ 6\ell & 2\ell^2 & -6\ell & 4\ell^2 \end{bmatrix}\bullet\begin{Bmatrix} v_1 \\ \theta_{z1} \\ v_2 \\ \theta_{z2} \end{Bmatrix}_{Local}$$

potential energy of deformation:

$$E_{pot. \atop element} = \frac{1}{2}\{d\}_{el}^T \bullet [k]_{el} \bullet \{d\}_{el}$$

$$\Updownarrow$$

$$E_{pot. \atop element} = \begin{Bmatrix} v_1 \\ \theta_{z1} \\ v_2 \\ \theta_{z2} \end{Bmatrix}_{Local}^T \bullet \frac{EI_z}{\ell^3}\begin{bmatrix} 12 & 6\ell & -12 & 6\ell \\ 6\ell & 4\ell^2 & -6\ell & 2\ell^2 \\ -12 & -6\ell & 12 & -6\ell \\ 6\ell & 2\ell^2 & -6\ell & 4\ell^2 \end{bmatrix}\bullet\begin{Bmatrix} v_1 \\ \theta_{z1} \\ v_2 \\ \theta_{z2} \end{Bmatrix}_{Local}$$

normal stress on any cross-section of the element ($\overline{x} = x / \ell$):

$$\sigma_x = -y \times E\left\{\frac{6}{\ell^2}(2\overline{x}-1)v_1 + \frac{4}{\ell}\left(\frac{3\overline{x}}{2}-1\right)\theta_{z1} - \frac{6}{\ell^2}(2\overline{x}-1)v_2 + \frac{2}{\ell}(3\overline{x}-1)\theta_{z2}\right\}$$

Shear stress on a cross-section of the element:
τ_{xy} and τ_{xz} cannot be calculated in the general case of any cross-sectional shapes.

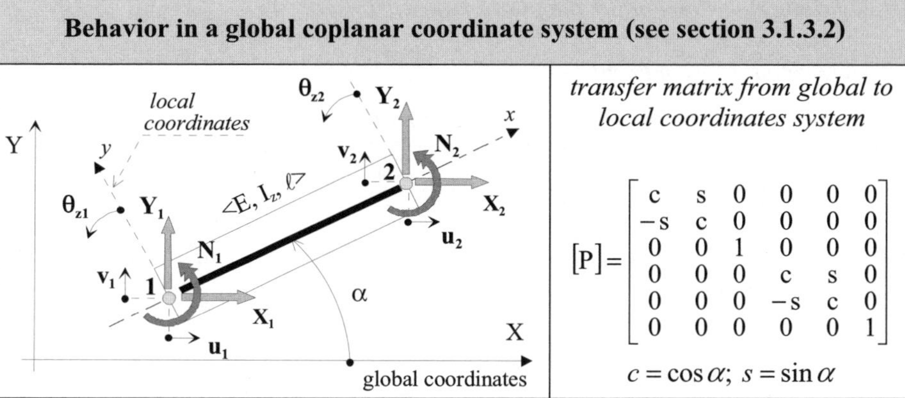

Behavior in a global coplanar coordinate system (see section 3.1.3.2)

$$[P] = \begin{bmatrix} c & s & 0 & 0 & 0 & 0 \\ -s & c & 0 & 0 & 0 & 0 \\ 0 & 0 & 1 & 0 & 0 & 0 \\ 0 & 0 & 0 & c & s & 0 \\ 0 & 0 & 0 & -s & c & 0 \\ 0 & 0 & 0 & 0 & 0 & 1 \end{bmatrix}$$

$$c = \cos\alpha; \quad s = \sin\alpha$$

$$\{F\}_{el}_{\,Local} = [P] \bullet \{F\}_{el}_{\,Global} \; ; \quad \{d\}_{el}_{\,Local} = [P] \bullet \{d\}_{el}_{\,Global}$$

$$[k]_{el}_{\,Global} = [P]^T \bullet [k]_{el}_{\,Local} \bullet [P]$$

behavior equation:

$$\{F\}_{el}_{\,Global} = [k]_{el}_{\,Global} \bullet \{d\}_{el}_{\,Global} \quad \hat{U} \quad \begin{Bmatrix} X_1 \\ Y_1 \\ N_1 \\ X_2 \\ Y_2 \\ N_2 \end{Bmatrix}_{Global} = [k]_{el}_{\,Global\,(6\times6)} \bullet \begin{Bmatrix} u_1 \\ v_1 \\ \theta_{z1} \\ u_2 \\ v_2 \\ \theta_{z2} \end{Bmatrix}_{Global}$$

potential energy of deformation:

$$E_{pot}_{\,element} = \frac{1}{2} \{d\}_{el}_{\,Global} \bullet [k]_{el}_{\,Global} \bullet \{d\}_{el}_{\,Global}$$

Figure 3.14. *Behavior of a plane-bending beam element*

3.2.4. *Triangular element for the plane state of stresses*

3.2.4.1. *Preliminary comment*

The plane state of stresses is described in Chapter 1 and its characteristics are summarized in [1.18]. The potential energy of deformation of an elementary volume $dx \times dy \times e$ has been calculated in Chapter 2, section 2.3.2 and its different expressions are summarized in [2.44]. These results are basic for defining a partition with geometrically simple elements of the structure outlined in Figure 1.14, and which is shown again below (Figure 3.15).

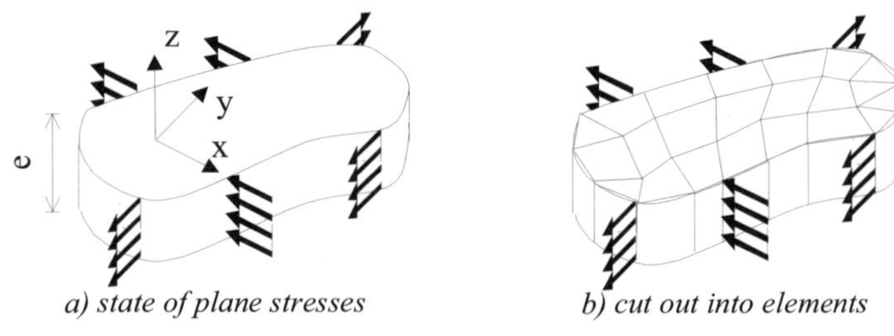

a) state of plane stresses   *b) cut out into elements*

Figure 3.15. *Meshing for a plane state of stresses*

It is clear from Figure 3.15b that the domain can be "covered as well as possible" with the help of prismatic elements with triangular or quadrangular base. On the other hand we have seen that for a plane state of stresses, stresses and deformations are the same in every (xy) plane perpendicular to the direction $\vec{z}$ in Figure 3.15. In such a case the prismatic elements will show up in the form of triangles or quadrilaterals (projection in (xy) plane)[24] and be so-called "membrane-elements".

3.2.4.2. *Definition of the element*

The element is represented in Figure 3.16. In the local system (*xyz*) where (*xy*) is the plane of the element, the nodes are placed at the three vertices marked (1), (2), (3). The displacements[25] as well as the corresponding loads (see [2.91]) are noted below in [3.27].

---

24 See Figure 2.13.
25 See [1.18].

node	coordinates of the node	(small) displacements of the node	corresponding loads
1	x_1, y_1	u_1 and v_1	X_1 and Y_1
2	x_2, y_2	u_2 and v_2	X_2 and Y_2
3	x_3, y_3	u_3 and v_3	X_3 and Y_3

$$[3.27]$$

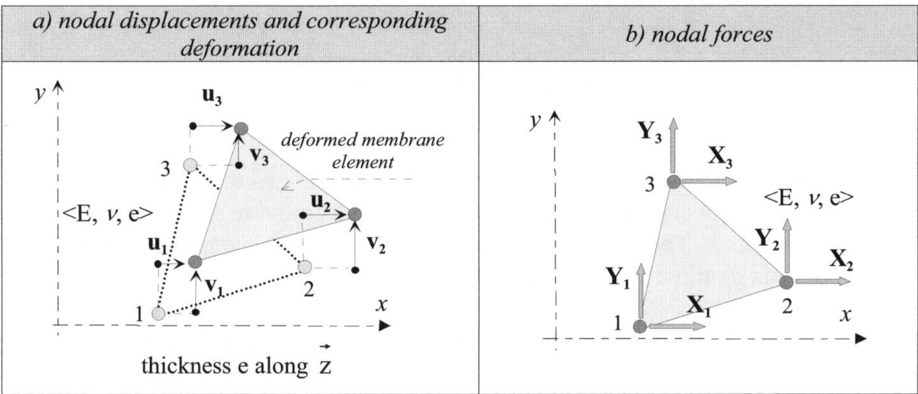

Figure 3.16. *Triangular element for plane state of stresses*

It can already be noted in Figure 3.16 that the element obtained is much more "artificial", i.e. it reflects reality much less than the previous beam element. The beam element can be "physically" loaded only at its two nodes by the resultant force and moment as described in the previous section. On the other hand, cohesion forces can be observed along common boundaries with other elements by examining Figure 3.15b. It is approximative for the model in Figure 3.16a to be loaded only at its nodes.

3.2.4.3. *Form of the displacement functions*

As a consequence of the previous comment, the analytical calculation of a realistic displacement field under the nodal forces is not even desirable because in any case it will not correspond to the material continuity, as is shown in Figure 3.17.

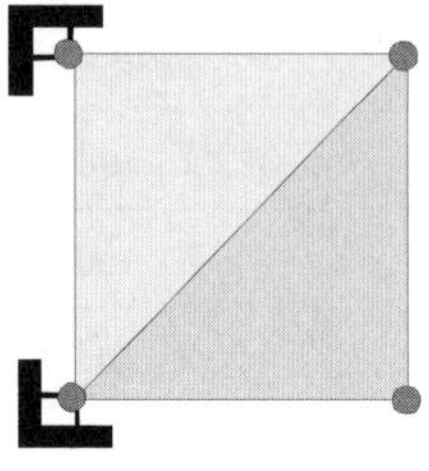

a) *assembly of two elements*

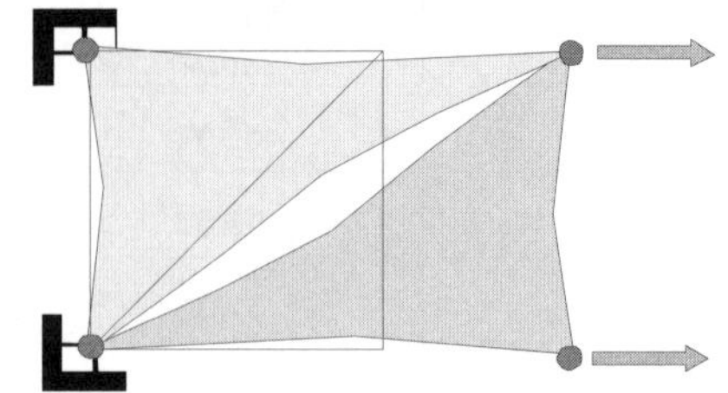

b) *real deformation of the loaded assembly: the material continuity is not ensured*

Figure 3.17. *Nodal loading*

For every element it is therefore necessary to define a rough displacement field which at least respects the material continuity while moving from one deformed element to the other. Then continuity of displacements along a common boundary of two elements is required.

The simplest approximate corresponding field is linear and written as:

$$u(x, y) = a_0 + a_1 x + a_2 y$$
$$v(x, y) = b_0 + b_1 x + b_2 y$$

[3.28]

In this case we can verify that the sides remain rectilinear after deformation. Coefficients a_j and b_j ($j = 0, 1, 2$) are determined by writing that u and v coincide with displacements u_i and v_i for every node i, thereby giving six equations:

$$u_i = a_0 + a_1 x_i + a_2 y_i \qquad i = 1, 2, 3$$
$$v_i = b_0 + b_1 x_i + b_2 y_i \qquad i = 1, 2, 3$$

[3.29]

Coefficients a_j and b_j appear as linear functions of displacements u_i and v_i. They are substituted in [3.28], and the displacement fields u(x,y) and v(x,y) take the following matrix form, where the interpolation matrix $[A(x, y)]$ shows up:

$$\left\{ \begin{array}{c} u \\ v \end{array} \right\} = [A(x, y)] \bullet \left\{ \begin{array}{c} u_1 \\ v_1 \\ u_2 \\ v_2 \\ u_3 \\ v_3 \end{array} \right\}$$

[3.30]

with:

$$[A(x,y)] = \frac{1}{2S}\begin{bmatrix} (\alpha_1 + \beta_1 x + \gamma_1 y) & 0 & (\alpha_2 + \beta_2 x + \gamma_2 y) & 0 & (\alpha_3 + \beta_3 x + \gamma_3 y) & 0 \\ 0 & (\alpha_1 + \beta_1 x + \gamma_1 y) & 0 & (\alpha_2 + \beta_2 x + \gamma_2 y) & 0 & (\alpha_3 + \beta_3 x + \gamma_3 y) \end{bmatrix}$$

$$[3.31]$$

where:

$$S: \text{area of the element}$$
$$\alpha_i = x_j y_k - x_k y_j; \quad \beta_i = y_j - y_k; \quad \gamma_i = -\left(x_j - x_k\right)$$

3.2.4.4. *Determination of the stiffness matrix*

Based on the form of potential energy of deformation in plane state of stresses [2.44], the following equation is obtained[26]:

$$E_{\substack{pot.\\ plane\\ stress}} = \frac{1}{2} e \int_S \begin{Bmatrix} \varepsilon_x \\ \varepsilon_y \\ \gamma_{xy} \end{Bmatrix}^T \bullet \begin{bmatrix} \dfrac{E}{1-v^2} & \dfrac{vE}{1-v^2} & 0 \\ \dfrac{vE}{1-v^2} & \dfrac{E}{1-v^2} & 0 \\ 0 & 0 & G \end{bmatrix} \bullet \begin{Bmatrix} \varepsilon_x \\ \varepsilon_y \\ \gamma_{xy} \end{Bmatrix} \times dS \qquad [3.32]$$

where we have seen that[27]:

$$\varepsilon_x = \frac{\partial u}{\partial x}; \qquad \varepsilon_y = \frac{\partial v}{\partial y}; \qquad \gamma_{xy} = \frac{\partial u}{\partial y} + \frac{\partial v}{\partial x}$$

It is therefore possible to calculate these deformations from the field [3.30]. In this way, it is necessary to differentiate [3.31] with respect to x or y. The terms of $[A]$ being first degree functions in x and y, their derivatives appear as constants, and the following equation is obtained:

$$\begin{Bmatrix} \varepsilon_x \\ \varepsilon_y \\ \gamma_{xy} \end{Bmatrix} = \frac{1}{2S} \begin{bmatrix} \beta_1 & 0 & \beta_2 & 0 & \beta_3 & 0 \\ 0 & \gamma_1 & 0 & \gamma_2 & 0 & \gamma_3 \\ \gamma_1 & \beta_1 & \gamma_2 & \beta_2 & \gamma_3 & \beta_3 \end{bmatrix} \bullet \begin{Bmatrix} u_1 \\ v_1 \\ u_2 \\ v_2 \\ u_3 \\ v_3 \end{Bmatrix} \qquad [3.33]$$

which can be replaced in [3.32]. After calculation:

---

26 It should be noted that: $dV = e \times dS$.
27 See [1.18].

$$E_{\substack{\text{pot.}\\\text{element}}} = \frac{1}{2}\begin{Bmatrix} u_1 \\ v_1 \\ u_2 \\ v_2 \\ u_3 \\ v_3 \end{Bmatrix}^{T} \bullet [k]_{el} \bullet \begin{Bmatrix} u_1 \\ v_1 \\ u_2 \\ v_2 \\ u_3 \\ v_3 \end{Bmatrix}$$

with:

$$[k]_{\acute{e}l} = \frac{E \times e}{4(1-v^2)S} \times \ldots\ldots$$

$$\ldots\begin{bmatrix} \beta_1^2 + \dfrac{1-v}{2}\gamma_1^2 & \dfrac{1+v}{2}\beta_1\gamma_1 & \beta_1\beta_2 + \dfrac{1-v}{2}\gamma_1\gamma_2 & v\beta_1\beta_2 + \dfrac{1-v}{2}\gamma_1\beta_2 & \beta_1\beta_3 + \dfrac{1-v}{2}\gamma_1\gamma_3 & v\beta_1\beta_3 + \dfrac{1-v}{2}\gamma_1\beta_3 \\[1em] & \gamma_1^2 + \dfrac{1-v}{2}\beta_1^2 & v\gamma_1\beta_2 + \dfrac{1-v}{2}\beta_1\gamma_2 & \gamma_1\gamma_2 + \dfrac{1-v}{2}\beta_1\beta_2 & v\gamma_1\beta_3 + \dfrac{1-v}{2}\beta_1\gamma_3 & \gamma_1\gamma_3 + \dfrac{1-v}{2}\beta_1\beta_3 \\[1em] S & & \beta_2^2 + \dfrac{1-v}{2}\gamma_2^2 & \dfrac{1+v}{2}\beta_2\gamma_2 & \beta_2\beta_3 + \dfrac{1-v}{2}\gamma_2\gamma_3 & v\beta_2\gamma_3 + \dfrac{1-v}{2}\gamma_2\beta_3 \\[1em] & Y & & \gamma_2^2 + \dfrac{1-v}{2}\beta_2^2 & v\gamma_2\beta_3 + \dfrac{1-v}{2}\gamma_3\beta_2 & \gamma_2\gamma_3 + \dfrac{1-v}{2}\beta_2\beta_3 \\[1em] & & M & & \beta_3^2 + \dfrac{1-v}{2}\gamma_3^2 & \dfrac{1+v}{2}\beta_3\gamma_3 \\[1em] & & & E & & \gamma_3^2 + \dfrac{1-v}{2}\beta_3^2 \end{bmatrix}$$

The behavior of this triangular plane element is therefore written in the coordinate system (xy)[28]:

$$\{F\}_{el} = [k]_{el} \bullet \{d\}_{el}$$

where the nodal forces correspond to the dof as represented in [3.27].

NOTES

❑ The triangular element in Figure 3.16 appears as "floating" in its plane. It is not properly linked in the lines stated in section 2.5.2. This absence of links leads to the possibility of two rigid-body movements in the local coordinate system which are translations along the two directions $\vec{x}$ and $\vec{y}$.

❑ Local and global coordinate systems: the triangular element has been defined in a local coordinate system (*xyz*). If we need to move to another coplanar coordinate (XYZ), we know that characteristic vectors (nodal displacement and nodal force)

---

28 See [2.12].

can show up in one or the other system according to equation [3.9] recalled below in Figure 3.18.

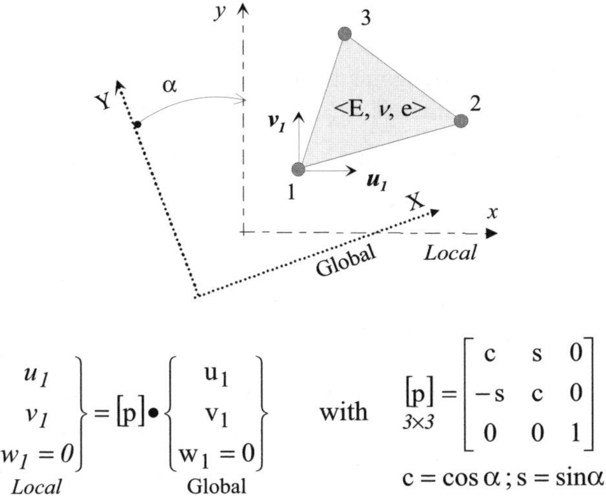

$$\left\{\begin{array}{c} u_1 \\ v_1 \\ w_1 = 0 \end{array}\right\}_{Local} = [p] \bullet \left\{\begin{array}{c} u_1 \\ v_1 \\ w_1 = 0 \end{array}\right\}_{Global} \qquad \text{with} \quad \underset{3\times3}{[p]} = \begin{bmatrix} c & s & 0 \\ -s & c & 0 \\ 0 & 0 & 1 \end{bmatrix}$$

$$c = \cos\alpha \,;\, s = \sin\alpha$$

Figure 3.18. *Local and global coordinates systems*

Thus, the totality of the dof and the totality of the nodal loads give rise to:

$$\underset{2\times2}{[p]} = \begin{bmatrix} c & s \\ -s & c \end{bmatrix}$$

$$\left\{\begin{array}{c} u_1 \\ v_1 \\ \dots \\ u_2 \\ v_2 \\ \dots \\ u_3 \\ v_3 \end{array}\right\}_{Local} = \begin{bmatrix} [p]_{2\times2} & 0 & 0 \\ 0 & [p]_{2\times2} & 0 \\ 0 & 0 & [p]_{2\times2} \end{bmatrix} \bullet \left\{\begin{array}{c} u_1 \\ v_1 \\ \dots \\ u_2 \\ v_2 \\ \dots \\ u_3 \\ v_3 \end{array}\right\}_{Global} \quad \text{and} \quad \left\{\begin{array}{c} X_1 \\ Y_1 \\ \dots \\ X_2 \\ Y_2 \\ \dots \\ X_3 \\ Y_3 \end{array}\right\}_{Local} = \begin{bmatrix} [p]_{2\times2} & 0 & 0 \\ 0 & [p]_{2\times2} & 0 \\ 0 & 0 & [p]_{2\times2} \end{bmatrix} \bullet \left\{\begin{array}{c} X_1 \\ Y_1 \\ \dots \\ X_2 \\ Y_2 \\ \dots \\ X_3 \\ Y_3 \end{array}\right\}_{Global} \qquad [3.34]$$

transfer matrix $[P](6\times6)$

A transfer matrix $[P](6\times6)$ appears, such that:

$$\{d\}_{el}^{Local} = [P] \bullet \{d\}_{el}^{Global} \quad ; \quad \{F\}_{el}^{Local} = [P] \bullet \{F\}_{el}^{Global}$$

Using this transfer matrix to write the behavior in the global coordinate system is done in accordance with the method described in [3.9] and the results are summarized in Figure 3.23.

❏ Once the dof of the triangular plane element are known in local coordinates (xyz), the plane state of stresses σ_x, σ_y, τ_{xy} inside the element can be arrived at. In fact the deformations ε_x, ε_y, γ_{xy} are known through [3.33]. The behavior equation in plane state of stresses [1.18] leads to:

$$\begin{Bmatrix} \sigma_x \\ \sigma_y \\ \tau_{xy} \end{Bmatrix} = \frac{1}{2S} \bullet \begin{bmatrix} \dfrac{E}{1-v^2} & \dfrac{vE}{1-v^2} & 0 \\ \dfrac{vE}{1-v^2} & \dfrac{E}{1-v^2} & 0 \\ 0 & 0 & G \end{bmatrix} \bullet \begin{bmatrix} \beta_1 & 0 & \beta_2 & 0 & \beta_3 & 0 \\ 0 & \gamma_1 & 0 & \gamma_2 & 0 & \gamma_3 \\ \gamma_1 & \beta_1 & \gamma_2 & \beta_2 & \gamma_3 & \beta_3 \end{bmatrix} \bullet \begin{Bmatrix} u_1 \\ v_1 \\ u_2 \\ v_2 \\ u_3 \\ v_3 \end{Bmatrix} \qquad [3.35]$$

β_i and γ_i being constants, these stresses do not depend on the x and y coordinates and remain constant inside the element. Thus when moving from one triangular element to its neighbor through a common boundary, we note a discontinuity of stresses which does not correspond to reality.

There is continuity of displacements but not of stresses

This leads to mediocre performances for this element, as illustrated in the following example.

3.2.4.5. *Example*

Figure 3.19a below shows a rectangular cantilever beam, fixed on the left and free on the right, bending in the main plane (xy) under the end force $\overrightarrow{F_2} = Y_2\overrightarrow{y}$. The results of the beam theory[29] show a state of plane stresses: σ_x, τ_{xy}.

---

29 See sections 1.5.2.2 and 1.5.2.6.

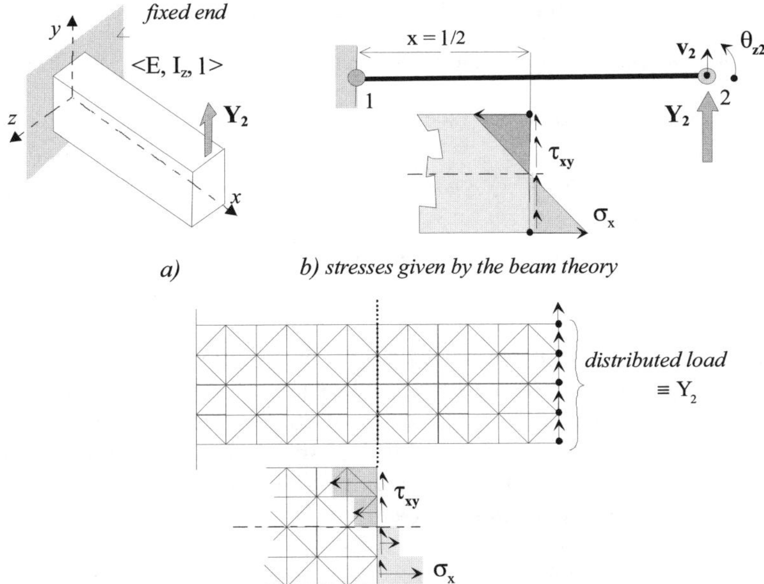

a)   b) stresses given by the beam theory

c) cross-section: stresses in triangular elements

Figure 3.19. *Comparison between two models*

Figure 3.19b shows a beam modeled by a single beam element (for example, the bending element of section 3.2.3). Considering the behavior relation of Figure 3.14 and the boundary conditions and loading corresponding to Figure 3.19a the following equation is obtained after deleting the lines and columns corresponding to the zero dof[30]:

$$\left\{ \begin{array}{c} Y_2 \\ 0 \end{array} \right\} = \left[\begin{array}{cc} \dfrac{12EI_z}{\ell^3} & -\dfrac{6EI_z}{\ell^2} \\ -\dfrac{6EI_z}{\ell^2} & \dfrac{4EI_z}{\ell} \end{array} \right] \bullet \left\{ \begin{array}{c} v_2 \\ \theta_{z2} \end{array} \right\}$$

which leads to:

$$v_2 = Y_2 \times \frac{\ell^3}{3EI_z}; \quad \theta_{z2} = \frac{3}{2\ell} \times v_2 = Y_2 \times \frac{\ell^2}{2EI_z} \qquad [3.36]$$

---

30 Here it is $v_1 = \theta_{z1} = 0$.

which are the "exact" values based on the theory of beams. Concerning normal stresses, for example, at the middle of the beam length $\left(x = \dfrac{\ell}{2} \right)$ (see Figure 3.14) the following equation is obtained (see [3.14]):

$$\sigma_x = -\,y \times E \times \frac{\theta_{z2}}{\ell} = -y \times Y_2 \times \frac{\ell}{2I_z}$$

which is the "exact" value given by the beam theory. Figure 3.19b shows the distribution.

Figure 3.19c shows the beam modeled by a large number of triangular elements. The stresses in a cross-section are obtained based on [3.35] and their distribution in the mid section is indicated. The inaccuracy of this discontinuous distribution with respect to the results of the beam model is evident. More precise information would require a prohibitive number of elements with a judicious meshing (see Figure 3.20) whereas one beam element should suffice[31].

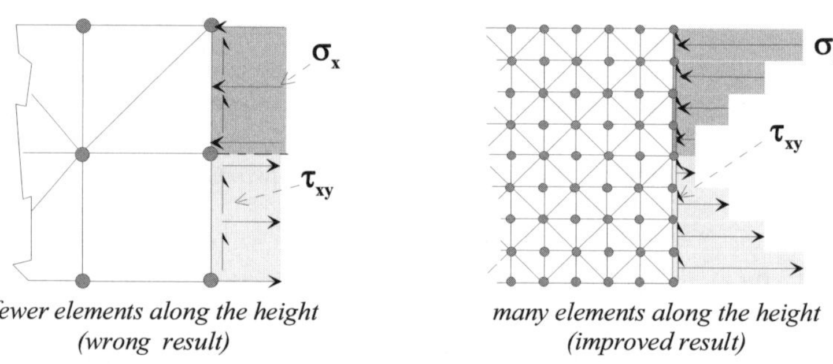

fewer elements along the height
(wrong result)   *many elements along the height*
(improved result)

Figure 3.20. *Need of fine discretization*

The same observation is valid for the deflection v due to bending, for example at the beam end. With Figure 3.19c as model, at the free extremity, the vertical displacements of nodes are of much lower value than the "exact" value [3.36]. Thus the defects of the previous triangular element can be summarized as follows:

♦ the displacements are continuous but the simplified corresponding field "forces" the element to deform in a rough manner. The element does not have its "real freedom" to move. As a result the modeled structure is more rigid compared to the real structure;

---

31 However, it should be noted that the triangular elements also give the values (not guaranteed) of the shear stress τ_{xy}, while the beam element gives only the normal stress (see section 3.2.3.2/*NOTES*).

◆ the stresses are constant in the element. Therefore, moving from one element to the other leads to discontinuities unlike the reality. As a consequence, the internal equilibrium of the model due to the nodal forces is different from the internal equilibrium of the real structure;

◆ this simple triangular element gives mediocre results.

3.2.4.6. *Performance improvement of the element*

It should be emphasized that the fundamental mediocrity of the previous element is due to stress that appears:

◆ discontinuous because the inter-forces between elements are limited to the nodal forces;

◆ constant in the element.

To improve the triangular element by attempting to correct these defects:

– the "points of contact" between the elements are multiplied;

– the degree of polynomials [3.28] representing the displacement field is necessarily increased. For example, for the element with six nodes of Figure 3.21a the displacement field is denoted as:

$$u(x, y) = a_0 + a_1 x + a_2 y + a_3 x^2 + a_4 xy + a_5 y^2$$

$$v(x, y) = b_0 + b_1 x + b_2 y + b_3 x^2 + b_4 xy + b_5 y^2$$

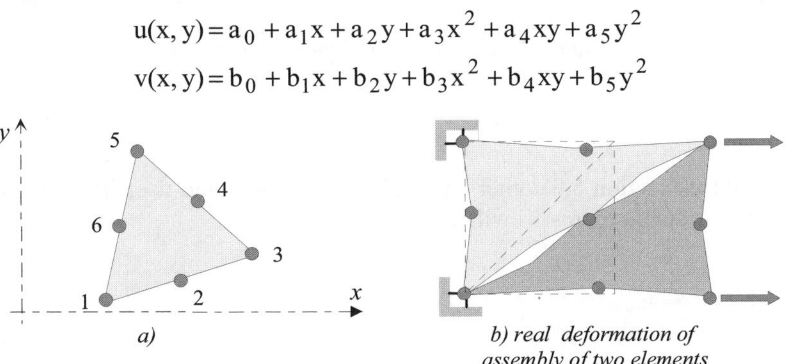

Figure 3.21. *Improving the triangular membrane-element*

By following a procedure similar to the one mentioned in section 3.2.4.6, a displacement field of type [3.30] with 12 dof and a second degree interpolation matrix $[A(x, y)]$ in x and y are obtained. Now the deformations, and along with them the stresses, vary linearly in the element and are first degree functions. The behavior of the assemblies of this type of elements is closer to reality[32]. We imagine that it could be possible, in this manner, to "manufacture" triangular membrane-elements which are more and more accurate, with polynomial displacement fields of higher degrees[33], based on Figure 3.22.

degree of polynomial of the displacement fields	*number of terms and nodes*	*Pascal's triangle*	*element*
1 linear	3	x y	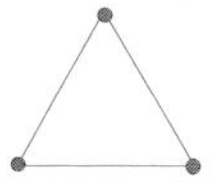
2 quadratic	6	x^2 xy y^2	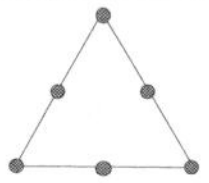
3 cubic	10	x^3 x^2y xy^2 y^3	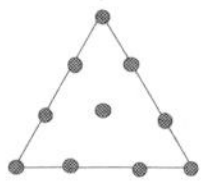
degree 4	15	x^4 x^3y x^2y^2 xy^3 y^4	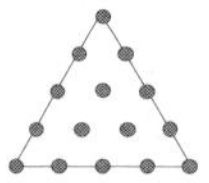

Figure 3.22. *Polynomial displacement fields*

NOTES

❐ In level II (section 5.3) other types of more reliable and user-friendly elements for the plane state of stresses will be seen.

---

32 Compare Figures 3.17b and Figure 3.21b in a qualitative manner.
33 Bibliographic references: J.F. Imbert.

3.2.4.7. *Summary*

The behavior of the triangular element working as membrane (plane stresses) can be summarized below.

<table>
<tr><td align="center">Behavior in the local coordinates system</td></tr>
</table>

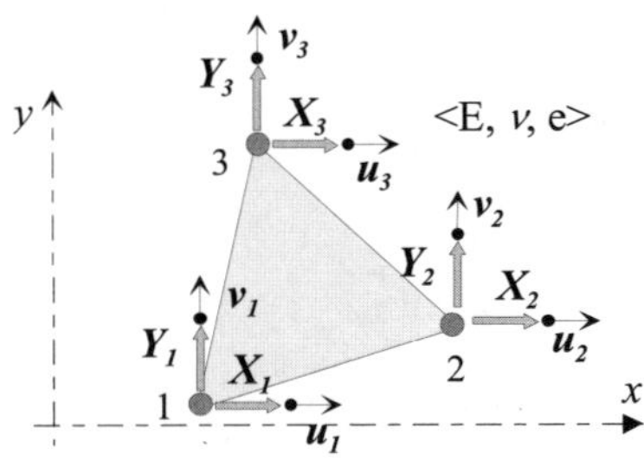

behavior equation:

$$\{F\}_{el} = [k]_{el} \bullet \{d\}_{el} \Leftrightarrow \begin{Bmatrix} X_1 \\ Y_1 \\ X_2 \\ Y_2 \\ X_3 \\ Y_3 \end{Bmatrix}_{Local} = [k]_{el}^{Local} \bullet \begin{Bmatrix} u_1 \\ v_1 \\ u_2 \\ v_2 \\ u_3 \\ v_3 \end{Bmatrix}_{Local}$$

$_{Local} \quad _{Local} \quad _{Local}$

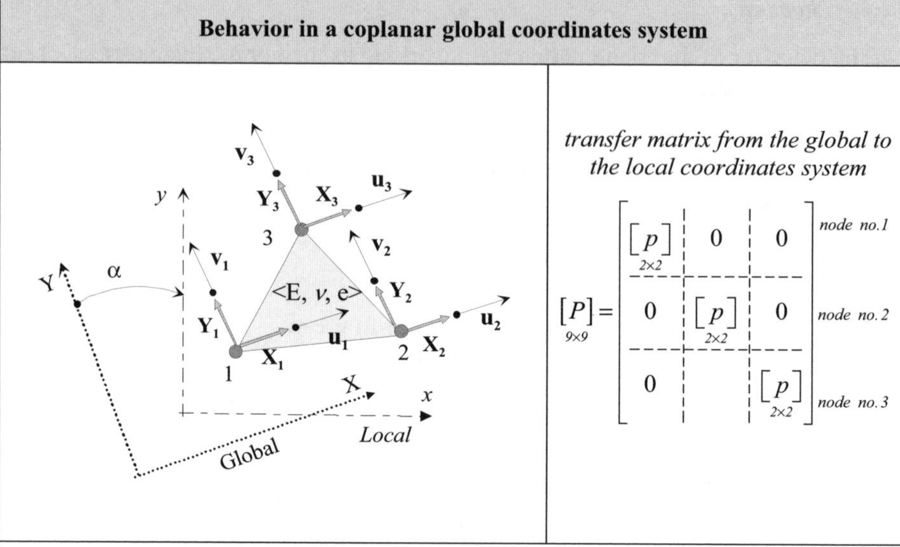

Behavior in a coplanar global coordinates system

transfer matrix from the global to the local coordinates system

$$[P]_{9\times9} = \begin{bmatrix} [p]_{2\times2} & 0 & 0 \\ 0 & [p]_{2\times2} & 0 \\ 0 & 0 & [p]_{2\times2} \end{bmatrix} \begin{array}{l} \textit{node no.1} \\ \\ \textit{node no.2} \\ \\ \textit{node no.3} \end{array}$$

$$\{F\}_{el}^{Local} = [P] \bullet \{F\}_{el}^{Global} \; ; \quad \{d\}_{el}^{Local} = [P] \bullet \{d\}_{el}^{Global}$$

$$[k]_{el}^{Global} = [P]^{T} \bullet [k]_{el}^{Local} \bullet [P]$$

behavior equation:

$$\{F\}_{el}^{Global} = [k]_{el}^{Global} \bullet \{d\}_{el}^{Global} \Leftrightarrow \begin{Bmatrix} X_1 \\ Y_1 \\ X_2 \\ Y_2 \\ X_3 \\ Y_3 \end{Bmatrix}_{Global} = [k]_{el}^{Global}{}_{(6\times6)} \bullet \begin{Bmatrix} u_1 \\ v_1 \\ u_2 \\ v_2 \\ u_3 \\ v_3 \end{Bmatrix}_{Global}$$

Figure 3.23. *Behavior of the triangular membrane-element*

3.3. Getting the global stiffness matrix of a structure

3.3.1. *Objective*

As indicated in section 3.1.2, the discretized elastic structure consists of a certain number of finite elements. These finite elements can either be identical or different. The simplest amongst these elements[34] have just been noticed. It is now suggested to go back to the construction of the global stiffness matrix $[K]_{str}$ of the structure with the help of the stiffness matrices of the elements with which it is formed.

3.3.2. *Mechanism of the assembly of elementary matrices*

3.3.2.1. *Example 1*

The structure represented by Figure 3.24, results in the assembly of three truss elements placed end to end. Thus a four node structure is obtained. The loading, as authorized for this type of element[35], is denoted as X_1, X_2, X_3, X_4. The associated degrees of freedom are: u_1, u_2, u_3, u_4[36].

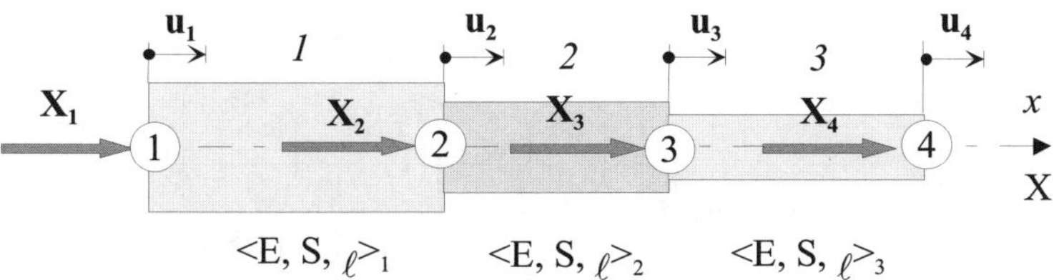

Figure 3.24. *Structure made of truss elements*

The deformation energy of a truss element, for example element 1, can be expressed as (see [3.11]):

$$E_{\substack{pot. \\ element\ 1}} = \frac{1}{2}\left(\frac{ES}{\ell}\right)_1 \left(u_1{}^2 + u_2{}^2 - 2u_2 u_1\right)$$

---

34 Chapter 5 gives other types of elements.
35 See Figure 3.8.
36 See [2.122].

Deformation energy of the complete structure can be considered as the sum of deformation energies of every element:

$$E_{\substack{\text{pot.} \\ \text{structure}}} = \sum_{i=1}^{3} E_{\substack{\text{pot.} \\ \text{element}}} = \frac{1}{2}\left\{ \left(\frac{ES}{\ell}\right)_1 u_1{}^2 + \left(\frac{ES}{\ell}\right)_1 u_2{}^2 - 2\left(\frac{ES}{\ell}\right)_1 u_1 u_2 \right\}.....$$

$$........+ \frac{1}{2}\left\{ \left(\frac{ES}{\ell}\right)_2 u_2{}^2 + \left(\frac{ES}{\ell}\right)_2 u_3{}^2 - 2\left(\frac{ES}{\ell}\right)_2 u_2 u_3 \right\}.......$$

$$.......+ \frac{1}{2}\left\{ \left(\frac{ES}{\ell}\right)_3 u_3{}^2 + \left(\frac{ES}{\ell}\right)_3 u_4{}^2 - 2\left(\frac{ES}{\ell}\right)_3 u_3 u_4 \right\}$$

$$E_{\substack{\text{pot.} \\ \text{structure}}} = \frac{1}{2}\left\{ \left(\frac{ES}{\ell}\right)_1 u_1{}^2 + \left[\left(\frac{ES}{\ell}\right)_1 + \left(\frac{ES}{\ell}\right)_2\right] u_2{}^2 + \left[\left(\frac{ES}{\ell}\right)_2 + \left(\frac{ES}{\ell}\right)_3\right] u_3{}^2 + \left(\frac{ES}{\ell}\right)_3 u_4{}^2 ... \right.$$

$$\left. ....- 2\left(\frac{ES}{\ell}\right)_1 u_1 u_2 - 2\left(\frac{ES}{\ell}\right)_2 u_2 u_3 - 2\left(\frac{ES}{\ell}\right)_3 u_3 u_4 \right\}$$

A quadratic form built on 4 dof u_1, u_2, u_3, u_4 is obtained. Using a known procedure we arrive at:

$$E_{\substack{\text{pot.} \\ \text{structure}}} = \frac{1}{2}\begin{bmatrix} u_1 & u_2 & u_3 & u_4 \end{bmatrix} \bullet \left\{ \begin{array}{l} \left(\dfrac{ES}{\ell}\right)_1 u_1 - \left(\dfrac{ES}{\ell}\right)_1 u_2 \\[2ex] -\left(\dfrac{ES}{\ell}\right)_1 u_1 + \left[\left(\dfrac{ES}{\ell}\right)_1 + \left(\dfrac{ES}{\ell}\right)_2\right] u_2 - \left(\dfrac{ES}{\ell}\right)_2 u_3 \\[2ex] -\left(\dfrac{ES}{\ell}\right)_2 u_2 + \left[\left(\dfrac{ES}{\ell}\right)_2 + \left(\dfrac{ES}{\ell}\right)_3\right] u_3 - \left(\dfrac{ES}{\ell}\right)_3 u4 \\[2ex] -\left(\dfrac{ES}{\ell}\right)_1 u_3 + \left(\dfrac{ES}{\ell}\right)_1 u_4 \end{array} \right\}$$

or in a form which is now well-known:

$$E_{\text{pot.}} = \frac{1}{2}\{d\}_{\text{str}}^{\text{T}} \bullet [K]_{\text{str}} \bullet \{d\}_{\text{str}}$$

$$\Updownarrow$$

$$E_{\substack{\text{pot.} \\ \text{structure}}} = \frac{1}{2}\left\{\begin{array}{c} u_1 \\ u_2 \\ u_3 \\ u_4 \end{array}\right\}^{T} \bullet \begin{bmatrix} \left(\dfrac{ES}{\ell}\right)_1 & -\left(\dfrac{ES}{\ell}\right)_1 & 0 & 0 \\[2ex] -\left(\dfrac{ES}{\ell}\right)_1 & \left[\left(\dfrac{ES}{\ell}\right)_1 + \left(\dfrac{ES}{\ell}\right)_2\right] & -\left(\dfrac{ES}{\ell}\right)_2 & 0 \\[2ex] 0 & -\left(\dfrac{ES}{\ell}\right)_2 & \left[\left(\dfrac{ES}{\ell}\right)_2 + \left(\dfrac{ES}{\ell}\right)_3\right] & -\left(\dfrac{ES}{\ell}\right)_3 \\[2ex] 0 & 0 & -\left(\dfrac{ES}{\ell}\right)_3 & \left(\dfrac{ES}{\ell}\right)_3 \end{bmatrix} \bullet \left\{\begin{array}{c} u_1 \\ u_2 \\ u_3 \\ u_4 \end{array}\right\}$$

where the stiffness matrix of the complete structure shows up. The behavior of this structure can be written[37] as:

$$\{F\}_{str} = [K]_{str} \bullet \{d\}_{str}$$

$$\Updownarrow$$

$$\begin{Bmatrix} X_1 \\ X_2 \\ X_3 \\ X_4 \end{Bmatrix} = \begin{bmatrix} \left(\dfrac{ES}{\ell}\right)_1 & -\left(\dfrac{ES}{\ell}\right)_1 & 0 & 0 \\ -\left(\dfrac{ES}{\ell}\right)_1 & \left[\left(\dfrac{ES}{\ell}\right)_1 + \left(\dfrac{ES}{\ell}\right)_2\right] & -\left(\dfrac{ES}{\ell}\right)_2 & 0 \\ 0 & -\left(\dfrac{ES}{\ell}\right)_2 & \left[\left(\dfrac{ES}{\ell}\right)_2 + \left(\dfrac{ES}{\ell}\right)_3\right] & -\left(\dfrac{ES}{\ell}\right)_3 \\ 0 & 0 & -\left(\dfrac{ES}{\ell}\right)_3 & \left(\dfrac{ES}{\ell}\right)_3 \end{bmatrix} \bullet \begin{Bmatrix} u_1 \\ u_2 \\ u_3 \\ u_4 \end{Bmatrix}$$

$$[3.37]$$

NOTES

❏ As represented in Figure 3.24 the structure appears to be "floating" along the $\vec{X}$ axis. Absence of a link with the environment gives rise to the possibility of a rigid-body movement (translation) along the $\vec{X}$ direction of loading. This structure is not properly linked in the lines specified in section 2.5.2. As a consequence, the inverse matrix of $[K]_{str}$ above (i.e. the flexibility matrix) does not exist. After verification it can be observed that the main determinant of matrix $[K]_{str}$ above is zero.

❏ The total energy can also be expressed in terms of the potential energy of elements with elementary stiffness matrix magnified to (4×4) size which is also that of the global stiffness matrix of the structure already studied. We obtain:

---

37 See [2.122].

$$E_{\substack{\text{pot.}\\\text{structure}}} = \frac{1}{2}\begin{Bmatrix} u_1 \\ u_2 \\ 0 \\ 0 \end{Bmatrix}^T \bullet \begin{bmatrix} \left(\dfrac{ES}{\ell}\right)_1 & -\left(\dfrac{ES}{\ell}\right)_1 & 0 & 0 \\[2mm] -\left(\dfrac{ES}{\ell}\right)_1 & \left(\dfrac{ES}{\ell}\right)_1 & 0 & 0 \\[2mm] 0 & 0 & 0 & 0 \\[1mm] 0 & 0 & 0 & 0 \end{bmatrix} \bullet \begin{Bmatrix} u_1 \\ u_2 \\ 0 \\ 0 \end{Bmatrix} \ldots\ldots$$

$$\ldots\ldots + \frac{1}{2}\begin{Bmatrix} 0 \\ u_2 \\ u_3 \\ 0 \end{Bmatrix}^T \bullet \begin{bmatrix} 0 & 0 & 0 & 0 \\[1mm] 0 & \left(\dfrac{ES}{\ell}\right)_2 & -\left(\dfrac{ES}{\ell}\right)_2 & 0 \\[2mm] 0 & -\left(\dfrac{ES}{\ell}\right)_2 & \left(\dfrac{ES}{\ell}\right)_2 & 0 \\[2mm] 0 & 0 & 0 & 0 \end{bmatrix} \bullet \begin{Bmatrix} 0 \\ u_2 \\ u_3 \\ 0 \end{Bmatrix} \ldots\ldots$$

$$\ldots\ldots + \frac{1}{2}\begin{Bmatrix} 0 \\ 0 \\ u_3 \\ u_4 \end{Bmatrix}^T \bullet \begin{bmatrix} 0 & 0 & 0 & 0 \\[1mm] 0 & 0 & 0 & 0 \\[1mm] 0 & 0 & \left(\dfrac{ES}{\ell}\right)_3 & -\left(\dfrac{ES}{\ell}\right)_3 \\[2mm] 0 & 0 & -\left(\dfrac{ES}{\ell}\right)_3 & \left(\dfrac{ES}{\ell}\right)_3 \end{bmatrix} \bullet \begin{Bmatrix} 0 \\ 0 \\ u_3 \\ u_4 \end{Bmatrix}$$

It is observed that the global stiffness matrix $[K]_{\text{str}}$ (4×4) is obtained directly by adding[38] the stiffness matrices of every element "reloaded" in a (4×4) table having the size of the global stiffness matrix. In other words, the global stiffness matrix $[K]_{\text{str}}$ appears as the sum of the elementary stiffness matrices redimensioned to the size of the global stiffness matrix, i.e.:

$$[K]_{\substack{\text{structure}}} = \begin{bmatrix} \left(\dfrac{ES}{\ell}\right)_1 & -\left(\dfrac{ES}{\ell}\right)_1 & 0 & 0 \\[2mm] -\left(\dfrac{ES}{\ell}\right)_1 & \left(\dfrac{ES}{\ell}\right)_1 & 0 & 0 \\[2mm] 0 & 0 & 0 & 0 \\[1mm] 0 & 0 & 0 & 0 \end{bmatrix} + \begin{bmatrix} 0 & 0 & 0 & 0 \\[1mm] 0 & \left(\dfrac{ES}{\ell}\right)_2 & -\left(\dfrac{ES}{\ell}\right)_2 & 0 \\[2mm] 0 & -\left(\dfrac{ES}{\ell}\right)_2 & \left(\dfrac{ES}{\ell}\right)_2 & 0 \\[2mm] 0 & 0 & 0 & 0 \end{bmatrix} + \begin{bmatrix} 0 & 0 & 0 & 0 \\[1mm] 0 & 0 & 0 & 0 \\[1mm] 0 & 0 & \left(\dfrac{ES}{\ell}\right)_3 & -\left(\dfrac{ES}{\ell}\right)_3 \\[2mm] 0 & 0 & -\left(\dfrac{ES}{\ell}\right)_3 & \left(\dfrac{ES}{\ell}\right)_3 \end{bmatrix}$$

$$[3.38]$$

---

38 See section 12.1.2.1.

The "reloading" as well as the operations of addition can be seen in Figure 3.25 where the three highlighted contours identify elementary stiffness matrices $[k]_1$, $[k]_2$ and $[k]_3$ respectively of elements 1, 2 and 3. Their location in the global stiffness matrix is linked to that of the dof of the corresponding elements that have been arranged here in the natural order of the nodes.

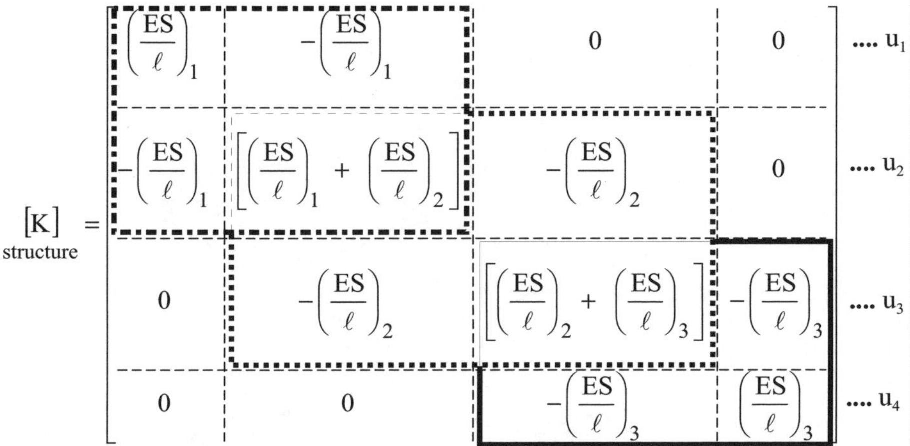

Figure 3.25. *Construction of the global stiffness matrix*

Another way of illustrating the assembly mechanism is graphically represented in Figure 3.26, where $A_i = \left(\dfrac{ES}{\ell}\right)_i$.

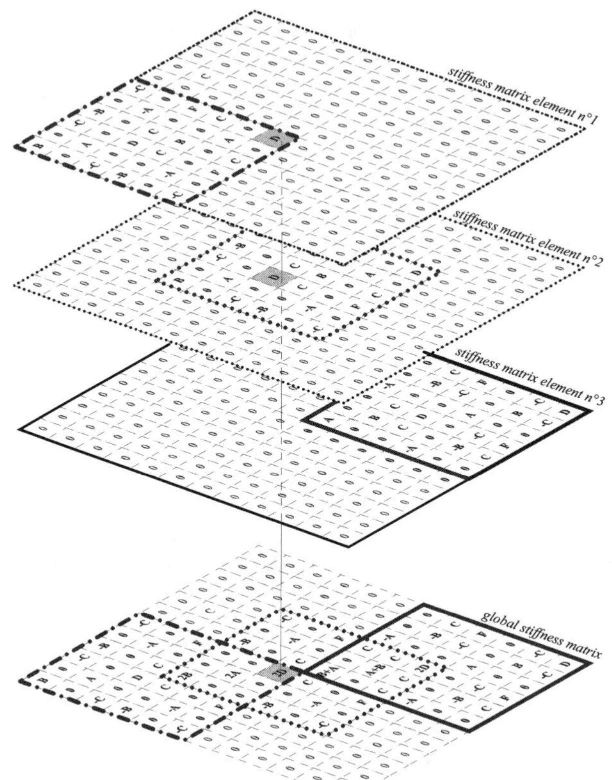

Figure 3.26. *Illustration of the assembly of the stiffness matrix*

❐ [3.38] can also be presented in the form:

$$
\begin{bmatrix}
\mathbf{K_{11}} & \mathbf{K_{12}} & \mathbf{K_{13}} & \mathbf{K_{14}} \\
\mathbf{K_{21}} & \mathbf{K_{22}} & \mathbf{K_{23}} & \mathbf{K_{24}} \\
\mathbf{K_{31}} & \mathbf{K_{32}} & \mathbf{K_{33}} & \mathbf{K_{34}} \\
\mathbf{K_{41}} & \mathbf{K_{42}} & \mathbf{K_{43}} & \mathbf{K_{44}}
\end{bmatrix}_{\text{structure}}
=
\sum_{\substack{\text{element no. }1 \\ (i=1)}}^{\substack{(i=3) \\ \text{element no. }3}}
\begin{bmatrix}
(k_{11})_i & (k_{12})_i & (k_{13})_i & (k_{14})_i \\
(k_{21})_i & (k_{22})_i & (k_{23})_i & (k_{24})_i \\
(k_{31})_i & (k_{32})_i & (k_{33})_i & (k_{34})_i \\
(k_{41})_i & (k_{42})_i & (k_{43})_i & (k_{44})_i
\end{bmatrix}_{\text{element no. }i}
$$

For example, here $\left(k_{11}\right)_1 = \left(\dfrac{ES}{\ell}\right)_1$ etc…, thus:

$$\mathbf{K_{11}} = (k_{11})_1 + (k_{11})_2 + (k_{11})_3 = (k_{11})_1 + 0 + 0 = (k_{11})_1$$

$$\mathbf{K_{13}} = (k_{13})_1 + (k_{13})_2 + (k_{13})_3 = 0 + 0 + 0 = 0$$

$$\mathbf{K_{33}} = (k_{33})_1 + (k_{33})_2 + (k_{33})_3 = 0 + (k_{33})_2 + (k_{33})_3 = \left[(k_{33})_2 + (k_{33})_3\right]$$

$\vdots$

This calculation method shows the formation of the terms $\mathbf{K_{IJ}}$, but it requires redimensioning of the elementary stiffness matrices so that they are equal in size to the global matrix. In practice, when the number of elements is large this method becomes too cumbersome.

☐ The degrees of freedom have been numbered in the natural order as in Figure 3.24. However, it is noted that this choice has been made *a priori*. For example, we can adopt another order, as in Figure 3.27.

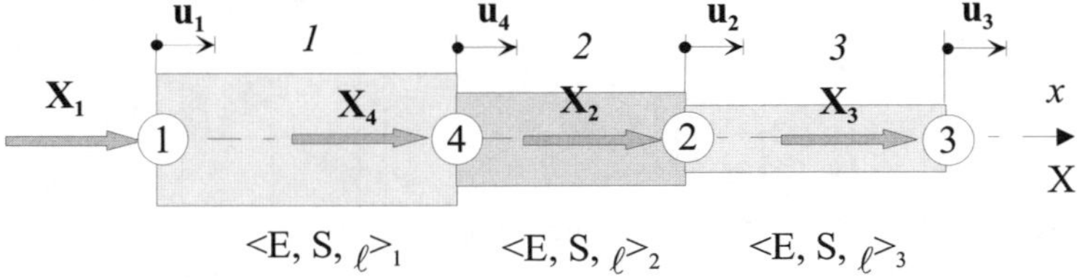

Figure 3.27. *Other possible numbering of the nodes*

The load and the dof being always denoted as $\begin{bmatrix} X_1 & X_2 & X_3 & X_4 \end{bmatrix}^T$ and $\begin{bmatrix} v_1 & v_2 & v_3 & v_4 \end{bmatrix}^T$ we ensure by following the same procedure that the equation below is obtained in place of [3.38]:

$$[K]_{str} = \begin{bmatrix} \left(\dfrac{ES}{\ell}\right)_1 & 0 & 0 & -\left(\dfrac{ES}{\ell}\right)_1 \\ 0 & 0 & 0 & 0 \\ 0 & 0 & 0 & 0 \\ -\left(\dfrac{ES}{\ell}\right)_1 & 0 & 0 & \left(\dfrac{ES}{\ell}\right)_1 \end{bmatrix} + \begin{bmatrix} 0 & 0 & 0 & 0 \\ 0 & \left(\dfrac{ES}{\ell}\right)_2 & 0 & -\left(\dfrac{ES}{\ell}\right)_2 \\ 0 & 0 & 0 & 0 \\ 0 & -\left(\dfrac{ES}{\ell}\right)_2 & 0 & \left(\dfrac{ES}{\ell}\right)_2 \end{bmatrix} + \begin{bmatrix} 0 & 0 & 0 & 0 \\ 0 & \left(\dfrac{ES}{\ell}\right)_3 & -\left(\dfrac{ES}{\ell}\right)_3 & 0 \\ 0 & -\left(\dfrac{ES}{\ell}\right)_3 & \left(\dfrac{ES}{\ell}\right)_3 & 0 \\ 0 & 0 & 0 & 0 \end{bmatrix}$$

i.e.:

$$[K]_{str} = \begin{bmatrix} \left(\dfrac{ES}{\ell}\right)_1 & 0 & 0 & -\left(\dfrac{ES}{\ell}\right)_1 \\[2em] 0 & \left[\left(\dfrac{ES}{\ell}\right)_2 + \left(\dfrac{ES}{\ell}\right)_3\right] & -\left(\dfrac{ES}{\ell}\right)_3 & -\left(\dfrac{ES}{\ell}\right)_2 \\[2em] 0 & -\left(\dfrac{ES}{\ell}\right)_3 & \left(\dfrac{ES}{\ell}\right)_3 & 0 \\[2em] -\left(\dfrac{ES}{\ell}\right)_1 & -\left(\dfrac{ES}{\ell}\right)_2 & 0 & \left[\left(\dfrac{ES}{\ell}\right)_1 + \left(\dfrac{ES}{\ell}\right)_2\right] \end{bmatrix}$$

Let us compare the configurations of the matrices $[K]_{st}$ corresponding to the two methods of numbering the nodes in Figure 3.28.

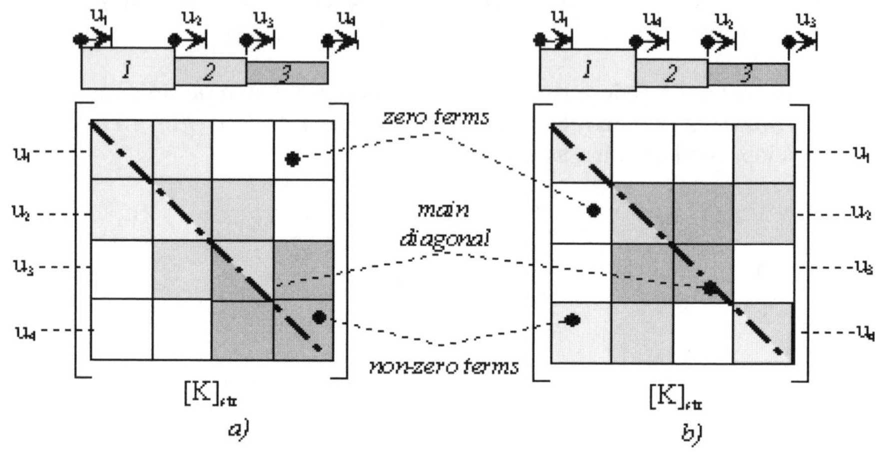

Figure 3.28. *Stiffness matrix: influence of the method of numbering the nodes*

It is observed that in the stiffness matrix the numbering order influences the geographical distribution of the non-zero terms. A "natural" order of the numbering regroups the terms around the main diagonal (Figure 3.28a). On the other hand, a "big" gap between the numbers of two consecutive nodes results in a "scattered" distribution of the non-zero terms in the table given in Figure 3.28b.

3.3.2.2. *Example 2*

Suppose that the previous structure is enlarged to 9 truss elements (Figure 3.29), the dof of the elements being numbered in the natural order.

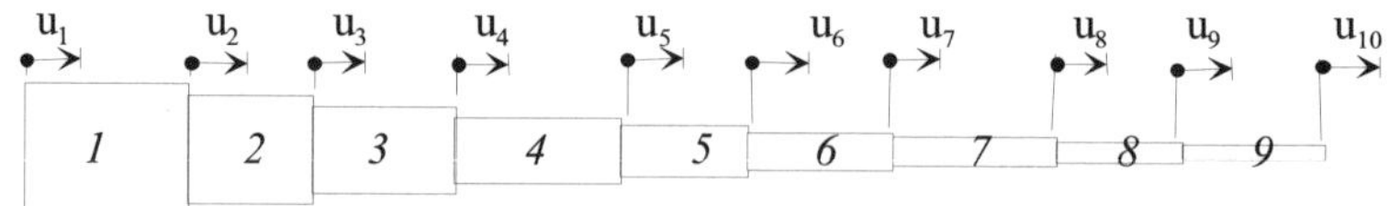

Figure 3.29. *Structure consisting of 9 truss elements*

Then the assembly mechanism considered in the previous section leads to $[K]_{str}$ *(10×10)* stiffness matrix similar to Figure 3.30 where only the non-zero terms are shaded.

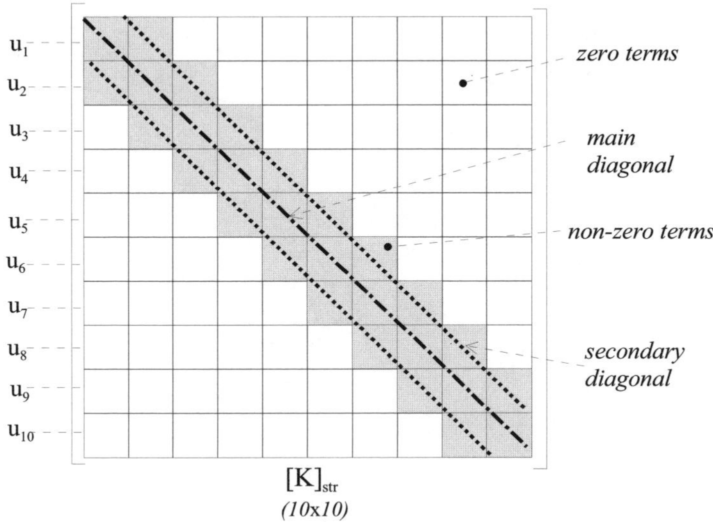

$$[K]_{str}$$
(10x10)

Figure 3.30. *Matrix [K]$_{st}$ for 9 truss elements*

NOTES

❏ The trend observed in Figure 3.28a is reinforced: the non-zero terms are regrouped on the main diagonal and on the two "secondary" diagonals. The (10×10) table in which the terms are arranged seem to be mainly "populated" with zero terms. This seems to be an important characteristic as it facilitates the numerical processing. This aspect of the "banded matrix" should be preferred and preserved. Thus, "jumping" when numbering the nodes should be avoided or at least kept to a minimum when there is no other option. Suppose the numbering of the nodes of Figure 3.29 is modified and represented as in Figure 3.31.

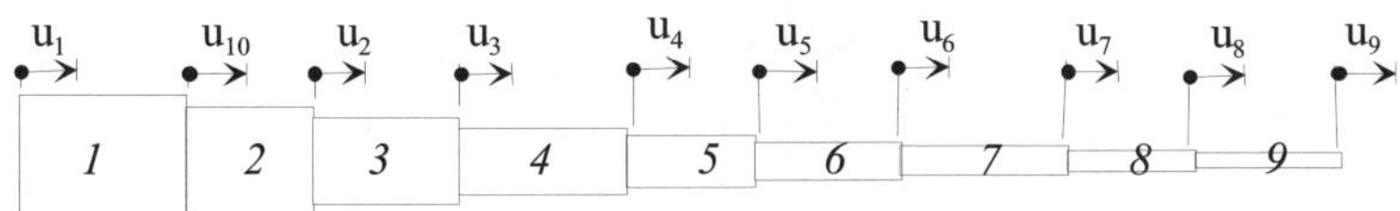

Figure 3.31. *Modification of the numbering of the nodes*

Then the stiffness matrix resembles that of Figure 3.32. It loses its "banded matrix" property. It is considered "badly disposed" for storing and numerical processing.

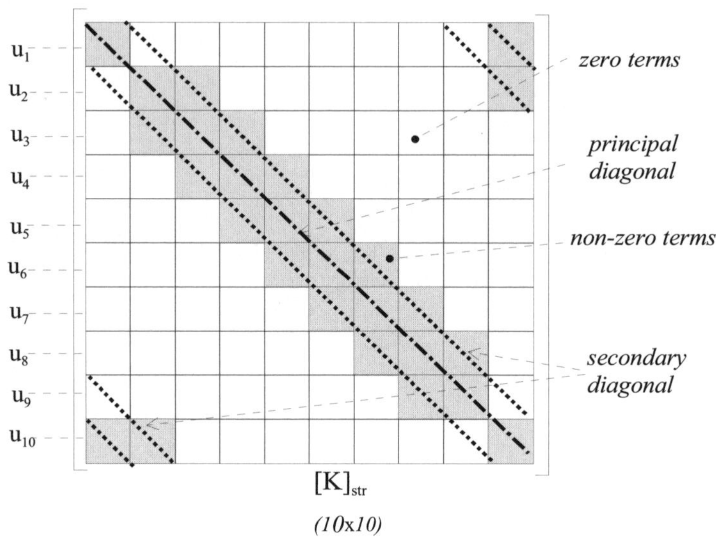

Figure 3.32. *Matrix [K]_sr for the numbering of Figure 3.31*

❏ The assembly mechanism of elementary matrices is not carried out by redimensioning them to the size of the global stiffness matrix because of the numerical calculations involved. The general [K]_str term can also be written as:

$$K_{IJ}=\sum_{i,j}\left(k_{ij}\right)_1 \times\delta_{iI}\times\delta_{jJ} + \sum_{i,j}\left(k_{ij}\right)_2 \times\delta_{iI}\times\delta_{jJ} +...+\sum_{i,j}\left(k_{ij}\right)_3 \times\delta_{iI}\times\delta_{jJ}$$

element no. 1   element no. 2   element no. 3

which enables "scrolling" of the k_{ij} terms of the elementary stiffness matrices. For the 9 elements forming the structure:

$$K_{IJ}=\sum_{m=1}^{9}\left(\sum_{i,j}\left(k_{ij}\right)_m \times\delta_{iI}\times\delta_{jJ}\right)$$

element no. m

where the (so-called Krönecker) symbol is used:

$$\delta_{iI} = 1 \text{ if } i = I$$

$$\delta_{iI} = 0 \text{ if } i \neq I$$

3.3.1. *Introduction*

Thus, an assembly rule for a structure discretized into M elements (m = 1,…M) and having "n" degrees of freedom appears. Every element has "n_e" degrees of freedom which form a sub-assembly of n dof of the complete structure, and which are noted in the global coordinate system as: $(d_{p+1}, d_{p+2},… d_i,… d_j,…d_{p+ne})$.

For example, Figure 3.33 represents a structure with "n" dof. A solid (or volumic) element with the form of a tetrahedron[39] has been traced. Its 12 dof indicated in the global system form the sub-assembly:

$$(d_{p+1},….d_{p+12}) \subset (d_1,….d_n)$$

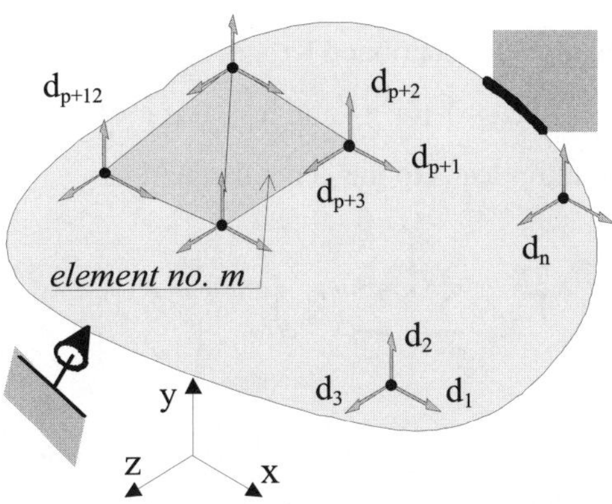

Figure 3.33. *dof-element as sub-assembly of dof structure*

Keeping in mind what is known up to now, the potential energy of an element is noted in the global system:

$$E_{\substack{pot.\\element}} = \frac{1}{2}\left\{\begin{matrix}d_{p+1}\\ \cdot \\ \cdot \\ d_{p+ne}\end{matrix}\right\}_G^T \bullet [k]_{\substack{element\\Global}} \bullet \left\{\begin{matrix}d_{p+1}\\ \cdot \\ \cdot \\ d_{p+ne}\end{matrix}\right\}_G = \frac{1}{2}\sum_{i=p+1}^{p+ne}\sum_{j=p+1}^{p+ne} k_{ij} d_i d_j$$

---

39 The "tetrahedron" volumic element is described in Chapter 5. It will be seen in section 5.5.2 that this element has four nodes and three dofs per node, that is, 12 dofs for the element.

The potential energy of the entire structure is written as:

$$E_{\substack{pot. \\ structure}} = \frac{1}{2}\begin{Bmatrix} d_1 \\ \cdot \\ \cdot \\ \cdot \\ d_n \end{Bmatrix}^{T} \bullet [K]_{structure} \bullet \begin{Bmatrix} d_1 \\ \cdot \\ \cdot \\ \cdot \\ d_n \end{Bmatrix} = \frac{1}{2}\sum_{I=1}^{n}\sum_{J=1}^{n} K_{IJ}d_I d_J$$

Therefore,[40] for the M elements forming the structure, we have:

$$\frac{1}{2}\sum_{I=1}^{n}\sum_{J=1}^{n} K_{IJ}d_I d_J = \sum_{m=1}^{M}\left\{\frac{1}{2}\sum_{i=p+1}^{p+ne}\sum_{j=p+1}^{p+ne}\left(k_{ij}\right)_m d_i d_j\right\}_{element\ no.\ m}$$

By introducing the already mentioned Krönecker symbol noted as δ_{iI}, where:

$$\delta_{iI} = 1 \text{ if } i = I \text{ and } \delta_{iI} = 0 \text{ if } i \neq I$$

The above identification leads to the general equation:

$$K_{IJ} = \sum_{m=1}^{M}\left(\sum_{i,j}\left(k_{ij}\right)_m \times \delta_{iI} \times \delta_{jJ}\right)_{element\ no.\ m} \tag{3.39}$$

NOTES

❏ In practice, the k_{ij} terms of the elementary stiffness matrix are scrolled only once, and they are "loaded" as and when necessary in a table with the dimensions of $[K]_{str}$. The values of the terms assigned to the same location are added up. When all the elements have been reviewed, the $[K]_{str}$ matrix is formed.

❏ For the structures formed with truss elements placed end to end and reviewed earlier, the non-zero terms of the stiffness matrix were located on the main and on the 2 secondary diagonals. This characteristic resulted in the numbering of the nodes following their natural order. In fact in this simple case the difference in the numbering between two adjacent nodes is 1 and this leads to the placing of the stiffness terms on the three diagonals previously mentioned (see Figures 3.28a and 3.30).

In the case of a complex structure discretized into numerous elements, it is normal that a same node belongs to a number of elements more than two. The total number of elements being large and the manual numbering being no longer

---

40 See [3.1].

possible, it is done using software so as to *minimize* the differences between the numbers assigned to the adjacent nodes.

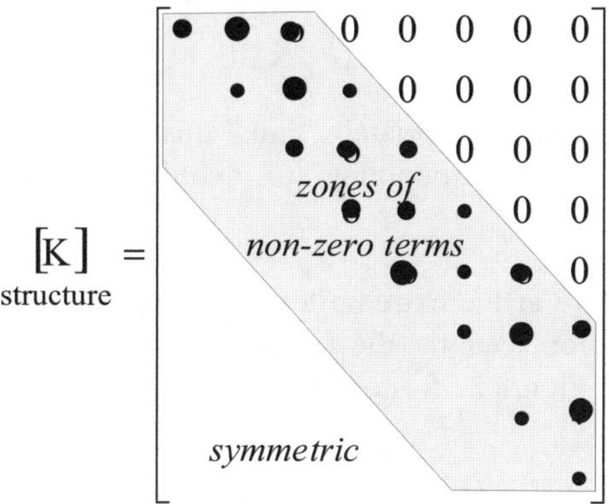

Figure 3.34. *Banded matrix*

This can create a stiffness matrix for the structure having its non-zero terms located on the main diagonal and on a *minimum* number of secondary diagonals. As already mentioned in the example in section 3.3.2.2, such a matrix is known by the name "banded matrix" (see Figure 3.34). The zero terms form a majority. To store the matrix, only the non-zero and the non-redundant terms (the symmetry of the matrix enables reduction of the number of terms to be retained) are stored in single file by the software. This type of storing is a necessity and an essential condition to be able to process the structures with a large number of dof (up to several hundreds of thousands).

3.4. Resolution of the system $\{F\} = [K] \bullet \{d\}$

3.4.1. *Linkage conditions*

Taking the case of assembly of three truss elements examined in the previous section (see Figure 3.24), it has been observed that behavior equation [3.37] was unworkable (singular stiffness matrix) because the structure of Figure 3.24 was not properly linked (it could have a rigid-body translation movement along $\vec{x}$). With the already used notations, behavior equation [3.37] can be written as:

$$\begin{Bmatrix} X_1 \\ X_2 \\ X_3 \\ X_4 \end{Bmatrix} = \begin{bmatrix} K_{11} & K_{12} & K_{13} & K_{14} \\ K_{21} & K_{22} & K_{23} & K_{24} \\ K_{31} & K_{32} & K_{33} & K_{34} \\ K_{41} & K_{42} & K_{43} & K_{44} \end{bmatrix} \bullet \begin{Bmatrix} u_1 \\ u_2 \\ u_3 \\ u_4 \end{Bmatrix} \qquad [3.40]$$

It is now necessary to appropriately "link" this structure, i.e., to position it in a complete manner in its surroundings to avoid any possibility of rigid-body movements.

Fixing this structure at the extreme left (node (1)) involves imposing the value $u_1 = 0$. Then X_1 is replaced by the linking action $X_{1\ell}$ (see section 2.6.4). The following diagram of Figure 3.35 is obtained.

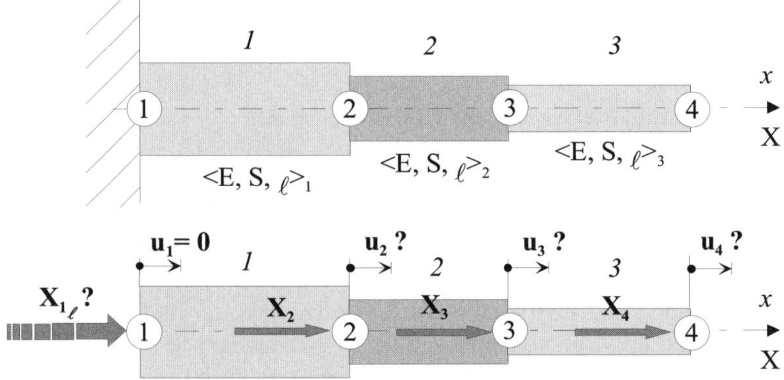

Figure 3.35. *Properly linked structure*

Equation [3.40] becomes:

$$\begin{Bmatrix} X_1 \\ X_2 \\ X_3 \\ X_4 \end{Bmatrix} = \begin{bmatrix} K_{11} & K_{12} & K_{13} & K_{14} \\ K_{21} & K_{22} & K_{23} & K_{24} \\ K_{31} & K_{32} & K_{33} & K_{34} \\ K_{41} & K_{42} & K_{43} & K_{44} \end{bmatrix} \bullet \begin{Bmatrix} u_1 \\ u_2 \\ u_3 \\ u_4 \end{Bmatrix} \qquad [3.41]$$

The stiffness matrix of the structure being singular it is known that this equation cannot be totally inverted to obtain the unknown values of the dof. Let us use the method already tried out in the examples of section 2.6.4:

a) in [3.41] by deleting the first row corresponding to the zero degree of freedom ($u_1 = 0$) and the column of the same row (first column) the following equation is obtained:

$$\begin{Bmatrix} X_2 \\ X_3 \\ X_4 \end{Bmatrix} = \begin{bmatrix} K_{22} & K_{23} & K_{24} \\ K_{32} & K_{33} & K_{34} \\ K_{42} & K_{43} & K_{44} \end{bmatrix} \bullet \begin{Bmatrix} u_2 \\ u_3 \\ u_4 \end{Bmatrix} \qquad [3.42]$$

b) by reverting to the first row of [3.40] that was deleted, the equation becomes:

$$X_{1\ell} = K_{12}u_2 + K_{13}u_3 + K_{14}u_4 \qquad [3.43]$$

Interpretation

At node (1), where $u_1 = 0$ (as a result of the fixed beam), it is not possible to impose a given value of the linking force $X_{1\ell}$. The value of this will be a result of the values of the solutions u_2, u_3, u_4 of [3.42] that is of the applied forces X_2, X_3, X_4. The force $X_{1\ell}$, mechanical action transmitted by the surroundings on the structure, appears as an additional unknown.

The steps of calculation are thus recalled:

➤ first, it is sufficient to delete the line and the column of the same row corresponding to the zero dof (i.e. known) in the complete stiffness matrix of structure [3.41] obtaining thereby the sub-matrix of stiffness [3.42]. Then the systems of equations [3.42] in which there are only the non-zero dof (i.e. the unknown) and the external imposed loads (i.e. the known) should be solved;

➤ second, the value $X_{1\ell}$ is calculated using equation [3.43] in which the values u_2, u_3, u_4 supplied by solving the systems of equations [3.42] are introduced.

3.4.2. *Generalization of the method*

The following applies to any structure. As described in the previous section, the structure's global matrix $[K]_{str}$ should first be constituted by assembling the elementary matrices $[k]_{el}$ expressed in the global system.

The complete system of equations giving the performance of the structure is written as:

$$\{F\}_{str} = [K]_{str} \bullet \{d\}_{str}$$

Stiffness matrix $[K]_{str}$ obtained is singular as the structure is still free since no linkage conditions have been specified. Secondly, the structure is properly linked (the rigid-body movements are no longer possible). A certain number of dof of this structure are then blocked i.e. their values are zero. Let these dof be denoted by

$\{d_o\}$ where ($\{d_o\} = \{0\}$). The dof that remained free are denoted $\{d^*\}$. These are the dof that need to be calculated. For the complete structure, this discrimination is represented by the notation: $\{d\}_{str} = \left\{\dfrac{d_o}{d^*}\right\}$. The complete system of equations, that is $\{F\}_{str} = [K]_{str} \bullet \{d\}_{str}$, is then rearranged in the following manner:

$$\left\{\frac{F_\ell}{F^*}\right\} = \left[\begin{array}{c|c}[K_\ell] & [K_\ell^*] \\ \hline [K_\ell^*]^T & [K^*]\end{array}\right] \bullet \left\{\frac{d_o = 0}{d^*}\right\}$$

where we observe the already identified[41] "partition" of the stiffness matrix in four "sub-matrices". This makes it possible for us to write two sub-systems:

$$\underset{\substack{\textit{applied forces}\\ \textit{known}}}{\{F^*\}} = [K^*] \bullet \underset{\textit{unknown}}{\{d^*\}} \tag{3.44}$$

$$\underset{\substack{\textit{linking actions}\\ \textit{unknown}}}{\{F_\ell\}} = [K_\ell^*] \bullet \underset{\textit{unknown}}{\{d^*\}} \tag{3.45}$$

The first of these equations is that which is obtained by the method described in the previous example. The lines and the columns of the same row corresponding to the zero dof in the initial global stiffness matrix were deleted.

➢ Firstly [3.44] should be solved by inversing matrix $[K^*]$ which is a sub-matrix of the complete initial stiffness matrix $[K]_{str}$ of the structure. This inversion leads[42] to the flexibility matrix of the discretized structure associated with its linkage conditions. This enables the calculation of the "free" dof that is $\{d^*\}$.

➢ The $\{d^*\}$ values found are then injected into system [3.45] which in turn gives linking actions $\{F_\ell\}$.

NOTES

❏ It should be remembered that this calculation procedure is independent of the redundancy degree due to the complete positioning of the structure. Here, the

---

41 See section 2.6.4.3.
42 See [2.122].

notion of an isostatic or hyperstatic system does not arise at any moment. Simply, a hypostatic[43] positioning should be avoided.

❏ In practice, the calculated values of the free dof $\left\{d^*\right\}$ mentioned above enable:

a) the calculation of the linking actions $\left\{F_\ell\right\}$ of the structure with its surroundings as indicated earlier

b) the calculation of the nodal forces of every element using:

$$\{F\}_{el} = [k]_{el} \bullet \{d\}_{el}$$

These calculations having been carried out in the global coordinate system, we can go back to the local system (see section 3.1.3). Knowing the dof values expressed in the local coordinate system helps us to go back to the stresses in the elements in view of controlling the dimensioning (mechanical strength of the structure).

Knowing the nodal forces in the local system makes possible the study of the behavior of external and internal links or joints of the structure[44], as well as the study of the stability for compressed zones.

3.5. Different types of finite elements available in industrial software

The finite elements checked in the previous sections do not allow the modeling of all types of structures. They are reserved for elongated structures composed of profiles or for structures where a plane state of stresses exists. According to the nature of the structure and the results desired, other finite elements are available in the calculation codes. Figure 3.36 gives an overview of the main finite element topologies that can be used in industrial software. These elements shall be described in Chapter 5. These days, software for computer aided design (CAD) contains what is called "solid modeling software". These tools enable an integrated transition to the calculations with solid finite elements with the help of automatic meshing software which discretizes the volume of the structure into elements whose geometries are generally tetrahedrons. As a step of the preliminary project, the pre-dimensioning of the part can be checked after applying the boundary conditions (links, loads) and after entry of material characteristics. However, the geometric models worked out with CAD do not always lead to the most appropriate calculation model using the finite element method. Thus, three typical examples of structures can be observed in Figure 3.36. The first two refer to elongated (beams) and thin

---

43 See section 2.5, Figure 2.39.
44 See Chapter 11.

(plates or shells) structures. For these kinds of structures, total integration of modeling choice in CAD software is not presently available. The designer has to intervene in the CAD model to change it into a model compatible with elements to be used[45]. For the time being, we shall only indicate the existence of other types of elements shown in the figure. These elements shall be dealt with in Chapter 5.

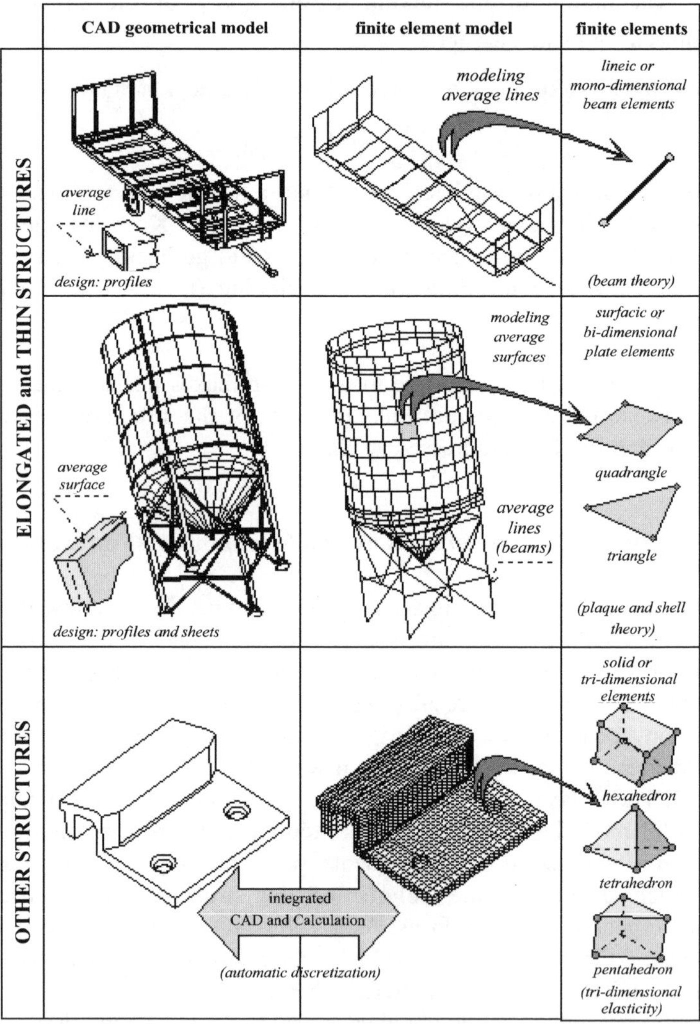

Figure 3.36. *Topology of the main types of finite elements*

45 It has already been indicated that beam elements, for example, were preferred for both precision of results and level of discretization to elements of plane stress (see section 3.2.4.5). They will also be preferred to solid elements (see Chapter 5). The same is valid for plate elements, of higher performance than solid – or volumic – elements.

Chapter 4

Applications: Discretization of Simple Structures

We are going to apply the results of the previous chapter to some simple structures.

4.1. Stiffness matrix of a spring

4.1.1. *Helical spring*

➢ *Object:*

Consider the helical spring of stiffness k_u of Figure 4.1 with elongations along the $\vec{x}\,\vec{x}$ axis. Its stiffness matrix has to be defined.

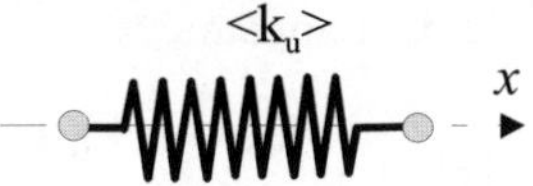

Figure 4.1. *Helical spring*

➢ *Solution:*

(1) and (2) are the ends of the spring (nodes). The loading comprises forces X_1 and X_2. The associated[1] dof are the nodal displacements along $\vec{x}$ marked as u_1 and u_2 in Figure 4.2.

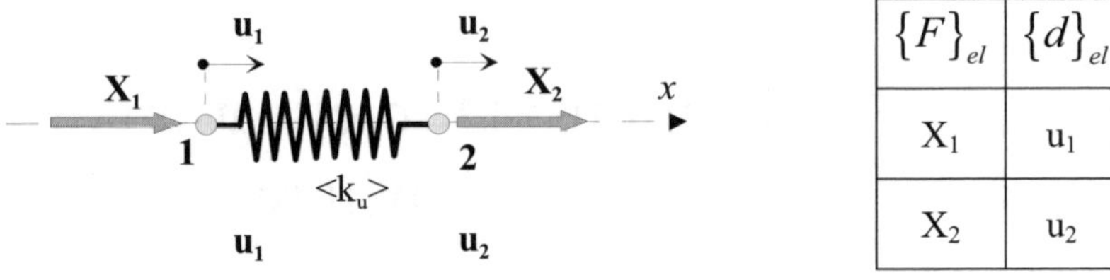

$\{F\}_{el}$	$\{d\}_{el}$
X_1	u_1
X_2	u_2

Figure 4.2. *Characteristics of the helical spring element*

The elongation of the spring is then written as $(u_2 - u_1)$, and the work done by the external forces becomes[2]:

$$W = \frac{1}{2}k_u(u_2 - u_1)^2 = E_{pot.}$$

that is a stored potential energy of deformation:

$$E_{pot.} = \frac{1}{2}k_u(u_1{}^2 + u_2{}^2 - 2u_1u_2)$$

The quadratic form of the nodal displacements is evident and it can also be written as:

$$E_{pot.} = \frac{1}{2}\begin{Bmatrix} u_1 \\ u_2 \end{Bmatrix}^T \bullet \begin{bmatrix} k_u & -k_u \\ -k_u & k_u \end{bmatrix} \bullet \begin{Bmatrix} u_1 \\ u_2 \end{Bmatrix} \qquad [4.1]$$

The stiffness matrix of the element appears:

$$[k]_{\acute{e}l} = \begin{bmatrix} k_u & -k_u \\ -k_u & k_u \end{bmatrix}$$

---

1 See [2.122].
2 See [2.6].

The behavior of this spring is summarized by[3]:

$$\{F\}_{\text{él}} = [k]_{\text{él}} \bullet \{d\}_{\text{él}} \Leftrightarrow \begin{Bmatrix} X_1 \\ X_2 \end{Bmatrix} = \begin{bmatrix} k_u & -k_u \\ -k_u & k_u \end{bmatrix} \bullet \begin{Bmatrix} u_1 \\ u_2 \end{Bmatrix}$$

As has been often indicated[4] for similar finite elements, the loaded element of Figure 4.2 "floats" in direction $\vec{x}$ of the load (rigid-body translation along $\vec{x}$). The stiffness matrix cannot[2] therefore be inverted and this can be checked by calculating the main determinant of this matrix, i.e.:

$$\Delta_{\text{principal}} = k_u{}^2 - k_u{}^2 = 0$$

4.1.2. *Spiral spring*

> *Object:*

Consider the spiral spring of stiffness k_θ of Figure 4.3 for the rotation around $\vec{z}$ axis of the associated pivot linkage. Its stiffness matrix is to be defined.

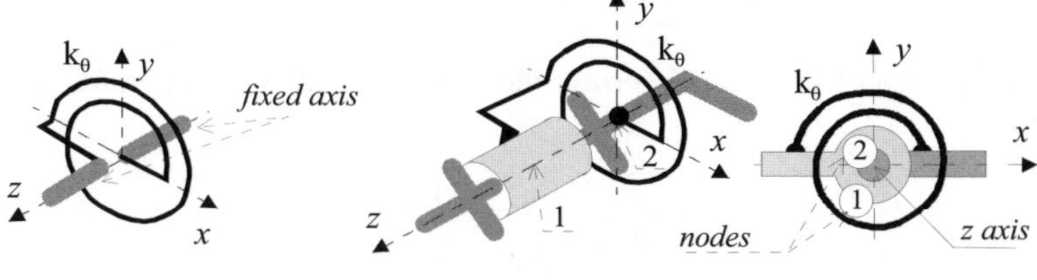

a) single spiral spring   *b) spiral spring associated a pivot linkage*

Figure 4.3. *Spiral spring*

> *Solution:*

Points (1) and (2) respectively indicate every part of the turning pair that schematically represents the pivot linkage (nodes of the spiral spring element). The load comprises moments N_1 and N_2. The associated[5] dof are the nodal rotations parallel to the $\vec{z}$ axis indicated as θ_{z1} and θ_{z2} of each of the constituent parts of the hinge (see Figure 4.4).

---

3 See [2.122].
4 See section 3.2.1.
5 See [2.122].

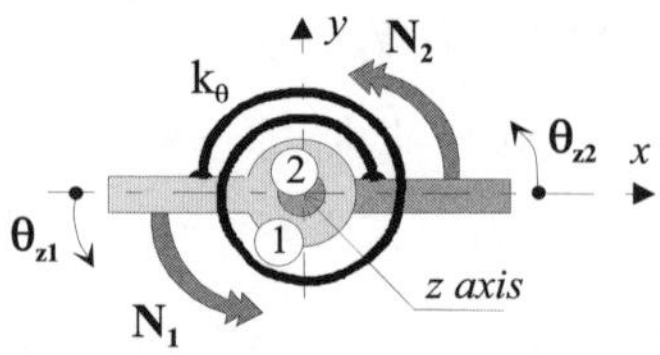

$\{F\}_{el}$	$\{d\}_{el}$
N_1	θ_{z1}
N_2	θ_{z2}

Figure 4.4. *Characteristics of the spiral spring element*

The angular deflected shape of the spring is expressed through $(\theta_{z2} - \theta_{z1})$. The work of the external moments become (see [2.7]).

$$W = \frac{1}{2} k_\theta (\theta_{z2} - \theta_{z1})^2 = E_{pot.}$$

that is a stored potential energy of deformation:

$$E_{pot.} = \frac{1}{2} k_\theta (\theta_{z1}^2 + \theta_{z2}^2 - 2\theta_{z1}\theta_{z2})$$

As in the case of the helical spring, the stiffness matrix of the spiral spring can be written as:

$$[k]_{él} = \begin{bmatrix} k_\theta & -k_\theta \\ -k_\theta & k_\theta \end{bmatrix} \tag{4.2}$$

As in the previous example, the behavior equation of this spiral spring can be derived from:

$$\{F\}_{él} = [k]_{él} \bullet \{d\}_{él} \Leftrightarrow \begin{Bmatrix} N_1 \\ N_2 \end{Bmatrix} = \begin{bmatrix} k_\theta & -k_\theta \\ -k_\theta & k_\theta \end{bmatrix} \bullet \begin{Bmatrix} \theta_{z1} \\ \theta_{z2} \end{Bmatrix}$$

In the configuration given in Figure 4.4, the same note can be made regarding the absence of a link of the spiral spring with its environment. This also leads to a singular stiffness matrix.

4.2. Assembly of elements

4.2.1. *Example 1*

> *Object:*

Consider the finite elements that are to be assembled and are outlined in Figure 4.5.

1) Write the stiffness matrix of the structure obtained after assembly as well as the corresponding behavior equation $\{F\}_{str} = [k]_{str} \cdot \{d\}_{str}$;

2) Fix the left end of the helical spring element and calculate the nodal displacements and the linking action under a tractional force exerted at the right end of the truss element.

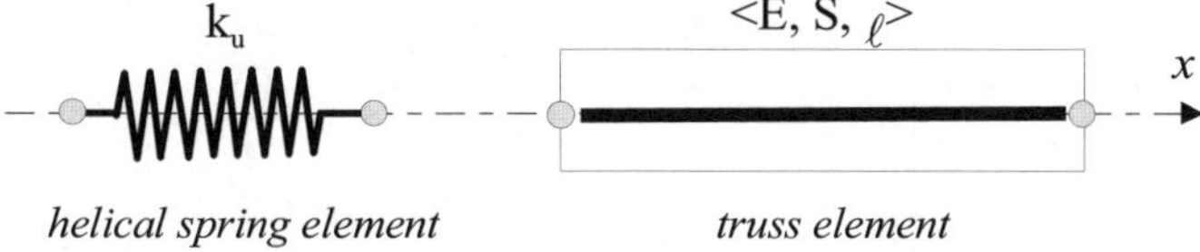

Figure 4.5. *Elements to be assembled*

> *Solution:*

1) The assembly results in a structure shown in Figure 4.6 having three nodes, three forces and three associated dof. Therefore, the stiffness matrix of this structure has three rows and three columns.

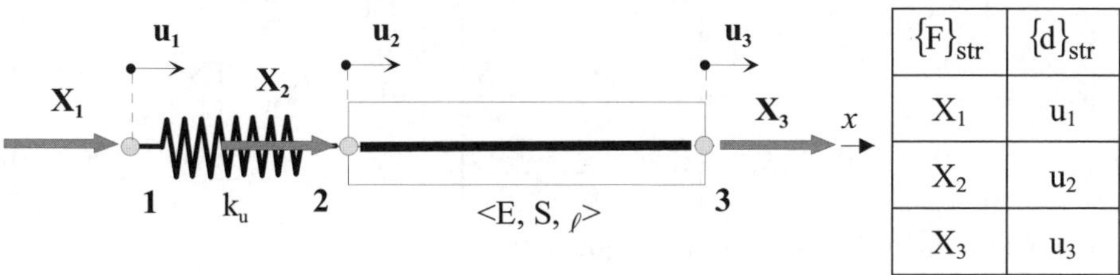

$\{F\}_{str}$	$\{d\}_{str}$
X_1	u_1
X_2	u_2
X_3	u_3

Figure 4.6. *Assembled structure*

The stiffness matrices of the helical spring element and of the truss element are known as are the equations of the corresponding potential energy of deformation (see [4.1] and [3.12]). Thus, we have:

$$E_{\substack{\text{pot.}\\ \text{structure}}} = \frac{1}{2} k_u (u_1{}^2 + u_2{}^2 - 2u_1 u_2) + \frac{1}{2}\frac{ES}{\ell}(u_2{}^2 + u_3{}^2 - 2u_2 u_3)$$

that is:

$$E_{\substack{\text{pot.}\\ \text{structure}}} = \frac{1}{2}\left[k_u u_1{}^2 + \left(k_u + \frac{ES}{\ell}\right)u_2{}^2 + \frac{ES}{\ell}u_3{}^2 - 2k_u u_1 u_2 - 2\frac{ES}{\ell}u_2 u_3 \right]$$

that can be expressed in the form:

$$E_{\substack{\text{pot.}\\ \text{structure}}} = \frac{1}{2}\{d\}_{str}^T \bullet [K]_{str} \bullet \{d\}_{str}$$

$$\Updownarrow$$

$$E_{\substack{\text{pot.}\\ \text{structure}}} = \frac{1}{2}\begin{Bmatrix} u_1 \\ u_2 \\ u_3 \end{Bmatrix}^T \bullet \begin{bmatrix} k_u & -k_u & 0 \\ -k_u & \left(k_u + \dfrac{ES}{\ell}\right) & -\dfrac{ES}{\ell} \\ 0 & -\dfrac{ES}{\ell} & \dfrac{ES}{\ell} \end{bmatrix} \bullet \begin{Bmatrix} u_1 \\ u_2 \\ u_3 \end{Bmatrix}$$

to show the stiffness matrix of the structure.

NOTE

❐ The stiffness matrix of each of the elements can be reloaded in a (3×3) table similar to $[K]_{str}$. Thus, from the stiffness matrices of each of the two elements[6], we can write:

$$[K]_{str} = \begin{bmatrix} k_u & -k_u & 0 \\ -k_u & k_u & 0 \\ 0 & 0 & 0 \end{bmatrix} + \begin{bmatrix} 0 & 0 & 0 \\ 0 & \dfrac{ES}{\ell} & -\dfrac{ES}{\ell} \\ 0 & -\dfrac{ES}{\ell} & \dfrac{ES}{\ell} \end{bmatrix} = \begin{bmatrix} k_u & -k_u & 0 \\ -k_u & \left(k_u + \dfrac{ES}{\ell}\right) & -\dfrac{ES}{\ell} \\ 0 & -\dfrac{ES}{\ell} & \dfrac{ES}{\ell} \end{bmatrix} \begin{matrix} \cdots u_1 \\[1.4em] \cdots u_2 \\[1.4em] \cdots u_3 \end{matrix}$$

and obtain the following behavior equation:

---

6 See [4.1] and [3.12].

$$\{F\}_{str} = [K]_{str} \bullet \{d\}_{str} \Leftrightarrow \begin{Bmatrix} X_1 \\ X_2 \\ X_3 \end{Bmatrix} = \begin{bmatrix} k_u & -k_u & 0 \\ -k_u & \left(k_u + \dfrac{ES}{\ell}\right) & -\dfrac{ES}{\ell} \\ 0 & -\dfrac{ES}{\ell} & \dfrac{ES}{\ell} \end{bmatrix} \bullet \begin{Bmatrix} u_1 \\ u_2 \\ u_3 \end{Bmatrix} \quad [4.3]$$

On calculating the main determinant of this matrix, the following is obtained:

$$k_u\left[\left(k_u + \frac{ES}{\ell}\right) \times \frac{ES}{\ell} - \left(\frac{ES}{\ell}\right)^2\right] + k_u \times \left(-k_u \frac{ES}{\ell}\right) = 0 \ \text{(unique matrix)}$$

This could be foreseen[7] because the structure in Figure 4.6 is not properly linked.

2) Let us fix the left end of the spring while the structure is subjected to a tractional force X_3 (here $X_2 = 0$) on the right. We then have the diagram given in Figure 4.7a.

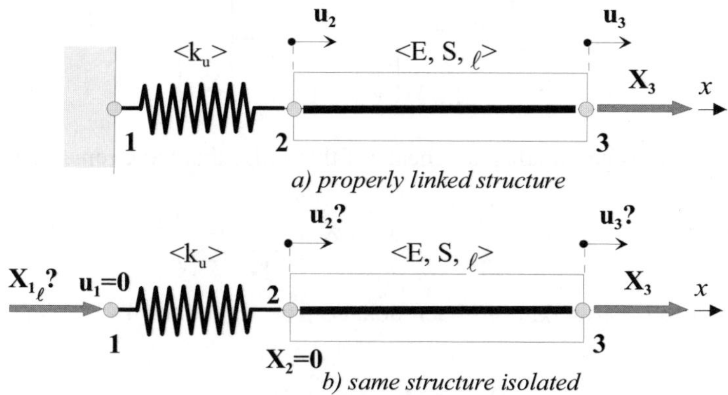

Figure 4.7. *Properly linked structure*

Based on the isolated structure in Figure 4.7b behavior equation [4.3] for these boundary conditions can be written as:

---

7 See [2.122].

$$\{F\}_{str} = [K]_{str} \bullet \{d\}_{str} \Leftrightarrow \begin{Bmatrix} X_{1\ell} \\ X_2 = 0 \\ X_3 \end{Bmatrix} = \begin{bmatrix} k_u & -k_u & 0 \\ -k_u & \left(k_u + \dfrac{ES}{\ell}\right) & -\dfrac{ES}{\ell} \\ 0 & -\dfrac{ES}{\ell} & \dfrac{ES}{\ell} \end{bmatrix} \bullet \begin{Bmatrix} u_1 = 0 \\ u_2 \\ u_3 \end{Bmatrix} \quad [4.4]$$

in which $X_{1\ell}$ is the force transmitted by the link. Following the procedure described in section 3.4.2, we delete the row and the column of the same row corresponding to the dof $u_1 = 0$, which gives us:

$$\begin{Bmatrix} 0 \\ X_3 \end{Bmatrix} = \begin{bmatrix} \left(k_u + \dfrac{ES}{\ell}\right) & -\dfrac{ES}{\ell} \\ -\dfrac{ES}{\ell} & \dfrac{ES}{\ell} \end{bmatrix} \bullet \begin{Bmatrix} u_2 \\ u_3 \end{Bmatrix}$$

By inverting this equation, we obtain:

$$\begin{Bmatrix} u_2 \\ u_3 \end{Bmatrix} = \begin{bmatrix} \dfrac{1}{k_u} & \dfrac{1}{k_u} \\ \dfrac{1}{k_u} & \left(\dfrac{1}{k_u} + \dfrac{\ell}{ES}\right) \end{bmatrix} \bullet \begin{Bmatrix} 0 \\ X_3 \end{Bmatrix}$$

which enables us to get the displacements of the nodes that have remained free:

$$u_2 = \frac{X_3}{k_u} \; ; \quad u_3 = X_3 \left(\frac{1}{k_u} + \frac{\ell}{ES}\right)$$

Reverting to the row that has been deleted from equation [4.4], it can be written as:

$$X_{1\ell} = -k_u \times u_2 + 0 \times u_3$$

i.e. by substituting the value of u_2 given above, we obtain:

$$X_{1\ell} = -X_3$$

This result, which is obvious in this simple case, helps in checking immediately the correct global equilibrium of the model.

4.2.2. *Example 2*

> *Object:*

Consider the finite elements that are to be assembled and are outlined in Figure 4.8:

1) Write the stiffness matrix of the structure obtained after assembly as well as the corresponding behavior equation $\{F\}_{str} = [k]_{str} \bullet \{d\}_{str}$.

2) The left end of the pivotal link of the spiral spring is fixed. Calculate the nodal displacements and the linking action under a transversal force exerted parallel to the $\vec{y}$ axis at the right end of the beam element under plane bending.

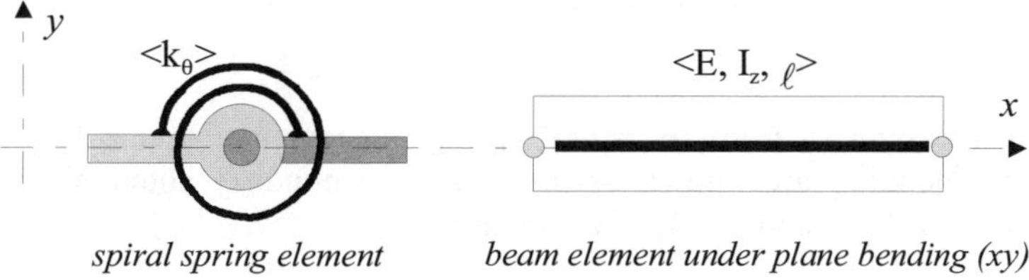

Figure 4.8. *Elements to be assembled*

> *Solution:*

1) Determining the global stiffness matrix of the structure.

The method underlying this assembly can be represented by the diagram in Figure 4.9.

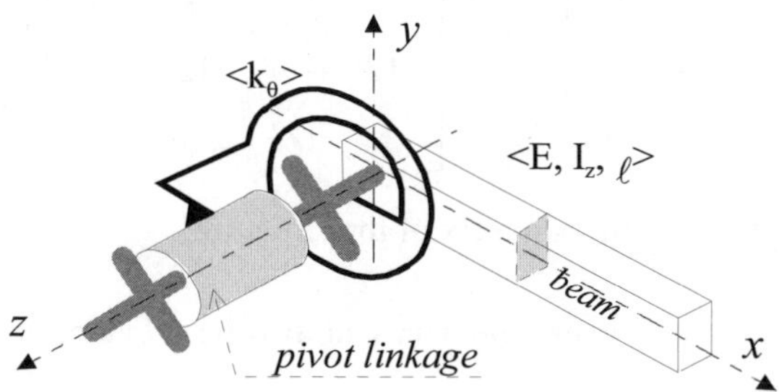

Figure 4.9. *Schematic representation of the method*

The assembly of these two elements leads to the structure with three nodes, five loads and five associated dof, as shown in Figure 4.10.

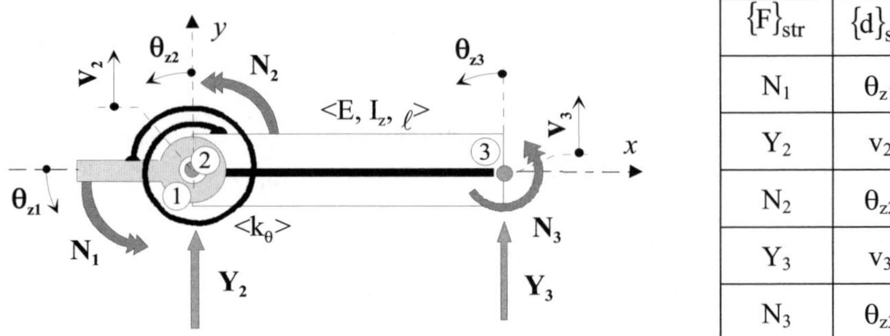

$\{F\}_{str}$	$\{d\}_{str}$
N_1	θ_{z1}
Y_2	v_2
N_2	θ_{z2}
Y_3	v_3
N_3	θ_{z3}

Figure 4.10. *Assembled structure with three nodes* ①, ②, ③

The stiffness matrices of the spiral spring element and of the beam element under plane bending are known as are the corresponding potential energies of deformation (see [4.2] and [3.24]). We obtain:

$$\underset{\text{structure}}{E_{pot.}} = \frac{1}{2}\{d\}_{str}^{T} \bullet [K]_{str} \bullet \{d\}_{str}$$

$$\Updownarrow$$

$$\underset{\text{structure}}{E_{pot.}} = \frac{1}{2}\begin{Bmatrix}\theta_{z1}\\\theta_{z2}\end{Bmatrix}^{T} \bullet \begin{bmatrix} k_\theta & -k_\theta \\ -k_\theta & k_\theta \end{bmatrix} \bullet \begin{Bmatrix}\theta_{z1}\\\theta_{z2}\end{Bmatrix} \ldots\ldots$$

$$\ldots + \frac{1}{2}\begin{Bmatrix}v_2\\\theta_{z2}\\v_3\\\theta_{z3}\end{Bmatrix}^{T} \bullet \frac{EI}{\ell^3}\begin{bmatrix} 12 & 6\ell & -12 & 6\ell \\ 6\ell & 4\ell^2 & -6\ell & 2\ell^2 \\ -12 & -6\ell & 12 & -6\ell \\ 6\ell & 2\ell^2 & -6\ell & 4\ell^2 \end{bmatrix} \bullet \begin{Bmatrix}v_2\\\theta_{z2}\\v_3\\\theta_{z3}\end{Bmatrix}$$

that leads to the (5×5) stiffness matrix of the structure.

The stiffness matrix of every element can also be reloaded in a (5×5) table of $[K]_{str}$, i.e.:

$$[K]_{str} = \begin{bmatrix} k_\theta & 0 & -k_\theta & 0 & 0 \\ 0 & 0 & 0 & 0 & 0 \\ -k_\theta & 0 & k_\theta & 0 & 0 \\ 0 & 0 & 0 & 0 & 0 \\ 0 & 0 & 0 & 0 & 0 \end{bmatrix} + \frac{EI}{\ell^3}\begin{bmatrix} 0 & 0 & 0 & 0 & 0 \\ 0 & 12 & 6\ell & -12 & 6\ell \\ 0 & 6\ell & 4\ell^2 & -6\ell & 2\ell^2 \\ 0 & -12 & -6\ell & 12 & -6\ell \\ 0 & 6\ell & 2\ell^2 & -6\ell & 4\ell^2 \end{bmatrix} \begin{matrix} \cdots\cdots\theta_{z1} \\ \cdots\cdots v_2 \\ \cdots\cdots\theta_{z2} \\ \cdots\cdots v_3 \\ \cdots\cdots\theta_{z3} \end{matrix}$$

$$[K]_{str} = \begin{bmatrix} k_\theta & 0 & -k_\theta & 0 & 0 \\ 0 & \dfrac{12EI}{\ell^3} & \dfrac{6EI}{\ell^2} & -\dfrac{12EI}{\ell^3} & \dfrac{6EI}{\ell^2} \\ -k_\theta & \dfrac{6EI}{\ell^2} & \left(k_\theta + \dfrac{4EI}{\ell}\right) & -\dfrac{6EI}{\ell^2} & \dfrac{2EI}{\ell} \\ 0 & -\dfrac{12EI}{\ell^3} & -\dfrac{6EI}{\ell^2} & \dfrac{12EI}{\ell^3} & -\dfrac{6EI}{\ell^2} \\ 0 & \dfrac{6EI}{\ell^2} & \dfrac{2EI}{\ell} & -\dfrac{6EI}{\ell^2} & \dfrac{4EI}{\ell} \end{bmatrix}$$

and the behavior equation obtained is:

$$\{F\}_{str} = [K]_{str} \bullet \{d\}_{str} \iff \begin{Bmatrix} N_1 \\ Y_2 \\ N_2 \\ Y_3 \\ N_3 \end{Bmatrix} = [K]_{str} \underset{(5\times5)}{\bullet} \begin{Bmatrix} \theta_{z1} \\ v_2 \\ \theta_{z2} \\ v_3 \\ \theta_{z3} \end{Bmatrix} \qquad [4.5]$$

It was observed that the main determinant of each of the two matrices $[k]_{el}$ was zero[8]. It can be easily verified that the main determinant $[K]_{str}$ is zero (unique matrix) as the structure in Figure 4.10 is not properly supported (possibility of rigid-body translation along $\vec{y}$ and rotation around $\vec{z}$)[9].

2) Application of the boundary conditions: let us fix the left end of the pivot linkage of the spiral spring while the structure is subjected to a transverse force Y_3 on the right. The diagram of Figure 4.11a is obtained.

---

8 See sections 3.2.1.2 and 4.1.1.
9 See [2.122].

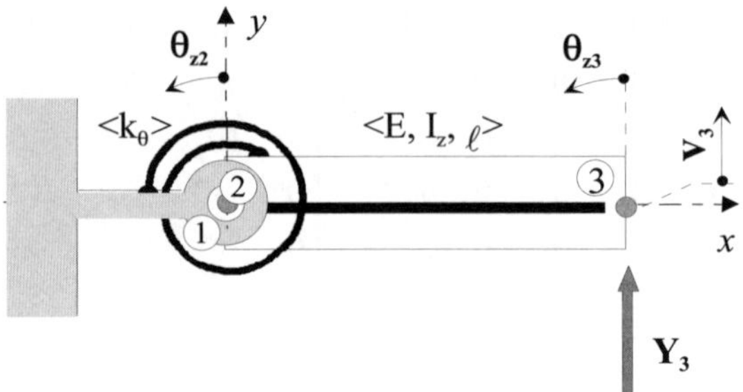

a) properly linked structure

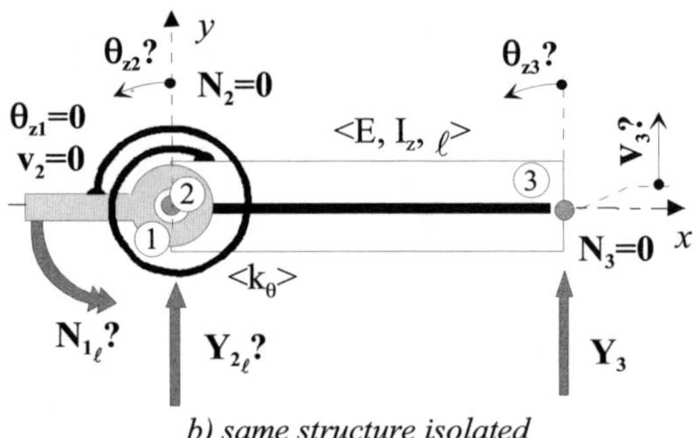

b) same structure isolated

Figure 4.11. *Properly linked structure*

Based on the isolated structure of Figure 4.11b, behavior equation [4.5] becomes:

$$\{F\}_{str} = [K]_{str} \bullet \{d\}_{str}$$

$$\Updownarrow$$

$$
\begin{Bmatrix} N_{1\ell} \\ Y_{2\ell} \\ N_2 = 0 \\ Y_3 \\ N_3 = 0 \end{Bmatrix}
=
\begin{bmatrix}
k_\theta & 0 & -k_\theta & 0 & 0 \\
0 & \dfrac{12EI}{\ell^3} & \dfrac{6EI}{\ell^2} & -\dfrac{12EI}{\ell^3} & \dfrac{6EI}{\ell^2} \\
-k_\theta & \dfrac{6EI}{\ell^2} & \left(k_\theta + \dfrac{4EI}{\ell}\right) & -\dfrac{6EI}{\ell^2} & \dfrac{2EI}{\ell} \\
0 & -\dfrac{12EI}{\ell^3} & -\dfrac{6EI}{\ell^2} & \dfrac{12EI}{\ell^3} & -\dfrac{6EI}{\ell^2} \\
0 & \dfrac{6EI}{\ell^2} & \dfrac{2EI}{\ell} & -\dfrac{6EI}{\ell^2} & \dfrac{4EI}{\ell}
\end{bmatrix}
\bullet
\begin{Bmatrix} \theta_{z1} = 0 \\ v_2 = 0 \\ \theta_{z2} \\ v_3 \\ \theta_{z3} \end{Bmatrix}
\qquad [4.6]
$$

in which $N_{1\ell}$ and $Y_{2\ell}$ are now mechanical linking actions transmitted by the left side of the pivot linkage to the axis of this link. By following the procedure described in section 3.4.2, we start by deleting the two rows corresponding to dof $\theta_{z1} = v_2 = 0$ and the two columns of the same row.

$$\left\{\begin{array}{c} 0 \\ Y_3 \\ 0 \end{array}\right\} = \left[\begin{array}{ccc} \left(k_\theta + \dfrac{4EI}{\ell}\right) & -\dfrac{6EI}{\ell^2} & \dfrac{2EI}{\ell} \\ -\dfrac{6EI}{\ell^2} & \dfrac{12EI}{\ell^3} & -\dfrac{6EI}{\ell^2} \\ \dfrac{2EI}{\ell} & -\dfrac{6EI}{\ell^2} & \dfrac{4EI}{\ell} \end{array}\right] \bullet \left\{\begin{array}{c} \theta_{z2} \\ v_3 \\ \theta_{z3} \end{array}\right\}$$

by inverting this "sub-matrix" we obtain:

$$\left\{\begin{array}{c} \theta_{z2} \\ v_3 \\ \theta_{z3} \end{array}\right\} = \left[\begin{array}{ccc} \dfrac{1}{k_\theta} & \dfrac{1}{k_\theta} & \dfrac{1}{k_\theta} \\ \dfrac{\ell}{k_\theta} & \left(\dfrac{\ell^3}{3EI} + \dfrac{\ell^2}{k_\theta}\right) & \left(\dfrac{\ell^2}{2EI} + \dfrac{\ell^2}{k_\theta}\right) \\ \dfrac{1}{k_\theta} & \left(\dfrac{\ell^2}{2EI} + \dfrac{\ell}{k_\theta}\right) & \left(\dfrac{\ell}{EI} + \dfrac{1}{k_\theta}\right) \end{array}\right] \bullet \left\{\begin{array}{c} 0 \\ Y_3 \\ 0 \end{array}\right\}$$

which helps in getting the displacements of the nodes that remained free, i.e.:

$$\theta_{z2} = \frac{Y_3 \times \ell}{k_\theta}; \quad v_3 = Y_3\left(\frac{\ell^3}{3EI} + \frac{\ell^2}{k_\theta}\right); \quad \theta_{z3} = Y_3\left(\frac{\ell^2}{2EI} + \frac{\ell}{k_\theta}\right)$$

On reverting to the rows that have been deleted in [4.6] we arrive at:

$$\left\{\begin{array}{c} N_{1\ell} \\ Y_{2\ell} \end{array}\right\} = \left[\begin{array}{ccc} -k_\theta & 0 & 0 \\ \dfrac{6EI}{\ell^2} & -\dfrac{12EI}{\ell^2} & \dfrac{6EI}{\ell^2} \end{array}\right] \bullet \left\{\begin{array}{c} \theta_{z2} \\ v_3 \\ \theta_{z3} \end{array}\right\}$$

The values found for the dof help in writing the equations:

$$N_{1\ell} = -k_\theta \times \theta_{z2} = -Y_3 \times \ell$$

$$Y_{2\ell} = \frac{6EI}{\ell^2}\theta_{z2} - \frac{12EI}{\ell^3}v_3 + \frac{6EI}{\ell^2}\theta_{z3} = -Y_3$$

It should be noted that these results make it possible to verify the global equilibrium of the model.

4.2.3. *Example 3*

> *Object:*

The finite elements outlined in Figure 4.12 are to be assembled.

1) Write the stiffness matrix of the structure after assembly in the global coordinates system (XY), as well as the corresponding behavior equation $\{F\}_{str} = [k]_{str} \bullet \{d\}_{str}$.

2) The beam element under simple plane bending is fixed at its left end as is the top end of the helical spring element. The structure is subjected to a moment $N_2 \vec{z}$ and a force $Y_2 \vec{y}$ at the right end of the beam element. Calculate the nodal displacements and the linking actions.

3) Special case: we suppose that the spring has a great stiffness noted $k = 10^6 \times \dfrac{EI}{\ell^3}$ and that the force along $\vec{y}$ is very strong such that $Y_2 \times \ell = 10^6 \times N_2$. By taking $\Delta = \dfrac{Y_2}{k}$, show that this special case is equivalent to the imposition of a displacement Δ at the right end of the beam element under bending.

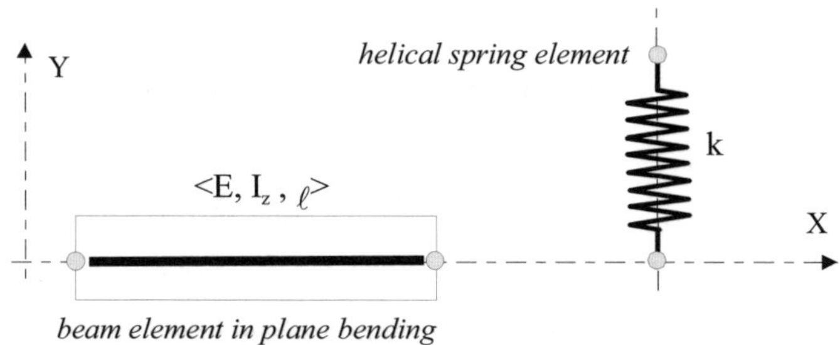

Figure 4.12. *Elements to be assembled*

> *Solution:*

1) The assembly leads to a structure with three nodes, five loads and five associated dof as shown in Figure 4.13.

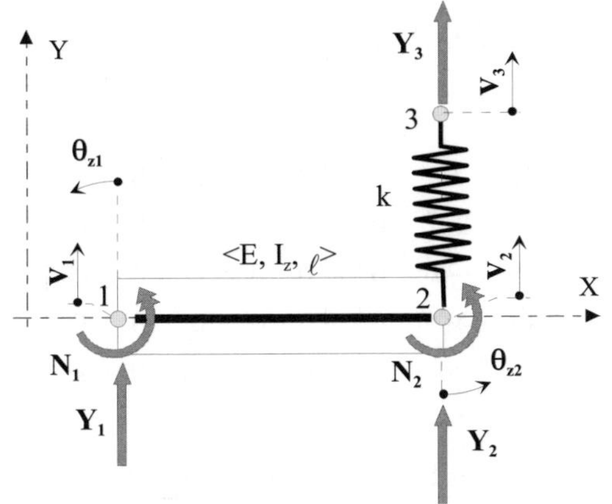

$\{F\}_{str}$	$\{d\}_{str}$
Y_1	v_1
N_1	θ_{z1}
Y_2	v_2
N_2	θ_{z2}
Y_3	v_3

Figure 4.13. *Assembled structure*

The stiffness matrices of the beam element under deflection and of the spring element are known (see [3.24] and [4.1]). We have:

$$E_{\substack{pot.\\structure}} = \frac{1}{2}\{d\}_{str}^{T} \bullet [K]_{str} \bullet \{d\}_{str}$$

$$\Updownarrow$$

$$E_{\substack{pot.\\structure}} = \frac{1}{2}\begin{Bmatrix} v_1 \\ \theta_{z1} \\ v_2 \\ \theta_{z2} \end{Bmatrix}^{T} \bullet \frac{EI}{\ell^3}\begin{bmatrix} 12 & 6\ell & -12 & 6\ell \\ 6\ell & 4\ell^2 & -6\ell & 2\ell^2 \\ -12 & -6\ell & 12 & -6\ell \\ 6\ell & 2\ell^2 & -6\ell & 4\ell^2 \end{bmatrix} \bullet \begin{Bmatrix} v_1 \\ \theta_{z1} \\ v_2 \\ \theta_{z2} \end{Bmatrix}$$

$$+ \frac{1}{2}\begin{Bmatrix} v_2 \\ v_3 \end{Bmatrix}^{T} \bullet \begin{bmatrix} k & -k \\ -k & k \end{bmatrix} \bullet \begin{Bmatrix} v_2 \\ v_3 \end{Bmatrix}$$

which leads to the (5×5) stiffness matrix of the structure[10]. This can also be obtained by reloading the stiffness matrix of each element in a (5×5)[11] table, that is:

---

10 See section 3.3.3.

11 See [3.38].

$$E_{\substack{pot.\\structure}} = \frac{1}{2}\{d\}_{str}^T \bullet [K]_{str} \bullet \{d\}_{str}$$

$$\Updownarrow$$

$$[K]_{str} = \frac{EI}{\ell^3}\begin{bmatrix} 12 & 6\ell & -12 & 6\ell & 0 \\ 6\ell & 4\ell^2 & -6\ell & 2\ell^2 & 0 \\ -12 & -6\ell & 12 & -6\ell & 0 \\ 6\ell & 2\ell^2 & -6\ell & 4\ell^2 & 0 \\ 0 & 0 & 0 & 0 & 0 \end{bmatrix} + \begin{bmatrix} 0 & 0 & 0 & 0 & 0 \\ 0 & 0 & 0 & 0 & 0 \\ 0 & 0 & k & 0 & -k \\ 0 & 0 & 0 & 0 & 0 \\ 0 & 0 & -k & 0 & k \end{bmatrix} \begin{matrix} \cdots\cdots v_1 \\ \cdots\cdots \theta_{z1} \\ \cdots\cdots v_2 \\ \cdots\cdots \theta_{z2} \\ \cdots\cdots v_3 \end{matrix}$$

$$[K]_{str} = \begin{bmatrix} \dfrac{12EI}{\ell^3} & \dfrac{6EI}{\ell^2} & -\dfrac{12EI}{\ell^3} & \dfrac{6EI}{\ell^2} & 0 \\ \dfrac{6EI}{\ell^2} & \dfrac{4EI}{\ell} & -\dfrac{6EI}{\ell^2} & \dfrac{2EI}{\ell} & 0 \\ -\dfrac{12EI}{\ell^3} & -\dfrac{6EI}{\ell^2} & \left(\dfrac{12EI}{\ell^3}+k\right) & -\dfrac{6EI}{\ell^2} & -k \\ \dfrac{6EI}{\ell^2} & \dfrac{2EI}{\ell} & -\dfrac{6EI}{\ell^2} & \dfrac{4EI}{\ell} & 0 \\ 0 & 0 & -k & 0 & k \end{bmatrix}$$

and the following behavior equation is obtained[12]:

$$\{F\}_{str} = [K]_{str} \bullet \{d\}_{str} \Leftrightarrow \begin{Bmatrix} Y_1 \\ N_1 \\ Y_2 \\ N_2 \\ Y_3 \end{Bmatrix} = [K]_{str} \bullet \begin{Bmatrix} v_1 \\ \theta_{z1} \\ v_2 \\ \theta_{z2} \\ v_3 \end{Bmatrix} \qquad [4.7]$$

$$(5\times5)$$

It is observed here that the structure "floats" (two rigid-body movements, a translation along $\vec{y}$ and a rotation around $\vec{z}$ are possible in the (XY) plane in the direction of the loads). The previous matrix $[K]_{str}$ is therefore singular and thus cannot be inverted[1].

2) Let us fix the beam element under bending at the left and the spring element at the top end. The structure is subjected to N_2 and Y_2. The diagram in Figure 4.14a is obtained.

---

12 See [2.122].

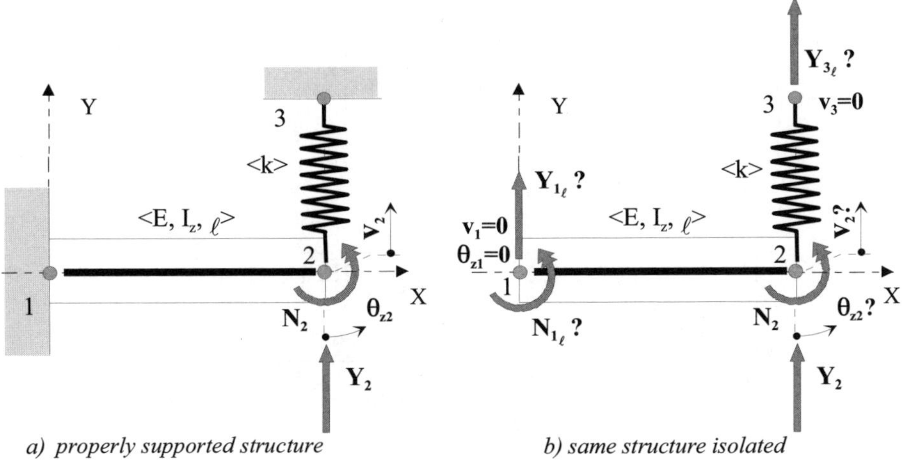

a) properly supported structure   b) same structure isolated

Figure 4.14. *Properly linked structure*

Based on the isolated structure of Figure 4.14b, behavior equation [4.7] becomes[13]:

$$\{F\}_{str} = [K]_{str} \bullet \{d\}_{str}$$

$$\Updownarrow$$

$$
\begin{Bmatrix} Y_{1\ell} \\ N_{1\ell} \\ Y_2 \\ N_2 \\ Y_{3\ell} \end{Bmatrix}
=
\begin{bmatrix}
\dfrac{12EI}{\ell^3} & \dfrac{6EI}{\ell^2} & -\dfrac{12EI}{\ell^3} & \dfrac{6EI}{\ell^2} & 0 \\[2ex]
\dfrac{6EI}{\ell^2} & \dfrac{4EI}{\ell} & -\dfrac{6EI}{\ell^2} & \dfrac{2EI}{\ell} & 0 \\[2ex]
-\dfrac{12EI}{\ell^3} & -\dfrac{6EI}{\ell^2} & \left(\dfrac{12EI}{\ell^3}+k\right) & -\dfrac{6EI}{\ell^2} & -k \\[2ex]
\dfrac{6EI}{\ell^2} & \dfrac{2EI}{\ell} & -\dfrac{6EI}{\ell^2} & \dfrac{4EI}{\ell} & 0 \\[2ex]
0 & 0 & -k & 0 & k
\end{bmatrix}
\bullet
\begin{Bmatrix} v_1 = 0 \\ \theta_{z1} = 0 \\ v_2 \\ \theta_{z2} \\ v_3 = 0 \end{Bmatrix}
\quad [4.8]
$$

in which $Y_{1\ell}$, $N_{1\ell}$ and $Y_{3\ell}$ are mechanical linking actions transmitted by the surroundings to the structure. By following the procedure given in section 3.4.2., the three rows corresponding to the zero dof and the three columns of the same row are deleted. We are left with:

---

13 See section 3.4.2.

$$\begin{Bmatrix} Y_2 \\ N_2 \end{Bmatrix} = \begin{bmatrix} \left(\dfrac{12EI}{\ell^3} + k \right) & -\dfrac{6EI}{\ell^2} \\ -\dfrac{6EI}{\ell^2} & \dfrac{4EI}{\ell} \end{bmatrix} \bullet \begin{Bmatrix} v_2 \\ \theta_{z2} \end{Bmatrix}$$

By inverting this we obtain:

$$\begin{Bmatrix} v_2 \\ \theta_{z2} \end{Bmatrix} = \begin{bmatrix} \dfrac{1}{\left(k + \dfrac{3EI}{\ell^3} \right)} & \dfrac{3}{2} \dfrac{1}{\left(k\ell + \dfrac{3EI}{\ell^2} \right)} \\ \dfrac{3}{2} \dfrac{1}{\left(k\ell + \dfrac{3EI}{\ell^2} \right)} & \dfrac{\ell}{4EI} + \dfrac{9}{4\left(k\ell^2 + \dfrac{3EI}{\ell} \right)} \end{bmatrix} \bullet \begin{Bmatrix} Y_2 \\ N_2 \end{Bmatrix}$$

that makes it possible to determine displacements of the nodes that have remained free, i.e.:

$$v_2 = \frac{Y_2}{\left(k + \dfrac{3EI}{\ell} \right)} + \frac{3}{2} \frac{N_2}{\left(k\ell + \dfrac{3EI}{\ell^2} \right)}$$

$$\theta_{z2} = \frac{3}{2} \frac{Y_2}{\left(k\ell + \dfrac{3EI}{\ell^2} \right)} + N_2 \left[\frac{\ell}{4EI} + \frac{9}{4\left(k\ell^2 + \dfrac{3EI}{\ell} \right)} \right] \qquad [4.9]$$

By reverting to the rows that have been deleted in [4.8], i.e. to the system:

$$\begin{Bmatrix} Y_{1\ell} \\ N_{1\ell} \\ Y_{3\ell} \end{Bmatrix} = \begin{bmatrix} -\dfrac{12EI}{\ell^3} & \dfrac{6EI}{\ell^2} \\ -\dfrac{6EI}{\ell^2} & \dfrac{2EI}{\ell} \\ -k & 0 \end{bmatrix} \bullet \begin{Bmatrix} v_2 \\ \theta_{z2} \end{Bmatrix}$$

the known values v_2 and θ_{z2} of the free dof help in writing:

$$Y_{1\ell} = -\frac{12EI}{\ell^3}v_2 + \frac{6EI}{\ell^2}\theta_{z2}$$

$$Y_{1\ell} = Y_2\left[\frac{9}{\left(\dfrac{k\ell^3}{EI}+3\right)} - \frac{1}{\left(\dfrac{k\ell^3}{12EI}+\dfrac{1}{4}\right)}\right] + 3N_2\left[\frac{1}{2\ell} + \frac{9}{\left(\dfrac{2k\ell^4}{EI}+6\ell\right)} - \frac{6}{\left(\dfrac{k\ell^4}{EI}+3\ell\right)}\right]$$

$$N_{1\ell} = -\frac{6EI}{\ell^2}v_2 + \frac{2EI}{\ell}\theta_{z2}$$

$$N_{1\ell} = Y_2\left[\frac{1}{\left(\dfrac{k\ell^2}{3EI}+\dfrac{1}{\ell}\right)} - \frac{1}{\left(\dfrac{k\ell^2}{6EI}+\dfrac{1}{2\ell}\right)}\right] + N_2\left[\frac{1}{2} + \frac{9}{\left(\dfrac{2k\ell^3}{EI}+6\right)} - \frac{1}{\left(\dfrac{k\ell^3}{9EI}+\dfrac{1}{3}\right)}\right]$$

$$Y_{3\ell} = -kv_2 = -\frac{Y_2}{\left(\dfrac{3EI}{k\ell^3}+1\right)} - \frac{3}{2}\frac{N_2}{\left(\dfrac{3EI}{k\ell^2}+\ell\right)}$$

3) Special case: $k = 10^6 \times \dfrac{EI}{\ell^3}$, $Y_2 \times \ell = 10^6 \times N_2$, $\Delta = \dfrac{Y_2}{k}$. Denoting negligible quantities by ε, ε', $\varepsilon'' \ll 1$, displacements [4.9] can be rewritten:

$$v_2' = \frac{Y_2}{k(1+\varepsilon)} + \varepsilon'\frac{Y_2}{k(1+\varepsilon)} \cong \frac{Y_2}{k} = \Delta$$

$$\theta_{z2}' = \frac{3}{2\ell}\frac{Y_2}{k(1+\varepsilon)} + N_2\frac{\ell}{4EI} + \frac{\varepsilon''}{\ell}\frac{Y_2}{k(1+\varepsilon)} \cong \frac{3}{2}\frac{\Delta}{\ell} + N_2\frac{\ell}{4EI}$$

By placing a spring of very high rigidity k and by exerting a very strong force Y_2 in order to obtain a required value $\Delta = \dfrac{Y_2}{k}$, it can be observed that everything happens as if a displacement Δ had been imposed at the end of the fixed beam element under bending before exerting a moment N_2. This is illustrated in Figure 4.15[14].

---

14 With the results of the example given in section 2.6.3.2 it can be verified that by applying a force at the end of a comparable cantilever beam, the resultant rotation and deflection at the extreme section are in the ratio: $\dfrac{\theta_z}{v} = \dfrac{3}{2\ell}$ (see [2.102]).

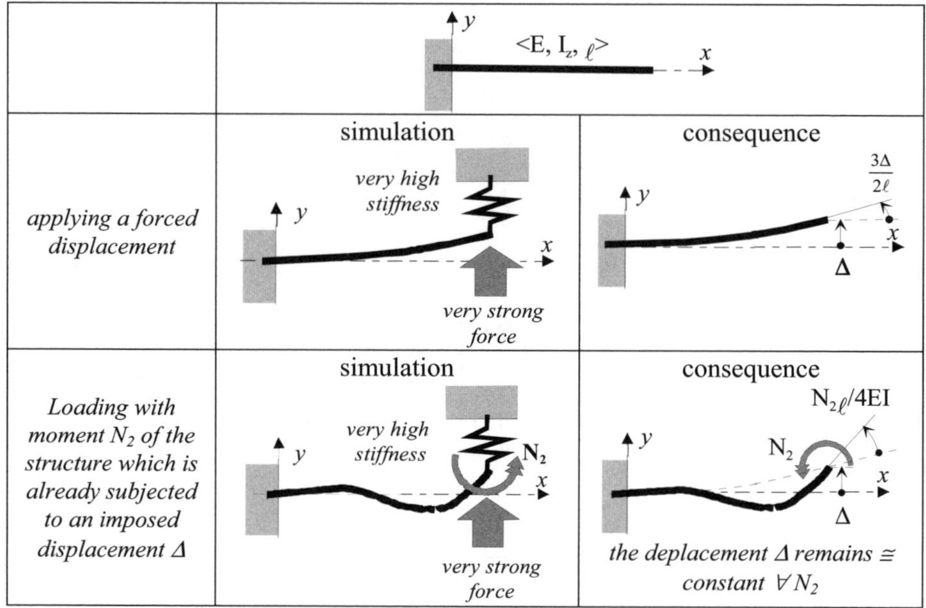

Figure 4.15. *Imposing a displacement*

NOTES

❏ This "stratagem" helps in imposing (linear or angular) displacements at specified places of the structures. These displacements remain unaffected by the intensity of the "real" loads that can be applied thereafter.

❏ This method is commonly used in calculation codes to impose non-zero boundary displacement-conditions.

4.2.4. *Assembly of a truss element and a beam element under simple plane bending*

➤ *Object:*

The finite elements for assembly are indicated in Figure 4.16.

Write the stiffness matrix of the structure obtained after assembly.

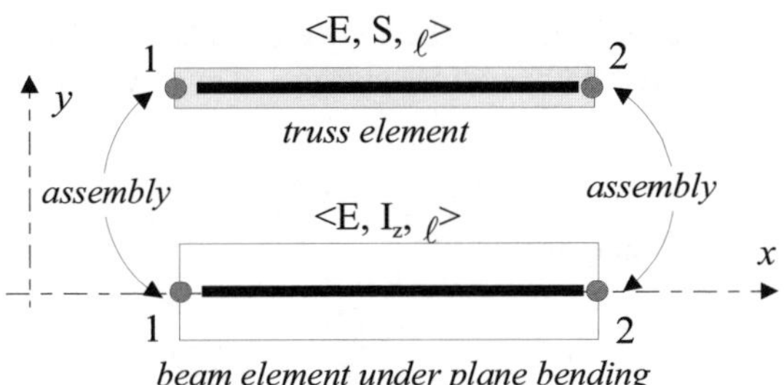

Figure 4.16. *Elements to be assembled*

➢ *Solution:*

The assembly leads to a structure with two nodes, six loads and six associated dof as in Figure 4.17. The stiffness matrix of this structure therefore has six rows and six columns.

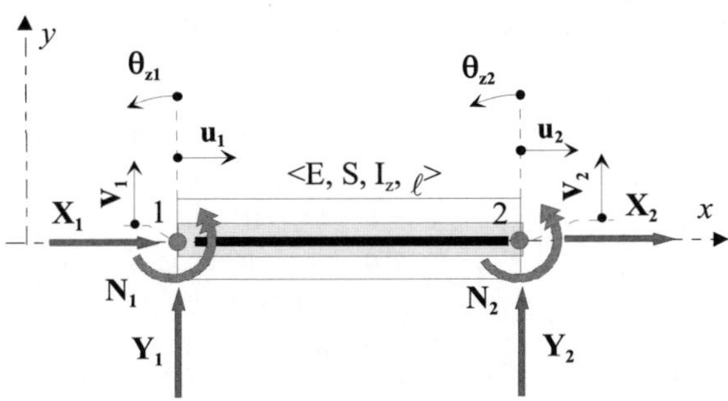

$\{F\}_{str}$	$\{d\}_{str}$
X_1	u_1
Y_1	v_1
N_1	θ_{z1}
X_2	u_2
Y_2	v_2
N_2	θ_{z2}

Figure 4.17. *Assembled structure*

The stiffness matrices of each of the two elements (see [3.12] and [3.24]) are known, as are the associated deformation energies. Thus, we have:

$$E_{\substack{pot. \\ structure}} = \frac{1}{2}\{d\}_{str}^{T} \bullet [K]_{str} \bullet \{d\}_{str}$$

$$\Updownarrow$$

$$E_{\substack{pot. \\ structure}} = \frac{1}{2}\begin{Bmatrix} u_1 \\ u_2 \end{Bmatrix}^{T} \bullet \begin{bmatrix} \dfrac{ES}{\ell} & -\dfrac{ES}{\ell} \\ -\dfrac{ES}{\ell} & \dfrac{ES}{\ell} \end{bmatrix} \bullet \begin{Bmatrix} u_1 \\ u_2 \end{Bmatrix}$$

$$+ \frac{1}{2}\begin{Bmatrix} v_1 \\ \theta_{z1} \\ v_2 \\ \theta_{z2} \end{Bmatrix}^{T} \bullet \frac{EI}{\ell^3}\begin{bmatrix} 12 & 6\ell & -12 & 6\ell \\ 6\ell & 4\ell^2 & -6\ell & 2\ell^2 \\ -12 & -6\ell & 12 & -6\ell \\ 6\ell & 2\ell^2 & -6\ell & 4\ell^2 \end{bmatrix} \bullet \begin{Bmatrix} v_1 \\ \theta_{z1} \\ v_2 \\ \theta_{z2} \end{Bmatrix}$$

which leads to the (6×6) stiffness matrix of the structure[15]. This can also appear by reloading the stiffness matrix of each element in a (6×6) table, i.e.[16]:

$$[K]_{str} = \begin{bmatrix} \dfrac{ES}{\ell} & 0 & 0 & -\dfrac{ES}{\ell} & 0 & 0 \\ 0 & 0 & 0 & 0 & 0 & 0 \\ 0 & 0 & 0 & 0 & 0 & 0 \\ -\dfrac{ES}{\ell} & 0 & 0 & \dfrac{ES}{\ell} & 0 & 0 \\ 0 & 0 & 0 & 0 & 0 & 0 \\ 0 & 0 & 0 & 0 & 0 & 0 \end{bmatrix} + \frac{EI}{\ell^3}\begin{bmatrix} 0 & 0 & 0 & 0 & 0 & 0 \\ 0 & 12 & 6\ell & 0 & -12 & 6\ell \\ 0 & 6\ell & 4\ell^2 & 0 & -6\ell & 2\ell^2 \\ 0 & 0 & 0 & 0 & 0 & 0 \\ 0 & -12 & -6\ell & 0 & 12 & -6\ell \\ 0 & 6\ell & 2\ell^2 & 0 & -6\ell & 4\ell^2 \end{bmatrix}$$

reloading of the (6×6) stiffness matrix
of the truss element

reloading of the (6×6) stiffness flexure matrix
of the beam element under deflection

---

15 See section 3.3.3.
16 See [3.38].

$$[K]_{str} = \begin{bmatrix} \dfrac{ES}{\ell} & 0 & 0 & -\dfrac{ES}{\ell} & 0 & 0 \\[2mm] 0 & \dfrac{12EI}{\ell^3} & \dfrac{6EI}{\ell^2} & 0 & -\dfrac{12EI}{\ell^3} & \dfrac{6EI}{\ell^2} \\[2mm] 0 & \dfrac{6EI}{\ell^2} & \dfrac{4EI}{\ell} & 0 & -\dfrac{6EI}{\ell^2} & \dfrac{2EI}{\ell} \\[2mm] -\dfrac{ES}{\ell} & 0 & 0 & \dfrac{ES}{\ell} & 0 & 0 \\[2mm] 0 & -\dfrac{12EI}{\ell^3} & -\dfrac{6EI}{\ell^2} & 0 & \dfrac{12EI}{\ell^3} & -\dfrac{6EI}{\ell^2} \\[2mm] 0 & \dfrac{6EI}{\ell^2} & \dfrac{2EI}{\ell} & 0 & -\dfrac{6EI}{\ell^2} & \dfrac{4EI}{\ell} \end{bmatrix} \qquad [4.10]$$

stiffness matrix (6×6) of the equivalent beam element

and we obtain the behavior equation[17]:

$$\{F\}_{str} = [K]_{str} \bullet \{d\}_{str} \iff \begin{Bmatrix} X_1 \\ Y_1 \\ N_1 \\ X_1 \\ Y_2 \\ N_2 \end{Bmatrix} = [K]_{str} \underset{(6\times6)}{\bullet} \begin{Bmatrix} u_1 \\ v_1 \\ \theta_{z1} \\ u_2 \\ v_2 \\ \theta_{z2} \end{Bmatrix}$$

NOTES

❏ Equation [4.10] shows that the stiffness matrices of the two basic elements are just added up without really being combined, because of the fact that their initial dof are different.

❏ This association provides an equivalent beam element (Figure 4.18) characterized by a simple plane bending behavior combined with a traction-compression behavior. It is therefore more efficient than each of its basic constituent elements.

---

17 See [2.122].

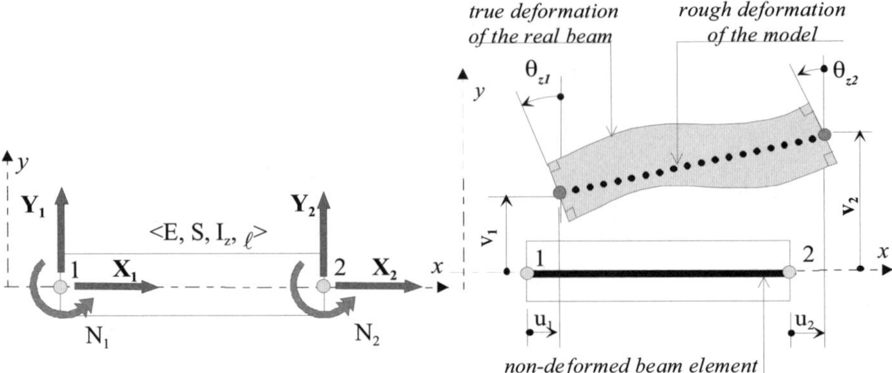

Figure 4.18. *Equivalent beam element (plane bending + traction-compression)*

4.3. Behavior in the global coordinate system

4.3.1. *Plane assembly of two truss elements*

> *Object:*

Consider the plane assembly of two truss elements (same material and same section) represented in Figure 4.19 giving a loaded structure as indicated. The two truss elements have $\vec{Z}$ axis pivot-linkage with the surroundings:

1) Write the dof and the stiffness matrices of each element in the global coordinate system (XY).

2) Write the stiffness matrix of the structure in the global coordinate system.

3) Calculate the nodal displacements and the linking actions in the global coordinate system by taking into account the links and the indicated load.

4) Calculate the displacements and the mechanical nodal actions in the local coordinate systems.

5) Thus, derive the normal stress in each of the elements.

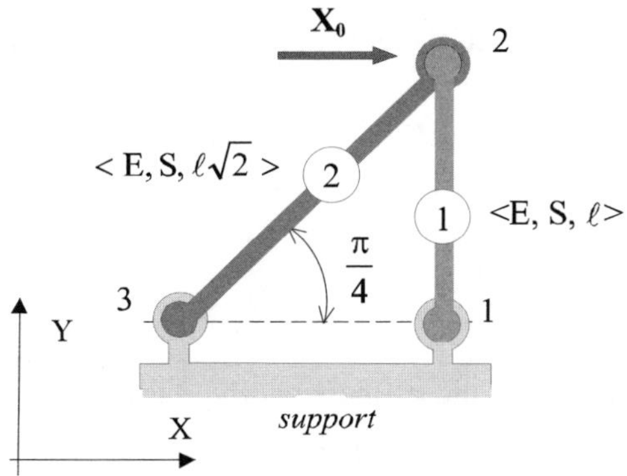

Figure 4.19. *Assembled structure under loading*

> *Solution:*

1) Global coordinate system:

♦ *Element 1*: from Figure 4.20, the following relation: $\vec{x_1} = \vec{Y}$, $\vec{y_1} = -\vec{X}$ between the local and the global coordinate system is observed.

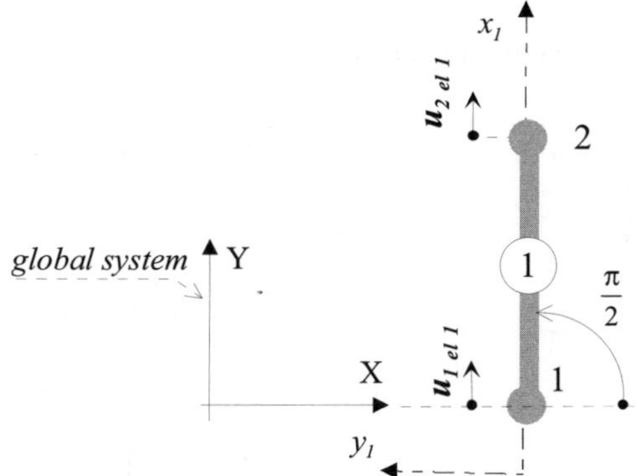

Figure 4.20. *Truss element 1*

i.e.:

$$\left\{ \begin{array}{c} \vec{x_1} \\ \vec{y_1} \end{array} \right\} = \left[\begin{array}{cc} 0 & 1 \\ -1 & 0 \end{array} \right] \bullet \left\{ \begin{array}{c} \vec{X} \\ \vec{Y} \end{array} \right\}$$

By reverting to the equation in Figure 3.8 with $\alpha = \dfrac{\pi}{2}$, we have:

$$\begin{Bmatrix} u_1 \\ v_1 \\ u_2 \\ v_2 \end{Bmatrix}_{\substack{el.1 \\ Local}} = \begin{bmatrix} 0 & 1 & 0 & 0 \\ -1 & 0 & 0 & 0 \\ 0 & 0 & 0 & 1 \\ 0 & 0 & -1 & 0 \end{bmatrix} \bullet \begin{Bmatrix} u_1 \\ v_1 \\ u_2 \\ v_2 \end{Bmatrix}_{\substack{el.1 \\ Global}} = [P]_{\acute{e}l1} \bullet \begin{Bmatrix} u_1 \\ v_1 \\ u_2 \\ v_2 \end{Bmatrix}_{\substack{el.1 \\ Global}} \qquad [4.11]$$

Let us take the equation of the stiffness matrix in the local system, i.e.[18]:

$$[k]_{\substack{\acute{e}l1 \\ Local}} = \begin{bmatrix} \dfrac{ES}{\ell} & -\dfrac{ES}{\ell} \\ -\dfrac{ES}{\ell} & \dfrac{ES}{\ell} \end{bmatrix} \begin{matrix} \cdots\, u_1 \\ \cdots\, u_2 \end{matrix} = \begin{bmatrix} \dfrac{ES}{\ell} & 0 & -\dfrac{ES}{\ell} & 0 \\ 0 & 0 & 0 & 0 \\ -\dfrac{ES}{\ell} & 0 & \dfrac{ES}{\ell} & 0 \\ 0 & 0 & 0 & 0 \end{bmatrix} \begin{matrix} \cdots\, u_1 \\ \cdots\, v_1 = 0 \\ \cdots\, u_2 \\ \cdots\, v_2 = 0 \end{matrix}$$

It has been observed that $[k]_{\substack{el \\ Global}}$ is obtained through the operation[19]:

$$[k]_{\substack{el.1 \\ Global}} = [P]_{el.1}^{T} \bullet [k]_{\substack{el.1 \\ Local}} \bullet [P]_{el.1}$$

from where using [4.11] we obtain successively:

$$[k]_{\substack{el.1 \\ Local}} \bullet [P]_{el.1} = \begin{bmatrix} \dfrac{ES}{\ell} & 0 & -\dfrac{ES}{\ell} & 0 \\ 0 & 0 & 0 & 0 \\ -\dfrac{ES}{\ell} & 0 & \dfrac{ES}{\ell} & 0 \\ 0 & 0 & 0 & 0 \end{bmatrix} \bullet \begin{bmatrix} 0 & 1 & 0 & 0 \\ -1 & 0 & 0 & 0 \\ 0 & 0 & 0 & 1 \\ 0 & 0 & -1 & 0 \end{bmatrix} = \begin{bmatrix} 0 & \dfrac{ES}{\ell} & 0 & -\dfrac{ES}{\ell} \\ 0 & 0 & 0 & 0 \\ 0 & -\dfrac{ES}{\ell} & 0 & \dfrac{ES}{\ell} \\ 0 & 0 & 0 & 0 \end{bmatrix}$$

$$[k]_{\substack{el.1 \\ Global}} = [P]_{el.1}^{T} \bullet [k]_{\substack{el.1 \\ Local}} \bullet [P]_{el.1} = \begin{bmatrix} 0 & -1 & 0 & 0 \\ 1 & 0 & 0 & 0 \\ 0 & 0 & 0 & -1 \\ 0 & 0 & 1 & 0 \end{bmatrix} \bullet \begin{bmatrix} 0 & \dfrac{ES}{\ell} & 0 & -\dfrac{ES}{\ell} \\ 0 & 0 & 0 & 0 \\ 0 & -\dfrac{ES}{\ell} & 0 & \dfrac{ES}{\ell} \\ 0 & 0 & 0 & 0 \end{bmatrix}$$

---

18 See Figure 3.8.
19 See Figure 3.8.

$$
[k]_{\substack{el.1 \\ Global}} =
\begin{bmatrix}
0 & 0 & 0 & 0 \\
0 & \dfrac{ES}{\ell} & 0 & -\dfrac{ES}{\ell} \\
0 & 0 & 0 & 0 \\
0 & -\dfrac{ES}{\ell} & 0 & \dfrac{ES}{\ell}
\end{bmatrix}
$$

◆ *Element 2*: the following relation between the local and the global coordinate system is observed (Figure 4.21): $\vec{x_2} = -\dfrac{\sqrt{2}}{2}\vec{X} - \dfrac{\sqrt{2}}{2}\vec{Y}$, $\vec{y_2} = \dfrac{\sqrt{2}}{2}\vec{X} - \dfrac{\sqrt{2}}{2}\vec{Y}$ i.e.:

$$
\begin{Bmatrix} \vec{x_2} \\ \vec{y_2} \end{Bmatrix} =
\begin{bmatrix} -\dfrac{\sqrt{2}}{2} & -\dfrac{\sqrt{2}}{2} \\ \dfrac{\sqrt{2}}{2} & -\dfrac{\sqrt{2}}{2} \end{bmatrix} \bullet
\begin{Bmatrix} \vec{X} \\ \vec{Y} \end{Bmatrix} =
\dfrac{\sqrt{2}}{2}\begin{bmatrix} -1 & -1 \\ 1 & -1 \end{bmatrix} \bullet
\begin{Bmatrix} \vec{X} \\ \vec{Y} \end{Bmatrix}
$$

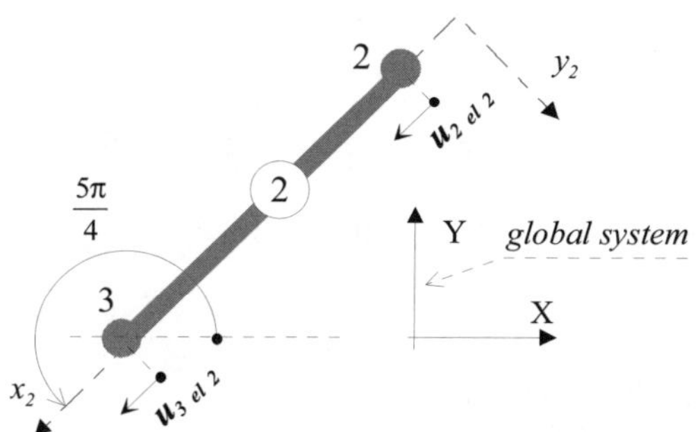

Figure 4.21. *Truss element 2*

We return to the equation in Figure 3.8, with $\alpha = \dfrac{5\pi}{4}$, i.e.:

$$
\begin{Bmatrix} u_2 \\ v_2 \\ u_3 \\ v_3 \end{Bmatrix}_{\substack{el.2 \\ Local}} =
\dfrac{\sqrt{2}}{2}
\left[\begin{array}{cc:cc}
-1 & -1 & 0 & 0 \\
1 & -1 & 0 & 0 \\ \hdashline
0 & 0 & -1 & -1 \\
0 & 0 & 1 & -1
\end{array}\right] \bullet
\begin{Bmatrix} u_2 \\ v_2 \\ u_3 \\ v_3 \end{Bmatrix}_{\substack{el.2 \\ Global}} =
[P]_{el.2} \bullet
\begin{Bmatrix} u_2 \\ v_2 \\ u_3 \\ v_3 \end{Bmatrix}_{\substack{el.2 \\ Global}}
\qquad [4.12]
$$

The stiffness matrix in the local system is written as:

$$\begin{bmatrix} k \end{bmatrix}_{\substack{\text{el.2} \\ \text{Local}}} = \begin{bmatrix} \dfrac{ES}{\ell} & -\dfrac{ES}{\ell} \\ -\dfrac{ES}{\ell} & \dfrac{ES}{\ell} \end{bmatrix} \begin{matrix} \cdots\, u_2 \\ \cdots\, u_3 \end{matrix} = \begin{bmatrix} \dfrac{ES}{\ell} & 0 & -\dfrac{ES}{\ell} & 0 \\ 0 & 0 & 0 & 0 \\ -\dfrac{ES}{\ell} & 0 & \dfrac{ES}{\ell} & 0 \\ 0 & 0 & 0 & 0 \end{bmatrix} \begin{matrix} \cdots\, u_2 \\ \cdots\, v_2 = 0 \\ \cdots\, u_3 \\ \cdots\, v_3 = 0 \end{matrix}$$

Let us calculate the stiffness matrix in the global system. With [4.12] we have successively:

$$\begin{bmatrix} k \end{bmatrix}_{\substack{\text{el.2} \\ \text{Local}}} \bullet \begin{bmatrix} P \end{bmatrix}_{\text{el.2}} = \begin{bmatrix} \dfrac{ES}{\ell} & 0 & -\dfrac{ES}{\ell} & 0 \\ 0 & 0 & 0 & 0 \\ -\dfrac{ES}{\ell} & 0 & \dfrac{ES}{\ell} & 0 \\ 0 & 0 & 0 & 0 \end{bmatrix} \bullet \dfrac{\sqrt{2}}{2} \begin{bmatrix} -1 & -1 & 0 & 0 \\ 1 & -1 & 0 & 0 \\ 0 & 0 & -1 & -1 \\ 0 & 0 & 1 & -1 \end{bmatrix}$$

$$= \dfrac{\sqrt{2}}{2} \begin{bmatrix} -\dfrac{ES}{\ell} & -\dfrac{ES}{\ell} & \dfrac{ES}{\ell} & \dfrac{ES}{\ell} \\ 0 & 0 & 0 & 0 \\ \dfrac{ES}{\ell} & \dfrac{ES}{\ell} & -\dfrac{ES}{\ell} & -\dfrac{ES}{\ell} \\ 0 & 0 & 0 & 0 \end{bmatrix}$$

Then[20]:

$$\begin{bmatrix} k \end{bmatrix}_{\substack{\text{el.2} \\ \text{Global}}} = \begin{bmatrix} P \end{bmatrix}_{\text{el.2}}^{\mathrm{T}} \bullet \begin{bmatrix} k \end{bmatrix}_{\substack{\text{el.2} \\ \text{Local}}} \bullet \begin{bmatrix} P \end{bmatrix}_{\text{el.2}}$$

$$= \dfrac{\sqrt{2}}{2} \begin{bmatrix} -1 & 1 & 0 & 0 \\ -1 & -1 & 0 & 0 \\ 0 & 0 & -1 & 1 \\ 0 & 0 & -1 & -1 \end{bmatrix} \bullet \dfrac{\sqrt{2}}{2} \begin{bmatrix} -\dfrac{ES}{\ell} & -\dfrac{ES}{\ell} & \dfrac{ES}{\ell} & \dfrac{ES}{\ell} \\ 0 & 0 & 0 & 0 \\ \dfrac{ES}{\ell} & \dfrac{ES}{\ell} & -\dfrac{ES}{\ell} & -\dfrac{ES}{\ell} \\ 0 & 0 & 0 & 0 \end{bmatrix}$$

---

20 See Figure 3.8.

$$[k]_{\substack{el.2 \\ Global}} = \frac{1}{2} \begin{bmatrix} \dfrac{ES}{\ell} & \dfrac{ES}{\ell} & -\dfrac{ES}{\ell} & -\dfrac{ES}{\ell} \\[2mm] \dfrac{ES}{\ell} & \dfrac{ES}{\ell} & -\dfrac{ES}{\ell} & -\dfrac{ES}{\ell} \\[2mm] -\dfrac{ES}{\ell} & -\dfrac{ES}{\ell} & \dfrac{ES}{\ell} & \dfrac{ES}{\ell} \\[2mm] -\dfrac{ES}{\ell} & -\dfrac{ES}{\ell} & \dfrac{ES}{\ell} & \dfrac{ES}{\ell} \end{bmatrix}$$

2) Stiffness matrix of the structure in the global system:

The total number of dof in the structure is $3 \times 2 = 6$ (three nodes, each having two dof in the global system); for example, we have the list:

$$\{d\}_{\substack{str \\ Global}}^{T} = \begin{bmatrix} u_1 & v_1 & u_2 & v_2 & u_3 & v_3 \end{bmatrix}_G$$

Let us reload the stiffness matrices written previously for each of the elements in a (6×6) table (dimensions of the global stiffness matrix).

The following[21] sum is obtained in the global system:

$$[k]_{\substack{str \\ Global}} = \begin{bmatrix} 0 & 0 & 0 & 0 & 0 & 0 \\ 0 & \dfrac{ES}{\ell} & 0 & -\dfrac{ES}{\ell} & 0 & 0 \\ 0 & 0 & 0 & 0 & 0 & 0 \\ 0 & -\dfrac{ES}{\ell} & 0 & \dfrac{ES}{\ell} & 0 & 0 \\ 0 & 0 & 0 & 0 & 0 & 0 \\ 0 & 0 & 0 & 0 & 0 & 0 \end{bmatrix} + \begin{bmatrix} 0 & 0 & 0 & 0 & 0 & 0 \\ 0 & 0 & 0 & 0 & 0 & 0 \\ 0 & 0 & \dfrac{ES}{2\ell} & \dfrac{ES}{2\ell} & -\dfrac{ES}{2\ell} & -\dfrac{ES}{2\ell} \\ 0 & 0 & \dfrac{ES}{2\ell} & \dfrac{ES}{2\ell} & -\dfrac{ES}{2\ell} & -\dfrac{ES}{2\ell} \\ 0 & 0 & -\dfrac{ES}{2\ell} & -\dfrac{ES}{2\ell} & \dfrac{ES}{2\ell} & \dfrac{ES}{2\ell} \\ 0 & 0 & -\dfrac{ES}{2\ell} & -\dfrac{ES}{2\ell} & \dfrac{ES}{2\ell} & \dfrac{ES}{2\ell} \end{bmatrix} \begin{matrix} \cdots u_1 \\ \cdots v_1 \\ \cdots u_2 \\ \cdots v_2 \\ \cdots u_3 \\ \cdots v_3 \end{matrix}$$

$$[k]_{\substack{str \\ Global}} = \begin{bmatrix} 0 & 0 & 0 & 0 & 0 & 0 \\ 0 & \dfrac{ES}{\ell} & 0 & -\dfrac{ES}{\ell} & 0 & 0 \\ 0 & 0 & \dfrac{ES}{2\ell} & \dfrac{ES}{2\ell} & -\dfrac{ES}{2\ell} & -\dfrac{ES}{2\ell} \\ 0 & -\dfrac{ES}{\ell} & \dfrac{ES}{2\ell} & \dfrac{3ES}{2\ell} & -\dfrac{ES}{2\ell} & -\dfrac{ES}{2\ell} \\ 0 & 0 & -\dfrac{ES}{2\ell} & -\dfrac{ES}{2\ell} & \dfrac{ES}{2\ell} & \dfrac{ES}{2\ell} \\ 0 & 0 & -\dfrac{ES}{2\ell} & -\dfrac{ES}{2\ell} & \dfrac{ES}{2\ell} & \dfrac{ES}{2\ell} \end{bmatrix} \begin{matrix} \cdots u_1 \\ \cdots v_1 \\ \cdots u_2 \\ \cdots v_2 \\ \cdots u_3 \\ \cdots v_3 \end{matrix}$$

---

21 See [3.38].

The behavior equation is written as[22]:

$$\{F\}_{str} = [k]_{str} \bullet \{d\}_{str} \iff \begin{Bmatrix} X_1 \\ Y_1 \\ X_2 \\ Y_2 \\ X_3 \\ Y_3 \end{Bmatrix} = [k]_{str} \bullet \begin{Bmatrix} u_1 \\ v_1 \\ u_2 \\ v_2 \\ u_3 \\ v_3 \end{Bmatrix}$$

Global   Global Global   Global   Global

3) The links imply: $u_1 = v_1 = u_3 = v_3 = 0$. The load is reduced to $\overrightarrow{X_0 X}$ at node (2). The previous behavior equation becomes[23]:

$$\begin{Bmatrix} X_{1\ell} \\ Y_{1\ell} \\ X_2 = X_0 \\ Y_2 = 0 \\ X_{3\ell} \\ Y_{3\ell} \end{Bmatrix}_{Global} = \frac{ES}{\ell} \begin{bmatrix} 0 & 0 & 0 & 0 & 0 & 0 \\ 0 & 1 & 0 & -1 & 0 & 0 \\ 0 & 0 & \dfrac{1}{2} & \dfrac{1}{2} & -\dfrac{1}{2} & -\dfrac{1}{2} \\ 0 & -1 & \dfrac{1}{2} & \dfrac{3}{2} & -\dfrac{1}{2} & -\dfrac{1}{2} \\ 0 & 0 & -\dfrac{1}{2} & -\dfrac{1}{2} & \dfrac{1}{2} & \dfrac{1}{2} \\ 0 & 0 & -\dfrac{1}{2} & -\dfrac{1}{2} & \dfrac{1}{2} & \dfrac{1}{2} \end{bmatrix} \bullet \begin{Bmatrix} u_1 = 0 \\ v_1 = 0 \\ u_2 \\ v_2 \\ u_3 = 0 \\ v_3 = 0 \end{Bmatrix}_{Global}$$

where $X_{1\ell}$, $Y_{1\ell}$, $X_{3\ell}$, $Y_{3\ell}$ are the mechanical linking actions in the global coordinate system.

By following the procedure in section 3.4.2, let us delete the rows corresponding to the blocked dof and the columns of the same row. We are left with:

$$\begin{Bmatrix} X_2 = X_0 \\ 0 \end{Bmatrix}_{Global} = \frac{ES}{\ell} \begin{bmatrix} \dfrac{1}{2} & \dfrac{1}{2} \\ \dfrac{1}{2} & \dfrac{3}{2} \end{bmatrix} \bullet \begin{Bmatrix} u_2 \\ v_2 \end{Bmatrix}_{Global}$$

which after inversion becomes:

$$\begin{Bmatrix} u_2 \\ v_2 \end{Bmatrix}_{Global} = \frac{\ell}{ES} \begin{bmatrix} 3 & -1 \\ -1 & 1 \end{bmatrix} \bullet \begin{Bmatrix} X_0 \\ 0 \end{Bmatrix}_{Global}$$

leading to the values:

---

22 See [2.122].
23 See section 3.4.2.

$$u_2 = 3\frac{X_0\ell}{ES} \; ; \; v_2 = -\frac{X_0\ell}{ES}$$

By reverting to the deleted rows and columns, i.e. to the system:

$$\begin{Bmatrix} X_{1\ell} \\ Y_{1\ell} \\ X_{3\ell} \\ Y_{3\ell} \end{Bmatrix}_{Global} = \frac{ES}{\ell}\begin{bmatrix} 0 & 0 \\ 0 & -1 \\ 1 & 1 \\ -\dfrac{1}{2} & -\dfrac{1}{2} \\ -\dfrac{1}{2} & -\dfrac{1}{2} \end{bmatrix} \bullet \begin{Bmatrix} u_2 \\ v_2 \end{Bmatrix}_{Global}$$

the calculated values u_2 and v_2 give us the linking forces:

$$X_{1\ell} = 0; \quad Y_{1\ell} = X_0; \quad X_{3\ell} = -X_0; \quad Y_{3\ell} = -X_0$$

these forces are shown in Figure 4.22. The correct global equilibrium of the model can thus be readily verified.

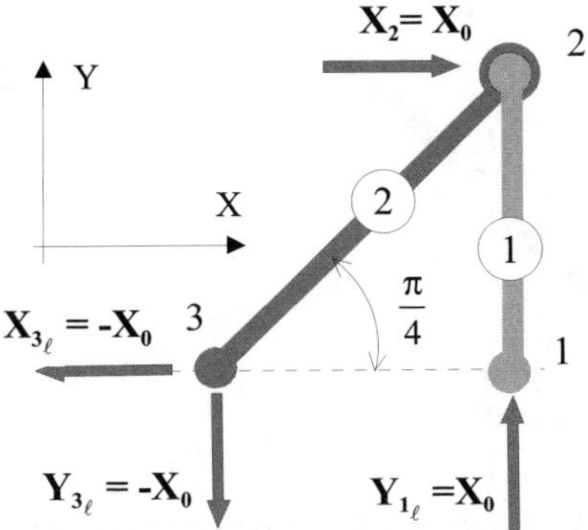

Figure 4.22. *Nodal forces on the structure*

4) Local coordinate systems

➢ Let us calculate the nodal displacements in the local coordinate systems of the elements:

◆ *element 1*: with [4.11]:

$$\underset{el.1}{u_1} = v_1 = 0 \; ; \; \underset{el.1}{v_1} = -u_1 = 0$$

$$\underset{el.1}{u_2} = v_2 = -\frac{X_0\ell}{ES} \; ; \; \underset{el.1}{v_2} = -u_2 = -3\frac{X_0\ell}{ES}$$

• *element 2*: with [4.12]:

$$u_2_{\text{el.2}} = \frac{\sqrt{2}}{2}\left(-u_2 - v_2\right) = -X_0 \frac{\ell\sqrt{2}}{ES} \; ;$$

$$v_2_{\text{el.2}} = \frac{\sqrt{2}}{2}\left(u_2 - v_2\right) = X_0 \frac{\ell 2\sqrt{2}}{ES}$$

$$u_3_{\text{el.2}} = \frac{\sqrt{2}}{2}\left(-u_3 - v_3\right) = 0 \; ; \quad v_3_{\text{el.2}} = \frac{\sqrt{2}}{2}\left(u_3 - v_3\right) = 0$$

The local coordinate systems of each truss element are represented in Figure 4.23.

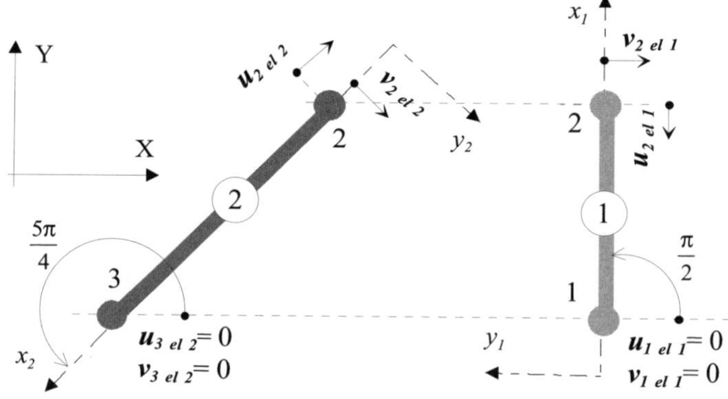

Figure 4.23. *Local coordinate systems of the truss elements*

The local displacements observed in these systems are illustrated (very much enlarged) in Figure 4.24. We can see that element 1 reaches its final position following a rigid-body rotation around the $\overrightarrow{z_1}$ axis and a contraction along its mean line. Only this contraction is taken into account in the potential energy of deformation, in line with the definition of such an element (see section 3.2.1.2). The rigid-body rotation does not result in deformation of the element. It is a general property: the "rigid-body" type movements do not deform the structures. Therefore their contribution to the potential energy of deformation is zero.

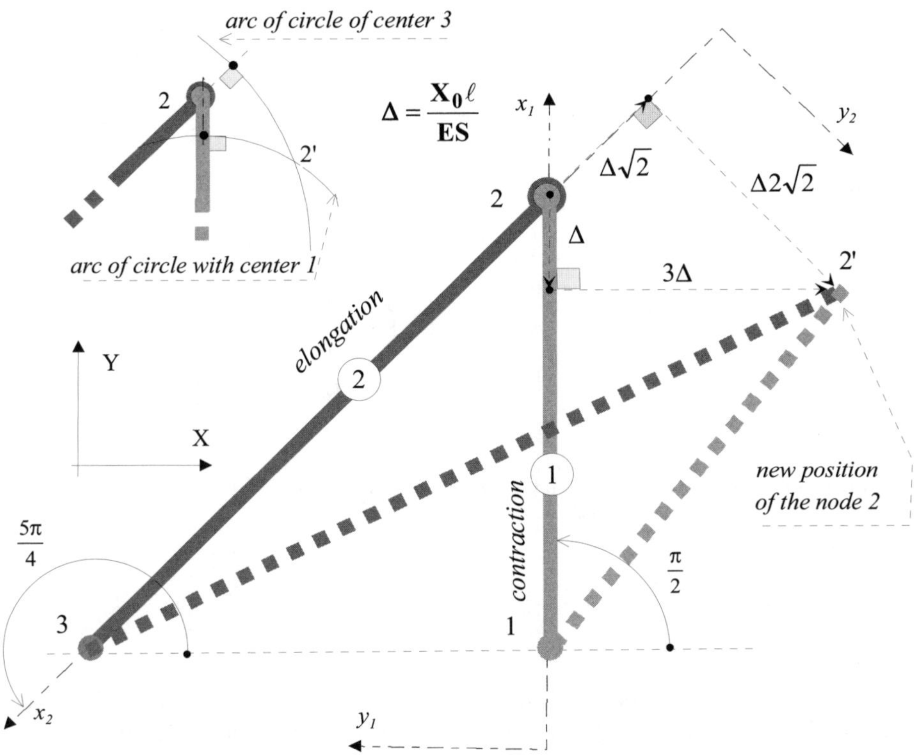

Figure 4.24. *Nodal displacements on the truss elements*
(the arcs of the circles are related to their tangents)

➢ Let us calculate the nodal forces:

The behavior equation of element 1 is written in its local system (see Figure 3.8):

$$
\{F\}_{el.1} = [k]_{el.1} \bullet \{d\}_{el.1} \Leftrightarrow
\begin{Bmatrix} X_1 \\ Y_1 \\ X_2 \\ Y_2 \end{Bmatrix}_{\substack{el.1 \\ Local}}
=
\begin{bmatrix}
\dfrac{ES}{\ell} & 0 & -\dfrac{ES}{\ell} & 0 \\
0 & 0 & 0 & 0 \\
-\dfrac{ES}{\ell} & 0 & \dfrac{ES}{\ell} & 0 \\
0 & 0 & 0 & 0
\end{bmatrix}_{\substack{el.1 \\ Local}}
\bullet
\begin{Bmatrix} u_1 = 0 \\ v_1 = 0 \\ u_2 \\ v_2 \end{Bmatrix}_{\substack{el.1 \\ Local}}
$$

Since the values u_2 and v_2 have already been calculated for element 1, we can
$\quad\quad\quad\quad\quad\quad$ el.1 $\quad\quad\quad$ el.1
derive:

$$X_1 = -\frac{ES}{\ell}u_2 = X_0 ; \quad Y_1 = 0 ; \quad X_2 = \frac{ES}{\ell}u_2 = -X_0 ; \quad Y_2 = 0$$
$$\text{el.1} \quad\quad\quad \text{el.1} \quad\quad\quad \text{el.1} \quad\quad\quad\quad \text{el.1} \quad \text{el.1} \quad\quad\quad \text{el.1}$$

Similarly for element 2 we have:

$$\{F\}_{\substack{\text{el.1}\\Local}} = [k]_{\substack{\text{el.1}\\Local}} \bullet \{d\}_{\substack{\text{el.1}\\Local}} \Leftrightarrow \begin{Bmatrix} X_1 \\ Y_1 \\ X_2 \\ Y_2 \end{Bmatrix}_{\substack{\text{el.1}\\Local}} = \begin{bmatrix} \dfrac{ES}{\ell} & 0 & -\dfrac{ES}{\ell} & 0 \\ 0 & 0 & 0 & 0 \\ -\dfrac{ES}{\ell} & 0 & \dfrac{ES}{\ell} & 0 \\ 0 & 0 & 0 & 0 \end{bmatrix}_{\substack{\text{el.1}\\Local}} \bullet \begin{Bmatrix} u_1 = 0 \\ v_1 = 0 \\ u_2 \\ v_2 \end{Bmatrix}_{\substack{\text{el.1}\\Local}}$$

Since the values u_2 and v_2 have already been calculated for element 2, we
$\quad\quad\quad\quad\quad\quad$ el2 $\quad\quad\quad$ el2
arrive at:

$$X_2 = \frac{ES}{\ell}u_2 = -X_0\sqrt{2} ; \quad Y_2 = 0 ; \quad X_3 = -\frac{ES}{\ell}u_3 = X_0\sqrt{2} ; \quad Y_3 = 0$$
$$\text{el 2} \quad\quad \text{el 2} \quad\quad\quad\quad \text{el 2} \quad\quad\quad\quad \text{el 2} \quad\quad \text{el 2} \quad\quad\quad\quad \text{el 2}$$

The nodal forces in the local systems thus calculated are illustrated in Figure
4.25. It should be ensured that element 1 is under compression and element 2 is
under tension.

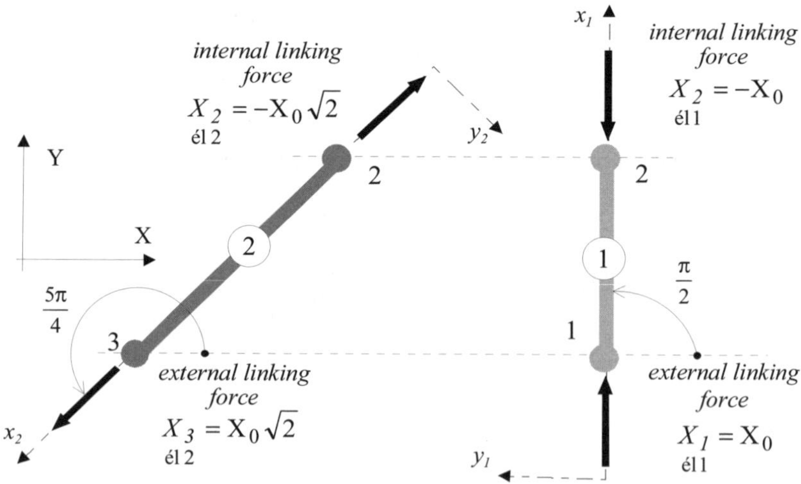

Figure 4.25. *Nodal forces on the truss elements*

5) Normal stresses in the elements

We have seen in section 3.2.1.2 how to express the displacement u(x) of any cross-section of a truss element with respect to the nodal displacements. Here we refer to the equation that was written in the local system; for example, for element 1:

$$u(\mathrm{x}) = \left[\left(1-\frac{\mathrm{x}}{\ell}\right) \quad \frac{\mathrm{x}}{\ell}\right] \bullet \left\{\begin{matrix} u_1 \\ u_2 \end{matrix}\right\}_{Local}$$

The behavior relation for the normal resultant is then written as[24]:

$$\mathcal{N}\Big|_{el.1} = ES\frac{du(\mathrm{x})}{d\mathrm{x}}\Big|_{el.1} = \frac{ES}{\ell}(u_2-u_1)\Big|_{el.1} = \frac{ES}{\ell}\times-\frac{\mathrm{X}_0\ell}{ES} = -\mathrm{X}_0$$

and for the normal stress in the truss we obtain[24]:

$$\sigma_\mathrm{x}\Big|_{el.1} = \frac{\mathcal{N}}{S}\Big|_{el.1} = -\frac{\mathrm{X}_0}{S} \quad \text{(compression)}$$

similarly for element 2:

$$\mathcal{N}\Big|_{el\,2} = ES\frac{du(\mathrm{x})}{d\mathrm{x}}\Big|_{el\,2} = \frac{ES}{\ell}(u_3-u_2)\Big|_{el\,2} = \frac{ES}{\ell}\times\frac{\mathrm{X}_0\ell\sqrt{2}}{ES} = \mathrm{X}_0\sqrt{2}$$

from where the normal stress in the truss can be written as:

$$\sigma_\mathrm{x}\Big|_{el2} = \frac{\mathcal{N}}{S}\Big|_{el2} = \frac{\mathrm{X}_0}{S}\sqrt{2} \quad \text{(traction)}$$

NOTES

❏ Figure 4.25 represents the two elements in equilibrium under the action of the nodal forces. It is readily verified in Figure 4.26 that the sum of the nodal forces at node (2) is equal to the external force $\mathrm{X}_0\overrightarrow{\mathrm{X}}$.

---

24 See section 1.5.2.1 and Figure 3.8.

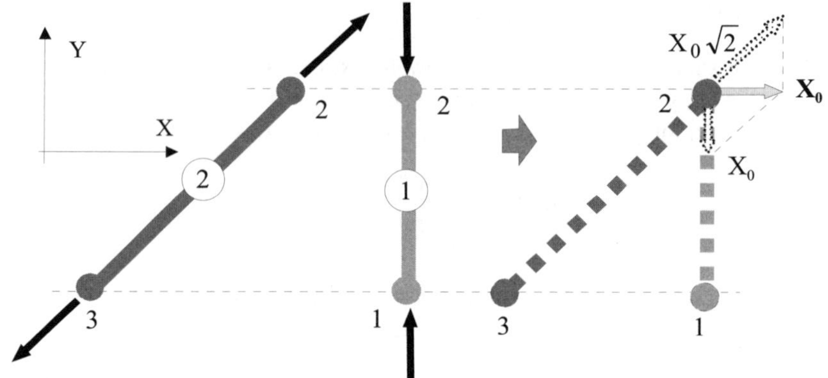

Figure 4.26. *Nodal forces*

❑ A structure consisting of assembled bars is called a truss-structure. It is often represented with basic triangular figures (triangular truss-structure systems) because, for articulated bars, the triangle is kinematically non-deformable, as shown in Figure 4.27.

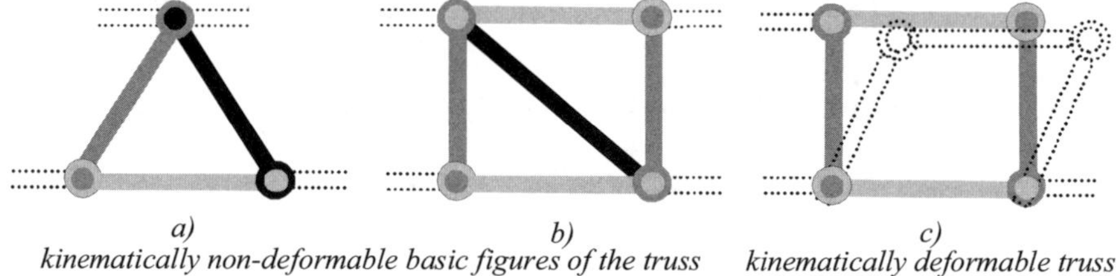

a)
kinematically non-deformable basic figures of the truss

b)

c)
kinematically deformable truss

Figure 4.27. *Plane truss-structures*

❑ Before using the finite elements method for calculation, special "manual" methods were used, suited mainly to the truss with patterns similar to those in Figure 4.27a. It was necessary to take into account internal isostatism or hyperstatism of the structure (based on whether or not it was possible to find the forces in the trusses without bringing in their strains[25]). As it has been already indicated[26] modeling through finite elements has "deleted" the notion of hyperstatism, that is the model is treated in the same manner whatever its external

---

25 Thus, it can be said that the truss in Figure 4.27a is internally isostatic, whereas that in Figure 4.27b is internally hyperstatic. The one in Figure 4.27c, which cannot retain its form, is known as internally hypostatic.
26 See section 2.6.4.3.

or internal hyperstatic nature. It is merely advisable to avoid the hypostatic configurations.

❒ However, at the design stage of numerous applications, the "isostatic" aspects should be kept in mind. When possible, a "sound" design can be obtained by reducing the hyperstatic character as much as possible. In fact when the structure is loaded, the smaller the degree of hyperstaticity at the design stage, the lower the superfluous stresses (links and cohesion) introduced by the external loads will be.

❒ This reduction of hyperstaticity is not always compatible with constructive solutions. Thus the trusses are more often built with bolted or welded gussets. If this is substituted by a model of calculation based on articulated trusses as shown in the example above, some components of the forces that can be transmitted by the links between the different parts of the structure are neglected. In the past, approximation with articulated trusses was practiced with the "manual" methods of calculation of trusses. It is still used in certain truss calculation codes using finite elements[27].

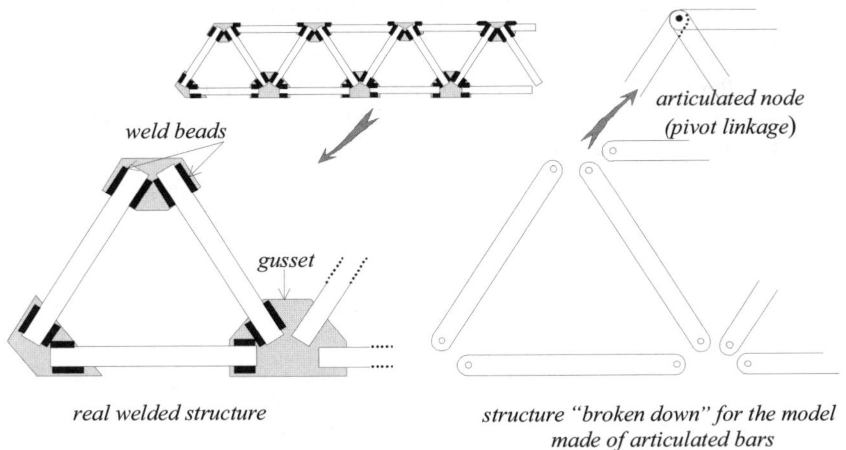

Figure 4.28. *Triangular truss structure*

____

27 It is however preferable to use beam elements in place of truss elements.

4.4. Bracket

4.4.1. *Objectives*

The bracket of Figure 4.29 is made of two steel beams with the same rectangular section welded at a right angle. Our intention is to:

➢ model the bracket by discretizing it into beam elements under plane bending "and" traction-compression, and by applying the boundary conditions (displacements and loading);

➢ obtain literal expressions of all nodal characteristics and clearly show the procedure.

NOTE

❏ This theme is particularly suited for carrying out a detailed study of educational value. In fact, in addition to the analytical approach done here, it allows:

– a numerical approach with the help of software for calculation using finite elements that can be used in parallel;

– an experimental approach: the bracket can be built and fitted with suitable instrumentation so as to enable the measurement of the displacements and the strains (respectively using dial and strain gauges)[28].

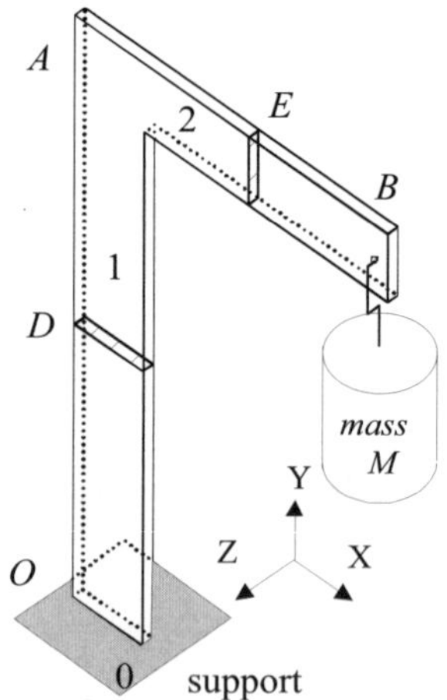

Figure 4.29. *Bracket*

---

28 An easily usable assembly is obtained by adopting the following dimensions for the steel beams (see Figure 4.31): $\ell = 195$ and $S = 30 \times 5$.

4.4.2. *Modeling*

4.4.2.1. *Definition of the beam element*

On analyzing the assembly in Figure 4.29 it can be foreseen that the horizontal part of the arm shall be loaded in plane bending (in the (XY) plane of the global system) whereas the vertical part shall be loaded in plane bending (in the (XY) plane) and in compression[29].

The combination made in section 4.2.4, and which consists of regrouping on one single element the properties of plane bending and that of traction-compression, meets this requirement. Figure 4.30 refers to the characteristics of such an element in the local coordinate system[30].

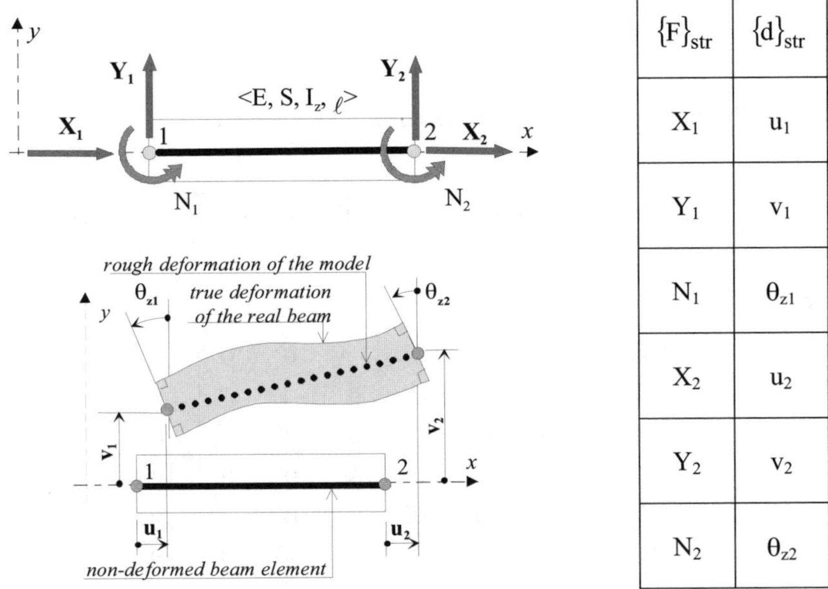

$\{F\}_{str}$	$\{d\}_{str}$
X_1	u_1
Y_1	v_1
N_1	θ_{z1}
X_2	u_2
Y_2	v_2
N_2	θ_{z2}

Figure 4.30. *Beam element in plane bending "+" traction-compression*

---

29 As an example this arm is referred to again in Chapter 9, section 9.4, which illustrates the systematic research of resultant forces and moments in the common sections of horizontal and vertical portion. We can then verify that the horizontal part is loaded in plane bending and the vertical part in plane bending and compression.

30 It should be remembered that in the entire chapter the simple structures studied are planes ((XY) plane) and they operate in this plane.

The behavior equation is recalled below (see [4.10]):

$$
\begin{Bmatrix} X_1 \\ Y_1 \\ N_1 \\ X_2 \\ Y_2 \\ N_2 \end{Bmatrix} =
\begin{bmatrix}
\dfrac{ES}{\ell} & 0 & 0 & -\dfrac{ES}{\ell} & 0 & 0 \\[2ex]
0 & \dfrac{12EI}{\ell^3} & \dfrac{6EI}{\ell^2} & 0 & -\dfrac{12EI}{\ell^3} & \dfrac{6EI}{\ell^2} \\[2ex]
0 & \dfrac{6EI}{\ell^2} & \dfrac{4EI}{\ell} & 0 & -\dfrac{6EI}{\ell^2} & \dfrac{2EI}{\ell} \\[2ex]
-\dfrac{ES}{\ell} & 0 & 0 & \dfrac{ES}{\ell} & 0 & 0 \\[2ex]
0 & -\dfrac{12EI}{\ell^3} & -\dfrac{6EI}{\ell^2} & 0 & \dfrac{12EI}{\ell^3} & -\dfrac{6EI}{\ell^2} \\[2ex]
0 & \dfrac{6EI}{\ell^2} & \dfrac{2EI}{\ell} & 0 & -\dfrac{6EI}{\ell^2} & \dfrac{4EI}{\ell}
\end{bmatrix}
\bullet
\begin{Bmatrix} u_1 \\ v_1 \\ \theta_{z1} \\ u_2 \\ v_2 \\ \theta_{z2} \end{Bmatrix}
$$

to simplify the notations, it can be assumed:

$$
A = \frac{ES}{\ell} \qquad B = \frac{12EI}{\ell^3} \qquad C = \frac{6EI}{\ell^2} \qquad D = \frac{4EI}{\ell} \qquad F = \frac{2EI}{\ell}
$$

i.e.:

$$
\begin{Bmatrix} X_1 \\ Y_1 \\ N_1 \\ X_2 \\ Y_2 \\ N_2 \end{Bmatrix} =
\begin{bmatrix}
A & 0 & 0 & -A & 0 & 0 \\
0 & B & C & 0 & -B & C \\
0 & C & D & 0 & -C & F \\
-A & 0 & 0 & A & 0 & 0 \\
0 & -B & -C & 0 & B & -C \\
0 & C & F & 0 & -C & D
\end{bmatrix}
\bullet
\begin{Bmatrix} u_1 \\ v_1 \\ \theta_{z1} \\ u_2 \\ v_2 \\ \theta_{z2} \end{Bmatrix}
\qquad [4.13]
$$

4.4.2.2. *Model using wires*

These elements are represented by a geometric "wire" model obtained by drawing the mean lines. The lengths of these mean lines are then dimensioned as represented in Figure 4.31.

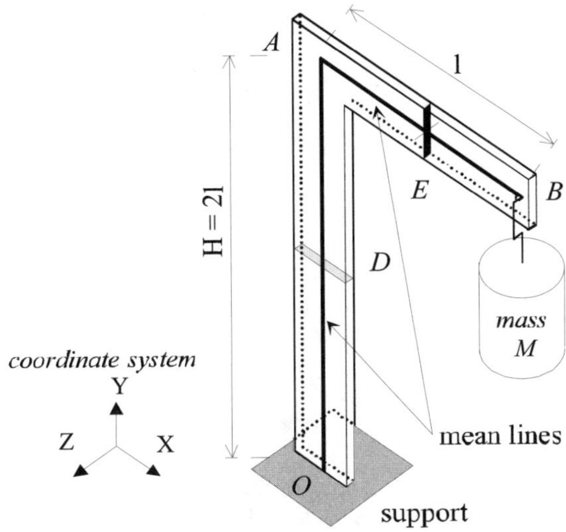

Figure 4.31. *Construction of the model using the mean lines*

On the geometric model let us choose nodes that will define the meshing in terms of beam finite elements. For example, three elements with four nodes represented in Figure 4.32 will be retained (the minimum meshing requires only two elements). For these elements (Figure 4.30), there are three dof at every node making a total of $3 \times 4 = 12$ dof. The bracket will have a (12×12) global stiffness matrix.

4.4.2.3. *Geometric properties of the beams*

Respecting proper orientation of the sections requires associating each mean line of an element to a local coordinate system composed of the main axes of the cross-sections of the beams as is illustrated in Figure 4.32.

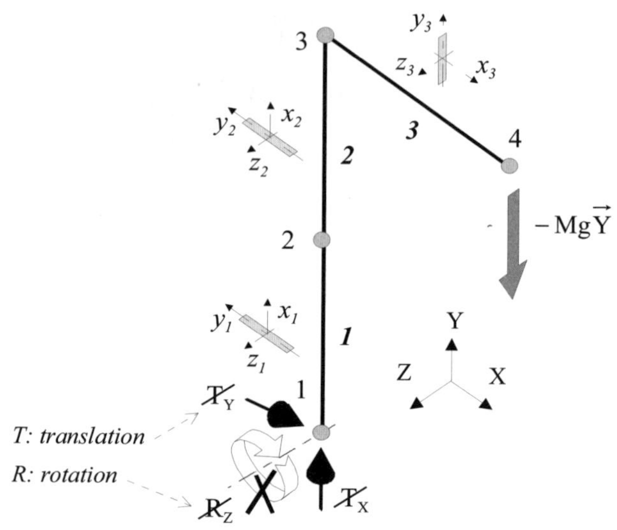

Figure 4.32. *Model in beam elements with support and loading conditions*

The following are configurations of the local coordinate systems.

element *1*	$\vec{x}_1 = \vec{Y}$	$\vec{y}_1 = -\vec{X}$	$\vec{z}_1 = \vec{Z}$
element *2*	$\vec{x}_2 = \vec{Y}$	$\vec{y}_2 = -\vec{X}$	$\vec{z}_2 = \vec{Z}$
element *3*	$\vec{x}_3 = \vec{X}$	$\vec{y}_3 = \vec{Y}$	$\vec{z}_3 = \vec{Z}$

$$[4.14]$$

The three elements have identical section characteristics which, for our requirement, are limited to: S (area) and I (quadratic moment around the local $\vec{z}$ axis).

Material: the three elements have identical material characteristics which are limited here to Young's modulus: E.

The above can be summarized in what is known as a connectivity table.

element n	node (i)	node (j)	material	area	quadratic moment
1	1	2	$E_1 = E$	$S_1 = S$	$I_1 = I$
2	2	3	$E_2 = E$	$S_2 = S$	$I_2 = I$
3	3	4	$E_3 = E$	$S_3 = S$	$I_3 = I$

4.4.2.4. *Support conditions*

The bracket is in a fixed link with the surroundings (embedded) (Figure 4.30). Therefore in the global coordinate system we have:

➤ *at node (1)*:

T_X translation blocked	T_Y translation blocked	R_Z rotation blocked
$u_1 = 0$	$v_1 = 0$	$\theta_{z1} = 0$

This results in unknown mechanical actions of link denoted as:

$X_{1\ell} = ?$	$Y_{1\ell} = ?$	$N_{1\ell} = ?$

The structure appears properly linked (rigid-body movement is not possible).

4.4.2.5. *Loading*

This is carried out by hooking the mass on to the bracket at B[31] (Figure 4.30), i.e.:

➤ *at node (4)* (Figure 4.32):

$X_4 = 0$	$Y_4 = -Mg$	$N_4 = 0$

The unknown displacements are denoted as:

translation T_X	translation T_Y	rotation R_Z
$u_4 = ?$	$v_4 = ?$	$\theta_{z4} = ?$

At the other free nodes (no. 2 and no. 3), the external loads are zero (no load).

---

31 To write this, the equilibrium of the mass M should have already been studied.

4.4.3. *Calculation of the elementary stiffness matrix in the global system*

Formula [4.13] characterizes the behavior of the beam element in the local system. It is therefore necessary to mention beforehand the behavior in the global system.

Reference is made to the procedure defined in [3.9]:

♦ The dof of an element expressed in the local and the global coordinate systems are geometrically connected by a formula of the type:

$$\{d\}_{Local} = [P] \bullet \{d\}_{Global}$$

♦ We have a similar one for the load:

$$\{F\}_{Local} = [P] \bullet \{F\}_{Global}$$

♦ It is known that the behavior equation of the element written in the local system is:

$$\{F\}_{Local} = [k]_{Local} \bullet \{d\}_{Local}$$

and it can be re-written in the global system as[32]:

$$\{F\}_{Global} = [k]_{Global} \bullet \{d\}_{Global}$$

with:

$$\{k\}_{Global} = [P]^{T} \bullet [k]_{Local} \bullet [P]$$

The following equations enable us to define the transfer matrix from the global to the local system:

♦ elements 1 and 2:

$$\left\{\begin{matrix} \vec{x} \\ \vec{y} \\ \vec{z} \end{matrix}\right\}_{Local} = \begin{bmatrix} 0 & 1 & 0 \\ -1 & 0 & 0 \\ 0 & 0 & 1 \end{bmatrix} \bullet \left\{\begin{matrix} \vec{X} \\ \vec{Y} \\ \vec{Z} \end{matrix}\right\}_{Global}$$

---

32 See [3.9].

matrix $[p]$ of formula [3.9] is obtained with $\alpha = \dfrac{\pi}{2}$, (see Figure 4.33). This leads to:

$$\left\{ \begin{array}{c} u \\ v \\ \theta_z \end{array} \right\} = \begin{bmatrix} 0 & 1 & 0 \\ -1 & 0 & 0 \\ 0 & 0 & 1 \end{bmatrix} \bullet \left\{ \begin{array}{c} u \\ v \\ \theta_z \end{array} \right\}$$

$$\qquad\quad Local \qquad\qquad\qquad\qquad Global$$

• element 3:

$$\left\{ \begin{array}{c} \vec{x} \\ \vec{y} \\ \vec{z} \end{array} \right\} = \begin{bmatrix} 1 & 0 & 0 \\ 0 & 1 & 0 \\ 0 & 0 & 1 \end{bmatrix} \bullet \left\{ \begin{array}{c} \vec{X} \\ \vec{Y} \\ \vec{Z} \end{array} \right\}$$

$$\qquad\quad Local \qquad\qquad\qquad\qquad Global$$

Here $\alpha = 0$, i.e.:

$$\left\{ \begin{array}{c} u \\ v \\ \theta_z \end{array} \right\} = \begin{bmatrix} 1 & 0 & 0 \\ 0 & 1 & 0 \\ 0 & 0 & 1 \end{bmatrix} \bullet \left\{ \begin{array}{c} u \\ v \\ \theta_z \end{array} \right\}$$

$$\qquad\quad Local \qquad\qquad\qquad\qquad Global$$

Transfer matrix from the global to the local coordinate system.

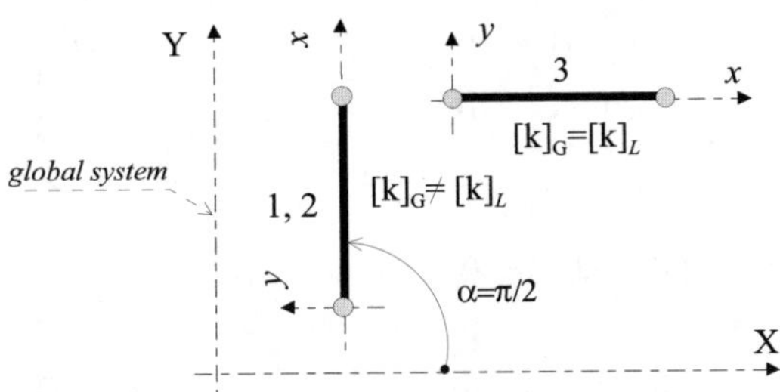

Figure 4.33. *Stiffness matrices global (G) and local (L)*

➢ *For elements 1 and 2:*

The nodes of these elements being noted as i and j, the transfer matrix is written as:

$$\begin{Bmatrix} u_i \\ v_i \\ \theta_{zi} \\ u_j \\ v_j \\ \theta_{zj} \end{Bmatrix} = \begin{bmatrix} 0 & 1 & 0 & 0 & 0 & 0 \\ -1 & 0 & 0 & 0 & 0 & 0 \\ 0 & 0 & 1 & 0 & 0 & 0 \\ 0 & 0 & 0 & 0 & 1 & 0 \\ 0 & 0 & 0 & -1 & 0 & 0 \\ 0 & 0 & 0 & 0 & 0 & 1 \end{bmatrix} \bullet \begin{Bmatrix} u_i \\ v_i \\ \theta_{zi} \\ u_j \\ v_j \\ \theta_{zj} \end{Bmatrix}$$

$$\text{Local} \qquad\qquad\qquad\qquad\qquad\qquad\qquad\qquad \text{Global}$$

Under these conditions let us calculate $[k]_{el}$ (Global) from $[k]_{el}$ (Local). We have successively:

$$[k]_{el}^{Local} \bullet [P] = \begin{bmatrix} A & 0 & 0 & -A & 0 & 0 \\ 0 & B & C & 0 & -B & C \\ 0 & C & D & 0 & -C & F \\ -A & 0 & 0 & A & 0 & 0 \\ 0 & -B & -C & 0 & B & -C \\ 0 & C & F & 0 & -C & D \end{bmatrix} \bullet \begin{bmatrix} 0 & 1 & 0 & 0 & 0 & 0 \\ -1 & 0 & 0 & 0 & 0 & 0 \\ 0 & 0 & 1 & 0 & 0 & 0 \\ 0 & 0 & 0 & 0 & 1 & 0 \\ 0 & 0 & 0 & -1 & 0 & 0 \\ 0 & 0 & 0 & 0 & 0 & 1 \end{bmatrix}$$

$$= \begin{bmatrix} 0 & A & 0 & 0 & -A & 0 \\ -B & 0 & C & B & 0 & C \\ -C & 0 & D & C & 0 & F \\ 0 & -A & 0 & 0 & A & 0 \\ B & 0 & -C & -B & 0 & -C \\ -C & 0 & F & C & 0 & D \end{bmatrix}$$

and then:

$$[k]_{el}^{Global} = [P]^T \bullet [k]_{el}^{Local} \bullet [P] = \begin{bmatrix} 0 & -1 & 0 & 0 & 0 & 0 \\ 1 & 0 & 0 & 0 & 0 & 0 \\ 0 & 0 & 1 & 0 & 0 & 0 \\ 0 & 0 & 0 & 0 & -1 & 0 \\ 0 & 0 & 0 & 1 & 0 & 0 \\ 0 & 0 & 0 & 0 & 0 & 1 \end{bmatrix} \bullet \begin{bmatrix} 0 & A & 0 & 0 & -A & 0 \\ -B & 0 & C & B & 0 & C \\ -C & 0 & D & C & 0 & F \\ 0 & -A & 0 & 0 & A & 0 \\ B & 0 & -C & -B & 0 & -C \\ -C & 0 & F & C & 0 & D \end{bmatrix}$$

$$\begin{bmatrix} k \end{bmatrix}_{\text{el}}^{\text{Global}} = \begin{bmatrix} B & 0 & -C & -B & 0 & -C \\ 0 & A & 0 & 0 & -A & 0 \\ -C & 0 & D & C & 0 & F \\ -B & 0 & C & B & 0 & C \\ 0 & -A & 0 & 0 & A & 0 \\ -C & 0 & F & C & 0 & D \end{bmatrix} \qquad [4.15]$$

Elements 1 and 2 being identical, their stiffness matrix in the global system can be expressed as:

$$\underset{\substack{\text{Global} \\ (6\times6)}}{[k]_1} = \underset{\substack{\text{Global} \\ (6\times6)}}{[k]_2} = \begin{bmatrix} B & 0 & -C & -B & 0 & -C \\ 0 & A & 0 & 0 & -A & 0 \\ -C & 0 & D & C & 0 & F \\ -B & 0 & C & B & 0 & C \\ 0 & -A & 0 & 0 & A & 0 \\ -C & 0 & F & C & 0 & D \end{bmatrix}$$

➢ *For element 3:*

The transfer matrix is written as:

$$\underset{\text{Local}}{\begin{Bmatrix} u_3 \\ v_3 \\ \theta_{z3} \\ u_4 \\ v_4 \\ \theta_{z4} \end{Bmatrix}} = \begin{bmatrix} 1 & 0 & 0 & 0 & 0 & 0 \\ 0 & 1 & 0 & 0 & 0 & 0 \\ 0 & 0 & 1 & 0 & 0 & 0 \\ 0 & 0 & 0 & 1 & 0 & 0 \\ 0 & 0 & 0 & 0 & 1 & 0 \\ 0 & 0 & 0 & 0 & 0 & 1 \end{bmatrix} \bullet \underset{\text{Global}}{\begin{Bmatrix} u_3 \\ v_3 \\ \theta_{z3} \\ u_4 \\ v_4 \\ \theta_{z4} \end{Bmatrix}}$$

Therefore, $[P] = [I]$ (identity matrix). It could be foreseen because coordinates $\left(\overrightarrow{x_3}, \overrightarrow{y_3}, \overrightarrow{z_3} \right)$ and $\left(\overrightarrow{X}, \overrightarrow{Y}, \overrightarrow{Z} \right)$ are parallel.

The stiffness matrix in the global coordinate system for element 3 is therefore the same as that in the local system. It is written as:

$$[k]_3 \underset{\substack{\text{Global} \\ (6\times6)}}{=} \begin{bmatrix} A & 0 & 0 & -A & 0 & 0 \\ 0 & B & C & 0 & -B & C \\ 0 & C & D & 0 & -C & F \\ -A & 0 & 0 & A & 0 & 0 \\ 0 & -B & -C & 0 & B & -C \\ 0 & C & F & 0 & -C & D \end{bmatrix} \qquad [4.16]$$

4.4.4. *Assembly of the global stiffness matrix [K]$_{str}$*

From now on we shall work in the global system. The assembly of the three basic matrices $[k]_1$, $[k]_2$ and $[k]_3$ of the three beam elements of the bracket makes it possible to obtain the 12×12 global matrix of the structure $[K]_{str}$.

By reverting to the method seen in section 3.3.3 the equation can be written as:

$$\underset{\substack{\text{pot.} \\ \text{structure}}}{E} = \frac{1}{2} \underset{(1\times12)}{\{d\}^{T}} \bullet \underset{(12\times12)}{[K]_{str}} \bullet \underset{(12\times1)}{\{q\}} \cdots$$

$$\cdots = \sum_{i=1}^{3} \underset{\text{el } i}{E_{\text{pot.}}} = \frac{1}{2}\underset{\text{el }1}{\{d\}^{T}} \bullet \underset{\text{el }1}{[k]} \bullet \underset{\text{el }1}{\{d\}} + \frac{1}{2}\underset{\text{el }2}{\{d\}^{T}} \bullet \underset{\text{el }2}{[k]} \bullet \underset{\text{el }2}{\{d\}} + \frac{1}{2}\underset{\text{el }3}{\{d\}^{T}} \bullet \underset{\text{el }3}{[k]} \bullet \underset{\text{el }3}{\{d\}}$$

The global stiffness matrix $\underset{12\times12}{[K]_{str}}$ can be obtained, for example, by adding the three basic stiffness matrices redimensioned to 12×12 (see [3.38]) in which the terms k_{ij}. are correctly positioned. To sequence the vector of the dof of the structure $\underset{(12\times1)}{\{d\}}$, let us adopt the list of nodal displacements read in the natural order of the nodes, i.e.:

$$\{d\}^{T} = [u_1 \quad v_1 \quad \theta_{z1} \quad u_2 \quad v_2 \quad \theta_{z2} \quad u_3 \quad v_3 \quad \theta_{z3} \quad u_4 \quad v_4 \quad \theta_{z4}]$$

Redimensioning each one of the three basic stiffness matrices in a (12×12) table leads to the reloading of equations [4.15] and [4.16] as specified in Figure 4.34, where the calculation of the term $\left(\dfrac{K}{\theta_{z2}\theta_{z2}}\right)_{str}$ is illustrated. It can be obtained in the following manner[33]:

$$\left(\frac{K}{\theta_{z2}\theta_{z2}}\right)_{str} = \left(\frac{k}{\theta_{z2}\theta_{z2}}\right)_{el\ 1} + \left(\frac{k}{\theta_{z2}\theta_{z2}}\right)_{el\ 2} = D + D = 2D$$

---

33 See also rule [3.39] for the assembly.

The reader can make sure that in this manner it is possible to obtain each of the terms of the global stiffness matrix as they are seen in the figure.

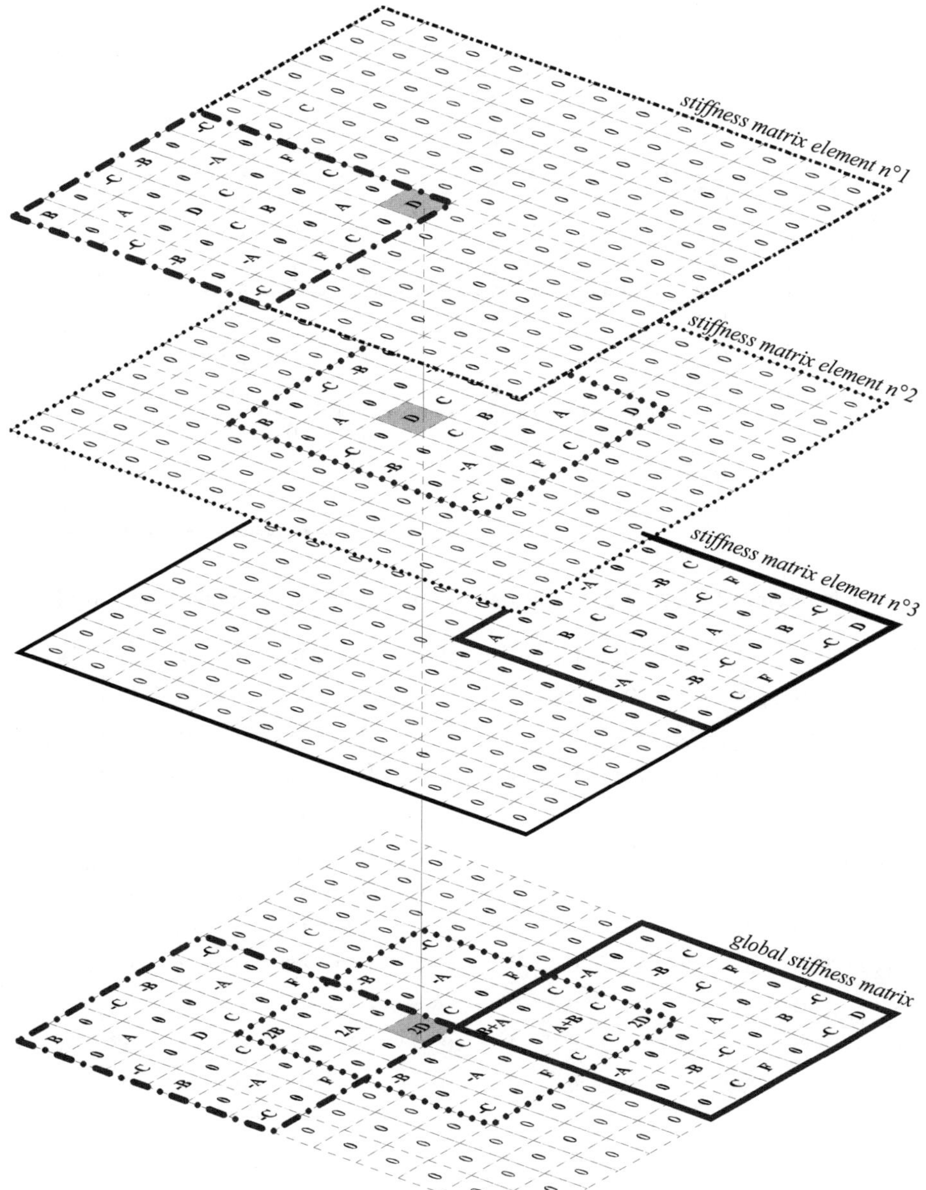

Figure 4.34. *Assembly method of the element stiffness matrices*

Global stiffness matrix $[K]_{\underset{12\times12}{str}}$ for the bracket is obtained below, where the display of the "projection" of the basic stiffness matrices of Figure 4.34 is retained.

$$\begin{bmatrix}
B & 0 & -C & -B & 0 & -C & 0 & 0 & 0 & 0 & 0 & 0 \\
0 & A & 0 & 0 & -A & 0 & 0 & 0 & 0 & 0 & 0 & 0 \\
-C & 0 & D & C & 0 & F & 0 & 0 & 0 & 0 & 0 & 0 \\
-B & 0 & C & 2B & 0 & 0 & -B & 0 & -C & 0 & 0 & 0 \\
0 & -A & 0 & 0 & 2A & 0 & 0 & -A & 0 & 0 & 0 & 0 \\
-C & 0 & F & 0 & 0 & 2D & C & 0 & F & 0 & 0 & 0 \\
0 & 0 & 0 & -B & 0 & 0 & B+A & 0 & C & -A & 0 & 0 \\
0 & 0 & 0 & 0 & 0 & 0 & 0 & A+B & C & 0 & -B & C \\
0 & 0 & 0 & -C & 0 & F & C & 0 & 2D & 0 & -C & F \\
0 & 0 & 0 & 0 & 0 & 0 & -A & 0 & 0 & A & 0 & 0 \\
0 & 0 & 0 & 0 & 0 & 0 & 0 & -B & -C & 0 & B & -C \\
0 & 0 & 0 & 0 & 0 & 0 & 0 & C & F & 0 & -C & D
\end{bmatrix}$$

The behavior of the assembled structure in the global system takes the form[34]:

$$\begin{Bmatrix} X_1 \\ Y_1 \\ N_1 \\ X_2 \\ Y_2 \\ N_2 \\ X_3 \\ Y_3 \\ N_3 \\ X_4 \\ Y_4 \\ N_4 \end{Bmatrix} = \begin{bmatrix}
B & 0 & -C & -B & 0 & -C & 0 & 0 & 0 & 0 & 0 & 0 \\
0 & A & 0 & 0 & -A & 0 & 0 & 0 & 0 & 0 & 0 & 0 \\
-C & 0 & D & C & 0 & F & 0 & 0 & 0 & 0 & 0 & 0 \\
-B & 0 & C & 2B & 0 & 0 & -B & 0 & -C & 0 & 0 & 0 \\
0 & -A & 0 & 0 & 2A & 0 & 0 & -A & 0 & 0 & 0 & 0 \\
-C & 0 & F & 0 & 0 & 2D & C & 0 & F & 0 & 0 & 0 \\
0 & 0 & 0 & -B & 0 & 0 & B+A & 0 & C & -A & 0 & 0 \\
0 & 0 & 0 & 0 & -A & 0 & 0 & A+B & C & 0 & -B & C \\
0 & 0 & 0 & -C & 0 & F & C & 0 & 2D & 0 & -C & F \\
0 & 0 & 0 & 0 & 0 & 0 & -A & 0 & 0 & A & 0 & 0 \\
0 & 0 & 0 & 0 & 0 & 0 & 0 & -B & -C & 0 & B & -C \\
0 & 0 & 0 & 0 & 0 & 0 & 0 & C & F & 0 & -C & D
\end{bmatrix} \bullet \begin{Bmatrix} u_1 \\ v_1 \\ \theta_{z1} \\ u_2 \\ v_2 \\ \theta_{z2} \\ u_3 \\ v_3 \\ \theta_{z3} \\ u_4 \\ v_4 \\ \theta_{z4} \end{Bmatrix}$$

$$[4.17]$$

---

34 See [2.122].

where care has been taken to maintain on every row the relation:

$$\text{"mechanical action} \Leftrightarrow \text{dof "}.$$

As indicated every time, this matrix $[K]_{\underset{12\times12}{str}}$ is singular because it is formed using "floating" elements.

4.4.5. *Establishing the linkage and loading conditions*

In section 4.4.2.4 it was noted that the fixed support of node (1) was expressed in the global system as $u_1 = v_1 = \theta_{z1} = 0$, the corresponding linking actions being $X_{1\ell}, Y_{1\ell}, N_{1\ell}$.

In section 4.4.2.5 it was noted that in the global system the external load is always reduced to a load at node (4): $Y_4 = -Mg$.

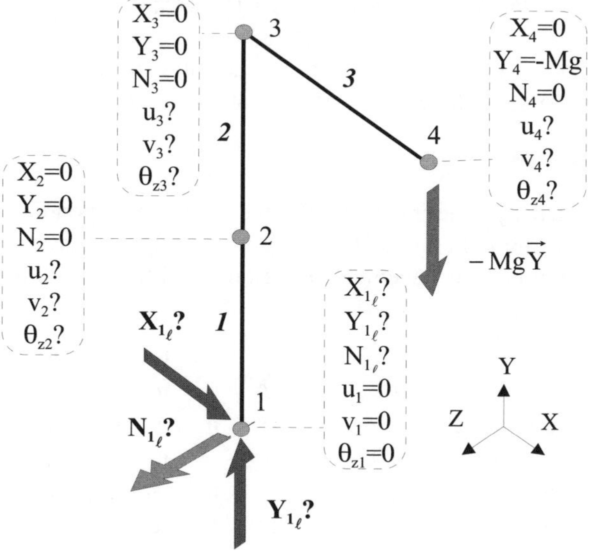

Figure 4.35. *Isolated bracket*

From the isolated bracket of Figure 4.35, behavior equation [4.17] is re-written as (see section 3.4.2):

$$
\begin{Bmatrix} X_{1\ell} = ? \\ Y_{1\ell} = ? \\ N_{1\ell} = ? \\ X_2 = 0 \\ Y_2 = 0 \\ N_2 = 0 \\ X_3 = 0 \\ Y_3 = 0 \\ N_3 = 0 \\ X_4 = 0 \\ Y_4 = -Mg \\ N_4 = 0 \end{Bmatrix}
=
\begin{bmatrix}
B & 0 & -C & -B & 0 & -C & 0 & 0 & 0 & 0 & 0 & 0 \\
0 & A & 0 & 0 & -A & 0 & 0 & 0 & 0 & 0 & 0 & 0 \\
-C & 0 & D & C & 0 & F & 0 & 0 & 0 & 0 & 0 & 0 \\
-B & 0 & C & 2B & 0 & 0 & -B & 0 & -C & 0 & 0 & 0 \\
0 & -A & 0 & 0 & 2A & 0 & 0 & -A & 0 & 0 & 0 & 0 \\
-C & 0 & F & 0 & 0 & 2D & C & 0 & F & 0 & 0 & 0 \\
0 & 0 & 0 & -B & 0 & 0 & B+A & 0 & C & -A & 0 & 0 \\
0 & 0 & 0 & 0 & -A & 0 & 0 & A+B & C & 0 & -B & C \\
0 & 0 & 0 & -C & 0 & F & C & 0 & 2D & 0 & -C & F \\
0 & 0 & 0 & 0 & 0 & 0 & -A & 0 & 0 & A & 0 & 0 \\
0 & 0 & 0 & 0 & 0 & 0 & 0 & -B & -C & 0 & B & -C \\
0 & 0 & 0 & 0 & 0 & 0 & 0 & C & F & 0 & -C & D
\end{bmatrix}
\bullet
\begin{Bmatrix} u_1 = 0 \\ v_1 = 0 \\ \theta_{z1} = 0 \\ u_2 = ? \\ v_2 = ? \\ \theta_{z2} = ? \\ u_3 = ? \\ v_3 = ? \\ \theta_{z3} = ? \\ u_4 = ? \\ v_4 = ? \\ \theta_{z4} = ? \end{Bmatrix}
$$

$$[4.18]$$

4.4.6. *Resolution of the linear system $\{F\}_{str} = [K]_{str} \bullet \{d\}_{str}$*

The resolution procedure of such a system has already been specified[35] and used several times. It results in the creation of two sub-systems:

➤ *Sub-system (I)[36]*

This is obtained by deleting in [4.18] the rows corresponding to the zero or "blocked" dof and the columns of the same row. In the sub-system thus obtained only the unknown nodal displacements (free dof) and the known nodal mechanical actions appear, i.e.:

$$
\begin{Bmatrix} 0 \\ 0 \\ 0 \\ 0 \\ 0 \\ 0 \\ 0 \\ -Mg \\ 0 \end{Bmatrix}
=
\begin{bmatrix}
2B & 0 & 0 & -B & 0 & -C & 0 & 0 & 0 \\
0 & 2A & 0 & 0 & -A & 0 & 0 & 0 & 0 \\
0 & 0 & 2D & C & 0 & F & 0 & 0 & 0 \\
-B & 0 & 0 & B+A & 0 & C & -A & 0 & 0 \\
0 & -A & 0 & 0 & A+B & C & 0 & -B & C \\
-C & 0 & F & C & C & 2D & 0 & -C & F \\
0 & 0 & 0 & -A & 0 & 0 & A & 0 & 0 \\
0 & 0 & 0 & 0 & -B & -C & 0 & B & -C \\
0 & 0 & 0 & 0 & C & F & 0 & -C & D
\end{bmatrix}
\bullet
\begin{Bmatrix} u_2 \\ v_2 \\ \theta_{z2} \\ u_3 \\ v_3 \\ \theta_{z3} \\ u_4 \\ v_4 \\ \theta_{z4} \end{Bmatrix}
$$

$$[4.19]$$

---

35 See section 3.4.2.
36 See [3.44].

We denote $\left[K^*\right]$ as the (9×9) sub-matrix of stiffness which characterizes the above equation.

It cannot be inverted reasonably using a standard manual procedure. The use of a formal calculating device or a pocket calculator becomes necessary to obtain its inverse. With values A, B, C, D, F defined in [4.13], the form of $\left[K^*\right]^{-1}$ as obtained from the formal calculating device is written as follows:

$$\left[K^*\right]^{-1} = \cdots$$

$$
\begin{bmatrix}
\frac{1}{(3\cdot E)}\cdot\frac{l^3}{I} & 0 & \frac{-1}{(2\cdot E)}\cdot\frac{l^2}{I} & \frac{5}{(6\cdot E)}\cdot\frac{l^3}{I} & 0 & \frac{-1}{(2\cdot E)}\cdot\frac{l^2}{I} & \frac{5}{(6\cdot E)}\cdot\frac{l^3}{I} & \frac{-1}{(2\cdot E)}\cdot\frac{l^3}{I} & \frac{-1}{(2\cdot E)}\cdot\frac{l^2}{I} \\[8pt]
0 & \frac{1}{E}\cdot\frac{1}{S} & 0 & 0 & \frac{1}{E}\cdot\frac{1}{S} & 0 & 0 & \frac{1}{E}\cdot\frac{1}{S} & 0 \\[8pt]
\frac{-1}{(2\cdot E)}\cdot\frac{l^2}{I} & 0 & \frac{1}{E}\cdot\frac{1}{I} & \frac{-3}{(2\cdot E)}\cdot\frac{l^2}{I} & 0 & \frac{1}{E}\cdot\frac{1}{I} & \frac{-3}{(2\cdot E)}\cdot\frac{l^2}{I} & \frac{1}{E}\cdot\frac{l^2}{I} & \frac{1}{E}\cdot\frac{1}{I} \\[8pt]
\frac{5}{(6\cdot E)}\cdot\frac{l^3}{I} & 0 & \frac{-3}{(2\cdot E)}\cdot\frac{l^2}{I} & \frac{8}{(3\cdot E)}\cdot\frac{l^3}{I} & 0 & \frac{-2}{E}\cdot\frac{l^2}{I} & \frac{8}{(3\cdot E)}\cdot\frac{l^3}{I} & \frac{-2}{E}\cdot\frac{l^3}{I} & \frac{-2}{E}\cdot\frac{l^2}{I} \\[8pt]
0 & \frac{1}{E}\cdot\frac{1}{S} & 0 & 0 & \frac{2}{E}\cdot\frac{1}{S} & 0 & 0 & \frac{2}{E}\cdot\frac{1}{S} & 0 \\[8pt]
\frac{-1}{(2\cdot E)}\cdot\frac{l^2}{I} & 0 & \frac{1}{E}\cdot\frac{1}{I} & \frac{-2}{E}\cdot\frac{l^2}{I} & 0 & \frac{2}{E}\cdot\frac{1}{I} & \frac{-2}{E}\cdot\frac{l^2}{I} & \frac{2}{E}\cdot\frac{l^2}{I} & \frac{2}{E}\cdot\frac{1}{I} \\[8pt]
\frac{5}{(6\cdot E)}\cdot\frac{l^3}{I} & 0 & \frac{-3}{(2\cdot E)}\cdot\frac{l^2}{I} & \frac{8}{(3\cdot E)}\cdot\frac{l^3}{I} & 0 & \frac{-2}{E}\cdot\frac{l^2}{I} & \frac{1}{(3\cdot(E\cdot(I\cdot S)))}\cdot(3\cdot I+8\cdot S\cdot l^2)\cdot1 & \frac{-2}{E}\cdot\frac{l^3}{I} & \frac{-2}{E}\cdot\frac{l^2}{I} \\[8pt]
\frac{-1}{(2\cdot E)}\cdot\frac{l^3}{I}\ \frac{1}{E}\cdot\frac{1}{S} & \frac{1}{E}\cdot\frac{l^2}{I} & \frac{-2}{E}\cdot\frac{l^3}{I}\ \frac{2}{E}\cdot\frac{1}{S} & \frac{2}{E}\cdot\frac{l^2}{I} & \frac{-2}{E}\cdot\frac{l^3}{I} & \frac{1}{(3\cdot(E\cdot(I\cdot S)))}\cdot(6\cdot I+7\cdot S\cdot l^2)\cdot1\ \frac{5}{(2\cdot E)}\cdot\frac{l^2}{I} \\[8pt]
\frac{-1}{(2\cdot E)}\cdot\frac{l^2}{I} & 0 & \frac{1}{E}\cdot\frac{1}{I} & \frac{-2}{E}\cdot\frac{l^2}{I} & 0 & \frac{2}{E}\cdot\frac{1}{I} & \frac{-2}{E}\cdot\frac{l^2}{I} & \frac{5}{(2\cdot E)}\cdot\frac{l^2}{I} & \frac{3}{E}\cdot\frac{1}{I}
\end{bmatrix}
$$

$$[4.20]$$

Equation [4.19] becomes:

$$
\begin{Bmatrix} u_2 \\ v_2 \\ \theta_{z2} \\ u_3 \\ v_3 \\ \theta_{z3} \\ u_4 \\ v_4 \\ \theta_{z4} \end{Bmatrix}
= \left[K^*\right]^{-1}_{9\times9} \bullet
\begin{Bmatrix} 0 \\ 0 \\ 0 \\ 0 \\ 0 \\ 0 \\ 0 \\ -Mg \\ 0 \end{Bmatrix}
$$

this enables us to obtain the values of the dof at nodes (2), (3) and (4), i.e.:

$u_2 = \dfrac{1}{2}\dfrac{Mg\ell^3}{EI}$	$v_2 = -\dfrac{Mg\ell}{ES}$	$\theta_{z2} = -\dfrac{Mg\ell^2}{EI}$
$u_3 = 2\dfrac{Mg\ell^3}{EI}$	$v_3 = -2\dfrac{Mg\ell}{ES}$	$\theta_{z3} = -2\dfrac{Mg\ell^2}{EI}$
$u_4 = 2\dfrac{Mg\ell^3}{EI}$	$v_4 = -2\dfrac{Mg\ell}{ES}\left(1+\dfrac{7}{6}\dfrac{S\ell^2}{I}\right)$	$\theta_{z4} = -\dfrac{5}{2}\dfrac{Mg\ell^2}{EI}$

$$[4.21]$$

➤ *Sub-system (II)*[37]

This is obtained by reverting to the previously deleted rows in system [4.17]. We arrive at (by eliminating the zero terms):

$$\begin{Bmatrix} X_{1\ell} \\ Y_{1\ell} \\ N_{1\ell} \end{Bmatrix} = \begin{bmatrix} -B & 0 & -C \\ 0 & -A & 0 \\ C & 0 & F \end{bmatrix} \bullet \begin{Bmatrix} u_2 \\ v_2 \\ \theta_{z2} \end{Bmatrix}$$

This sub-system can then be written as:

$$\begin{cases} X_{1\ell} = -Bu_2 - C\theta_{z2} \\ Y_{1\ell} = -Av_2 \\ N_{1\ell} = Cu_2 + F\theta_{z2} \end{cases}$$

With the previous values [4.13] and [4.21] of the dof u_2, v_2, θ_{z2}, this leads to:

$X_{1\ell} = 0$	$Y_{1\ell} = Mg$	$N_{1\ell} = Mg\ell$

The global equilibrium of the model is then readily verified.

4.4.7. *Additional study of the behavior of the bracket*

4.4.7.1. *Internal linking forces on each of the isolated elements*

The nodal displacements in the global systems being known, it is possible to know the nodal forces acting on each element. For this it is necessary to go back to

---

37 See [3.45].

the behavior equation in the local system of the element. The transfer matrix defined in section 4.4.3 enables us to find the values of the dof mentioned in the local system of each element. We can then write:

$$\{F\}_{el3} = [k]_{el3} \bullet \{d\}_{el3}$$
$$\underset{Local}{} \quad \underset{Local}{} \quad \underset{Local}{}$$

which shall give us the values of the nodal forces in the local system. These are represented on Figure 4.36. They consist of:

– internal linking actions between adjacent elements;

– external loading;

– external linking actions.

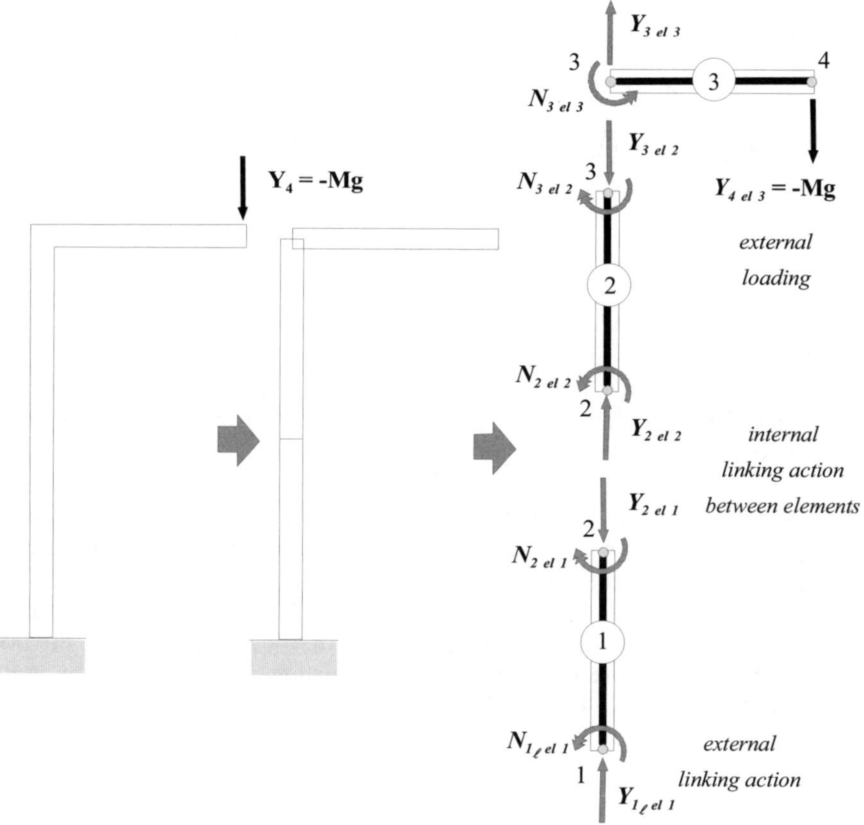

Figure 4.36. *Nodal forces (local coordinates-system): equilibrium of beam elements*

For example, for element 3 (same local and global coordinate systems), we will have:

$$\{F\}_{el3} = [k]_{el3} \bullet \{d\}_{el3}$$
$$Local \qquad Local \qquad Local$$

i.e. (see [4.10]):

$$
\begin{Bmatrix} X_3 = ? \\ Y_3 = ? \\ N_3 = ? \\ X_4 = 0 \\ Y_4 = -Mg \\ N_4 = 0 \end{Bmatrix}_{el3,\,Local}
=
\begin{bmatrix}
\dfrac{ES}{\ell} & 0 & 0 & -\dfrac{ES}{\ell} & 0 & 0 \\[2mm]
0 & \dfrac{12EI}{\ell^3} & \dfrac{6EI}{\ell^2} & 0 & -\dfrac{12EI}{\ell^3} & \dfrac{6EI}{\ell^2} \\[2mm]
0 & \dfrac{6EI}{\ell^2} & \dfrac{4EI}{\ell} & 0 & -\dfrac{6EI}{\ell^2} & \dfrac{2EI}{\ell} \\[2mm]
-\dfrac{ES}{\ell} & 0 & 0 & \dfrac{ES}{\ell} & 0 & 0 \\[2mm]
0 & -\dfrac{12EI}{\ell^3} & -\dfrac{6EI}{\ell^2} & 0 & \dfrac{12EI}{\ell^3} & -\dfrac{6EI}{\ell^2} \\[2mm]
0 & \dfrac{6EI}{\ell^2} & \dfrac{2EI}{\ell} & 0 & -\dfrac{6EI}{\ell^2} & \dfrac{4EI}{\ell}
\end{bmatrix}
\bullet
\begin{Bmatrix} u_3 \\ v_3 \\ \theta_{z3} \\ u_4 \\ v_4 \\ \theta_{z4} \end{Bmatrix}_{el3,\,Local}
$$

with the values found in [4.21], the internal linking actions on element 3 are obtained:

$$\begin{Bmatrix} X_3 = 0 \\ Y_3 = Mg \\ N_3 = Mg\ell \end{Bmatrix}_{Local}$$

NOTES

❏ The search for the nodal linking forces is necessary to dimension the assemblies (bolts, rivets, welding[38]) when building the structure or to verify the stability of the compressed zones (buckling).

❏ These nodal forces are used in the local coordinate system. Here, for element 3 having its local coordinate system blended with the global system, the use of a transfer matrix is not necessary.

In the case of elements 1 and 2 where the local coordinate system is different from the global system the displacements should first be expressed in the local coordinate system before using the behavior equation. Thus it should be written successively (see section 4.4.3):

$$\{d\}_{el} = [P] \bullet \{d\}_{el}$$
$$Local \qquad\qquad Global$$

then

$$\{F\}_{el} = [k]_{el} \bullet \{d\}_{el}$$
$$Local \qquad Local \qquad Local$$

---

38 See Chapter 11.

4.4.7.2. *Normal stresses*

The normal stresses are generated here by normal force $\mathcal{N}$ and bending moment $\mathcal{M}f_z$. We refer to the equations[39]:

- normal force: $\left(\sigma_x\right)_{\mathcal{N}} = \dfrac{\mathcal{N}}{S}$

- bending moment: $\left(\sigma_x\right)_{\mathcal{M}f_z} = -\dfrac{\mathcal{M}f_z}{I} \times y$

corresponding to the distributions in Figure 4.37 below.

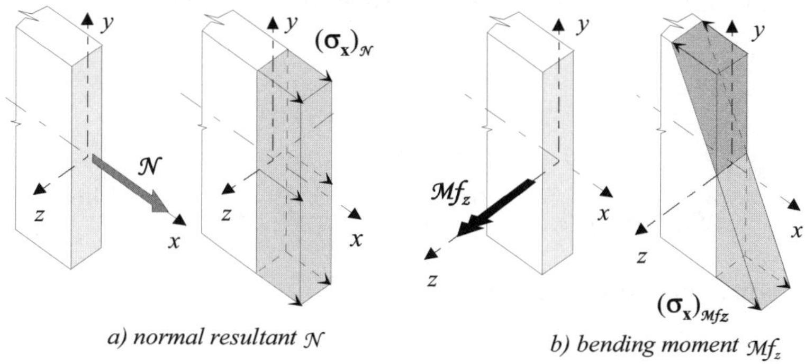

a) normal resultant $\mathcal{N}$ *b) bending moment $\mathcal{M}f_z$*

Figure 4.37. *Normal stresses*

We have seen that knowing the local dof helped in obtaining these normal stresses in the traction-compression elements for $\left(\sigma_x\right)_{\mathcal{N}}$ and plane bending elements for $\left(\sigma_x\right)_{\mathcal{M}f_z}$. Corresponding equations[40] are:

$$\left(\sigma_x\right)_{\mathcal{N}} = \frac{E}{\ell}\left(u_4 - u_3\right)$$

$$\left(\sigma_x\right)_{\mathcal{M}f_z} = -y \times E\left\{\frac{6}{\ell^2}\left(2\overline{x}-1\right)v_3 + \frac{4}{\ell}\left(\frac{3\overline{x}}{2}-1\right)\theta_{z3} - \frac{6}{\ell^2}\left(2\overline{x}-1\right)v_4 + \frac{2}{\ell}\left(3\overline{x}-1\right)\theta_{z4}\right\}$$

---

39 See sections 1.5.2.1 and 1.5.2.6.
40 See Figures 3.8 and 3.14.

with $\overline{x} = \dfrac{x}{\ell}$, where the nodal values u_3, u_4, v_3, v_4, θ_{z3}, θ_{z4} are known (equations [4.21]).

The total normal stress is the sum of the previous equations:

$$\left(\sigma_x\right)_{\mathcal{N}} + \left(\sigma_x\right)_{\mathcal{M}f_z}$$

Thus, the calculation of normal stresses can be made in every cross-section by giving an appropriate value, in the x abscissa, of the current section that we wish to study.

For example, for element 3 and in its associated local coordinate system we shall have at the nodes of this element with the values of the dof obtained in [4.21]:

◆ node (3)[41]: $x = 0$

$$\left(\sigma_x\right)_{\mathcal{N}} = \frac{E}{\ell}(u_4 - u_3) = 0$$

$$\left(\sigma_x\right)_{\mathcal{M}f_z} = -y \times E\left\{\frac{6}{\ell^2}(v_4 - v_3) - \frac{2}{\ell}(\theta_{z4} + 2\theta_{z3})\right\} = \frac{Mg \times \ell}{I} \times y$$

The form of Figure 4.38 is obtained.

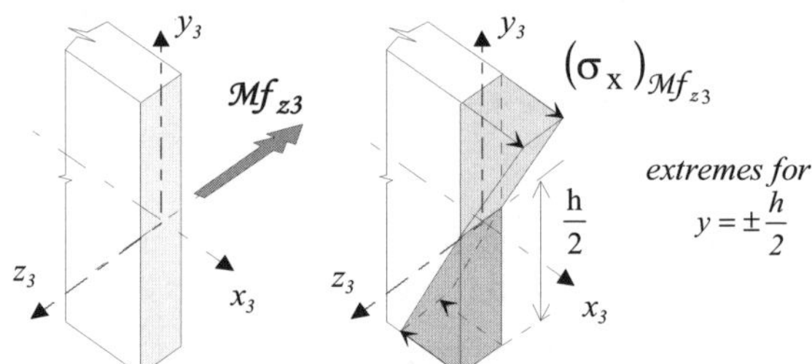

Figure 4.38. *Distribution of the normal stresses at node no. 3*

◆ node (4): $x = \ell$

$$\left(\sigma_x\right)_{\mathcal{N}} = \frac{E}{\ell}(u_4 - u_3) = 0$$

$$\left(\sigma_x\right)_{\mathcal{M}f_z} = -y \times E\left\{\frac{6}{\ell^2}(v_4 - v_3) - \frac{2}{\ell}(2\theta_{z4} + \theta_{z3})\right\} = 0$$

---

41 See Figures 4.32 and 4.36.

4.4.8. *Using computing software*

We have dealt with literal calculus by choosing three elements for modeling the bracket. It could be thought that when a software is used, an increase in the number of elements (increase in the meshing density, which is easy to achieve) will result in an increasing accuracy of the results obtained. In fact, if the load is limited to the force $Y_4 = -Mg$ taken into account here, a greater meshing density is not required. On the contrary, the number of elements can even be reduced to two (a horizontal and a vertical element instead of three). Then, the "exact" solution of the beam-theory is again found. Here three elements have been taken not to increase the accuracy of the model but to illustrate the assembly mechanism of the stiffness matrix. On the contrary, the cases of more complex loading, like the loads in Figure 4.39a for example, would require a greater number of elements. It is the same case when we want to take into account the weights of the elements that are distributed at each of the nodes (see Figure 4.39b).

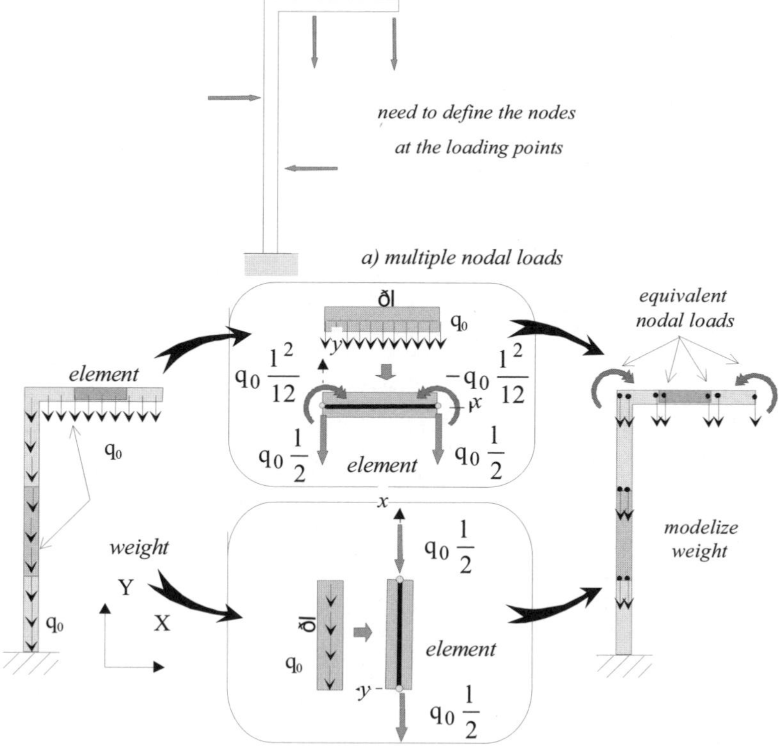

Figure 4.39. *Other types of loads*

NOTES

❐ Taking into account the distributed loads.

Figure 4.39 represents a horizontal element and a vertical element subjected to a distributed load q_0 (N/m). This distributed load can be replaced by equivalent nodal loads that produce the same work as the distributed force when the element deforms under bending and under traction-compression[42]. The method involves obtaining the same potential energy of deformation for the real load and for the equivalent nodal load. It is shown that these nodal forces consist of a force and a moment.

---

42 The calculation to prove this equivalent loading is not carried out as a part of this work. Let us just note that the displacement field $u(x)$, $v(x)$, $\theta_z(x)$ in every section of the x abscissa is known with respect to the nodal displacements. This helps in the calculation mentioned here (see section 3.2.1.2 and [3.21]).

PART 2

Level 2

In order to approach Part 2, it is better to be familiar with the results of Part 1, and especially with text highlighted by a vertical shaded strip, as well as with the summaries at the end of sections and chapters.

As the term "Level 2" suggests, this part of the book is meant for an educational program of higher level in industrial design, i.e. corresponding to a 3-year undergraduate degree or Licence (L3) or bachelor's degree, if necessary completed by a one-year Master's (M1) diploma (EU studies).

Chapter 5 provides the characteristics of the main finite elements found in computing softwares. Next is Chapter 6 which helps with the understanding of the principle of calculation in dynamics. The criteria for designing and the techniques of modeling (Chapters 7 and 8) will be particularly useful in the area of industrial applications.

Chapter 5

Other Types of Finite Elements

We mentioned in Chapter 3[1] that the discretization of a structure by finite elements required an assortment of elements, each one characterized by a known basic stiffness matrix $[k]_{el}$. Some simple elements were examined in section 3.2, and the corresponding principles of calculation of such matrices were shown. Nevertheless, the elements in Chapter 3 enable the study of only a limited category of structures. The software for industrial calculation involves a much wider range of elements that enable adaptation to geometries and the most varied boundary conditions. These elements were only illustrated in Figure 3.36. The main characteristics of these elements are shown in the following sections.

5.1. Return to local and global coordinate systems

5.1.1. *Transfer matrix*

In Chapter 3, [3.9], a coplanar local and global coordinate system having in common a direction $\vec{z} \equiv \vec{Z}$ were used.

We specified the manner in which the components of a vector expressed in one of these systems could be deduced from the components in the other system through a linear relation. We thus defined a so-called "transfer" matrix $[p]$ in summary [3.9], that we will illustrate again in Figure 5.1a below.

---

1 See section 3.1.2.

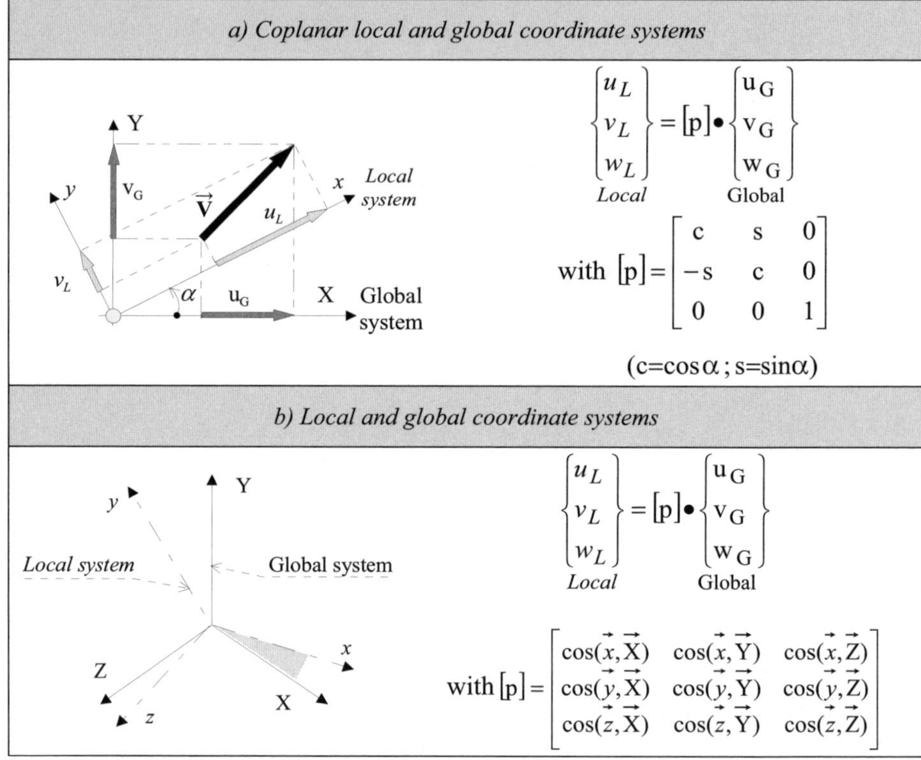

$$\begin{Bmatrix} u_L \\ v_L \\ w_L \end{Bmatrix} = [p] \bullet \begin{Bmatrix} u_G \\ v_G \\ w_G \end{Bmatrix}$$

$$\text{with } [p] = \begin{bmatrix} c & s & 0 \\ -s & c & 0 \\ 0 & 0 & 1 \end{bmatrix}$$

$$\begin{Bmatrix} u_L \\ v_L \\ w_L \end{Bmatrix} = [p] \bullet \begin{Bmatrix} u_G \\ v_G \\ w_G \end{Bmatrix}$$

$$\text{with } [p] = \begin{bmatrix} \cos(\vec{x},\vec{X}) & \cos(\vec{x},\vec{Y}) & \cos(\vec{x},\vec{Z}) \\ \cos(\vec{y},\vec{X}) & \cos(\vec{y},\vec{Y}) & \cos(\vec{y},\vec{Z}) \\ \cos(\vec{z},\vec{X}) & \cos(\vec{z},\vec{Y}) & \cos(\vec{z},\vec{Z}) \end{bmatrix}$$

Figure 5.1. *Transfer matrix*

We can easily extend this concept of a transfer matrix to cases where the two coordinate systems (*Local* and Global) have any direction in space[2] with respect to each other. We thus obtain the form $[p]$ in Figure 5.1b.

While observing that $\cos(\vec{i},\vec{J}) = \cos(\vec{J},\vec{i})$, we note that this transfer matrix verifies the characteristic[3] relation: $[p]^{-1} = [p]^{T}$.

---

2 To establish this transfer matrix, see Chapter 12.

3 In fact, if we invert the relation in Figure 5.1b we have:

$$\begin{Bmatrix} u_G \\ v_G \\ w_G \end{Bmatrix}_{Global} = \begin{bmatrix} \cos(\vec{X},\vec{x}) & \cos(\vec{X},\vec{y}) & \cos(\vec{X},\vec{z}) \\ \cos(\vec{Y},\vec{x}) & \cos(\vec{Y},\vec{y}) & \cos(\vec{Y},\vec{z}) \\ \cos(\vec{Z},\vec{x}) & \cos(\vec{Z},\vec{y}) & \cos(\vec{Z},\vec{z}) \end{bmatrix} \bullet \begin{Bmatrix} u_L \\ v_L \\ w_L \end{Bmatrix}_{Local} = \begin{bmatrix} \cos(\vec{x},\vec{X}) & \cos(\vec{y},\vec{X}) & \cos(\vec{z},\vec{X}) \\ \cos(\vec{x},\vec{Y}) & \cos(\vec{y},\vec{Y}) & \cos(\vec{z},\vec{Y}) \\ \cos(\vec{x},\vec{Z}) & \cos(\vec{y},\vec{Z}) & \cos(\vec{z},\vec{Z}) \end{bmatrix} \bullet \begin{Bmatrix} u_L \\ v_L \\ w_L \end{Bmatrix}_{Local} = [p]^{T} \bullet \begin{Bmatrix} u_L \\ v_L \\ w_L \end{Bmatrix}_{Local}$$

We thus obtain: $[p]^{-1} = [p]^{T}$.

5.1.2. *Summary*

Summary [3.9], already written for coplanar coordinate systems, may thus be immediately generalized for coordinate systems oriented in any direction in space.

<table>
<tr><td colspan="2" align="center">***Cases of any local and global coordinate systems***</td></tr>
<tr><td colspan="2">

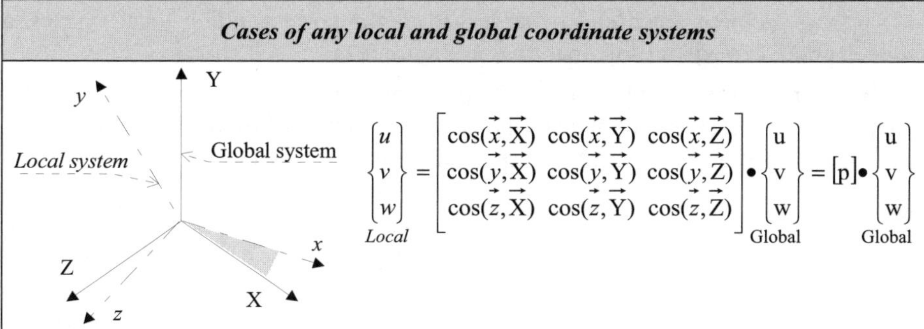

$$\begin{Bmatrix} u \\ v \\ w \end{Bmatrix}_{Local} = \begin{bmatrix} \cos(\vec{x},\vec{X}) & \cos(\vec{x},\vec{Y}) & \cos(\vec{x},\vec{Z}) \\ \cos(\vec{y},\vec{X}) & \cos(\vec{y},\vec{Y}) & \cos(\vec{y},\vec{Z}) \\ \cos(\vec{z},\vec{X}) & \cos(\vec{z},\vec{Y}) & \cos(\vec{z},\vec{Z}) \end{bmatrix} \bullet \begin{Bmatrix} u \\ v \\ w \end{Bmatrix}_{Global} = [p] \bullet \begin{Bmatrix} u \\ v \\ w \end{Bmatrix}_{Global}$$

</td></tr>
<tr><td colspan="2">

Degrees of freedom of an element:

$$\{d\}_{el}^{Local} = [P] \bullet \{d\}_{el}^{Global} \quad \text{with} \quad [P]_{\substack{transfer \\ matrix}} = \begin{bmatrix} [p] & 0 & 0 \\ 0 & [p] & 0 \\ 0 & 0 & \ddots \end{bmatrix} \begin{matrix} node\ 1 \\ node\ 2 \\ \cdots\cdots \end{matrix}$$

</td></tr>
<tr><td colspan="2">

Nodal force on an element:

$$\{F\}_{el}^{Local} = [P] \bullet \{F\}_{el}^{Global}$$

</td></tr>
<tr><td colspan="2" align="center">***Behavior relation of the element***</td></tr>
<tr><td>

Local system:

$$\{F\}_{el}^{Local} = [k]_{el}^{Local} \bullet \{d\}_{el}^{Local}$$

</td><td>

Global system:

$$\{F\}_{el}^{Global} = [k]_{el}^{Global} \bullet \{d\}_{el}^{Global}$$

</td></tr>
<tr><td colspan="2" align="center">

$$[k]_{el}^{Global} = [P]^{T} \bullet [k]_{el}^{Local} \bullet [P]$$

</td></tr>
<tr><td colspan="2">

Potential energy of deformation of the element:

$$E_{\substack{pot. \\ element}} = \frac{1}{2}\{d\}_{el}^{T} \bullet [k]_{el} \bullet \{d\}_{el} \Big|_{Local} = \frac{1}{2}\{d\}_{el}^{T} \bullet [k]_{el} \bullet \{d\}_{el} \Big|_{Global}$$

</td></tr>
</table>

$$[5.1]$$

5.2. Complete beam element (any loading case)

5.2.1. *Preliminary comments*

In Chapter 3, section 3.2 we studied:

– a truss element: each end or node had in the corresponding local system, a single degree of freedom (dof) specified in Figure 5.2a;

– a beam element in plane bending, deflecting in a principal plane, shown for example by (xy). Each end or node of the beam had in the corresponding local system, 2 dof: one translation and one rotation shown as v and θ_z in Figure 5.2b.

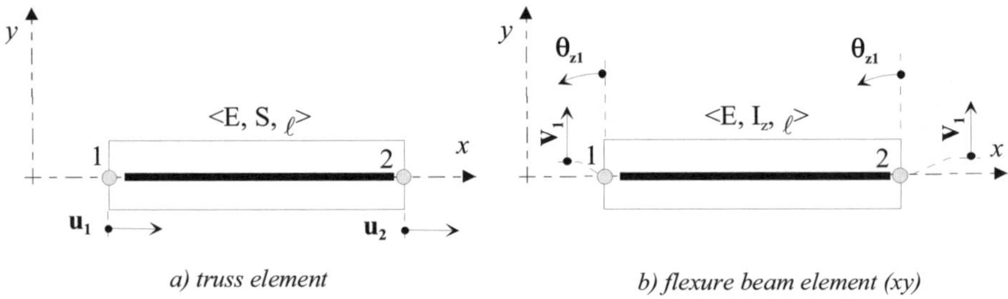

a) truss element   *b) flexure beam element (xy)*

Figure 5.2. *Elements already studied*

In Chapter 4, section 4.2.4, we associated these two elements in order to obtain a single element with better performance, which would work both in plane bending and in traction-compression. Adopting the same approach, we will now try to complete this element and use it in space, so as to be able to use its most general performance that includes all the properties of beams (traction-compression, bending, torsion).

The beam element to be defined will be able to bend in each of the principal planes: (xy) and (xz). We therefore deduce the following degree of freedoms at one node and in the local coordinate system of the element:

– traction-compression: u;

– bending: (xy): v, θ_z;

– bending (xz): w, θ_y;

– torsional: θ_x,

i.e. 6 dof in the local coordinate system. There are small displacements of the cross-section whose geometric center coincides with the node in question. The beam element possessing an original node and an end node shall therefore be characterized by:

$$2 \times 6 = 12 \, \text{dof}$$

These degrees of freedom are illustrated in Figure 5.3. Each degree of freedom is associated with its corresponding resultant force obtained from the projections of the resultant force and the moment exerted on the element at each of its ends by its surroundings. We thus obtain $2 \times 6 = 12$ "associated forces" represented in Figure 5.3. The stiffness matrix of this element will thus have 12 rows and 12 columns.

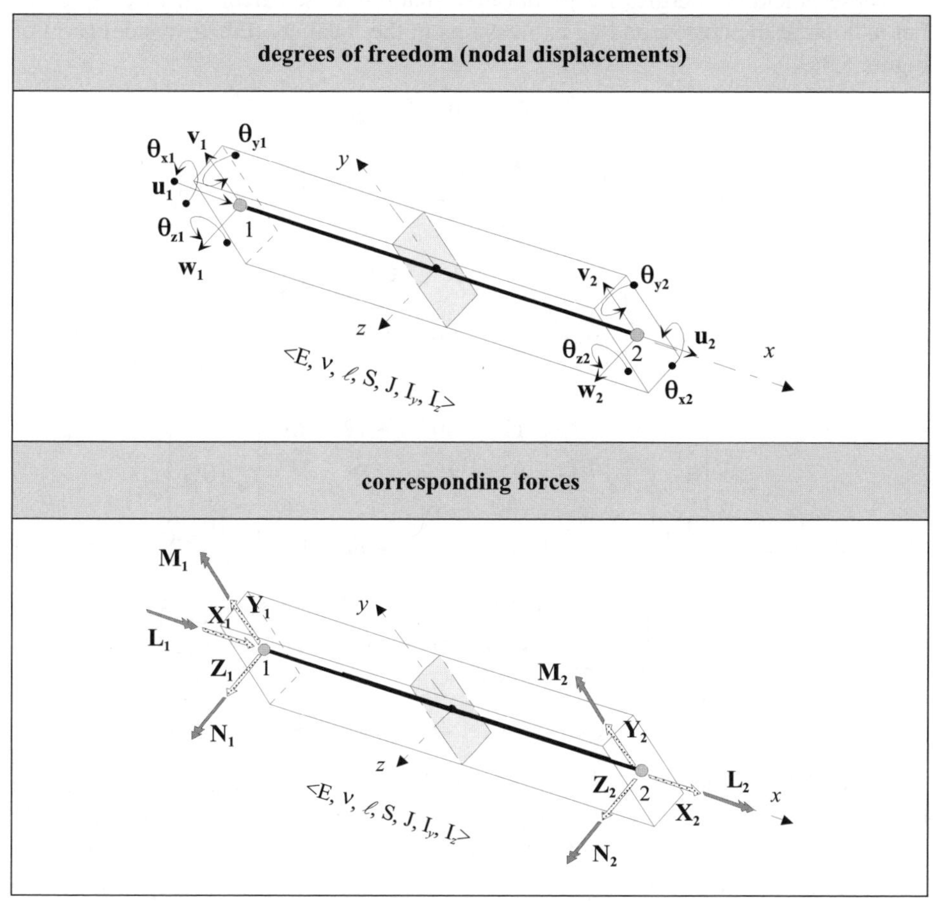

Figure 5.3. *Complete beam element*

5.2.2. *Obtaining the stiffness matrix in the local coordinate system*

Let us note the potential deformation energy of the element that is calculated for a set of 12 dof written as $\{d\}_{el}$. We will obtain:

$$\underset{element}{E_{pot.}} = \frac{1}{2}\{d\}_{el}^{T} \bullet [k]_{el} \bullet \{d\}_{el} = \underset{\substack{traction\text{-}compression \\ element}}{E_{pot.}} + \underset{\substack{torsional \\ element}}{E_{pot.}} + \underset{\substack{bending \\ element\,(xy)}}{E_{pot.}} + \underset{\substack{bending \\ element(xz)}}{E_{pot.}}$$

These potential energies for partial deformations were calculated in Chapter 3[4]. Let us look at the corresponding expressions in the local coordinate system (xyz) of Figure 5.3.

$$\underset{element}{E_{pot.}} = \frac{1}{2}\begin{Bmatrix} u_1 \\ u_2 \end{Bmatrix}^{T} \bullet \begin{bmatrix} \dfrac{ES}{\ell} & -\dfrac{ES}{\ell} \\ -\dfrac{ES}{\ell} & \dfrac{ES}{\ell} \end{bmatrix} \bullet \begin{Bmatrix} u_1 \\ u_2 \end{Bmatrix} \cdots$$

$$\cdots + \frac{1}{2}\begin{Bmatrix} \theta_{x1} \\ \theta_{x2} \end{Bmatrix}^{T} \bullet \begin{bmatrix} \dfrac{GJ}{\ell} & -\dfrac{GJ}{\ell} \\ -\dfrac{GJ}{\ell} & \dfrac{GJ}{\ell} \end{bmatrix} \bullet \begin{Bmatrix} \theta_{x1} \\ \theta_{x2} \end{Bmatrix} \cdots$$

$$\cdots + \frac{1}{2}\begin{Bmatrix} v_1 \\ \theta_{z1} \\ v_2 \\ \theta_{z2} \end{Bmatrix}^{T} \bullet \frac{EI_z}{\ell^3}\begin{bmatrix} 12 & 6\ell & -12 & 6\ell \\ 6\ell & 4\ell^2 & -6\ell & 2\ell^2 \\ -12 & -6\ell & 12 & -6\ell \\ 6\ell & 2\ell^2 & -6\ell & 4\ell^2 \end{bmatrix} \bullet \begin{Bmatrix} v_1 \\ \theta_{z1} \\ v_2 \\ \theta_{z2} \end{Bmatrix} \cdots$$

$$\cdots + \frac{1}{2}\begin{Bmatrix} w_1 \\ \theta_{y1} \\ w_2 \\ \theta_{y2} \end{Bmatrix}^{T} \bullet \frac{EI_y}{\ell^3}\begin{bmatrix} 12 & -6\ell & -12 & -6\ell \\ -6\ell & 4\ell^2 & 6\ell & 2\ell^2 \\ -12 & 6\ell & 12 & 6\ell \\ -6\ell & 2\ell^2 & 6\ell & 4\ell^2 \end{bmatrix} \bullet \begin{Bmatrix} w_1 \\ \theta_{y1} \\ w_2 \\ \theta_{y2} \end{Bmatrix}$$

Following the method already used[5], we can obtain the stiffness matrix of the beam element (12×12) by reloading the partial stiffness matrixes in a (12×12) table. Let us define the column of the 12 dof of the element in the following order:

$$\{d\}_{el} = \begin{bmatrix} u_1 & v_1 & w_1 & \theta_{x1} & \theta_{y1} & \theta_{z1} & u_2 & v_2 & w_2 & \theta_{x2} & \theta_{y2} & \theta_{z2} \end{bmatrix}^{T} \quad [5.2]$$

---

4 See Figures 3.8, 3.10 and 3.14.
5 See sections 3.3.2 or 4.2.4.

We will thus obtain:

$$
[k]_{el}^{Local} =
\begin{bmatrix}
\frac{ES}{\ell} & 0 & 0 & 0 & 0 & 0 & -\frac{ES}{\ell} & 0 & 0 & 0 & 0 & 0 \\
 & 0 & 0 & 0 & 0 & 0 & 0 & 0 & 0 & 0 & 0 & 0 \\
 & & 0 & 0 & 0 & 0 & 0 & 0 & 0 & 0 & 0 & 0 \\
 & S & & 0 & 0 & 0 & 0 & 0 & 0 & 0 & 0 & 0 \\
 & & Y & & 0 & 0 & 0 & 0 & 0 & 0 & 0 & 0 \\
 & & & M & & 0 & 0 & 0 & 0 & 0 & 0 & 0 \\
 & & & & M & & \frac{ES}{\ell} & 0 & 0 & 0 & 0 & 0 \\
 & & & & & E & & 0 & 0 & 0 & 0 & 0 \\
 & & & & & & T & & 0 & 0 & 0 & 0 \\
 & & & & & & & & & 0 & 0 & 0 \\
 & & & & & & & R & & & 0 & 0 \\
 & & & & & & & & Y & & & 0
\end{bmatrix}
+
\begin{bmatrix}
0 & 0 & 0 & 0 & 0 & 0 & 0 & 0 & 0 & 0 & 0 & 0 \\
 & 0 & 0 & 0 & 0 & 0 & 0 & 0 & 0 & 0 & 0 & 0 \\
 & & 0 & 0 & 0 & 0 & 0 & 0 & 0 & 0 & 0 & 0 \\
 & S & & \frac{GJ}{\ell} & 0 & 0 & 0 & 0 & 0 & -\frac{GJ}{\ell} & 0 & 0 \\
 & & Y & & 0 & 0 & 0 & 0 & 0 & 0 & 0 & 0 \\
 & & & M & & 0 & 0 & 0 & 0 & 0 & 0 & 0 \\
 & & & & & & 0 & 0 & 0 & 0 & 0 & 0 \\
 & & & & M & & & 0 & 0 & 0 & 0 & 0 \\
 & & & & & E & & & 0 & 0 & 0 & 0 \\
 & & & & & & T & & & \frac{GJ}{\ell} & 0 & 0 \\
 & & & & & & & R & & & 0 & 0 \\
 & & & & & & & & Y & & & 0
\end{bmatrix}
+ \dots
$$

$$
\dots +
\begin{bmatrix}
0 & 0 & 0 & 0 & 0 & 0 & 0 & 0 & 0 & 0 & 0 & 0 \\
 & \frac{12EI_z}{\ell^3} & 0 & 0 & 0 & \frac{6EI_z}{\ell^2} & 0 & -\frac{12EI_z}{\ell^3} & 0 & 0 & 0 & \frac{6EI_z}{\ell^2} \\
 & & 0 & 0 & 0 & 0 & 0 & 0 & 0 & 0 & 0 & 0 \\
 & S & & 0 & 0 & 0 & 0 & 0 & 0 & 0 & 0 & 0 \\
 & & Y & & 0 & 0 & 0 & 0 & 0 & 0 & 0 & 0 \\
 & & & M & & \frac{4EI_z}{\ell} & 0 & -\frac{6EI_z}{\ell^2} & 0 & 0 & 0 & \frac{2EI_z}{\ell} \\
 & & & & M & & 0 & 0 & 0 & 0 & 0 & 0 \\
 & & & & & E & & 0 & 0 & 0 & 0 & 0 \\
 & & & & & & T & & 0 & 0 & 0 & 0 \\
 & & & & & & & & 0 & 0 & 0 & 0 \\
 & & & & & & & R & & 0 & 0 & 0 \\
 & & & & & & & & Y & & & \frac{4EI_z}{\ell}
\end{bmatrix}
+ \dots
$$

$$
\dots +
\begin{bmatrix}
0 & 0 & 0 & 0 & 0 & 0 & 0 & 0 & 0 & 0 & 0 & 0 \\
 & 0 & 0 & 0 & 0 & 0 & 0 & 0 & 0 & 0 & 0 & 0 \\
 & & \frac{12EI_y}{\ell^3} & 0 & -\frac{6EI_y}{\ell^2} & 0 & 0 & 0 & -\frac{12EI_y}{\ell^3} & 0 & -\frac{6EI_y}{\ell^2} & 0 \\
 & S & & 0 & 0 & 0 & 0 & 0 & 0 & 0 & 0 & 0 \\
 & & Y & & \frac{4EI_y}{\ell} & 0 & 0 & 0 & \frac{6EI_y}{\ell^2} & 0 & \frac{2EI_y}{\ell} & 0 \\
 & & & M & & 0 & 0 & 0 & 0 & 0 & 0 & 0 \\
 & & & & & & 0 & 0 & 0 & 0 & 0 & 0 \\
 & & & & M & & & 0 & 0 & 0 & 0 & 0 \\
 & & & & & E & & & \frac{12EI_y}{\ell^3} & 0 & \frac{6EI_y}{\ell^2} & 0 \\
 & & & & & & T & & & 0 & 0 & 0 \\
 & & & & & & & R & & & \frac{4EI_y}{\ell} & 0 \\
 & & & & & & & & Y & & & 0
\end{bmatrix}
$$

After addition, the stiffness matrix of the complete beam element in its local coordinate system is written as:

$$[k]_{el}^{Local} = \begin{bmatrix}
\dfrac{ES}{\ell} & 0 & 0 & 0 & 0 & 0 & -\dfrac{ES}{\ell} & 0 & 0 & 0 & 0 & 0 \\[2mm]
& \dfrac{12EI_z}{\ell^3} & 0 & 0 & 0 & \dfrac{6EI_z}{\ell^2} & 0 & -\dfrac{12EI_z}{\ell^3} & 0 & 0 & 0 & \dfrac{6EI_z}{\ell^2} \\[2mm]
& & \dfrac{12EI_y}{\ell^3} & 0 & -\dfrac{6EI_y}{\ell^2} & 0 & 0 & 0 & -\dfrac{12EI_y}{\ell^3} & 0 & -\dfrac{6EI_y}{\ell^2} & 0 \\[2mm]
& & & \dfrac{GJ}{\ell} & 0 & 0 & 0 & 0 & 0 & -\dfrac{GJ}{\ell} & 0 & 0 \\[2mm]
& & & & \dfrac{4EI_y}{\ell} & 0 & 0 & 0 & \dfrac{6EI_y}{\ell^2} & 0 & \dfrac{2EI_y}{\ell} & 0 \\[2mm]
& & & & & \dfrac{4EI_z}{\ell} & 0 & -\dfrac{6EI_z}{\ell^2} & 0 & 0 & 0 & \dfrac{2EI_z}{\ell} \\[2mm]
& & & & & & \dfrac{ES}{\ell} & 0 & 0 & 0 & 0 & 0 \\[2mm]
& & & & & & & \dfrac{12EI_z}{\ell^3} & 0 & 0 & 0 & -\dfrac{6EI_z}{\ell^2} \\[2mm]
& & & \text{SYMMETRY} & & & & & \dfrac{12EI_y}{\ell^3} & 0 & \dfrac{6EI_y}{\ell^2} & 0 \\[2mm]
& & & & & & & & & \dfrac{GJ}{\ell} & 0 & 0 \\[2mm]
& & & & & & & & & & \dfrac{4EI_y}{\ell} & 0 \\[2mm]
& & & & & & & & & & & \dfrac{4EI_z}{\ell}
\end{bmatrix}$$

$$[5.3]$$

Thus, the behavior of this beam element in the local coordinate system[6] is:

$$\{F\}_{el} = [k]_{el} \bullet \{d\}_{el}$$

where the nodal forces correspond to the dof as shown below:

$$\{d\}_{el} = \begin{bmatrix} u_1 & v_1 & w_1 & \theta_{x1} & \theta_{y1} & \theta_{z1} & u_2 & v_2 & w_2 & \theta_{x2} & \theta_{y2} & \theta_{z2} \end{bmatrix}^T$$
$$\updownarrow \quad \updownarrow \quad \updownarrow \quad \updownarrow \quad \updownarrow \quad \updownarrow \quad \updownarrow \quad \updownarrow \quad \updownarrow \quad \updownarrow \quad \updownarrow \quad \updownarrow$$
$$\{F\}_{el} = \begin{bmatrix} X_1 & Y_1 & Z_1 & L_1 & M_1 & N_1 & X_2 & Y_2 & Z_2 & L_2 & M_2 & N_2 \end{bmatrix}^T$$

$$[5.4]$$

NOTES

❏ When discretizing a structure, this finite element shall be modeled by the length of its mean line ℓ and its two end nodes. We will observe, on stiffness matrix [5.3], that it will also be necessary to provide:

---

6 See [2.121].

– properties of the material: modulus of longitudinal elasticity E and Poisson's ratio v^7;

– characteristics of the cross-section: area S, torsion constant J, quadratic moments with respect to axes $\vec{y}$ and $\vec{z}$ written respectively as I_y and I_z.

❏ The beam element shown in Figure 5.4 appears to be floating in space, i.e. it is not properly linked as defined in section 2.5.2. This absence of linkage allows rigid-body motions. Here, there are six possible rigid-body motions:

– three translations along the three directions $\vec{x}$, $\vec{y}$, $\vec{z}$;

– three rotations around the same directions.

Stiffness matrix [5.3] is thus singular.

❏ *Local* and global coordinate system: in Figure 5.4, a global coordinate system (XYZ) that is distinct from the *local* system (*xyz*) is shown.

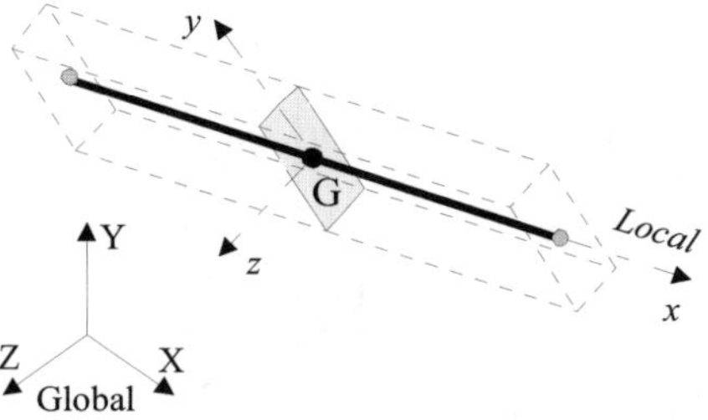

Figure 5.4. *Local and global coordinate systems*

We must recall that the projections of the same vector in each of these two coordinate systems are deduced from the relations in [5.1]. We can see for the vectors in question several groups, each one having three components:

– for the dof:

$$\{d\}_{el} = \begin{bmatrix} \underbrace{u_1 \quad v_1 \quad w_1}_{\substack{\textit{vector translation} \\ \textit{of node 1}}} & \vdots & \underbrace{\theta_{x1} \quad \theta_{y1} \quad \theta_{z1}}_{\substack{\textit{vector rotation} \\ \textit{of section 1} \\ \textit{to node 1}}} & \vdots & \underbrace{u_2 \quad v_2 \quad w_2}_{\substack{\textit{vector translation} \\ \textit{of node 2}}} & \vdots & \underbrace{\theta_{x2} \quad \theta_{y2} \quad \theta_{z2}}_{\substack{\textit{vector rotation} \\ \textit{of section 2} \\ \textit{to node 2}}} \end{bmatrix}^T$$

---

7 For the value of G, the property: $G = \dfrac{E}{2(1+v)}$ is used (see Chapter 10, section 10.1.5.2).

– for the internal forces:

$$\{F\}_{el} = [\; \underbrace{X_1 \quad Y_1 \quad Z_1}_{\substack{\textit{force} \\ \textit{at node 1}}} \; \vdots \; \underbrace{L_1 \quad M_1 \quad N_1}_{\substack{\textit{moment} \\ \textit{at node 1} \\ \textit{on section 1}}} \; \vdots \; \underbrace{X_2 \quad Y_2 \quad Z_2}_{\substack{\textit{force} \\ \textit{at node 2}}} \; \vdots \; \underbrace{L_2 \quad M_2 \quad N_2}_{\substack{\textit{moment} \\ \textit{at node 2} \\ \textit{on section 2}}} \;]^T$$

each of these vectors can be written in both the coordinate systems through matrix $[p]$ appearing in [5.1].

We thus obtain:

$$\begin{Bmatrix} u_1 \\ v_1 \\ w_1 \\ \cdots \\ \theta_{x1} \\ \theta_{y1} \\ \theta_{z1} \\ \cdots \\ u_2 \\ v_2 \\ w_2 \\ \cdots \\ \theta_{x2} \\ \theta_{y2} \\ \theta_{z2} \end{Bmatrix}_{Local} = \begin{bmatrix} [p]_{3\times3} & 0 & 0 & 0 \\ 0 & [p]_{3\times3} & 0 & 0 \\ 0 & 0 & [p]_{3\times3} & 0 \\ 0 & 0 & 0 & [p]_{3\times3} \end{bmatrix} \bullet \begin{Bmatrix} u_1 \\ v_1 \\ w_1 \\ L \\ \theta_{x1} \\ \theta_{y1} \\ \theta_{z1} \\ L \\ u_2 \\ v_2 \\ w_2 \\ L \\ \theta_{x2} \\ \theta_{y2} \\ \theta_{z2} \end{Bmatrix}_{Global}$$

where the matrix is the *transfer matrix* $[P]\,(12\times12)$.

$$\Leftrightarrow \quad \{d\}_{el}^{Local} = [P] \bullet \{d\}_{el}^{Global} \qquad [5.5]$$

$$\begin{Bmatrix} X_1 \\ Y_1 \\ Z_1 \\ \cdots \\ L_1 \\ M_1 \\ N_1 \\ \cdots \\ X_2 \\ Y_2 \\ Z_2 \\ \cdots \\ L_2 \\ M_2 \\ N_2 \end{Bmatrix}_{Local} = \begin{bmatrix} [p]_{3\times3} & 0 & 0 & 0 \\ 0 & [p]_{3\times3} & 0 & 0 \\ 0 & 0 & [p]_{3\times3} & 0 \\ 0 & 0 & 0 & [p]_{3\times3} \end{bmatrix} \bullet \begin{Bmatrix} X_1 \\ Y_1 \\ Z_1 \\ \cdots \\ L_1 \\ M_1 \\ N_1 \\ \cdots \\ X_2 \\ Y_2 \\ Z_2 \\ \cdots \\ L_2 \\ M_2 \\ N_2 \end{Bmatrix}_{Global}$$

where the matrix is the *transfer matrix* $[P]\,(12\times12)$.

$$\Leftrightarrow \quad \{F\}_{el}^{Local} = [P] \bullet \{F\}_{el}^{Global} \qquad [5.6]$$

$[P]$ is thus the transfer matrix between the global and the local coordinate system. We can therefore use it following the method described in [5.1]. The result is summarized in Figure 5.8.

❏ When the dof of the beam element in the local coordinate system are known, we can deduce the normal stress. We will thus obtain[8]:

$$\sigma_x = \frac{\mathcal{N}}{S} + \frac{\mathcal{M}f_y \times z}{I_y} - \frac{\mathcal{M}f_z \times y}{I_z} \qquad [5.7]$$

or:

$$\sigma_x = E\frac{du}{dx} - z \times E\frac{d^2w}{dx^2} - y \times E\frac{d^2v}{dx^2} \qquad [5.8]$$

Remember that the interpolation functions defined for the displacements[9] allow the expression of u, v and in an identical manner that of w in any cross-section, from the knowledge of the dof in the local coordinate system. This will enable access to the distribution of the normal stress σ_x for any abscissa $0 \le x \le \ell$.

❏ In practice, finite element softwares recognize only the "globalized" characteristics of the cross-section of the element, that is S, J, I_y, I_z. Details of the geometry of the section "escape" the postprocessor. Therefore, the software is generally programmed to calculate only the normal stress:

– the calculation is done for particular values of x: for example in the two end sections ($x = 0$ and $x = \ell$)

– the calculation is based on relation [5.7] where the normal resultant and bending moments $\mathcal{N}$, $\mathcal{M}f_y$ and $\mathcal{M}f_z$ are previously calculated from the behavior relation $\{F\}_{el} = [k]_{el} \bullet \{d\}_{el}$.

In fact, we can observe in Figure 5.5 that very close to node 1, the resultant cohesion force and moment can be referred to as $\{Coh_{/1}\} = \left\{ \begin{array}{c} \overrightarrow{\mathcal{R}_1} \\ \overrightarrow{\mathcal{M}_{1/G}} \end{array} \right\}$. The equilibrium of the infinitesimal segment, as illustrated, results in the following: X_1, Y_1, Z_1 are the projections of $-\overrightarrow{\mathcal{R}_1}$ and L_1, N_1, M_1 are the projections of $-\overrightarrow{\mathcal{M}_{1/G}}$.

---

8 See sections 3.2.1.2 (*NOTES*) and 3.2.3.2 (*NOTES*). See also sections 1.5.2.1, 1.5.2.5 and 1.5.2.6.
9 See [3.10] and [3.21].

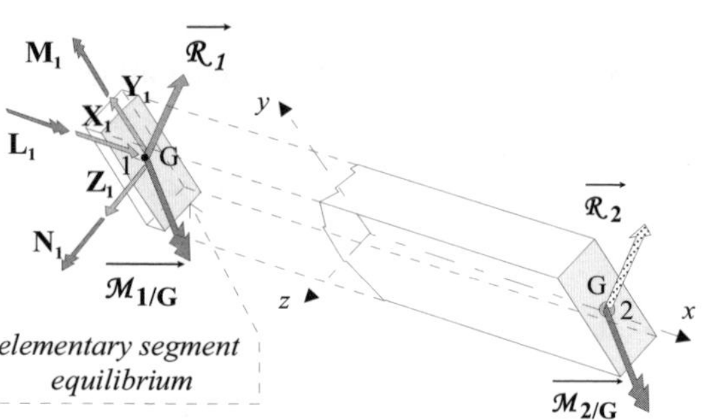

Figure 5.5. *Resultant cohesion force and moment, and nodal forces*

In node 2, the resultant cohesion forces and moments are written as

$$\{Coh_{/2}\}=\left\{\frac{\overrightarrow{\mathcal{R}_2}}{\overrightarrow{\mathcal{M}_{2/G}}}\right\}.$$

Here X_2, Y_2, Z_2 are the projections of $\overrightarrow{\mathcal{R}_2}$ and L_2, N_2, M_2 are the projections of $\overrightarrow{\mathcal{M}_{2/G}}$ [10]. We can thereby deduce the normal resultant and bending moments necessary for the calculation of the normal stress in 1 and 2:

– at node 1: $\mathcal{N}_1 = -X_1$; $\mathcal{M}f_{y_1} = -M_1$; $\mathcal{M}f_{z1} = -N_1$;

– at node 2: $\mathcal{N}_2 = X_2$; $\mathcal{M}f_{y_2} = M_2$; $\mathcal{M}f_{z2} = N_2$.

❑ In relations [5.7] the ratios $\dfrac{z}{I_y}$ and $\dfrac{y}{I_z}$ appear. We must therefore provide the software with the maximum values $\dfrac{z}{I_y}$ and $\dfrac{y}{I_z}$ of these quotients for the cross-section of the element as data input. They are entered in the following forms called "section modulus":

---

10 We will remember that the two shaded cross-sections where the resultant cohesion forces and moments are found are directed by the same normal axis $\vec{x}$. The software assigns an axis $\vec{x}$ to each beam element, which can be recognized by querying the postprocessor. It can be useful if we want to use the nodal forces given above. This is especially the case when it is necessary to control the resistance of structural joints: screws, rivets, welded joints (see Chapter 11).

$$\frac{I_y}{z_{max}} \text{ and } \frac{I_z}{y_{max}} \qquad\qquad [5.9]$$

On the basis of formula [5.7], the software retains in each end section the combination that results in maximum stress in absolute value. Note that this presupposes that the geometry of the section verifies:

$$-y_{max} \le y \le y_{max} \; ; \quad -z_{max} \le z \le z_{max}$$

This is not always the case. It is thus, in general, an approximate pessimistic calculation, unless the section is able to accurately verify this geometric property.

❑ As we mentioned earlier, in the case of a beam element in plane bending[11], the knowledge of dof values is not sufficient to access the shear stress distribution induced by the shear force in the plane of cross-sections[12]. This is why we do not find this type of information in software postprocessors. The same comment is valid for shear stresses due to torsional[13] couple.

5.2.3. *Improvement in performances of this beam element*

5.2.3.1. *Supplementary deformation due to the shear*

We mentioned in Chapter 2, section 2.3.1.4, that for plane bending in principal plane (xy), the beam element deformation resulted from (see Figure 2.16):

– an elementary rotation $d\theta_z$;

– an average distortion due to the shear force, and written as $\left(\dfrac{dv}{dx} - \theta_z\right)$.

Leading to the elementary potential energy of deformation written as[14]:

$$E_{pot.bending} = \frac{1}{2}\int_0^\ell E \times I_z \times \left(\frac{d\theta_z}{dx}\right)^2 \times dx + \frac{1}{2}\int_0^\ell G \times k_y \times S \times \left(\frac{dv}{dx} - \theta_z\right)^2 \times dx$$

where coefficient of reduced section – or shear coefficient – k_y appeared[15].

---

11 See section 3.2.3.2 (*NOTES*).

12 Except in specific cases of standard simple sections for which we can express shear stress in analytical form. In the case of any sections, it is possible to calculate the shear stresses numerically by referring to the detailed description in Chapter 9 (see [9.60]). This involves making available a specific additional software working from the input of the geometry of the section.

13 See sections 3.2.2.2 (*NOTES*) and 9.3.2.5 and [9.27].

14 See [2.31].

Since the beginning of our study of bending beam elements, we have placed ourselves within the framework of Bernoulli's[16] hypothesis that neglects the average distortion $\left(\dfrac{dv}{dx}-\theta_z\right)$. In this case, the second integral above disappears. In fact, except in specific cases mentioned a little later in the notes, the contribution of this second integral (written as Φ_y in stiffness matrix [5.10]) is negligible in comparison to that of the first. Let us remember that it would be the same if we were to examine the bending in the principal plane (xz), where a complementary energy due to the shear effort $\mathcal{T}_z$ appears, and is linked to a shear correction factor k_z (the contribution is written as Φ_z in stiffness matrix [5.10]). When these complementary energies[17] are taken into account, a stiffness matrix modified as follows is obtained:

$$
[k]_{el}^{Local} =
\begin{bmatrix}
\dfrac{ES}{\ell} & 0 & 0 & 0 & 0 & 0 & -\dfrac{ES}{\ell} & 0 & 0 & 0 & 0 & 0 \\[2ex]
 & \dfrac{12EI_z}{\ell^3(1+\Phi_y)} & 0 & 0 & 0 & \dfrac{6EI_z}{\ell^2(1+\Phi_y)} & 0 & -\dfrac{12EI_z}{\ell^3(1+\Phi_y)} & 0 & 0 & 0 & \dfrac{6EI_z}{\ell^2(1+\Phi_y)} \\[2ex]
 & & \dfrac{12EI_y}{\ell^3(1+\Phi_z)} & 0 & \dfrac{6EI_y}{\ell^2(1+\Phi_z)} & 0 & 0 & 0 & -\dfrac{12EI_y}{\ell^3(1+\Phi_z)} & 0 & -\dfrac{6EI_y}{\ell^2(1+\Phi_z)} & 0 \\[2ex]
S & & & \dfrac{GJ}{\ell} & 0 & 0 & 0 & 0 & 0 & -\dfrac{GJ}{\ell} & 0 & 0 \\[2ex]
 & Y & & & \dfrac{(4+\Phi_z)EI_y}{\ell(1+\Phi_z)} & 0 & 0 & 0 & \dfrac{6EI_y}{\ell^2(1+\Phi_z)} & 0 & \dfrac{(2-\Phi_z)EI_y}{\ell(1+\Phi_z)} & 0 \\[2ex]
 & & M & & & \dfrac{(4+\Phi_y)EI_z}{\ell(1+\Phi_y)} & 0 & -\dfrac{6EI_z}{\ell^2(1+\Phi_y)} & 0 & 0 & 0 & \dfrac{(2-\Phi_y)EI_z}{\ell(1+\Phi_y)} \\[2ex]
 & & & M & & & \dfrac{ES}{\ell} & 0 & 0 & 0 & 0 & 0 \\[2ex]
 & & & & E & & & \dfrac{12EI_z}{\ell^3(1+\Phi_y)} & 0 & 0 & 0 & -\dfrac{6EI_z}{\ell^2(1+\Phi_y)} \\[2ex]
 & & & & & T & & & \dfrac{12EI_y}{\ell^3(1+\Phi_z)} & 0 & \dfrac{6EI_y}{\ell^2(1+\Phi_z)} & 0 \\[2ex]
 & & & & & & R & & & \dfrac{GJ}{\ell} & 0 & 0 \\[2ex]
 & & & & & & & Y & & & \dfrac{(4+\Phi_z)EI_y}{\ell(1+\Phi_z)} & 0 \\[2ex]
 & & & & & & & & & & & \dfrac{(4+\Phi_y)EI_z}{\ell(1+\Phi_y)}
\end{bmatrix}
$$

$$[5.10]$$

---

15 The principle of calculation of the two shear coefficients is described in Chapter 9 and summarized in Table [9.58].

16 See [2.32].

17 The calculation is not described in this book. For such calculation, we can no longer use approximation $\theta_z \cong \dfrac{dv}{dx}$. We should use the equilibrium and performance relations summarized in Chapter 9, Figure 9.59.

in which $\quad \Phi_y = \dfrac{12EI_z}{\ell^2 Gk_y S} \; ; \; \Phi_z = \dfrac{12EI_y}{\ell^2 Gk_z S}$

NOTES

❏ Numerical values of shear coefficients k_y and k_z are known in the case of geometrically simple sections[18].

For geometrically complex sections, it is necessary to have a specific additional software that will be provided with the geometry of the section[19].

❏ It is recommended to use the shear correction factors k_y and k_z (or even, depending on the equations used, the shear correction factors $k_y S$ and $k_z S$) in the following cases:

– for static problems: when we have to use very short beams or composite beams (particularly sandwich beams);

– for dynamic problems[20]: when we wish to consider a large number of frequencies and vibration modes that are specific to the structure.

❏ In general cases (other than specific cases mentioned here above), we seldom influence the result of the calculations by providing the values by default:

$$k_y = k_z = 1 \text{ (or } k_y S = k_z S = S)$$

❏ If, in expression [5.10], we delete the terms Φ_y and Φ_z characterizing the shear force, we revert to the simplified form [5.3] that suffices in general cases.

5.2.3.2. *Combination of bending and torsion*

In the figures of the previous section, we showed a beam element with a rectangular section (it had a center of symmetry). Indeed, in section 3.2.2 we had restricted our study of torsion to beams whose cross-section exhibited a center of symmetry. In such a case, the center of symmetry is blended together with the geometric center through which the local-coordinate axis $\bar{x}$ passes. This obviously concerns specific cases: rectangular, circular, tubular, IPN beams, etc. When the section does not exhibit a center of symmetry, the torsional rotation takes place around a distinct point of the geometric center as we can see in Figure 5.6. This

---

18 See Chapter 9, Table [9.58].
19 See Chapter 9, Table [9.59].
20 See Chapter 6.

point is the center of torsion. A moment $\vec{\mathcal{M}} = L\vec{x}$ thus creates a rotation θ_x, but also a linear displacement of the geometric center G written as v in Figure 5.6.

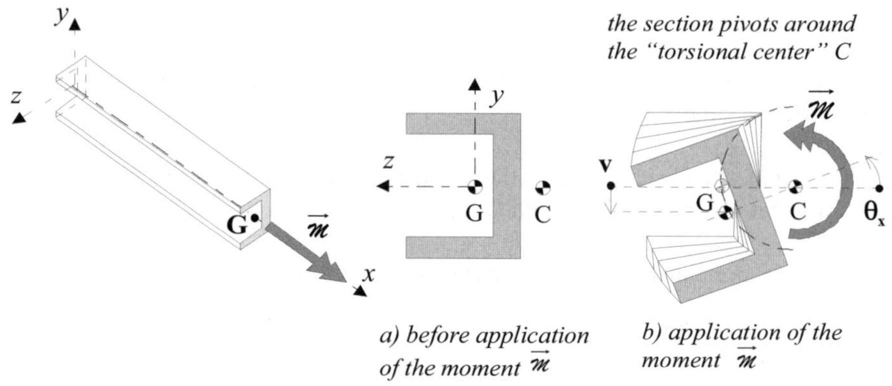

Figure 5.6. *Coupling between bending and torsion*

In addition in Figure 5.7, the application of a nodal force in G such that $\vec{F} = Y\vec{y}$ parallel to the principal axis y causes on the end section, not only a nodal displacement v colinear to Y, but also a rotation θ_x as indicated in the figure.

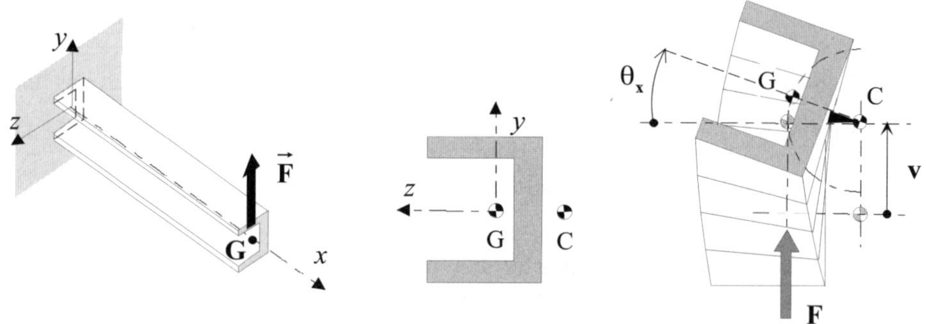

Figure 5.7. *Coupling between bending and torsion*

Let us simultaneously apply $\vec{F} = Y\vec{y}$ and $\vec{\mathcal{M}} = L\vec{x}$. We obtain from the previous equation:

$$\begin{array}{l} v \;\; = \alpha_{11}Y + \alpha_{12}L \\ \theta_x = \alpha_{21}Y + \alpha_{22}L \end{array} \quad \text{that is} \quad \left\{\begin{array}{c} v \\ \theta_x \end{array}\right\} = \left[\begin{array}{cc} \alpha_{11} & \alpha_{12} \\ \alpha_{21} & \alpha_{22} \end{array}\right] \bullet \left\{\begin{array}{c} Y \\ L \end{array}\right\}$$

Using an approach that has been used several times in Chapter 2, section 2.4, we can show that the flexibility matrix $[\alpha]$ that appears here is symmetric ($\alpha_{12} = \alpha_{21}$). From there by inversion, we would obtain a behavior-relation of the type:

$$\left\{\begin{array}{c} Y \\ L \end{array}\right\} = \left[\begin{array}{cc} k_{11} & k_{12} \\ k_{21} & k_{22} \end{array}\right] \bullet \left\{\begin{array}{c} v \\ \theta_x \end{array}\right\}$$

We can observe here coefficients k_{12} and k_{21} that characterize the coupling of the phenomena of bending and torsion and disappear when the section exhibits a center of symmetry. When we examine the behavior of a beam element possessing such an asymmetric section, we must write a stiffness matrix $[k]_{el}$ where non-zero
$$\qquad\qquad\qquad\qquad\qquad\qquad\qquad\qquad\qquad\qquad\qquad\qquad Local$$
coefficients appear that characterize this coupling. A large majority of industrial computing softwares do not include this coupling and use a stiffness matrix such as [5.10][21]. That is why we will limit ourselves here to the description of the beam element whose section shows a center of symmetry.

---

21 The beam element with a center of torsion that is distinct from the geometric center of the section can be found as an option in some industrial computing codes.

5.2.4. *Summary*

The behavior of the beam element is summarized in the following table. It will be illustrated and modeled at the time of calculation by its center line joining the geometric centers (or nodes) of the two end sections.

<table>
<tr><td align="center">Beam element</td></tr>
<tr><td align="center">in the local coordinate system</td></tr>
</table>

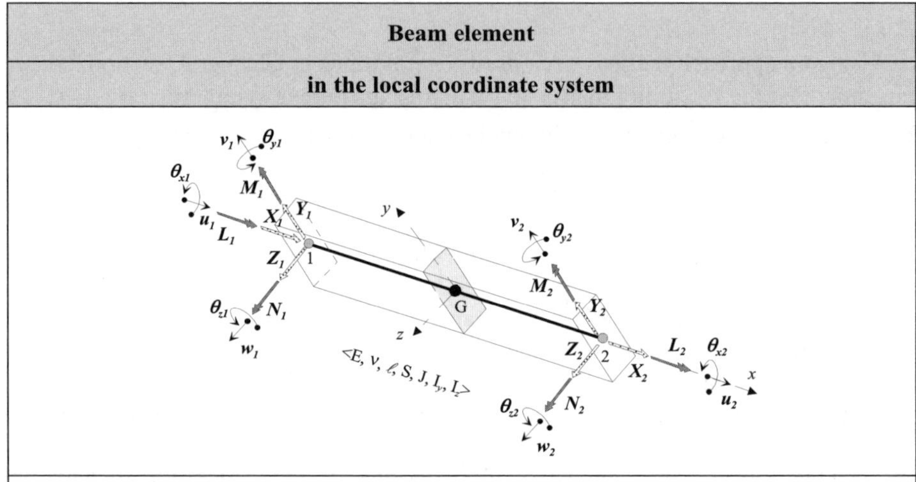

Properties that are necessary to define the element:

⇨ nature of the material: E, ν[22]

⇨ geometry of the element:

   ◆ length: ℓ

   ◆ area of cross-section: S

   ◆ section moduli: $k_y S$ (bending in plane (xy)) and $k_z S$ (bending in plane (xz))

   ◆ torsion constant: J

   ◆ principal quadratic moments: I_y (with respect to axis Gy) and I_z (with respect to axis Gz)

   ◆ flexure moduli: $\dfrac{I_y}{z_{max}}$ (bending in the plane (xz)) and $\dfrac{I_z}{y_{max}}$ (bending in the plane(xy))

---

22 Usually, for isotropic materials, we provide as input for the software, Poisson's ratio ν rather than shear modulus G. The software calculates G from relation $G = \dfrac{E}{2(1 + \nu)}$. This last relation is justified in Chapter 10 (see [10.11]).

Behavior relation: $\{F\}_{el} = [k]_{el} \bullet \{d\}_{el}$
Local   Local   Local

$$
\begin{Bmatrix} X_1 \\ Y_1 \\ Z_1 \\ L_1 \\ M_1 \\ N_1 \\ X_2 \\ Y_2 \\ Z_2 \\ L_2 \\ M_2 \\ N_2 \end{Bmatrix}_{Local}
=
\begin{bmatrix}
\frac{ES}{\ell} & 0 & 0 & 0 & 0 & 0 & -\frac{ES}{\ell} & 0 & 0 & 0 & 0 & 0 \\
 & \frac{12EI_z}{\ell^3(1+\Phi_y)} & 0 & 0 & 0 & \frac{6EI_z}{\ell^2(1+\Phi_y)} & 0 & -\frac{12EI_z}{\ell^3(1+\Phi_y)} & 0 & 0 & 0 & \frac{6EI_z}{\ell^2(1+\Phi_y)} \\
 & & \frac{12EI_y}{\ell^3(1+\Phi_z)} & 0 & \frac{6EI_y}{\ell^2(1+\Phi_z)} & 0 & 0 & 0 & -\frac{12EI_y}{\ell^3(1+\Phi_z)} & 0 & -\frac{6EI_y}{\ell^2(1+\Phi_z)} & 0 \\
 & S & & \frac{GJ}{\ell} & 0 & 0 & 0 & 0 & 0 & -\frac{GJ}{\ell} & 0 & 0 \\
 & Y & & & \frac{(4+\Phi_z)EI_y}{\ell(1+\Phi_z)} & 0 & 0 & 0 & \frac{6EI_y}{\ell^2(1+\Phi_z)} & 0 & \frac{(2-\Phi_z)EI_y}{\ell(1+\Phi_z)} & 0 \\
 & & M & & & \frac{(4+\Phi_y)EI_z}{\ell(1+\Phi_y)} & 0 & -\frac{6EI_z}{\ell^2(1+\Phi_y)} & 0 & 0 & 0 & \frac{(2-\Phi_y)EI_z}{\ell(1+\Phi_y)} \\
 & & & M & & & \frac{ES}{\ell} & 0 & 0 & 0 & 0 & 0 \\
 & & & & E & & & \frac{12EI_z}{\ell^3(1+\Phi_y)} & 0 & 0 & 0 & -\frac{6EI_z}{\ell^2(1+\Phi_y)} \\
 & & & & & T & & & \frac{12EI_y}{\ell^3(1+\Phi_z)} & 0 & \frac{6EI_y}{\ell^2(1+\Phi_z)} & 0 \\
 & & & & & & R & & & \frac{GJ}{\ell} & 0 & 0 \\
 & & & & & & & Y & & & \frac{(4+\Phi_z)EI_y}{\ell(1+\Phi_z)} & 0 \\
 & & & & & & & & & & & \frac{(4+\Phi_y)EI_z}{\ell(1+\Phi_y)}
\end{bmatrix}
\bullet
\begin{Bmatrix} u_1 \\ v_1 \\ w_1 \\ \theta_{x1} \\ \theta_{y1} \\ \theta_{z1} \\ u_2 \\ v_2 \\ w_2 \\ \theta_{x2} \\ \theta_{y2} \\ \theta_{z2} \end{Bmatrix}_{Local}
$$

where $\Phi_y = \dfrac{12EI_z}{k_y GS\ell^2}$ and $\Phi_z = \dfrac{12EI_y}{k_z GS\ell^2}$

Potential energy of deformation:

$$
\underset{\text{element}}{E_{pot}} = \frac{1}{2}\{d\}^T_{el} \bullet [k]_{el} \bullet \{d\}_{el}
$$
$$
Local \quad Local \quad Local
$$

$$
\underset{\text{element}}{E_{pot.}} = \frac{1}{2}
\begin{Bmatrix} u_1 \\ v_1 \\ w_1 \\ \theta_{x1} \\ \theta_{y1} \\ \theta_{z1} \\ u_2 \\ v_2 \\ w_2 \\ \theta_{x2} \\ \theta_{y2} \\ \theta_{z2} \end{Bmatrix}^T_{Local}
\bullet
\underset{\substack{Local \\ (12\times12)}}{[k]_{el}}
\bullet
\begin{Bmatrix} u_1 \\ v_1 \\ w_1 \\ \theta_{x1} \\ \theta_{y1} \\ \theta_{z1} \\ u_2 \\ v_2 \\ w_2 \\ \theta_{x2} \\ \theta_{y2} \\ \theta_{z2} \end{Bmatrix}_{Local}
$$

In the global coordinate system

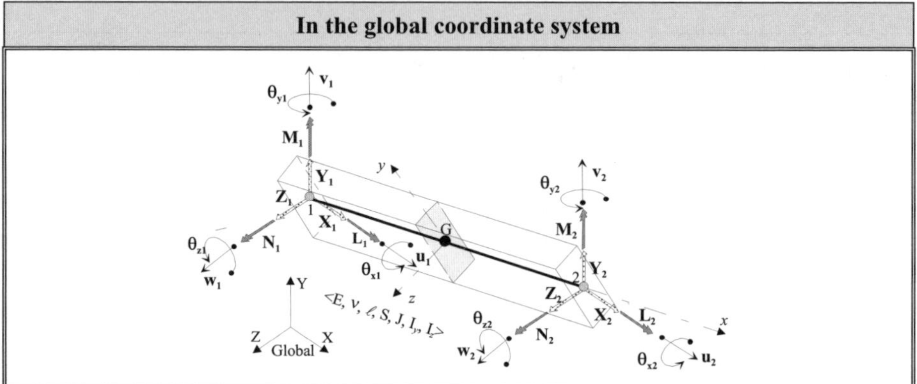

Transfer matrix from the global coordinate system to the local system:

$$\underset{12\times12}{[P]} = \begin{bmatrix} \underset{3\times3}{[p]} & 0 & 0 & 0 \\ 0 & \underset{3\times3}{[p]} & 0 & 0 \\ 0 & 0 & \underset{3\times3}{[p]} & 0 \\ 0 & 0 & 0 & \underset{3\times3}{[p]} \end{bmatrix} \quad with \quad \underset{3\times3}{[p]} = \begin{bmatrix} \cos(\vec{x},\vec{X}) & \cos(\vec{x},\vec{Y}) & \cos(\vec{x},\vec{Z}) \\ \cos(\vec{y},\vec{X}) & \cos(\vec{y},\vec{Y}) & \cos(\vec{y},\vec{Z}) \\ \cos(\vec{z},\vec{X}) & \cos(\vec{z},\vec{Y}) & \cos(\vec{z},\vec{Z}) \end{bmatrix}$$

$$\underset{Local}{\{F\}_{el}} = [P]\bullet\underset{Global}{\{F\}_{el}} \; ; \quad \underset{Local}{\{d\}_{el}} = [P]\bullet\underset{Global}{\{d\}_{el}} \; ; \quad \underset{Global}{[k]_{el}} = [P]^{T}\bullet\underset{Local}{[k]_{el}}\bullet[P]$$

Behavior relation:

$$\underset{Global}{\{F\}_{el}} = \underset{Global}{[k]_{el}}\bullet\underset{Global}{\{d\}_{el}} \Leftrightarrow \underset{Global}{\begin{Bmatrix} X_1 \\ Y_1 \\ Z_1 \\ L_1 \\ M_1 \\ N_1 \\ X_2 \\ Y_2 \\ Z_2 \\ L_2 \\ M_2 \\ N_2 \end{Bmatrix}} = \underset{\substack{Global \\ (12\times12)}}{[k]_{el}}\bullet\underset{Global}{\begin{Bmatrix} u_1 \\ v_1 \\ w_1 \\ \theta_{x1} \\ \theta_{y1} \\ \theta_{z1} \\ u_2 \\ v_2 \\ w_2 \\ \theta_{x1} \\ \theta_{y1} \\ \theta_{z2} \end{Bmatrix}}$$

Potential energy of deformation:

$$\underset{element}{E_{pot}} = \frac{1}{2}\underset{Global}{\{d\}_{el}^{T}}\bullet\underset{Global}{[k]_{el}}\bullet\underset{Global}{\{d\}_{el}}$$

Figure 5.8. *The beam element*

5.3. Elements for the plane state of stress

5.3.1. *Triangular element*

5.3.1.1. *Preliminary notes*

Figure 5.9 below repeats Figure 3.15 of Chapter 3 that showed, in the case of plane state of stress, the partitioning of the structure into triangular and quadrilateral elements. A triangular element was defined in section 3.2.4. We had restricted ourselves in the case of a global coordinate system coplanar to a local system. The main results concerning the triangular element were summarized in [3.23].

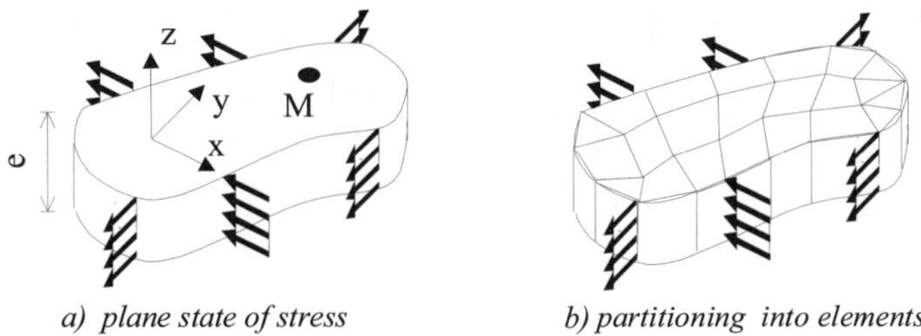

Figure 5.9. *Meshing for the plane state of stresses*

When we wish to go from a local coordinate system (*xyz*) to a global system (XYZ) with any direction in space, we know that each of the vectors characteristic of the triangular element (nodal displacement and nodal force) can manifest themselves in each of the systems on the basis of the transformation that involves matrix $[p]$ of [5.1], referred to hereafter in Figure 5.10.

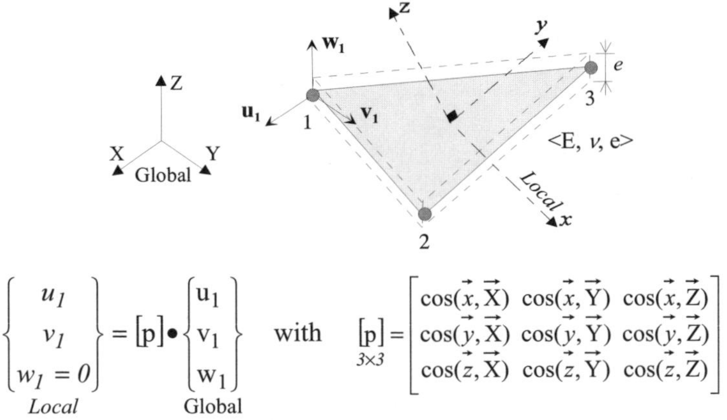

$$\left\{ \begin{array}{c} u_1 \\ v_1 \\ w_1 = 0 \end{array} \right\}_{Local} = [p] \bullet \left\{ \begin{array}{c} u_1 \\ v_1 \\ w_1 \end{array} \right\}_{Global} \quad \text{with} \quad [p]_{3\times3} = \begin{bmatrix} \cos(\vec{x},\vec{X}) & \cos(\vec{x},\vec{Y}) & \cos(\vec{x},\vec{Z}) \\ \cos(\vec{y},\vec{X}) & \cos(\vec{y},\vec{Y}) & \cos(\vec{y},\vec{Z}) \\ \cos(\vec{z},\vec{X}) & \cos(\vec{z},\vec{Y}) & \cos(\vec{z},\vec{Z}) \end{bmatrix}$$

Figure 5.10. *Local and global coordinate systems*

We will thus obtain for all dof and all nodal forces:

$$
\begin{Bmatrix} u_1 \\ v_1 \\ 0 \\ \cdots \\ u_2 \\ v_2 \\ 0 \\ \cdots \\ u_3 \\ v_3 \\ 0 \end{Bmatrix}_{Local} = \begin{bmatrix} [\mathbf{p}]_{3\times3} & 0 & 0 \\ 0 & [\mathbf{p}]_{3\times3} & 0 \\ 0 & & [\mathbf{p}]_{3\times3} \end{bmatrix}_{\substack{transfer\ matrix \\ [P]\,(9\times9)}} \bullet \begin{Bmatrix} u_1 \\ v_1 \\ w_1 \\ \cdots \\ u_2 \\ v_2 \\ w_2 \\ \cdots \\ u_3 \\ v_3 \\ w_3 \end{Bmatrix}_{Global}
\quad \text{and} \quad
\begin{Bmatrix} X_1 \\ Y_1 \\ 0 \\ \cdots \\ X_2 \\ Y_2 \\ 0 \\ \cdots \\ X_3 \\ Y_3 \\ 0 \end{Bmatrix}_{Local} = \begin{bmatrix} [\mathbf{p}]_{3\times3} & 0 & 0 \\ 0 & [\mathbf{p}]_{3\times3} & 0 \\ 0 & & [\mathbf{p}]_{3\times3} \end{bmatrix}_{\substack{transfer\ matrix \\ [P]\,(9\times9)}} \bullet \begin{Bmatrix} X_1 \\ Y_1 \\ Z_1 \\ \cdots \\ X_2 \\ Y_2 \\ Z_2 \\ \cdots \\ X_3 \\ Y_3 \\ Z_3 \end{Bmatrix}_{Global}
$$

$$[5.11]$$

Corresponding to a transfer matrix $[P]\ (9\times9)$ such that:

$$\{d\}_{el}^{Local} = [P] \bullet \{d\}_{el}^{Global} \ ; \quad \{F\}_{el}^{Local} = [P] \bullet \{F\}_{el}^{Global}$$

We can thus use this transfer matrix to express the behavior relation in the global coordinate system in conformity with [5.1]. The result is summarized in Figure 5.11.

5.3.1.2. *Summary*

We can summarize hereafter the general performance of the triangular element for plane elasticity (state of plane stress)[23].

<table>
<tr><td align="center">Triangular element for plane state of stress</td></tr>
<tr><td align="center">in the local coordinate system</td></tr>
</table>

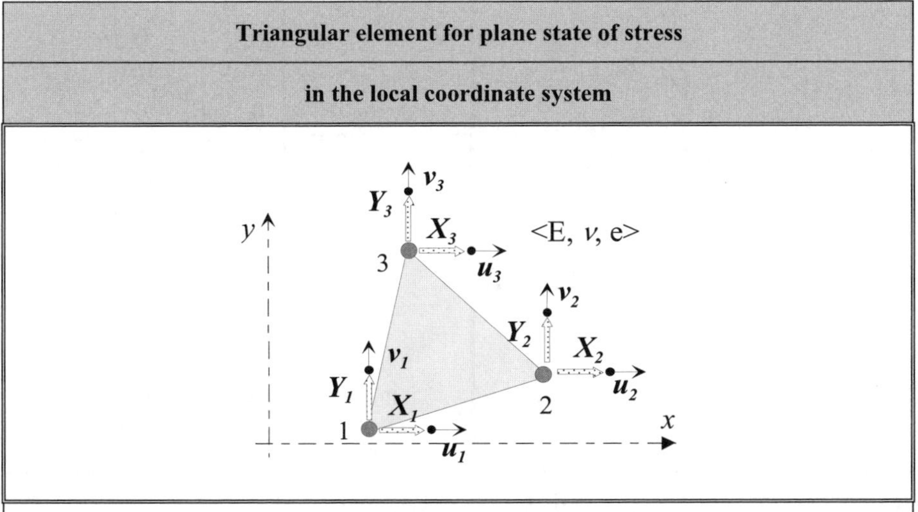

Behavior relation[24]:

$$\{F\}_{el} = [k]_{el} \bullet \{d\}_{el} \Leftrightarrow \begin{Bmatrix} X_1 \\ Y_1 \\ X_2 \\ Y_2 \\ X_3 \\ Y_3 \end{Bmatrix}_{Local} = [k]_{el} \bullet \begin{Bmatrix} u_1 \\ v_1 \\ u_2 \\ v_2 \\ u_3 \\ v_3 \end{Bmatrix}_{Local}$$
$$\text{Local} \quad \text{Local} \quad \text{Local} \qquad\qquad\qquad \text{Local}$$

---

23 The calculation of the potential energy of deformation has been described in Chapter 3, section 3.2.4.4. It will not be recalled in summaries.

24 The detailed expression of $[k]_{el}$ can be found in section 3.2.4.4.

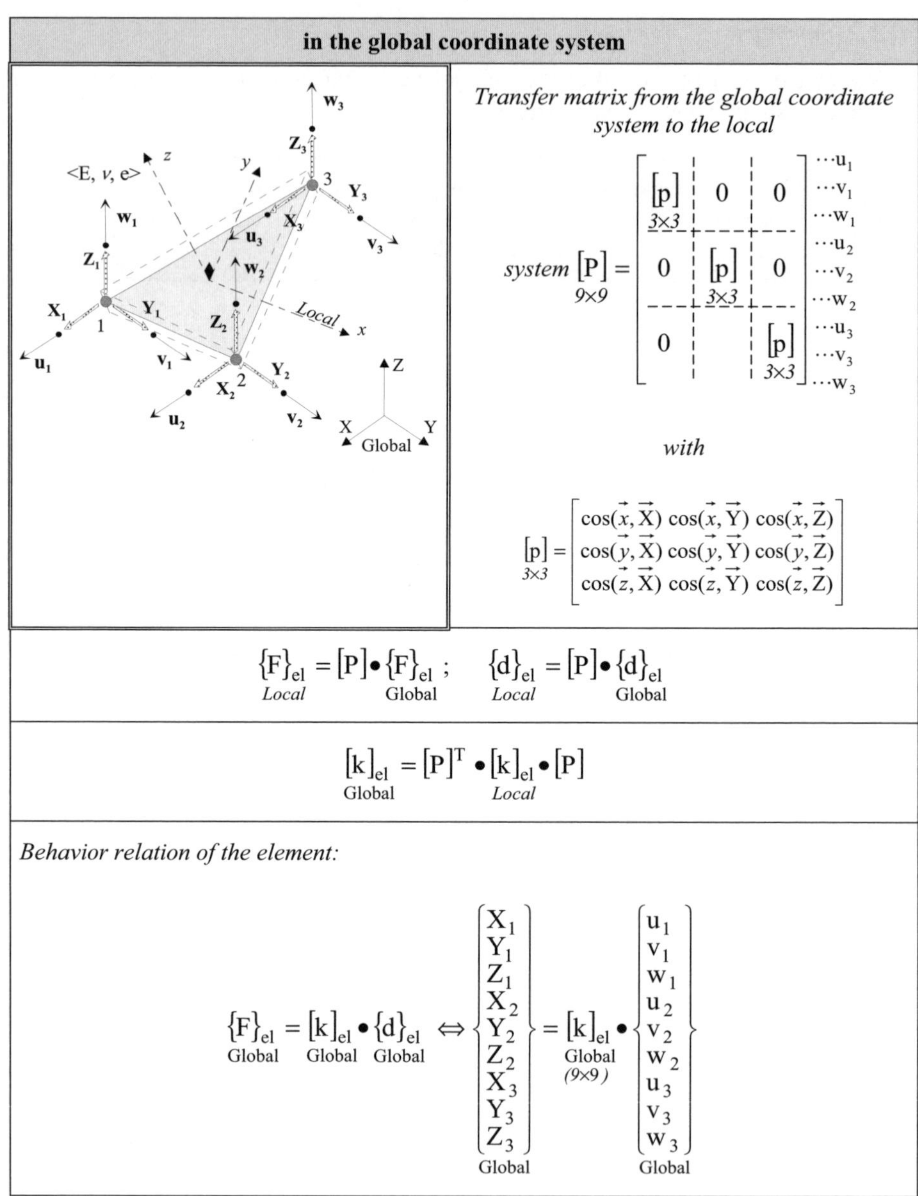

$$system\ \begin{bmatrix}P\end{bmatrix}_{9\times9} = \begin{bmatrix} \begin{bmatrix}p\end{bmatrix}_{3\times3} & 0 & 0 \\ 0 & \begin{bmatrix}p\end{bmatrix}_{3\times3} & 0 \\ 0 & 0 & \begin{bmatrix}p\end{bmatrix}_{3\times3} \end{bmatrix} \begin{matrix}\cdots u_1\\ \cdots v_1\\ \cdots w_1\\ \cdots u_2\\ \cdots v_2\\ \cdots w_2\\ \cdots u_3\\ \cdots v_3\\ \cdots w_3\end{matrix}$$

with

$$\begin{bmatrix}p\end{bmatrix}_{3\times3} = \begin{bmatrix} \cos(\vec{x},\vec{X}) & \cos(\vec{x},\vec{Y}) & \cos(\vec{x},\vec{Z}) \\ \cos(\vec{y},\vec{X}) & \cos(\vec{y},\vec{Y}) & \cos(\vec{y},\vec{Z}) \\ \cos(\vec{z},\vec{X}) & \cos(\vec{z},\vec{Y}) & \cos(\vec{z},\vec{Z}) \end{bmatrix}$$

$$\{F\}_{el}^{Local} = [P]\bullet\{F\}_{el}^{Global}\ ; \quad \{d\}_{el}^{Local} = [P]\bullet\{d\}_{el}^{Global}$$

$$[k]_{el}^{Global} = [P]^{T}\bullet[k]_{el}^{Local}\bullet[P]$$

Behavior relation of the element:

$$\{F\}_{el}^{Global} = [k]_{el}^{Global}\bullet\{d\}_{el}^{Global} \Leftrightarrow \begin{Bmatrix} X_1\\ Y_1\\ Z_1\\ X_2\\ Y_2\\ Z_2\\ X_3\\ Y_3\\ Z_3 \end{Bmatrix}_{Global} = [k]_{el}^{Global}_{(9\times9)}\bullet\begin{Bmatrix} u_1\\ v_1\\ w_1\\ u_2\\ v_2\\ w_2\\ u_3\\ v_3\\ w_3 \end{Bmatrix}_{Global}$$

Figure 5.11. *Triangular element for the plane state of stress*

5.3.2. *Quadrilateral element in plane state of stress*

5.3.2.1. *Rectangular element*

In Figure 5.12, a rectangular element[25] in its local coordinate system (xy) is showed. The nodes are the four apexes. There are therefore $4 \times 2 = 8$ local dof.

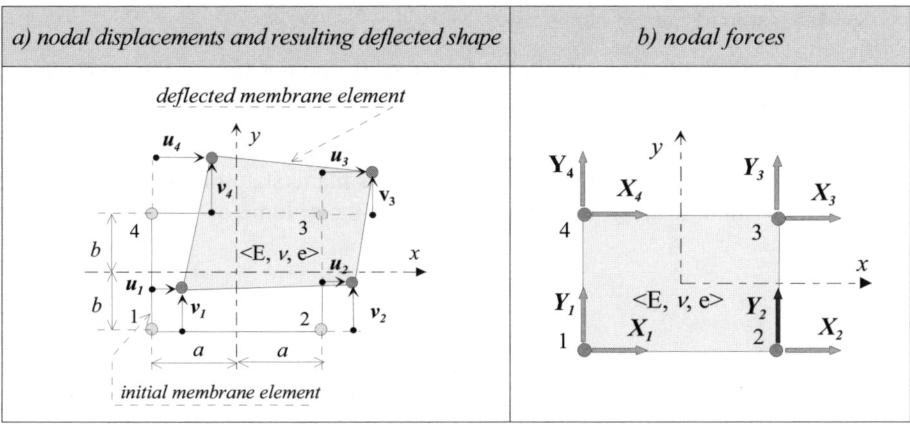

Figure 5.12. *Rectangular element for plane state of stress*

Following the same study plan as that of the triangular element, we must successively define the following:

♦ a displacement field, here:

$$u(x, y) = a_0 + a_1 x + a_2 y + a_3 xy$$
$$v(x, y) = b_0 + b_1 x + b_2 y + b_3 xy$$

that leads to an interpolation matrix $[A(x, y)]$ such that:

$$\left\{ \begin{matrix} u \\ v \end{matrix} \right\} = [A(x, y)] \bullet \left\{ \begin{matrix} u_1 \\ v_1 \\ u_2 \\ v_2 \\ u_3 \\ v_3 \\ u_4 \\ v_4 \end{matrix} \right\}$$

---

25 It should be noted that we always consider the case of plane state of stress (see Figure 5.9).

with the form:

$$[A(x,y)] = \frac{1}{S}\begin{bmatrix} \alpha_1 & 0 & \alpha_2 & 0 & \alpha_3 & 0 & \alpha_4 & 0 \\ 0 & \alpha_1 & 0 & \alpha_2 & 0 & \alpha_3 & 0 & \alpha_4 \end{bmatrix} \qquad [5.12]$$

where we wrote:

$S = 4ab$ (area of the element);

$\alpha_i = (a \pm x)(b \pm y)$ (the signs being functions of the position of the nodes under consideration).

• the deformations ε_x, ε_y, γ_{xy}[26]. They are always obtained by derivation of $[A(x,y)]$. The potential energy of deformation for plane state of stress (see [2-43]) can then be written, leading after integration over the volume of the element, to the form:

$$E_{\substack{pot.\\element}} = \frac{1}{2}\begin{Bmatrix} u_1 \\ v_1 \\ u_2 \\ v_2 \\ u_3 \\ v_3 \\ u_4 \\ v_4 \end{Bmatrix}^T \bullet [k]_{el} \bullet \begin{Bmatrix} u_1 \\ v_1 \\ u_2 \\ v_2 \\ u_3 \\ v_3 \\ u_4 \\ v_4 \end{Bmatrix}$$

• the behavior of the element according to [2.121]:

$$\{F\}_{el} = [k]_{el} \bullet \{d\}_{el}$$

NOTES

❏ The rectangular element illustrated in Figure 5.12 is not linked to its surroundings. We can make the same comment in this case as the one already made for all the elements previously studied: its stiffness matrix is singular (possibility of two rigid-body translations in the plane).

❏ Local coordinate system and global system: as for the previous triangular element, we can go to a global coordinate system (XYZ) by means of the transformation [5.1] shown in Figure 5.13 hereafter.

---

26 See [1.18].

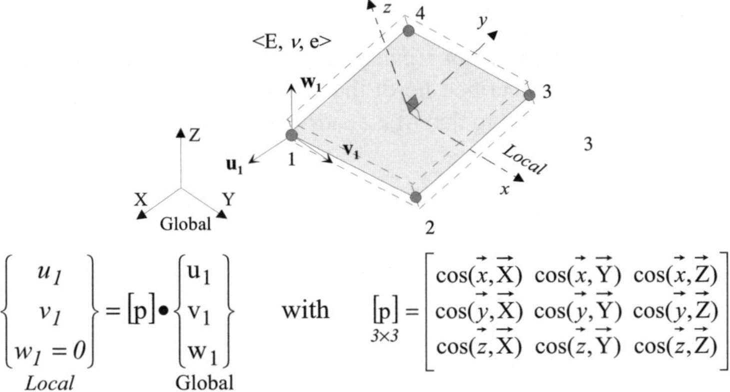

$$\left\{\begin{array}{c} u_1 \\ v_1 \\ w_1 = 0 \end{array}\right\} = [\mathrm{p}] \bullet \left\{\begin{array}{c} u_1 \\ v_1 \\ w_1 \end{array}\right\} \quad \text{with} \quad \underset{3\times3}{[\mathrm{p}]} = \begin{bmatrix} \cos(\vec{x},\vec{X}) & \cos(\vec{x},\vec{Y}) & \cos(\vec{x},\vec{Z}) \\ \cos(\vec{y},\vec{X}) & \cos(\vec{y},\vec{Y}) & \cos(\vec{y},\vec{Z}) \\ \cos(\vec{z},\vec{X}) & \cos(\vec{z},\vec{Y}) & \cos(\vec{z},\vec{Z}) \end{bmatrix}$$

$$\textit{Local} \qquad\qquad \textit{Global}$$

Figure 5.13. *Local and global coordinate systems*

We thus obtain for all dof and all nodal forces:

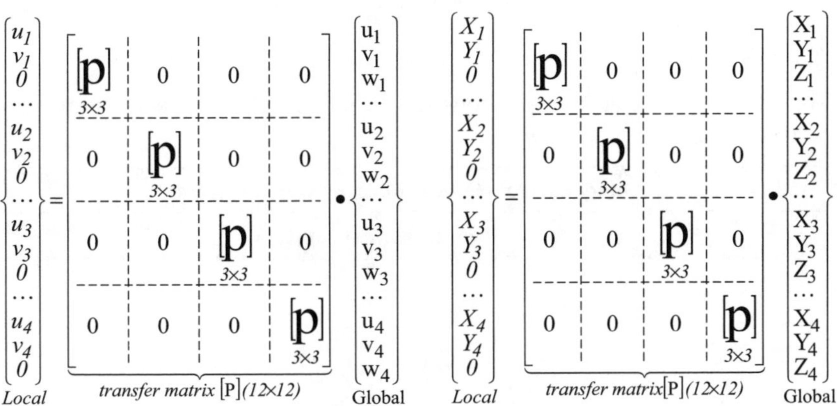

$$[5.13]$$

where appears the transfer matrix $[P](12\times12)$ such that:

$$\{d\}_{el} = [P] \bullet \{d\}_{el} \;\; ; \;\; \{F\}_{el} = [P] \bullet \{F\}_{el}$$
$$\textit{Local} \qquad \textit{Global} \quad \textit{Local} \qquad \textit{Global}$$

When the local dof are known, we can deduce the components σ_x, σ_y, τ_{xy} of plane stresses from the known deformations[27]. The interpolation matrix [5.12] being of second degree with respect to x and y, the deformations and therefore the stresses vary linearly through the element.

---

27 See identical note on the triangular element in section 3.2.4.4 (relation [3.35]).

5.3.2.2. *Quadrilateral element*

When the plane element is not rectangular, but only quadrilateral (Figure 5.14b), we use procedures (not described here) that allow the transfer of properties of the previous rectangular element to that of a quadrilateral element.

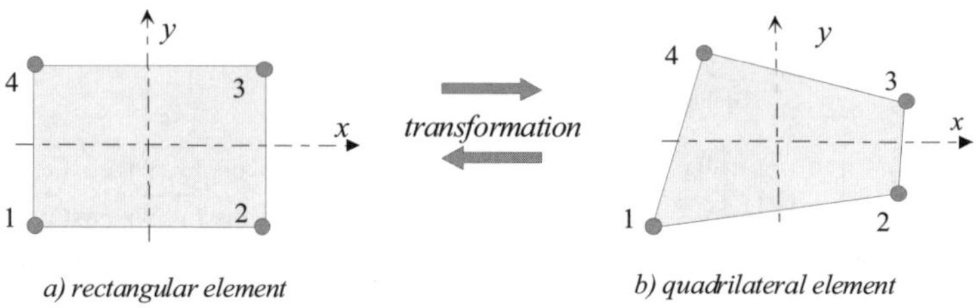

Figure 5.14. *Quadrilateral element for plane state of stress*

We can then mesh the structure by adapting the quadrilateral form so as to recreate as best as possible the real boundaries of the structure (see Figure 5.9).

5.3.2.3. *Summary*

We shall focus on the previous quadrilateral element and summarize its properties in the following figure.

<table>
<tr><td align="center">Quadrilateral element in plane state of stress</td></tr>
<tr><td align="center">in the local coordinate system</td></tr>
</table>

Behavior relation:

$$\{F\}_{el} = [k]_{el} \bullet \{d\}_{el} \Leftrightarrow \begin{Bmatrix} X_1 \\ Y_1 \\ X_2 \\ Y_2 \\ X_3 \\ Y_3 \\ X_4 \\ Y_4 \end{Bmatrix} = [k]_{el} \bullet \begin{Bmatrix} u_1 \\ v_1 \\ u_2 \\ v_2 \\ u_3 \\ v_3 \\ u_4 \\ v_4 \end{Bmatrix}$$

Local   Local   Local   Local   Local   Local

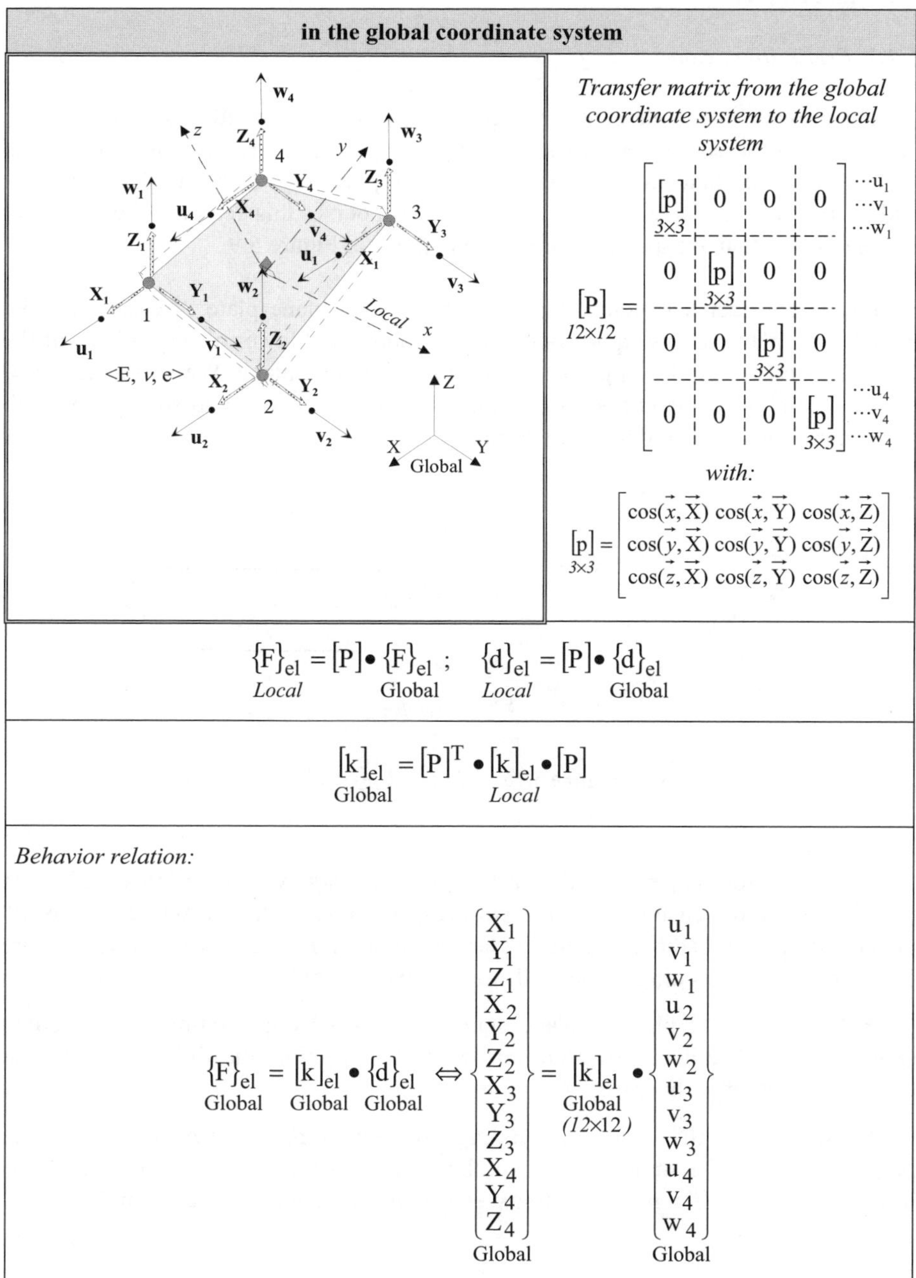

$$\{F\}_{el}^{Local} = [P] \bullet \{F\}_{el}^{Global} \; ; \quad \{d\}_{el}^{Local} = [P] \bullet \{d\}_{el}^{Global}$$

$$[k]_{el}^{Global} = [P]^T \bullet [k]_{el}^{Local} \bullet [P]$$

Behavior relation:

$$\{F\}_{el}^{Global} = [k]_{el}^{Global} \bullet \{d\}_{el}^{Global} \Leftrightarrow \begin{Bmatrix} X_1 \\ Y_1 \\ Z_1 \\ X_2 \\ Y_2 \\ Z_2 \\ X_3 \\ Y_3 \\ Z_3 \\ X_4 \\ Y_4 \\ Z_4 \end{Bmatrix}^{Global} = [k]_{el}^{Global \; (12\times12)} \bullet \begin{Bmatrix} u_1 \\ v_1 \\ w_1 \\ u_2 \\ v_2 \\ w_2 \\ u_3 \\ v_3 \\ w_3 \\ u_4 \\ v_4 \\ w_4 \end{Bmatrix}^{Global}$$

Figure 5.15. *Quadrilateral element in plane state of stress*

5.4. Plate element

5.4.1. *Preliminary notes*

The previous plane elements characterized plane states of stresses, limited in the plane (xy) to the components of stress: σ_x, σ_y, τ_{xy}. We recall that according to the definition of plane stresses, the latter vary in plane (xy) (they are functions of x and y) but do not depend on coordinate z. Of course, this requires the action of internal forces that are specific, for example, those in Figure 5.9.

Let us consider a structure that is often seen: a plane plate. We call "middle plane"[28] the plane that is equidistant from the upper and lower plane surfaces of the plate. The axis $\vec{z}$ is placed perpendicular to this middle plane. We will qualify this plate as "thin" if its thickness e according to $\vec{z}$ is relatively low in comparison with the dimensions of the middle plane Figure 5.16.

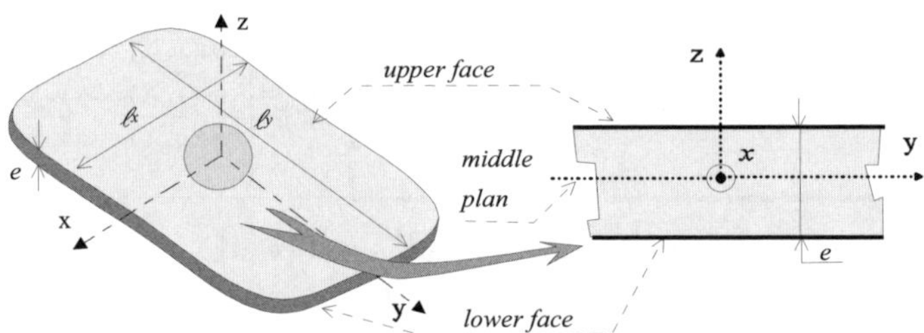

Figure 5.16. *Thin plate:* $\dfrac{e}{\ell} \ll 1$

We know from experience that a thin plate appears even more "flexible" when its longitudinal dimensions are large with respect to its thickness. When it is "bent" under appropriate loading, its middle plane becomes a middle curved surface. Figure 5.17 shows that according to the loads that have been applied on this plate, we can:

◆ "work on" the plate in plane state of stresses by generating plane loading (Figure 5.17a). We have just seen, in the previous section, the finite elements that are representative of this case;

◆ "work on" the plate in bending by generating displacements that are perpendicular to the middle plane of the plate (Figure 5.17b). In this case, the loadings are of a different nature: transversal force, moments of axes $\vec{x}$ and $\vec{y}$.

---

28 The notion of a middle plane is not necessary for the plane elements defined previously for a plane state of stress.

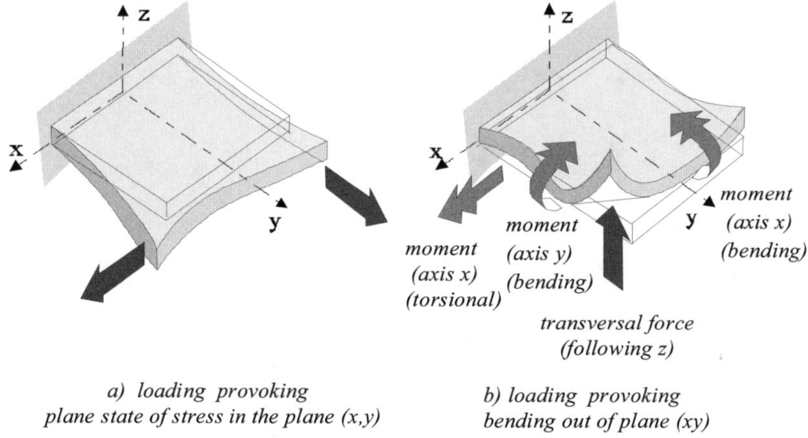

a) loading provoking
plane state of stress in the plane (x,y)

b) loading provoking
bending out of plane (xy)

Figure 5.17. *Behaviors of a plane plate*

The study of bending will not be described in this book. Let us only quote some results.

We thus isolate in Figure 5.18 a small plate element $dx \times dy \times e$ from a plate in bending and on which the stress components used in the plate theory are shown:

♦ we find components σ_x, σ_y, τ_{xy} in Figure 5.18b, which vary linearly across the thickness[29];

♦ we now observe the other components τ_{xz} and τ_{yz} which are called transverse shear stresses that we generally tend to neglect when the plates are thin;

♦ we note the absence of the normal component σ_z. Outside the zones of application of the loading forces, the plate being thin and its upper and lower surfaces being free, the stress σ_z is necessarily very weak or zero. That is why it is not represented[30].

---

29 We can remember a similar stress evolution for the bending of beams with a unique normal stress σ_x (see sections 1.5.2.5 and 1.5.2.6).

30 We also tend to neglect σ_z when pressure is applied on the plate. It is then thought that the values of σ_z due to the pressure are negligible in comparison to other stresses.

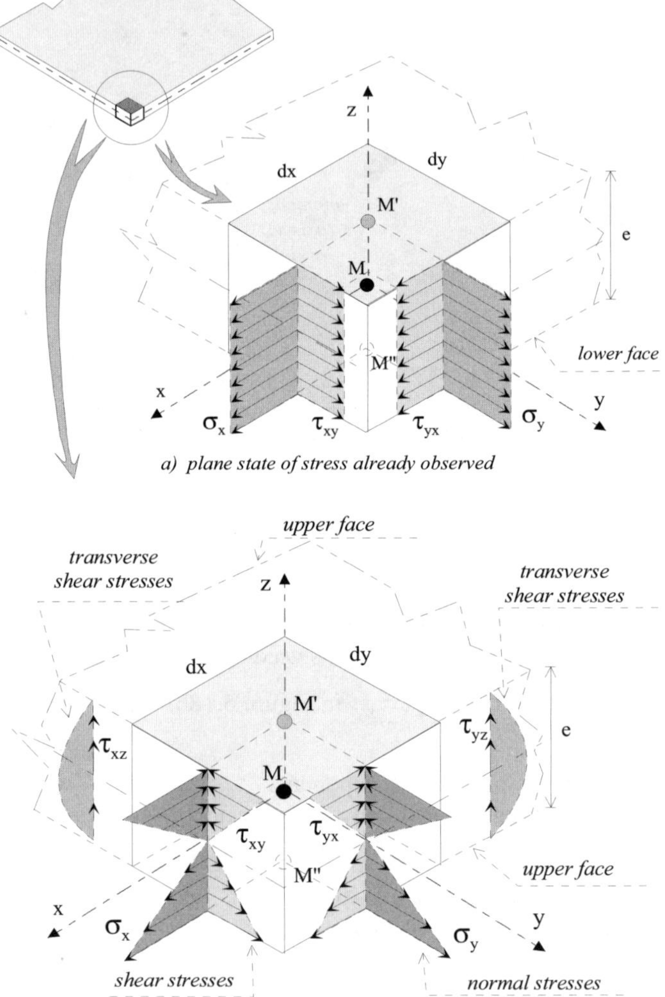

a) *plane state of stress already observed*

b) *bending stresses*

Figure 5.18. *Stresses in a bending plate*

5.4.2. *Resultant forces and moments for cohesion forces*

On the basic domain shown in Figure 5.18, we can define elementary resultant forces and moments for cohesion forces acting on the faces normal to $\vec{x}$ and $\vec{y}$. In order to do this, we use a similar approach to that in Chapter 1, section 1.5.1. The elementary cohesion forces $\overrightarrow{df_1}$ and $\overrightarrow{df_2}$ acting respectively on faces dS_1 and dS_2 are represented in Figure 5.19 and the resultant forces and moments are deduced[31]. In the same figure, these resultant forces and moments have also been projected on axes $\vec{x},\ \vec{y}$ and $\vec{z}$.

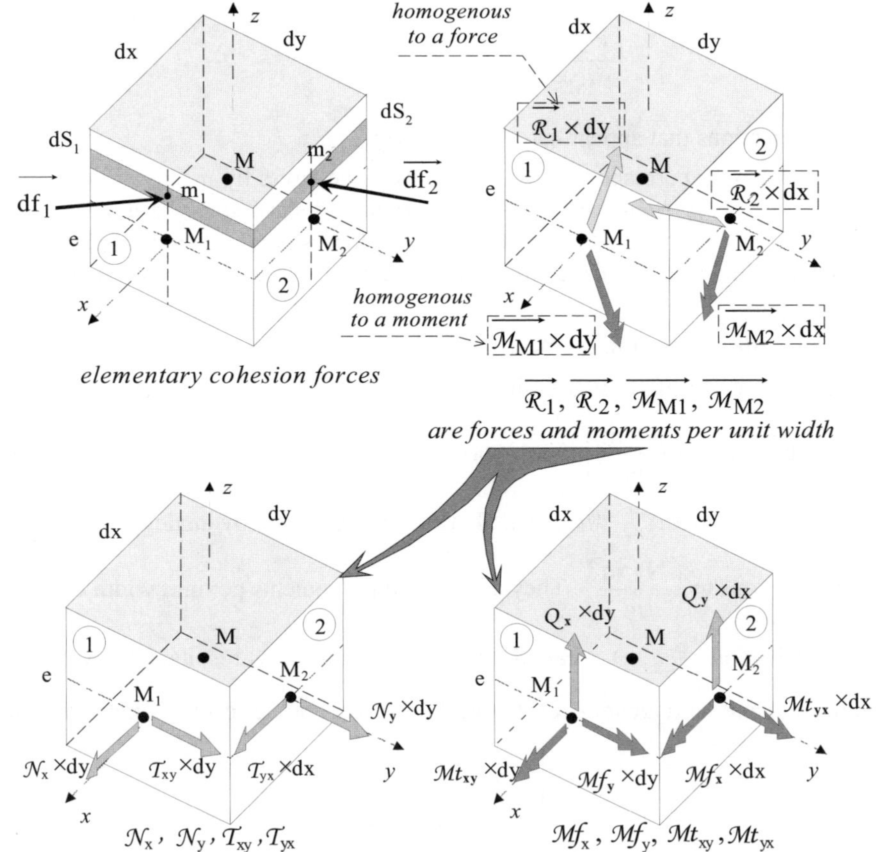

Figure 5.19. *Resultant forces and moments for cohesion forces*

---

31 For example, on surface element 1, the resultant force and moment for cohesion forces $\overrightarrow{\mathcal{R}_1}$ (N/m) and $\overrightarrow{\mathcal{M}_{M1}}$ (N.m/m) are such that:

$$\overrightarrow{\mathcal{R}_1}\times dy = \int\overrightarrow{df_1}\ \text{ and }\ \overrightarrow{\mathcal{M}_{M1}}\times dy = \int\overrightarrow{M_1m_1}\wedge\overrightarrow{df_1}$$

We thus observe on face element 1 with normal $\vec{x}$ and after dividing by dy[32]:

$$\overrightarrow{\mathcal{R}_1} = \mathcal{N}_x \vec{x} + \mathcal{T}_{xy} \vec{y} + Q_x \vec{z}$$

$$\overrightarrow{\mathcal{M}_{M1}} = \mathcal{M}t_{xy} \vec{x} + \mathcal{M}f_y \vec{y}$$

$$[5.14]$$

and on face element 2 with normal $\vec{y}$ and after dividing by dx:

$$\overrightarrow{\mathcal{R}_2} = \mathcal{T}_{yx} \vec{x} + \mathcal{N}_y \vec{y} + Q_y \vec{z}$$

$$\overrightarrow{\mathcal{M}_{M2}} = \mathcal{M}f_x \vec{x} + \mathcal{M}t_{yx} \vec{y}$$

$$[5.15]$$

The projections that appear here characterize:

➤ the behavior corresponding to a state of plane stress by means of projections $\mathcal{N}_x$, $\mathcal{N}_y$, $\mathcal{T}_{xy}$, $\mathcal{T}_{yx}$[33] and known as the "normal and shear stress resultants". They are homogenous to $(N \times m^{-1})$, corresponding to forces per unit width of plate. They correspond to the state of plane stress in Figure 5.18a. This aspect has been studied in the previous section;

➤ the bending behavior, by means of:

– projections $\mathcal{M}f_x$, $\mathcal{M}f_y$ called "bending moments" (by analogy with beams)

and homogenous to $\left(\dfrac{N \times m}{m} \right)$. They correspond to moments per unit width of plate,

– projections $\mathcal{M}t_{xy}$, $\mathcal{M}t_{yx}$ called "torsion moments" (by analogy with beams)

and homogenous to $\left(\dfrac{N \times m}{m} \right)$. They correspond to moments per unit width of plate,

– projections Q_x, Q_y called "transverse shear stress resultants" (by analogy

with beams) and homogenous to $(N \times m^{-1})$, corresponding to forces per unit width

of plate.

---

32 We would successively obtain the following for the resultant force acting on surface element ①:

$$\overrightarrow{\mathcal{R}_1}dy = \mathcal{N}_x dy\vec{x} + \mathcal{T}_{xy}dy\vec{y} + Q_x dy\vec{z} = \vec{x}\int_e \sigma_x \times dz \times dy + \vec{y}\int_e \tau_{xy} \times dz \times dy + \vec{z}\int_e \tau_{xz} \times dz \times dy$$

which corresponds to the following definitions of the resultant force projections, also called the normal and shear stress resultant:

$$\mathcal{N}_x = \int_e \sigma_x \times dz \; ; \quad \mathcal{T}_{xy} = \int_e \tau_{xy} \times dz \; ; \quad Q_x = \int_e \tau_{xz} \times dz$$

33 The behavior of the state of plane stresses was dealt with in section 5.3 without needing definition of these plane normal and shear stress resultant.

NOTES

❏ We will recollect the property of reciprocity of the shear forces ($\tau_{xy} = \tau_{yx}$). On examination of Figure 5.18a, we can easily establish the relation:

$$\mathcal{T}_{xy} = \tau_{xy} \times e = \tau_{yx} \times e = \mathcal{T}_{yx}$$

❏ Based on the same property, on examination of Figure 5.18b, we can easily establish that:

$$\mathcal{M}t_{xy} = \int_e \tau_{xy} \times z \times dz = -\int_e \tau_{yx} \times z \times dz = -\mathcal{M}t_{yx}$$

❏ We note that the projections on $\vec{z}$ of moments $\overrightarrow{\mathcal{M}_{M1}} \times dy$ or $\overrightarrow{\mathcal{M}_{M2}} \times dx$ in Figure 5.19 are zero, since all elementary cohesion forces $\overrightarrow{df_1}$ or $\overrightarrow{df_2}$ pass by supports $M_1\vec{z}$ or $M_2\vec{z}$ We can thus conclude that vector moments $\overrightarrow{\mathcal{M}_{M1}}$ and $\overrightarrow{\mathcal{M}_{M2}}$ are found in the middle plane (xy) of the plate.

❏ We can observe that the loadings inducing bending in Figure 5.17b correspond to the projections of Figure 5.19, and include:

– transverse forces following $\vec{z}$
– bending and torsional moments

[5.16]

5.4.3. *Plate element in bending*

5.4.3.1. *Rectangular element*

This can be seen on Figure 5.20. The nodal loads in (a) are defined at the four apexes on the basis of the loads described previously and summarized in [5.16].

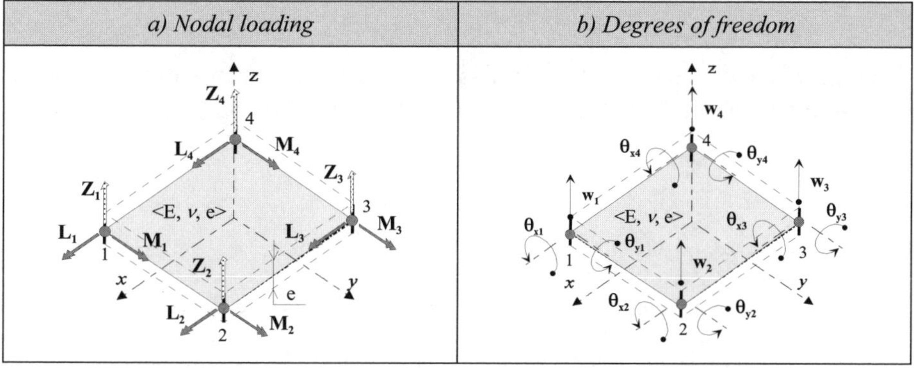

Figure 5.20. *Rectangular plate element in bending*

The 12 corresponding degrees of freedom[34] are shown in Figure 5.20b.

We will restrict ourselves to a brief description of the procedure that will allow us to determine the stiffness matrix in the local coordinate system, without calculating it. Let us simply say that the stages are similar to those used for the elements that have already been defined. We shall begin by writing a displacement field, characteristic of the bending. For example, in the case of thin plates, we will write the transverse displacement w(x,y) in a polynomial form:

$$w(x, y) = a_0 + a_1 x + a_2 y + a_3 x^2 + a_4 xy + a_5 y^2 + a_6 x^3 + a_7 x^2 y + a_8 xy^2 + a_9 y^3 ...$$
$$..... + a_{10} x^3 y + a_{11} xy^3$$

Assuming: $\theta_x \cong \dfrac{\partial w}{\partial y}$, $\theta_y \cong -\dfrac{\partial w}{\partial x}$ [35], we obtain a matrix of interpolation $\left[A(x, y)\right]$ such that:

$$\begin{Bmatrix} w \\ \theta_x \\ \theta_y \end{Bmatrix} = \underset{(3\times 12)}{\left[A(x,y)\right]} \bullet \begin{Bmatrix} w_1 \\ \theta_{x1} \\ \theta_{y1} \\ w_2 \\ ..... \\ \theta_{y4} \end{Bmatrix} = \underset{(3\times 12)}{\left[A(x,y)\right]} \bullet \underset{(12\times 1)}{\{d\}_{el}}$$

that allows access to specific plate deformations, and thus to the potential energy of deformation of any element in the form[36]:

$$E_{\substack{pot. \\ \textit{bending} \\ \textit{element}}} = \frac{1}{2} \{d\}_{el}^T \bullet [k]_{el} \bullet \{d\}_{el}$$

From which we can derive the stiffness matrix. The performance of the element in the local coordinate system (x, y, z) is written with the notations of Figure 5.20a[37]:

---

34 See [2.90].
35 This approximation constitutes in the case of plates, the equivalent of Bernoulli's hypothesis for beams (see sections 1.5.2.5 and 1.5.2.6).
36 The deformations that are specific to plates appear in elastic behavior relations that are not dealt with in this book.
37 See [2.121].

$$\{F\}_{el} = [k]_{el} \bullet \{d\}_{el} \Leftrightarrow \begin{Bmatrix} Z_1 \\ L_1 \\ M_1 \\ Z_2 \\ L_2 \\ M_2 \\ \vdots \\ M_4 \end{Bmatrix} = [k]_{el} \underset{(12\times12)}{\bullet} \begin{Bmatrix} w_1 \\ \theta_{x1} \\ \theta_{y1} \\ w_2 \\ \theta_{x2} \\ \theta_{y2} \\ \vdots \\ \theta_{y4} \end{Bmatrix}$$

NOTES

❏ The element in Figure 5.20 appears to be floating in space. It is not properly linked, as seen in section 2.5.2. This absence of connection results in the three possible rigid-body motions:

  – a translation following direction $\vec{z}$;

  – a rotation around an axis parallel to $\vec{x}$;

  – a rotation around an axis parallel to $\vec{y}$.

❏ The nodes of element in Figure 5.20 appear as the centers of the four small vertical intersection lines show with length e (thickness of the plate). The nodal dof effectively show the possible displacements of such intersection lines. We observe that a nodal dof of type θ_z did not exist as it characterized the rotation of the nodal intersection line around itself, which does not constitute a possible physical displacement. This is in keeping with the loadings seen above, in which, in particular, there did not appear any moment with axis z (see Figure 5.19).

❏ The plate element described here is one amongst other plate elements in bending with higher performance that are available in computing softwares.

❏ As in the case of rectangular elements for states of plane stresses, it is possible to go from the characterization of a rectangular bending element to that of a quadrilateral bending element.

❏ *Local* coordinate system and global system: the passage from the local coordinate system (*xyz*) to the distinct global system (XYZ) has already been described for preceding elements and on the basis of summary [5.1]. For node 1 for example, we will obtain:

$$\begin{Bmatrix} 0 \\ 0 \\ w_1 \\ \cdots \\ \theta_{x1} \\ \theta_{y1} \\ 0 \end{Bmatrix} = \begin{bmatrix} [\mathbf{p}]_{3\times3} & \vdots \\ \cdots & \cdots \\ \vdots & [\mathbf{p}]_{3\times3} \end{bmatrix} \bullet \begin{Bmatrix} u_1 \\ v_1 \\ w_1 \\ \cdots \\ \theta_{x1} \\ \theta_{y1} \\ \theta_{z1} \end{Bmatrix}$$

Local   transfer matrix   Global
$[P]\,(6\times6)$

and for all dof:

$$\begin{Bmatrix} 0 \\ 0 \\ w_1 \\ \cdots \\ \theta_{x1} \\ \theta_{y1} \\ 0 \\ \cdots \\ \vdots \\ \cdots \\ \theta_{x4} \\ \theta_{y4} \\ 0 \end{Bmatrix} = [P]_{24\times24} \bullet \begin{Bmatrix} u_1 \\ v_1 \\ w_1 \\ \cdots \\ \theta_{x1} \\ \theta_{y1} \\ \theta_{z1} \\ \cdots \\ \vdots \\ \cdots \\ \theta_{x4} \\ \theta_{y4} \\ \theta_{z4} \end{Bmatrix} \quad \Leftrightarrow \quad \{d\}_{el} = [P]_{24\times24} \bullet \{d\}_{el}$$

Local   Global   Local   Global
24×1   24×1

and the same for the nodal forces: $\quad \{F\}_{el} = [P]_{24\times24} \bullet \{F\}_{el}$

Local   Global

❏ When after calculation, the global and then the local dof are obtained, we can calculate the bending stresses components σ_x, σ_y, τ_{xy} that appear in Figure 5.18b. This is always explained in the same manner: the displacement field everywhere in the element is function of the dof. Therefore, deformations are also expressed according to the dof. We observe that in Figure 5.18b, stresses σ_x, σ_y, τ_{xy} attain their extreme values on the upper and lower faces (or sheets) of the plate. It is therefore in these two zones that the postprocessor of the software will calculate their values.

5.4.3.2. *Triangular element*

Many triangular bending elements have been developed with definitions that try to get them to recreate with as much precision as possible, the real behavior of a plate thus meshed. From the point of view of the software user, they manifest themselves with dof and nodal strains similar to those above. Figure 5.21 shows the triangular element with its local coordinate system.

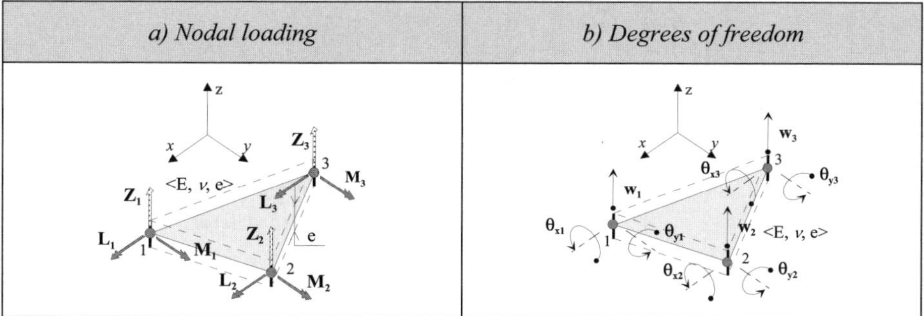

| *a) Nodal loading* | *b) Degrees of freedom* |

Figure 5.21. *Triangular bending element*

With the illustration in Figure 5.21, we obtain the following behavior from the method described for the rectangular element:

$$\{F\}_{el} = [k]_{el} \bullet \{d\}_{el} \Leftrightarrow \begin{Bmatrix} Z_1 \\ L_1 \\ M_1 \\ Z_2 \\ L_2 \\ M_2 \\ \vdots \\ M_3 \end{Bmatrix} = [k]_{el} \bullet \begin{Bmatrix} w_1 \\ \theta_{x1} \\ \theta_{y1} \\ w_2 \\ \theta_{x2} \\ \theta_{y2} \\ \vdots \\ \theta_{y3} \end{Bmatrix}$$
$$\underset{(9\times9)}{}$$

NOTES

❏ We can review the comments relating to the preceding rectangular bending element so as to indicate:

– the possibility of three rigid body motions in space;

– the non-existence of a rotation dof around axis $\vec{z}$, such that θ_{zi} in node i, that has no physical significance;

– the possibility of finding stresses σ_x, σ_y, τ_{xy} when the dof are known, with their display on the upper and lower plate faces;

– the description of the behavior in the global coordinate system (XYZ) that is distinct from the local system (*xyz*) and used for the definition of the element, by means of a transfer matrix $[P]$ such that:

$$
\left\{
\begin{array}{c}
0 \\
0 \\
w_1 \\
\cdots \\
\theta_{x1} \\
\theta_{y1} \\
0 \\
\cdots \\
\vdots \\
\cdots \\
\theta_{x3} \\
\theta_{y3} \\
0
\end{array}
\right\}
=
[P] \bullet
\left\{
\begin{array}{c}
u_1 \\
v_1 \\
w_1 \\
\cdots \\
\theta_{x1} \\
\theta_{y1} \\
\theta_{z1} \\
\cdots \\
\vdots \\
\cdots \\
\theta_{x3} \\
\theta_{y3} \\
\theta_{z3}
\end{array}
\right\}
\quad \Leftrightarrow \quad
\{d\}_{el} = [P] \bullet \{d\}_{el}
$$

$$
\underset{Local}{} \quad \underset{18\times18}{} \quad \underset{Global}{}
$$

$$
\underset{\substack{Local \\ 18\times1}}{} \qquad \underset{\substack{Global \\ 18\times1}}{}
$$

and the same for the nodal forces:
$$
\{F\}_{el} = [P] \bullet \{F\}_{el}
$$
$$
\underset{Local}{} \quad \underset{18\times18}{} \quad \underset{Global}{}
$$

5.4.4. *Complete plate element*

Figure 5.17 illustrated the two behaviors of a plate structure:

– the behavior in state of plane stress: we have defined the corresponding elements in the preceding section (see section 5.3);

– the behavior in bending (subject of the present section).

It is evident that the real loadings on plate structures are generally such that these two behaviors "coexist", hence the need to include in the same element the properties of "state of plane stress" as well as those of "bending". The procedure is identical to the one that had allowed us to define the complete beam element in space in section 5.2. Let us review the principle. The potential energy of deformation of a complete plate element should appear as the sum:

$$
\underset{\substack{complete \\ element}}{E_{pot.}} = \underset{\substack{plane\ stress \\ element}}{E_{pot.}} + \underset{\substack{bending \\ element}}{E_{pot.}}
$$

$$
\underset{\substack{complete \\ element \\ (plane\ stress + bending)}}{E_{pot.}} = \frac{1}{2} \underset{\substack{plane \\ stress}}{\{d\}_{el}^{T}} \bullet \underset{\substack{plane \\ stress}}{[k]_{el}} \bullet \underset{\substack{plane \\ stress}}{\{d\}_{el}} + \frac{1}{2} \underset{bending}{\{d\}_{el}^{T}} \bullet \underset{bending}{[k]_{el}} \bullet \underset{bending}{\{d\}_{el}}
$$

The stiffness matrix of the complete plate element is obtained by updating the "partial" stiffness matrices here above in a table that has the dimension of the complete stiffness matrix. For example:

○ The complete rectangular plate element shown in Figure 5.22 possesses in all $8+12=20$ dof in the local coordinate system.

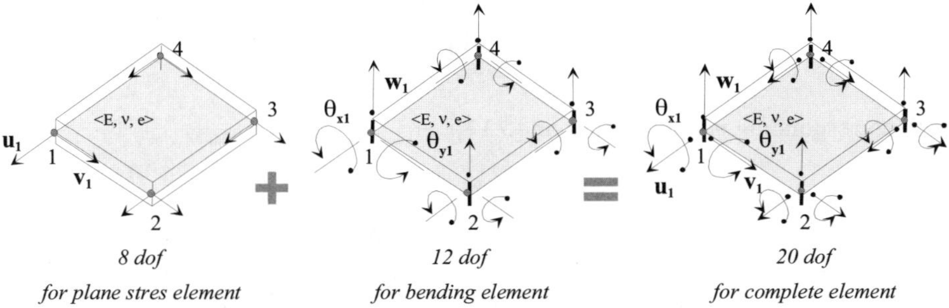

8 dof　　　　　　12 dof　　　　　　20 dof

for plane stres element　　for bending element　　for complete element

Figure 5.22. Complete rectangular plate element

We thus obtain:

$$\underset{\substack{\text{\textit{complete rectangular element}}\\ \text{\textit{(plane stress + bending)}}}}{E_{pot.}} = \frac{1}{2}\underset{1\times20}{\{d\}_{el}}{}^{T}\bullet\underset{20\times20}{[k]_{el}}\bullet\underset{20\times1}{\{d\}_{el}}$$

The behavior of this element in the local coordinate system (xyz) is written as[38]:

$$\underset{20\times1}{\{F\}_{el}} = \underset{20\times20}{[k]_{el}}\bullet\underset{20\times1}{\{d\}_{el}}$$

where we note the following correspondence for the nodal forces:

$$\{d\}_{el} = \begin{bmatrix} u_1 & v_1 & w_1 & \theta_{x1} & \theta_{y1} & u_2 & v_2 & w_2 & \theta_{x2} & \theta_{y2} & \cdots\cdots & \theta_{y4} \end{bmatrix}^{T}$$
$$\quad\;\updownarrow\;\;\;\updownarrow\;\;\;\updownarrow\;\;\;\updownarrow\;\;\;\updownarrow\;\;\;\updownarrow\;\;\;\updownarrow\;\;\;\updownarrow\;\;\;\updownarrow\;\;\;\updownarrow\;\;\cdots\cdots\;\;\updownarrow$$
$$\{F\}_{el} = \begin{bmatrix} X_1 & Y_1 & Z_1 & L_1 & M_1 & X_2 & Y_2 & Z_2 & L_2 & M_2 & \cdots\cdots & M_4 \end{bmatrix}^{T}$$

$$[5.17]$$

○ The complete triangular plate element shown in Figure 5.23 possesses $6+9=15$ dof in the local system.

We thus obtain:

$$E_{pot.} = \frac{1}{2}\{d\}_{el}^{\,T} \bullet [k]_{el} \bullet \{d\}_{el}$$

$$\underset{\substack{complete\ element \\ (plane\ stress\ +\ bending)}}{} \qquad \underset{1\times15 \quad 15\times15 \quad 15\times1}{}$$

Leading to behavior in local coordinate system (xyz) written as:

$$\{F\}_{el} = [k]_{el} \bullet \{d\}_{el}$$

$$\underset{15\times1 \quad\quad 15\times15 \quad 15\times1}{}$$

with corresponding nodal forces as in [5.17].

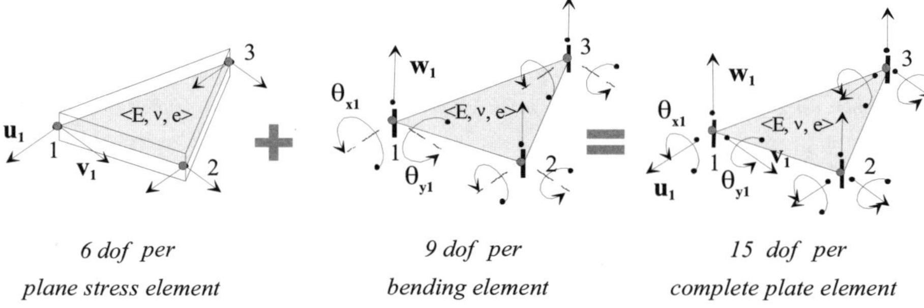

6 dof per

plane stress element

9 dof per

bending element

15 dof per

complete plate element

Figure 5.23. *Complete triangular plate element*

NOTES

❏ The notes made earlier for the plane stress and bending elements are valid here. It can be mentioned for the "floating" elements in Figures 5.22 and 5.23:

– the possibility of five rigid body motions in the local co-ordinate system (xyz);

– the non-existence of a rotation dof θ_{zi} at node i;

– the knowledge of stresses σ_x, σ_y, τ_{xy} when the dof are known, and the display of their values on the upper and lower faces (bending) and on the middle plane (plane stress) of the element;

– the behavior in a global coordinate system (XYZ) that is distinct from the local system (xyz) is obtained by means of a transfer matrix $[P]$ such that:

$$\begin{Bmatrix} u_1 \\ v_1 \\ w_1 \\ \theta_{x1} \\ \theta_{y1} \\ 0 \\ u_2 \\ \vdots \end{Bmatrix} = [P] \bullet \begin{Bmatrix} u_1 \\ v_1 \\ w_1 \\ \theta_{x1} \\ \theta_{y1} \\ \theta_{z1} \\ u_2 \\ \vdots \end{Bmatrix} \qquad \Leftrightarrow \qquad \{d\}_{el} = [P] \bullet \{d\}_{el}$$

$$\text{\textit{Local}} \qquad\qquad \text{Global} \qquad\qquad\qquad \text{\textit{Local}} \qquad\qquad \text{Global}$$

and the same for the nodal forces: $\{F\}_{el} = [P] \bullet \{F\}_{el}$

$$\qquad\qquad\qquad\qquad\qquad\qquad \text{\textit{Local}} \qquad\quad \text{Global}$$

with dimension of $[P]$:

- 24×24 for the rectangular element,

- 18×18 for the triangular element;

– the possibility of defining a complete quadrilateral plate element from a rectangular one.

SUMMARY

We will thus summarize the complete plate element with the two forms:

– quadrilateral, that we can derive from the rectangular plate element (previous note);

– triangular.

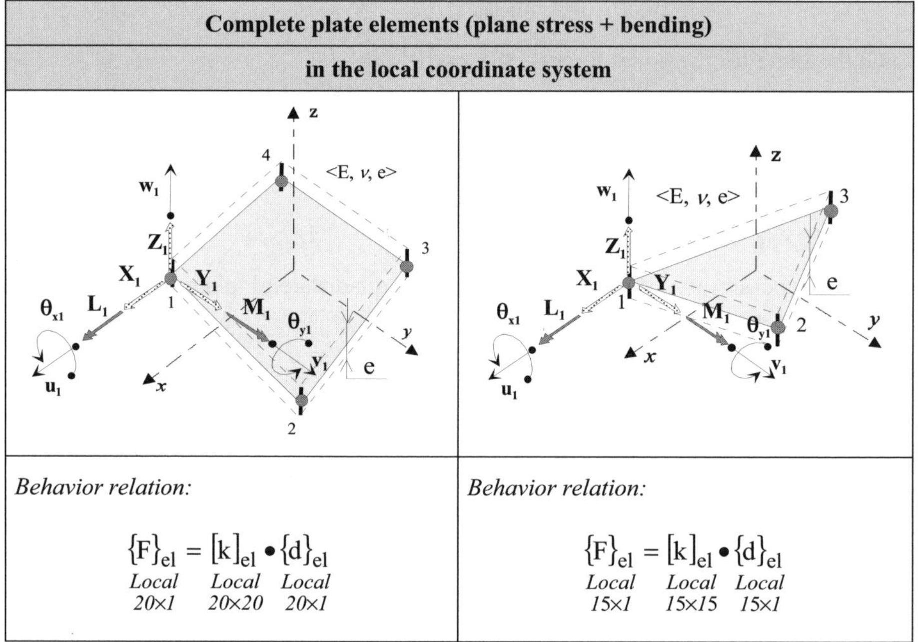

Complete plate elements (plane stress + bending)

in the local coordinate system

Behavior relation:

$$\{F\}_{el} = [k]_{el} \bullet \{d\}_{el}$$
Local   Local   Local
20×1   20×20   20×1

Behavior relation:

$$\{F\}_{el} = [k]_{el} \bullet \{d\}_{el}$$
Local   Local   Local
15×1   15×15   15×1

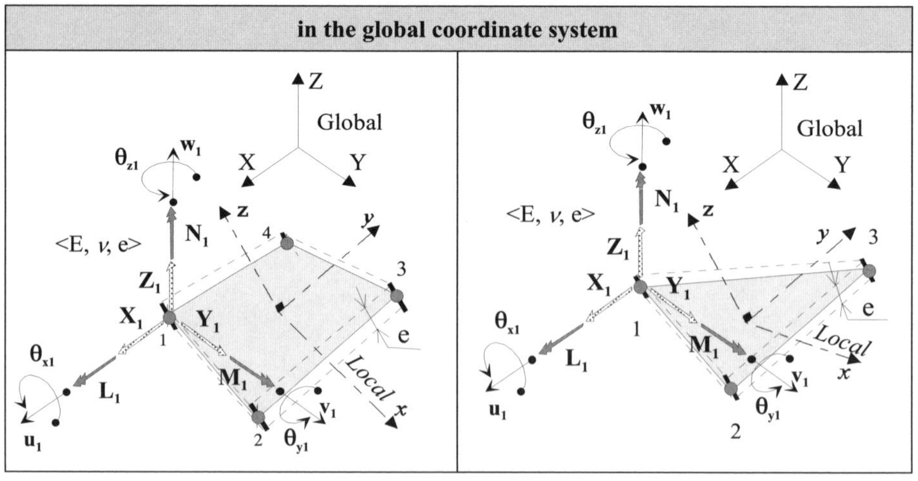

in the global coordinate system

$\{F\}_{el} = [P] \bullet \{F\}_{el}$ *Local* *24×24* Global $\{d\}_{el} = [P] \bullet \{d\}_{el}$ *Local* *24×24* Global	$\{F\}_{el} = [P] \bullet \{F\}_{el}$ *Local* *18×18* Global $\{d\}_{el} = [P] \bullet \{d\}_{el}$ *Local* *18×18* Global
$[k]_{el} = [P]^T \bullet [k]_{el} \bullet [P]$ Global *24×24* *Local* *24×24* *24×24*	$[k]_{el} = [P]^T \bullet [k]_{el} \bullet [P]$ Global *18×18* *Local* *18×18* *18×18*
Behavior relation: $\{F\}_{el} = [k]_{el} \bullet \{d\}_{el}$ Global Global Global *24×1* *24×24* *24×1*	*Behavior relation:* $\{F\}_{el} = [k]_{el} \bullet \{d\}_{el}$ Global Global Global *18×1* *18×18* *18×1*

Figure 5.24. *Complete plate elements*

5.5. Elements for complete states of stresses

5.5.1. *Preliminary notes*

In Chapter 1 (Figure 1.1) we isolated in a structure of any geometry a small domain, initially spherical, surrounding a point M. When the structure was subjected to any loading, we indicated that the small sphere changed in an ellipsoid whose three orthogonal axes were the principal directions (noted as $\vec{1}$, $\vec{2}$, $\vec{3}$) for the stresses. On the elementary faces, perpendicular to these principal directions, acted the principal stresses $\overrightarrow{C_{1(M,\vec{1})}}$, $\overrightarrow{C_{2(M,\vec{2})}}$, $\overrightarrow{C_{3(M,\vec{3})}}$ that we will henceforth call $\vec{\sigma_1}$, $\vec{\sigma_2}$, $\vec{\sigma_3}$. Let us remember their specificity: they are perpendicular to the elementary faces in question. The small domain, initially spherical, expands – or contracts – following directions $\vec{1}$, $\vec{2}$, $\vec{3}$. In Figure 5.25a we showed the six faces above of the ellipsoid perpendicular to the principal directions. They form a small parallelipiped on which we do not observe any tangential strain.

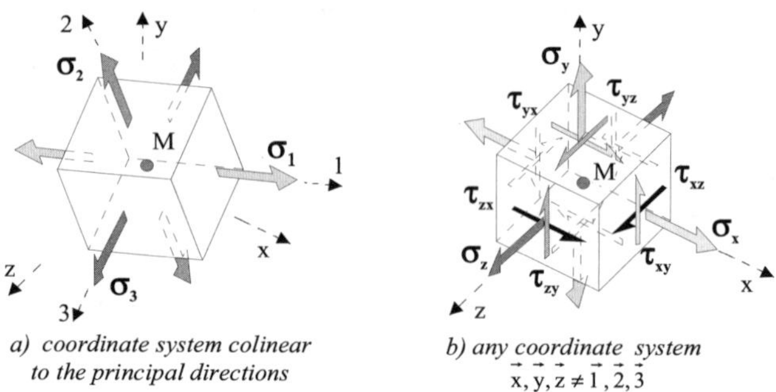

a) coordinate system colinear
to the principal directions

b) any coordinate system
$\vec{x}, \vec{y}, \vec{z} \neq \vec{1}, \vec{2}, \vec{3}$

Figure 5.25. *Complete state of stress*

When the local coordinate system used, i.e. $\vec{x}$, $\vec{y}$, $\vec{z}$, is distinct from system $\vec{1}$, $\vec{2}$, $\vec{3}$ formed by the principal directions, we consider a small parallelepiped with plane faces perpendicular to these new axes. The state of stress is then complete. It includes normal stresses but also tangential – or shear – stresses. Projections of these stresses on axes $\vec{x}$, $\vec{y}$, $\vec{z}$ are shown in Figure 5.25b.

There are six in total[39]:

• three normal components: σ_x, σ_y, σ_z ;

• three tangential components: τ_{xy}, τ_{yz}, τ_{zx} .

A specific deformation is associated with each of these components. The latter are written as[40]:

expansions or contractions   *angular distorsions*

$\varepsilon_x \quad \varepsilon_y \quad \varepsilon_z \quad \vdots \quad \gamma_{xy} \quad \gamma_{yz} \quad \gamma_{zx}$

$\updownarrow \quad \updownarrow \quad \updownarrow \qquad \updownarrow \quad \updownarrow \quad \updownarrow$

$\sigma_x \quad \sigma_y \quad \sigma_z \quad \vdots \quad \tau_{xy} \quad \tau_{yz} \quad \tau_{zx}$

normal stresses   *tangential stresses*

---

39 For more details, see Chapter 10, section 10.2 and relations [10.27].

40 In fact, due to σ_x for example, there is a main linear deformation ε_x along x and two secondary linear deformations along y and z, proportional to ε_x , respectively $\varepsilon_y = -\nu\varepsilon_x$ and $\varepsilon_z = -\nu\varepsilon_x$, where ν is the Poisson's coefficient already seen (see section 1.4). The two other normal stresses σ_y and σ_z have analogous effects.

It can be shown (section 10.2) that the properties already mentioned for the state of plane stress[41] generalize as follows:

• the elastic displacement field, that is the small displacement of any point M of the structure, has the following components in axes $\vec{x}$, $\vec{y}$, $\vec{z}$:

$$u(x, y, z) ; \qquad v(x, y, z) ; \qquad w(x, y, z)$$

• the deformations are expressed as:

$$\varepsilon_x = \frac{\partial u}{\partial x}; \quad \varepsilon_y = \frac{\partial v}{\partial y}; \quad \varepsilon_z = \frac{\partial w}{\partial z}$$
$$\gamma_{xy} = \frac{\partial u}{\partial y}+\frac{\partial v}{\partial x}; \quad \gamma_{yz} = \frac{\partial v}{\partial z}+\frac{\partial w}{\partial y}; \quad \gamma_{xz} = \frac{\partial w}{\partial x}+\frac{\partial u}{\partial z} \qquad [5.18]$$

• these deformations and stresses are linked by a "behavior relation" that is written in the behavior domain that interests us (elastic and linear) as:

$$\begin{Bmatrix} \sigma_x \\ \sigma_y \\ \sigma_z \\ \tau_{xy} \\ \tau_{yz} \\ \tau_{zx} \end{Bmatrix} = \frac{E}{(1+v)(1-2v)} \begin{bmatrix} 1-v & v & v & 0 & 0 & 0 \\ v & 1-v & v & 0 & 0 & 0 \\ v & v & 1-v & 0 & 0 & 0 \\ 0 & 0 & 0 & \dfrac{1-2v}{2} & 0 & 0 \\ 0 & 0 & 0 & 0 & \dfrac{1-2v}{2} & 0 \\ 0 & 0 & 0 & 0 & 0 & \dfrac{1-2v}{2} \end{bmatrix} \bullet \begin{Bmatrix} \varepsilon_x \\ \varepsilon_y \\ \varepsilon_z \\ \gamma_{xy} \\ \gamma_{yz} \\ \gamma_{zx} \end{Bmatrix} \quad [5.19]$$

• the potential energy of deformation for an element with volume dV takes the following form:

$$dE_{pot.} = \frac{1}{2} \begin{Bmatrix} \varepsilon_x \\ \varepsilon_y \\ \varepsilon_z \\ \gamma_{xy} \\ \gamma_{yz} \\ \gamma_{zx} \end{Bmatrix}^T \bullet \frac{E}{(1+v)(1-2v)} \begin{bmatrix} 1-v & v & v & 0 & 0 & 0 \\ v & 1-v & v & 0 & 0 & 0 \\ v & v & 1-v & 0 & 0 & 0 \\ 0 & 0 & 0 & \dfrac{1-2v}{2} & 0 & 0 \\ 0 & 0 & 0 & 0 & \dfrac{1-2v}{2} & 0 \\ 0 & 0 & 0 & 0 & 0 & \dfrac{1-2v}{2} \end{bmatrix} \bullet \begin{Bmatrix} \varepsilon_x \\ \varepsilon_y \\ \varepsilon_z \\ \gamma_{xy} \\ \gamma_{yz} \\ \gamma_{zx} \end{Bmatrix} \times dV$$

$$[5.20]$$

---

41 See Chapter 1, [1.18].

It thus becomes necessary to examine the "volume" of the material constituting the structure, that requires the definition of finite elements called "three-dimensional", "3D", "volumic" or "solid". Expressions [5.18], [5.19] and [5.20] enable us to define these elements on the same principles as those used for previous elements.

5.5.2. *Solid tetrahedric element*

This type of element allows us to close a given volume of material with a minimum number of plane faces (4). This is shown in Figure 5.26. The nodes are the four vertices of the tetrahedron. The element has 12 dof associated with 12 nodal forces. Our experience of the triangular element in plane state of stress allows us to describe this new element[42] more rapidly.

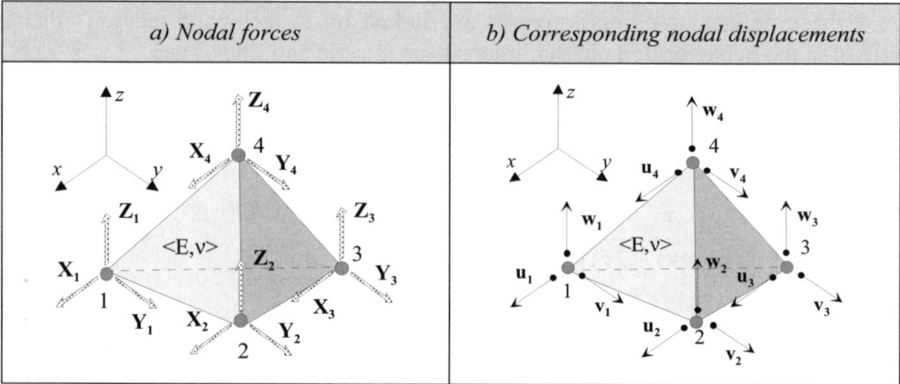

Figure 5.26. *Tetrahedric 3D element*

It presents similarities with the triangular element. Particularly, we can make the same preliminary comment regarding the absence here of cohesion forces on the faces since the linkage forces between neighboring elements are located at the nodes. The simplest solid tetrahedric element shall be characterized by a displacement field inspired by field [3.28], i.e.:

$$
\begin{aligned}
u(x, y) &= a_0 + a_1 x + a_2 y + a_3 z \\
v(x, y) &= b_0 + b_1 x + b_2 y + b_3 z \\
w(x, y) &= c_0 + c_1 x + c_2 y + c_3 z
\end{aligned}
\qquad [5.21]
$$

---

42 See sections 3.2.4 and 5.3.1.

Following the same method, (coefficients a_j, b_j, c_j are obtained by stating that u, v, w coincide with the dof in each node i of coordinates x_i, y_i, z_i) we obtain the interpolated form that is now familiar to us:

$$\begin{Bmatrix} u \\ v \\ w \end{Bmatrix} = \underset{(3\times12)}{[A(x,y,z)]} \bullet \begin{Bmatrix} u_1 \\ v_1 \\ w_1 \\ u_2 \\ v_2 \\ w_2 \\ u_3 \\ v_3 \\ w_3 \\ u_4 \\ v_4 \\ w_4 \end{Bmatrix} \qquad [5.22]$$

where the terms of interpolation matrix $[A]$ are of first degree in x, y, z. We can thus calculate the deformations [5.18] that appear as constants in the entire volume of the element. We obtain:

$$\begin{Bmatrix} \varepsilon_x \\ \varepsilon_y \\ \varepsilon_z \\ \gamma_{xy} \\ \gamma_{yz} \\ \gamma_{zx} \end{Bmatrix} = [A'] \bullet \begin{Bmatrix} u_1 \\ v_1 \\ w_1 \\ u_2 \\ v_2 \\ w_2 \\ u_3 \\ v_3 \\ w_3 \\ u_4 \\ v_4 \\ w_4 \end{Bmatrix} \qquad [5.23]$$

These values are introduced in expression [5.20] of the potential energy of deformation. After calculations we obtain:

$$E_{\substack{\text{pot.}\\\text{element}}} = \frac{1}{2} \underset{1\times12}{\begin{Bmatrix} u_1 \\ v_1 \\ w_1 \\ u_2 \\ v_2 \\ w_2 \\ u_3 \\ v_3 \\ w_3 \\ u_4 \\ v_4 \\ w_4 \end{Bmatrix}^T} \bullet \underset{12\times12}{[k]_{el}} \bullet \underset{12\times1}{\begin{Bmatrix} u_1 \\ v_1 \\ w_1 \\ u_2 \\ v_2 \\ w_2 \\ u_3 \\ v_3 \\ w_3 \\ u_4 \\ v_4 \\ w_4 \end{Bmatrix}}$$

where the stiffness matrix of the element appears in local coordinate system. We shall not describe this matrix in detail here. The behavior of this solid element is thus noted in the local coordinate system (xyz)[43]:

$$\{F\}_{el} = [k]_{el} \bullet \{d\}_{el}$$

where the nodal forces correspond to their respective dof as shown in Figure 5.26.

NOTES

All the notes made in the triangular element case can be taken up once again here.

❏ The solid element shown in Figure 5.26 appears to be floating. The definition of the dof of the element leads to three rigid body displacements in the local coordinate system which are the three translations following axes $\vec{x}$, $\vec{y}$, $\vec{z}$. The stiffness matrix mentioned here above is therefore singular.

❏ When the dof are known in the local coordinate system (xyz), we obtain the deformations by [5.23]. Then expression [5.19] allows us to deduce the stresses in the form:

$$\begin{Bmatrix} \sigma_x \\ \sigma_y \\ \sigma_z \\ \tau_{xy} \\ \tau_{yz} \\ \tau_{zx} \end{Bmatrix} = \frac{E}{(1+\nu)(1-2\nu)} \begin{bmatrix} 1-\nu & \nu & \nu & 0 & 0 & 0 \\ \nu & 1-\nu & \nu & 0 & 0 & 0 \\ \nu & \nu & 1-\nu & 0 & 0 & 0 \\ 0 & 0 & 0 & \frac{1-2\nu}{2} & 0 & 0 \\ 0 & 0 & 0 & 0 & \frac{1-2\nu}{2} & 0 \\ 0 & 0 & 0 & 0 & 0 & \frac{1-2\nu}{2} \end{bmatrix} \bullet [A'] \bullet \begin{Bmatrix} u_1 \\ v_1 \\ w_1 \\ u_2 \\ v_2 \\ w_2 \\ u_3 \\ v_3 \\ w_3 \\ u_4 \\ v_4 \\ w_4 \end{Bmatrix} \quad [5.24]$$

❏ We can see in this relation that stresses in an element are constants as for deformations. By crossing a face element common to two adjacent elements, we thus observe, in general, a discontinuity of stresses. The internal forces are transmitted, not in a continuous manner in the material, but only in concentrated points that are the nodes. Displacements are continuous, but not stresses.

❏ We can go from the local coordinate system (xyz) to the global system (XYZ) by means of the transfer matrix $[P]$ that can be easily defined thanks to [5.1]:

---

43 See [2.121].

$$
\left\{\begin{array}{c} u_1 \\ v_1 \\ w_1 \\ u_2 \\ v_2 \\ w_2 \\ u_3 \\ v_3 \\ w_3 \\ u_4 \\ v_4 \\ w_4 \end{array}\right\}_{\substack{Local \\ 12\times 1}} = \underset{(12\times 12)}{[P]} \bullet \left\{\begin{array}{c} u_1 \\ v_1 \\ w_1 \\ u_2 \\ v_2 \\ w_2 \\ u_3 \\ v_3 \\ w_3 \\ u_4 \\ v_4 \\ w_4 \end{array}\right\}_{\substack{Global \\ 12\times 1}} \Leftrightarrow \{d\}_{el}^{\,Local} = \underset{(12\times 12)}{[P]} \bullet \{d\}_{el}^{\,Global}
$$

❏ As for the triangular element in plane state of stress, it is possible to improve the behavior of such elements with similar concerns as those mentioned in section 3.2.4, that we shall not discuss further here.

5.5.3. *Solid parallelepipedic element*

This is shown in Figure 5.27 in its local coordinate system.

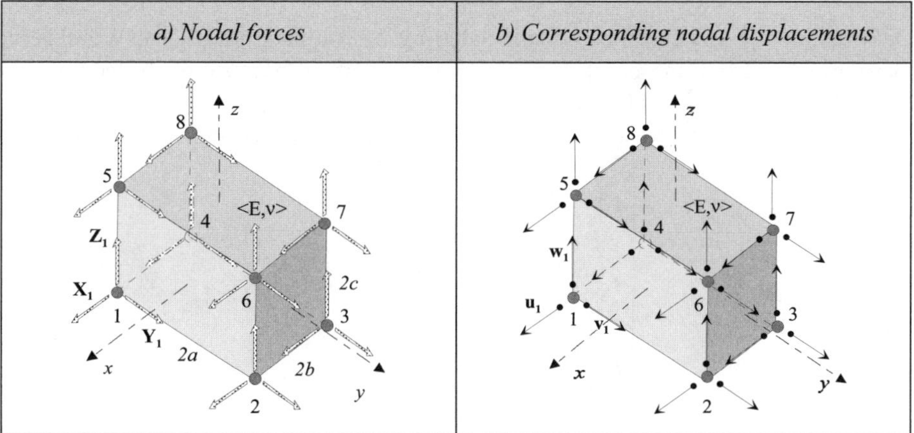

Figure 5.27. *Solid parallelepipedic element*

It possesses 8 nodes and therefore $3\times 8 = 24$ dof, with which are associated 24 nodal forces, following the now well known principle. Only the forces and the dof at node 1 have been named in Figure 5.27. We will rapidly describe, as for the element in the previous section, the procedure that defines this element. The displacement field is stated in the form:

$$u(x, y) = a_0 + a_1 x + a_2 y + a_3 z + a_4 xy + a_5 yz + a_6 zx + a_7 xyz$$
$$v(x, y) = b_0 + b_1 x + b_2 y + b_3 z + b_4 xy + b_5 yz + b_6 zx + b_7 xyz \qquad [5.25]$$
$$w(x, y) = c_0 + c_1 x + c_2 y + c_3 z + c_4 xy + c_5 yz + c_6 zx + c_7 xyz$$

a_j, b_j, c_j are known by writing that u, v, w coincide with the dof in each node i of coordinates x_i, y_i, z_i. This leads to:

$$\left\{ \begin{matrix} u \\ v \\ w \end{matrix} \right\}_{3\times 1} = \left[A(x, y, z) \right]_{3\times 24} \bullet \left\{ \begin{matrix} u_1 \\ v_1 \\ w_1 \\ \vdots \\ \vdots \\ u_8 \\ v_8 \\ w_8 \end{matrix} \right\}_{24\times 1} \qquad [5.26]$$

with
$$\left[A(x, y, z) \right] = \frac{1}{V} \begin{bmatrix} \alpha_1 & 0 & 0 & \alpha_2 & 0 & 0 & \cdots & 0 \\ 0 & \alpha_1 & 0 & 0 & \alpha_2 & 0 & \cdots & 0 \\ 0 & 0 & \alpha_1 & 0 & 0 & \alpha_2 & \cdots & \alpha_8 \end{bmatrix}$$

in which: $V = 8abc$, volume of the element and $\alpha_i = (a \pm x)(b \pm y)(c \pm z)$. Deformations [5.18] can thus be calculated by derivation from [5.26]. We observe that they vary from one point to another in the element because of the polynomial form of expressions [5.25].

$$\left\{ \begin{matrix} \varepsilon_x \\ \varepsilon_y \\ \varepsilon_z \\ \gamma_{xy} \\ \gamma_{yz} \\ \gamma_{zx} \end{matrix} \right\}_{6\times 1} = \left[A'(x, y, z) \right]_{6\times 24} \bullet \left\{ \begin{matrix} u_1 \\ v_1 \\ w_1 \\ \vdots \\ \vdots \\ u_8 \\ v_8 \\ w_8 \end{matrix} \right\}_{24\times 1} \qquad [5.27]$$

These values are introduced in expression [5.20] of the potential energy of deformation and lead, after calculations, to:

$$E_{pot.} = \frac{1}{2}\begin{Bmatrix} u_1 \\ v_1 \\ w_1 \\ \vdots \\ \vdots \\ u_8 \\ v_8 \\ w_8 \end{Bmatrix}^{T} \bullet [k]_{el} \bullet \begin{Bmatrix} u_1 \\ v_1 \\ w_1 \\ \vdots \\ \vdots \\ u_8 \\ v_8 \\ w_8 \end{Bmatrix}$$

$$\underset{1\times24}{} \qquad \underset{24\times24}{} \qquad \underset{24\times1}{}$$

where the stiffness matrix of the element appears in its local coordinate system. This matrix is not described here. The behavior of this solid element is thus written in the local system (xyz)[44]:

$$\{F\}_{el} = [k]_{el} \bullet \{d\}_{el}$$

where the nodal forces correspond to their respective dof as shown in Figure 5.27.

NOTES

The same notes can be made as for the previous tetrahedric element.

❏ The solid element shown in Figure 5.27 appears to be floating. It could have three possible rigid body displacements corresponding to the three translations following each of axes $\vec{x}$, $\vec{y}$, $\vec{z}$ colinear to the nodal dof. The stiffness matrix mentioned above is therefore singular.

❏ When the dof are known in the local coordinate system, we obtain the deformations by [5.27] and expression [5.19] allows the calculation of the stresses. The values of these stresses vary in the element.

❏ We can go from the local coordinate system (xyz) to a distinct global one (XYZ) by transfer matrix $[P]$ (built according to [5.1]) as follows:

$$\begin{Bmatrix} u_1 \\ v_1 \\ w_1 \\ \vdots \\ u_8 \\ v_8 \\ w_8 \end{Bmatrix} = [P] \bullet \begin{Bmatrix} u_1 \\ v_1 \\ w_1 \\ \vdots \\ u_8 \\ v_8 \\ w_8 \end{Bmatrix} \quad \Leftrightarrow \quad \{d\}_{el} = [P] \bullet \{d\}_{el}$$

$$\underset{\substack{Local \\ 24\times1}}{} \qquad \underset{(24\times24)}{} \qquad \underset{\substack{Global \\ 24\times1}}{} \qquad\qquad \underset{Local}{} \quad \underset{(24\times24)}{} \quad \underset{Global}{}$$

---

44 See [2.121].

❏ The behavior of such an element can be improved on the basis of the notes made in section 3.2.4.6, by using supplementary nodes placed on the 12 edges of the parallelepiped (element with $8 + 12 = 20$ nodes). This aspect is not dealt with here.

❏ It is possible to define from the current parallelepipedic element a solid element with 8 nodes, called a hexahedric element, thus allowing a better adaptation to the geometric boundaries of the structure that we are proposing to discretize (Figure 5.28a). Nevertheless, we must be aware that the vertices of a quadrilateral perimeter are not necessarily coplanar. It is therefore not possible to exceed some out-of-flatness limits with this type of element (Figure 5.28b).

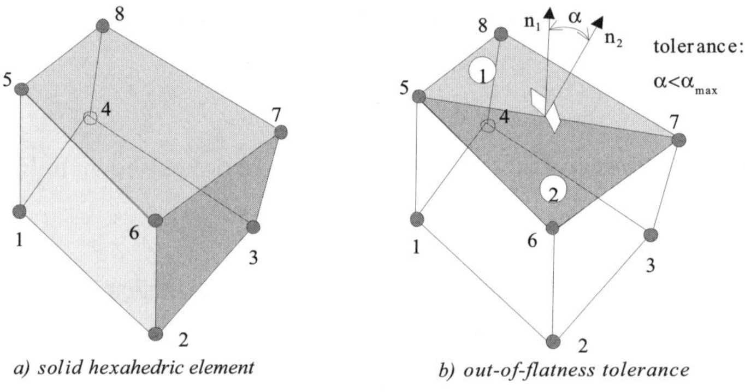

Figure 5.28. *Hexahedric element with 8 nodes*

❏ Another topology of solid element is available in finite elements softwares. This is the pentahedric element with 6 nodes (5 faces) shown in Figure 5.29. It possesses nodal properties that are identical to those of the solid elements that have already been studied.

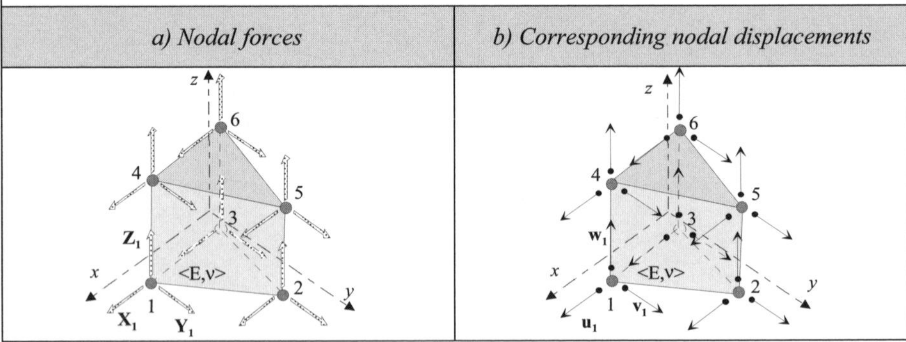

Figure 5.29. *Pentahedric element with 6 nodes*

◻ Let us also mention that we find in data libraries solid axisymmetric elements characterized by:

– an axis of revolution;

– a triangle or a quadrilateral located in a meridian plane.

The element represents the behavior of the axisymmetric-solid generated by the rotation of angle 2π of the triangle or the quadrilateral around the axis (Figure 5.30). Such elements allow the rapid modeling of thick axisymmetric structures, while avoiding the numerous dof that would result from a discretization with the preceding solid elements. We can observe in Figure 5.30c, for nodal forces, the same properties of axisymmetry as for the dof in Figure 5.30b: these characteristic quantities are identical in any meridian plane.

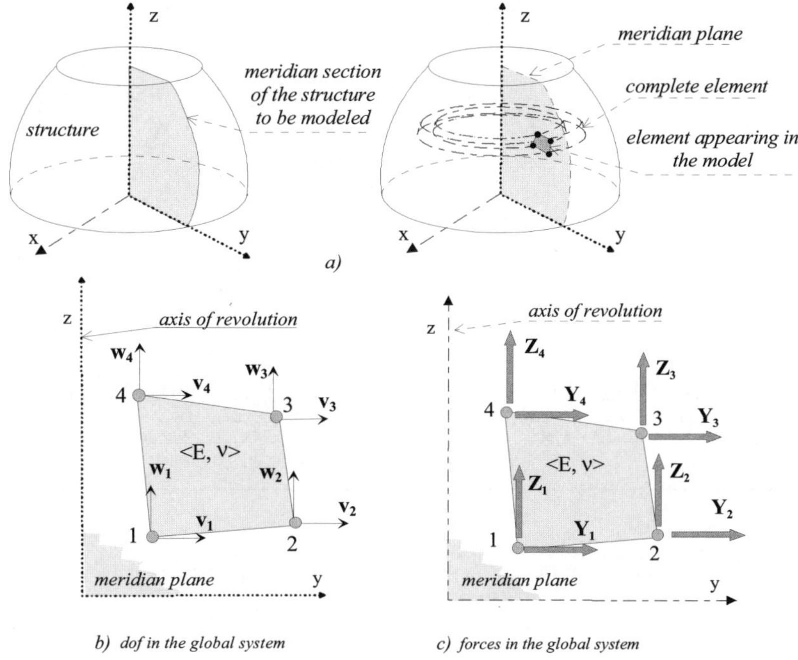

Figure 5.30. *Solid axisymmetric quadrilateral element*

SUMMARY

We can summarize the properties of the two main solid elements above in the following figure.

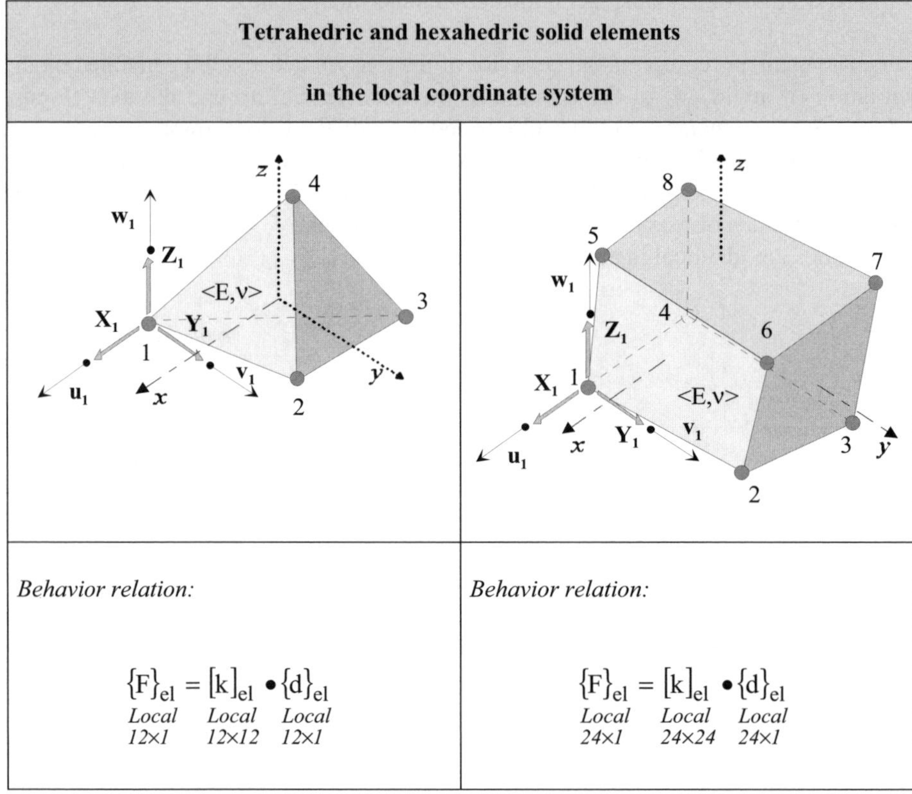

Tetrahedric and hexahedric solid elements	
in the local coordinate system	

Behavior relation:

$$\{F\}_{el} = [k]_{el} \bullet \{d\}_{el}$$

Local   Local   Local
12×1   12×12   12×1

Behavior relation:

$$\{F\}_{el} = [k]_{el} \bullet \{d\}_{el}$$

Local   Local   Local
24×1   24×24   24×1

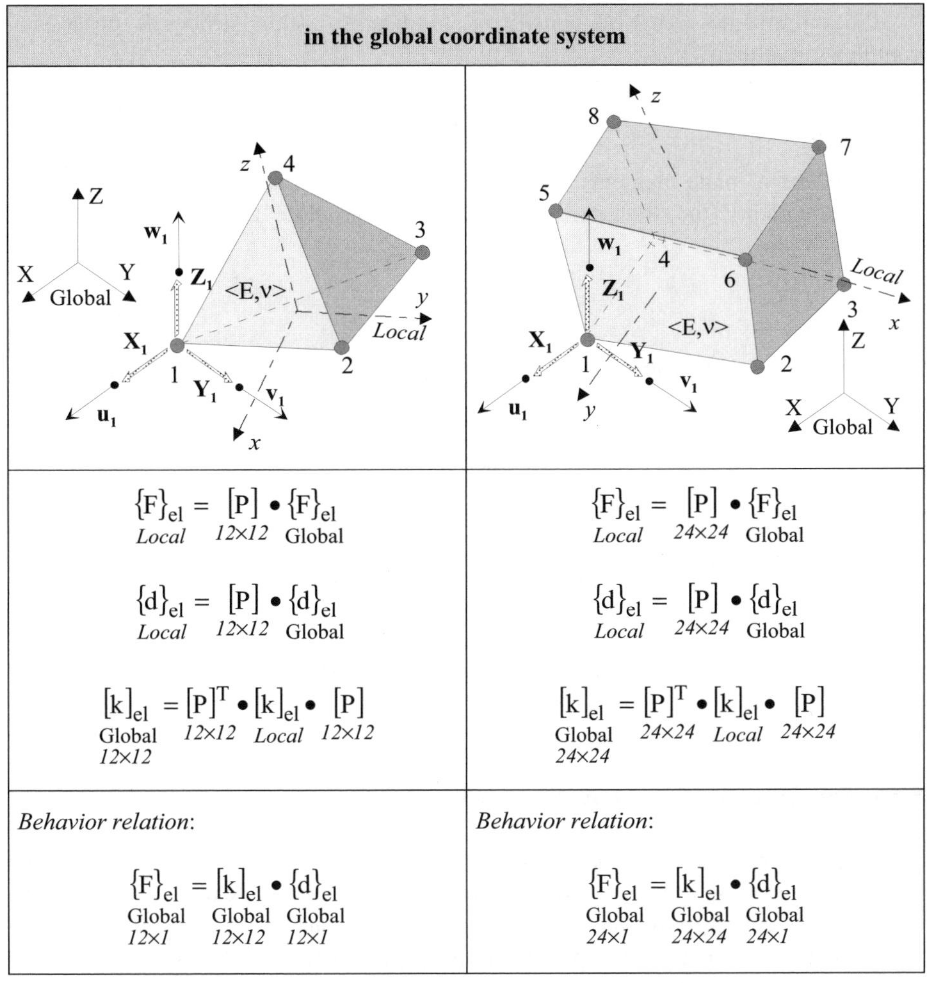

Figure 5.31. *Tetrahedric and hexahedric solid elements*

5.6. Shell elements

5.6.1. *Preliminaries*

The plate elements in section 5.4 were planes (see, for example, Figure 5.24). We can also find in data libraries elements that are "curved plates", the so-called shells. In these elements, the middle *plane* of the plate is replaced by a middle *surface* characterized by *curvatures*, as shown in Figure 5.32.

Two approaches may be considered to discretize thin structures possessing complex curvatures:

– the use of curved shell elements which coincide with the complex geometry of the structure; or

– the use of plate elements ensuring only an approximate representation of the structure geometry (the structure is called a "faceted structure").

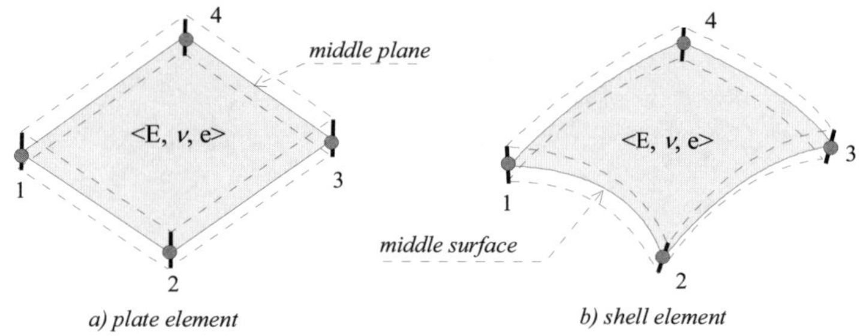

Figure 5.32. *Plate and shell elements*

Without spending time on thin and thick shell elements, we will only describe one specific case.

5.6.2. *Specific case of axisymmetric shells*

Amongst the structures that possess shell geometry, a frequent category is one that exhibits symmetry of revolution around an axis. These are called "axisymmetric shells". We can impose on these structures boundary limits (attachments and loadings) that are themselves axisymmetric (Figure 5.33a), or boundary limits that do not possess any particular symmetry (Figure 5.33b).

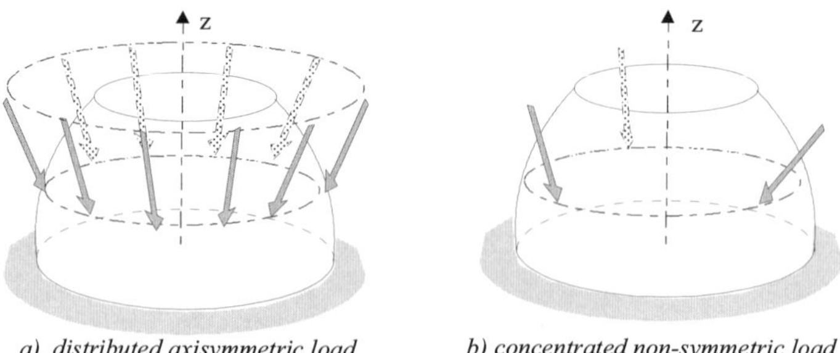

Figure 5.33. *Axisymmetric shells*

5.6.3. *Axisymmetric shell element with axisymmetric boundaries*

This element is cut out in the axisymmetric shell, shown for example in Figure 5.34. In this case of geometry and loading, mechanical characteristic (displacements, deformations, stresses) are independent on the polar angle θ[45]. This element is characterized by:

– an axis of revolution;

– an curved segment or arc representing the middle surface cut by the meridian plane (Figure 5.34b).

The nodes are the ends of this arc. The nodal dof are the displacements in the meridian plane of the small intersection-line that has a node as its center point (Figure 5.34b). We can thus observe three dof per node:

– two translations;

– one rotation;

and according to usual rules, we note the corresponding nodal forces in Figure 5.34c.

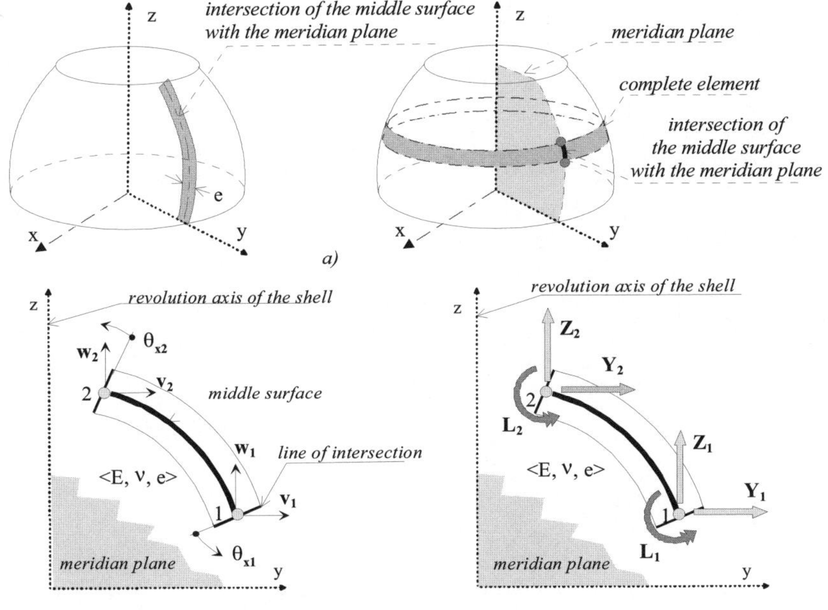

Figure 5.34. *Axisymmetric shell element*

---

45 θ is the angle of the standard cylindrical coordinates r, θ, z.

NOTES

❏ When the loading is not axisymmetric, we can use particular moduli available in finite element softwares. These moduli are based upon methods that allow the use of a preceding axisymmetric element where a supplementary nodal dof, perpendicular to the plane of the figure (meridian plane), appears.

❏ The axisymmetric shell element allows us to model more rapidly and with less degrees of freedom thin axisymmetric structures that should have been discretized with the plate or curved shell elements defined earlier.

Chapter 6

Introduction to Finite Elements
for Structural Dynamics

Until now, the structures discussed were motionless and the loads were independent of time. We have been dealing with "static" structural cases. Nevertheless, we can find a large variety of real cases where structures, having complete positioning, can be animated by movements of low magnitudes. These magnitudes are sufficiently low so as not to modify the geometry of the structure in a perceptible way. For example, this is the case for structures subjected to quickly variable mechanical forces with short durations. While applying these forces, small movements of the structure occur. Once these forces cease, these small movements persist but ease down gradually. It is said that the structure vibrates freely. We also find cases where structures are permanently subjected to variable but persistent loads. It is hence said that the vibrations of the structure are "permanent" or "forced". All these cases correspond to dynamic behaviors of the structure.

In this chapter, we will give an outline of the extension of the finite element method to the study of structural dynamics. We will see how to adapt the static formulation already seen to the study of dynamic behavior, by limiting ourselves to a freely vibrating structure. In this view, we will avoid speaking in this chapter about Lagrange's equations and linearization. On the one hand, this is to respect the level objectives of this book, and on the other to reveal the motion equations in a more physical manner such as resulting from a loading that is no longer static, but dynamic, i.e. including inertial forces.

6.1. Principles and characteristics of dynamic study

6.1.1. *Example 1*

6.1.1.1. *Description of motion*

Let us consider the beam in Figure 6.1 clamped at its left end (1). It has the vertical plane (xy) as symmetry plane and can bend in this plane. We fix on the right end (2) of this beam a *lumped* mass of value M_2. The beam is assumed to be massless. When we move this mass M_2 away vertically from its stable position and then release it, we observe that the beam-mass system oscillates in the vertical plane. These oscillations are said to be "free". They decrease gradually in magnitude as time increases and end up dying out. They are qualified as "damped". This means that all the initially potential deformation energy gradually disappears, is dissipated, and this happens in general in several ways[1]. We will neglect these energy losses here. For small magnitudes, displacements of a current cross-section of x axis of this beam are always noted as v and θ_z [2]. Here they are functions of x *and* of the time parameter t that is $v(x, t)$ and $\theta_z(x, t)$. Due to the small values of v, the movement of mass M_2 consists almost of a vertical translation v_2. To simplify, let us suppose at this stage that we are in a state of weightlessness ($g = 0$).

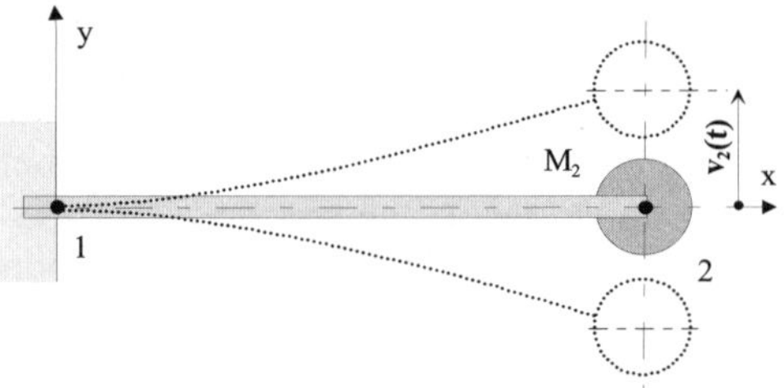

Figure 6.1.

Let us consider the vibrating system at moment t. The acceleration of the mass is noted as $\ddot{v}_2$. In Figure 6.2, we dissociated this mass and thus subjected to cohesion forces[3], noted as $\overrightarrow{F_2}$ on the beam and as $-\overrightarrow{F_2}$ on the mass.

---

1 The origins of energy loss are multiple. They are here due to "frictions" of various natures: friction of contact in the clamped support, internal friction (within material which gets deformed continuously), friction of air on the moving structure.

2 See Chapter 3, Figure 3.11.

3 Strictly speaking, the cohesion resultant forces and moments acting on the right end (2) of the beam in Figure 6.2a are written as: $\left\{Coh._{\text{beam/mass}}\right\}_2 = \left\{\begin{array}{c}\overrightarrow{F_2}\\\overrightarrow{\mathcal{M}_2}\end{array}\right\}_2$. This would

6.1.1.2. *Dynamic behavior relation*

Let us consider the "lumped" mass M_2 subjected to force $-\overrightarrow{F_2}$. Since its motion direction is parallel to axis $\overrightarrow{y}$, the fundamental equation of dynamics or "dynamic behavior relation" is written as:

$$-\overrightarrow{F_2} = M_2 \ddot{v}_2 \overrightarrow{y}$$

$\overrightarrow{F_2}$ is thus colinear with axis $\overrightarrow{y}$. We will write down in Figure 6.2b $\overrightarrow{F_2} = Y_2 \overrightarrow{y}$.

From where it follows:

$$-Y_2 = M_2 \ddot{v}_2 \qquad\qquad [6.1]$$

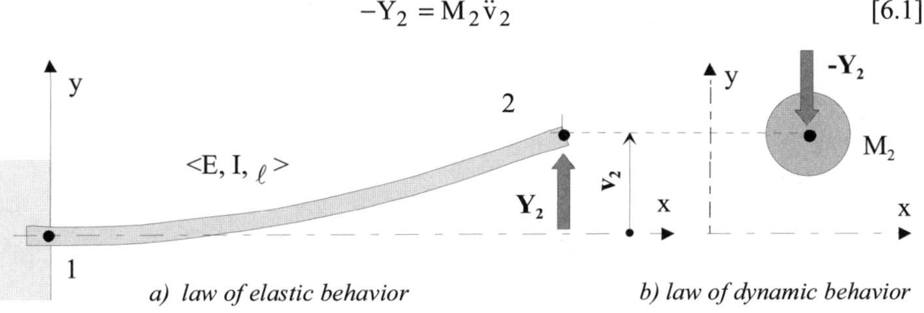

a) *law of elastic behavior*   b) *law of dynamic behavior*

Figure 6.2. *Dissociated mass and beam*

6.1.1.3. *Elastic behavior relation*

In Figure 6.2a, the beam thus appears loaded by a single force at the end:

$$\overrightarrow{F_2} = Y_2 \overrightarrow{y} = -M_2 \ddot{v}_2 \overrightarrow{y}$$

Let us use the behavior relation characterizing this beam that is considered as a single finite element in plane bending (xy).

---

correspond to the application on mass M_2 of a reciprocal system $\left\{ \begin{array}{c} -\overrightarrow{F_2} \\ -\overrightarrow{\mathcal{m}_2} \end{array} \right\}_2$. However, mass M_2 being assumed to be lumped, its inertia moments are zero and the verification of the dynamic moment equation leads to the nullity of $-\overrightarrow{\mathcal{m}_2}$.

It is written as[4]:

$$\{F\}=[K]\bullet\{d\}\Rightarrow\begin{Bmatrix}Y_1\\N_1\\Y_2\\N_2\end{Bmatrix}=\frac{EI}{\ell^3}\begin{bmatrix}12 & 6\ell & -12 & 6\ell\\6\ell & 4\ell^2 & -6\ell & 2\ell^2\\-12 & -6\ell & 12 & -6\ell\\6\ell & 2\ell^2 & -6\ell & 4\ell^2\end{bmatrix}\bullet\begin{Bmatrix}v_1\\\theta_{z1}\\v_2\\\theta_{z2}\end{Bmatrix}\qquad[6.2]$$

Let us adapt this relation to the links (clamping in (1)) and to the load. It becomes:

$$\begin{Bmatrix}Y_{1\ell}\\N_{1\ell}\\Y_2\\0\end{Bmatrix}=\frac{EI}{\ell^3}\begin{bmatrix}12 & 6\ell & -12 & 6\ell\\6\ell & 4\ell^2 & -6\ell & 2\ell^2\\-12 & -6\ell & 12 & -6\ell\\6\ell & 2\ell^2 & -6\ell & 4\ell^2\end{bmatrix}\bullet\begin{Bmatrix}v_1=0\\\theta_{z1}=0\\v_2\\\theta_{z2}\end{Bmatrix}\qquad[6.3]$$

where $Y_{1\ell}$ and $N_{1\ell}$ are the linking forces. Following the standard procedure[5], we delete the lines and columns of the same row corresponding to the clamped dof ($v_1=\theta_{z1}=0$). It remains:

$$\begin{Bmatrix}Y_2\\0\end{Bmatrix}=\frac{EI}{\ell^3}\begin{bmatrix}12 & -6\ell\\-6\ell & 4\ell^2\end{bmatrix}\bullet\begin{Bmatrix}v_2\\\theta_{z2}\end{Bmatrix}\qquad[6.4]$$

i.e.[6]:

$$\begin{Bmatrix}v_2\\\theta_{z2}\end{Bmatrix}=\begin{bmatrix}\dfrac{\ell^3}{3EI} & \dfrac{\ell^2}{2EI}\\[2mm]\dfrac{\ell^2}{2EI} & \dfrac{\ell}{EI}\end{bmatrix}\bullet\begin{Bmatrix}Y_2\\0\end{Bmatrix}$$

6.1.1.4. *Equation of motion*

Now let us "put back" together the beam and the mass that we had dissociated in Figure 6.2. Taking into account [6.1], the relation above becomes:

$$\begin{Bmatrix}v_2\\\theta_{z2}\end{Bmatrix}=\begin{bmatrix}\dfrac{\ell^3}{3EI} & \dfrac{\ell^2}{2EI}\\[2mm]\dfrac{\ell^2}{2EI} & \dfrac{\ell}{EI}\end{bmatrix}\bullet\begin{Bmatrix}-M_2\ddot{v}_2\\0\end{Bmatrix}\qquad[6.5]$$

---

4 See Figure 3.14.
5 See section 3.4.2.
6 See, for example, [2.119].

i.e.:

$$v_2 = -\frac{M_2 \ell^3}{3EI}\ddot{v}_2 \; ; \; \theta_{z2} = -\frac{M_2 \ell^2}{2EI}\ddot{v}_2$$

We can see that the dynamic values of the two degrees of freedom will be deduced from the solution of the differential equation:

$$\ddot{v}_2 + \frac{3EI}{M_2 \ell^3}v_2 = 0 \qquad\qquad [6.6]$$

This equation is standard. By rewriting it as $\ddot{v}_2 + \omega^2 v_2 = 0$, its solution can be put in the form: $v_2(t) = v_{02} \times \cos(\omega t + \varphi)$ in which:

- v_{02} (homogenous to a length) represents the maximum value of motion elongation;

- ω (radians/sec) is the pulsation or the "circular frequency" of the motion[7];

- φ (radians) is the phase angle.

We have here $\omega = \sqrt{\dfrac{3EI}{M_2 \ell^3}}$ (which gives the frequency motion f (Hertz) such as

$f = \dfrac{1}{2\pi}\sqrt{\dfrac{3EI}{M_2 \ell^3}}$). Constants v_{02} and φ are determined from the initial conditions

of motion, i.e. from values $v_2(t = 0)$ and $\dot{v}_2(t = 0)$.

After complete resolution of [6.6], and quantities v_{02}, φ, ω being known, we will have:

$$v_2(t) = v_{02} \times \cos(\omega t + \varphi) \quad ; \quad \theta_{z2}(t) = -\frac{\ell^2}{2EI}M_2 \ddot{v}_2(t) = \frac{3}{2}\frac{v_{02}}{\ell}\times\cos(\omega t + \varphi)$$

We can then again consider the lines and columns of same row in system [6.3] which had initially been removed to calculate the linking actions. We obtain:

$$\begin{Bmatrix}Y_{1\ell} \\ N_{1\ell}\end{Bmatrix} = \frac{EI}{\ell^3}\begin{bmatrix}-12 & 6\ell \\ -6\ell & 2\ell^2\end{bmatrix}\bullet\begin{Bmatrix}v_2 \\ \theta_{z2}\end{Bmatrix}$$

---

7 It should be noted that the free oscillation frequency of the motion (the so-called natural frequency or eigenfrequency) is linked to the pulsation or circular frequency ω by the relation:

$f(Hertz) = \dfrac{\omega}{2\pi}$.

where we must replace $v_2(t)$ and $\theta_{z2}(t)$ by their values now known. We thus obtain "dynamic" linking actions.

NOTES

❑ In this approach, we have "replaced" the concentrated mass M_2 in motion by the action which it exerts on the structure beam, that is the mass product and the acceleration with changed sign $(-M_2\ddot{v}_2)$ that we also call "inertia force".

❑ We had assumed the absence of gravity in the earlier study. Let us now consider the gravity field. Then mass M_2 has a weight $-M_2 g\vec{y}$. By using the behavior relation $\{F\}=[k]\bullet\{d\}$ adapted to this static load applied to the beam, we can calculate the corresponding degrees of freedom, that is: $v_{2_{sta}}$ and $\theta_{z2_{sta}}$. For this we start from [6.4]:

$$\begin{Bmatrix} -M_2 g \\ 0 \end{Bmatrix} = \frac{EI}{\ell^3} \begin{bmatrix} 12 & -6\ell \\ -6\ell & 4\ell^2 \end{bmatrix} \bullet \begin{Bmatrix} v_{2_{sta}} \\ \theta_{z2_{sta}} \end{Bmatrix} \qquad [6.7]$$

If we now consider at the same time the framework of the previously described motion *and* the gravity field, system [6.4] becomes[8]:

$$\begin{Bmatrix} -M_2 g - M_2\ddot{v}_2 \\ 0 \end{Bmatrix} = \frac{EI}{\ell^3} \begin{bmatrix} 12 & -6\ell \\ -6\ell & 4\ell^2 \end{bmatrix} \bullet \begin{Bmatrix} v_2 \\ \theta_{z2} \end{Bmatrix}$$

we note that we can break up the degrees of freedom into a "static" part (not depending on time) and a "dynamic" part evolving with time:

$$v_2 = v_{2_{sta}} + v_{2_{dyn}}$$

$$\theta_{z2} = \theta_{z2_{sta}} + \theta_{z2_{dyn}}$$

The earlier relation can thus be written as:

$$\begin{Bmatrix} -M_2 g \\ 0 \end{Bmatrix} + \begin{Bmatrix} -M_2\ddot{v}_2 \\ 0 \end{Bmatrix} = \frac{EI}{\ell^3} \begin{bmatrix} 12 & -6\ell \\ -6\ell & 4\ell^2 \end{bmatrix} \bullet \begin{Bmatrix} v_{2_{sta}} \\ \theta_{z2_{sta}} \end{Bmatrix} + \frac{EI}{\ell^3} \begin{bmatrix} 12 & -6\ell \\ -6\ell & 4\ell^2 \end{bmatrix} \bullet \begin{Bmatrix} v_{2_{dyn}} \\ \theta_{z2_{dyn}} \end{Bmatrix}$$

---

8 Force $\vec{F_2}$ applied at the right end of the beam is now written as:

$$\vec{F_2} = Y_2\vec{y} = (-M_2\ddot{v}_2 - M_2 g)\vec{y}.$$

Taking into account static equilibrium relation [6.7], and also noting that $\ddot{v}_2 = \ddot{v}_{2_{dyn}}$, there remains the following relation in which $v_{2_{dyn}}$ and $\theta_{z2_{dyn}}$ represent, by definition, the dynamic elongations *whose origin is at the equilibrium configuration*:

$$\left\{ \begin{array}{c} -M_2 \ddot{v}_{2_{dyn}} \\ 0 \end{array} \right\} = \frac{EI}{\ell^3} \begin{bmatrix} 12 & -6\ell \\ -6\ell & 4\ell^2 \end{bmatrix} \bullet \left\{ \begin{array}{c} v_{2_{dyn}} \\ \theta_{z2_{dyn}} \end{array} \right\}$$

Equation [6.5] is thus formally maintained as:

$$\left\{ \begin{array}{c} v_{2_{dyn}} \\ \\ \theta_{z2_{dyn}} \end{array} \right\} = \begin{bmatrix} \dfrac{\ell^3}{3EI} & \dfrac{\ell^2}{2EI} \\ \dfrac{\ell^2}{2EI} & \dfrac{\ell}{EI} \end{bmatrix} \bullet \left\{ \begin{array}{c} -M_2 \ddot{v}_{2_{dyn}} \\ \\ 0 \end{array} \right\}$$

The influence of gravity on the motion disappears if magnitudes
of the motion start from the static equilibrium configuration.

❏ It also follows that from the earlier relation in system [6.3], the linking actions $Y_{1\ell}$ and $N_{1\ell}$ seem to be the superposition of static linking actions (related to the presence of gravity) and dynamic linking actions (primarily due to the existence of the motion). In fact from [6.3] we obtain:

$$\left\{ \begin{array}{c} Y_{1\ell} \\ N_{1\ell} \end{array} \right\} = \frac{EI}{\ell^3} \begin{bmatrix} -12 & 6\ell \\ -6\ell & 2\ell^2 \end{bmatrix} \bullet \left\{ \begin{array}{c} v_{2_{sta}} + v_{2_{dyn}} \\ \theta_{z2_{sta}} + \theta_{z2_{dyn}} \end{array} \right\}$$

$$\left\{ \begin{array}{c} Y_{1\ell} \\ N_{1\ell} \end{array} \right\} = \frac{EI}{\ell^3} \begin{bmatrix} -12 & 6\ell \\ -6\ell & 2\ell^2 \end{bmatrix} \bullet \left\{ \begin{array}{c} v_{2_{sta}} \\ \theta_{z2_{sta}} \end{array} \right\} + \frac{EI}{\ell^3} \begin{bmatrix} -12 & 6\ell \\ -6\ell & 2\ell^2 \end{bmatrix} \bullet \left\{ \begin{array}{c} v_{2_{dyn}} \\ \theta_{z2_{dyn}} \end{array} \right\}$$

from where the possibility of writing:

$$\left\{ \begin{array}{c} Y_{1\ell_{sta}} \\ N_{1\ell_{sta}} \end{array} \right\} = \frac{EI}{\ell^3} \begin{bmatrix} -12 & 6\ell \\ -6\ell & 2\ell^2 \end{bmatrix} \bullet \left\{ \begin{array}{c} v_{2_{sta}} \\ \theta_{z2_{sta}} \end{array} \right\} \; ; \; \left\{ \begin{array}{c} Y_{1\ell_{dyn}} \\ N_{1\ell_{dyn}} \end{array} \right\} = \frac{EI}{\ell^3} \begin{bmatrix} -12 & 6\ell \\ -6\ell & 2\ell^2 \end{bmatrix} \bullet \left\{ \begin{array}{c} v_{2_{dyn}} \\ \theta_{z2_{dyn}} \end{array} \right\}$$

*static
linking
action*

*dynamic
linking
actions*

$$[6.8]$$

6.1.2. *Example 2*

6.1.2.1. *Dynamic behavior relation*

Still looking at the massless beam of Figure 6.1, the "lumped" mass M_2 fixed at the right end is replaced here by a solid occupying a non-neglectible domain in space, as shown in Figure 6.3. This solid has point (2) as center of gravity. As particular inertial properties, we consider its mass M_2 and inertia moment $\mathcal{I}_2$ around an axis parallel to $\vec{z}$ passing through (2)[9].

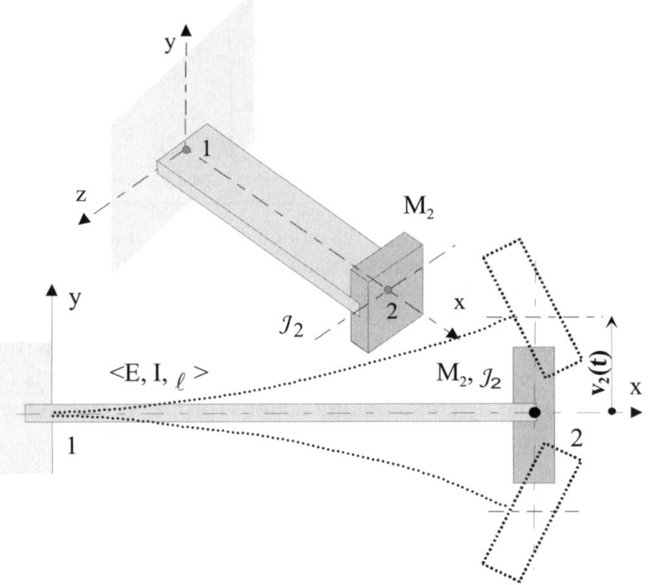

Figure 6.3. *Cantilever beam with non-lumped mass at its free end*

The influence of gravity will no longer be taken into account. We saw in fact (see NOTE above) that the gravity brought out static displacements which thus define a configuration of equilibrium. However, it does not influence the motion provided that we locate dynamic displacements with their origin at equilibrium configuration. By simply denoting the dynamic displacements as v_2 and θ_{z2}[10], the equivalence of Figure 6.2 becomes Figure 6.4.

---

9 For this solid, axes $\vec{x}$, $\vec{y}$, $\vec{z}$ from (2) are the principal axes of inertia.

10 We will quickly write the notations v_2 and θ_{z2} instead of $v_{2\mathrm{dyn}}$ and $\theta_{z2\mathrm{dyn}}$ (see [6.8]).

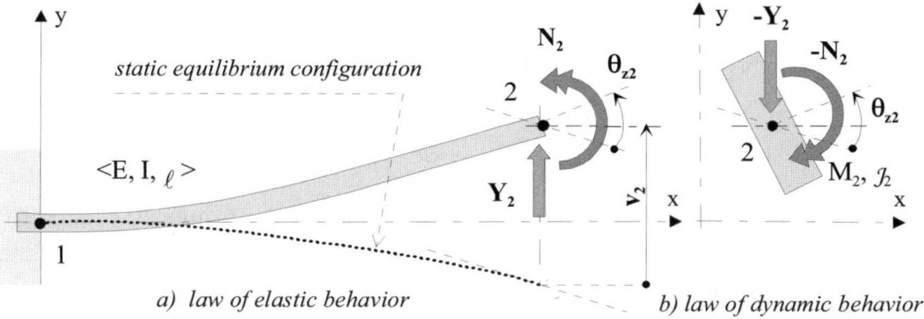

Figure 6.4. *Dissociated solid and beam*

In Figure 6.4 we have dissociated the solid from the beam. The cohesion force and moment on the right end (2) of the beam (Figure 6 4a) are written as:

$$\{Coh._{\text{solid/beam}}\}_2 = \left\{ \frac{\overrightarrow{F_2}}{\mathcal{m}_2} \right\}_2$$

which corresponds to the application on the solid of Figure 6.4b of a reciprocal force and moment $\left\{ \begin{array}{c} -\overrightarrow{F_2} \\ -\overrightarrow{\mathcal{m}_2} \end{array} \right\}_2$. Since translation v_2 is parallel to axis $\overrightarrow{y}$ and rotation θ_{z2} is parallel to axis $\overrightarrow{z}$, the fundamental principle of dynamics for this solid is written as:

– equation of resultant: $-\overrightarrow{F_2} = M_2 \ddot{v}_2 \overrightarrow{y}$
– equation of moment: $-\overrightarrow{\mathcal{m}_2} = \mathcal{J}_2 \ddot{\theta}_{z2} \overrightarrow{z}$

Force $\overrightarrow{F_2}$ appears colinear to axis $\overrightarrow{y}$. We will set down: $\overrightarrow{F_2} = Y_2 \overrightarrow{y}$.

The moment $\overrightarrow{\mathcal{m}_2}$ appears colinear to axis $\overrightarrow{z}$. We will set down: $\overrightarrow{\mathcal{m}_2} = N_2 \overrightarrow{z}$, (see Figure 6.4) i.e.:

$$-Y_2 = M_2 \ddot{v}_2$$
$$-N_2 = \mathcal{J}_2 \ddot{\theta}_{z2}$$

[6.9]

6.1.2.2. *Elastic behavior relation*

In Figure 6.4a, the beam is loaded by a force and a moment on its end. Following an approach similar to that of the earlier section, the behavior relation of this beam that is reduced to a unique and single element is written as[11]:

---

11 See Figure 3.13.

$$\begin{Bmatrix} Y_{1\ell} \\ N_{1\ell} \\ Y_2 \\ N_2 \end{Bmatrix} = \frac{EI}{\ell^3} \begin{bmatrix} 12 & 6\ell & -12 & 6\ell \\ 6\ell & 4\ell^2 & -6\ell & 2\ell^2 \\ -12 & -6\ell & 12 & -6\ell \\ 6\ell & 2\ell^2 & -6\ell & 4\ell^2 \end{bmatrix} \bullet \begin{Bmatrix} v_1 = 0 \\ \theta_{z1} = 0 \\ v_2 \\ \theta_{z2} \end{Bmatrix} \qquad [6.10]$$

where $Y_{1\ell}$ and $N_{1\ell}$ are the "dynamic" linking actions, i.e. primarily related to the existence of the motion. After removing the rows and columns corresponding to zero degrees of freedom[12]:

$$\begin{Bmatrix} Y_2 \\ N_2 \end{Bmatrix} = \frac{EI}{\ell^3} \begin{bmatrix} 12 & -6\ell \\ -6\ell & 4\ell^2 \end{bmatrix} \bullet \begin{Bmatrix} v_2 \\ \theta_{z2} \end{Bmatrix} \qquad [6.11]$$

6.1.2.3. *Equations of motion*

Now let us "put back" together the beam and solid, which we have dissociated in Figure 6.4. After taking into account [6.9], the above relation [6.11] is rewritten as:

$$\begin{Bmatrix} -M_2\ddot{v}_2 \\ -\mathcal{J}_2\ddot{\theta}_{z2} \end{Bmatrix} = \frac{EI}{\ell^3} \begin{bmatrix} 12 & -6\ell \\ -6\ell & 4\ell^2 \end{bmatrix} \bullet \begin{Bmatrix} v_2 \\ \theta_{z2} \end{Bmatrix}$$

We thus obtain two differential equations each containing v_2 and θ_{z2} and which for this reason are called "coupled" equations:

$$\begin{bmatrix} M_2\ddot{v}_2 + \dfrac{12EI}{\ell^3} v_2 - \dfrac{6EI}{\ell^2}\theta_{z2} = 0 \\[2em] \mathcal{J}_2\ddot{\theta}_{z2} - \dfrac{6EI}{\ell^2} v_2 + \dfrac{4EI}{\ell}\theta_{z2} = 0 \end{bmatrix} \qquad [6.12]$$

6.1.2.4. *Eigenmodes of vibration*

We must now solve this linear second order differential system with constant coefficients. Let us remember the steps of such a solution:

♦ we write down as solutions:

$$v_2 = v_{02} \times \cos(\omega t + \varphi)$$
$$\theta_{z2} = \theta_{0z2} \times \cos(\omega t + \varphi)$$

where, as previously, ω (rad/s) is the natural pulsation or circular frequency, and φ (rad) is the phase of the motion.

---

12 See section 3.4.2.

♦ we introduce these expressions in system [6.12], which after simplification leads to[13]:

$$\left[\begin{array}{l} -M_2\omega^2 v_{02} + \dfrac{12EI}{\ell^3} v_{02} - \dfrac{6EI}{\ell^2}\theta_{0z2} = 0 \\[2mm] -\mathcal{J}_2\omega^2\theta_{0z2} - \dfrac{6EI}{\ell^2} v_{02} + \dfrac{4EI}{\ell}\theta_{0z2} = 0 \end{array} \right.$$

or:

$$\left[\begin{array}{l} \left(\dfrac{12EI}{\ell^3} - M_2\omega^2\right) v_{02} - \dfrac{6EI}{\ell^2}\theta_{0z2} = 0 \\[3mm] -\dfrac{6EI}{\ell^2} v_{02} + \left(\dfrac{4EI}{\ell} - \mathcal{J}_2\omega^2\right)\theta_{0z2} = 0 \end{array} \right. \qquad [6.13]$$

which can be further written as:

$$\left[\begin{array}{cc} \left(\dfrac{12EI}{\ell^3} - M_2\omega^2\right) & -\dfrac{6EI}{\ell^2} \\[3mm] -\dfrac{6EI}{\ell^2} & \left(\dfrac{4EI}{\ell} - \mathcal{J}_2\omega^2\right) \end{array} \right] \bullet \left\{ \begin{array}{c} v_{02} \\[2mm] \theta_{0z2} \end{array} \right\} = \left\{ \begin{array}{c} 0 \\[2mm] 0 \end{array} \right\} \qquad [6.14]$$

We note a linear and homogenous system with non-zero solutions v_{02} and θ_{0z2} (since there is movement). Thus, the principal determinant of this system must be zero[14]. The nullity of this determinant leads to:

$$\left(\frac{12EI}{\ell^3} - M_2\omega^2\right)\left(\frac{4EI}{\ell} - \mathcal{J}_2\omega^2\right) - \left(-\frac{6EI}{\ell^2}\right)^2 = 0$$

or:

$$\omega^4 - \omega^2\left(\frac{4EI}{\mathcal{J}_2\ell} + \frac{12EI}{M_2\ell^3}\right) + \frac{12}{\mathcal{J}_2 M_2}\left(\frac{EI}{\ell^2}\right)^2 = 0 \qquad [6.15]$$

We obtain the so-called equation for circular eigenfrequencies. Solving it provides two positive roots ω_1^2 and ω_2^2 (for example, $\omega_1^2 < \omega_2^2$). We verify in fact in equation [6.15] that their sum and products are positive.

---

13 We note in fact that $\ddot{v}_2 = -\omega^2 v_2$ and $\ddot{\theta}_2 = -\omega^2\theta_2$.

14 This signifies also that the two relations constituting system [6.13] are related by a simple ratio of proportionality.

• any of relations [6.13] provide for each root ω_i^2 found (i = 1 or 2) a ratio of the values v_{02} and θ_{0z2}. For example with the first relation [6.13]:

$$\frac{(\theta_{0z})_i}{(v_0)_i} = \frac{\dfrac{12EI}{\ell^3} - M_2\omega_i^2}{\dfrac{6EI}{\ell^2}} = \frac{2}{\ell} - \frac{M_2\ell^2}{6EI}\omega_i^2 = f(\omega_i^2) \qquad [6.16]$$

The general solution of [6.12] is thus written as:

$$v_2(t) = (v_{02})_1 \times \cos(\omega_1 t + \varphi_1) + (v_{02})_2 \times \cos(\omega_2 t + \varphi_2)$$
$$\theta_{z2}(t) = (v_{02})_1 f(\omega_1^2) \times \cos(\omega_1 t + \varphi_1) + (v_{02})_2 f(\omega_2^2) \times \cos(\omega_2 t + \varphi_2) \qquad [6.17]$$

The general motion is the superposition of two sinusoidal motions (also known as harmonic), each one characterized by an eigenfrequency $\dfrac{\omega_i}{2\pi}$ and a phase φ_i (i = 1, 2). These are called "eigenmodes of vibration". In concrete terms, each eigenmode of vibration is a particular movement where all the points of the concerned structure "vibrate" with the same frequency and the same phase (for example, displacements of all the points of the structure reach their extreme values at the same time).

The intensity of each of these eigenmodes of vibration depends on the initial conditions of generating the motion, i.e. the values:

$$v_2(t=0); \quad \dot{v}_2(t=0); \quad \theta_{z2}(t=0); \quad \dot{\theta}_{z2}(t=0)$$

These four initial values allow the calculation of the four quantities having remained unknown until now in [6.17], i.e.:

$$(v_{02})_1 \qquad (v_{02})_2 \qquad \varphi_1 \qquad \varphi_2$$

Here, a numerical application would show a general form of each eigenmode of vibration as described in Figure 6.5.

Figure 6.5a represents the mode known as "low frequency", where all the points of the structure move at the same frequency $\dfrac{\omega_1}{2\pi}$. The phase angle φ_1 being single for a given mode, all the points reach an extreme magnitude at the same moment. All the intermediate deflected shapes are homothetic of the extreme deformation. Any non-zero intermediate deflected shape is sufficient to characterize the deformation corresponding to this mode ("modal deflected shape" or "eigenshape"). The same notes hold true for the "high frequency" mode in Figure 6.5b.

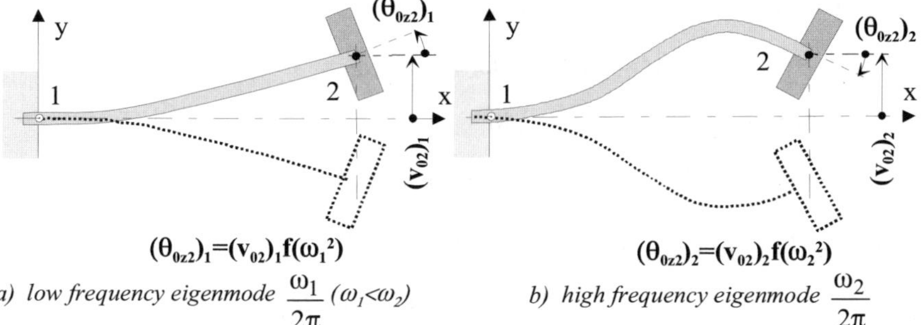

$$(\theta_{0z2})_1 = (v_{02})_1 f(\omega_1{}^2) \qquad\qquad (\theta_{0z2})_2 = (v_{02})_2 f(\omega_2{}^2)$$

a) low frequency eigenmode $\dfrac{\omega_1}{2\pi}$ *($\omega_1 < \omega_2$)*   *b) high frequency eigenmode* $\dfrac{\omega_2}{2\pi}$

Figure 6.5. *Modal deflected shapes (eigenshapes): extreme values (highly exaggerated)*

At this stage, displacements $v_2(t)$ and $\theta_{z2}(t)$ are known. We can thus return to relation [6.10] to calculate the unknown linking actions, i.e.:

$$\begin{Bmatrix} Y_{1\ell}(t) \\ N_{1\ell}(t) \end{Bmatrix} = \frac{EI}{\ell^3} \begin{bmatrix} 12 & -6\ell \\ -6\ell & 4\ell^2 \end{bmatrix} \bullet \begin{Bmatrix} v_2(t) \\ \theta_{z2}(t) \end{Bmatrix}$$

in which we replace $v_2(t)$ and $\theta_{z2}(t)$ by their development [6.17].

NOTES

❏ Returning to system [6.12], we note that it can be rewritten in the form:

$$\underbrace{\begin{bmatrix} M_2 & 0 \\ 0 & \mathcal{J}_2 \end{bmatrix}}_{mass\ matrix} \bullet \underbrace{\begin{Bmatrix} \ddot{v}_2 \\ \ddot{\theta}_{z2} \end{Bmatrix}}_{free\ dof} + \underbrace{\frac{EI}{\ell^3} \begin{bmatrix} 12 & -6\ell \\ -6\ell & 4\ell^2 \end{bmatrix}}_{stiffness\ matrix} \bullet \underbrace{\begin{Bmatrix} v_2 \\ \theta_{z2} \end{Bmatrix}}_{free\ dof} = \begin{Bmatrix} 0 \\ 0 \end{Bmatrix}$$

Then appears what can be called a diagonal "mass matrix" by the side of the stiffness matrix of the properly linked structure. By noting this matrix as $[m]$, this differential system can be written as:

$$[m] \bullet \{\ddot{d}\} + [k] \bullet \{d\} = \{0\} \tag{6.18}$$

The diagonal nature of this mass matrix is due to the fact that we have fixed mass and inertia moment at the free node.

❏ The procedure above for solving differential system [6.12] can be taken again [6.18] in the following way:

    ◆ the unknown dynamic free dof being $\{d\}$, we note the solution as an eigenmode:

$$\{d\} = \cos(\omega t + \varphi) \times \{d_0\}$$

i.e. by deriving twice with respect to time:

$$\{\ddot{d}\} = -\omega^2 \cos(\omega t + \varphi) \times \{d_0\} = -\omega^2 \{d\}$$

that we substitute in [6.18]. We obtain:

$$-\omega^2 [m] \bullet \{d_0\} + [k] \bullet \{d_0\} = \{0\} \qquad [6.19]$$

by putting $\{d_0\}$ in factor we obtain the equivalent of [6.14]:

$$\big([k] - \omega^2 [m]\big) \bullet \{d_0\} = \{0\} \qquad [6.20]$$

The equation for "circular eigenfrequencies" [6.15] is written as:

$$\det\big([k] - \omega^2 [m]\big) = 0$$

♦ let us write solution [6.17] under the form:

$$\begin{Bmatrix} v_2 \\ \theta_{z2} \end{Bmatrix} = \left(v_{02}\right)_1 \cos\left(\omega_1 t + j_1\right) \times \begin{Bmatrix} 1 \\ f(\omega_1{}^2) \end{Bmatrix} + \left(v_{02}\right)_2 \cos\left(\omega_2 t + j_2\right) \times \begin{Bmatrix} 1 \\ f(\omega_2{}^2) \end{Bmatrix}$$

eigenshape no. 1   eigenshape no. 2

where it should be noted that $v_{02} \times f(\omega^2) = \theta_{0z2}$ (see Figure 6.5). In this relation appear shapes of modes (modal shapes or eigenshapes) characterized by column matrices: $\{\Phi_i\} = \begin{Bmatrix} 1 \\ f(\omega_i{}^2) \end{Bmatrix}$ (with i = 1 or 2).

The solution can thus take a more general form:

$$\{d\} = \cos\left(\omega_1 t + j_1\right) \times a_1 \times \{\Phi_1\} + \cos\left(\omega_2 t + j_2\right) \times a_2 \times \{\Phi_2\} \qquad [6.21]$$

eigenshape no. 1   eigenshape no. 2

We note that there are *as many eigenshapes as there are degrees of freedom* (two in this case, see Figure 6.5).

❒ Let us replace $\{d_0\}$ by the solution $a_1 \times \{\Phi_1\}$ in relation [6.20]. We have:

$$\big([k] - \omega_1{}^2 [m]\big) \bullet \{\Phi_1\} = \{0\}$$

i.e.:

$$[k] \bullet \{\Phi_1\} = \omega_1{}^2 [m] \bullet \{\Phi_1\}$$

let us multiply by the inverse $[m]^{-1}$ of the mass matrix:

$$[m]^{-1} \bullet [k] \bullet \{\Phi_1\} = \omega_1^2 \times \{\Phi_1\}$$

It is said that $\{\Phi_1\}$ is an "eigenvector" of matrix $[m]^{-1} \bullet [k]$ and that ω_1^2 is its associated eigenvalue[15]. This allows systematic research of eigenfrequencies and modes of vibration in the following way:

- we start with system [6.19] rewritten as:

$$[m]^{-1} \bullet [k] \bullet \{d_0\} = \omega^2 \times \{d_0\}$$

- we numerically calculate the eigenvalues ω_i^2 and associated eigenvectors $\{\Phi_i\}$ of matrix $[m]^{-1} \bullet [k]$. There are as many eigenvalues and eigenvectors as free dof. The calculated eigenvectors are homothetic to eigenmodes, and then appear sufficient to define "eigenshapes" already noted by carrying out direct calculation [6.16].

- n being the number of free dof, the general solution can be written as:

$$\{d\} = \sum_{i=1}^{n} \cos(\omega_i t + \varphi_i) \times a_i \times \{\Phi_i\} \tag{6.22}$$

In this expression, the constants a_i and φ_i still remain unknown as long as we do not use the initial conditions. Nevertheless, only knowing the eigenvectors (or eigenshapes) $\{\Phi_i\}$ is enough to describe the corresponding modal deflected shape. For the example discussed here, Figure 6.6 is defined only from $\{\Phi_1\}$ and $\{\Phi_2\}$. We can compare it to Figure 6.5.

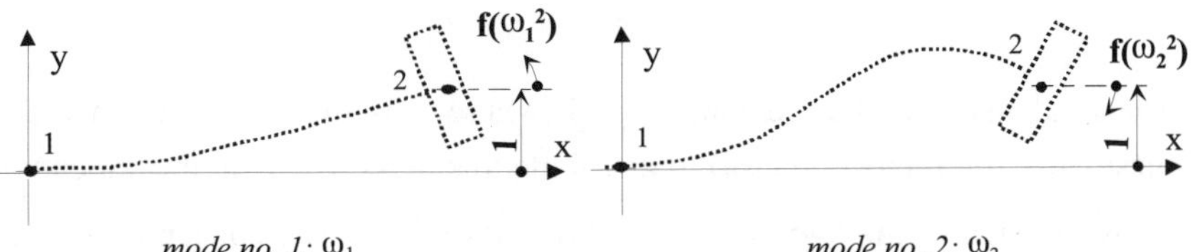

mode no. 1: ω_1 mode no. 2: ω_2

eigenvector or eigenshape: $\{\Phi_1\} = \left\{ \begin{matrix} 1 \\ f(\omega_1^2) \end{matrix} \right\}$ eigenvector or eigenshape: $\{\Phi_2\} = \left\{ \begin{matrix} 1 \\ f(\omega_2^2) \end{matrix} \right\}$

Figure 6.6. *Eigenvectors describe the eigenmodes of vibration*

---

15 See Chapter 12, section 12.1.4.

❏ The eigenvectors or shapes in Figure 6.6 are defined with a homothety factor since, in relation [6.22], the true value of product $a_i \times \{\Phi_i\}$ can be obtained with an infinite number of pairs of values $(a_i, \{\Phi_i\})$. For example, in Figure 6.6, the first component of $\{\Phi_1\}$ and $\{\Phi_2\}$ was arbitrarily selected as equal to 1. It is still said that the two eigenshapes are "normed" with respect to their first component. We could of course have chosen to norm with respect to the second component, or even by both components[16].

6.2. Mass properties of beams

6.2.1. *Finite beam element in dynamic bending plane*

We will now take into account the mass of the element. For this purpose, we consider a beam element with the characteristics indicated in Figure 6.7 for plane bending (plane (xy)). The mass per unit volume of the element is noted as ρ. We wish to study, by as simple a method as possible, the bending vibration (plane (xy)) of a beam using this element.

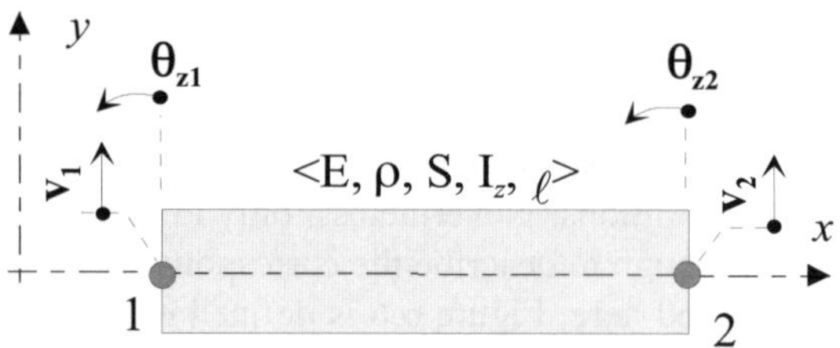

Figure 6.7. *Beam element in plane bending*

The total mass of the element is $\rho\ell S$ where S is the cross-section area. The simplest way (if not most precise) to distribute this mass results in the allocation of a half mass at each of the two nodes, i.e. $M_1 = M_2 = \dfrac{\rho\ell S}{2}$. The concentration of mass properties in nodes also implies the definition of *an equivalent moment of inertia*

---

16 This method used in the numerical calculation algorithms of eigenvalues and eigenvectors can, from a practical point of view, facilitate the physical interpretation of the eigenmodes because it allows the comparison of extreme values of the degrees of freedom.

around the axis parallel to $\vec{z}$ defined by analogy with that of an elementary cross-section of beam of thickness dx[17]. This is written for each node as:

$$\mathcal{I}_1 = \mathcal{I}_2 = \int_0^{\ell/2} d\mathcal{I}_z = \int_0^{\ell/2} \rho \times dx \times I_z = \rho \frac{I_z \times \ell}{2}$$

Each node i is thus fitted with:

– a lumped mass $M_i = \dfrac{\rho \times S \times \ell}{2}$;

– a moment of inertia $\mathcal{I}_i = \dfrac{\rho \times I_z \times \ell}{2}$ [18].

This is illustrated in Figure 6.8.

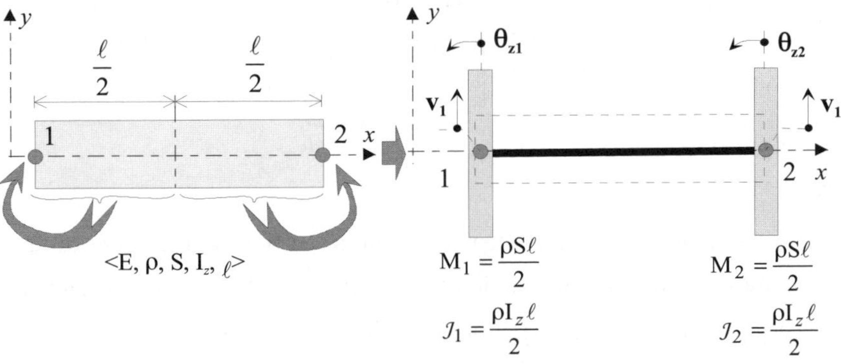

Figure 6.8. *Beam element in dynamic plane bending*

---

17 The moment of inertia of an elementary cross-section of beam of thickness dx around the axis passing through the elastic center and parallel to $\vec{z}$ is written as:

$$d\mathcal{I}_z = \int_S \rho dV y^2 = \int_S \rho dx dS y^2 = \rho dx \int_S y^2 dS = \rho dx I_z$$

18 It will be noted that the definition of the equivalent moment of inertia is not very precise here. In particular, we can no longer talk about moment of inertia around an axis, as for a thin cross-section of beam. It is instead a moment of inertia referred to the middle plane (x,z) of the beam.

Taking again the approach of the precedent section (movement in plane (xy)), the nodes are loaded by inertial loads which we have already expressed (see [6.9]). The behavior relation of the beam element is thus written as[19]:

$$\begin{Bmatrix} Y_1 \\ N_1 \\ Y_2 \\ N_2 \end{Bmatrix} = \begin{Bmatrix} -M_1\ddot{v}_1 \\ -\mathcal{J}_1\ddot{\theta}_{Z1} \\ -M_2\ddot{v}_2 \\ -\mathcal{J}_2\ddot{\theta}_{Z2} \end{Bmatrix} = \frac{EI}{\ell^3} \begin{bmatrix} 12 & 6\ell & -12 & 6\ell \\ 6\ell & 4\ell^2 & -6\ell & 2\ell^2 \\ -12 & -6\ell & 12 & -6\ell \\ 6\ell & 2\ell^2 & -6\ell & 4\ell^2 \end{bmatrix} \bullet \begin{Bmatrix} v_1 \\ \theta_{z1} \\ v_2 \\ \theta_{z2} \end{Bmatrix}$$

The notes of the earlier section always remaining valid, this behavior relation can be written in the form [6.18]:

$$[m]\bullet\{\ddot{d}\}+[k]\bullet\{d\}=\{0\}$$

i.e.:

$$\underbrace{\begin{bmatrix} M_1 & 0 & 0 & 0 \\ 0 & \mathcal{J}_1 & 0 & 0 \\ 0 & 0 & M_2 & 0 \\ 0 & 0 & 0 & \mathcal{J}_2 \end{bmatrix}}_{\substack{\textit{mass matrix} \\ \textit{of element}}} \bullet \begin{Bmatrix} \ddot{v}_1 \\ \ddot{\theta}_{Z1} \\ \ddot{v}_2 \\ \ddot{\theta}_{Z2} \end{Bmatrix} + \underbrace{\frac{EI}{\ell^3} \begin{bmatrix} 12 & 6\ell & -12 & 6\ell \\ 6\ell & 4\ell^2 & -6\ell & 2\ell^2 \\ -12 & -6\ell & 12 & -6\ell \\ 6\ell & 2\ell^2 & -6\ell & 4\ell^2 \end{bmatrix}}_{\substack{\textit{stiffness matrix} \\ \textit{of element}}} \bullet \begin{Bmatrix} v_1 \\ \theta_{z1} \\ v_2 \\ \theta_{z2} \end{Bmatrix} = \begin{Bmatrix} 0 \\ 0 \\ 0 \\ 0 \end{Bmatrix} \qquad [6.23]$$

NOTES

❏ As often indicated, such an element is not properly linked[20]. There can be two rigid body movements (a translation along axis $\vec{y}$ and a rotation around an axis parallel to $\vec{z}$). Relation [6.23] is thus not exploitable in this state. We could, for example, clamp the element at its left end ($v_1 = \theta_{z1} = 0$). This would brings us back to the same case as in the earlier section (see Figure 6.3), which we have just analyzed, and for which we can simply peform $M_2 = \dfrac{\rho S\ell}{2}$ and $\mathcal{J}_2 = \dfrac{\rho I_z\ell}{2}$. The results that we thus obtain (frequencies and modes) are not sufficient to reflect the real dynamic behavior[21].

---

19 See Figure 3.13.

20 See section 2.5.2.

21 Indeed, the complete study of plane dynamic bending of such a beam regarded as a one-dimensional continuous medium shows that its movement results in the superposition of an infinite succession of eigenmodes. Each of them is characterized by a circular eigenfrequency associated with an eigenshape. Compared to that, the movement of the beam element fixed at one end is characterized by only two eigenmodes (see section 6.1.2) and can thus reflect very approximately the first two modes found when studying the beam as a continuous medium.

❏ In relation [6.23] the inertial dynamic load appears via a diagonal matrix as we had already indicated in notes on the earlier section. It is called the "mass" matrix of the element.

> For the dynamic case, a finite element is at least[22] characterized by a stiffness matrix and a mass matrix.

❏ We can easily write the kinetic energy of the element corresponding to the mass concentration of Figure 6.8. We obtain (Koenig's theorem):

$$E_{\underset{\text{element}}{\text{kin.}}} = \frac{1}{2}M_1\dot{v}_1{}^2 + \frac{1}{2}\mathcal{J}_1\dot{\theta}_{z1}{}^2 + \frac{1}{2}M_2\dot{v}_2{}^2 + \frac{1}{2}\mathcal{J}_2\dot{\theta}_{z2}{}^2$$

It is known that such a quadratic form can be rewritten as[23]:

$$E_{\underset{\text{element}}{\text{kin.}}} = \frac{1}{2}\begin{Bmatrix}\dot{v}_1\\\dot{\theta}_{z1}\\\dot{v}_2\\\dot{\theta}_{z2}\end{Bmatrix}^T \bullet \underbrace{\begin{bmatrix}M_1 & 0 & 0 & 0\\0 & \mathcal{J}_1 & 0 & 0\\0 & 0 & M_2 & 0\\0 & 0 & 0 & \mathcal{J}_2\end{bmatrix}}_{\substack{\textit{mass matrix}\\\textit{element}}} \bullet \begin{Bmatrix}\dot{v}_1\\\dot{\theta}_{z1}\\\dot{v}_2\\\dot{\theta}_{z2}\end{Bmatrix}$$

$$\Downarrow$$

$$E_{\underset{\text{element}}{\text{kin.}}} = \frac{1}{2}\{\dot{d}\}_{el}^T \bullet [m]_{el} \bullet \{\dot{d}\}_{el}$$

The kinetic energy appears as a quadratic form of "velocities" of dof, built on the diagonal mass matrix. Here, movement is due to the action of an inertial mass load and damping is not taken into account. The material system thus considered does not dissipate energy. It is called the conservative, i.e. the sum of the kinetic energy and the potential elastic energy remains constant in the course of time.

$$E_{\underset{\text{element}}{\text{kin}}} + E_{\underset{\text{element}}{\text{pot}}} = Cte$$

❏ Local and global coordinate systems

The dynamic elongations in question have very low magnitudes, i.e. comparable with those of elastic displacements which appear in static problems. The local coordinate systems for finite elements, in a similar way, are thus

---

22 It is shown that a third matrix (damping matrix) appears when we take into account the damping properties of material.
23 See section 12.1.3.

regarded as invariable, and therefore fixed in the global coordinate system. We saw several times how to pass from the static dof expressed in the local coordinate system to the dof expressed in the global system[24]:

$$\underset{Local}{\{d\}_{el}} = [P] \bullet \underset{Global}{\{d\}_{el}}$$

We immediately verify that such a relation is maintained when it concerns the velocities:

$$\underset{Local}{\{\dot{d}\}_{el}} = [P] \bullet \underset{Global}{\{\dot{d}\}_{el}}$$

Consequently, the kinetic energy represented as higher in the local coordinate system of the beam element thus also takes the form:

$$\underset{element}{E_{kin.}} = \frac{1}{2} \underset{Global}{\{\dot{d}\}_{el}}^{T} \bullet [P]^{T} \bullet \underset{Local}{[m]_{el}} \bullet [P] \bullet \underset{Global}{\{\dot{d}\}_{el}}$$

A mass matrix of the element is therefore written in the global coordinate system as:

$$\underset{Global}{[m]_{el}} = [P]^{T} \bullet \underset{Local}{[m]_{el}} \bullet [P]$$

We can note that we obtain it by a transformation identical to that which we had already used for the stiffness matrix.

6.2.2. Discretization of a beam for dynamic bending

Let us suppose that we wish to know in a more precise way the dynamic behavior of the cantilever beam represented in Figure 6.9 bending in its middle plane (xy).

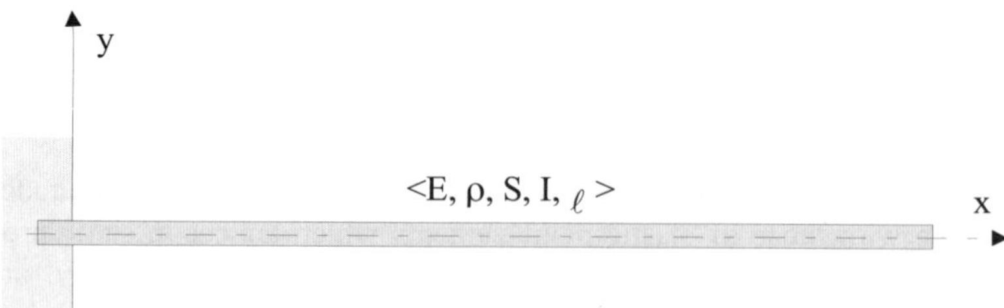

Figure 6.9. *Cantilever continuous beam before discretization*

---

24 See [5.1].

As noted previously, reducing this beam to only one element does not give very significant results compared to those of continuum mechanics. We can then consider a partition, or discretization of the beam into a number M of identical elements, for example M = 3 in Figure 6.10a.

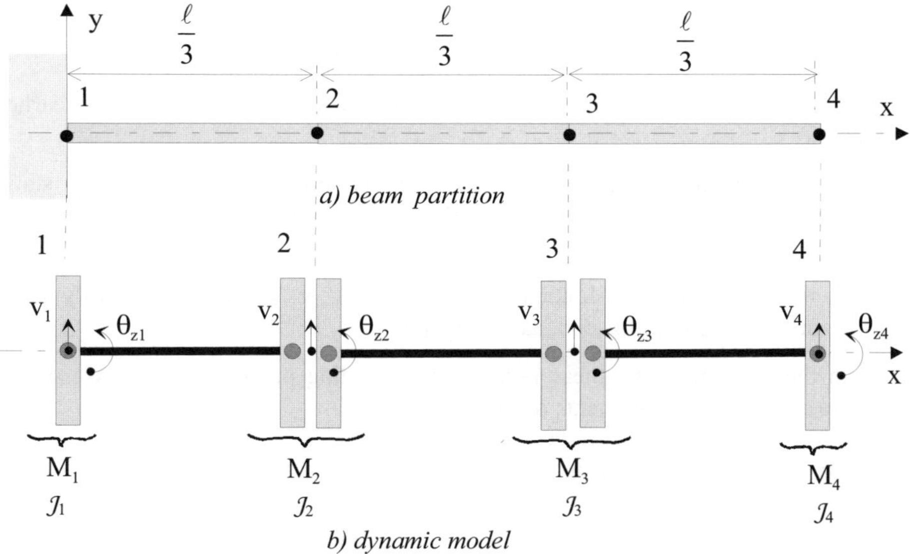

Figure 6.10. *Discretization into three beam-elements*

The "inertial forces and moments" will be created by the masses and moments of inertia noted as in Figure 6.10b, where we have:

$$M_1 = M_4 = \rho S \frac{\ell}{6} \quad ; \quad M_2 = M_3 = \rho S \frac{\ell}{6} + \rho S \frac{\ell}{6} = \rho S \frac{\ell}{3}$$

$$J_1 = J_4 = \rho I_z \frac{\ell}{6} \quad ; \quad J_2 = J_3 = \rho I_z \frac{\ell}{6} + \rho I_z \frac{\ell}{6} = \rho I_z \frac{\ell}{3}$$

[6.24]

It is known that the elastic behavior of this structure bending in its middle plane (xy) would be written as[25]:

---

25 See section 3.4.2.

$$\begin{Bmatrix} Y_1 \\ N_1 \\ Y_2 \\ N_2 \\ Y_3 \\ N_3 \\ Y_4 \\ N_4 \end{Bmatrix} = \begin{bmatrix} K \end{bmatrix}_{\substack{\text{structure} \\ (8\times8)}} \bullet \begin{Bmatrix} v_1 \\ \theta_{z1} \\ v_2 \\ \theta_{z2} \\ v_3 \\ \theta_{z3} \\ v_4 \\ \theta_{z4} \end{Bmatrix} \qquad\qquad [6.25]$$

We will complete this writing in accordance with the approach of the earlier sections:

♦ in the dynamic case, the magnitudes of dynamic dof start from the static equilibrium configuration consecutive to the action of gravity, which is not studied here;

♦ there is an embedded linkage on the left end: $v_1 = \theta_{z1} = 0$. Then dynamic linking actions appear at this end, as $Y_{1\ell}$ and $N_{1\ell}$;

♦ in the dynamic case, there is an inertial loading on free dof because of the presence of masses and moments of inertia (see [6.9]).

Relation [6.25] thus takes the form:

$$\begin{Bmatrix} Y_{1\ell} \\ N_{1\ell} \\ -M_2\ddot{v}_2 \\ -J_2\ddot{\theta}_{z2} \\ -M_3\ddot{v}_3 \\ -J_3\ddot{\theta}_{z3} \\ -M_4\ddot{v}_4 \\ -J_4\ddot{\theta}_{z4} \end{Bmatrix} = \begin{bmatrix} K \end{bmatrix}_{\substack{\text{structure} \\ (8\times8)}} \bullet \begin{Bmatrix} v_1 = 0 \\ \theta_{z1} = 0 \\ v_2 \\ \theta_{z2} \\ v_3 \\ \theta_{z3} \\ v_4 \\ \theta_{z4} \end{Bmatrix}$$

The standard procedure is as follows[26]: lines as well as columns of the same row corresponding to zero-values of dof are removed. The system reduces to:

---

26 See section 3.4.2.

$$-\begin{bmatrix} M_2 & 0 & 0 & 0 & 0 & 0 \\ 0 & \mathcal{J}_2 & 0 & 0 & 0 & 0 \\ 0 & 0 & M_3 & 0 & 0 & 0 \\ 0 & 0 & 0 & \mathcal{J}_3 & 0 & 0 \\ 0 & 0 & 0 & 0 & M_4 & 0 \\ 0 & 0 & 0 & 0 & 0 & \mathcal{J}_4 \end{bmatrix} \bullet \begin{Bmatrix} \ddot{v}_2 \\ \ddot{\theta}_{z2} \\ \ddot{v}_3 \\ \ddot{\theta}_{z3} \\ \ddot{v}_4 \\ \ddot{\theta}_{z4} \end{Bmatrix} = \begin{bmatrix} K^* \end{bmatrix}_{\substack{\text{structure} \\ (6\times6)}} \bullet \begin{Bmatrix} v_2 \\ \theta_{z2} \\ v_3 \\ \theta_{z3} \\ v_4 \\ \theta_{z4} \end{Bmatrix}$$

or:

$$\begin{bmatrix} M^* \end{bmatrix}_{str} \bullet \left\{ \ddot{d}^* \right\}_{str} + \begin{bmatrix} K^* \end{bmatrix}_{str} \bullet \left\{ d^* \right\}_{str} = \{0\} \qquad [6.26]$$

There remain six free dynamic dof. The solution of this linear differential system is obtained following the diagram described in section 6.1.2. Let us point out the successive stages:

* the solution is written as (see [6.22]):

$$\left\{ d^* \right\} = \sum_{i=1}^{6} \cos\left(\omega_i t + \phi_i\right) \times a_i \times \{\Phi_i\} \qquad [6.27]$$

where six circular eigenfrequencies and six eigenshapes appear. We determine them by setting in [6.26]:

$$\left\{ d^* \right\} = \cos\left(\omega t + \phi\right) \times a \times \{\Phi\}$$

this leads to:

$$-\omega^2 \begin{bmatrix} M^* \end{bmatrix}_{str} \bullet \{\Phi\} + \begin{bmatrix} K^* \end{bmatrix}_{str} \bullet \{\Phi\} = 0$$

i.e.:

$$\left(\begin{bmatrix} M^* \end{bmatrix}_{str}^{-1} \bullet \begin{bmatrix} K^* \end{bmatrix}_{str} \right) \bullet \{\Phi\} = \omega^2 \{\Phi\}$$

- the circular eigenfrequencies ω_i^2 ($i = 1,..6$) are the six eigenvalues of matrix $\left[M^*\right]_{str}^{-1} \bullet \left[K^*\right]_{str}$, and the eigenshapes $\{\Phi_i\}$ ($i = 1,..6$) are the six associated
(6×6)
eigenvectors[27].

NOTES

❏ The numerical resolution of system [6.26] on the above principle thus provides: six circular eigenfrequencies and six eigenshapes. At this stage, we are not able to complete the calculation of the general solution [6.27] where six constants a_i and six constants φ_i still remain unknown. For the determination of these constants it would be necessary to know the (6×2) initial conditions of motion, i.e. six $\left\{d^*\right\}_{t=0}$ values and six $\left\{\dot{d}^*\right\}_{t=0}$ values.

❏ We can see that the kinetic energy of the complete structure before installation of links can be written as the sum of the kinetic energies of each element in Figure 6.10, i.e.:

$$E_{\substack{kin. \\ structure}} = \sum_{i=1}^{3} E_{\substack{kin. \\ element\ i}}$$

we obtain:

$$E_{\substack{kin. \\ structure}} = \frac{1}{2}\left(\frac{\rho\ell S}{6}\dot{v}_1^2 + \frac{\rho I_z \ell}{6}\dot{\theta}_{z1}^2 + \frac{\rho\ell S}{6}\dot{v}_2^2 + \frac{\rho I_z \ell}{6}\dot{\theta}_{z2}^2\right)....$$
$$+ \frac{1}{2}\left(\frac{\rho\ell S}{6}\dot{v}_2^2 + \frac{\rho I_z \ell}{6}\dot{\theta}_{z2}^2 + \frac{\rho\ell S}{6}\dot{v}_3^2 + \frac{\rho I_z \ell}{6}\dot{\theta}_{z3}^2\right)....$$
$$+ \frac{1}{2}\left(\frac{\rho\ell S}{6}\dot{v}_3^2 + \frac{\rho I_z \ell}{6}\dot{\theta}_{z3}^2 + \frac{\rho\ell S}{6}\dot{v}_4^2 + \frac{\rho I_z \ell}{6}\dot{\theta}_{z4}^2\right)$$

or again with the values [6.24]:

$$E_{\substack{kin. \\ structure}} = \frac{1}{2}\left(M_1\dot{v}_1^2 + J_1\dot{\theta}_{z1}^2 + M_2\dot{v}_2^2 + J_2\dot{\theta}_{z2}^2 ...\right.$$
$$\left. + M_3\dot{v}_3^2 + J_3\dot{\theta}_{z3}^2 + M_4\dot{v}_4^2 + J_4\dot{\theta}_{z4}^2\right)$$

---

[27] The computing cores for finite element software integrate algorithms for the calculus of eigenvalues and eigenvectors mentioned here. It is pointed out (see NOTE in section 6.1.2) that the eigenvectors are defined with a homothetic factor.

or in matrix form:

$$E_{kin.\ structure} = \sum_{element\ no.\ 1}^{element\ no.\ 3} \frac{1}{2}\{\dot{d}\}_{el}^{T} \bullet [m]_{el} \bullet \{\dot{d}\}_{el} = \frac{1}{2}\{\dot{d}\}_{str}^{T} \bullet [M]_{str} \bullet \{\dot{d}\}_{str}$$

$$\Downarrow$$

$$E_{kin.\ structure} = \frac{1}{2}\begin{Bmatrix} \dot{v}_1 \\ \dot{\theta}_{z1} \\ \dot{v}_2 \\ \dot{\theta}_{z2} \\ \dot{v}_3 \\ \dot{\theta}_{z3} \\ \dot{v}_4 \\ \dot{\theta}_{z4} \end{Bmatrix}^{T} \bullet \underbrace{\begin{bmatrix} M_1 & 0 & 0 & 0 & 0 & 0 & 0 & 0 \\ 0 & \mathcal{J}_1 & 0 & 0 & 0 & 0 & 0 & 0 \\ 0 & 0 & M_2 & 0 & 0 & 0 & 0 & 0 \\ 0 & 0 & 0 & \mathcal{J}_2 & 0 & 0 & 0 & 0 \\ 0 & 0 & 0 & 0 & M_3 & 0 & 0 & 0 \\ 0 & 0 & 0 & 0 & 0 & \mathcal{J}_3 & 0 & 0 \\ 0 & 0 & 0 & 0 & 0 & 0 & M_4 & 0 \\ 0 & 0 & 0 & 0 & 0 & 0 & 0 & \mathcal{J}_4 \end{bmatrix}}_{diagonal\ mass\ matrix} \bullet \begin{Bmatrix} \dot{v}_1 \\ \dot{\theta}_{z1} \\ \dot{v}_2 \\ \dot{\theta}_{z2} \\ \dot{v}_3 \\ \dot{\theta}_{z3} \\ \dot{v}_4 \\ \dot{\theta}_{z4} \end{Bmatrix}$$

We can note here that the assembly rule of the mass matrix is similar to that used for the assembly of stiffness matrix[28]. We can also illustrate the assembly rule of the mass matrix by reloading the mass matrix (4×4) of an element in a table having the size of the mass matrix of the complete structure (8×8). For example, for element 1:

$$\begin{matrix} \dot{v}_1 \rightarrow \\ \dot{\theta}_{z1} \rightarrow \\ \dot{v}_2 \rightarrow \\ \dot{\theta}_{z2} \rightarrow \end{matrix} \begin{bmatrix} \dfrac{\rho S\ell}{6} & 0 & 0 & 0 \\ 0 & \dfrac{\rho I_z \ell}{6} & 0 & 0 \\ 0 & 0 & \dfrac{\rho S\ell}{6} & 0 \\ 0 & 0 & 0 & \dfrac{\rho I_z \ell}{6} \end{bmatrix}$$

diagonal mass matrix of element no. 1, (4×4)

$$\xrightarrow{resetting}$$

$$\begin{bmatrix} \dfrac{\rho S\ell}{6} & 0 & 0 & 0 & 0 & 0 & 0 & 0 \\ 0 & \dfrac{\rho I_z \ell}{6} & 0 & 0 & 0 & 0 & 0 & 0 \\ 0 & 0 & \dfrac{\rho S\ell}{6} & 0 & 0 & 0 & 0 & 0 \\ 0 & 0 & 0 & \dfrac{\rho I_z \ell}{6} & 0 & 0 & 0 & 0 \\ 0 & 0 & 0 & 0 & 0 & 0 & 0 & 0 \\ 0 & 0 & 0 & 0 & 0 & 0 & 0 & 0 \\ 0 & 0 & 0 & 0 & 0 & 0 & 0 & 0 \\ 0 & 0 & 0 & 0 & 0 & 0 & 0 & 0 \end{bmatrix} \begin{matrix} \leftarrow \dot{v}_1 \\ \leftarrow \dot{\theta}_{z1} \\ \leftarrow \dot{v}_2 \\ \leftarrow \dot{\theta}_{z2} \\ \leftarrow \dot{v}_3 \\ \leftarrow \dot{\theta}_{z3} \\ \leftarrow \dot{v}_4 \\ \leftarrow \dot{\theta}_{z4} \end{matrix}$$

*diagonal mass matrix of element no. 1
reset in table (8×8)
corresponding to complete structure*

❐ It has been noted above that the magnitudes of dynamic dof start from the static equilibrium configuration. This makes it possible to avoid the static study. Let us again verify this property here.

---

28 See section 3.3.3.

In the gravity field, the mass discretization of Figure 6.10 yields to gravity efforts according to $\vec{y}$. The behavior relation for this loading is obtained from [6.25] after clamping the left end beam, in the form:

$$
\begin{Bmatrix}
Y_{1\ell_{sta}} \\
N_{1\ell_{sta}} \\
-M_2 g \\
0 \\
-M_3 g \\
0 \\
-M_4 g \\
0
\end{Bmatrix}
= \begin{bmatrix} K \end{bmatrix}_{\substack{structure \\ (8\times8)}} \bullet
\begin{Bmatrix}
v_{1_{sta}} = 0 \\
\theta_{z1_{sta}} = 0 \\
v_{2_{sta}} \\
\theta_{z2_{sta}} \\
v_{3_{sta}} \\
\theta_{z3_{sta}} \\
v_{4_{sta}} \\
\theta_{z4_{sta}}
\end{Bmatrix}_{static\ dof}
\qquad [6.28]
$$

By noting that $\ddot{v}_i = \ddot{v}_{i_{dyn}}$ and $\ddot{\theta}_{zi} = \ddot{\theta}_{zi_{dyn}}$ since the static dof do not depend on time, the behavior under the total loading (static and inertial forces) of the structure becomes:

$$
\begin{Bmatrix}
Y_{1\ell_{sta}} + Y_{1\ell_{dyn}} \\
N_{1\ell_{sta}} + N_{1\ell_{dyn}} \\
-M_2 g - M_2 \ddot{v}_{2_{dyn}} \\
0 - \mathcal{J}_2 \ddot{\theta}_{z2_{dyn}} \\
-M_3 g - M_3 \ddot{v}_{3_{dyn}} \\
0 - \mathcal{J}_3 \ddot{\theta}_{z3_{dyn}} \\
-M_4 g - M_4 \ddot{v}_{4_{dyn}} \\
0 - \mathcal{J}_4 \ddot{\theta}_{z4_{dyn}}
\end{Bmatrix}
= \begin{bmatrix} K \end{bmatrix}_{\substack{structure \\ (8\times8)}} \bullet
\begin{Bmatrix}
v_1 = 0 \\
\theta_{z1} = 0 \\
v_{2_{sta}} + v_{2_{dyn}} \\
\theta_{z2_{sta}} + \theta_{z2_{dyn}} \\
v_{3_{sta}} + v_{3_{dyn}} \\
\theta_{z3_{sta}} + \theta_{z3_{dyn}} \\
v_{4_{sta}} + v_{4_{dyn}} \\
\theta_{z4_{sta}} + \theta_{z4_{dyn}}
\end{Bmatrix}_{static\ dof}
$$

That can be further broken down into:

$$
\left\{ \begin{array}{c} Y_{1\ell_{sta}} \\ N_{1\ell_{sta}} \\ -M_2 g \\ 0 \\ -M_3 g \\ 0 \\ -M_4 g \\ 0 \end{array} \right\} + \left\{ \begin{array}{c} Y_{1\ell_{dyn}} \\ N_{1\ell_{dyn}} \\ -M_2 \ddot{v}_{2_{dyn}} \\ -J_2 \ddot{\theta}_{z2_{dyn}} \\ -M_3 \ddot{v}_{3_{dyn}} \\ -J_3 \ddot{\theta}_{z3_{dyn}} \\ -M_4 \ddot{v}_{4_{dyn}} \\ -J_4 \ddot{\theta}_{z4_{dyn}} \end{array} \right\} = [K]_{\substack{structure \\ (8\times8)}} \bullet \left\{ \begin{array}{c} 0 \\ 0 \\ v_{2_{sta}} \\ \theta_{z2_{sta}} \\ v_{3_{sta}} \\ \theta_{z3_{sta}} \\ v_{4_{sta}} \\ \theta_{z4_{sta}} \end{array} \right\} + [K]_{\substack{structure \\ (8\times8)}} \bullet \left\{ \begin{array}{c} 0 \\ 0 \\ v_{2_{dyn}} \\ \theta_{z2_{dyn}} \\ v_{3_{dyn}} \\ \theta_{z3_{dyn}} \\ v_{4_{dyn}} \\ \theta_{z4_{dyn}} \end{array} \right\}
$$

static dof   dynamic dof

We can verify here that we find again systems [6.25] and [6.28] which can be decoupled. We can thus independently study static and dynamic behaviors.

6.2.3. *Other types of dynamic behaviors of a beam*

The principle of concentration of mass properties used in the earlier sections for the bending of beams also applies when studying the dynamic behavior in tension-compression and torsion. Hereafter the corresponding elements are quickly given.

6.2.3.1. *Truss element in dynamic tension-compression*

The truss element is represented with its properties in Figure 6.11a.

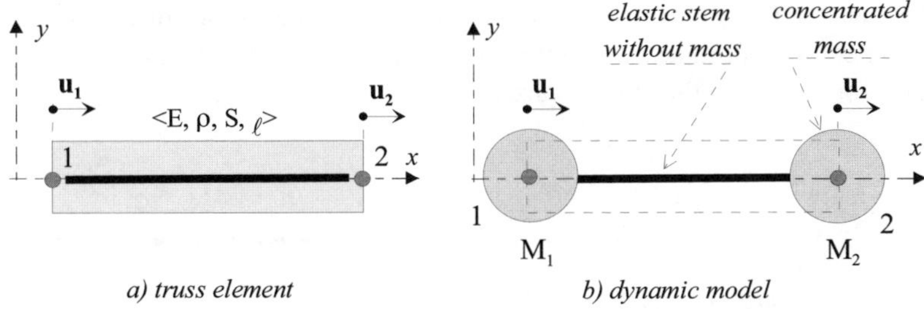

a) truss element   b) dynamic model

Figure 6.11. Truss element for dynamic study of tension-compression

By following the same principle of concentration of mass properties to the nodes, the dynamic model of the element is characterized here by a middle line with elastic properties, which can be stretched in tension-compression, and a lumped mass at each node as represented in Figure 6.11b with the value:

$$M_1 = M_2 = \frac{\rho S \ell}{2}$$

where ρ is the mass per unit volume and $\rho S \ell$ represents the total mass of the element.

NOTES

❏ In the dynamic field, each node of such an element is the seat of an "inertial effort":

$$-M_i \ddot{u}_i$$

❏ The kinetic energy of such an element is (see [6.1]):

$$E_{kin.} = \frac{1}{2}\left(M_1 \dot{u}_1{}^2 + M_2 \dot{u}_2{}^2\right)$$

We can already rewrite this expression as:

$$E_{kin.} = \frac{1}{2}\begin{Bmatrix}\dot{u}_1 \\ \dot{u}_2\end{Bmatrix}^T \bullet \begin{bmatrix} M_1 & 0 \\ 0 & M_2 \end{bmatrix} \bullet \begin{Bmatrix}\dot{u}_1 \\ \dot{u}_2\end{Bmatrix}$$

where the diagonal "mass matrix" of the finite element appears:

$$[m]_{el} = \begin{bmatrix} \dfrac{\rho S \ell}{2} & 0 \\ 0 & \dfrac{\rho S \ell}{2} \end{bmatrix}$$

❏ By considering again the methodology of earlier sections 6.1 to 6.2.2, the reader will be easily able to establish the dynamic equations and the form of the solution (frequencies, eigenmodes of vibrations) for simple structures built with this element, for example, those in Figure 6.12.

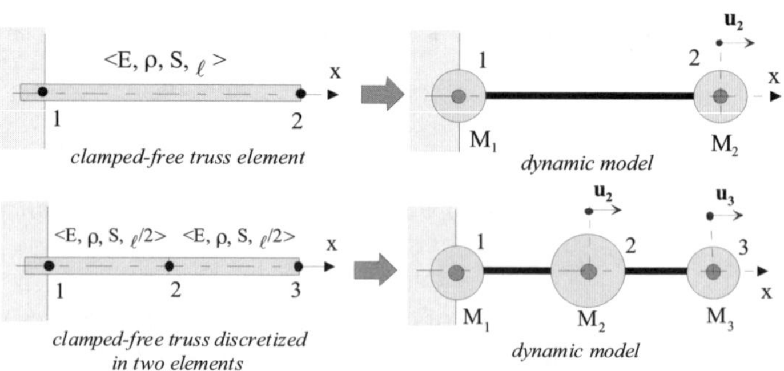

Figure 6.12. *Examples*

6.2.3.2. *Beam element with circular cross-section in dynamic torsion*

The beam element (here a shaft of circular section) is represented with its properties, in Figure 6.13a.

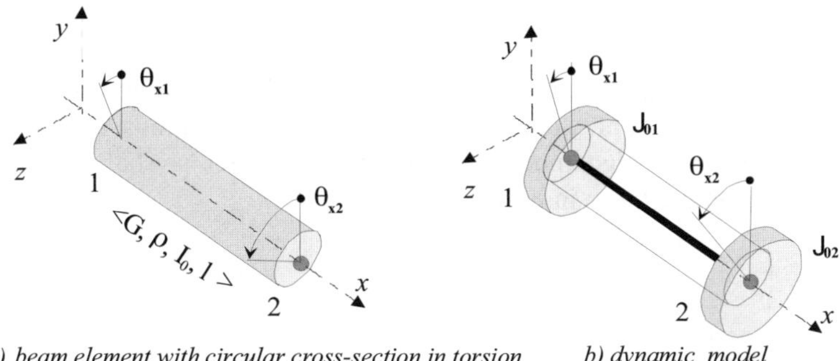

a) beam element with circular cross-section in torsion   *b) dynamic model*

Figure 6.13. *Shaft element for dynamic torsion*

The dynamic model of the element is characterized here by an elastic center line, deformable by torsion and two moments of inertia around the longitudinal axis $\vec{x}$ lumped at the nodes, such that (see Figure 6.13b):

$$\mathcal{J}_{o1} = \mathcal{J}_{o2} = \frac{\rho I_o \ell}{2}$$

where ρ is the mass per unit volume and $\rho I_o \ell$ represents the total inertia moment of the element around axis $\vec{x}$ (remember that I_o is the quadratic polar moment of a circular cross-section).

NOTES

❑ In the dynamic domain, each node of such an element is the seat of an "inertial moment":

$$- \mathcal{J}_{oi} \times \ddot{\theta}_{xi}$$

❑ The kinetic energy of such an element is written as:

$$E_{kin.} = \frac{1}{2}\left(\mathcal{J}_{o1}\dot{\theta}_{x1}{}^2 + \mathcal{J}_{o2}\dot{\theta}_{x2}{}^2\right)$$

We can easily rewrite this expression as:

$$E_{kin.} = \frac{1}{2}\left\{\begin{matrix}\dot{\theta}_{x1}\\\dot{\theta}_{x2}\end{matrix}\right\}^T \bullet \begin{bmatrix}\mathcal{J}_{o1} & 0\\0 & \mathcal{J}_{o1}\end{bmatrix} \bullet \left\{\begin{matrix}\dot{\theta}_{x1}\\\dot{\theta}_{x2}\end{matrix}\right\}$$

where the diagonal "mass matrix" of the finite element appears:

$$[m]_{el} = \begin{bmatrix} \dfrac{\rho I_o \ell}{2} & 0 \\ 0 & \dfrac{\rho I_o \ell}{2} \end{bmatrix}$$

❏ By again applying the methodology of the earlier sections 6.1 to 6.2.2, the reader will easily be able to establish the dynamic equations and form of the solution (frequencies, eigenmodes of vibrations) for circular shafts built with this element, for example those in Figure 6.14.

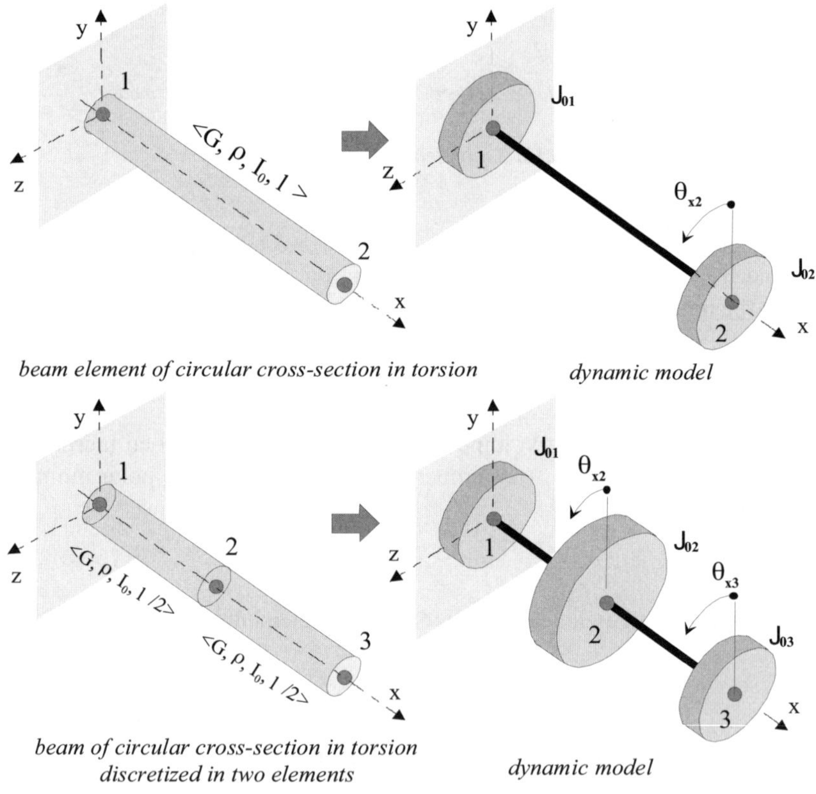

beam element of circular cross-section in torsion   *dynamic model*

*beam of circular cross-section in torsion
discretized in two elements*   *dynamic model*

Figure 6.14. *Examples*

❑ We saw in the earlier chapter how a beam element could combine the elastic properties of bending, tension-compression and torsion. The stiffness matrices relating to each of these properties made it possible to constitute a stiffness matrix for the complete beam element [5.3][29]. We can do the same with the mass matrices, since the partial kinetic energy can add up as for the potential strain energy. For example, the mass properties above (bending in plane (xy), tension-compression, torsion) combine in the following way, by further integrating bending in the other principal plane (xz):

$$
\underset{\substack{\text{element}}}{E_{\text{kin.}}} = \underset{\substack{\textit{tension–compression} \\ \textit{element}}}{E_{\text{kin.}}} + \underset{\substack{\textit{torsion} \\ \textit{element}}}{E_{\text{kin.}}} + \underset{\substack{\textit{bending (xy)} \\ \textit{element}}}{E_{\text{kin.}}} + \underset{\substack{\textit{bending (xz)} \\ \textit{element}}}{E_{\text{kin.}}}
$$

i.e.:

$$
\underset{\substack{\text{element}}}{E_{\text{kin.}}} = \frac{1}{2} \begin{Bmatrix} \dot{u}_1 \\ \\ \dot{u}_2 \end{Bmatrix}^T \bullet \begin{bmatrix} \dfrac{\rho S \ell}{2} & 0 \\ 0 & \dfrac{\rho S \ell}{2} \end{bmatrix} \bullet \begin{Bmatrix} \dot{u}_1 \\ \\ \dot{u}_2 \end{Bmatrix} + \frac{1}{2} \begin{Bmatrix} \dot{\theta}_{x1} \\ \\ \dot{\theta}_{x2} \end{Bmatrix}^T \bullet \begin{bmatrix} \dfrac{\rho I_o \ell}{2} & 0 \\ 0 & \dfrac{\rho I_o \ell}{2} \end{bmatrix} \bullet \begin{Bmatrix} \dot{\theta}_{x1} \\ \\ \dot{\theta}_{x2} \end{Bmatrix} + \dots
$$

$$
\dots \frac{1}{2} \begin{Bmatrix} \dot{v}_1 \\ \dot{\theta}_{z1} \\ \dot{v}_2 \\ \dot{\theta}_{z2} \end{Bmatrix}^T \bullet \begin{bmatrix} \dfrac{\rho S \ell}{2} & 0 & 0 & 0 \\ 0 & \dfrac{\rho I_z \ell}{2} & 0 & 0 \\ 0 & 0 & \dfrac{\rho S \ell}{2} & 0 \\ 0 & 0 & 0 & \dfrac{\rho I_z \ell}{2} \end{bmatrix} \bullet \begin{Bmatrix} \dot{v}_1 \\ \dot{\theta}_{z1} \\ \dot{v}_2 \\ \dot{\theta}_{z2} \end{Bmatrix} + \frac{1}{2} \begin{Bmatrix} \dot{w}_1 \\ \dot{\theta}_{y1} \\ \dot{w}_2 \\ \dot{\theta}_{y2} \end{Bmatrix}^T \bullet \begin{bmatrix} \dfrac{\rho S \ell}{2} & 0 & 0 & 0 \\ 0 & \dfrac{\rho I_y \ell}{2} & 0 & 0 \\ 0 & 0 & \dfrac{\rho S \ell}{2} & 0 \\ 0 & 0 & 0 & \dfrac{\rho I_y \ell}{2} \end{bmatrix} \bullet \begin{Bmatrix} \dot{w}_1 \\ \dot{\theta}_{y1} \\ \dot{w}_2 \\ \dot{\theta}_{y2} \end{Bmatrix}
$$

We readjust the partial mass matrices appearing above in a dimension table (12×12) corresponding to the column of the 12 dof of the element written in the following order:

$$
\{d\}_{el} = \begin{bmatrix} u_1 & v_1 & w_1 & \theta_{x1} & \theta_{y1} & \theta_{z1} & u_2 & v_2 & w_2 & \theta_{x2} & \theta_{y2} & \theta_{z2} \end{bmatrix}^T
$$

---

29 See section 5.2.2.

362   Modeling and Dimensioning of Structures

we obtain:

$$
[m]_{el}^{Local} = \begin{bmatrix}
\dfrac{\rho S\ell}{2} & 0 & 0 & 0 & 0 & 0 & \dfrac{\rho S\ell}{2} & 0 & 0 & 0 & 0 & 0 \\
 & 0 & 0 & 0 & 0 & 0 & 0 & 0 & 0 & 0 & 0 & 0 \\
 & & 0 & 0 & 0 & 0 & 0 & 0 & 0 & 0 & 0 & 0 \\
 & S & & 0 & 0 & 0 & 0 & 0 & 0 & 0 & 0 & 0 \\
 & & Y & & 0 & 0 & 0 & 0 & 0 & 0 & 0 & 0 \\
 & & & M & & 0 & 0 & 0 & 0 & 0 & 0 & 0 \\
 & & & & M & & \dfrac{\rho S\ell}{2} & 0 & 0 & 0 & 0 & 0 \\
 & & & & & E & & 0 & 0 & 0 & 0 & 0 \\
 & & & & & & T & & 0 & 0 & 0 & 0 \\
 & & & & & & & & 0 & 0 & 0 & 0 \\
 & & & & & & & R & & 0 & 0 & 0 \\
 & & & & & & & & Y & & & 0
\end{bmatrix}
+ \begin{bmatrix}
0 & 0 & 0 & 0 & 0 & 0 & 0 & 0 & 0 & 0 & 0 & 0 \\
 & 0 & 0 & 0 & 0 & 0 & 0 & 0 & 0 & 0 & 0 & 0 \\
 & & 0 & 0 & 0 & 0 & 0 & 0 & 0 & 0 & 0 & 0 \\
 & S & & \dfrac{\rho I_0\ell}{2} & 0 & 0 & 0 & 0 & 0 & \dfrac{\rho I_0\ell}{2} & 0 & 0 \\
 & & Y & & 0 & 0 & 0 & 0 & 0 & 0 & 0 & 0 \\
 & & & M & & 0 & 0 & 0 & 0 & 0 & 0 & 0 \\
 & & & & M & & 0 & 0 & 0 & 0 & 0 & 0 \\
 & & & & & E & & 0 & 0 & 0 & 0 & 0 \\
 & & & & & & T & & \dfrac{\rho I_0\ell}{2} & 0 & 0 & 0 \\
 & & & & & & & R & & 0 & 0 & 0 \\
 & & & & & & & & Y & & & 0 \\
 & & & & & & & & & & & 0
\end{bmatrix}
+ \ldots
$$

$$
\ldots + \begin{bmatrix}
0 & 0 & 0 & 0 & 0 & 0 & 0 & 0 & 0 & 0 & 0 & 0 \\
\dfrac{\rho S\ell}{2} & 0 & 0 & 0 & 0 & 0 & 0 & 0 & 0 & 0 & 0 & 0 \\
 & & 0 & 0 & 0 & 0 & 0 & 0 & 0 & 0 & 0 & 0 \\
 & S & & 0 & 0 & 0 & 0 & 0 & 0 & 0 & 0 & 0 \\
 & & Y & & 0 & 0 & 0 & 0 & 0 & 0 & 0 & 0 \\
 & & & M & & \dfrac{\rho I_z\ell}{2} & 0 & 0 & 0 & 0 & 0 & 0 \\
 & & & & M & & 0 & 0 & 0 & 0 & 0 & 0 \\
 & & & & & E & & \dfrac{\rho S\ell}{2} & 0 & 0 & 0 & 0 \\
 & & & & & & T & & 0 & 0 & 0 & 0 \\
 & & & & & & & R & & 0 & 0 & 0 \\
 & & & & & & & & Y & & 0 & 0 \\
 & & & & & & & & & & & \dfrac{\rho I_z\ell}{2}
\end{bmatrix}
+ \begin{bmatrix}
0 & 0 & 0 & 0 & 0 & 0 & 0 & 0 & 0 & 0 & 0 & 0 \\
 & 0 & 0 & 0 & 0 & 0 & 0 & 0 & 0 & 0 & 0 & 0 \\
\dfrac{\rho S\ell}{2} & 0 & 0 & 0 & 0 & 0 & 0 & 0 & 0 & 0 & 0 \\
 & S & & 0 & 0 & 0 & 0 & 0 & 0 & 0 & 0 & 0 \\
 & Y & & \dfrac{\rho I_y\ell}{2} & 0 & 0 & 0 & 0 & 0 & 0 & 0 & 0 \\
 & & M & & 0 & 0 & 0 & 0 & 0 & 0 & 0 & 0 \\
 & & & M & & 0 & 0 & 0 & 0 & 0 & 0 & 0 \\
 & & & E & & 0 & 0 & 0 & 0 & 0 & 0 \\
 & & & & T & & \dfrac{\rho S\ell}{2} & 0 & 0 & 0 \\
 & & & & R & & 0 & 0 & 0 \\
 & & & & & Y & & \dfrac{\rho I_y\ell}{2} & 0 \\
 & & & & & & & 0
\end{bmatrix}
$$

leading to the mass matrix of the element in its local coordinates:

$$
[m]_{el}^{Local} = \begin{bmatrix}
\dfrac{\rho S\ell}{2} & 0 & 0 & 0 & 0 & 0 & 0 & 0 & 0 & 0 & 0 & 0 \\
 & \dfrac{\rho S\ell}{2} & 0 & 0 & 0 & 0 & 0 & 0 & 0 & 0 & 0 & 0 \\
 & & \dfrac{\rho S\ell}{2} & 0 & 0 & 0 & 0 & 0 & 0 & 0 & 0 & 0 \\
 & S & & \dfrac{\rho I_0\ell}{2} & 0 & 0 & 0 & 0 & 0 & 0 & 0 & 0 \\
 & & Y & & \dfrac{\rho I_y\ell}{2} & 0 & 0 & 0 & 0 & 0 & 0 & 0 \\
 & & & M & & \dfrac{\rho I_z\ell}{2} & 0 & 0 & 0 & 0 & 0 & 0 \\
 & & & & M & & \dfrac{\rho S\ell}{2} & 0 & 0 & 0 & 0 & 0 \\
 & & & & & E & & \dfrac{\rho S\ell}{2} & 0 & 0 & 0 & 0 \\
 & & & & & & T & & \dfrac{\rho S\ell}{2} & 0 & 0 & 0 \\
 & & & & & & & R & & \dfrac{\rho I_0\ell}{2} & 0 & 0 \\
 & & & & & & & & Y & & \dfrac{\rho I_y\ell}{2} & 0 \\
 & & & & & & & & & & & \dfrac{\rho I_z\ell}{2}
\end{bmatrix}
$$

6.3. Generalization

What is given above only constitutes one approach in treating freely vibrating structures. The mass and inertial properties are here concentrated on the nodes, and we have limited ourselves to conservative systems. By using these simple principles, the making of dynamic models can thus be extended with the help of other types of finite elements.

NOTES

❏ There is a more general approach to calculate mass matrices, applicable to all types of finite elements: it starts from the writing of kinetic energy of an elementary domain taken in the element, using the displacement fields defined by the laws of interpolation already seen on several occasions[30]. We then integrate elementary kinetic energy to find the kinetic energy of the whole element considered as a quadratic form of dof velocities. The mass matrix which then appears from this quadratic form is no longer diagonal but only symmetric. It is calculated with a velocity field defined in the same form as the field of displacements. The mass matrix is thus obtained on a principle identical to that which allows the calculation of rigidity matrix $[k]$. It is said that we obtain a "consistent" mass matrix $[m]^{31}$.

❏ We saw that the most significant results of free vibration study (see [6.29]) consist of obtaining the eigenfrequencies and associated eigenshapes. This information is basic in view of studying more general dynamic cases at designing or verification stages of a structure. For example, finite element softwares use eigenmodes and frequencies to rebuild the actual movement of a structure with imposed forces or displacements depending on time (transitory vibrations, shocks). As we know, calculating displacements or dof then makes it possible to go back to strains and stresses. We will thus obtain the "dynamic" stresses, i.e. essentially due to movement, which are to be further added to static stresses.

---

30 See [3.10], [3.21], [3.30], [5.12], [5.22], [5.26].
31 We will not detail this approach here, referring the reader to specialized works.

6.4. Summary

We summarize below the procedure which leads to equations of the dynamic behavior of structures, and most significant elements of the solution.

<table>
<tr><td colspan="2" align="center">Dynamic behavior of a structure
(free vibrations; without damping; structure properly linked)</td></tr>
<tr><td colspan="2" align="center">"n" free "dynamic" dof $\{d\}$, magnitudes of which start
from the static equilibrium configuration</td></tr>
<tr>
<td>Kinetic energy:

$$E_{kin.} = \frac{1}{2}\{\dot{d}\}^T \bullet [M]_{structure} \bullet \{\dot{d}\}$$

where $[M]$ is the symmetric mass matrix
$(n \times n)$</td>
<td>Potential energy:

$$E_{pot.} = \frac{1}{2}\{d\}^T \bullet [K]_{structure} \bullet \{d\}$$

where $[K]$ is the symmetric stiffness matrix
$(n \times n)$</td>
</tr>
<tr><td colspan="2">Conservative system:

$$E_{kin.} + E_{pot.} = C^{te}$$</td></tr>
<tr><td colspan="2">Equations of dynamic behavior:

$$[M]_{structure} \bullet \{\ddot{d}\} + [K]_{structure} \bullet \{d\} = \{0\}$$</td></tr>
<tr><td colspan="2">Solution:

$$\{d\}_{(n\times1)} = \sum_{i=1}^{n} \cos(\omega_i t + \varphi_i) \times a_i \times \{\Phi_i\}_{(n\times1)}$$

characterized by:

• n circular eigenfrequencies ω_i (i = 1,...n)

• n eigenshapes $\{\Phi_i\}$ (i = 1,...n) defined with a homothetic factor</td></tr>
<tr><td colspan="2">Calculation of ω_i and $\{\Phi_i\}$:

– the ω_i^2 are the eigenvalues of $[M]_{structure}^{-1} \bullet [K]_{structure}$

– the $\{\Phi_i\}$ are the eigenvectors associated with these eigenvalues</td></tr>
<tr><td colspan="2" align="center">constants a_i and φ_i depend upon the initial conditions
of motion; in general they are not calculated</td></tr>
</table>

$$[6.29]$$

Chapter 7

Criteria for Dimensioning

7.1. Designing and dimensioning

Any design engineer who studies the behavior of a structure under loading with the objective of designing it foresees two aspects of the problem:

– *dimensioning regarding rigidity criteria*: the structure must not suffer deflection higher than the limits fixed by the specifications. A very high flexibility must be avoided as it would lead to lower performances in service;

– *dimensioning regarding resistance criteria*: the structure should remain "intact" and continue its normal work under the loads that are applied to it.

According to circumstances, one or both of these factors may predominate. For example, in normal conditions, a structure should be sufficiently rigid. On the other hand, in extreme conditions, it should resist destruction. Figure 7.1 provides an example of this.

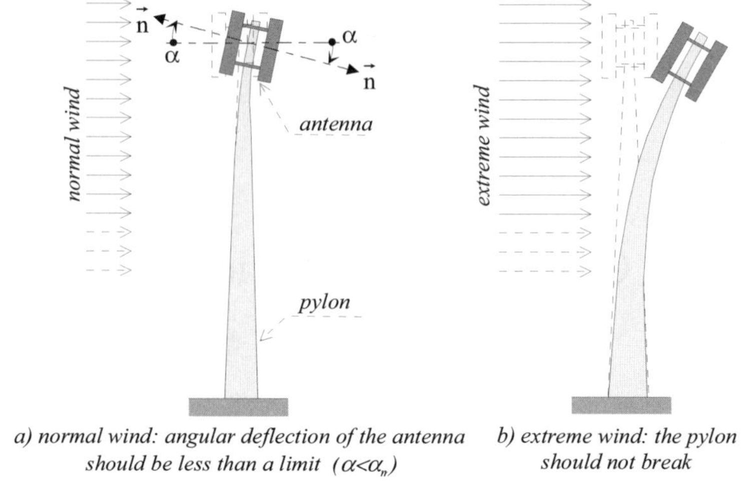

*a) normal wind: angular deflection of the antenna
should be less than a limit ($\alpha<\alpha_n$)*   *b) extreme wind: the pylon
should not break*

Figure 7.1. *Pylon supporting antenna for cellular telephones*

The specifications are illustrated in this figure. It should be noted that the specifications take into account rigidity *and* resistance.

For us, dimensioning will mean checking that a designed structure, which enables us to prepare a finite element model, will respect the limits in resistance and in rigidity according to the specifications.

Considering what has been seen until now, dimensioning regarding rigidity is carried out "naturally" with the finite element analysis. In fact the solution of the system $\{F\}_{\text{str}} = [K]_{\text{str}} \bullet \{d\}_{\text{str}}$ gives us the displacements of free nodes of the modeled structure. We can immediately test if these displacements remain within the limits prescribed by the specifications. Therefore, in this chapter, we are going to essentially examine the aspects related to *dimensioning regarding resistance.*

Dimensioning regarding resistance consists of checking that the structure can support the loads prescribed by the specifications. Under these loads, the model made up of finite elements will supply a distribution of stresses in the structure. This distribution has to be identified, interpreted and compared to the possible performances of the material making up this structure. This comparison generally takes the form of the following ratio:

$$C_S = \frac{\text{maximum allowable stress for the material}}{\text{stress calculated from the model}}$$

The maximum allowable stress for the material comes from results of tests carried out on a test specimen. The calculated stress is noted on the model after calculation. "C_s" is the safety factor. Its value depends on a large number of parameters related to the application. It is known through procedures detailed in specific national or international standards or regulations of the concerned application categories. In addition, a number of regulations or direct studies are carried out by manufacturers. This makes it possible to determine the increase, combinations and weightings to be applied to the loading which corresponds to a normal service of the structure (weight of snow, wind action, vibrations, seismic effects, different types of accidental temporary overloading). Under these "weighted" loads, the check on the structure can be performed in a practical form deduced from the earlier definition of C_s:

$$\text{calculated stress on the model} \leq \frac{\text{maximum allowable stress for the material}}{C_S}$$

NOTES

❏ A variant to express the resistance condition consists of evaluating the margin of security denoted as m in the form:

$$m = \frac{\text{maximum allowable stress for the material} - \text{computed stress on the model}}{\text{computed stress on the model}}$$

that is: $$m = C_s - 1$$

❏ The standardized tests (tension, compression, shear) enable us to define the elastic limit and ultimate, or rupture strength, of materials involved, by producing simple states of stresses in test specimen with standardized geometries. On the other hand, real work conditions of an industrial structure lead to complex states of stresses that can be determined thanks to modeling with finite elements. Therefore we can evaluate how complex stresses can be recognized as admissible compared to available limits obtained by tests with simple states of stresses.

❏ Dimensioning regarding resistance could take into account:

– loads that are nearly invariable, or with little variation in time or even with less repetition. These are thus known as static loads leading to static dimensioning;

– repeated loads or cyclic loads. Beyond a thousand cycles, the dimensioning will call upon another notion called "fatigue" (dimensioning regarding fatigue).

❏ In addition to the aims regarding rigidity and resistance in the different parts of the structure, we also have to check the behavior of the joints or assemblies of these different parts, by means of specific considerations[1].

❏ The designer follows one of the two approaches in a practical manner:

– starting from the *dimensions of the structure measured in the drawings*, the designer verifies that the margin (or the safety factor) is higher than the one fixed by the regulation;

or:

– the designer calculates the *minimum dimensions* of the structure to be designed so that the rigidity and resistance conditions in the common parts and joints are verified[2].

For a static load, the general method can be hereafter summarized assuming that the structure is working in the linear elastic domain.

---

1 See Chapter 11.

2 These minimum dimensions of course depend on previously chosen materials. They will serve as a basis for the design department in view of final dimensioning.

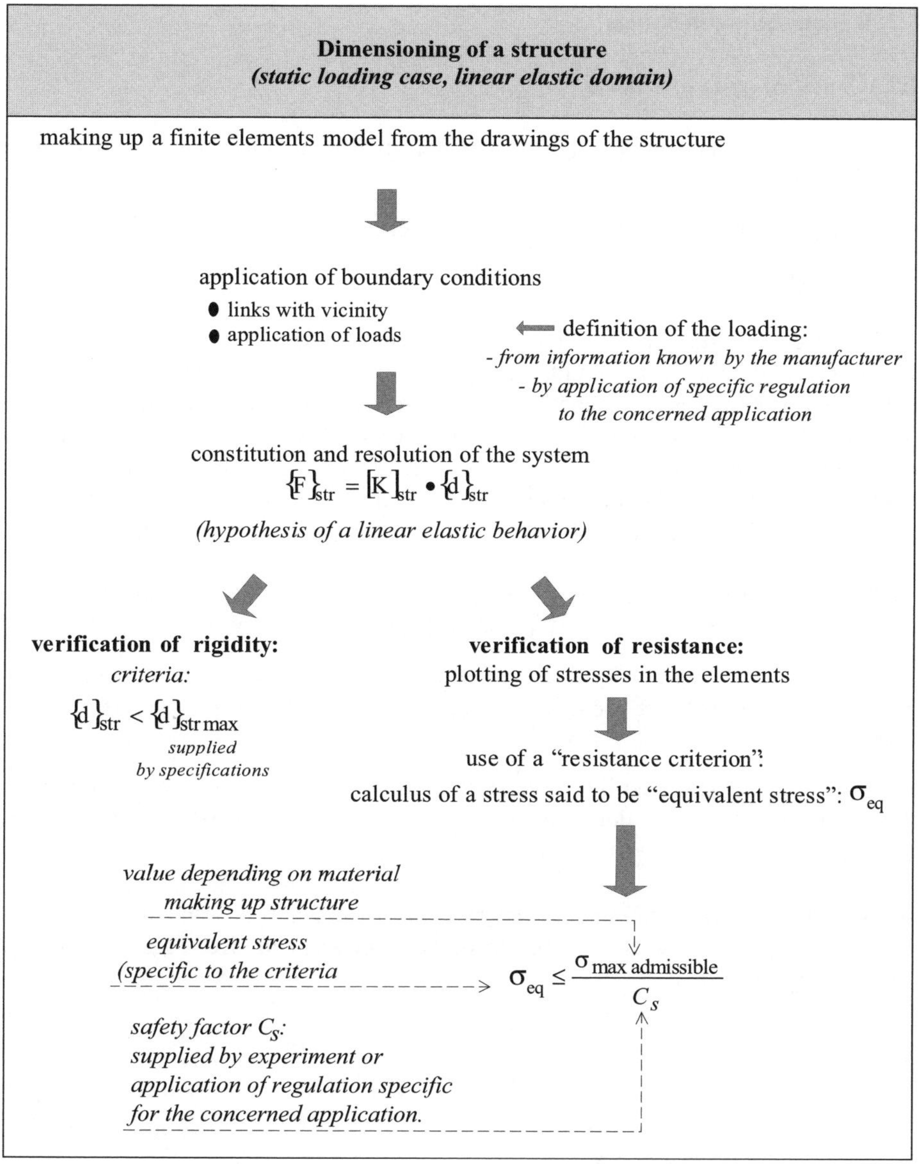

Dimensioning of a structure
(static loading case, linear elastic domain)

making up a finite elements model from the drawings of the structure

application of boundary conditions
- links with vicinity
- application of loads

definition of the loading:
- from information known by the manufacturer
- by application of specific regulation
to the concerned application

constitution and resolution of the system

$$\{F\}_{str} = [K]_{str} \bullet \{d\}_{str}$$

(hypothesis of a linear elastic behavior)

verification of rigidity:
criteria:

$$\{d\}_{str} < \{d\}_{str\,max}$$

supplied
by specifications

verification of resistance:
plotting of stresses in the elements

use of a "resistance criterion":

calculus of a stress said to be "equivalent stress": σ_{eq}

value depending on material
making up structure

equivalent stress
(specific to the criteria

safety factor C_S:
supplied by experiment or
application of regulation specific
for the concerned application.

$$\sigma_{eq} \leq \frac{\sigma_{max\,admissible}}{C_S}$$

Figure 7.2. *General dimensioning procedure for static loading*

7.2. Dimensioning in statics

7.2.1. *The two types of criteria*

According to the type of damage targeted, two types of criteria for maximum stresses can be used:

– the non-plastification criterion (elastic limit);

– the non-rupture criterion (rupture limit).

The structures should often be designed keeping these two aspects in mind. Thus the structure of an aircraft should, during its actual service, remain in a field where none of its structural parts reaches the elastic limit beyond which yielding, leading to plastic deformation, begins. Besides this normal use, the specifications define exceptional loads due to some incidents, breakdowns, local failures, atmospheric conditions, and particular maneuvers. None of the essential structural parts of the aircraft should collapse under these loads, and in any case not immediately.

The behaviors of structure materials are generally known through the simple tension test given in Chapter 1[3]. Remember that we obtained there:

– an area characterized by a reversible strain, defining the elastic domain (as far as we are concerned, we are only dealing with the linear elastic behavior in this book);

– an area characterized by irreversible deformations, typical of the elastoplastic domain (the test sample does not return to its original form when removing the loading).

Nevertheless, considering the materials taken into account and the treatments that they undergo, these two areas can be very different in size. Figure 7.3 shows some examples of this:

---

3 See section 1.3.1, Figure 1.9.

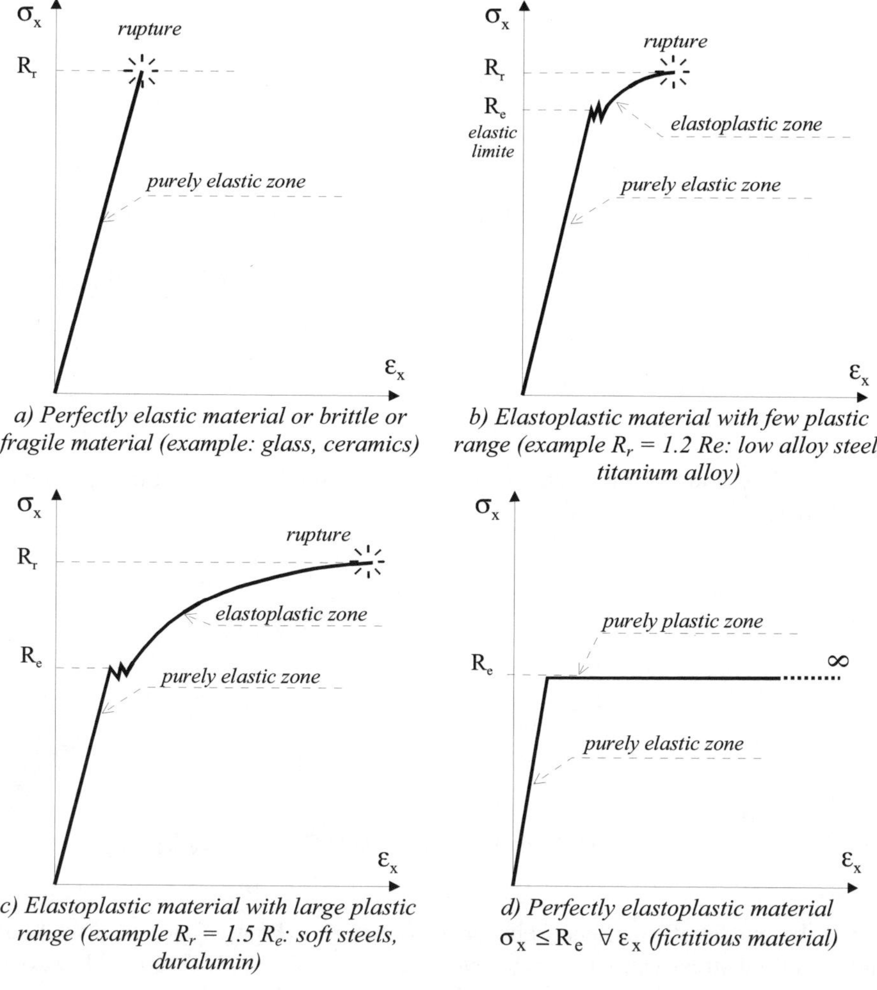

a) Perfectly elastic material or brittle or fragile material (example: glass, ceramics)

b) Elastoplastic material with few plastic range (example $R_r = 1.2\ R_e$: low alloy steel, titanium alloy)

c) Elastoplastic material with large plastic range (example $R_r = 1.5\ R_e$: soft steels, duralumin)

d) Perfectly elastoplastic material $\sigma_x \leq R_e\ \forall\, \varepsilon_x$ (fictitious material)

Figure 7.3. *Different types of materials*

The resistance criterion should, therefore, be linked to the nature of the material. For example, considering a part under mechanical loading, it is known that the brutal variations of shapes and cross-sections provoke internal "overstress", also called "stress concentration" (see Figure 7.4).

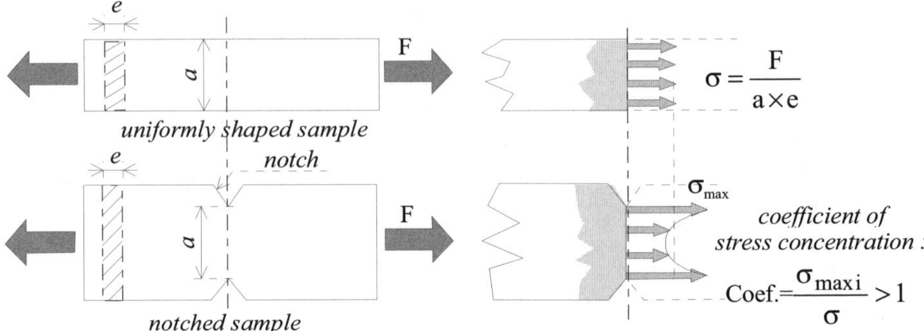

Figure 7.4. *Internal overstress (stress concentration)*

Figure 7.5 shows the location of this phenomenon for some geometric singularities:

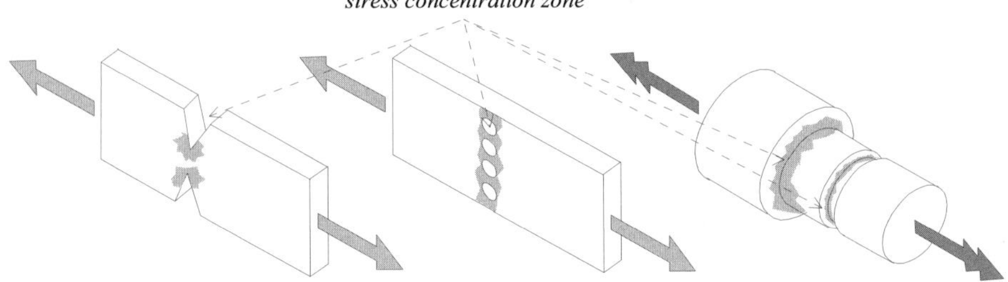

Figure 7.5. *Geometric singularities*

● if the part is of a perfectly elastoplastic material (Figure 7.3d) then we know that the highest stress value is R_e whatever the level of the applied forces. The zones of concentration, initially characterized by considerably increasing gradients of stress[4], gradually see these gradients reducing when the rate of the applied forces increases: the stresses tend to become "uniform" simultaneously as the "plastified" zones become larger (local plastic adaptation);

● if the material used is of a brittle type (Figure 7.3a) or has a weak plastic range, (Figure 7.3b) then the stress concentrations continue and the maximum local

---

4 We will say without giving more details that the "stress gradient" is high in a certain domain when a high stress variation is noted between neighboring points of the domain.

stress increases dangerously with the applied forces. The failure will start in the localized zones shown in Figure 7.5[5].

NOTES

❏ When a structure part is affected by local stress concentration, it is important to reach the values of stresses in the concerned domain by the most precise way possible. This should be the more accurately modeled part in the domain in question as the material used has a weak plastic range. This could result in locally increasing the finite element density (meshing) or using higher performance elements. This is not always easy in practice.

7.2.2. *Elasticity limit criterion*

7.2.2.1. *Intrinsic surface*

Several criteria are available that mark the passage from the purely elastic behavior domain to the elastoplastic behavior domain. Yielding occurs, i.e. irreversible relative creeping of the grains that make up the fine structure of the material. Such creeping produces an angular variation phenomenon or "distortion" already mentioned several times earlier[6]. In Figure 7.6a a loaded structure S and a partition of this structure are shown. Let us consider on the cross-section consecutive to this partition an elementary facet dS around a point M, its normal axis $\vec{n}$ and the stress vector $\overrightarrow{C}_{(M,\vec{n})}$[7].

For a given loading, let us increase all the components of this loading in the same proportions. The vector $\overrightarrow{C}_{(M,\vec{n})} = \sigma\vec{n} + \tau\vec{t}$ then maintains the same support, and the irreversible creeping at M will begin when the intensity of $\overrightarrow{C}_{(M,\vec{n})}$ reaches a maximum value. This limit stress vector will be defined by an orientation and a maximum modulus:

– the orientation depends on the components of the loading;

– the maximum modulus depends on these components and also on the nature of the material.

---

5 We can find in specialized books values of the increase in stress, or "coefficients of stress concentration" for standard geometric singularities (see bibliography: J.P. Faurie).
6 See for example Figure 1.18.
7 See section 1.2.

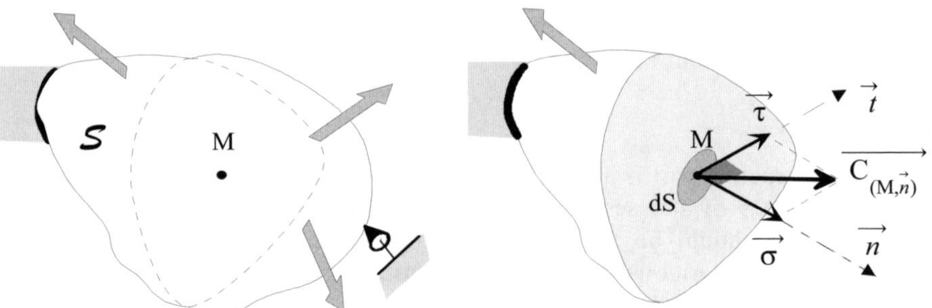

a)   Stress vector in M corresponding to a given loading

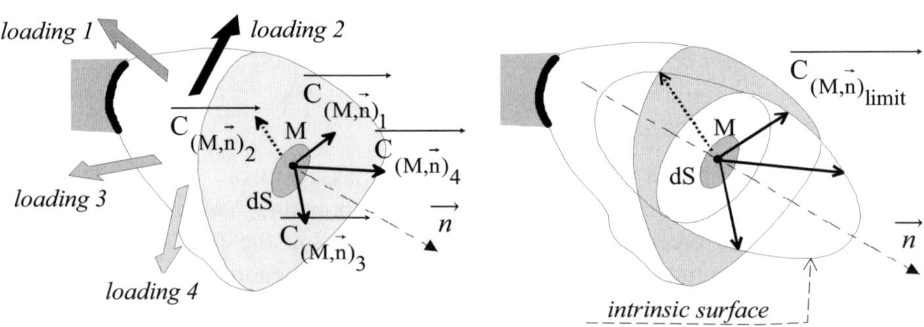

b) stress vectors in M corresponding to four distinct loadings 1, 2, 3, 4

c) envelope of the limit stress vectors at elastic limit: intrinsic surface

Figure 7.6. *Intrinsic surface*

Now let us suppose that we can vary the loadings (nature and magnitude) on this structure as we wish. Thus the stress vector $\overrightarrow{C_{(M,\vec{n})}} = \sigma\vec{n} + \tau\vec{t}$ varies in size and direction (Figure 7.6b). For each maximum allowable loading, we note the corresponding limit stress vector. Then the extremities of all these stress limit vectors are found on a surface called the *intrinsic surface* (Figure 7.6c).

NOTE

❑ For isotropic materials this intrinsic surface is a surface of revolution. Let us refer for example to Figure 7.6a and let us show on the same plane passing through the normal $\vec{n}$ all the limit stress vectors, each one marking the limit of elastic behavior. Their extremities generate a curve denoted by $\mathcal{C}$ in Figure 7.7. Due to the isotropy of the material:

– this curve has normal $\vec{n}$ for axis of symmetry

– we obtain an identical curve in all the planes passing through $\vec{n}$.

Thus, we create a surface of revolution of axis $\vec{n}$.

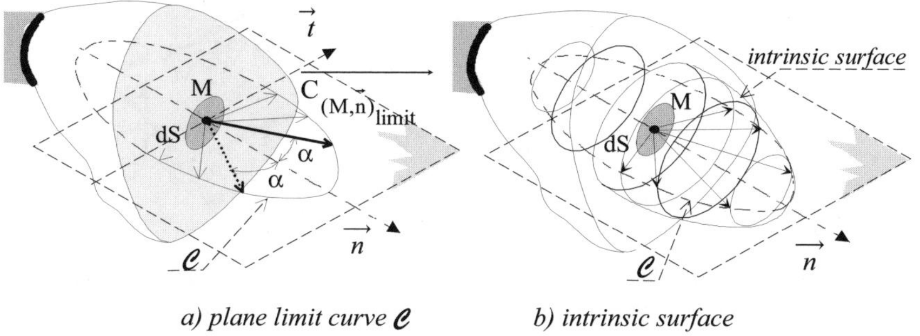

a) plane limit curve $\mathcal{C}$ b) intrinsic surface

Figure 7.7. *Generation of intrinsic surface*

The experiments carried out on the plastic flow, in the second half of the 19th century, led to the formulation of criteria enabling us – in a more or less precise manner – to note quantitatively the fact that the extremity of the stress vector reaches the intrinsic surface. The common criteria usually deal with the creep phenomenon within the material (distortion) by considering the following factors:

– the shear stresses, leading to the Tresca criterion;

– the distortion energy, leading to the Von Mises criterion.

These two criteria provide values for the elasticity limits with relatively small differences. They are both integrated with postprocessors of finite element software. We will examine the Von Mises criterion in the following.

7.2.2.2. *Complete state of stresses*

When the local coordinate system used is different from system $\vec{1}$, $\vec{2}$, $\vec{3}$ of main stress directions, a complete state of stresses consists of normal as well as tangential stress components as described in Chapter 5[8], components which are shown in Figure 7.8 given below.

---

8 See section 5.5 and also [10.21].

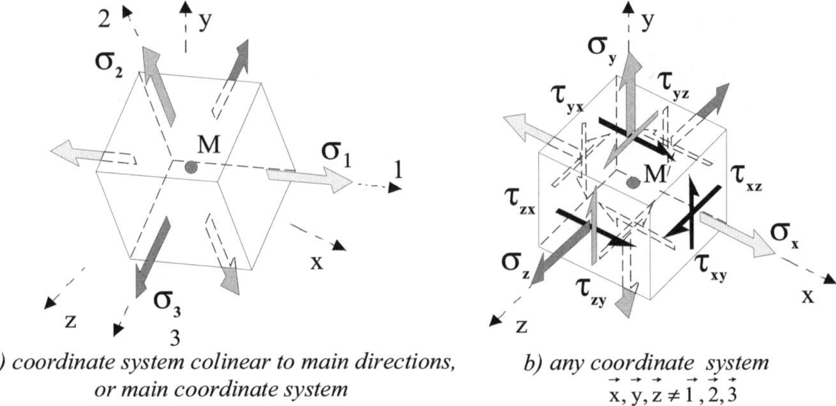

a) coordinate system colinear to main directions,
or main coordinate system

b) any coordinate system
$$\vec{x}, \vec{y}, \vec{z} \neq \vec{1}, \vec{2}, \vec{3}$$

Figure 7.8. *Stress components of complete state of stress*

7.2.2.3. *Von Mises criterion*

This criterion postulates that a material locally reaches its elastic limit and begins to yield when the distortion energy of an elementary domain of the material reaches a higher limit or critical threshold, experimentally determined by a uniaxial tension test on a test sample. This energy distortion is described in Chapter 10[9], and we then find for a complete state of stress:

$$dE_{pot.\ distor.} = \frac{1+\nu}{6E}\left[\left(\sigma_x - \sigma_y\right)^2 + \left(\sigma_y - \sigma_z\right)^2 + \left(\sigma_z - \sigma_x\right)^2 + 6\left(\tau_{xy}{}^2 + \tau_{yz}{}^2 + \tau_{zx}{}^2\right)\right] \times dV \quad [7.1]$$

○ Equivalent Von Mises stress (normal stress)

If we adapt what is written above to a uniaxial tension test case (axis $\vec{x}$), then the only non-zero stress component is limited to σ_x. We have with [7.1]:

$$dE_{pot.\ distortion} = \frac{1+\nu}{6E} \times 2\sigma_x{}^2 \times dV$$

The critical value of this energy will be reached when:

$$\sigma_x = R_e \text{ (elastic limit)}$$

i.e.:

$$dE_{pot.\ critical\ distortion} = \frac{1+\nu}{6E} \times 2R_e{}^2 \times dV \qquad [7.2]$$

---

9 See section 10.2.7, [10.36].

We will avoid irreversible yielding, or plastification, if we avoid "saturation" of the elastic limit criterion, thus:

$$dE_{pot.\ distortion} < dE_{pot.\ critical\ distortion}$$

This brings us, by comparing [7.1] and [7.2], to write for the complete state of stresses:

$$\sqrt{\frac{1}{2}\left[(\sigma_x - \sigma_y)^2 + (\sigma_y - \sigma_z)^2 + (\sigma_z - \sigma_x)^2 + 6\left(\tau_{xy}^2 + \tau_{yz}^2 + \tau_{zx}^2\right)\right]} < R_e$$

The first member is homogenous to a stress which we will call equivalent normal Von Mises stress[10]:

$$\sigma_{eq\ V\ Mises} = \sqrt{\frac{1}{2}\left[(\sigma_x - \sigma_y)^2 + (\sigma_y - \sigma_z)^2 + (\sigma_z - \sigma_x)^2 + 6\left(\tau_{xy}^2 + \tau_{yz}^2 + \tau_{zx}^2\right)\right]} \qquad [7.3]$$

and the criterion to be verified is written as:

$$\sigma_{eq\ V.Mises} < R_e \qquad [7.4]$$

NOTES

❏ The Von Mises stress is said to be:

– normal: because we compare it to R_e resulting from the uniaxial tension test[11];

– equivalent: because it combines the components of the real state of stresses. The Von Mises stress is not a real stress present in the structure, but only a convenient entity given by the postprocessor with the objective of evaluating the risk of plastification.

❏ In fact, the postprocessor of a finite element software calculates the stress components in the elements.

---

10 The equivalent normal stress can also be expressed in the main coordinate system with the main stresses σ_1, σ_2, σ_3. We immediately obtain from [7.3]:

$$\sigma_{eq\ V\ Mises} = \sqrt{\frac{1}{2}\left[(\sigma_1 - \sigma_2)^2 + (\sigma_2 - \sigma_3)^2 + (\sigma_3 - \sigma_1)^2\right]}$$

11 See section 1.3.1.

These stress values:

– correspond to hypotheses which have enabled us to define the elements. In Chapter 5 we were able to see how stresses could be expressed inside the different types of elements[12];

– are calculated in the local coordinate system of each element, which is distinct from that of the main directions (main coordinate system).

It is possible to display the magnitude levels of each of the six components of stresses all over the model of the structure under study, or even several components if we have a "windowing system" facility on the work station screen. However, each image will only give a partial pictoral description of the complete state of stresses. Studying only one image may prove to be misleading. The equivalent Von Mises stress has the advantage of combining all the components of stresses.

❐ The equivalent Von Mises stress [7.3], which can be calculated as an option in postprocessors of finite element software, should remain less than the elastic limit in order to respect the criteria [7.4] i.e.:

$$\sigma_{eq\ V.Mises} < R_e$$

❐ When a highly loaded structural part has geometric singularities, especially re-entrant angles or sudden variations of sections, the presence of local stress concentrations (see Figure 7.5) means that the display of values of $\sigma_{eq\ V.Mises}$ can rapidly go beyond R_e locally. This can have several causes (possibly cumulative):

– the real behavior of the structure is purely elastic, leading to the limit being exceeded. Thus, the method of calculation is not the only cause of these high values;

– the degree of discretization (density of elements) is insufficient, finite element covering a local domain where the stresses evolve too fast from one point to another;

– the elements used do not have the capacity of adapting to rapid variations of stresses.

Therefore, the interpretation of results is difficult. We must *consider the behavior of the structure against the behavior of the finite elements*. In reality, as far as the behavior of the structure is concerned, and if the material allows it, there will be local plastic adaptations[13]. Thus, the stresses no longer increase in

---

12 See for example [5.19].
13 See section 7.2.1.

proportion to the load; their growth is slower. A structure can thus generally remain globally in the elastic domain (reversible) whereas locally the elastic limit has been exceeded.

❑ In case of plane state of stress in plane (xy), the stress is denoted σ_x, σ_y, τ_{xy}[14], and expression [7.3] becomes:

$$\sigma_{eq\,V.Mises} = \sqrt{\sigma_x{}^2 + \sigma_y{}^2 - \sigma_x\sigma_y + 3\tau_{xy}{}^2} \qquad [7.5]$$

❑ Elastic limit in shear: let us suppose that we recreate on a test sample a plane state of stress corresponding to pure shear[15], which is reduced to:

$$\sigma_x = \sigma_y = 0 \; ; \; \tau_{xy} \neq 0$$

Relation [7.3] is written as:

$$\sigma_{eq\,V.Mises} = \sqrt{3}\,\tau_{xy}$$

and criterion [7.4] is saturated when τ_{xy} reaches the elastic shear limit or distortion limit noted as R_{eg}:

$$\sqrt{3}\,R_{eg} = R_e$$

The expression of the elastic shear limit is derived from the resulting Von Mises criterion:

$$R_{eg} = 0.58\,R_e \qquad [7.6]$$

❑ In the case of beams, the typical components of stresses are shown in Figure 7.9, which are σ_x, τ_{xy}, τ_{xz}[16]. Thus [7.3] leads to the following form of equivalent stress of Von Mises:

$$\sigma_{eq\,V.Mises} = \sqrt{\sigma_x{}^2 + 3\left(\tau_{xy}{}^2 + \tau_{xz}{}^2\right)} \Leftrightarrow \sigma_{eq\,V.Mises} = \sqrt{\sigma_x{}^2 + 3\tau^2} \qquad [7.7]$$

---

14 See [1.18].
15 See Chapter 1, section 1.4.2.3.
16 See Figure 1.29.

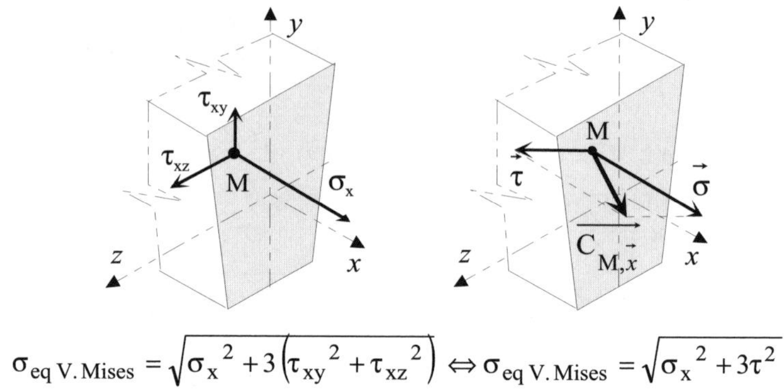

$$\sigma_{eq\,V.Mises} = \sqrt{\sigma_x{}^2 + 3\left(\tau_{xy}{}^2 + \tau_{xz}{}^2\right)} \Leftrightarrow \sigma_{eq\,V.Mises} = \sqrt{\sigma_x{}^2 + 3\tau^2}$$

Figure 7.9. *Case of beams*

It must be noted that when a calculation is processed from beam elements, we cannot display the equivalent normal stress $\sigma_{eq\,V.Mises}$. In fact the tangential stresses τ_{xy} and τ_{xz} are not taken into account in the software. Evaluation of these stresses requires specific programs that can carry out extra calculations in the cross-section domain[17]. In fact, these stresses depend on the shape of the cross-section, and this shape is not known as an input by the preprocessor, which only recognizes the median line of the beam element.

❑ In the case of plates, stresses on one element $dx \times dy \times e$ are given in Figure 7.10[18]. Thus, [7.3] leads to the following expression of equivalent Von Mises[19] stress:

$$\sigma_{eq\,V.Mises} = \sqrt{\sigma_x{}^2 + \sigma_y{}^2 - \sigma_x\sigma_y + 3\tau_{xy}{}^2} \qquad [7.8]$$

In spite of the formal resemblance of this expression with [7.5], here we do not deal with plane state of stresses, as seen in Figure 7.10, where the components of bending stresses, and thus the value of $\sigma_{eq\,V.Mises}$, evolves with altitude z, i.e. has a proper value at each point on the perpendicular in M(xy) to the middle plane (xy) of the plate.

---

17 These calculations are carried out on the basis given in detail in Chapter 9; see [9.27], [9.59] and [9.60].

18 See section 5.4.

19 When the plates studied are "thin" (see section 5.4.1), The shear stresses τ_{xz} and τ_{yz}, still called "transversal shear", are left out for the evaluation of $\sigma_{eq\,V.Mises}$.

In Figure 7.10 it is seen that the highest values of $\sigma_{eq\,V.Mises}$ will be located on higher or lower surfaces or skins of the plate (points M' and M"). In addition to the mid-surface, it is also in these zones that the software postprocessor evaluates the levels of equivalent Von Mises stresses [20].

❏ In case of solid elements (see section 5.3), the state of stresses is complete, and the equivalent normal stress preserves the general form [7.3] i.e.:

$$\sigma_{eq\,V.Mises} = \frac{1}{\sqrt{2}}\sqrt{\left(\sigma_x - \sigma_y\right)^2 + \left(\sigma_y - \sigma_z\right)^2 + \left(\sigma_z - \sigma_x\right)^2 + 6\left(\tau_{xy}^{\,2} + \tau_{yz}^{\,2} + \tau_{xz}^{\,2}\right)}$$

$$[7.9]$$

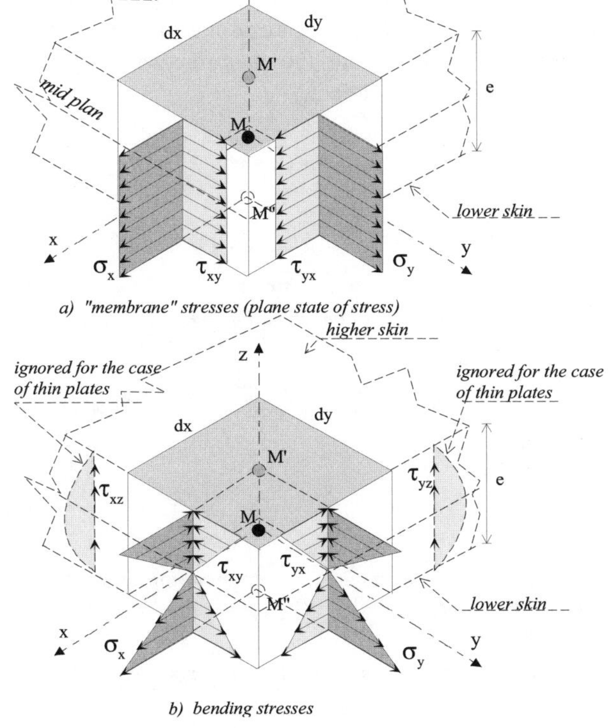

a) "membrane" stresses (plane state of stress)

b) bending stresses

Figure 7.10. *Case of plates*

---

20 The post-processor overwrites the distributions of stresses represented in Figures 7.10a and 7.10b to give:

$$\sigma_x = \sigma_x{}_{\text{membrane}} + \sigma_x{}_{\text{bending}} \quad ; \quad \sigma_y = \sigma_y{}_{\text{membrane}} + \sigma_y{}_{\text{bending}} \quad ; \quad \tau_{xy} = \tau_{xy}{}_{\text{membrane}} + \tau_{xy}{}_{\text{bending}}$$

7.2.3. *Non-rupture criterion*

7.2.3.1. *Brittle materials*

It is relatively easy to define a criterion for non-rupture for materials said to be brittle, or fragile, such as glass or ceramics which do not yield. These materials remain elastic until break point (see Figure 7.3a), so the earlier criterion which marked the passage to the elastic limit becomes a criterion for rupture limit since there is failure at the limit of elastic behavior. The rupture limits (merged with the elastic limits) are even higher as the number of internal defects is low. In fact, these defects can be considered geometric singularities, and we have indicated that the increase of stresses (concentrations) developed in such singularities (Figure 7.5). We shall thus find such "stress concentrations" in the crack initiation zones shown in Figure 7.11. The size of the parts and the manufacturing process influence the number and the direction of internal defects. For example, for the strain given in the figure, the bulky part in Figure 7.11a is more prone to developing cracks than the long thin part in Figure 7.11b, where the manufacturing process (rolling, cold drawing, extrusion) has directed the defects.

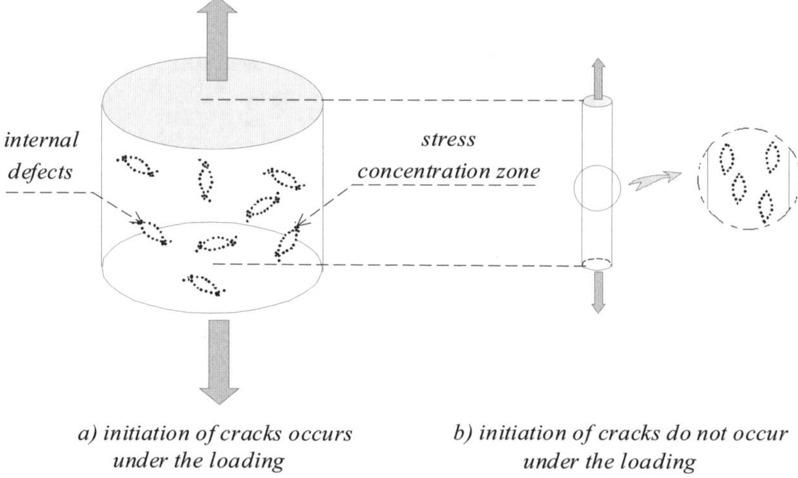

Figure 7.11. *The scale effect: resistance to tension is greater in b) than in a)*

NOTES

❏ For these brittle materials, the intrinsic surface characterizing the elastic limit in Figure 7.6c becomes an intrinsic surface for the rupture limit.

7.2.3.2. *Elastoplastic materials*

It is more difficult to define a non-rupture criterion when the structure part or the structure itself has an elastoplastic behavior (Figures 7.3b and 7.3c). Let us study the following examples.

○ *Example 1: current zone of a structure*

We consider a material with an elastoplastic behavior illustrated in Figure7.12a. With this material we make the beam in Figure 7.12b which deflects in the main plane (xy) under action of a pure bending moment $\overrightarrow{\mathcal{M}} = \mathcal{M}\overrightarrow{z}$ which we increase gradually.

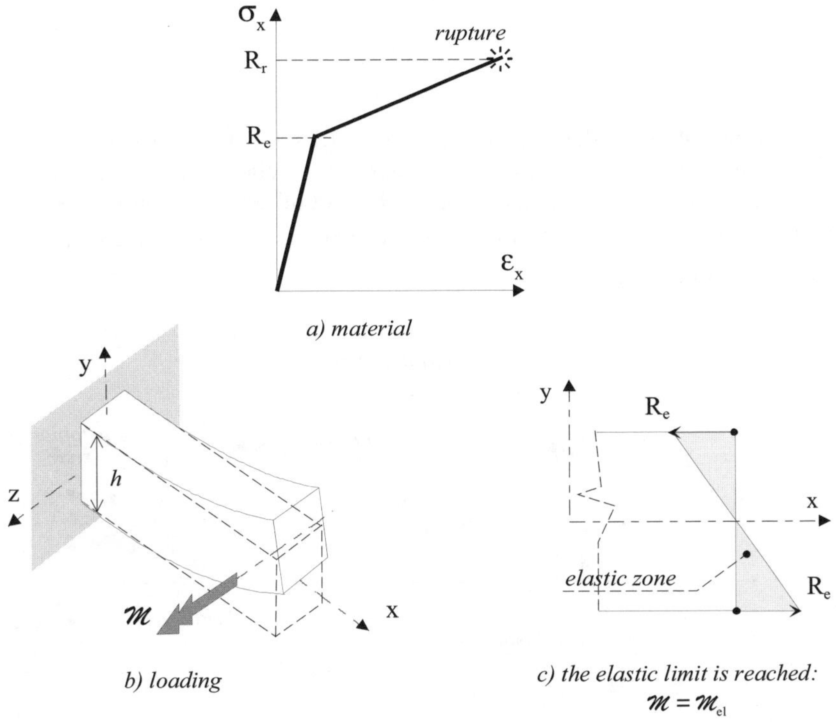

Figure 7.12. *Elastic limit on each cross-section*

In this case we know the distribution of normal bending stress. It is linear in the elastic domain, with the form recalled here[21]:

---

21 See section 1.5.2.6.

$$\sigma_x = -\frac{\mathcal{M}f_z \times y}{I_z} \tag{7.10}$$

with I_z quadratic moment around the axis $\vec{z}$, and here $\mathcal{M}f_z = \mathcal{M}$. In Figure 7.12c, the elastic limit is reached when $\mathcal{M}$ reaches a value $\mathcal{M}_{el}$ such that:

$$R_e = \frac{\mathcal{M}_{el} \times \dfrac{h}{2}}{I_z}$$

Let us increase the moment beyond value $\mathcal{M}_{el}$. The cross-sections remaining plane[22], the dilatation ε_x is always given by[23]:

$$\varepsilon_x = -y\frac{d\theta_z}{dx}$$

The normal stresses develop as in Figure 7.13. In an increasing portion of the considered cross-section, these stresses exceed the elastic limit. We can see that it is possible to increase the moment $\mathcal{M}$ at the end of the beam until a break value denoted as $\mathcal{M}_{rupt}$ is reached, in which the extreme stresses reach the rupture limit R_r; see Figure 7.13c.

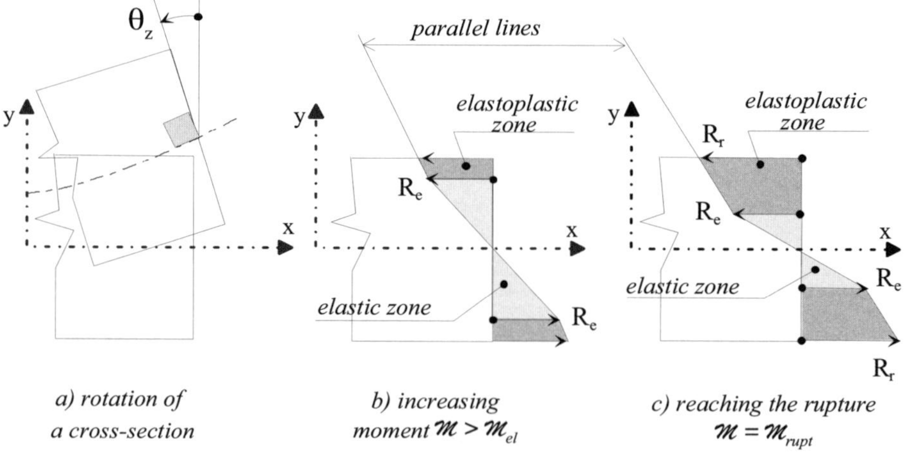

a) rotation of
a cross-section

b) increasing
moment $\mathcal{M} > \mathcal{M}_{el}$

c) reaching the rupture
$\mathcal{M} = \mathcal{M}_{rupt}$

Figure 7.13. *Rupture limit*

---

22 See Figure 9.36.
23 See section 1.5.2.6.

In Figure 7.14 we can compare the stress distributions for the bending moments corresponding to the elastic limit $\mathcal{M}_{el}$ and to the rupture limit $\mathcal{M}_{rupt}$, as well as the (fictitious) distribution that we would obtain by keeping relation [7.10] until the rupture stress R_r is reached. We can then note that such a relation, based on the elastic calculation rules, no longer corresponds to reality.

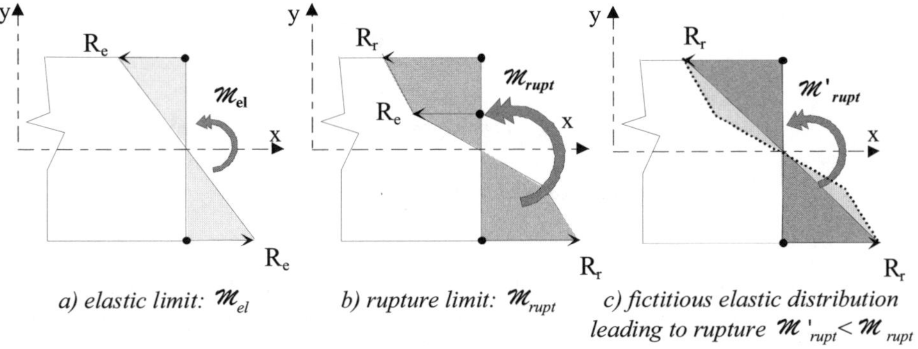

a) elastic limit: $\mathcal{M}_{el}$ b) rupture limit: $\mathcal{M}_{rupt}$ c) fictitious elastic distribution leading to rupture $\mathcal{M}'_{rupt} < \mathcal{M}_{rupt}$

Figure 7.14. *Normal stress distributions*

We obtain a (fictitious) bending moment denoted as $\mathcal{M}'_{rupt}$ in Figure 7.14c where we can see without calculation that it is lower than the real rupture moment $\mathcal{M}_{rupt}$ of Figure 7.14b. This moment $\mathcal{M}'_{rupt}$ is thus a pessimistic evaluation enabling reliable dimensioning toward rupture:

$$\mathcal{M}'_{rupt} < \mathcal{M}_{rupt} \qquad [7.11]$$

NOTES

❏ Considering the characteristic curve of the material[24], when we follow the strain evolution corresponding to the loading in Figure 7.14, we obtain in the elastoplastic domain, before the rupture moment of Figure 7.14b, the shaded part shown hereafter in Figure 7.15b. The maximum normal stress σ_B is then higher than the elastic limit, as we can see in Figure 7.15a.

---

24 See Figure 7.12a.

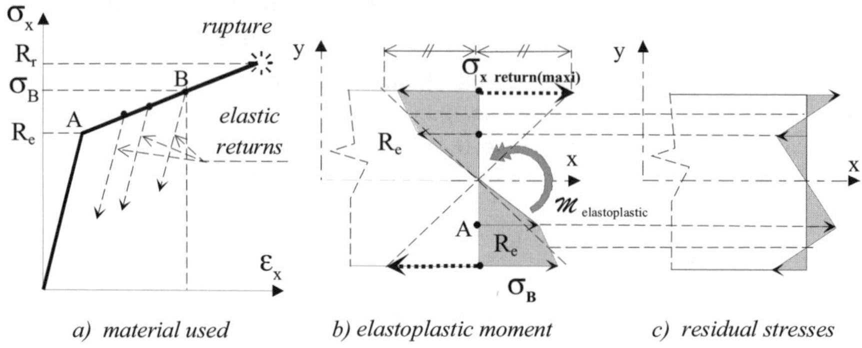

a) material used   *b) elastoplastic moment*   *c) residual stresses*

Figure 7.15. *Removing elastoplastic moment*

Now if we remove the moment at the end of the beam, the plane sections will undergo an inverse rotation corresponding to a variation of ε_x in the sense of "elastic return" given in Figure 7.15a with[25]:

$$\varepsilon_{x\,return} = -y\left(\frac{d\theta_z}{dx}\right)_{return}$$

The final moment at the end of the beam being zero, the elastic returns correspond to a moment: $\mathcal{M}_{return} = -\mathcal{M}_{elastoplastic}$. The returns indicated in Figure 7.15a are "linear" with a slope identical to the "elastic" one, that means that the evolution of the "return stress" is proportionate to $\varepsilon_{x\,return}$. Thus, as per Hooke's law:

$$\sigma_{x\,return} = E \times \varepsilon_{x\,return} = -y \times E\left(\frac{d\theta_z}{dx}\right)_{return}$$

or even:

$$\mathcal{M}_{return} = \int_{section} -y \times \sigma_x \times dS = E \times I_z \left(\frac{d\theta_z}{dx}\right)_{retunr} = -\mathcal{M}_{elastoplastic}$$

which enables us to know $\left(\dfrac{d\theta_z}{dx}\right)_{return}$ and thus the stress distribution, i.e. with the earlier relations:

---

25 See section 1.5.2.6.

$$\sigma_{x\,return} = -\frac{\mathcal{M}_{elastoplastic}}{I_z} \times y$$

This distribution is represented in Figure7.15b in dotted lines. Thus, when the beam is no longer loaded, we obtain a residual state of normal stresses resulting from the superposition of successive loads $\mathcal{M}_{elastoplastic}$ then $\mathcal{M}_{return} = -\mathcal{M}_{elastoplastic}$. These residual stresses are shown in Figure 7.15c.

Therefore:

• if we again load the beam with an increasing moment up to the earlier value $\mathcal{M}_{elastoplastic}$, we will apply an *elastic* moment, as opposed to $\mathcal{M}_{return}$ which was a purely elastic moment. Thus, we will not create permanent new (plastic) strains. The plastic strains created earlier now allow a load remaining in the elastic domain higher than the one that the beam could support in its primitive state. We say that we have increased the elastic limit by "*strain hardening*";

• if we load the beam with an opposite sign moment, we can see that the residual state of stresses of Figure 7.15c will aggravate the tensions and compressions which are going to develop. We shall reach the elastic limit faster than if we started from the primitive state before strain hardening. The strain hardening is advantageous if the load always acts in the same direction.

○ *Example 2*: *attachment area of a structure*

Let us consider the assembly of two strips of identical metal sheets denoted (a) and (b), riveted together with three identical rivets (material with characteristics R_e, R_r) represented in Figure 7.16. Here we deem that the failure of the assembly is due to the breaking of rivets under the load $\vec{F}$ acting on the structure.

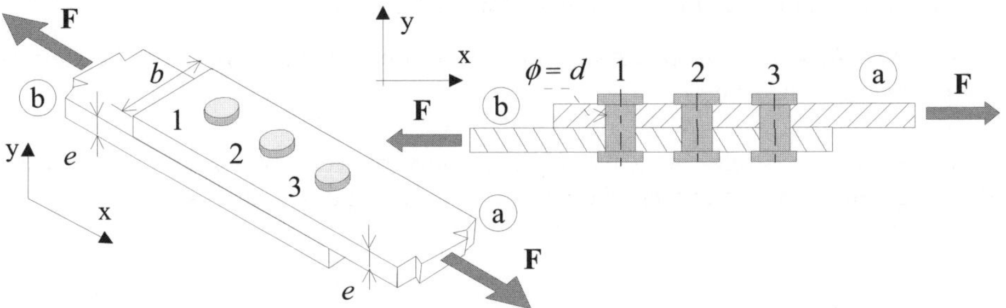

Figure 7.16. *Assembly with three rivets*

When we submit the assembly to an increasing force denoted by F, we observe in Figure 7.17 that each rivet transmits a force F_i (i = 1, 2, 3) such that:

$$F = F_1 + F_2 + F_3 \qquad\qquad [7.12]$$

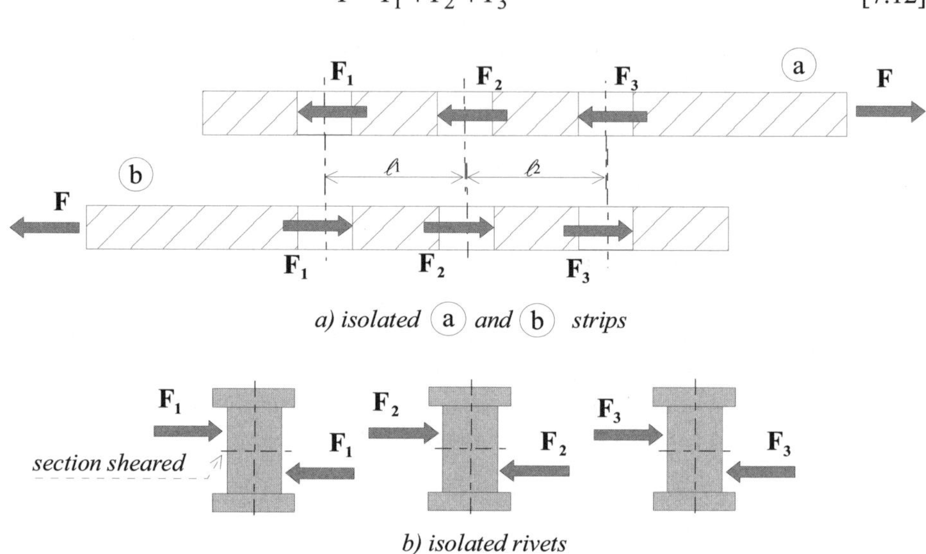

a) isolated ⓐ and ⓑ strips

b) isolated rivets

Figure 7.17. *Isolated strips and rivets*

The rods of the rivets are assumed to be perfectly fitted to their holes so that the assembly is strictly without clearance[26].

In such an assembly, the sheets and rivets are made up of elastoplastic materials which are strained simultaneously. Thus, an exact study of the phenomenon appears immediately complex. Let us examine two behavior hypotheses:

♦ *First hypothetical behavior*: the flexibility of the assembled materials overrides that of the rivets.

The sheets are stretched in the same quantity between two consecutive rivets. This lengthening is given by Hooke's law (see [1.2]), and is written as:

---

26 Nevertheless, we assume that the rivets are not tightened to the sheets, which leads to zero contact pressure between the rivet heads and the sheets.

$$\frac{\Delta\ell_1}{\ell_1} = \underbrace{\frac{F_1}{E \times b \times e}}_{\textit{(higher sheet)}} = \underbrace{\frac{F - F_1}{E \times b \times e}}_{\textit{(lower sheet)}} \quad ; \quad \frac{\Delta\ell_2}{\ell_2} = \underbrace{\frac{F - F_3}{E \times b \times e}}_{\textit{(higher sheet)}} = \underbrace{\frac{F_3}{E \times b \times e}}_{\textit{(lower sheet)}} \qquad [7.13]$$

i.e.:
$$F_1 = F - F_1 \; ; \quad F - F_3 = F_3$$

from there we deduce:

$$F_1 = F_3 = \frac{F}{2}$$

and therefore, with [7.12]:

$$F_2 = 0$$

Thus, the second rivet does not exercise any force. When F increases, F_1 and F_3 increase, each of the rivets (1) and (3) receive opposite strains from the two sheets, which "shear" their circular section located at the interface between the sheets (Figure 7.17b). Due to the friction-fit, we consider[27] the vicinity of this interface to be a state of pure uniform shear with the value:

$$\tau_i = \frac{F_i}{\pi \dfrac{d^2}{4}} \qquad (i = 1,2,3)$$

When the elastic shear limit R_{eg} of the rivets is reached (see [7.6]), the force F being exerted on the sheets has a value of:

$$F_{el} = 2 \times R_{eg} \times \pi \frac{d^2}{4} \qquad [7.14]$$

Then irreversible yielding begins to form. The compatibility of the elongations [7.13] is no longer valid, and rivet 2 begins to transmit a force $F_2 \neq 0$. With force F continuing to increase, forces F_1, F_2, F_3 transmitted by the rivets become uniform until rupture, where it can be considered that each transmits the same force (when the end of the elastoplastic zone of the rivet material is reached, the forces applied are clearly equal, as shown in Figure 7.18 below).

---

27 For further information and development, see Chapter 11, section 11.1.2.2.

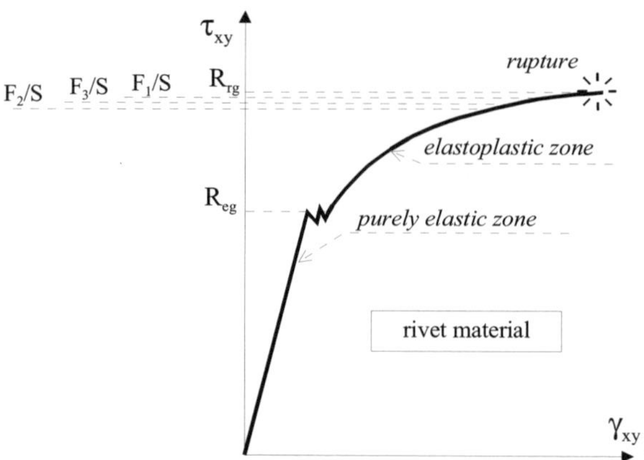

Figure 7.18. *Behavior of rivet material*

The maximum force being exerted on the sheets at the rupture moment of the rivets is then:

$$F_{rupt} = 3 \times R_{rg} \times \pi \frac{d^2}{4} \tag{7.15}$$

where R_{rg} represents the rupture shear stress of the rivets.

NOTES

❏ We can note in this example that it is easy to find the maximum force [7.15] that can be transmitted at rivet failure, whilst below the elastic limit of the rivets, a more detailed study of the loading of rivets in the elastic domain should be made.

❏ We are placed in the hypothetical case where the flexibility of assembled materials is significant. When the riveting pattern is more general, the study of the force transmitted by the rivets below their elastic limit becomes a very difficult problem. It requires a specific delicate modeling by finite elements whose results should be handled with caution. It is therefore more convenient to evaluate the rupture force than the force corresponding to elastic shear limit reached in one or several rivets of the assembly.

❏ As given in the earlier example, if for the assembled parts, we preserve the rules of elastic calculation until the rupture failure of the rivets[28] [7.14] we would obtain for the maximum transmissible force:

---

28 Such a calculation is not valid since the material of the rivets is plastified before breaking. Thus, the compatibility of elastic deformations of the assembled parts is disturbed.

$$F'_{rupt} = 2 \times R_{rg} \times \pi \frac{d^2}{4}$$

Then comparing with [7.15], here as well we see a pessimistic evaluation:

$$F'_{rupt} < F_{rupt} \qquad [7.16]$$

♦ *Second hypothetical behavior*: the flexibility of rivets exceeds that of the sheet strips.

Now the sheet strips are supposed to be much more rigid than the rivets, which means practically unstretchable. The rivets behave as "springs" of stiffness k[29] illustrated in Figure 7.19.

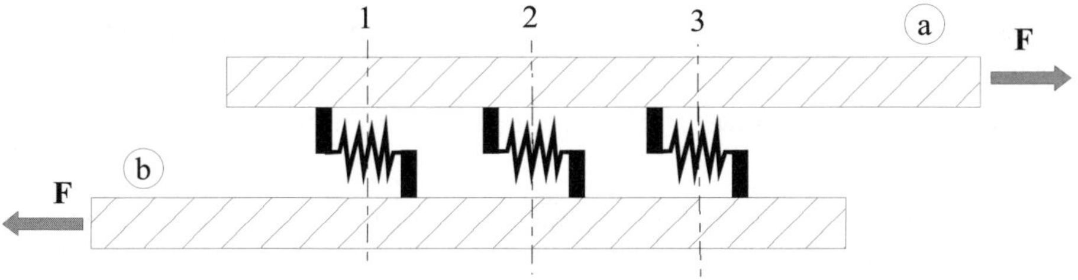

Figure 7.19. *Exceeding flexibility of the three rivets.*

When the assembly is loaded, the contractions denoted as Δ_i of the three springs are identical.

We have on each spring $F_i = k \times \Delta_i$, from which we immediately obtain: $F_1 = F_2 = F_3$.

The strains of the springs result from the strains of the rivet's rods and of their contact zones. They increase with F. We can thus exceed elastic shear limit R_{eg} in the interface cross-sections and reach the rupture limit in shear R_{rg} simultaneously for the three rivets. We thus arrive at the expression:

$$F_{rupt} = 3 \times R_{rg} \times \pi \frac{d^2}{4} \qquad [7.17]$$

---

29 See section 2.2.2 for the definition of the spring stiffness.

NOTES

❒ In the two earlier behavior hypotheses (flexible assembled parts or flexible rivets) the second appears more realistic. In fact, the interface of the assembled parts is of large dimensions as compared to the shear sections of the rivets. For the transmission area of the force being located in the sheared cross-sections and their vicinity, we believe that stresses will be increased in the rivets and immediate vicinity of the contact zones of the rivet rods. Therefore, the largest deformations will be formed here.

❒ Taking into consideration the flexibility of the assembly components (here the rivets), we see that [7.17] enables us to access more quickly their simultaneous behavior at rupture failure[30].

○ As a conclusion the two examples above (current zone of a structure and attachment area) show that:

♦ it is difficult to follow the development of strains in the elastoplastic domain (examples 1 and 2);

♦ the rupture loads are not always easy to characterize (difficult in example 1 and easy in example 2);

♦ if we exploit the elastic stress distributions by prolonging them (in an unjustified manner) until rupture, we underestimate the maximum force that a structure can support (pessimistic evaluation which is oriented towards the safety aspect). In fact:

– in example 1, we saw [7.11] $\mathcal{M}'_{rupt} < \mathcal{M}_{rupt}$,

– in example 2, we saw [7.16] $F'_{rupt} \leq F_{rupt}$.

There we have, for want of anything better, a method of evaluating (of under-evaluating) the maximum breaking loads on a structure[31].

---

30 This approach is again dealt with in Chapter 11.
31 According to the nature of the application, such an evaluation can be judged "too" pessimistic if it leads to maximum loads under-evaluated in an exaggerated manner.

7.3. Dimensioning in fatigue[32]

7.3.1. *Fatigue phenomenon*

Here we will limit ourselves to the damage on a structure part due to a large number of load cycles (higher than 1,000) creating stresses lower than the elastic limit. When the number of cycles is lower, the fatigue strength tends to blend with the rupture strength R_r and we go back to the earlier dimensioning in static. The fatigue break begins in zones where stress concentrations exist, which explains that the break occurs even when the level of stresses in the current zone is lower than the elastic limit. These zones can be (Figure 7.20):

– geometric shape singularities;

– internal singularities (defects) and/or surface texture singularities (machining roughness).

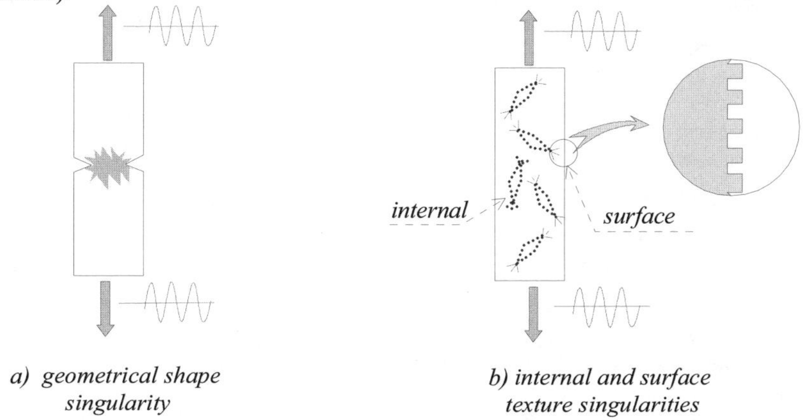

Figure 7.20. *Initiation of fatigue failure*

When fatigue cracks develop, the damaging development can be rapid (breaking after a small number of cycles), slow (breaking after a large number of cycles) or can halt (non-breakage whatever the number of cycles). This phenomenon is illustrated in Figure 7.21.

---

32 In the following we shall restrict ourselves to a simple approach of the fatigue phenomenon.

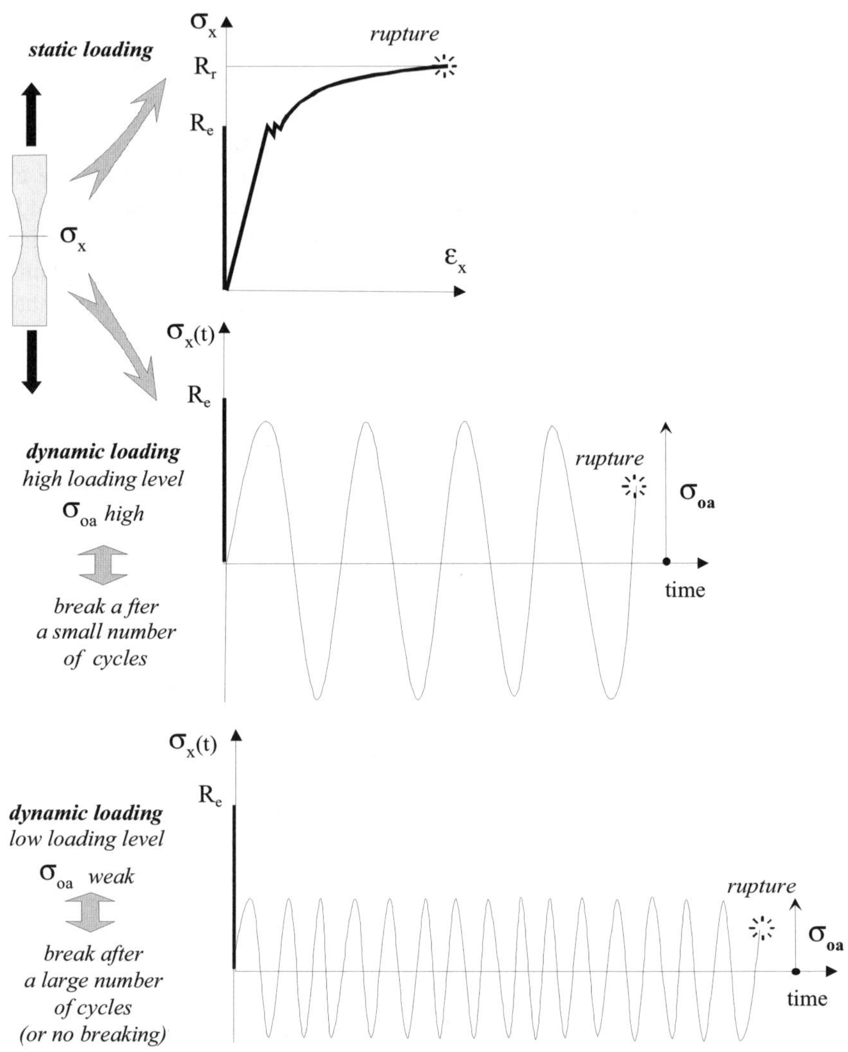

Figure 7.21. *Fatigue breaking*

7.3.2. *Fatigue test*

Specific machines for fatigue tests are used, which enable the exertion of sinusoidal loadings at low frequencies with homogenous states of stresses in the test samples[33]. By varying the peak value of the alternate stress denoted as σ_{oa} on Figure 7.21, we note that the break occurs for a variable number of cycles N. This test sample provides a measure point (σ_{oa} , N). This test is thus costly as it requires a number of test samples. We obtain the fatigue curve also called Wöhler's curve in Figure 7.22.

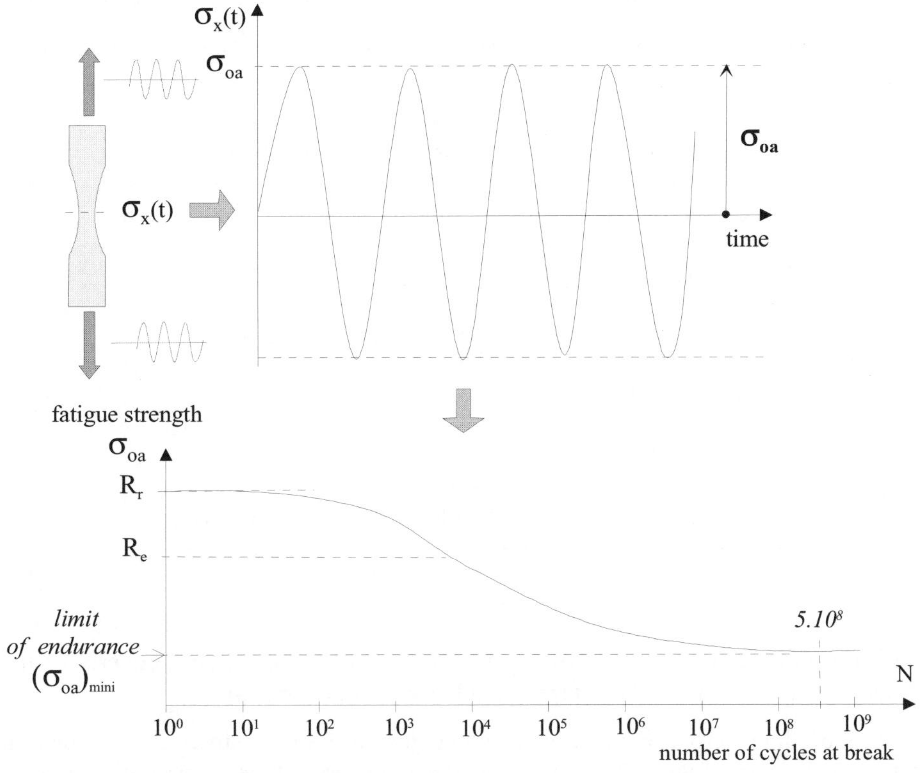

Figure 7.22. *Curve of the fatigue test or Wöhler's curve*

NOTES

❏ When the number of cycles becomes very large, this fatigue curve takes on an asymptotic appearance of horizontal type. For ferrous alloys, we can find a veritable horizontal asymptote, defined by an ordinate $(\sigma_{oa})_{mini}$ constituting a limit of endurance. This is of particular interest for the life cycles of rotating

---

33 Some applications require specific fatigue tests with "non-homogenous" states of stresses. This is typically the case for rotating machine shafts.

machine shafts (in steel) for which the number of load cycles is the product of the rotation speed in revolutions per minute multiplied by the life cycle (in minutes).

❏ For non-ferrous alloys (light alloys for example), the asymptote does not exist, and there is always a break point after a high number of cycles. By convention, we define the endurance limit stress $(\sigma_{oa})_{min}$ for a corresponding cycles value $N = 5 \times 10^8$. The following table gives the orders of magnitude for endurance[34] limits.

alloy	Endurance limit (stress)
Cast iron and steel	$(\sigma_{oa})_{min} \cong 0.4 \times R_r$
Flat rolled steel	$(\sigma_{oa})_{min} \cong 0.5 \times R_r$
High tensile steel	$0.3 \times R_r \leq (\sigma_{oa})_{min} \leq 0.4 \times R_r$
Cast light alloys	$(\sigma_{oa})_{min} \cong 0.3 \times R_r$
Light alloys	$(\sigma_{oa})_{min} \cong 0.4 \times R_r$
Titanium alloys	$0.3 \times R_r \leq (\sigma_{oa})_{min} \leq 0.6 \times R_r$

$$[7.18]$$

❏ The fatigue curve seen above does not provide an "intrinsic" property of the material. It depends on the nature of the latter of course, but also:

♦ on the macroscopic and microscopic geometry of the structure, as shown in Figure 7.11 and Figure 7.20. Thus, a part with large dimensions will present a higher probability of internal defects than a part of small dimensions: a steam turbine shaft for electrical energy that is voluminous and forged is more sensitive to fatigue than an automobile piston shaft, small and machined in a wiredrawn rod;

♦ on the surface roughness and of the degree of corrosion;

♦ on the surface treatment: shot blasting treatment, heat treatment, physico-chemical treatment, which create residual compression stresses that close up the micro-cracks on the surface;

---

34 See bibliography: J. Drouin.

♦ on the temperature, which develops the elastoplastic zone. An elastoplastic material, Figure 7.3, becomes fragile when the temperature reduces[35].

Thus, the fatigue test represented in Figure 7.22 can serve as a reference only if it is done in accordance with standardized procedures.

7.3.3. *Modeling of the fatigue*

7.3.3.1. *Modeling of dynamic loading*

The real variable strains on a structure (handling machine, airplane, rail chassis, etc.) lead to irregular stress levels as given in Figure 7.23a.

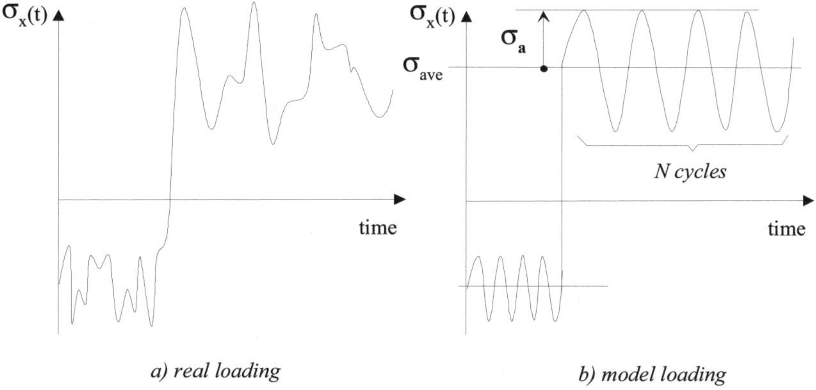

Figure 7.23. *Modeling of dynamic loading*

These evolutions are schematized in a diagram as subsets or "sequences" of cyclical or "wavy" loads which act successively on the structure, each one characterized by:

♦ σ_a : alternate stress around average value;

♦ σ_{ave} : average stress;

♦ N: the number of cycles.

---

35 The temperature can actually contribute to the stress distribution within a loaded structure. It is said to concern a "thermo-mechanical" loading. Such an influence of the temperature is not taken into account in this work.

7.3.3.2. *Corresponding fatigue test*

The fatigue test is modified. An average stress value appears, which influences the number of cycles to failure as shown in Figure 7.24. For a given average stress, the fatigue strength is the alternate stress σ_a versus the number of cycles N at failure.

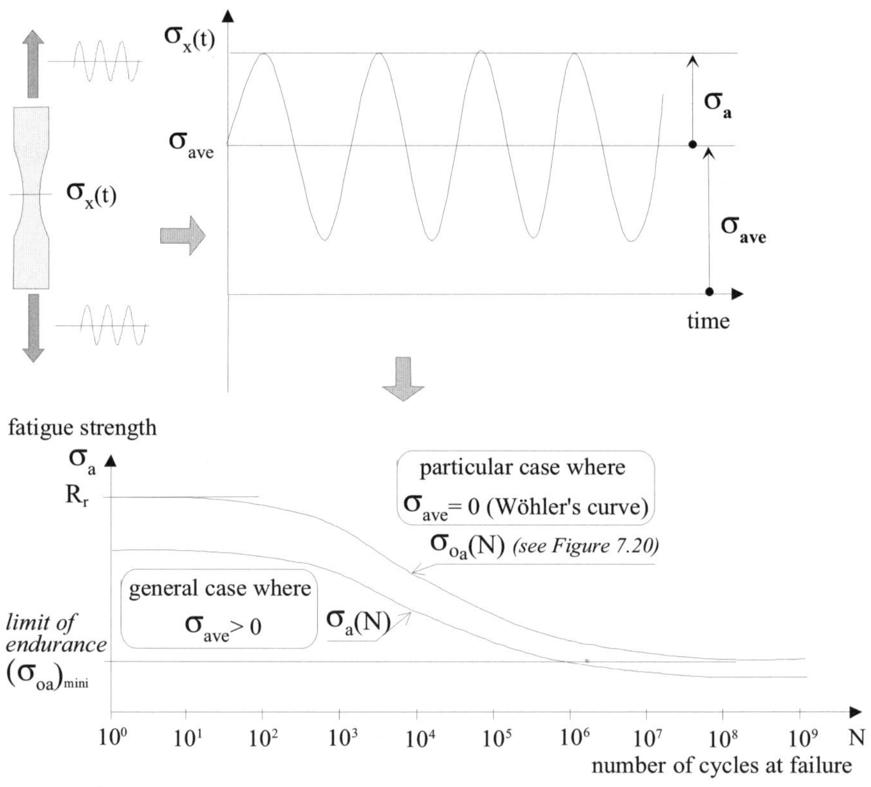

Figure 7.24. *Evolution of the fatigue strength when average stress varies*

❏ Real structures result in the lay-out and assembly of parts characterized by specific geometric shapes, such as beams, plates or shells, or thick localized parts. The states of stresses depend on the mechanical loadings as well as on the geometry of these zones. Except for particular cases, (test samples for example) states of stresses are rarely simple, i.e., only reduced to one component: normal stress, shear stress. When it concerns a plane state of stresses, or a state of stresses characteristic of a beam, or of a plate, or a general one, the stress $\sigma_x(t)$ in Figure 7.24 should be replaced by the Von Mises equivalent stress, i.e. $\sigma_{V.Mises}(t)$ with the forms [7.5], [7.7], [7.8] or [7.9] as the case may be.

❏ With the same concern, when in a zone of the structure under study a geometric singularity exists where a stress concentration develops, we must evaluate the equivalent stress in this zone.

7.3.4. *Estimation of fatigue strength*

7.3.4.1. *Case of a simple wavy load*

We dispose of empirical formulae enabling us to inter-relate the characteristics of the fatigue test in Figure 7.24. We quote below the one which gives the most "pessimistic" results[36], called the Soderberg formula:

$$\frac{\sigma_a}{\sigma_{oa}} + \frac{\sigma_{ave}}{R_e} = 1 \qquad [7.19]$$

in which:

– σ_a is the alternate stress of Figure 7.24;

– σ_{oa} is the fatigue strength shown by Wohler's curve in Figure 7.22 (in the absence of average stress);

– σ_{ave} is the average stress;

– R_e is the elastic tension strength on a test sample.

The practical problem that occurs is the following: knowing the characteristics (σ_a, σ_{ave}) of Von Mises equivalent stress in a given zone of the structure as well as the elastic limit R_e of the material used, what is the number N_0 of load cycles that the structure can support in this zone.

The method to obtain the solution is illustrated below:

---

36 This formula is more pessimistic than the better known Goodman's formula, written as:

$\dfrac{\sigma_a}{\sigma_{oa}} + \dfrac{\sigma_{ave}}{R_r} = 1$ where R_r is the rupture tension strength in the test sample.

data	Soderberg's formula	result	Fatigue test curve (Figure 7.22)
(σ_a, σ_{ave}) R_e	$\dfrac{\sigma_a}{\sigma_{oa}} + \dfrac{\sigma_{ave}}{R_e} = 1$	$\sigma_{oa} \rightarrow$	*possible number of cycles* N_0

$$[7.20]$$

NOTES

❏ This method presupposes that we know the fatigue test curve, which is not always the case. Nevertheless we can obtain estimation of the possible number of cycles by using an empirical relation which reflects Wölher's fatigue curve in Figure 7.22 and which corresponds to the approximate evolutions given hereafter[37]. Thus, in step [7.20], we can replace the experimental curve of the fatigue test by the approximate curve [7.21].

---

37 See bibliography: Bazergui *et al.*

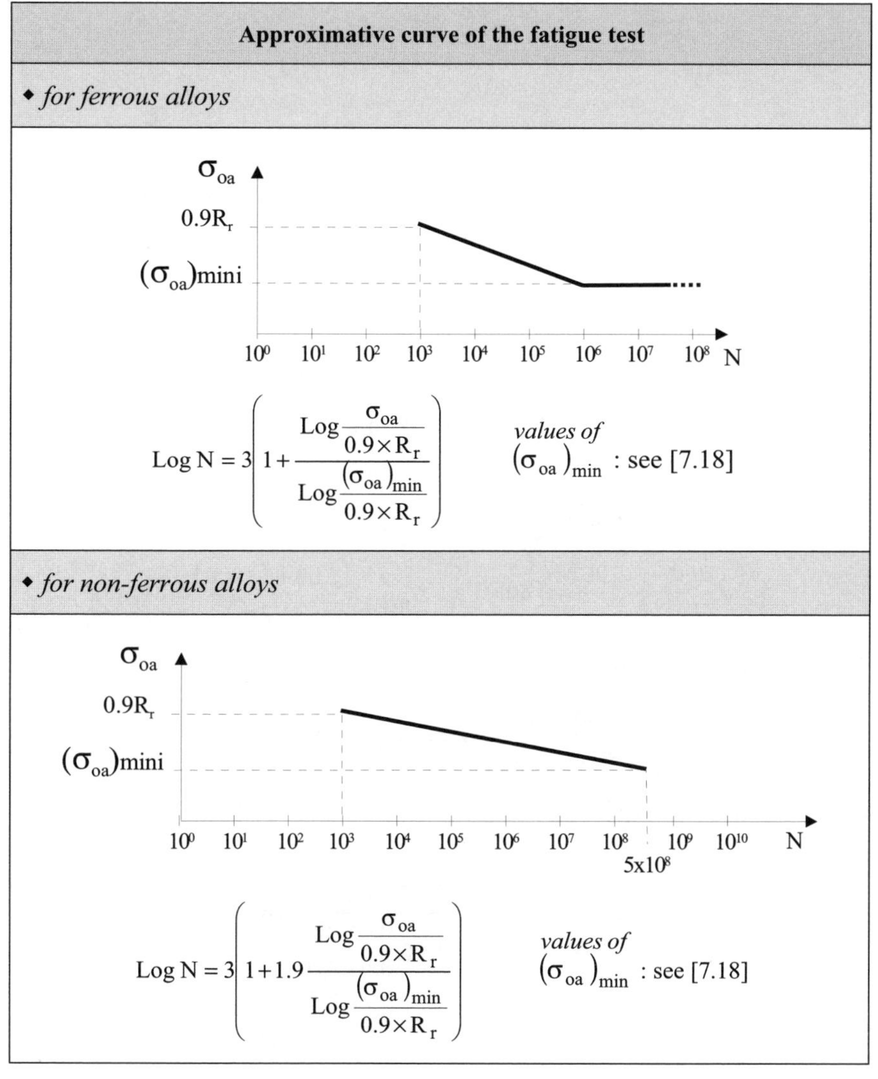

Approximative curve of the fatigue test

• *for ferrous alloys*

$$\text{Log N} = 3\left(1 + \dfrac{\text{Log}\dfrac{\sigma_{oa}}{0.9 \times R_r}}{\text{Log}\dfrac{(\sigma_{oa})_{min}}{0.9 \times R_r}}\right)$$

values of
$(\sigma_{oa})_{min}$: see [7.18]

• *for non-ferrous alloys*

$$\text{Log N} = 3\left(1 + 1.9\dfrac{\text{Log}\dfrac{\sigma_{oa}}{0.9 \times R_r}}{\text{Log}\dfrac{(\sigma_{oa})_{min}}{0.9 \times R_r}}\right)$$

values of
$(\sigma_{oa})_{min}$: see [7.18]

[7.21]

○ Example:

We consider a structure part made in duraluminum (EN AW-2024: R_e = 280 MPa; R_r = 420 MPa) subjected to a wavy load as in Figure 7.25, and we wish to estimate the number of possible load cycles without failure.

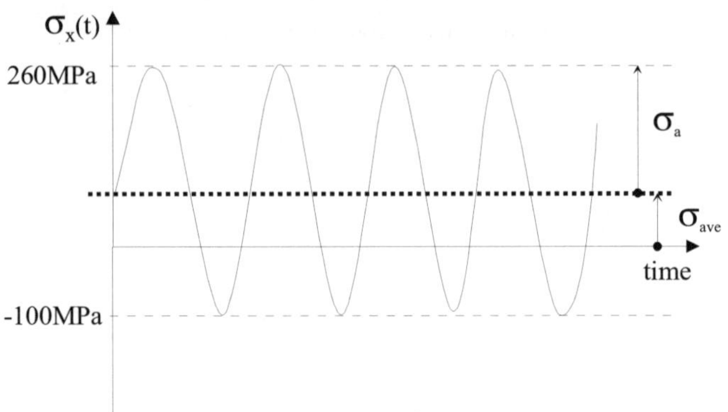

Figure 7.25. *Wavy load*

From Figure 7.25, we infer:

$$\sigma_a = \frac{260 - (-100)}{2} = 180\text{MPa} \; ; \; \sigma_{ave} = \frac{260 + (-100)}{2} = 80\text{MPa}$$

With [7.20] we obtain:

$$\frac{180}{\sigma_{oa}} + \frac{80}{280} = 1 \Rightarrow \sigma_{oa} = 252\text{MPa}$$

and with [7.18] and [7.21]:

$$\text{Log N}_0 = 3\left(1 + 1.9 \frac{\text{Log} \dfrac{252}{0.9 \times 420}}{\text{Log} \dfrac{0.4 \times 420}{0.9 \times 420}}\right) = 5.85$$

i.e. $\text{N}_0 = 7\,\text{E5 cycles}$.

7.3.4.2. *Case of multiple wavy loads*

Figure 7.23 represented a loading modeled with two successive wavy load sequences. It is possible to extend this modeling by taking into account a general case where the structure is strained by n sequences of wavy loads i (i = 1, ..., n), each load sequence with the characteristics: $(\sigma_a)_i$, $(\sigma_{ave})_i$, $(N)_i$ as given in Figure 7.26.

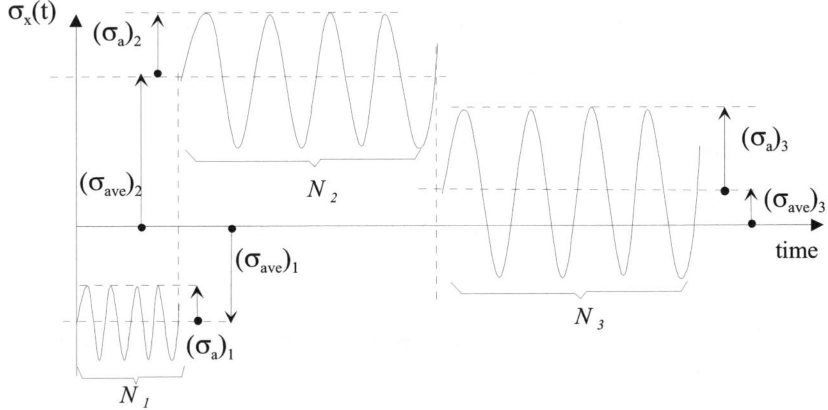

Figure 7.26. *General case (n wavy load sequences)*

We then examine the influence of each of the wavy loads as per the method given in the previous section. Thus, for load i according to Figure 7.26 and with [7.20]:

data:

$$\left.\begin{array}{r}(\sigma_a)_i \\ (\sigma_{ave})_i \\ (N)_i \\ R_e\end{array}\right\} \Rightarrow \frac{(\sigma_a)_i}{(\sigma_{oa})_i}+\frac{(\sigma_{ave})_i}{R_e}=1 \Rightarrow (\sigma_{oa})_i \Rightarrow \begin{array}{c}\textit{experimental} \\ \textit{or} \\ \textit{approximate} \\ \textit{fatigue curve} \\ [7.21]\end{array} \Rightarrow N_{oi}\ \begin{array}{l}\textit{number of} \\ \textit{cycles} \\ \textit{before} \\ \textit{failure}\end{array}$$

We then compare the total number of cycles possible N_{oi} before failure to the number of cycles N_i decided for this load:

– if $N_i \geq N_{oi}$: there is fatigue failure. The structure cannot support this wavy load sequence, and *a fortiori* all the wavy loads sequences given in Figure 7.26;

– if $N_i < N_{oi}$: we form the quotient $\dfrac{N_i}{N_{oi}}$

We next repeat this step by successively examining each load sequence in Figure 7.26, $i = 1,\ldots, n$. Then we consider the sum of the obtained quotients:

$$\frac{N_1}{N_{o1}} + \frac{N_2}{N_{o2}} + \frac{N_3}{N_{o3}} + \ldots\ldots + \frac{N_n}{N_{on}} = \sum_{i=1}^{n} \frac{N_i}{N_{oi}} \qquad [7.22]$$

○ Palmgren-Miner's rule

– if $\displaystyle\sum_{i=1}^{n} \frac{N_i}{N_{oi}} \geq 1$: the structure cannot support the multiple wavy load sequences;

– if $\displaystyle\sum_{i=1}^{n} \frac{N_i}{N_{oi}} < 1$: the structure supports the multiple wavy load sequences.

NOTES

❒ In the case where $\displaystyle\sum_{i=1}^{n} \frac{N_i}{N_{oi}} < 1$, we can estimate the margin of increase of the number of cycles of one or several partial wavy loads until the rupture limit is reached. For example, we wish to know the increase ΔN_2 of the number of cycles N_2 in Figure 7.26 which would lead to failure. We then write, by putting: $N_2' = N_2 + \Delta N_2$:

$$\frac{N_1}{N_{o1}} + \frac{N'_2}{N_{o2}} + \frac{N_3}{N_{o3}} + \ldots\ldots + \frac{N_n}{N_{on}} = 1$$

i.e. with [7.22]:

$$\frac{N'_2}{N_{o2}} - \frac{N_2}{N_{o2}} + \sum_{i=1}^{n} \frac{N_i}{N_{oi}} = 1$$

leading to the desired increase ΔN_2 :

$$\frac{\Delta N_2}{N_{o2}} = 1 - \sum_{i=1}^{n} \frac{N_i}{N_{oi}} \qquad [7.23]$$

❒ The rule above is approximate. We can present the following points against it:

– it does not take into consideration the application order of the partial wavy load sequences;

– the partial wavy loads leading to a stress σ_{oa} lower than the endurance limit stress (see Figure 7.22) appear to bring a zero contribution in the damage procedure as the number of cycles at failure $N_{oi} \rightarrow \infty$ and so $\dfrac{N_i}{N_{oi}} \rightarrow 0$.

There are more elaborate rules for evaluation of damages in order to cure these anomalies[38].

❒ Let us again mention that in practice, the national and international standards for designing codes enable us to define the critical stress values allowable under static or cyclical loads according to application domains.

❒ Cyclical loads can, according to the frequency of loading, be accompanied or not by significant "inertial" forces related to the accelerations. These are for example the inertial forces which appear in the earlier chapter (Chapter 6). If these forces have to be considered, we should remember that the "dynamic stresses" generated by the masses in movement will add themselves to the static stresses.

---

38 See bibliography: C. Bleuzen.

Chapter 8

Practical Aspects of Finite Element Modeling

8.1. Use of finite element software

8.1.1. *Introduction*

Chapter 5 is devoted to the definition of standard elements available in finite element software. We have mainly distinguished, for a wide area of applications:

– beam elements;

– plate elements;

– solid elements.

The characteristics of these elements appear in software as summarized in the following figures. In view of modeling, first of all the geometry of the structure must be determined with the help of CAD software. Then on the basis of this geometry the finite element mesh generation or "discretization" of the structure will be carried out. This mesh generation can be done:

– without any operator intervention to clarify one or several types of elements. The mesh generation is said to be automatic. This is the case of solid modeling with CAD software. The generated elements will be of solid type, for example tetrahedrons such as those shown in section 5.5.2. In this case, it must be kept in mind that the apparent convenience of the mesh generation can mask the insufficiencies pointed out in the notes in section 5.5.2;

– controlled by the operator. The latter should distinguish beforehand on the geometric CAD definition parts that are assimilable for instance to beams, others to plates or shells, or others that are definitely thicker. Then, the operator must modify the CAD representation in a way that will make it more compatible with the graphic representation of corresponding elements.

Thus, it follows that:

– parts assimilable to beams must be reduced to centerlines which need to be defined and positioned;

– parts to be modeled by plates must be reduced to their middle surfaces which need to be defined and positioned;

– thick voluminous parts should not, in theory, be transformed since they immediately lead to a discretization in solid elements.

Finally, in the case where different types of elements need to be linked, specific precautions can prove to be necessary to ensure a satisfactory connection, as explained later in section 8.1.3.

8.1.2. *Summary tables of the properties of elements*

We summarize in the following tables the main types of elements seen in Chapter 5:

♦ input data to be provided relating to the element and relating to the model;

♦ results available after calculation, relating to the element and relating to the model.

Beam element	
NODES	**ELEMENT**
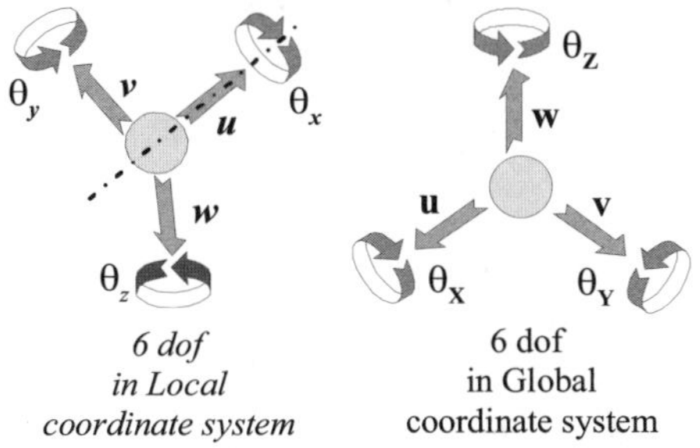	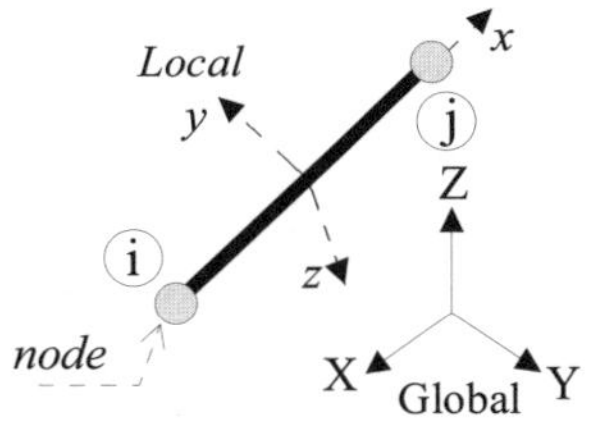
model data	***element data***
boundary conditions: ◆ **external linkings**: traditional joints (simple pointwise support, cylindrical joint, revolute joint, spherical joint, spherical-and-sliding joint, clamped joint) ◆ **internal linkings**: *by default:* clamped linking *by imposed "relaxation" inside internal linkings (in Local coordinate system):* pointwise support, cylindrical joint, revolute joint, spherical-and-sliding joint, spherical joint, etc. ◆ **nodal loads**: *forces:* X, Y, Z *moments*: L, M, N	***orientation of the Local coordinate system*** (planes *xy* and *xz*) ***cross-section properties***: area S, shear sections S_{ry}, S_{rz}, torsion constant J, quadratic moments I_y, I_z, flexure moduli I_y/z_{max}, I_z/y_{max} ***properties of isotropic material***: Young's modulus E, Poisson's ratio ν, mass and (or) weight per unit volume. ***linear distributed loads:*** forces and moments
output for the model	***output for the element***
nodal displacements (*Global coordinate system):* ◆ **linears**: u, v, w (translations of cross-section) ◆ **angulars**: θ_X, θ_Y, θ_Z (rotations of cross-section) ***mechanical actions on external linkings*** (Global coordinate system): *forces:* X, Y, Z *moments*: L, M, N	***plotting of the deflected shape*** ***shading*** with colors: iso-displacement, and normal iso-stress zones along center-lines ***normal stresses*** due to $\mathcal{N}$, $\mathcal{M}f_y$, $\mathcal{M}f_z$, combination of the three stresses ***nodal forces on the element*** (*Local coordinate system*): X_i, Y_i, Z_i, L_i, M_i, N_i, X_j, Y_j, Z_j, L_j, M_j, N_j

Figure 8.1. *Use of characteristics of the beam element in finite element software*

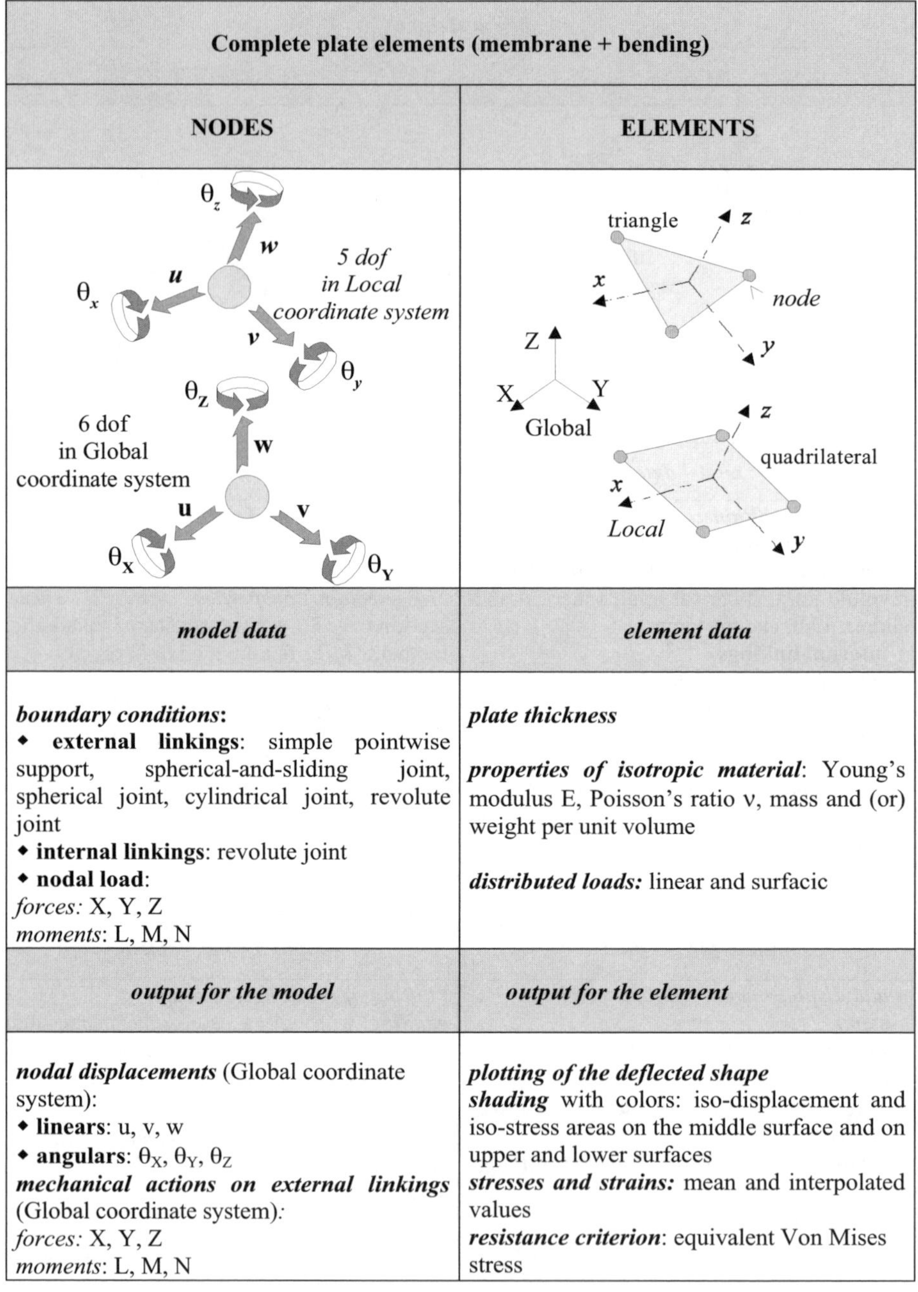

model data	element data
boundary conditions: ♦ **external linkings**: simple pointwise support, spherical-and-sliding joint, spherical joint, cylindrical joint, revolute joint ♦ **internal linkings**: revolute joint ♦ **nodal load**: *forces:* X, Y, Z *moments:* L, M, N	*plate thickness* *properties of isotropic material*: Young's modulus E, Poisson's ratio ν, mass and (or) weight per unit volume *distributed loads:* linear and surfacic
output for the model	***output for the element***
nodal displacements (Global coordinate system): ♦ **linears**: u, v, w ♦ **angulars**: θ_X, θ_Y, θ_Z *mechanical actions on external linkings* (Global coordinate system): *forces:* X, Y, Z *moments:* L, M, N	*plotting of the deflected shape* *shading* with colors: iso-displacement and iso-stress areas on the middle surface and on upper and lower surfaces *stresses and strains:* mean and interpolated values *resistance criterion*: equivalent Von Mises stress

Figure 8.2. *Use of characteristics of plate elements in finite element software*

<table>
<tr><td colspan="2" align="center">Solid 3D elements</td></tr>
<tr><td align="center">NODES</td><td align="center">ELEMENTS</td></tr>
<tr>
<td> 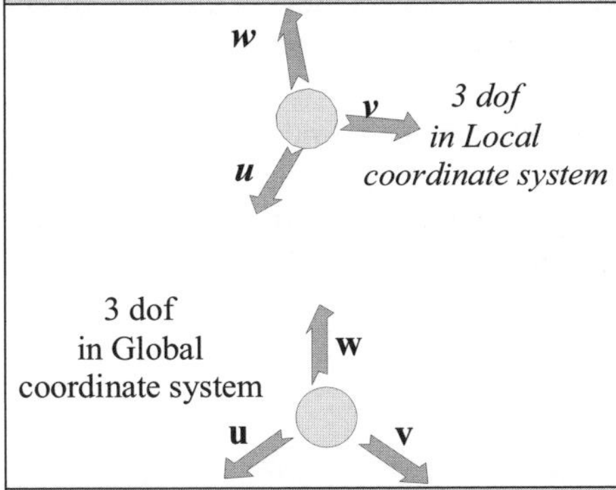

</td>
<td> 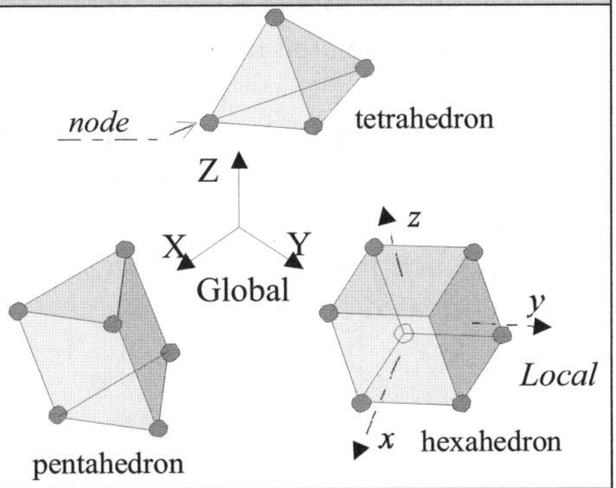

</td>
</tr>
<tr><td align="center">model data</td><td align="center">element data</td></tr>
<tr>
<td>

boundary conditions:
♦ external linkings: simple pointwise support, spherical-and-sliding joint, spherical joint
♦ internal linkings: spherical joint
♦ nodal loads:
forces: X, Y, Z

</td>
<td>

properties of isotropic material: Young's modulus E, Poisson's ratio ν, mass and (or) weight per unit volume

distributed loads: linear and surfacic

</td>
</tr>
<tr><td align="center">output for the model</td><td align="center">output for the element</td></tr>
<tr>
<td>

nodal displacements (Global coordinate system):
♦ linears: u, v, w
mechanical actions on external linkings (Global coordinate system):
forces: X, Y, Z

</td>
<td>

plotting of the deflected shape
shading with colors: iso-displacement and iso-stress zones
stresses and strains: mean and interpolated values
resistance criterion: equivalent Von Mises stress

</td>
</tr>
</table>

Figure 8.3. *Use of characteristics of solid 3D elements in finite element software*

NOTES

❏ When we examine a structure for modeling, the problem arises at the beginning to retain one or several types of elements capable of the best compromise between the simplicity of the model and its expected performances.

The table in Figure 8.4 presents for the same example a view of advantages and disadvantages related to the use of the three types of elements above.

❏ The global model and local modeling: we can also find a compromise by creating the simplest possible global model, so as to find out the behavior in current zones. Thereafter, we make a local model in the particular zone where more detailed information is necessary. This process is illustrated in Figure 8.5 in a weld fabricated chassis.

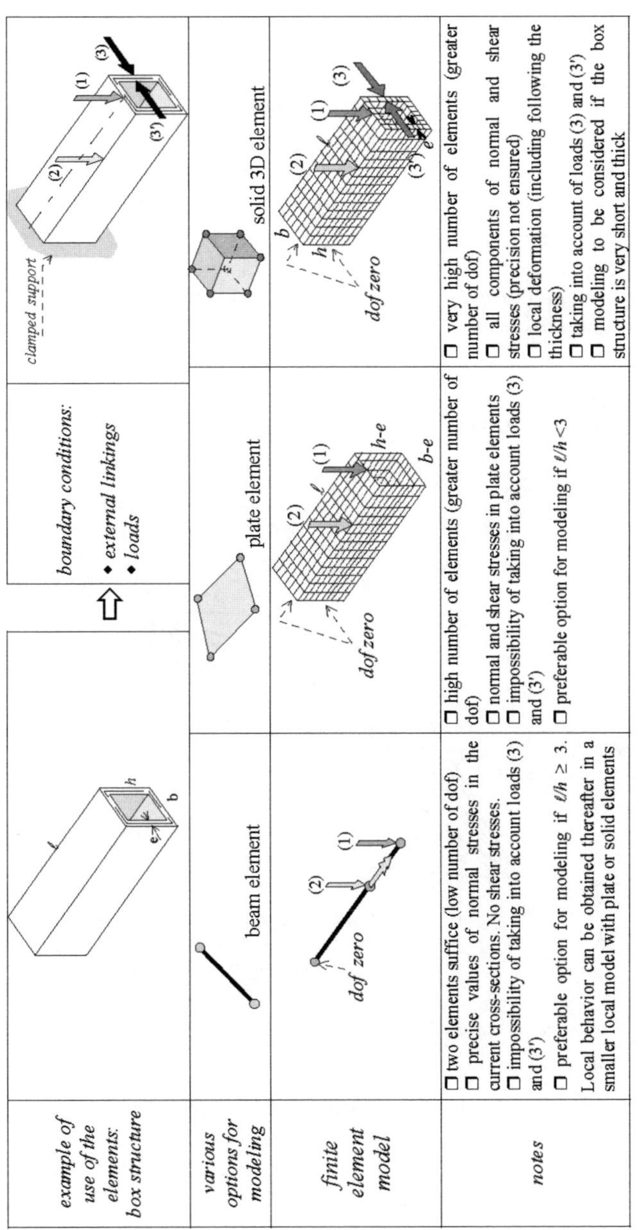

Figure 8.4. *The same structure modeled by each of the three element types*

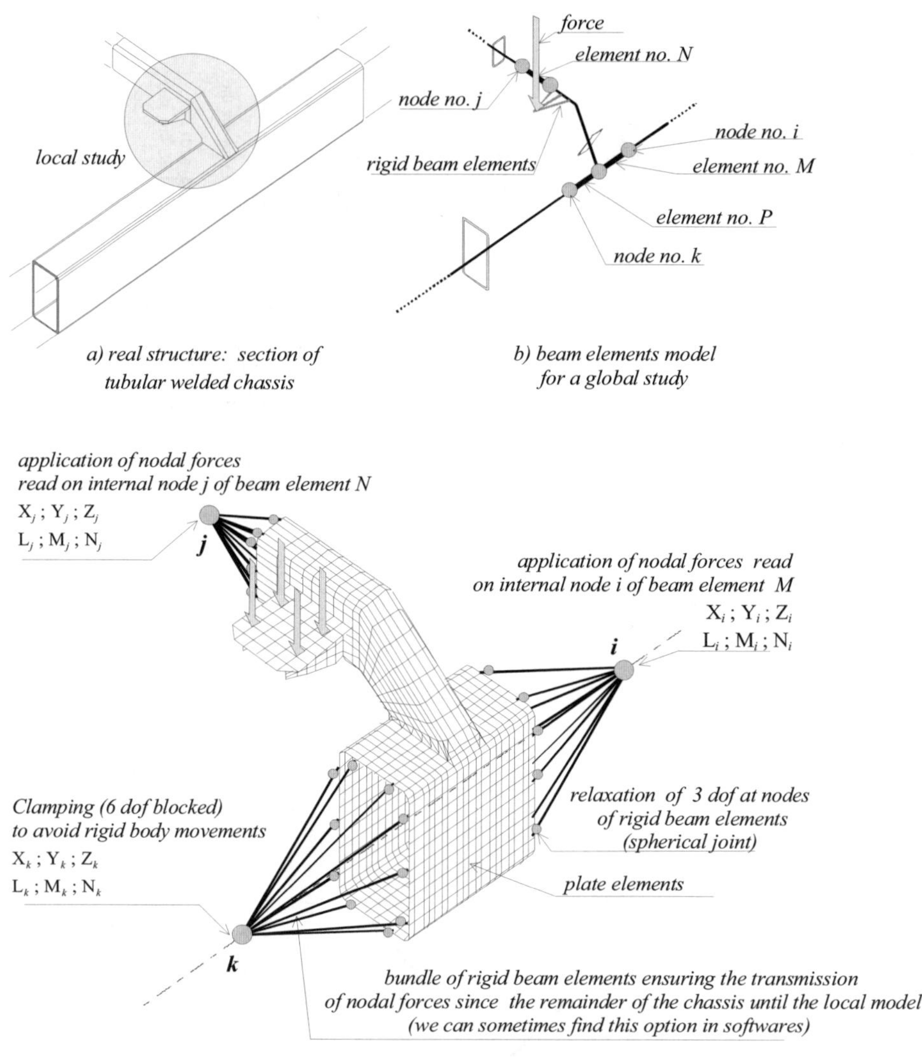

Figure 8.5. *Global model and local model*

8.1.3. *Connection between elements of different types*

8.1.3.1. *Introduction*

In Chapter 5, we have reviewed an assortment of elements sufficient to allow the discretization of various structures. Each type of element is characterized by a certain number of dof at each node. Figure 8.6 recalls these dof for the main elements which we have already seen. In this figure, we have considered a single node and adopted parallel local coordinate systems for each of the elements.

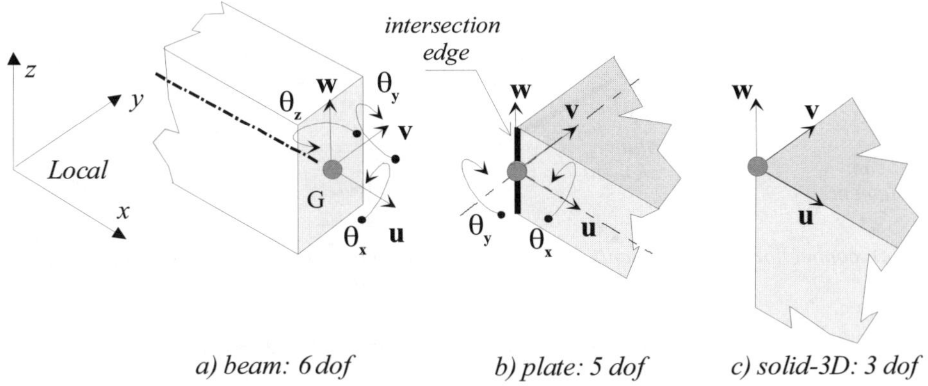

Figure 8.6. *Nodal degrees of freedom of main element types*

It is noted immediately that the beam element possesses the maximum number of dof (6) making it possible to mark the position of a solid in space, in the occurrence of the cross-section which has the node as geometric center. The plate element has 5 nodal dof as it lacks rotation θ_z in comparison to the beam in Figure 8.6b. In fact, as we have already pointed out[1], the five nodal dof here characterize the position in space of the vertical intersection edge of the plate which has the said node as its center. Thus a rotation of θ_z does not have physical significance for this intersection edge which cannot turn around itself as it is geometrically recognizable as a line without thickness. Finally, the three nodal dof of the solid 3D-element characterize the position in space of the geometric point constituted by the node itself. Here, rotations such that θ_x, θ_y or θ_z do not have any physical value.

From this, there are differences in nodal "mobility" between elements of different types. For instance, if we connect the elements of Figure 8.6, we will

---

1 See section 5.4.3/*NOTES*.

observe an incomplete internal linking at the common node. This particularity is shown on the following examples.

8.1.3.2. *Example 1*

Let us consider the connection of beam and plate elements in Figure 8.6. It can be seen in Figure 8.7 that there is no rotation linking around $\vec{z}$ axis. The beam and the plate have a relative degree of freedom which is the angle of rotation θ_z. These two elements are connected with a revolute joint with $\vec{z}$ axis on the connection node (Figure 8.7c).

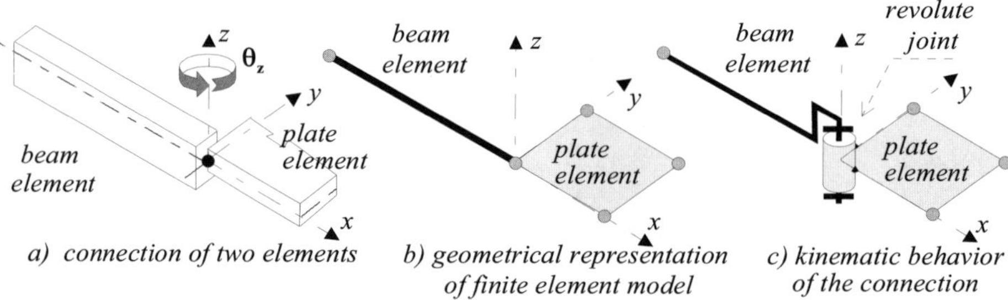

Figure 8.7. *Connection of a beam and a plate element*

Consequence: consider models as represented in the next figure (Figure 8.8). They will be able to present a rigid-body motion of a part of the structure, in spite of a proper linkage of the whole structure with its vicinity. It follows that the stiffness sub-matrix of the structure, deduced from the complete stiffness matrix after suppression of corresponding lines with zero[2] dof and columns of the same row, is still singular.

---

2 See application of support conditions in section 3.4.

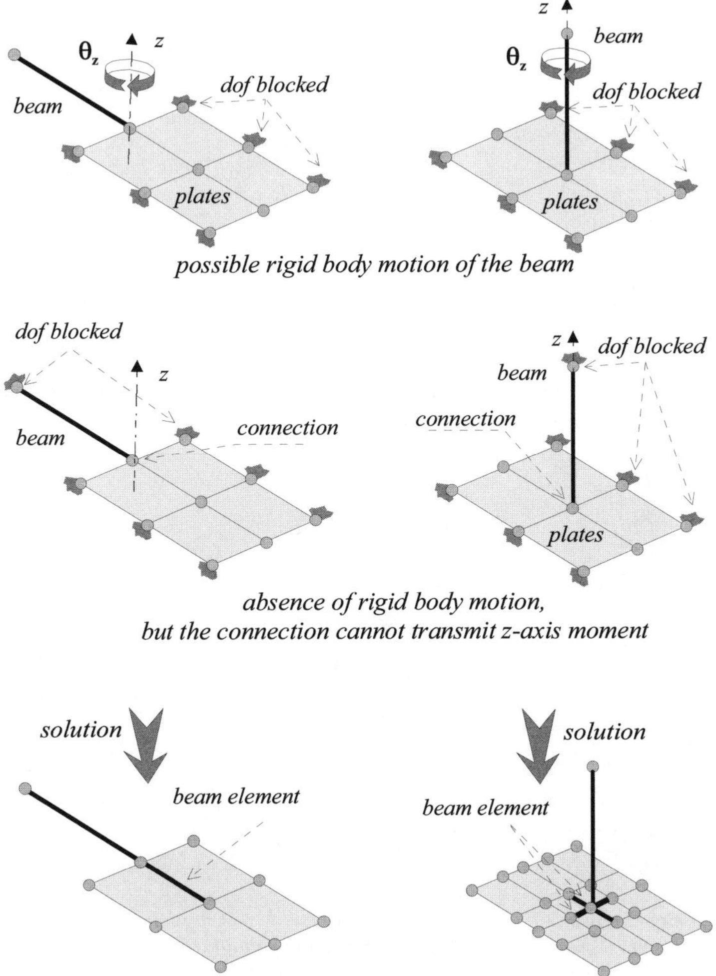

Figure 8.8. *How to carry out a correct link-up*

8.1.3.3. *Example 2*

We encounter the same particularity when we connect the plate element and the solid 3D-element in Figure 8.6. This link-up is represented in Figure 8.9. The two elements have three relative dof of rotation at the linking node. They are linked with a spherical joint.

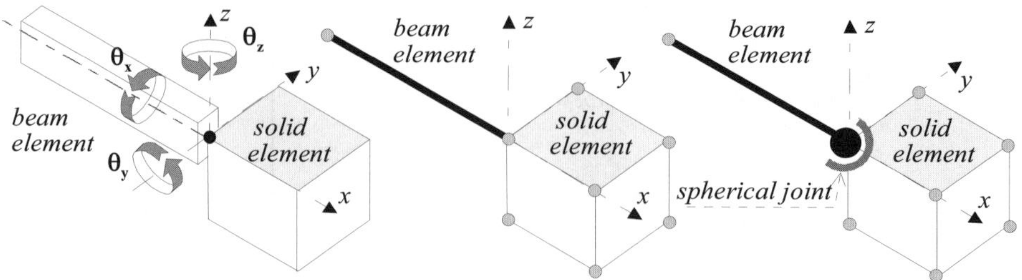

Figure 8.9. *Connection of a beam and a solid 3D-element*

The same type of note can be made regarding the possible motion of a part of the entire modeled structure in Figure 8.10.

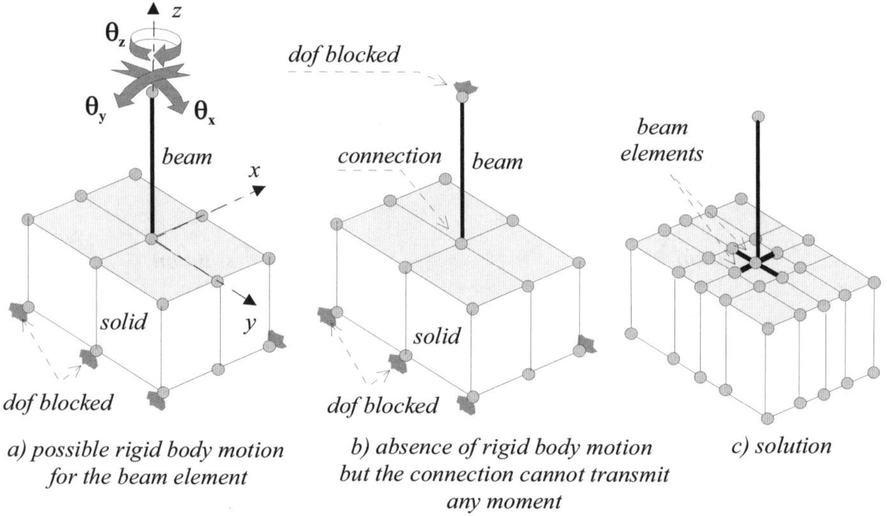

Figure 8.10. *How to carry out a correct link-up*

8.1.3.4. *Example 3*

Finally, the association of a plate and a solid 3D-element in Figure 8.11 will necessitate analogous precautions as we encounter the same type of problem at the connecting joints.

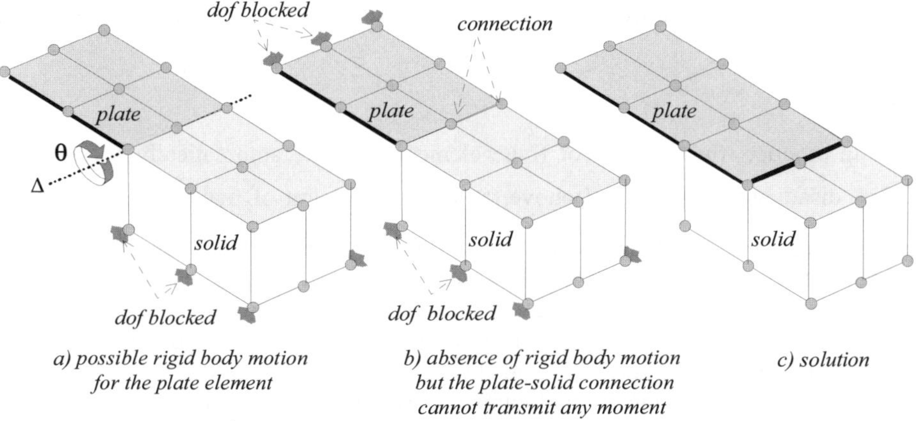

a) possible rigid body motion
for the plate element

b) absence of rigid body motion
but the plate-solid connection
cannot transmit any moment

c) solution

Figure 8.11. *Connection of plates and solid elements*

8.1.3.5. *Conclusion*

These simple examples show that assembling different types of finite elements can pose two types of problems:

– Even if we believe we have properly linked the structure with its vicinity, some parts of the model are still capable of rigid-body motion(s) due to incomplete internal linking connections. The finite element software will calculate the stiffness matrix of the whole structure and then apply the calculation process already mentioned[3] to eliminate rows and columns corresponding to zero dof at the supports. We thus obtain the "sub-system"

$$\{F^*\} = \left[K^*\right] \bullet \{d^*\} \tag{8.1}$$

in which $\{d^*\}$ represents the free dof. The software will then try to invert $\left[K^*\right]$. The possible rigid-body motion of localized parts of the model make this inversion impossible. The sub-matrix $\left[K^*\right]$ is singular. System [8.1] cannot be solved. This is what is illustrated on the previous figures.

– We should therefore take the precaution of eliminating possibilities of rigid-body motion of substructure parts by blocking supplementary appropriate dof. However, after loading the structure and calculation, the results may be locally

---

3 See section 3.4.

misleading as the transmissible forces through the internal joints in question do not reflect reality.

NOTES

❏ In practice, the majority of finite element software automatically corrects the first disadvantage identified above, to allow inversion of matrix $\left[K^* \right]$. For this, plates and solid 3D-elements are equipped with "springs" at the connecting joints. Then it becomes possible to "rigidify" these joints toward clamped junctions. This can be illustrated in Figure 8.12 for a plate element. Thus internal linkings are completed, and in this way the rigid-body motions of parts of the model are eliminated. However, "artificial" stiffness that is added is "speculative". Thus the forces transmitted by the internal joints thus rigidified do not conform.

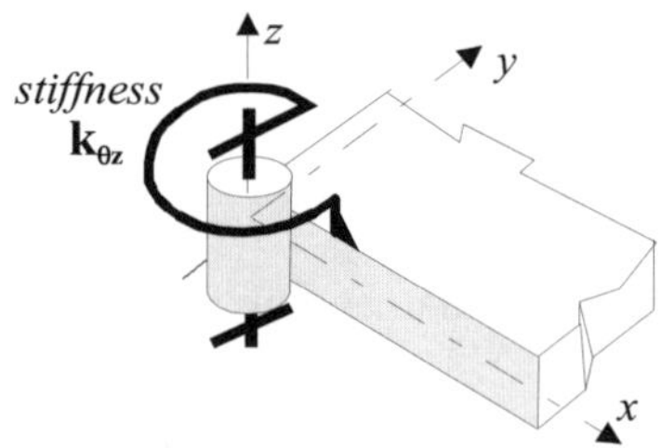

Figure 8.12. *Addition of stiffness* k_{θ_z} *at a plate element node*

❏ Consequently from the earlier part, we can generalize in the following way the rule that we had mentioned in section 2.5.2.2 regarding a "properly" linked structure:

> A structure is properly linked when rigid-body motion of the entire structure or of any *substructure part* is not possible.

[8.2]

8.1.4. *Other practical aspects*

It is certainly impossible to review all the structure particularities that can influence the design of a model through finite elements. We can only cite frequently observed cases.

8.1.4.1. *Symmetric structures*

Consider a structure with one (or several) geometric symmetry plane(s). If the linkage conditions at boundaries and the loading present the same symmetric properties, then we can create a fractional model. Figure 8.13 illustrates this principle which leads to an important dof saving for the model.

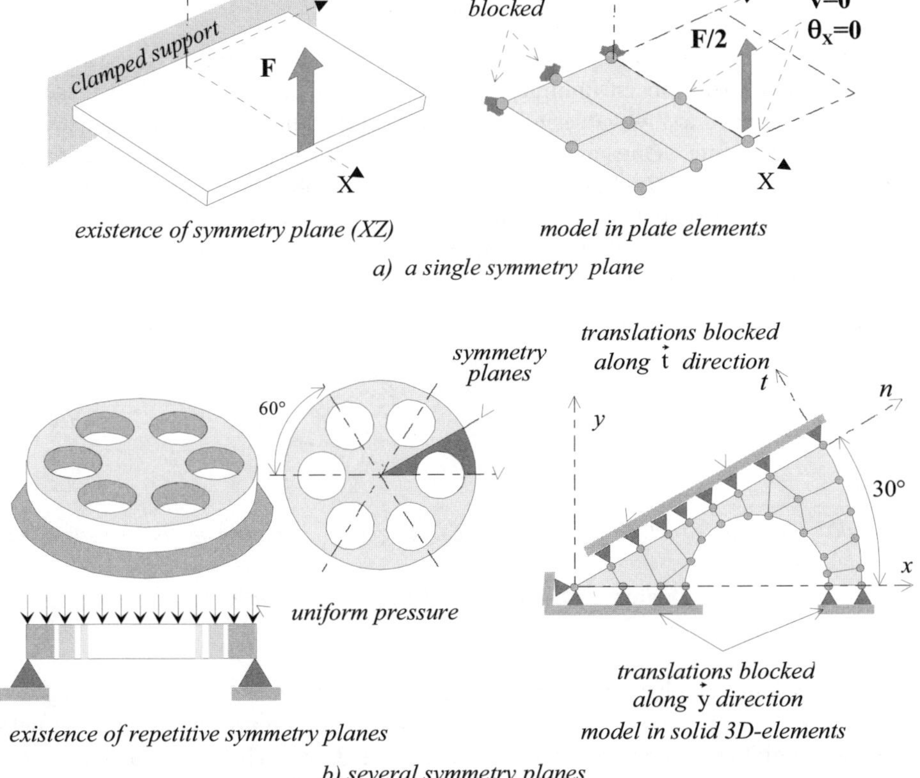

existence of symmetry plane (XZ)   *model in plate elements*

a) a single symmetry plane

existence of repetitive symmetry planes   *model in solid 3D-elements*

b) several symmetry planes

Figure 8.13. *Use of symmetry properties*

8.1.4.2. *"Floating" models*

We have already pointed out repeatedly that a structure not properly linked with its surroundings presents one or several possibilities of rigid-body motion, making it impossible to solve the behavior equation:

$$\{F\}_{str} = \underset{structure}{[K]} \bullet \{d\}_{str}$$

This is because the stiffness matrix is singular. Nevertheless, we can find loaded structures in equilibrium even without being completely connected to their surroundings. This is for instance the case of an aircraft in flight. The problem remains however: the structure is in equilibrium under a set of external forces, and it is desirable to know its mechanical behavior, i.e., to be able to solve a relation of the type:

$$\{F\}_{str} = \underset{structure}{[K]} \bullet \{d\}_{str}$$

For this it is necessary to prevent any possible rigid-body motion (there are six in space), which results in an unlimited increase in dof values. Therefore these dof must be maintained at low values without interfering with the real strain and stresses inside the structure. One method consists of "artificially" creating pointwise supports with carefully chosen directions. The number of supports should be necessary and sufficient to eliminate all rigid-body motions. As a simplified example, let us look at a structure "floating" in the plane of Figure 8.14a, loaded by a set of balanced external forces.

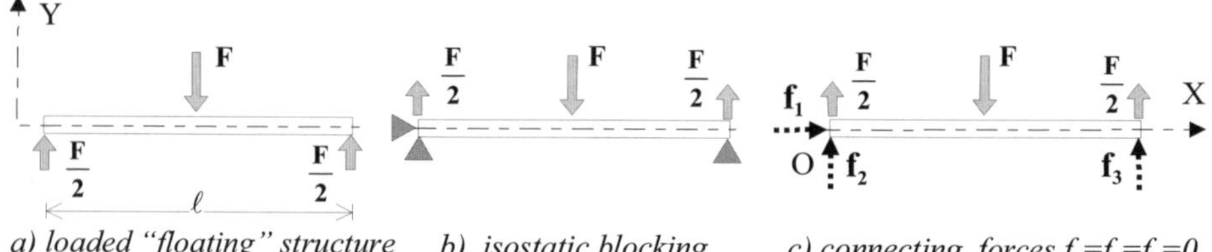

a) loaded "floating" structure   b) isostatic blocking   c) connecting forces $f_1{=}f_2{=}f_3{=}0$

Figure 8.14. *Isostatic blocking on a structure*

This structure can have three possible rigid-body motions: translation along $\vec{X}$, translation along $\vec{Y}$, and rotation around $\vec{Z}$. These can be prevented with the help of simple pointwise supports as shown in Figure 8.14b. Thus, in principle, three components of connecting forces are created: f_1, f_2, f_3 in the global coordinate system (XY). These forces must themselves check equilibrium relations since the external loading is already balanced. This is represented as follows:

– sum of forces projected on $\vec{X}$: $f_1 = 0$;

– sum of forces projected on $\vec{Y}$: $f_2 + f_3 = 0$;

– moment zero on O, projected on $\vec{Z}$: $\ell \times f_3 = 0$

or even in matrix form:

$$\begin{bmatrix} 1 & 0 & 0 \\ 0 & 1 & 1 \\ 0 & 0 & \ell \end{bmatrix} \bullet \begin{Bmatrix} f_1 \\ f_2 \\ f_3 \end{Bmatrix} = \begin{Bmatrix} 0 \\ 0 \\ 0 \end{Bmatrix}$$

geometrical
matrix
3×3

The determinant of the geometric matrix is written as:

$$\Delta = \ell \neq 0$$

It is non-zero, and therefore the solution of the equation system reduces to:

$$f_1 = f_2 = f_3 = 0$$

In conclusion, the rigid-body motions have been eliminated. Nevertheless the initial external load and its effects on the structure (stresses, strains) are not by any means modified since no additional force acts on this structure. It is said that we have carried out an "isostatic blocking"[4]. The operator must carefully choose the supports to be used so as to obtain an equilibrium which is truly statically determinate. In the previous example the positioning was done without any ambiguity. Let us now consider a somewhat more complex "floating" structure in Figure 8.15.

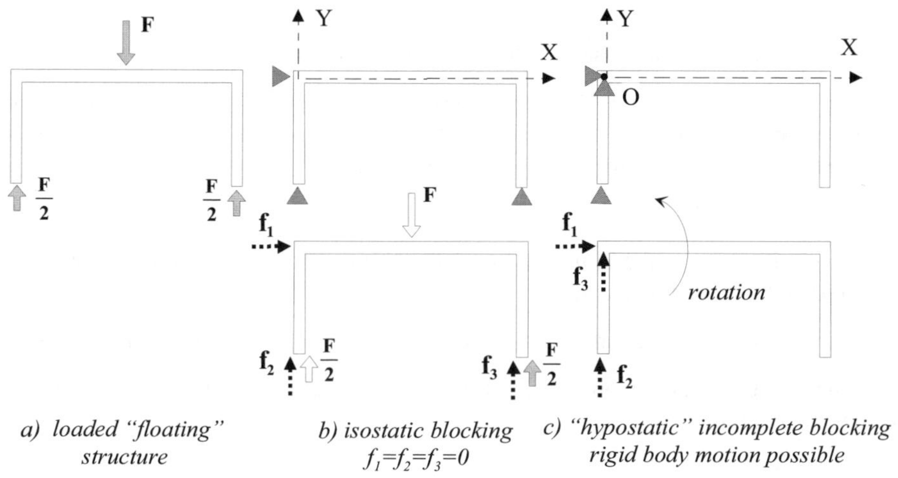

a) loaded "floating"
structure

b) isostatic blocking
$f_1=f_2=f_3=0$

c) "hypostatic" incomplete blocking
rigid body motion possible

Figure 8.15. *Complete and incomplete blocking*

---

4 See Figure 2.39.

When the simple pointwise supports are placed as in Figure 8.15b, we are brought back to the previous case. Now, if the supports are placed as in Figure 8.15c (colinear f_2 and f_3), we obtain as equilibrium relations:

– sum of forces projected on $\vec{X}$: $f_1 = 0$;

– sum of forces projected on $\vec{Y}$: $f_2 + f_3 = 0$;

– moment zero on O, projected on $\vec{Z}$: $0 = 0$

or even in matrix form:

$$\begin{bmatrix} 1 & 0 & 0 \\ 0 & 1 & 1 \\ 0 & 0 & 0 \end{bmatrix} \bullet \begin{Bmatrix} f_1 \\ f_2 \\ f_3 \end{Bmatrix} = \begin{Bmatrix} 0 \\ 0 \\ 0 \end{Bmatrix}$$

geometrical matrix
3×3

We observe that the geometric matrix is evidently singular (its principal determinant is zero). The structure can still have a rigid-body movement which is a rotation around axis $\vec{Z}$. The incomplete blocking is said to be unstable or "hypostatic"[5].

◯ Let us generalize the case of a "floating" structure in space.

There are six possible rigid-body movements. Let us "carefully" install six simple pointwise supports to eliminate these six motions. Thus we obtain six components of force, one for each pointwise support:

$$f_1, \ldots, f_6$$

The equilibrium of the structure under these six components can be projected on the three global coordinate axes as:

– three linear relations on $f_1, \ldots, f_6$ expressing that the resultant force is zero;

– three linear relations on $f_1, \ldots, f_6$ expressing that the resultant moment is zero on any given point.

These six linear relations can be regrouped in a matrix form as follows:

---

5 See also Figure 2.39.

$$[G] \bullet \left\{ \begin{matrix} f_1 \\ f_2 \\ \vdots \\ f_6 \end{matrix} \right\} = \left\{ \begin{matrix} 0 \\ 0 \\ \vdots \\ 0 \end{matrix} \right\} \qquad [8.3]$$

where spatial positions and directions of pointwise supports appear in matrix $[G]$. It can thus be qualified as a "geometric matrix".

♦ If matrix $[G]$ is "regular" (principal determinant $\neq 0$), the solution of system [8.3] is reduced to $f_1 = f_2 = \ldots = f_6 = 0$.

The complete blocking thus achieved is so-called statically determinate or "isostatic". The structure behavior under the initial external loads is not disturbed since no additional force acts on the structure. In practice, in the behavior relation of the free structure:

$$\{F\}_{ext} = \underset{structure}{[K]} \bullet \{d\}_{str}$$

so that isostatic blocking will result, according to the discussed procedure[6], by the removal of six rows and columns corresponding to dof which are canceled as soon as the six pointwise supports are created. We then obtain the sub-system:

$$\{F^*\} = \underset{structure}{\left[K^* \right]} \bullet \{d^*\}_{str}$$

and now inversion of sub-matrix $\left[K^* \right]_{str}$ is possible.

♦ If matrix $[G]$ is singular (principal determinant zero) then there remains one, or several, possibilities of rigid-body movements. The blocking is incomplete (unstable state, or "hypostatic" positioning). It is then necessary to reconsider the positions as well as the directions of the simple pointwise supports.

NOTES

❏ A statically determinate positioning or "isostatic" positioning can be achieved in different ways. The operator generally has a large number of possible choices. Thus, if we again take the example of Figure 8.15, we have several options for isostatic blocking, as illustrated in Figure 8.16.

---

6 See [2.122] and section 3.4.

Figure 8.16. *Different options for isostatic blocking*

❐ Some finite element software establish by themselves an isostatic positioning if the model appears to be "floating" in space (not properly supported). This operation is transparent for the operator but can involve some problems since the operator loses control of the initial blocking.

8.1.4.3. *Modeling of fabricated welded structures*

Structures built on this principle generally include assemblies of thin-walled components:

– thin-walled profiles with I, U, L, T open cross-sections;

– thin-walled profiles with rectangular or circular closed cross-sections;

– thin-walled extruded profiles with complex section shapes;

– flame-cut plates.

These types of parts involve problems of modeling at their junctions. These problems are mentioned hereafter, as well as their solutions.

◯ Profiles linked by welded joints.

Modeling with beam elements is carried out using centerlines as indicated in Figure 8.17 given below.

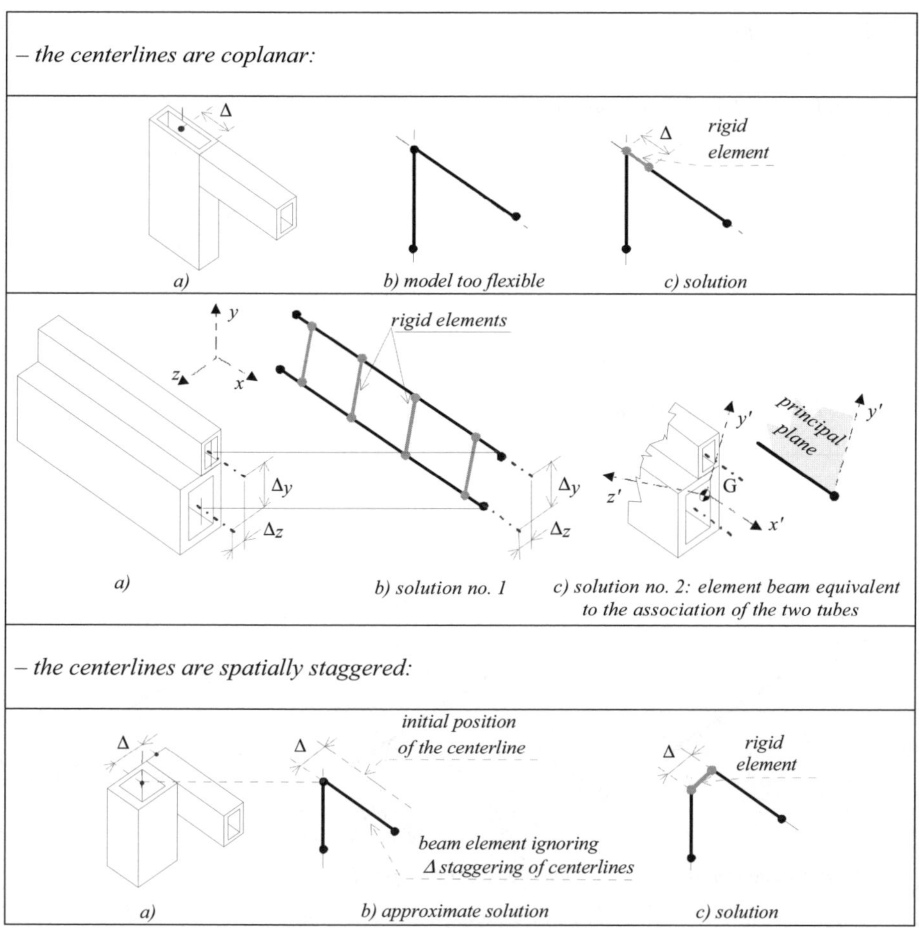

Figure 8.17. *Profile modeling at welded joints*

○ Plate welded to a profile

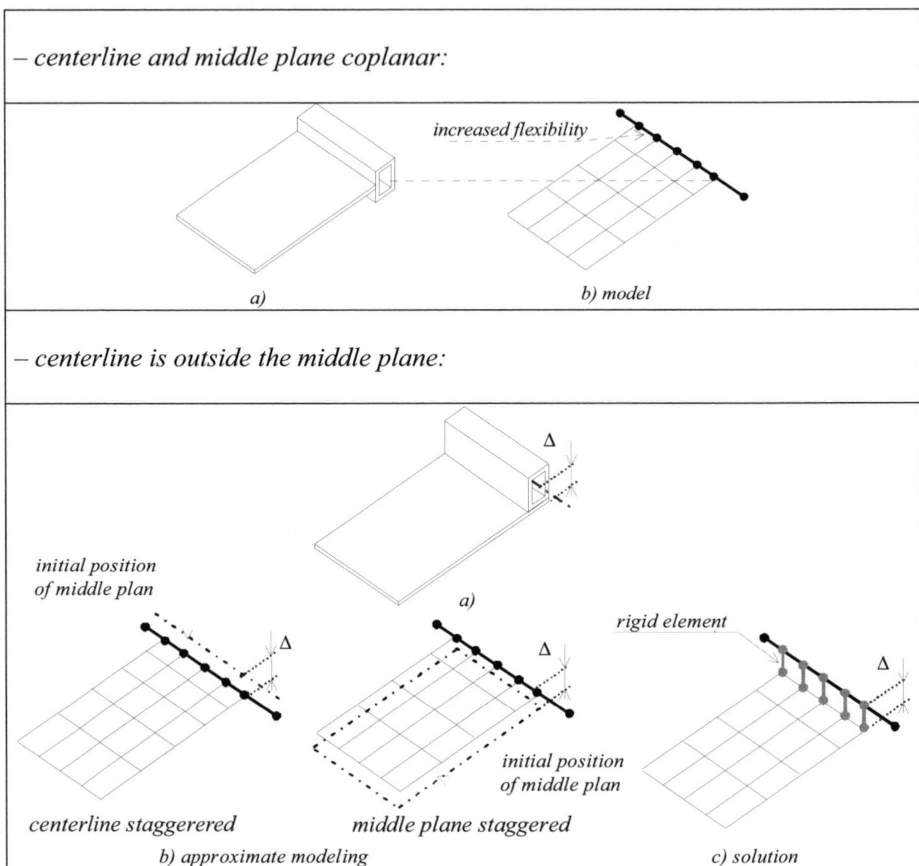

Figure 8.18. *Welding of a plate and a profile*

○ Welding of two plates

– spot welding:

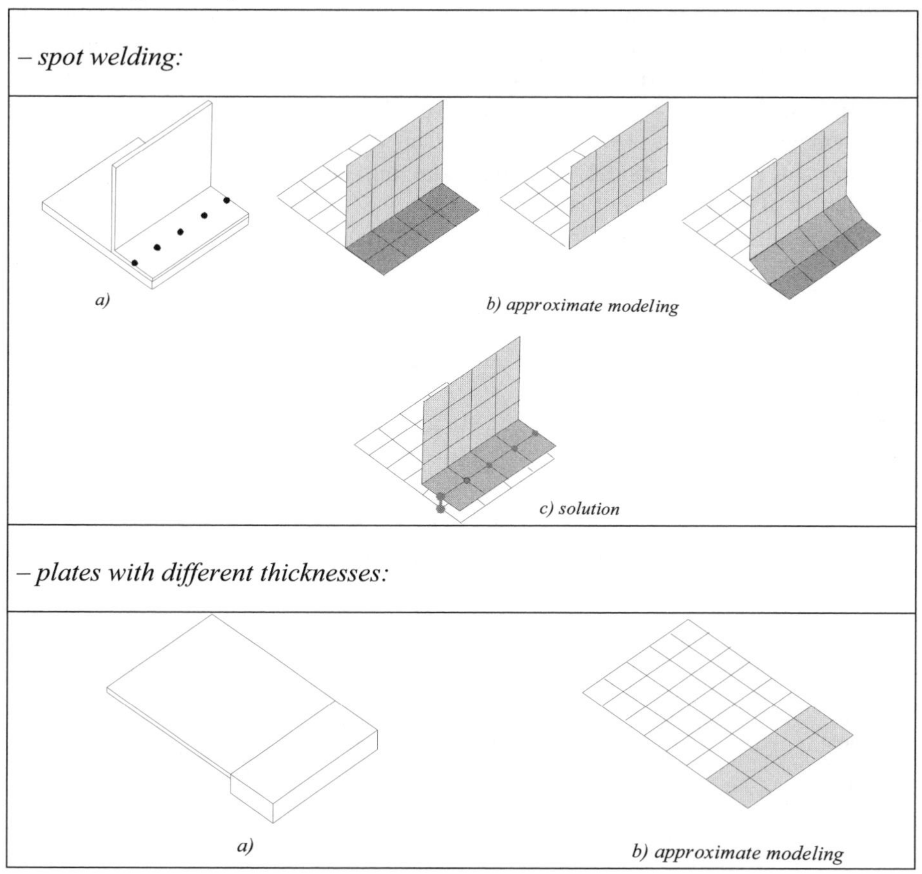

– plates with different thicknesses:

Figure 8.19. *Weld-joined plates*

8.1.4.4. *"Nonlinear geometric" behavior*

Let us consider in Figure 8.20a a beam connected to its vicinity through a revolute joint at each extremity. A force is applied at beam mid-length as shown. We model this structure with two elements in Figure 8.20b. The boundary displacement conditions are noted in the same figure. We know that the bending of such a beam is characterized by a curvature of the centerline, initially rectilinear and blended with the $\vec{x}$-axis. Here the distance ℓ between points (1) and (3) remains constant, whereas such a curvature corresponds to an increase in the length of the centerline. A normal tensile force is thus produced in the beam related to this lengthening, but it is not proportional to the displacement v_2 of the mid-node.

We say that we are dealing with "nonlinear geometric"[7] behavior. In reality, the joints at extremities render the beam less deformable (or more stiff) in bending (the work of force F is converted into bending *and* tensile strain energy, and not only into bending energy): the v_{2real} displacement of Figure 8.20c is weaker than the displacement v_2 of Figure 8.20d calculated on the basis of the model with finite elements. In fact, the model does not take into account this stiffening as the formulation of beam element seen in Chapter 5 is carried out without considering lengthening effects of the centerline[8].

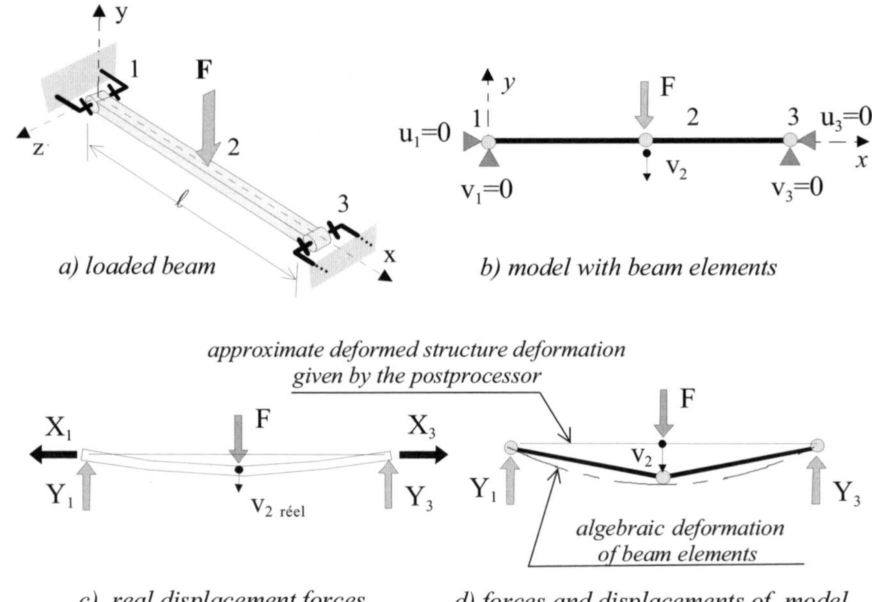

Figure 8.20. *Behavior differences between the model and the real structure: "nonlinear geometric behavior"*

7 There is another type of nonlinear behavior which is the material nonlinear behavior when we enter the elastoplastic zone. The finite element software can integrate such a behavior from the input data of the elastoplastic law. This aspect is not studied in this book.
8 See section 5.2.

NOTES

❐ Finite element software includes specific modules to take into account this type of behavior, but which must be used with care as the verification of the calculational results then becomes difficult.

❐ When such nonlinear forces appear, the results obtained with linear models (that is in areas where the strains and dof are linked by a linear law of the type $\{F\} = [K] \bullet \{d\}$ are misleading. For example, Figure 8.21 shows a welded steel nacelle. We want to know the maximum of force F that this nacelle can tolerate at rupture.

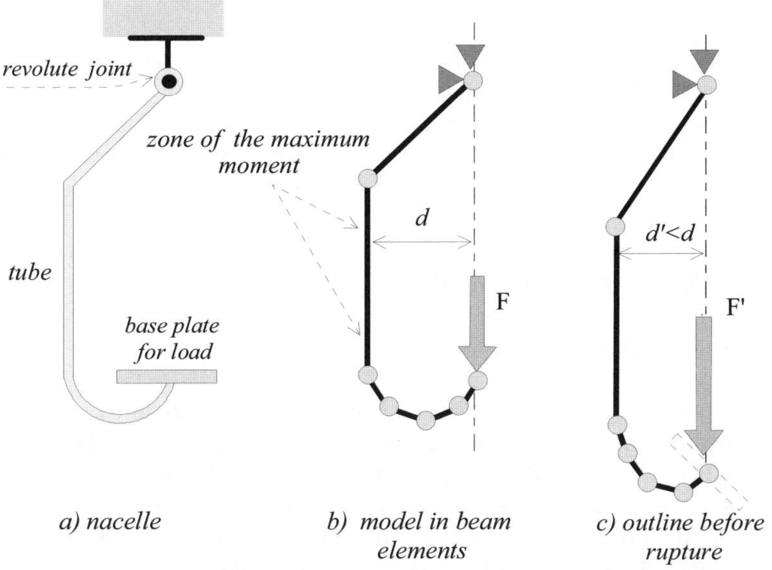

Figure 8.21. *Another example of nonlinear geometric behavior*

We can see in Figure 8.21b that the model presents a bending moment in the vertical tube which is proportional to the "moment arm" *d* whatever the calculated displacements under the load. On the other hand, the appearance of the real deformation of the structure at failure (Figure 8.21c) brings about a "moment arm" d'<d. For the same maximum rupture moment, the real force F' will be greater than force F calculated by means of the model. The analysis with the model will give a very pessimistic result as compared to reality.

❑ In the preceding part we have been able to observe over and over again that the linking forces and internal forces in a structure depend on:

    – the structure geometry;

    – the materials used;

    – the loading;

    – the linkings at boundaries.

When these linking forces and internal forces also depend (greatly) on the *deformed configuration* of the structure we say that we are dealing with nonlinear geometric behavior.

❑ Another category of nonlinear effects concerns the compression forces which, on slender parts or thin structures, can bring about an unstable behavior. When these applied forces exceed certain limits, all or part of the free dof increase in an uncontrollable manner. We say that the phenomenon of buckling has begun. The study of this standard phenomenom is not addressed in this book.

8.2. Example 1: machine-tool shaft

8.2.1. *Simulation exercise*

In this example we propose a preliminary modeling work allowing the use, in a second stage, of finite element software. The structure is the machine shaft of (Figure 8.22). The assembly includes mechanical components such as bearings, toothed gears, etc. The creation of a finite element model necessitates a preliminary technical training on these components. In fact, considering the performances of the finite elements chosen, it will be necessary to restitute in the model a mechanical behavior conforming to that of the real component. Thereafter the use of the computational code will make it possible to check that the machine shaft resists the loading and that the strains, particularly the cross-section rotations under bending, remain less than the swivelling angles allowed for by the bearings.

The drawing of the whole unit in Figure 8.22 partially shows the intermediate shaft of a machine-tool gear box. Shaft (2) receives its rotation motion thanks to gear transmission: toothed wheel (4) (ϕ_{pitch}: 225), sliding through splines (ϕ_{pitch}: 65), meshing at point K^9 with a pinion gear (9) not shown. A second sliding gear (3) with two pinions, (for the largest, ϕ_{pitch}: 150) sliding through splines (ϕ_{pitch}: 55) allowing

---

9 See Figure 8.23.

transmission of rotation motion to another parallel shaft (not shown) through a toothed wheel (5), point of meshing J. All the gears are straight bevel gears.

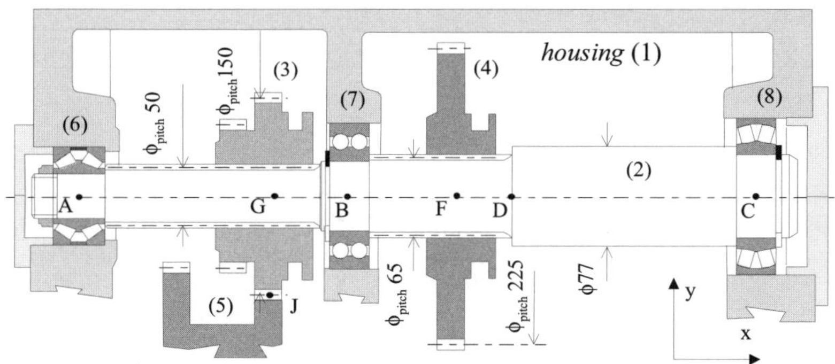

Figure 8.22. *Shaft of machine-tool*

The revolute joint of the shaft (2) in housing (1) is carried out by 3 bearings:

– a double-row of conical rollers bearing (6), (type TDI TIMKEN®) composed of a single inner race, two outer races and an external adjusting spacer. This bearing ensures the location of the axis (2) in the housing. Moreover, we note that such a bearing cannot swivel;

– a double-row ball bearing (7), which cannot swivel, but which can slide in the housing;

– a double-row barrel bearing (8), which is a self-aligning bearing.

8.2.2. *Data*

– global coordinate system: xyz;

– units: forces in N, moments in N.mm, lengths in mm;

– dimensions (in mm): AG = 140, GB = 60, BF = 75, FD = 60, DC = 165;

– material: steel S45C (R_e = 355 MPa), Young's modulus = 210,000 MPa;

– loading:

- the weight of the shaft itself (2) is not taken into account,

- the meshing forces on the toothed wheels are represented in Figure 8.23.

force and moment of (9)/(4) modeled in K	force and moment of (5)/(3) modeled in J
$\{\mathcal{F}_{9/4}\}_K = \left\{ \begin{array}{l} \overrightarrow{K_{9/4}} = 2100\vec{y} - 765\,\vec{z} \\ \hline \overrightarrow{\mathcal{M}_{K9/4}} = \vec{0} \end{array} \right\}$	$\{\mathcal{F}_{5/3}\}_J = \left\{ \begin{array}{l} \overrightarrow{J_{5/3}} = 1146\vec{y} - 3150\vec{z} \\ \hline \overrightarrow{\mathcal{M}_{J5/3}} = \vec{0} \end{array} \right\}$

$$[8.4]$$

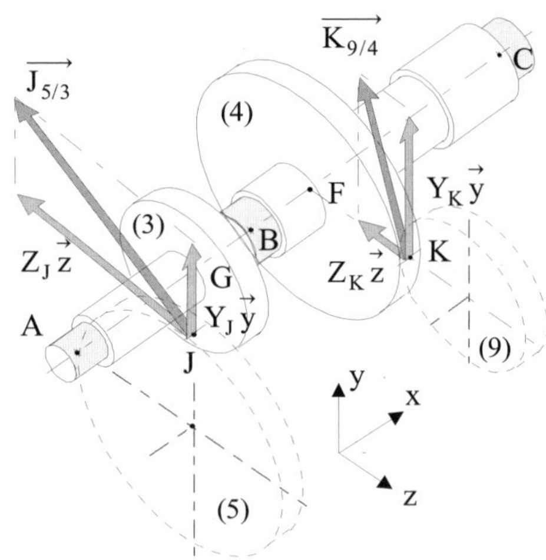

Figure 8.23. *Meshing forces on the wheels (3) and (4)*

8.2.3. *Successive steps of modeling*

As defined previously, shaft (2) constitutes the structure to be modeled.

8.2.3.1. *Definition of the loading*

The resultant forces and moments are deduced from [8.4] at points F and G on the centerline of shaft (2).

resultant force and moment (9)/(4) written at point F	resultant force and moment (5)/(3) written at point G
$$\{\mathcal{F}_{9/4}\}_F = \left\{ \frac{\overrightarrow{F_{9/4}} = \overrightarrow{K_{9/4}}}{\overrightarrow{\mathcal{M}_{F9/4}} = \overrightarrow{FK} \wedge \overrightarrow{F_{9/4}}} \right\}_F$$ with $\overrightarrow{FK} = 112.5\vec{z}$ i.e.: $$\{\mathcal{F}_{9/4}\}_F = \left\{ \frac{\overrightarrow{F_{9/4}} = Y_F\,\vec{y} + Z_F\,\vec{z}}{\overrightarrow{\mathcal{M}_{F9/4}} = L_F\,\vec{x}} \right\}_F$$ with: $Y_F = 2100N$; $Z_F = -765N$ $L_F = -236250Nmm$	$$\{\mathcal{F}_{5/3}\}_G = \left\{ \frac{\overrightarrow{G_{5/3}} = \overrightarrow{J_{5/3}}}{\overrightarrow{\mathcal{M}_{G5/3}} = \overrightarrow{GJ} \wedge \overrightarrow{J_{5/3}}} \right\}_G$$ with $\overrightarrow{GJ} = -75\vec{y}$ i.e.: $$\{\mathcal{F}_{5/3}\}_G = \left\{ \frac{\overrightarrow{G_{5/3}} = Y_G\,\vec{y} + Z_G\,\vec{z}}{\overrightarrow{\mathcal{M}_{G5/3}} = L_G\,\vec{x}} \right\}_G$$ with: $Y_G = 1146N$; $Z_G = -3150N$ $L_G = +236250Nmm$

$$[8.5]$$

The corresponding resultant forces and moments are represented in Figure 8.24.

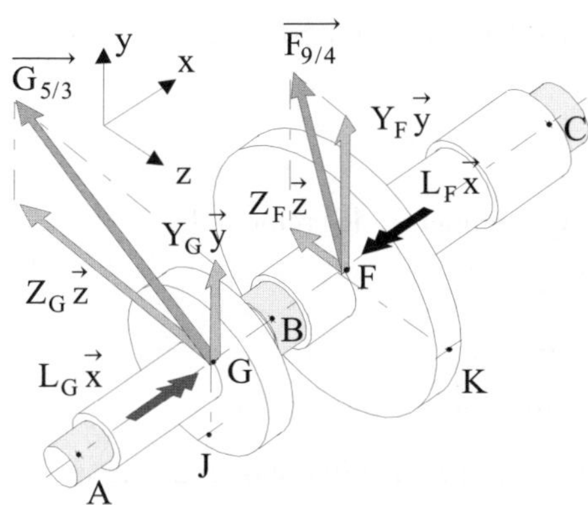

Figure 8.24. *Mechanical actions on the centerline of the shaft*

8.2.3.2. *Linkings of the structure at boundaries*

Shaft (2) with its gears and bearings is linked to its surroundings (housing) through outer races of the three bearings. If we suppose an infinite value for internal contact rigidity of these bearings, the joint corresponding to each bearing is kinematically perfect and conforms to standardized pictorial diagrams of Figure 8.25. These joints are going to influence the discretization and reduce to zero a certain number of dof (dof blocked).

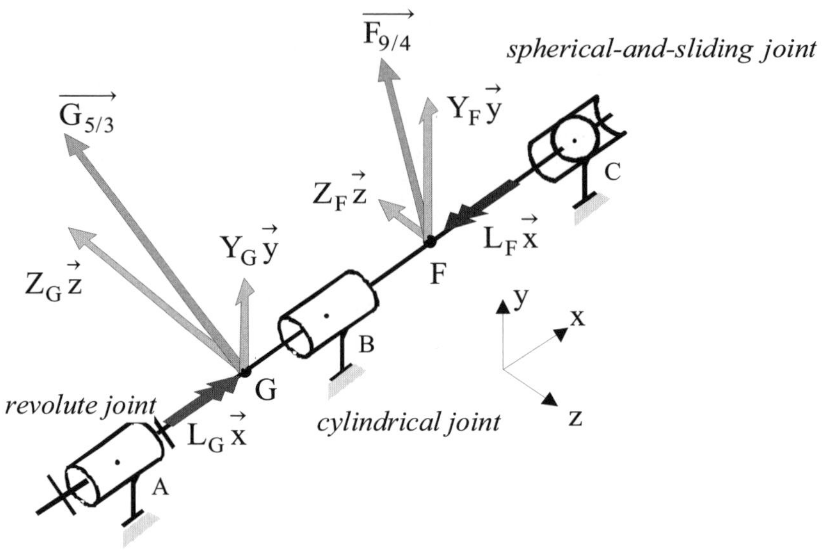

Figure 8.25. *Linkings of shaft (2) with the housing*

NOTES

❐ We can note the high number of dof removed:

– revolute joint: 5 dof;

– cylindrical joint: 4 dof;

– spherical-and-sliding joint: 2 dof;

that means a total of 11 dof removed as shown in Figure 8.27. However, shaft (2) can still have a rigid-body movement (rotation around axis $\vec{x}$). Its blocking is incomplete ("hypostatic positioning"[10]) and this possibility of motion must be removed without disturbing the mechanical behavior, according to the method in section 8.1.4.2. In Figure 8.27, this rotation can be prevented by canceling θ_{x2}, as is explained further.

❐ Internal linkings: the minimal kinematic pictorial diagram of the shaft equipped with sliding-through-splines gears (Figure 8.26) shows the prismatic slider joints between the shaft and the two units (3) and (4).

---

10 See Figure 2.39 and section 8.1.4.2.

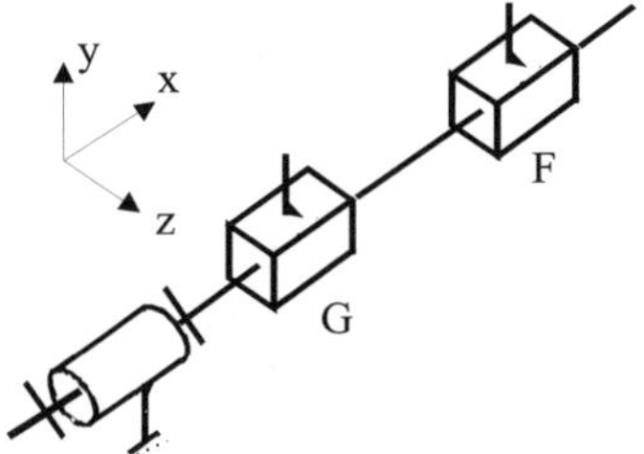

Figure 8.26. *Minimal kinematic pictorial diagram of the shaft*

In the geometric configuration of Figure 8.22, the toothed wheels occupy a well defined position. The prismatic slider joints can be replaced by clamped joints F and G for the loading application.

❑ In rotating shaft applications implying dynamic behavior problems, it is advisable to consider a radial stiffness (finite) due to elasticity of internal contacts in the bearings. It is possible for such cases to be put it into practice through spring elements placed according to the dof directions which would be normally blocked in Figures 8.25 and 8.27.

❑ The double-row of conical roller bearings and the double-row of ball bearings can be modeled more precisely when we know their swivelling rigidity from the technical manufacturer's data. This can be done through spiral springs already defined and used earlier, for which the operator gives the corresponding stiffnesses. We can then check if the calculated deflected shape of the shaft is compatible with the maximum swivel allowed for these bearings.

8.2.3.3. *Finite element discretization of shaft (2)*

Figure 8.25 leads to the choice of beam elements[11]. We can note that shaft (2) of Figure 8.22 is characterized by three distinct circular cross-sections (neglected splines). In the same figure, observing the diagram leads to a minimum of five elements and six nodes i.e.: A, G, B, F, D, C. The corresponding model is represented in 8.27 with earlier nodes numbered. We made the canceled degrees of freedom at linking joints apparent. Thus at node 1 (pivot joint A) we blocked:

– 3 translational degrees of freedom u_1, v_1, w_1 (symbolized here by T_x, T_y, T_z);

– 2 rotational degrees of freedom θ_{y1}, θ_{z1} (symbolized here by R_y, R_z);

– etc.

---

11 See section 5.2.

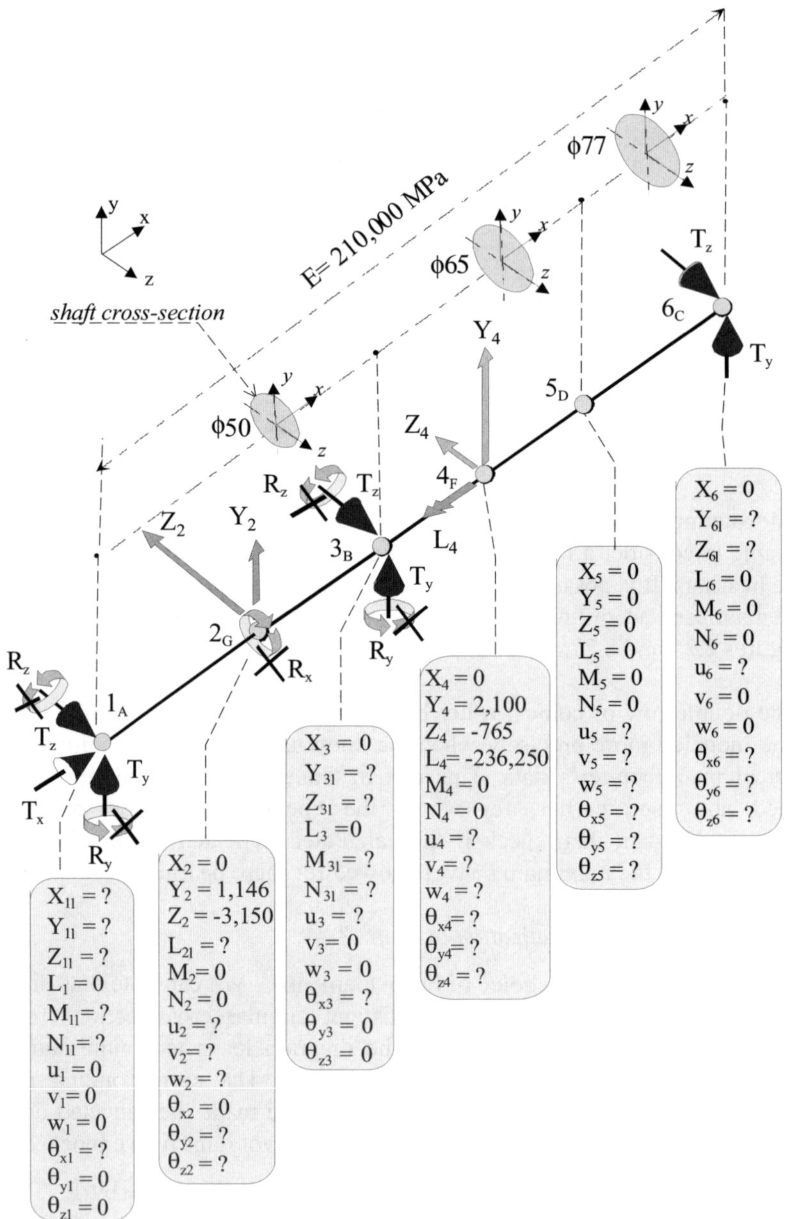

Figure 8.27. *The beam element model*

NOTES

❏ It will be noted that the operating mode of the rotating shaft (uniform rotation) corresponds to rigid-body motion. As we saw in section 8.1.4.2, we must remove the latter by creating on this structure (already balanced under the meshing forces and the linking forces) a complete "isostatic" positioning. This is done by blocking the rotation θ_x of an ordinary node. Here we have removed the rigid-body rotation around axis $\vec{x}$ by blocking θ_{x2} (denoted R_x) at node 2 (point G). We can then continue to exert on this point a moment $L_2 = L_G = 236,250$ Nmm, or indifferently remove it (which has been done here) since this will not modify the linking moment $L_{2\ell}$ [12]. We could also have chosen to cancel rotation θ_{x4} of node 4 (point F). The design engineer can thus obtain the deflected configuration that he finds the most convenient at the time of interpretation of computational results.

❏ The local coordinate systems of beam elements are all parallel to the global coordinate system (xyz).

❏ With six dof at each of six nodes, the behavior of the complete structure after assembly takes the form:

$$\{F\}_{str} = [K]_{str} \bullet \{d\}_{str}$$
$$\phantom{\{F\}_{str}} {}_{36\times1} \quad {}_{36\times36} \quad {}_{36\times1}$$

The structure still not properly linked to its surroundings, we know that the stiffness matrix is singular since shaft (2) can have six rigid-body motions (3 translations parallel to axes $\vec{x}$, $\vec{y}$, $\vec{z}$ and three rotations around these same axes). After canceling 12 linking dof as described in Figure 8.27 the system is thus "partitioned":

$$\left\{ \begin{array}{c} F_\ell \\ {}_{12\times1} \\ F^* \\ {}_{24\times1} \end{array} \right\} = \left[\begin{array}{c|c} K_\ell & K_\ell^* \\ {}_{12\times12} & {}_{12\times24} \\ \hline K_\ell^{*T} & K^* \\ {}_{24\times12} & {}_{24\times24} \end{array} \right] \bullet \left\{ \begin{array}{c} d_\ell = 0 \\ {}_{12\times1} \\ d^* \\ {}_{24\times1} \end{array} \right\} \qquad [8.6]$$

Following the known method [13], we should solve the system:

$$\left\{ \begin{array}{c} F^* \\ {}_{24\times1} \end{array} \right\} = \left[\begin{array}{c} K^* \\ {}_{24\times24} \end{array} \right] \bullet \left\{ \begin{array}{c} d^* \\ {}_{24\times1} \end{array} \right\}$$

which provides after inversion of sub-matrix $\left[K^* \right]$ the values of the free dof $\{d^*\}$.

Thus, by recovering the lines and columns removed from system [8.6], we come to:

---

12 See section 2.6.4.
13 See section 3.4.2.

$$\left\{ \begin{matrix} F_\ell \\ {\scriptstyle 12\times1} \end{matrix} \right\} = \left[\begin{matrix} K_\ell^* \\ {\scriptstyle 12\times24} \end{matrix} \right] \bullet \left\{ \begin{matrix} d^* \\ {\scriptstyle 24\times1} \end{matrix} \right\}$$

The known values of $\{d^*\}$ enable us to calculate the 12 components of the linking forces $\{F_\ell\}$. We can verify once again in this example a particularity already mentioned of the finite element method[14]: it does not depend on the degree of hyperstatism of the structure[15] under study.

❏ The free dof $\{d^*\}$ being known, we obtain as a result the normal stresses – due here to bending – in the five elements making up the shaft[16].

8.3. Example 2: thin-walled structures

The suggested example of Figure 8.28 is a rotating support arm (the revolute joint will not be considered). This fabricated welded structure will lead us to several finite element models which we will compare qualitatively.

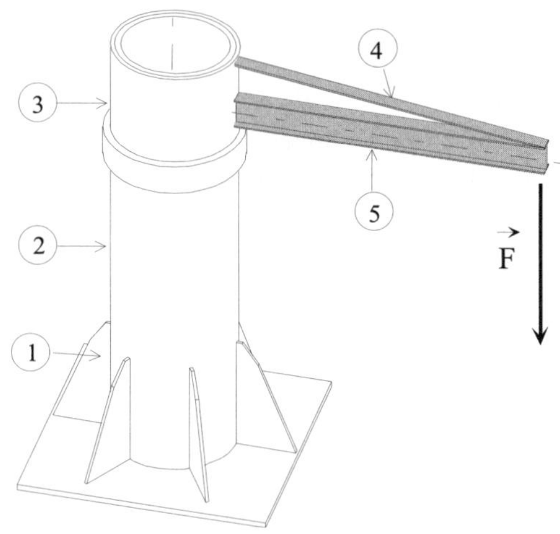

Figure 8.28. *Rotating support arm*

---

14 See section 4.3.1, *NOTES*.

15 Here shaft (2) has hyperstatic linking conditions (after canceling rotation θ_{z2} as specified above). It is in fact subject to 12 linking forces whereas we can write only 6 equilibrium equations in the coordinate system (xyz). The hyperstatic degree is then: $12 - 6 = 6$. We also say that this structure is six times "externally hyperstatic".

16 See [5.8].

Starting from a design with profiles and tubes, we obtain here a thin-walled structure. In view of finite element modeling, we essentially use beam and/or plate elements.

8.3.1. *Model based on beam elements*

8.3.1.1. *Methodology*

From Figure 8.28 we trace over the lengths and positions of the centerlines of all the profiles as shown in Figure 8.29.

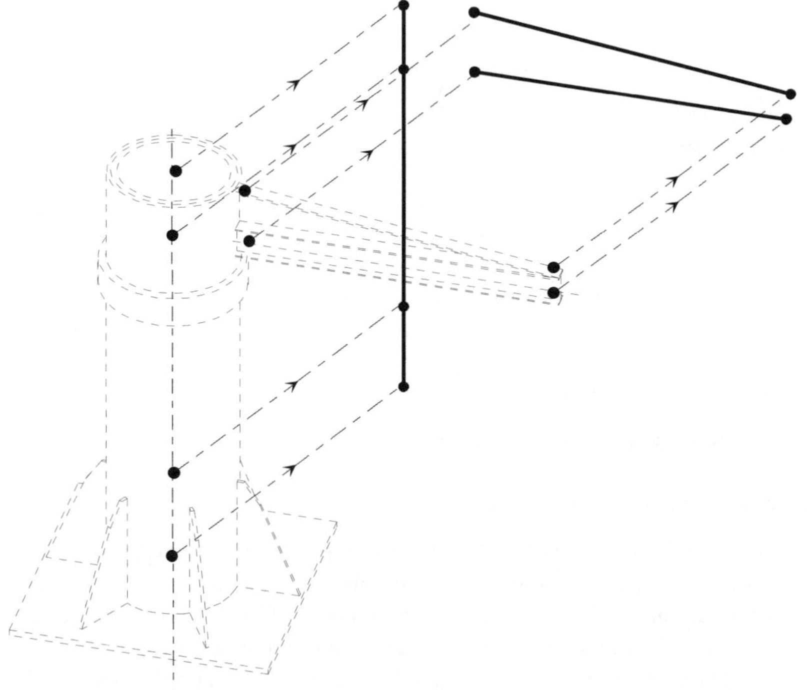

Figure 8.29. *Definition of the profile centerlines of beams elements*

We immediately encounter the characteristic difficulty of modeling thin-walled structures as already indicated in section 8.1.4.3, i.e. the problem of profile centerlines junction as shown in Figure 8.30.

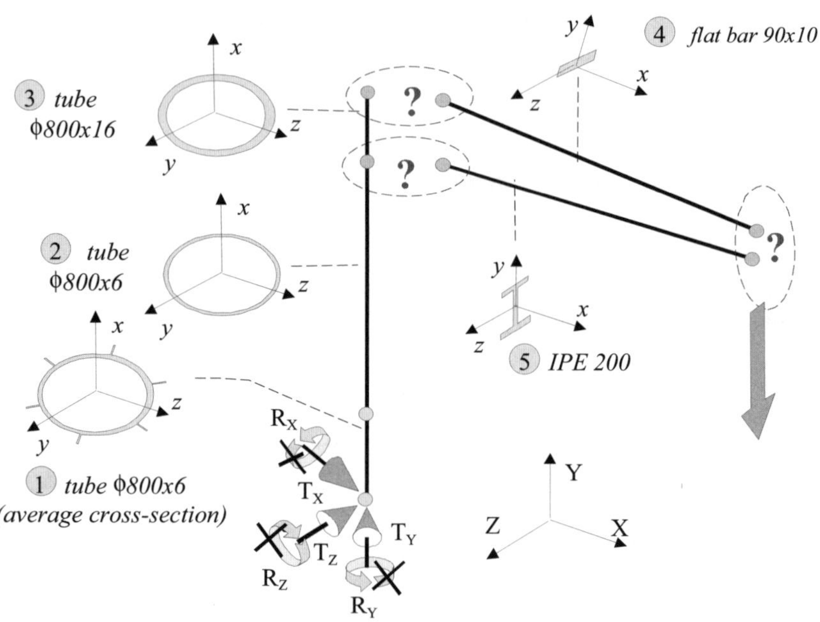

Figure 8.30. *"Commonplace" beam element model*

8.3.1.2. *Junctions of centerlines*

The problem can be handled in two ways:

○ First solution (see Figure 8.31)

Whilst preserving the earlier tracing for the centerline lengths and positions, junctions of centerlines are created by so-called "kinematic" or "rigid" elements. They enable an integral transmission of forces and moments without becoming deformed. If the kinematic element is not available in the software, we can replace it by a very stiff beam element (for example by according a Young's modulus 10 times higher than that of the material of the elements thus joined).

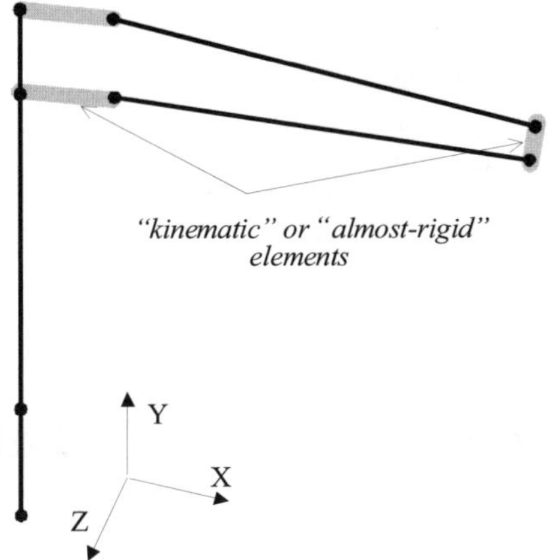

Figure 8.31. *Model drawn up from beam and "kinematic" elements*

○ Second solution (see Figure 8.32)

From the drawing of the structure, we elaborate a new model by preserving the centerlines of the most important elements and by "manipulating" the centerline lengths and positions of the remaining profiles. When the concerned profiles are sufficiently slender, the geometric modifications are relatively minor. Thus, this constitutes a general practice.

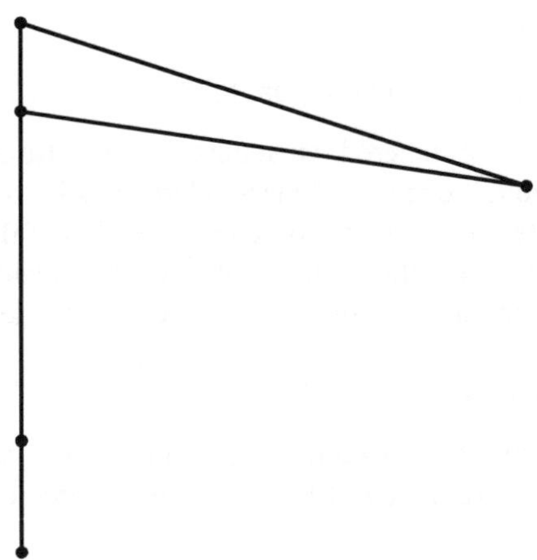

Figure 8.32. *Model drawn up from beam elements only*

NOTES

❏ The earlier model and the real structure are compared by a view of the top in Figure 8.33. In Figure 8.33b, the centerline junction of the IPE section is on the centerline of the cylindrical tube. The difference in size of cross-sections for the elements thus connected will notably modify the mechanical behavior of the modeled support arm. The model becomes more flexible than the real support arm.

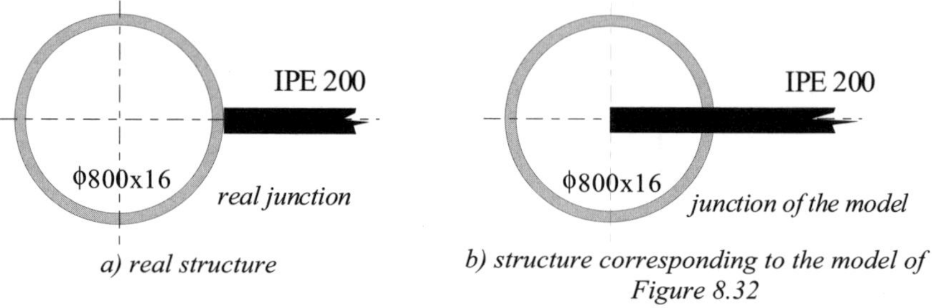

a) real structure

b) structure corresponding to the model of Figure 8.32

Figure 8.33. *Very different cross-section sizes for the tube and the IPE*

❏ Density of meshing for a beam element model

The number of elements is conditioned by the dimensions of cross-sections, the boundary and loading conditions, the materials and the geometry of the structure. It can be reduced here by taking advantage of the fact that there is no distributed load [17].

❏ Progressive sections in zone (1), Figure 8.28

In this zone we note tapered stiffeners issuing from the base-plate. The equivalent cross-section, "tube + stiffeners" changes with height. Certain software automatically calculate the cross-section characteristics of this type of profile. The operator can also break up the profile into several beam elements, each with characteristics corresponding to cross-dimensions at mean length.

8.3.1.3. *Relieving of linkings*

The most complete linking between two beam elements is a clamped joint at the junction node. Thus, the 6 *relative* dof between the two elements are removed at this

---

17 See Figure 4.39. When the load is distributed on the profiles, the meshing density influences the results. For example, the results obtained with one single element of length ℓ affected by a distributed load q (N/mm) is different from the one with two elements of half length affected by the same distributed load q.

place. It is possible to partially or totally "relieve" this linking by allowing one or several relative dof, i.e. by installing, at the exact place of the considered node, a standardized classic joint rather than the clamped joint[18]. Such an operation can bring the model closer to reality. Furthermore, using two different models built on this principle, we can also approach the technological reality of a linking which is only "semi-clamped".

We propose to clarify this point for the linkings of flat bar (4) (Figure 8.28). This bar is welded at its extremities, which thus act as clamped joints. The modeling, mentioned earlier, is taken up again in Figure 8.34.

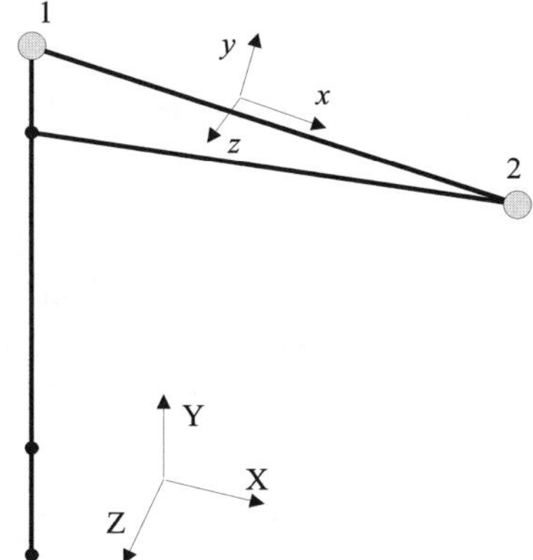

Figure 8.34. *Model with internal clamped joints*

The nodal forces which balance the beam element modeling flat bar (4) in its *local coordinate system* (*xyz*) are illustrated in Figure 8.35 (bending in plane (*xy*)).

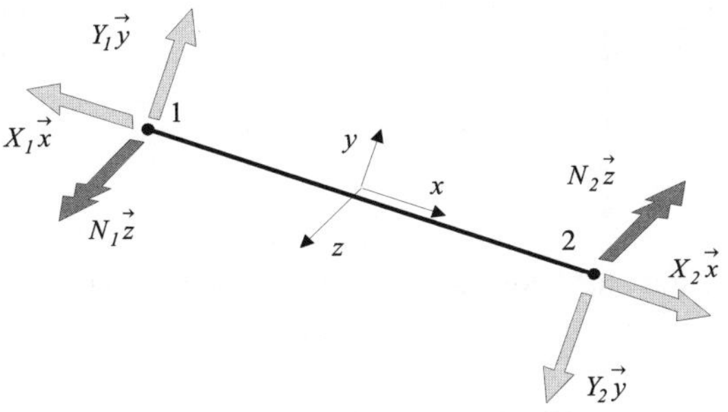

Figure 8.35. *Balancing of beam element (4)*

---

18 See Appendices.

We can note in Figure 8.28 that flat bar (4) plays the structural role of a tie bar, i.e., it is stretched. Let us now suppose that this plate is not welded at its extremities but equipped with a hinge (revolute joint) as represented in Figure 8.36.

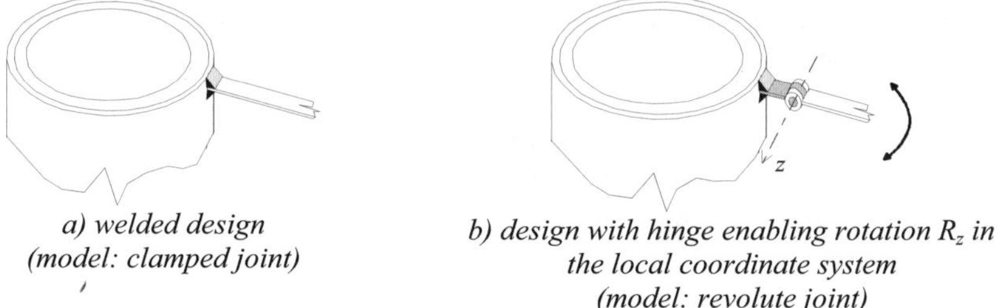

a) welded design
(model: clamped joint)

b) design with hinge enabling rotation R_z in
the local coordinate system
(model: revolute joint)

Figure 8.36. *Relieving of an internal linking*

Thus it is necessary to relieve the rotation R_z on nodes 1 and 2 in the local coordinate system of the beam element. A revolute joint is thus created at each of these junctions. The model of the support arm with the flat bar thus equipped is represented in Figure 8.37. By relieving the two nodes of the beam element, the stiffness bending around local axis $\vec{z}$ disappears.

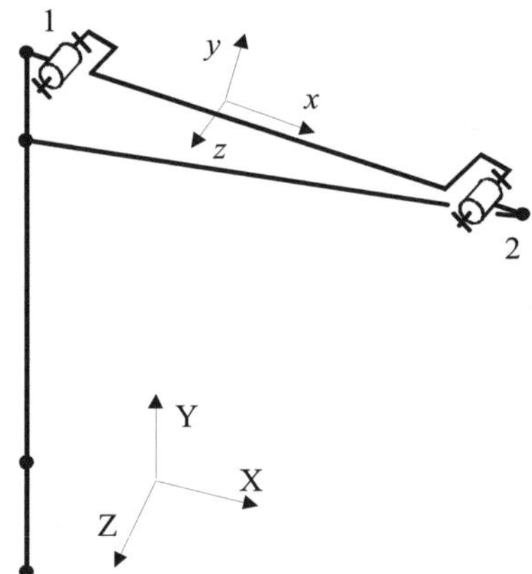

Figure 8.37. *Relieved model: internal revolute joints*

The nodal forces which ensure balancing of beam element (4) are represented in Figure 8.38, in the local coordinate system. Strains characterizing bending have disappeared, only the traction force remains.

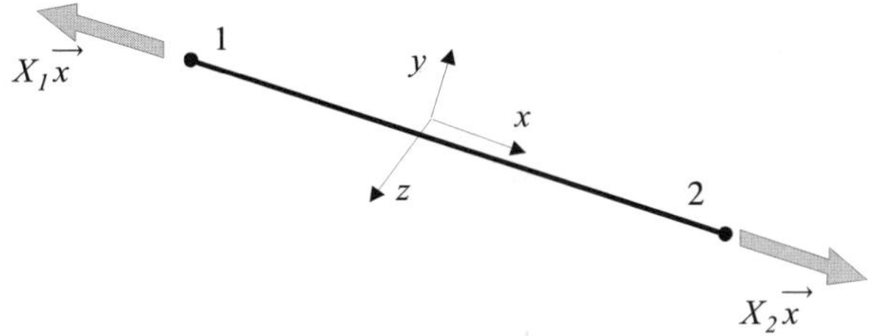

Figure 8.38. *Balancing of relieved beam element (4)*

NOTES

❐ The flat bar will still work in the same way (tension) if supplementary relieving is achieved at preceding nodes:

– for example, let us relieve at each node the three rotations R_x, R_y, R_z around local axes. We obtain at each extremity a spherical joint. The beam element has become a truss element[19]. It now presents a possible rigid-body rotational motion around local axis $\vec{x}$, as represented in Figure 8.39. Therefore the general stiffness matrix of the structure is singular and cannot be inverted[20];

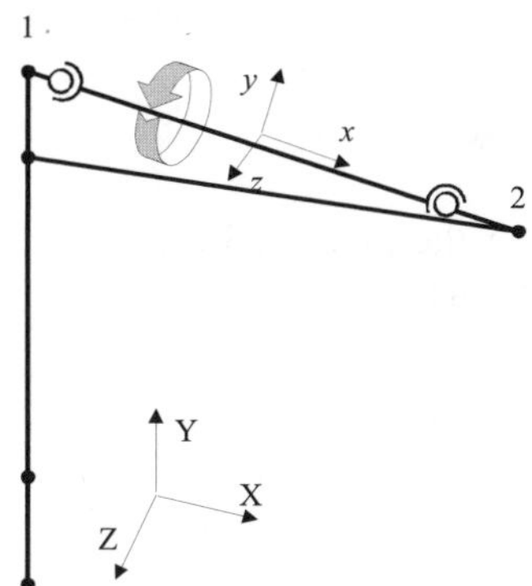

Figure 8.39. *Relieving model: internal spherical joint*

– for each node, let us relieve the two rotations R_y, R_z around the local axes. In Figure 8.40, we obtain at each extremity a slotted spherical joint[21]. This is equivalent to a Cardan joint. In this case there is no rigid-body motion.

---

19 See section 3.2.1.
20 See [8.2].
21 See Appendices.

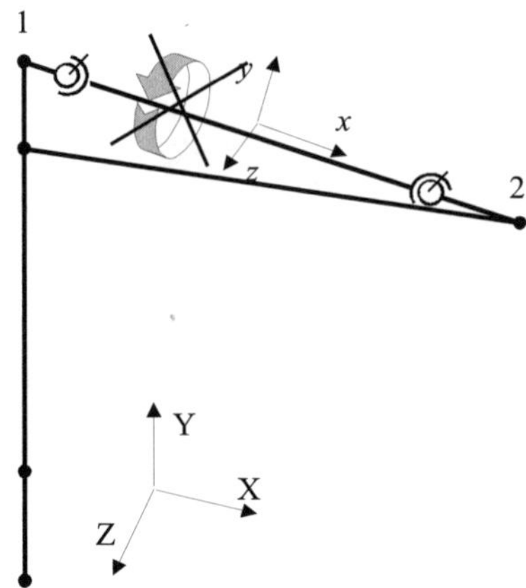

Figure 8.40. *Relieved model: internal slotted spherical joints*

❏ According to what precedes, it appears possible to model the main standardized types of structural joints: external joints between elements and surroundings as well as internal joints between elements. This is essentially valid for beams because of the presence of the six nodal dof.

8.3.2. *Model in plate elements*

From the *mid-surfaces* of the tubes and cross-sections, we can create the plate elements model as in Figure 8.41.

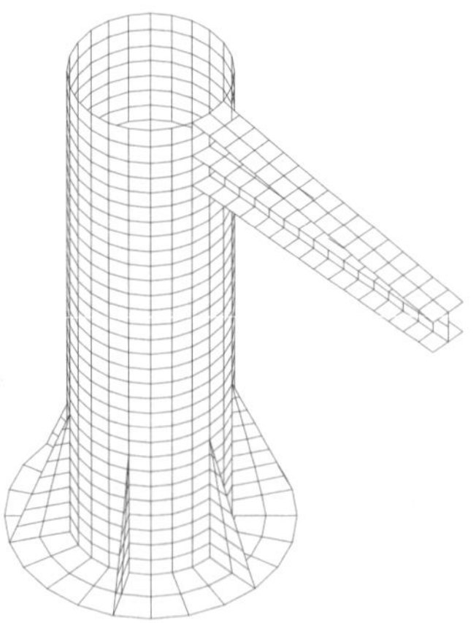

Figure 8.41. *Model in plate elements*

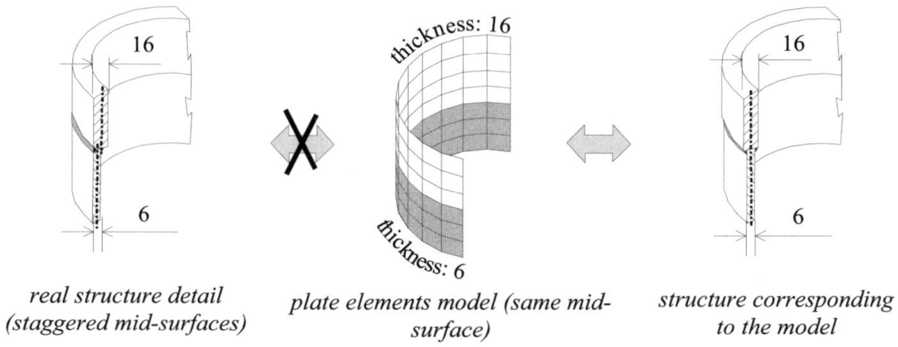

real structure detail
(staggered mid-surfaces)

plate elements model (same mid-surface)

structure corresponding to the model

Figure 8.42. *Processing of mid-surfaces*

Here Figure 8.42 shows that the staggering of mid-surfaces has been ignored.

8.3.3. *Model in beam and plate elements*

In Figure 8.43, the operator has chosen to model the slender parts of the structure in beam elements. These elements are associated with plate elements. In section 8.1.3.2 we have emphasized the incomplete compatibility between the nodal dof of the beam elements (6 dof) and those of the plate elements (5 dof). Some software overlooks this difficulty. If the software does not automatically solve this problem, the operator can use the modeling artifice shown in Figure 8.8, and shown again in Figure 8.43 given below.

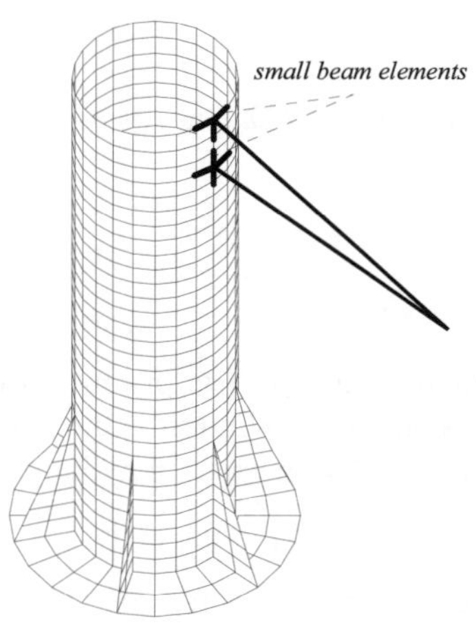

Figure 8.43. *Mixed model with beams and plate elements*

It is to be noted that such an association of elements enables us to simplify and lighten the earlier model (Figure 8.41), whilst ensuring a good accuracy.

8.4. Example 3: modeling of a massive structure

We propose to examine the case of a self-balanced massive structure under heavy loading. We will also illustrate in this example how to take into account the geometric and boundary symmetries.

8.4.1. *Problem*

Figure 8.44 represents the pictorial view of a hydraulic portable punching machine used on site for stamping out and crimping operations.

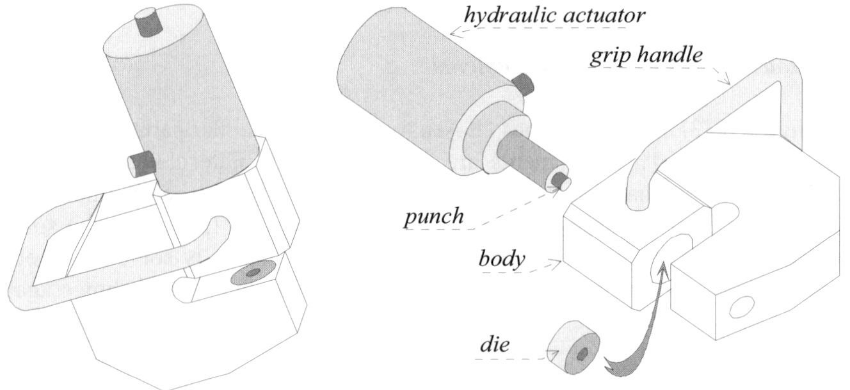

Figure 8.44. *Hydraulic punching machine*

It is made up of:

– a goose-neck shaped thick body;

– a double-acting hydraulic actuator. The piston-rod is equipped at its end with an inter-changeable punch stem. The cylinder of the actuator is fitted and screwed into the body. This actuator is fed by an external hydraulic group and flexible pipes;

– a die fitted in a cylindrical housing in the body;

– a grip handle welded to the body for the operator to hold the punching machine.

8.4.2. *Steps of modeling*

Preparatory work is carried out in view of a static linear behavior study of the body when the actuator exerts a significant force on the die.

8.4.2.1. *Structural parts*

Only the body will be considered. We ignore the weight and the gripping force exerted by the operator on the grip handle. The latter will thus not be modeled.

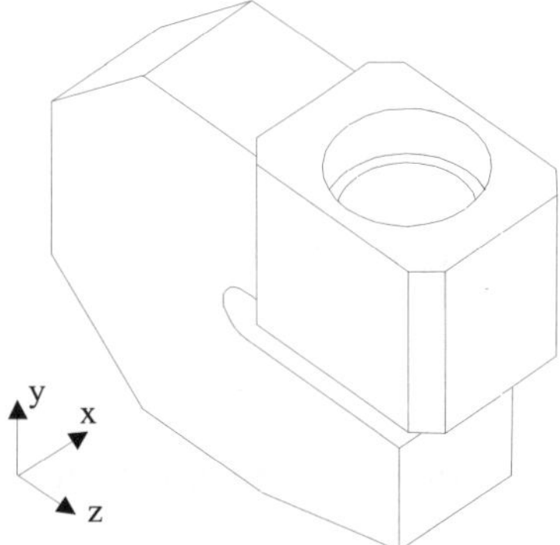

Figure 8.45. *The body as structural part*

8.4.2.2. *Choosing the type of finite element*

In Figure 8.45, we observe that the body is a bulky component, stretching in the three space directions. It will therefore be modeled in 3D-solid elements[22].

8.4.2.3. *Forces applied on the body*

They are reduced to:

– the actions of the threaded extension of the cylinder barrel on the internal thread in the body;

– the actions of the die on its surface bearing in the body.

The structure being balanced, the resultant forces for these actions will reduce to two directly opposite forces along the axis of the actuator as represented in Figure 8.46.

---

22 See Figures 5.31 and 8.3.

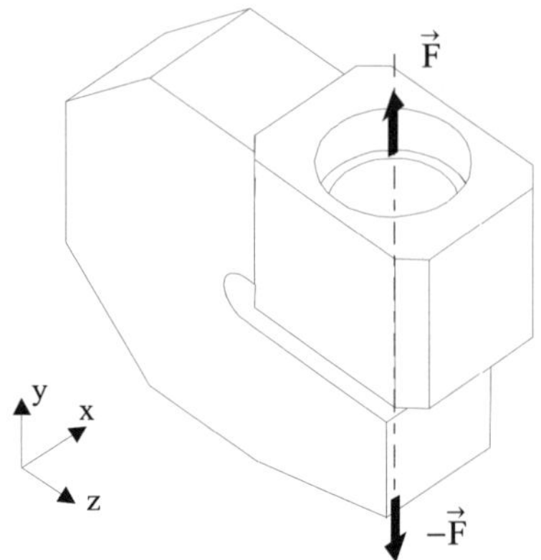

Figure 8.46. *Balancing of the body*

8.4.2.4. *Boundary conditions*

Given below are the two ways of envisaging the application of boundary forces and displacements on the model.

○ **1st case**: balanced (or self-balanced) "floating" structure

⇨ loading: we have to apply uniformly distributed loads (see Figure 8.47) upon the bearing surfaces of the "actuator-body" interface and the "die-body" interface. As mentioned earlier, these loads should correspond to the two opposite forces of Figure 8.46.

⇨ external linkings: here the structure is not properly linked, it is floating in space as discussed in section 8.1.4.2.

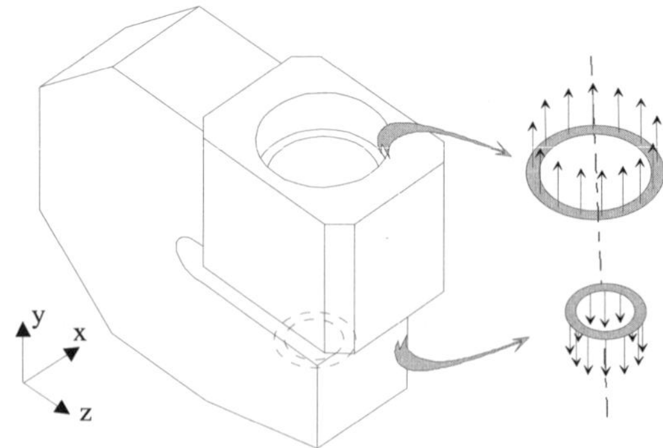

Figure 8.47. *Floating structure*

We have indicated[23] that some software can set a statically determinate position (6 dof blocked) of the model in the global coordinate system. We have seen that such an operation did not create parasite stresses on the structure. Nevertheless, it is better that the operator himself sets the complete isostatic blocking that it estimates has adapted best to the further development of results. For example, searching for the deflected shape of the goose neck implies carefully choosing a node where the only dof T_y will be blocked, as represented in Figure 8.48.

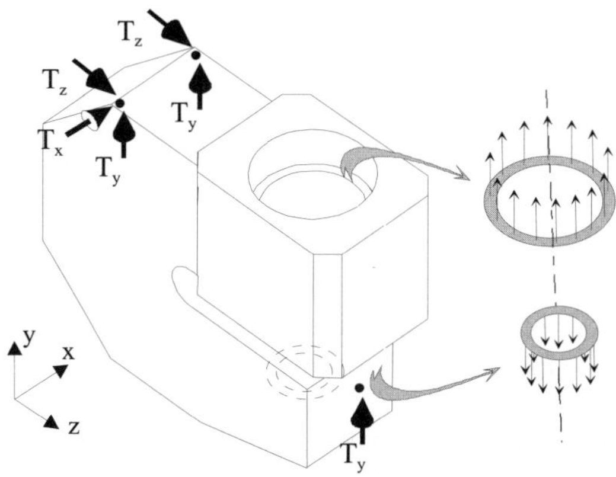

Figure 8.48. *Complete isostatic blocking set by the operator*

○ **2nd case**: properly linked structure

⇨ loading: here we can conserve the force exerted on the body by the actuator;

⇨ external linkings: thus, we will cancel the dof on the "die-body" bearing zone as shown in Figure 8.49[24].

---

23 See section 8.1.4.2.

24 In Figure 8.49 the canceling of translations Tx or Tz of *"one other node"* must be done with the precautions mentioned in Figure 8.15. Canceling the translation Ty on all nodes of the "die-body" contact zone means assuming that in reality the small vertical displacements of these nodes are identical. This is not quite true. Reality is thus forced, and we locally "stiffen" the structure. However, there we touch upon the limit in the possibilities of the operator. In fact, in the earlier case regarding an isostatic positioning, we assumed that the force exerted by the die on the body was uniformly distributed, which was not true either.

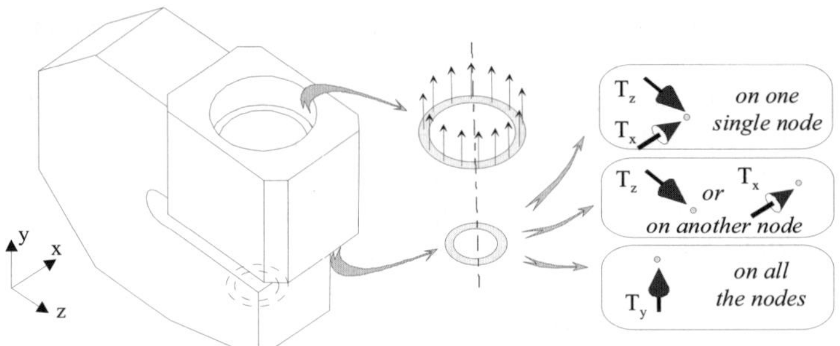

Figure 8.49. *Properly linked structure*

8.4.2.5. *Taking symmetries into account*

In order to respect boundary symmetries and geometric symmetries, the mechanical behavior of a partial model reduced by means of symmetry considerations must be identical to that of the complete model.

Here the median plan of the body (Figure 8.50), parallel to plane (yz), is a symmetry plane for geometry and for the force and displacement boundary conditions.

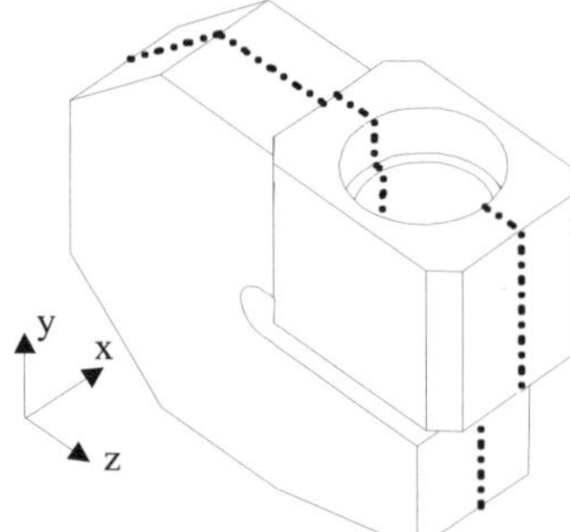

Figure 8.50. *Mid-plan of the body*

We can reduce the model to half of the body, divided as shown in Figure 8.50, by giving it modified boundary conditions in the following way:

⇨ loading: in the "actuator-body" contact zone we shall apply a uniformly distributed surface load identical to the earlier one (Figure 8.51);

⇨external linkings: external joints: we must force the mid-plane of the half-structure to remain plane and parallel to plane (yz). This is represented in Figure

8.51: all the translation dof T_x must be blocked. The structure then conserves a possibility of two plane rigid-body motions (middle plane (yz)). Translation T_y and rotation around axis $\vec{x}$ will be removed by canceling all the translation dof T_y of nodes on the "die-body" bearing zone.

The remaining motion of translation T_z will be eliminated by blocking any node following this direction.

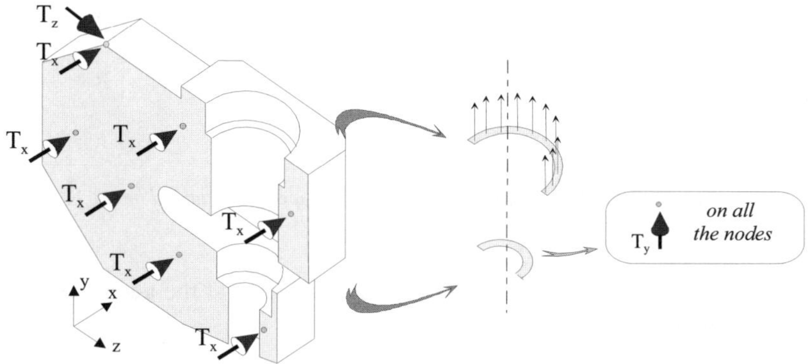

Figure 8.51. *Boundary conditions for the half body model*

8.4.2.6. *Other aspects of the modeling*

O Material properties

All the solid elements are in the same material (Young's modulus and Poisson's ratio).

O Mesh generation

The mesh generation of the body model in solid elements, done for example through integrated CAD-Finite element software, does not pose a problem. Given the weak radius of the curve (geometric singularity) and the significant effort values, the zone in the hollow of the goose neck will show large stress concentration. The operator will have to make the meshing more dense in this zone.

8.4.3. *Comments on the validity of the model*

The general shape is that of a deeply curved beam. The extension of beam theory to high curvatures provides[25] the stresses in the cross-section represented in Figure 8.52, with the appearance of Figure 8.53.

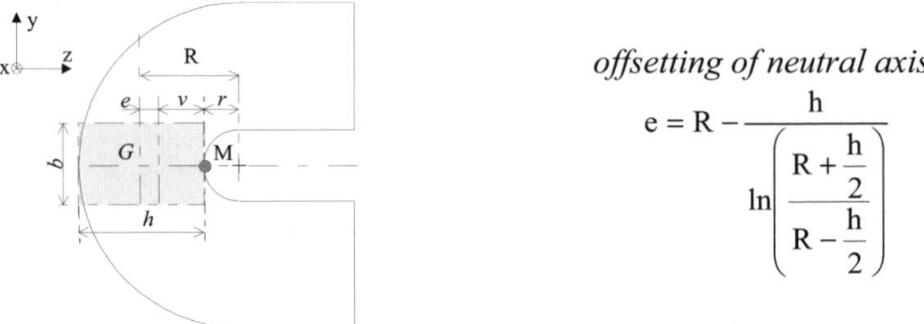

$$e = R - \frac{h}{\ln\left(\dfrac{R + \dfrac{h}{2}}{R - \dfrac{h}{2}}\right)}$$

Figure 8.52. *Off-setting of neutral axis in a deeply curved beam*

Mf_x is the bending moment and $\mathcal{N}$ is the normal force at the geometric center G of the cross-section.

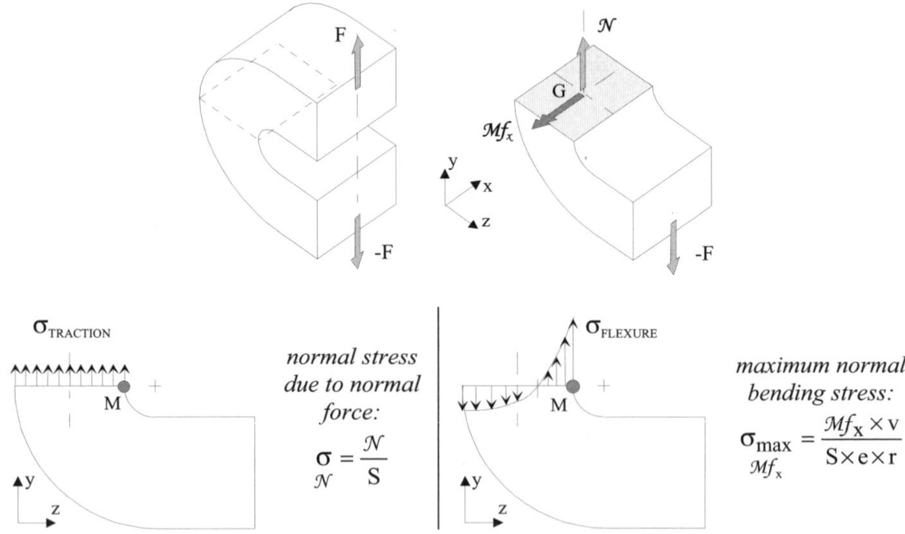

$$\sigma_{\mathcal{N}} = \frac{\mathcal{N}}{S}$$

$$\sigma_{max} = \frac{Mf_x \times v}{S \times e \times r}$$

Figure 8.53. *Normal stresses on the cross-section passing through M (deeply curved beam)*

---

25 See bibliography: J. Courbon.

We can thus estimate the maximum normal stress at point M: $\sigma_{max} = \sigma_{max} + \sigma$.
$$\phantom{\sigma_{max} = \sigma}{}_{Mf_x}{}_{N}$$

In Figure 8.54 this stress is compared with the calculated distribution obtained with the model. Thus, we can verify the order of magnitude of values provided by the model. In the model we observe that near the free lateral faces the stress reduces whereas at the center it is clearly higher than the "theoretical" value.

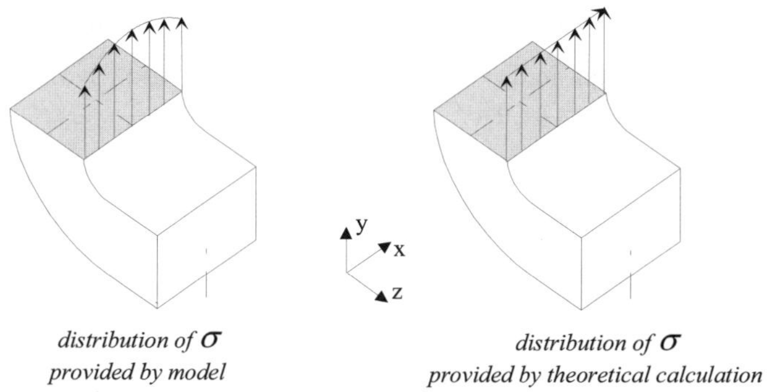

distribution of σ
provided by model

distribution of σ
provided by theoretical calculation

Figure 8.54. *Qualitative comparison with the model results*

8.5. Summary of the successive modeling steps

8.5.1. *Preliminary analysis*

Let us summarize the successive practical stages which lead to the construction of a model.

○ Definition of the study objectives

We first have to check that the specifications and requirements of the contractor are quite clear and compatible with the use that can be made with a finite element model. Will a linear static analysis be sufficient? With what kinds of elements? Should we have to densify the meshing in certain zones? Does the use of the structure described in the specifications also extend to nonlinear behavior cases (plastification, stability, etc.), or dynamic overloads (shocks), etc.? Thus, many aspects make the choice of modeling highly dependent on the specifications.

○ Preliminary study from the engineering drawings:

⇨ analysis of the structure drawings and its surroundings;

⇨ choosing the type of analysis (static, dynamic[26], etc.);

⇨ study of total or only local mechanical behavior[27];

⇨ types of structure parts: thin, slender, bulky;

⇨ choosing finite element types to be used as per the structure's characteristics;

⇨ definition of the geometry of the model (determination of centerlines or mid-surfaces) or directly taking the CAD model in view of modeling in solid elements;

⇨ existence of exploitable symmetries (a geometric symmetry with a dissymmetric loading cannot be used [28]);

⇨ boundary conditions:

  - nodal loads,

  - external linkings (note that the structure has to be properly linked),

  - internal linkings;

⇨ material properties;

⇨ geometric properties of beam cross-sections for beam elements;

⇨ geometric properties of other elements: thicknesses of plate and shell elements.

8.5.2. *Model verification and validation*

The creation of a satisfactory finite element model is not immediate. A critical attitude should be preserved before and after calculation.

8.5.2.1. *Before calculation*

The graphic and interactive software tools enable the checking of:

• the model geometry (testing of the nodes properties, various image displays of the model (projections and perspectives));

• the quality of the mesh generation (coloring of elements, exploded views to isolate elements):

---

26 There are other kinds of finite element analysis which have not been described in this book: permanent or transient thermal analysis, nonlinear analysis (nonlinear geometric behavior, see section 8.1.4.4, or elastoplastic behavior), analysis with modules dedicated to composite materials, etc.

27 See Figure 8.5.

28 Except if we have an adapted software module.

- continuity of the mesh generation in a given zone or between different linked meshed zones,

- distortion of plate and solid elements (they should not have differences between length, width and height that are too large);

+ boundary conditions (visible on displayed nodes through testing);

+ material properties and cross-sections of beam elements (shaded colored images when testing elements).

8.5.2.2. *After calculation*

O Geometric and inertial general checks:

– area, volume, mass or weight of the structure[29], location of the center of gravity of the model.

O Equilibrium checking:

– the structure's weight should be balanced by the vertical actions at external linkings.

O Displacements checking:

– amplification of the displayed deflected shape of the structure in order to detect the "badly attached" elements and comparison with an intuitive approach of the mechanical behavior;

– if possible, checking the magnitude of displacements with the help of simple expressions and formulae (remember that displacements are very weak compared to the dimensions of the structure)

O Checking the stress values:

– on external boundaries of the model, the normal stresses should be equal to the applied external pressure. They should be zero on free boundaries. In the opposite case the meshing density should be increased so that real stress gradients can be better reflected;

– we must be careful with stress concentrations (high stress gradients) when analyzing results in abrupt shape or cross-section variations;

– it is better to select colored isostresses in each element rather than general smoothed colored isostresses extending on the whole structure, which may hide stress peaks.

---

29 Contractors usually know the weight of the structure to be studied.

O Using automatic error estimation:

– local precision indicator: this indicator is based on the difference of stress values between elements adjacent to (or around) common nodes. It reveals the anomalies of gradients of calculated stresses. It can also be expressed in terms of energy. It gives information on the zones to be re-meshed more finely but in no case on the value of the calculation error;

– global error indicator: it evaluates potential energy of deformation. The most precise model to be retained is one which minimizes this energy. The density of potential energy is also used to observe the way in which the elements take part in the calculation of the potential energy of the whole structure.

8.5.3. *Corresponding use of the software*

Here, we point out the chronology of the stages previously described while repositioning them in the main parts of finite element software.

<table>
<tr><td align="center">preprocessor</td></tr>
<tr><td>

modeling of the structure:
   ⇨ geometric description of the model in CAD;
   ⇨ discretization of the structure in finite elements:
     - mesh generation
     - data input
   ⇨ verification of the geometric modeling and of the discretization in finite elements

</td></tr>
<tr><td align="center">processor</td></tr>
<tr><td>

computation
   according to the procedure already described[30]

</td></tr>
<tr><td align="center">postprocessor</td></tr>
<tr><td>

analysis of results:
   ⇨ verification of the finite elements model issuing from the postprocessing phase,
   ⇨ viewing: displacements, stresses (eigenfrequencies and eigenshapes, etc. for dynamic analysis),
   ⇨ validation of results with (as per case):
       - a theoretical analysis whenever possible
       - using formulae[31]
       - eventually a stripped finite element model
       - a similar model
       - results of tests on similar structures
   ⇨ resetting of the model if required

</td></tr>
</table>

---

30 See for example section 3.4.2.
31 See bibliography: Warren C. Young.

PART 3

Supplements

In this part we have included elements intended to supplement the practical aspects of dimensioning, but which can also be consulted independently. Chapter 9 deals with straight beams. We find there an elementary approach which corresponds to 1-year or 2-year undergraduate degrees, as well as a thorough study accessible for the students of a BEng degree or for students following a Master's degree. Chapter 10 provides some additional complements of elasticity thereby helping to give a better understanding of states of stresses. However, since it avoids discussing the evolution of the deformation according to the orientation of the corresponding facet, it can be accessed at the 3-year undergraduate degree level. Chapter 11 deals with structural joints. We have strived to highlight the principles of checking the main modes of structural joints while trying to keep to a simple approach (3-year undergraduate degree) here too. Finally, Chapter 12 summarizes some standard basic mathematical properties and rules on which this book rests.

Chapter 9

Behavior of Straight Beams

In Chapter 1 we mentioned for a straight beam the main results concerning stresses and deformations produced by the resultant forces and moments coming from cohesion forces[1]. These results, though useful for the following chapters, were given without justification. The theory of straight beams is well-known since it relates to a structural part which plays a basic role in innumerable applications. Moreover, it forms a matter for study by itself which can be analyzed independently of the rest of the book.

This chapter provides the supporting data for the results given earlier without proof in section 1.5.2. As far as possible we shall strive to maintain the same simplicity – for example, torsion and bending will be described in a very basic manner. However, we shall also analyze some special interesting phenomena regarding the behavior of such structural elements even though they are more complex. Finally, despite the traditional nature of the notions related to straight beams and the various descriptions that have been given in literature, certain developments in this chapter will be treated in an original way. This is especially the case for the calculus of shear stresses due to the shear resultants.

9.1. The "straight beam" model

9.1.1. *Definition*

In what follows, we shall define a straight beam as a solid made up of a homogenous isotropic[1] material:

– *of cylindrical shape:* the corresponding cylinder is generated by a set of parallel generators which follow one or more director curves with *a priori* any shape. To create straight beams, it is convenient to adopt plane director curves and generators that are orthogonal to the plane of the curve as represented in Figure 9.1. The director curves delimit a domain called the cross-section of the beam;

– *slender:* large longitudinal dimension ℓ of the beam as against the largest transversal dimension h in a cross-section. The beam slenderness is then represented by the ratio $\dfrac{\ell}{h}$. We observe in practice a beam behavior as described below for cases where $\dfrac{\ell}{h} > 3$. This limit value can also depend on the manner in which a beam is loaded.

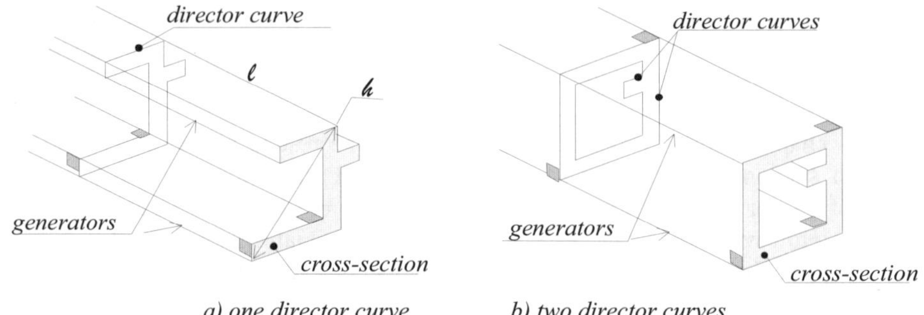

a) one director curve   b) two director curves

Figure 9.1. *Straight beams*

9.1.2. *Main or "principal" axis of a cross-section*

Necessity to define a particular coordinate system and an origin from an infinite number of systems of axes and origins possible in a plane cross-section is justified below. Figure 9.2a shows a beam cross-section S (for example, as in Figure 9.1a), having orthonormal axes $\vec{y}$ and $\vec{z}$ with origin G.

---

1 See section 1.5.2.

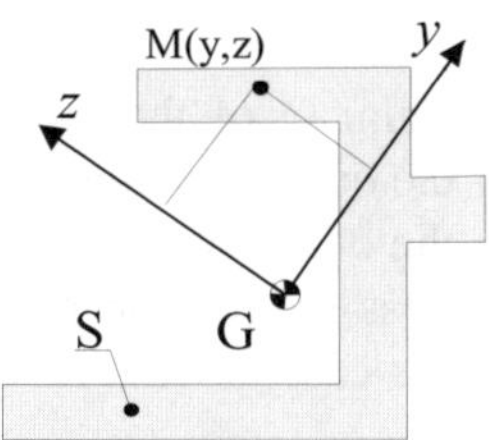

a) principal quadratic axes
(second moments of area maximum and minimum
and zero product moment of area)

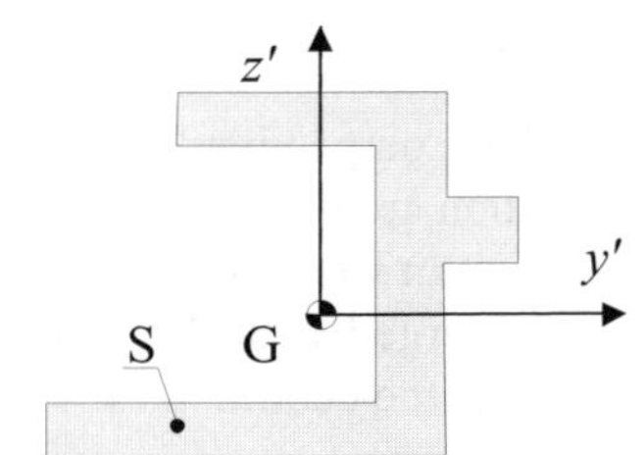

b) non-principal quadratic axes
(non-zero product moment of area)

Figure 9.2. *Principal and non-principal central axes (G is the geometric center)*

These are the "central" axes having the barycenter or the geometric center of the cross-section as origin. In what follows, we shall retain the specific axes with underlined properties of Figure 9.2a. If, in these axes, M(y, z) is a given point on the cross-section with y and z coordinates, relations characterizing these axes and the geometric center of the section are summarized in the table below in which dS represents an elementary surface in the cross-section whose total area is S.

Coordinate system linked to a current cross-section	
origin: geometric center G of the section	$\displaystyle\int_S y\,dS = \int_S z\,dS = 0$
orientation of the orthogonal axes $\vec{y}$ and $\vec{z}$:	$\displaystyle I_{Gyz} = \int_S yz\,dS = 0$ (I_{Gyz} : product moment of area)
the so-called "main" second moments of area, or main quadratic moments	$\displaystyle I_{G\vec{y}} = I_y = \int_S z^2\,dS$ $\displaystyle I_{G\vec{z}} = I_z = \int_S y^2\,dS$

$$[9.1]$$

NOTES

❑ Two perpendicular axes passing through any point O of the cross-section in question are described as main or principal axes if the quadratic moment assumes a maximum value for one and a minimum value for the other. Added to that, if point O is merged with the geometric center G of the cross-section the main axes also become the central axes. We will always consider this case in what follows.

❑ Specific additional software working from input of the geometry of the section based on CAD make it possible to locate the "main" axes and calculating the quadratic moments I_y and I_z. It is also possible to refer to formulae (see bibliography)[2].

❑ The locus of geometric centers G of all cross-sections form what is known as the "centerline" of the beam.

9.1.3. *Applied loadings*

The beam in Figure 9.3 is subjected to:

– forces exerted on the current zone: they can be distributed on a certain domain (surface, linear), or concentrated at particular points. They can include moments;

– forces exerted by other structures at the boundaries of the beam under study (structures S_1 and S_2 in Figure 9.3). The geometric centers of end sections of the beam are noted as G_1 and G_2. The mechanical actions exerted by S_1 and S_2 on the beam reduce to applied resultant force and moment[3] in G_1 and G_2.

---

2 These results should be known for the following standard sections:

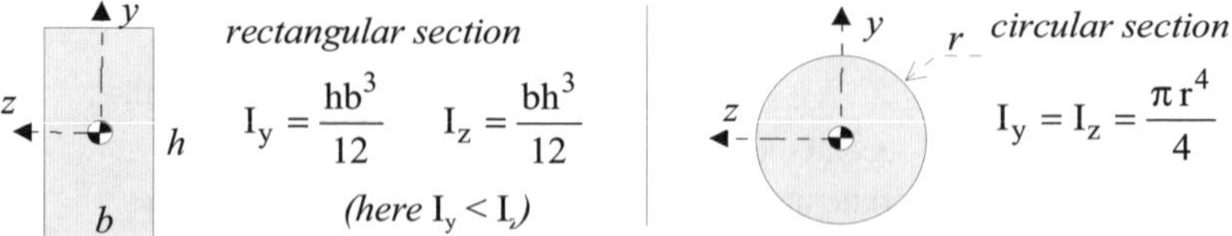

3 It should be remembered that the transmissible mechanical actions from structure S_1 to the beam can be modeled by resultant force and moment at point G_1 summarized in condensed writing as $\left\{ F_{S_1 \,/\, \text{beam}} \right\}_{G_1}$ i.e.:

– the resultant of mechanical actions $\overrightarrow{\mathcal{R}}_{S_1 \,/\, \text{beam}}$

– the resultant moment $\overrightarrow{\mathcal{M}}_{G_1 \, S_1 \,/\, \text{beam}}$ at G_1 of mechanical actions

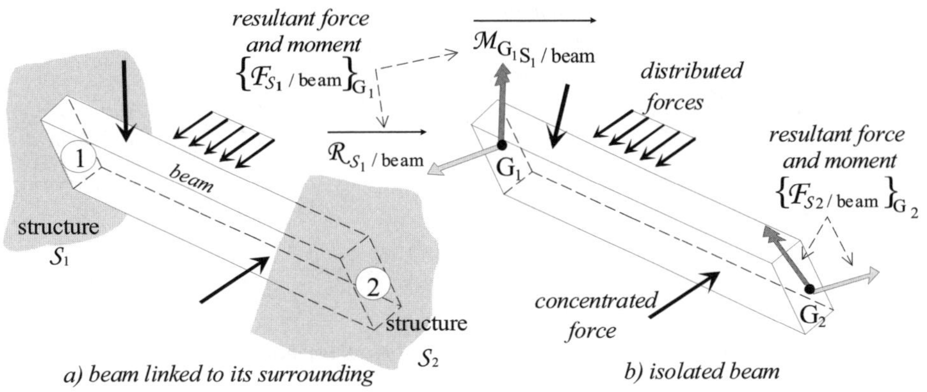

Figure 9.3. *Forces applied on a beam*

9.1.4. *Cohesion force and moment on a current cross-section*

In the particular case of beams, the search for modeling cohesion forces (see section 1.1) is particularly important because it helps in going back to the stress distributions on the cross-sections.

Figure 9.4a takes the case of the loaded beam within a structure in equilibrium (which is in addition also loaded). Figure 9.4b shows an isolated beam element limited by sections ① and ② with centers G_1 and G_2.

It should be recalled that the mechanical actions of the remaining structure on the considered beam reduce to resultant force and moment $\{F_{S_1 / \text{beam}}\}_{G_1}$ at point G_1 and $\{F_{S_2 / \text{beam}}\}_{G_2}$ at point G_2. Let us consider a cross-section with G as geometric center. Thus, we can consider the beam as the union of two parts marked I and II on both sides of the section in question.

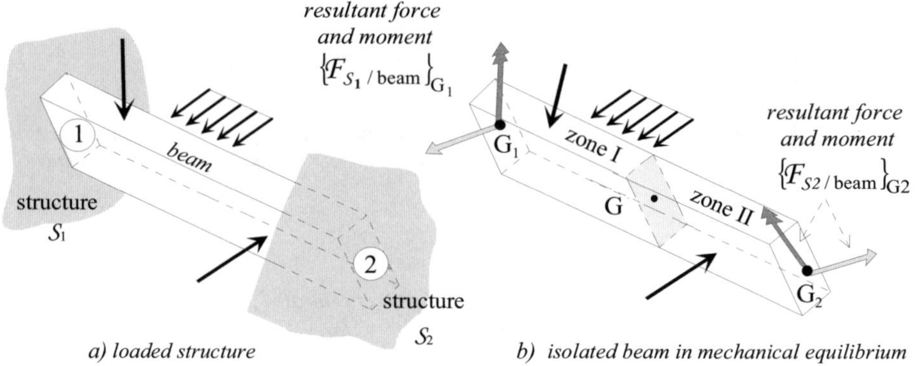

a) loaded structure   b) isolated beam in mechanical equilibrium

Figure 9.4. *Beam as element of structure*

9.1.4.1. *Equilibrium of the beam*

Applying the fundamental principle of statics at the center G let us write the equilibrium of the whole beam under study (union of zones I and II[4]) (see Figure 9.5a).

$$\{\mathcal{F}_{\text{ext/I}}\}_G + \{\mathcal{F}_{\text{ext/II}}\}_G \equiv \{0\} \qquad [9.2]$$

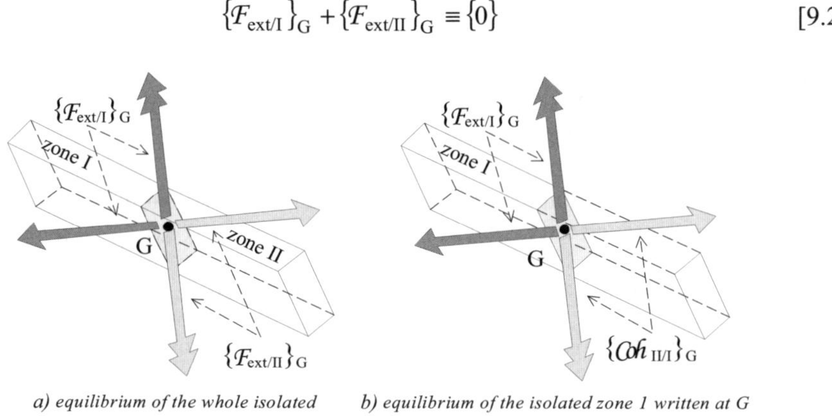

a) equilibrium of the whole isolated   b) equilibrium of the isolated zone 1 written at G
beam written at G   (here appear the cohesive resultant force and moment
                                    on cross-section)

Figure 9.5. *Two equilibrium configurations: whole beam and portion of beam*

---

4 It should be noted that for the calculation of resultant force and moment $\{\mathcal{F}_{\text{ext/I}}\}_G$, resultant force and moment $\{\mathcal{F}_{S_1/\text{beam}}\}_{G_1}$ as well as the external mechanical actions applied on zone I of the beam take place. The same approach is applied for the calculation of the couple $\{\mathcal{F}_{\text{ext/II}}\}_G$.

9.1.4.2. *Defining the resultant force and moment for cohesion forces*

Let us isolate zone I by deleting zone II Figure 9.5b. The deleted part should be replaced by the actions that it exerts on the remaining part located on the cross-section with center G in order to balance the remaining part. These actions are the cohesion forces (see section 1.1). Their distribution, illustrated in Figure 9.6a is not yet known, but they can be modeled at point G by a resultant force and a resultant moment summarized by the writing: $\{Coh_{\mathrm{II/I}}\}_G$

Balancing of zone I imposes:

$$\{\mathcal{F}_{\text{ext/I}}\}_G + \{Coh_{\mathrm{II/I}}\}_G \equiv \{0\} \qquad [9.3]$$

by comparing equations [9.2] and [9.3], the following is observed:

$$\{Coh_{\mathrm{II/I}}\}_G = \{\mathcal{F}_{\text{ext/II}}\}_G \qquad [9.4]$$

NOTES

❑ Taking into account the equilibrium equation [9.3], we can write:

$$\{Coh_{\mathrm{II/I}}\}_G = -\{\mathcal{F}_{\text{ext/I}}\}_G \qquad [9.5]$$

❑ Here zone I is isolated but the reasoning is the same if zone II is isolated [5].

9.1.4.3. *Projections of cohesive resultant force and moment on local axis*

$\{Coh_{\mathrm{II/I}}\}$ summarizes the following elements at G (see Figure 9.6a):

➢ the resultant of elementary cohesive forces $\overrightarrow{df_{i\,(\mathrm{II/I})}}$, that is $\overrightarrow{\mathcal{R}_{\mathrm{II/I}}}$;

---

5 In an identical manner, we could have isolated the zone II by deleting zone I. In that case the cohesion forces shall be modeled by a cohesion resultant force and moment at G as: $\{Coh_{\mathrm{I/II}}\}_G$. Balancing of zone II would then give: $\{\mathcal{F}_{\text{ext/II}}\}_G + \{Coh_{\mathrm{I/II}}\}_G \equiv \{0\}$. By comparing the previous equation and [9.2], we would arrive at: $\{Coh_{\mathrm{I/II}}\}_G = \{\mathcal{F}_{\text{ext/I}}\}_G$.

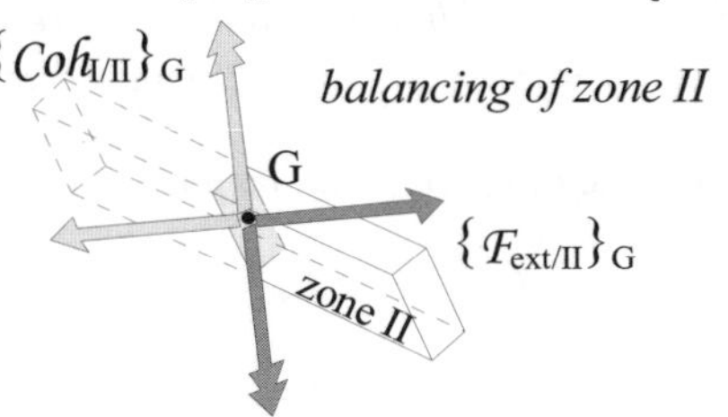

➤ the resultant moment, sum of the moments of elementary cohesive forces $\overrightarrow{df_{i\,(II/I)}}$ with reference to point G, that is $\overrightarrow{\mathcal{M}_{G(II/I)}}$:

$$\{Coh._{II/I}\}_G = \left\{ \begin{array}{l} \overrightarrow{\mathcal{R}_{II/I}} = \displaystyle\int_S \overrightarrow{df_{i\,(II/I)}} \\[2ex] \overrightarrow{\mathcal{M}_{G(II/I)}} = \displaystyle\int_S \overrightarrow{GM_i} \wedge \overrightarrow{df_{i\,(II/I)}} \end{array} \right\}_G$$

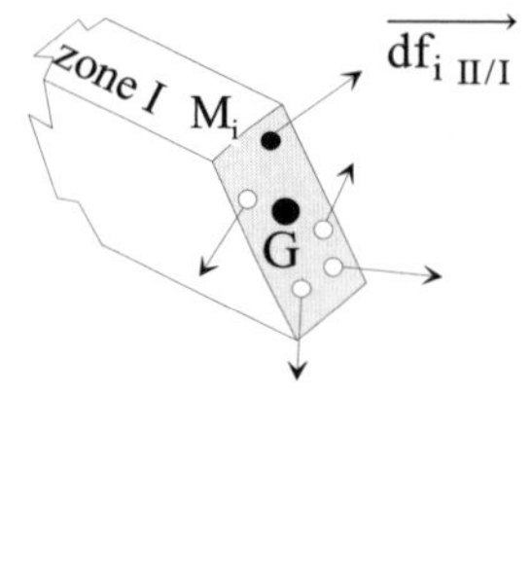

a) cohesive forces

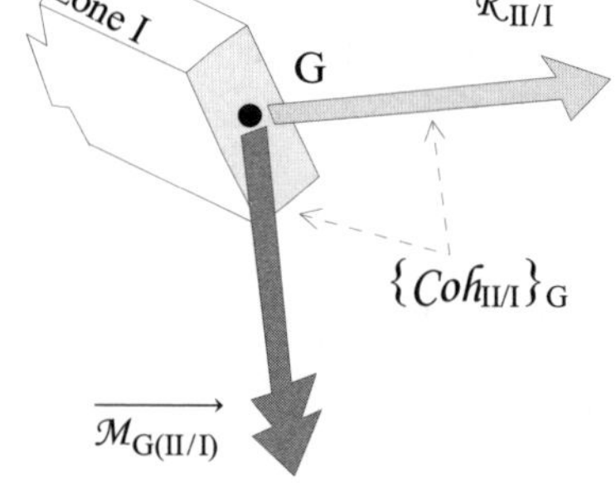

b) resultant force and moment at G

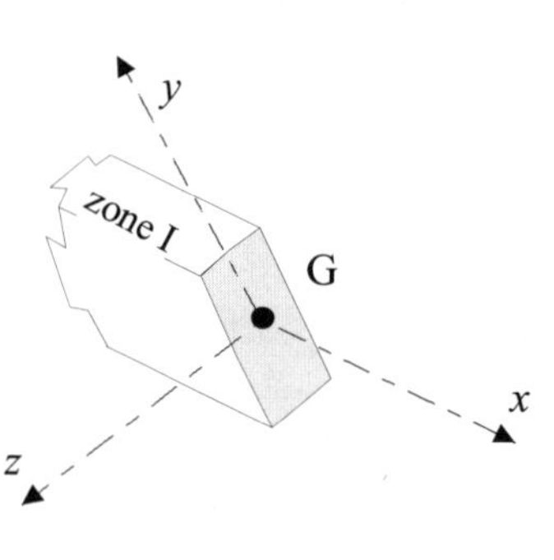

c) local coordinate system

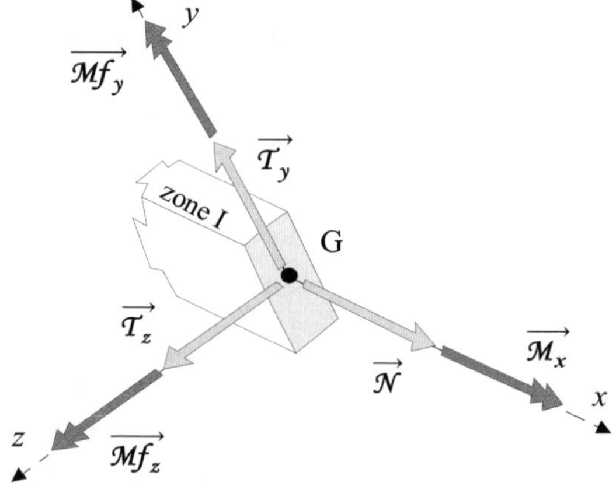

d) 6 projections of the resultant force and moment in the local coordinate system

Figure 9.6. *Cohesion resultant force and moment*

These vectors are projected on a coordinate system related to the isolated cross-section denoted the "local coordinate system" in Figure 9.6c. It is defined based on a method already mentioned in section 1.2.1. One of the axes forms the outward normal to the isolated cross-section passing by G and thus colinear to the centerline, i.e. the $\bar{x}$ axis. The other two axes are colinear to the main axes (see [9.1]) of the concerned section (this is $\vec{y}$ and $\vec{z}$ in Figure 9.6).

The six projections of the resultant force and moment in the local coordinate system marked "r" are then defined as follows:

$$\{Coh_{II/I}\}_G = \left\{ \begin{array}{l} \overrightarrow{\mathcal{R}_{II/I}} = \mathcal{N}\,\vec{x} + \mathcal{T}_y\,\vec{y} + \mathcal{T}_z\,\vec{z} \\[2ex] \hline \overrightarrow{\mathcal{M}_{G(II \to I)}} = \mathcal{M}_x\,\vec{x} + \mathcal{M}f_y\,\vec{y} + \mathcal{M}f_z\,\vec{z} \end{array} \right\}_{G,\,r}$$

[9.6]

NOTES

❏ Each one of these projections means the existence of a specific distribution of cohesive forces. We have seen in section 1.2, that the stresses can be defined based on the cohesion forces. In other words, in terms of stresses, it will be seen that every projection appearing in [9.6] is associated with a specific stress distribution.

❏ These projections are fundamental for the study of the beam behavior. The following terminology is used to designate them:

$\mathcal{N}$ or $\mathcal{N}_x$: normal resultant force $\mathcal{M}_x$: longitudinal moment [6]

$\mathcal{T}_y$: shear resultant force $\mathcal{M}f_y$: bending moment

$\mathcal{T}_z$: shear resultant force $\mathcal{M}f_z$: bending moment

---

6 When the cross-section of a beam has a center of symmetry (for example the rectangular section represented in Figure 9.6), the longitudinal moment $\mathcal{M}_x$ is also called the "torsional moment" $\mathcal{M}t$ (see section 9.3.2.4 and [9.21]).

9.1.5. *Hypothesis of the beam theory*

9.1.5.1. *Hypothesis on stresses*

○ Saint Venant's hypothesis

The cohesion resultant force and moment $\{Coh_{II/I}\}_G$ acts on a current cross-section (see Figure 9.6). As indicated earlier, we shall see that each of the projections of the resultant force and moment [9.6] reflects the existence of a corresponding stress distribution on the section. The total stresses on this section thus seem directly connected to the cohesive resultant force and moment, and not to local characteristics (point of application, line of action, etc.) of the real forces applied all along the beam. In other words, it is possible to imagine an infinite number of distributions of different forces on the deleted zone II of Figure 9.5 and they will produce the same stresses on the current cross-section under study, provided they have the same resultant force and moment at G. This is illustrated by the example in Figure 9.7 where on the concerned cross-section the stresses are identical for loadings 1 and 2. This assertion simplifies the reality. This can be observed, for example, in Figure 9.8, where the concentrated opposite forces applied on the beam result in zero resultant forces on the current section.

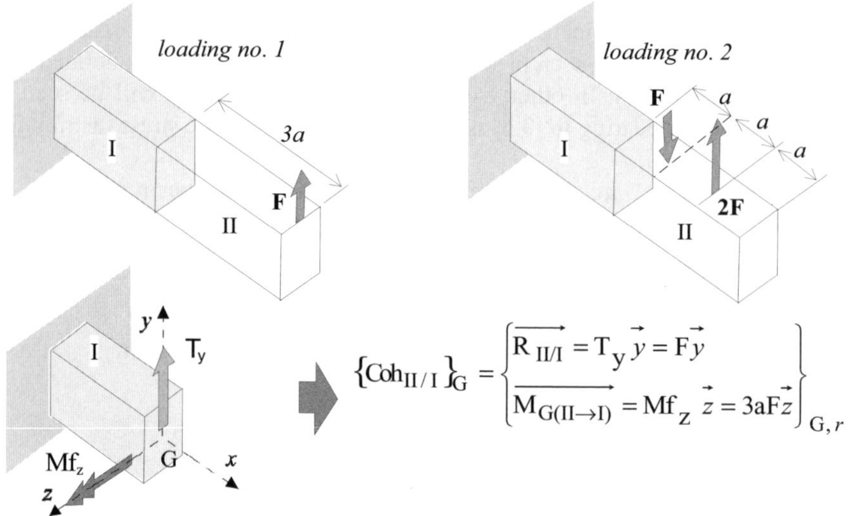

$$\{Coh_{II/I}\}_G = \left\{ \begin{array}{l} \overrightarrow{R_{II/I}} = T_y\,\vec{y} = F\vec{y} \\ \hline \overrightarrow{M_{G(II\to I)}} = Mf_z\,\vec{z} = 3aF\vec{z} \end{array} \right\}_{G,r}$$

Figure 9.7. *Distinct loadings 1 and 2 with same cohesive resultant force and moment at G*

If we go by the previous hypothesis, these forces should produce zero stresses on this same cross-section. This seems possible on the current "distanced" cross-section of Figure 9.8a, but it is not possible for the cross-section in Figure 9.8b which is very close to the zone of application of the forces.

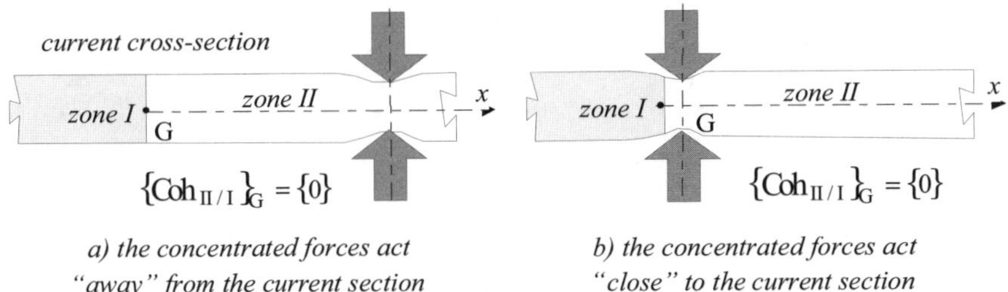

a) the concentrated forces act
"away" from the current section

b) the concentrated forces act
"close" to the current section

Figure 9.8. *Limit of Saint Venant's hypothesis*

Consequently this hypothesis can be stated as follows:

"away"[7] from the zone of application of localized forces, the stress distribution on a current cross-section depends only on the cohesive resultant force and moment for this section.

This is known as "Saint Venant's hypothesis".

○ Stresses on a current cross-section

Figure 9.9 shows a current cross-section of a beam subjected to a certain load, and the local coordinate system as defined previously in 9.1.4.3.

---

7 The term "away" here is purposely left imprecise. In reality, as an indicative value, the minimum "critical" distance between localized forces and the cross-section in question should be the largest dimension h that can be found in the cross-section.

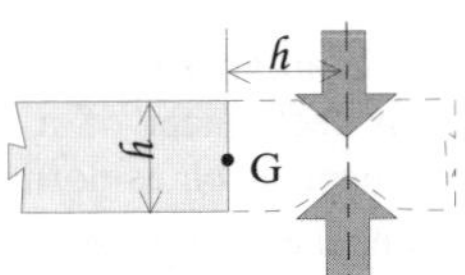

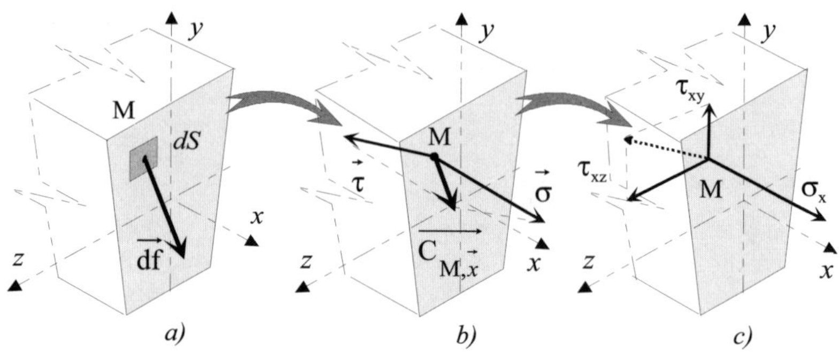

Figure 9.9. *Stresses at point M of a cross-section*

Let us represent a cohesion force marked $\overrightarrow{df}$ (see Figure 9.9a) acting on an elementary domain dS of this section. To this corresponds a stress $\overrightarrow{C}_{(M,\vec{x})}$ (see section 1.2) having in general, a normal component parallel to $\vec{x}$ and a tangential component in the plane of the cross-section.

In Figure 9.9b we have:

$$\overrightarrow{C}_{(M,\vec{x})} = \vec{\sigma} + \vec{\tau}$$

that is in the local coordinate system:

$$\overrightarrow{C}_{(M,\vec{x})} = \sigma_x\,\vec{x} + \tau_{xy}\,\vec{y} + \tau_{xz}\,\vec{z}$$

σ_x is a normal stress acting on the cross-section, τ_{xy} and τ_{xz} are the shear stresses acting in the plane of the cross-section.

we suppose that there are no stresses other than σ_x ; τ_{xy} ; τ_{xz} [8]

[9.7]

---

[8] This is an approximation. In section 10.2.2, it is seen that in a three-dimensional medium and on each facet of a small cubic element, it is possible to define other similar stress components such as normal stresses and tangential stresses. In the case of beams that interest us here, the components other than σ_x, τ_{xy} and τ_{xz} are ignored as they are supposed much weaker and negligible in comparison.

NOTES

❏ We should pay attention to the fact that stresses σ_x, τ_{xy}, and τ_{xz} are very different from the three stresses characterizing a plane state of stresses as described in section 1.4.

❏ This simplification of the stresses matches Saint Venant's hypothesis. It can be easily guessed from Figure 9.8b that due to compression the beam is laterally deformed at the places where concentrated forces act. It is obvious that such a deformation cannot be due only to the effect of stress σ_x. Near the points of application of forces there are normal stresses perpendicular to direction $\vec{x}$. They become weaker once we move away from this zone. Thus the two previous hypotheses on stresses become coherent.

❏ Deformations corresponding to stresses σ_x, τ_{xy}, and τ_{xz}:

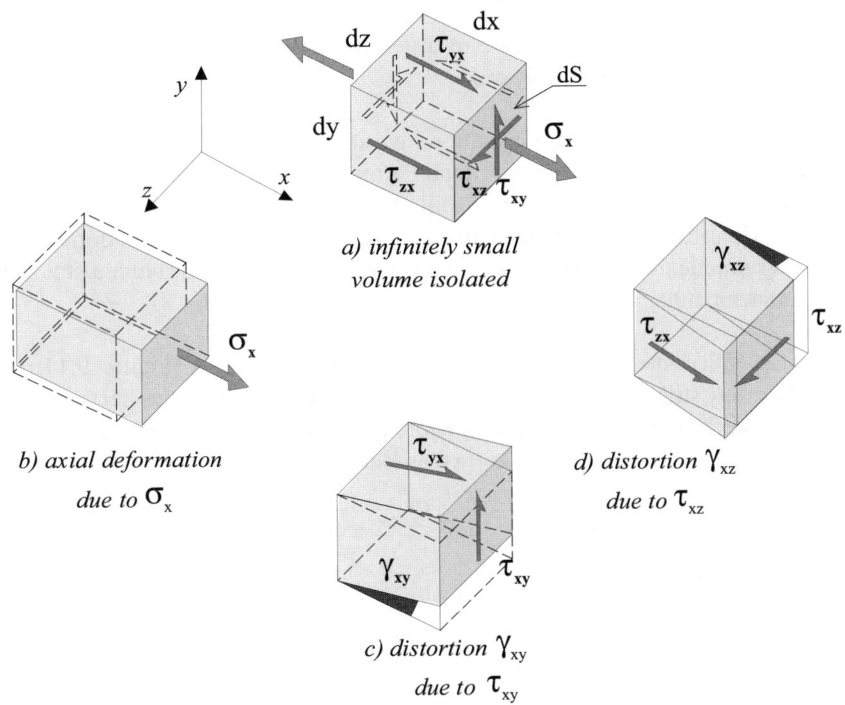

Figure 9.10. *Deformations due to stresses*

If the infinitely small cubic element defined from the facet dS in Figure 9.9a is isolated, it can be seen that the reciprocity property of the shear stresses[9] introduces stresses τ_{yx} and τ_{zx} in Figure 9.10a (it is known that $\tau_{yx} = \tau_{xy}$ and $\tau_{zx} = \tau_{xz}$). Figure 9.10b represents the deformed shape of the small element (axial deformation or dilatation along $\vec{x}$) due to the only stress σ_x. Figure 9.10c shows the angular distortion γ_{xy} due to the shear stress τ_{xy} and Figure 9.10d the angular distortion γ_{xz} due to the shear stress τ_{xz}.

We can revert to the forms [1.6] and [1.12] of the behavior relations where E and G are respectively the longitudinal elasticity and the shear modulus:

$$\varepsilon_x = \frac{\sigma_x}{E} \; ; \qquad \gamma_{xy} = \frac{\tau_{xy}}{G} \; ; \qquad \gamma_{xz} = \frac{\tau_{xz}}{G} {}^{10} \tag{9.8}$$

9.1.5.2. *Hypothesis on deformations*

The beam in Figure 9.3 is made of an elastic and linear material[11] and deforms under the action of the applied forces.

As per the hypothesis on deformation , we state that initially plane cross-sections (see Figure 9.11a) remain "almost" plane after deformation, i.e. the displacements outside the initial plane of the section still appear very small compared to the size of the latter. Therefore, if we consider two infinitely close deformed sections it is then possible to practically superimpose one section over the other through a small rigid-body displacement.

This is called the generalized Navier-Bernoulli principle (Figure 9.11b). The reality is often still further simplified by assuming that the cross-sections remain plane after deformation and also perpendicular to the deformed centerline (Figure 9.11c). This is what is commonly known as Bernoulli's hypothesis.

---

9 See section 1.2.3 and 10.2.2.

10 Referring to the linear elastic nature of the material constituting the beam, when the three stresses σ_x, τ_{xy} and τ_{xz} act simultaneously, the total deformation of the small element appears as the superimposition of one on the other of the three deformations illustrated in Figure 9.10.

11 See section 1.3.3.

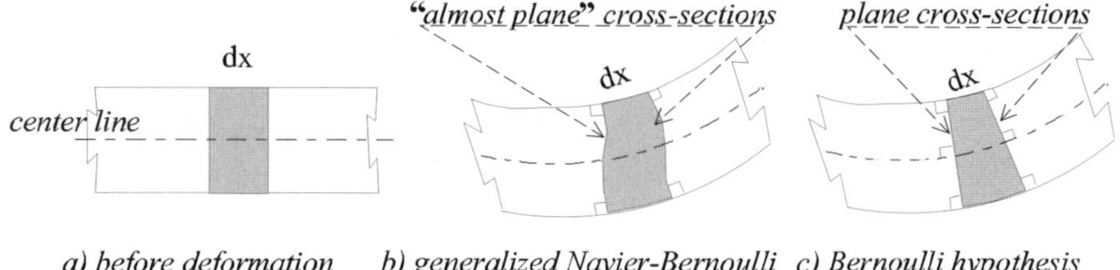

a) before deformation   b) generalized Navier-Bernoulli   c) Bernoulli hypothesis
principle

Figure 9.11. *Hypothesis on deformations*

NOTES

❐ Bernoulli's hypothesis can all the more be verified as concerned cross-sections are in the current zone of the beam (for example away from the concentrated forces indicated in Figure 9.8). Such a hypothesis seems to go with Saint Venant's hypothesis, and thus with the distribution [9.7] retained for the stresses on a current cross-section.

❐ Recall that in the entire book, the deformation of structures under study appear as small displacements that do not modify the initial geometry in a noticeable way. Especially for beams, the previous hypothesis on cross-section deformation causes very small displacements of any cross-section in its own plane and outside this plane. All these small displacements are the result of very small local deformations [9.8][12].

9.1.6. *Microscopic equilibrium*

Let us isolate an elementary slice of a beam subjected to any load (Figure 9.12a). As indicated in the previous section, the hypothesis in [9.7] on stresses expresses the existence of normal stress σ_x and shear stresses τ_{xy} and τ_{xz} on the cross-sections that limit this slice.

---

12 See section 2.4 and Figure 2.20.

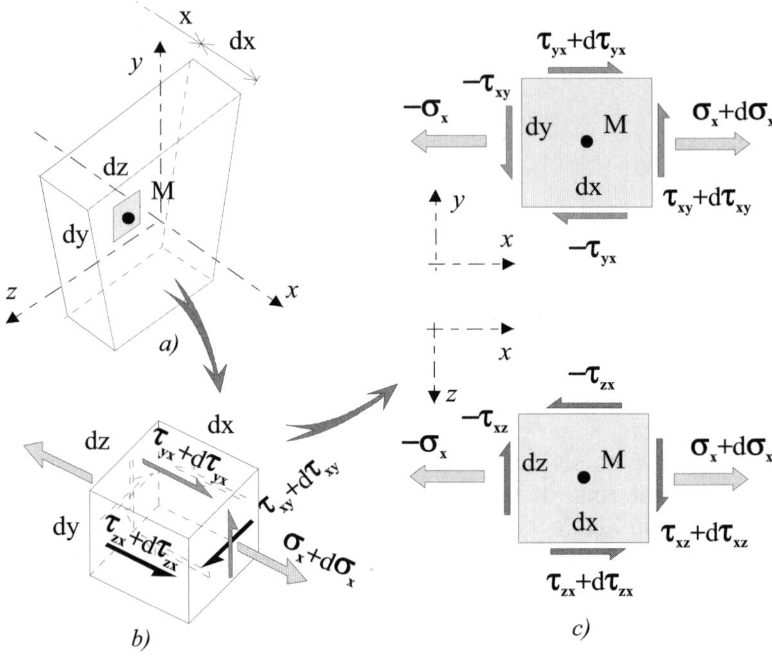

Figure 9.12. *Equilibrium of an infinitely small domain*

Starting from this slice, let us isolate an infinitely small parallelepipedic volume $dx \times dy \times dz$ (Figure 9.12b). As already seen in Figure 9.10a, let us position the stresses on the two facets perpendicular to $\vec{x}$ and on the other facets[13]. Figure 9.12c represents the domain seen in the (xy) and (xz) planes where the stresses σ_x, τ_{xy} and τ_{xz} appear. In the general case of any loading, these stresses vary from one point to another of the beam. They undergo weak algebraic increases, denoted by $d\sigma_x$, $d\tau_{xy}$ and $d\tau_{xz}$, when we move from one facet to the other parallel facet covering the distance dx along $\vec{x}$, or dy along $\vec{y}$ or dz along $\vec{z}$.

Let us project on $\vec{x}$ axis the equilibrium of the small parallelepiped, that is the sum of projected elementary forces applied on it is zero:

---

13 Referring again to the reciprocity property $\tau_{xy} = \tau_{yx}$ and $\tau_{xz} = \tau_{zx}$.

$$-\sigma_x dydz + (\sigma_x + d\sigma_x)dydz - \tau_{xy} dxdz + (\tau_{xy} + d\tau_{xy})dxdz.......$$
$$.......-\tau_{xz} dxdy + (\tau_{xz} + d\tau_{xz})dxdy = 0$$

i.e.:

$$d\sigma_x dydz + d\tau_{xy} dxdz + d\tau_{xz} dxdy = 0$$

According to the previous comment, when we move from the center of the facet of abscissa x to that of abscissa $x + dx$, there is an increase:

$$\sigma_x = \sigma_x(x, y, z) \Rightarrow d\sigma_x = \frac{\partial\sigma_x}{\partial x} dx + \frac{\partial\sigma_x}{\partial y} dy + \frac{\partial\sigma_x}{\partial z} dz = \frac{\partial\sigma_x}{\partial x} dx \;^{14}$$

because $dy = dz = 0$ (the path covered has a single component dx).

Similarly, when we move along $\vec{y}$ of dy:

$$\tau_{xy} = \tau_{xy}(x, y, z) \Rightarrow d\tau_{xy} = \frac{\partial\tau_{xy}}{\partial y} dy$$

because $dx = dz = 0$ (the path covered has a single component dy).

In addition, when we move dz along $\vec{z}$:

$$\tau_{xz} = \tau_{xz}(x, y, z) \Rightarrow d\tau_{xz} = \frac{\partial\tau_{xz}}{\partial z} dz$$

because $dx = dy = 0$ (the path covered has a single component dz).

The equilibrium relation written above becomes:

$$\frac{\partial\sigma_x}{\partial x} dxdydz + \frac{\partial\tau_{xy}}{\partial y} dydxdz + \frac{\partial\tau_{xz}}{\partial z} dzdxdy = 0$$

where the volume of the basic parallelepiped appears $dV = dx \times dy \times dz$. On simplification it becomes:

$$\frac{\partial\sigma_x}{\partial x} + \frac{\partial\tau_{xy}}{\partial y} + \frac{\partial\tau_{xz}}{\partial z} = 0 \tag{9.9}$$

---

14 Remembering the total differential of a function f(x,y,z):
$$df = \frac{\partial f}{\partial x} dx + \frac{\partial f}{\partial y} dy + \frac{\partial f}{\partial z} dz$$

NOTE

❐ This equation is denoted the "local longitudinal equilibrium equation". It defines the equilibrium along the $\vec{x}$ direction on a *microscopic scale.*

9.2. Mesoscopic equilibrium or equilibrium extended to a whole cross-section

Figure 9.13 brings back the loaded beam of Figure 9.3. Two infinitely close cross-sections having geometric centers G and G' (separated by dx) are studied.

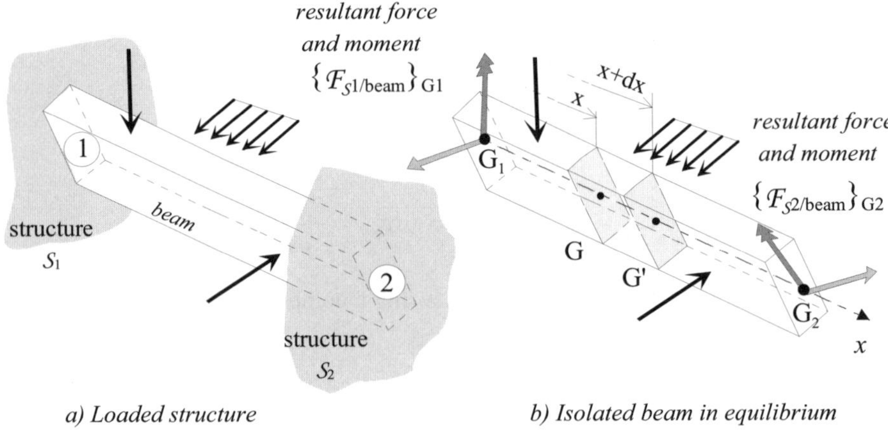

Figure 9.13. *Two infinitely near cross-sections*

Let us carry out a first separation at G on the abscissa x. The action of x^+ on x^- can be modeled by the cohesive resultant force and moment at G (Figure 9.14a):

$$\{Coh_{x^+/x^-}\}_G = \left\{ \begin{array}{c} \vec{\mathcal{R}} \\ \vec{\mathcal{M}} \end{array} \right\}_G$$

Let us carry out a second separation at G' on the abscissa $x' = x + dx$. The action of x'^+ on x'^- can be modeled by the cohesive resultant force and moment at G' (Figure 9.14b):

$$\{Coh_{x'^+/x'^-}\}_{G'} = \left\{ \begin{array}{c} \vec{\mathcal{R}} + \vec{d\mathcal{R}} \\ \vec{\mathcal{M}} + \vec{d\mathcal{M}} \end{array} \right\}_{G'}$$

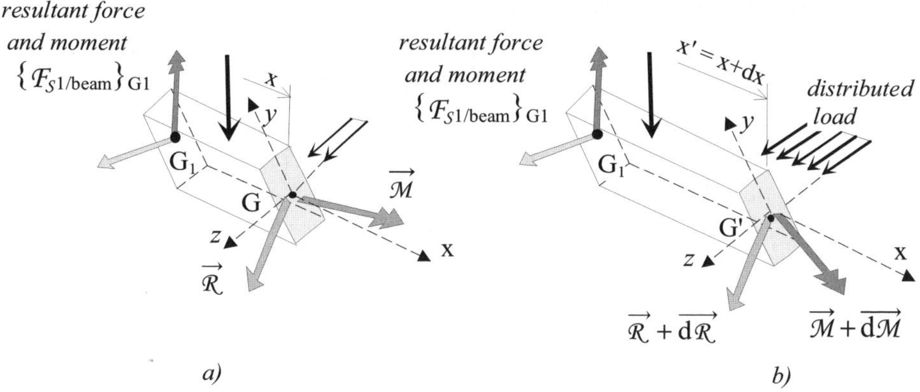

Figure 9.14. *Two infinitely near partitions of the beam in equilibrium*

In Figure 9.15 we have isolated the beam slice of thickness dx with the mechanical actions that it subjected to:

• a distributed load $\vec{q}$ (N/m) supposed to act on the mean line GG' of the beam, which projects in the local coordinate system as follows:

$$\vec{q} = q_x\,\vec{x} + q_y\,\vec{y} + q_z\,\vec{z}$$

It produces on slice dx a resulant force and resultant moment at G' marked as:

$$\left\{\mathcal{F}_{\text{distributed load}/dx}\right\}_{G'} = \left\{\begin{array}{c}\vec{q}dx \\ \vec{0}\end{array}\right\}_{G'}{}^{15}$$

• two cohesive resultant forces and moments: $\left\{Coh_{x^-/x^+}\right\}_G$ and $\left\{Coh_{x'^+/x'^-}\right\}_{G'}$.

It is noted that $\left\{Coh_{x^-/x^+}\right\}_G = -\left\{Coh_{x^+/x^-}\right\}_G = \left\{\begin{array}{c}-\vec{\mathcal{R}} \\ -\vec{\mathcal{M}}\end{array}\right\}_G$.

---

15 The moment at G' of distributed load is written as: $\dfrac{1}{2}\overrightarrow{G'G} \wedge \vec{q}dx = -\dfrac{1}{2}\vec{x} \wedge \vec{q}(dx)^2$ and it introduces an infinitely small quantity of the second order that we ignore here.

The fundamental principle of statics enables us to write:

$$\left\{ Coh_{x^-/x^+} \right\}_{G'} + \left\{ Coh_{x'^+/x'^-} \right\}_{G'} + \left\{ \mathcal{F}_{\text{distributed load}/dx} \right\}_{G'} = \{0\}$$

$$\left\{ Coh_{x^-/x^+} \right\}_{G'} + \left\{ Coh_{x'^+/x'^-} \right\}_{G'} + \left\{ \begin{array}{c} \vec{q}dx \\ \vec{0} \end{array} \right\}_{G'} = \{0\} \qquad [9.10]$$

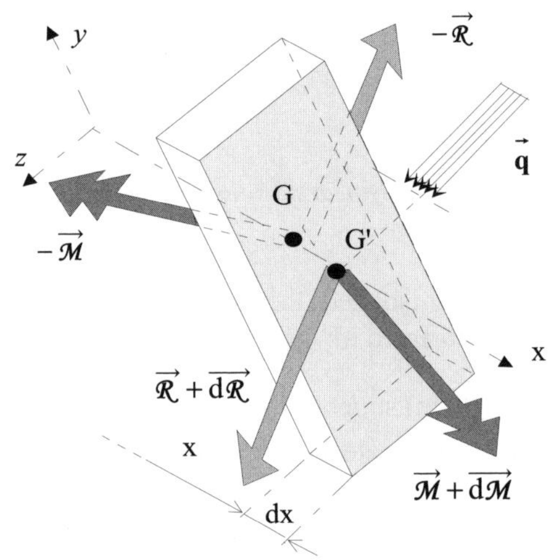

Figure 9.15. *Equilibrium of a slice of beam*

From [9.10] the following vectorial equations of equilibrium are derived:

$$\left[\begin{array}{l} -\vec{\mathcal{R}} + \vec{\mathcal{R}} + \overrightarrow{d\mathcal{R}} + \vec{q}dx = \vec{0} \\ -\vec{\mathcal{M}} + \vec{\mathcal{M}} + \overrightarrow{d\mathcal{M}} + \overrightarrow{G'G} \wedge -\vec{\mathcal{R}} = \vec{0} \end{array} \right.$$

i.e.:

$$\left[\begin{array}{l} \overrightarrow{d\mathcal{R}} + \vec{q}dx = \vec{0} \\ \overrightarrow{d\mathcal{M}} + dx\,\vec{x} \wedge \vec{\mathcal{R}} = \vec{0} \end{array} \right. \qquad [9.11]$$

Let us project the cohesive resultant force and moment at G in the local coordinate system. Based on [9.6] we have:

$$\left\{Coh_{(x^+ / x^-)}\right\}_G = \left\{\begin{array}{l} \overrightarrow{\mathcal{R}} = \mathcal{N}\,\vec{x} + \mathcal{T}_y\,\vec{y} + \mathcal{T}_z\,\vec{z} \\[2mm] \overrightarrow{\mathcal{M}_G} = \mathcal{M}_x\,\vec{x} + \mathcal{M}f_y\,\vec{y} + \mathcal{M}f_z\,\vec{z} \end{array}\right\}_{G,r}$$

Relations [9.11] are then rewritten as:

$$\left[\begin{array}{l} d\mathcal{N}\,\vec{x} + d\mathcal{T}_y\,\vec{y} + d\mathcal{T}_z\,\vec{z} + q_x\,dx\,\vec{x} + q_y dx\,\vec{y} + q_z dx\,\vec{z} = \vec{0} \\[2mm] d\mathcal{M}_x\,\vec{x} + d\mathcal{M}f_y\,\vec{y} + d\mathcal{M}f_z\,\vec{z} - \mathcal{T}_z\,dx\,\vec{y} + \mathcal{T}_y dx\,\vec{z} = \vec{0} \end{array}\right.$$

or even as:

$$\left[\begin{array}{l} \left(\dfrac{d\mathcal{N}}{dx} + q_x\right)\vec{x} + \left(\dfrac{d\mathcal{T}_y}{dx} + q_y\right)\vec{y} + \left(\dfrac{d\mathcal{T}_z}{dx} + q_z\right)\vec{z} = \vec{0} \\[4mm] \dfrac{d\mathcal{M}_x}{dx}\,\vec{x} + \left(\dfrac{d\mathcal{M}f_y}{dx} - \mathcal{T}_z\right)\vec{y} + \left(\dfrac{d\mathcal{M}f_z}{dx} + \mathcal{T}_y\right)\vec{z} = \vec{0} \end{array}\right.$$

Thus, equilibrium equations extended to a cross-section are obtained:

$$\left[\begin{array}{ll} \dfrac{d\mathcal{N}}{dx} + q_x = 0 \qquad & \dfrac{d\mathcal{M}_x}{dx} = 0 \\[4mm] \dfrac{d\mathcal{T}_y}{dx} + q_y = 0 & \dfrac{d\mathcal{M}f_y}{dx} - \mathcal{T}_z = 0 \\[4mm] \dfrac{d\mathcal{T}_z}{dx} + q_z = 0 & \dfrac{d\mathcal{M}f_z}{dx} + \mathcal{T}_y = 0 \end{array}\right. \qquad [9.12]$$

NOTE

❏ Until now we have studied the equilibrium of the structure at different "scales". The previous section dealt with "microscopic" equilibrium. Here, the cross-section domain is considerably larger because it is finite, but the isolated slice of beam is still characterized by an infinitely small thickness dx. Such an equilibrium will be called "mesoscopic". If the equilibrium of the complete beam, as represented in Figure 9.13b is examined now, the corresponding equilibrium will be called "macroscopic". These scale levels are indicated in Figure 9.16.

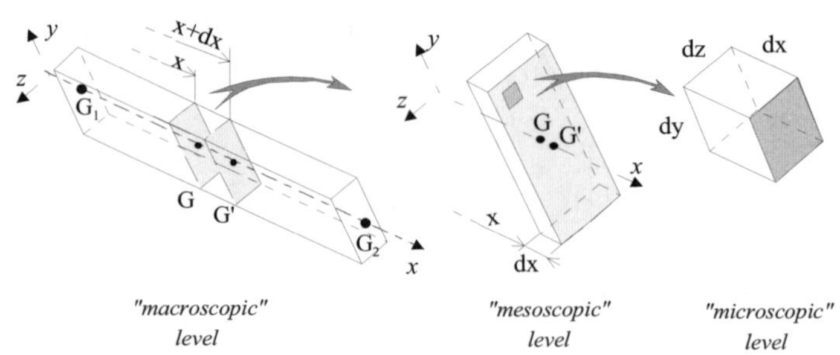

Figure 9.16. *Different scale levels*

9.3. Behavior relations and stresses

9.3.1. *Normal resultant*

9.3.1.1. *Definition*

Figure 9.17 shows a beam of any cross-section. The $\vec{x}$ axis is the centerline passing through the geometric centers G. The left end of the beam is clamped. It is subjected to a force $\vec{F} = X\vec{x}$ at the right end. The beam elongates or contracts along the direction $\vec{F}$.

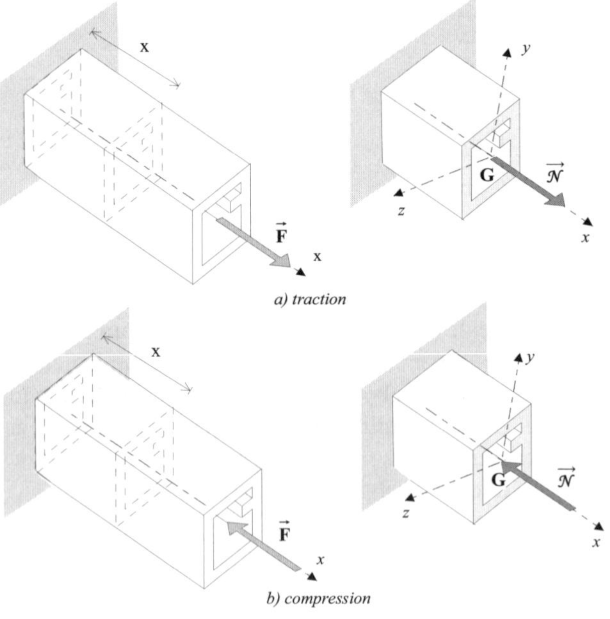

Figure 9.17. *Loading in traction or compression*

Then on the current cross-section of abscissa x, the resultant force and moment of cohesive forces are written as:

$$\left\{ Coh_{(x^+/x^-)} \right\}_G = \left\{ \begin{array}{l} \overrightarrow{\mathcal{R}_{(x^+/x^-)}} = X\vec{x} \\ \hline \overrightarrow{\mathcal{M}_{G(x^+/x^-)}} = \vec{0} \end{array} \right\}_{G,r}$$

From [9.6], it is observed that the only non-zero component of resultant force which acts on the cross-section is the normal resultant $\mathcal{N}$, i.e.:

$$\mathcal{N} = X$$

9.3.1.2. *Deformation of an elementary slice of beam*

The stresses corresponding to a resultant positive or negative normal are assumed to be constant[16] on a current cross-section which is sufficiently far from the end where force $\vec{F}$[17] is applied. According to Hooke's law[18] the dilatation $\varepsilon_x = \dfrac{du(x)}{dx}$ of the beam slice is also constant on the cross-section domain.

Under these conditions, if Hooke's law $\varepsilon_x = \dfrac{\sigma_x}{E}$ is integrated on the cross-section domain of area S, we have:

$$\int_S \varepsilon_x \, dS = \int_S \frac{\sigma_x}{E} dS$$

which gives:

$$\varepsilon_x \times S = \frac{1}{E} \sigma_x \times S$$

Whereas it can be noted that the normal resultant force is just the sum of the elementary cohesive forces $\sigma_x \, dS$:

$$\mathcal{N} = \int_S df_n = \int_S \sigma_x \, dS = \sigma_x \times S \qquad [9.13]$$

---

16 It can be shown that this uniform distribution is exact because it is the solution to the problem of elasticity which is raised.
17 see section 9.1.5.1.
18 See [1.6].

from which:

$$\frac{du(x)}{dx} \times S = \frac{1}{E}\, \mathcal{N}$$

Thus, the behavior relation due to the normal resultant $\mathcal{N}$ for the beam slice of length dx is obtained:

$$\mathcal{N} = ES\frac{du(x)}{dx} \qquad [9.14]$$

Figure 9.18 shows the deformation of such an elementary slice for a rectangular cross-section beam.

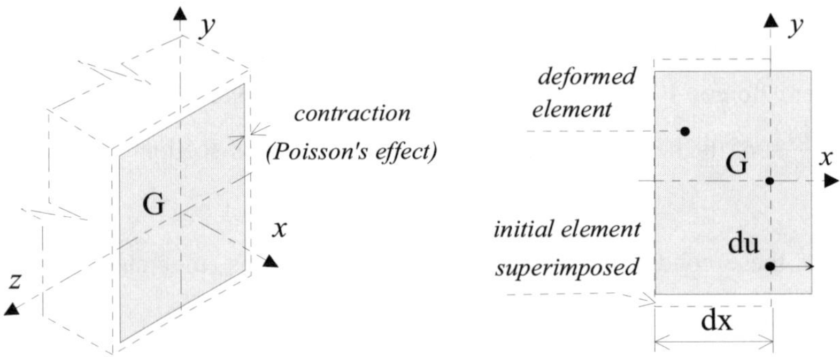

Figure 9.18. *Evolution of the shape of a rectangular section and deformation of an elementary beam slice under traction*

9.3.1.3. *Stresses on a cross-section*

From [9.13] we obtain:

$$\sigma_x = \frac{\mathcal{N}}{S}$$

For example, for a rectangular cross-section the distribution of normal stresses is represented in 9.19.

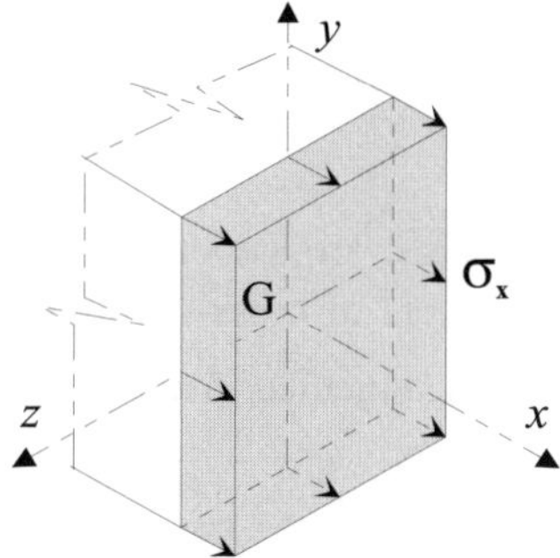

Figure 9.19. *Normal stress distribution due to normal resultant*

A resultant positive normal results in traction stresses $\sigma_x > 0$.

A resultant negative normal results in compression stresses $\sigma_x < 0$.

In brief:

Traction-compression	
resultant force of the cohesive forces on a cross-section: $\mathcal{N}$: normal resultant along $\vec{x}$	*displacement of a section:* $u(x)$ translation along $\vec{x}$
behavior relation due to a normal resultant: $$\mathcal{N} = ES\frac{du(x)}{dx}$$	
normal stress: $$\sigma_x = \frac{\mathcal{N}}{S}$$	

$$[9.15]$$

9.3.2. *Torsional loading*

9.3.2.1. *Definition*

Figure 9.20a shows a beam of any cross-section. The $\vec{x}$ axis is the centerline that passes through the geometric centers of cross-sections. The left end of the beam is clamped and the right extremity is subjected to a moment $\overrightarrow{\mathcal{M}} = \mathrm{L}\vec{x}$.

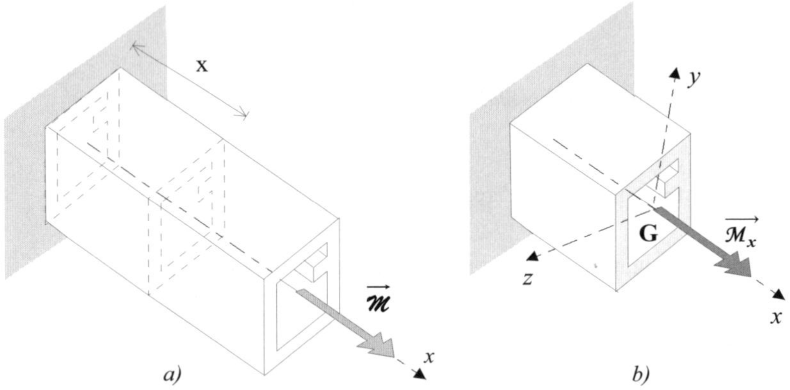

Figure 9.20. *Torsional loading*

Then in the current cross-section of abscissa x, the resultant force and moment of cohesion forces are written as:

$$\left\{ \mathcal{C}oh_{(x^+/x^-)} \right\}_G = \left\{ \begin{array}{c} \overrightarrow{\mathcal{R}_{(x^+/x^-)}} = \vec{0} \\ \hline \overrightarrow{\mathcal{M}_{G(x^+/x^-)}} = \mathrm{L}\vec{x} \end{array} \right\}_{G,r}$$

As per [9.6] the only non-zero resultant force of the elementary cohesion forces acting on this cross-section is the longitudinal moment $\mathcal{M}_x$ (see Figure 9.20b), i.e.:

$$\mathcal{M}_x = \mathrm{L}$$

9.3.2.2. *Deformation of an elementary beam slice*

Away from the clamped end[19], the current cross-sections rotate around an axis parallel to $\vec{x}$. The longitudinal moment $\mathcal{M}_x$ being constant here, angle $\theta_x(x)$ by which the cross-section rotates along abscissa x evolves along the $\vec{x}$ axis in such a way that between two cross-sections separated by dx (elementary slice of beam) the same relative rotation $\theta_x(x)$ is observed irrespective of x (see Figure 9.21). This can also be written by introducing the quotient $\dfrac{d\theta_x}{dx}$ or rate of rotation:

$$\frac{d\theta_x(x)}{dx} = C^{te}$$

The corresponding phenomenon shall be termed as *"uniform torsion"*.

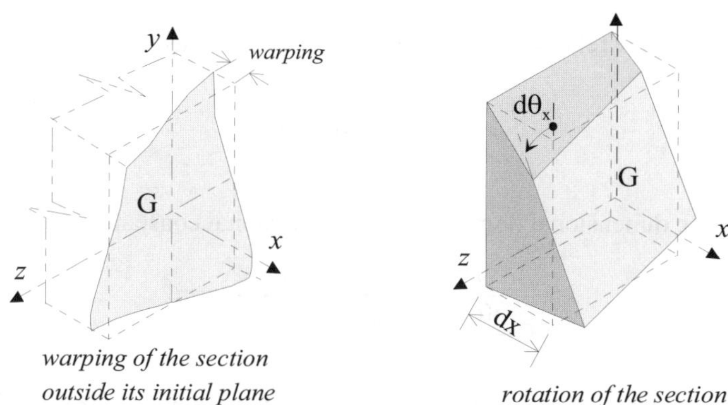

warping of the section
outside its initial plane

rotation of the section

Figure 9.21. *Evolution of the initial rectangular shape of a cross-section and deformation of an elementary slice of beam under torsion*

The greater the longitudinal moment the greater is this "rate of rotation". Thus we arrive at:

$$\frac{d\theta_x(x)}{dx} = A \times \mathcal{M}_x$$

---

19 When the cross-sectional shape of a beam is anything other than circular geometry, the standard term "uniform" given to the torsion is not valid "near" a clamped support. In fact, section 9.3.2.4 will show that the sections "warp" during uniform torsion (a schematic representation of this is given in Figure 9.21). Since warping does not take place at the clamped support ("constrained" torsion), the torsion obeys other laws locally.

the constant A depending on:

– the nature of the material. Further detailed study shall show the shear stresses on the cross-sections, that produce distortions (see [9.8]). The rotation of cross-sections is caused by these distortions. It is known that the corresponding behavior relation introduces the shear modulus G^{20} that will be taken here as the characteristic constant of the material;

– the shape of the section. Certain section shapes show a better rigidity against torsion (see section 9.3.2.5). A detailed study of torsion will lead to a characteristic constant related to the geometry of the section shape, and known as torsional constant J (or Saint Venant's constant).

The proportionality relation above can then be written in the form[21]:

$$\frac{d\theta_x(x)}{dx} = \frac{1}{G \times J} \mathcal{M}_x$$

It is the behavior relation due to the longitudinal moment $\mathcal{M}_x$, which is written as:

$$\mathcal{M}_x = G \times J \frac{d\theta_x(x)}{dx} \qquad [9.16]$$

We note the dimensional homogenity of the torsional constant:

$$J = \frac{\mathcal{M}_x}{G \dfrac{d\theta_x(x)}{dx}} \equiv \frac{N.m}{N.m^{-2} \times m^{-1}} = m^4$$

this constant having the dimension of a second moment of area, or quadratic moment.

9.3.2.3. *Simple case of a circular section*

The corresponding beam is shown in Figure 9.22a. The only non-zero resultant force of cohesion forces that act on the cross-section is the longitudinal moment $\mathcal{M}_x$ which we will call here "torsional moment" $\mathcal{M}t^{22}$ (see Figure 9.22b), i.e.:

$$\mathcal{M}_x = \mathcal{M}t = L$$

---

20 See [9.8].

21 G and J are placed at denominator because their increase improves the rigidity against torsion and leads to a lowering of the rotation rate $\dfrac{d\theta_x(x)}{dx}$.

22 For cross-sections of any shape the longitudinal moment $\mathcal{M}_x$ differs in general from the torsional moment $\mathcal{M}t$. This aspect is developed in section 9.3.2.3.

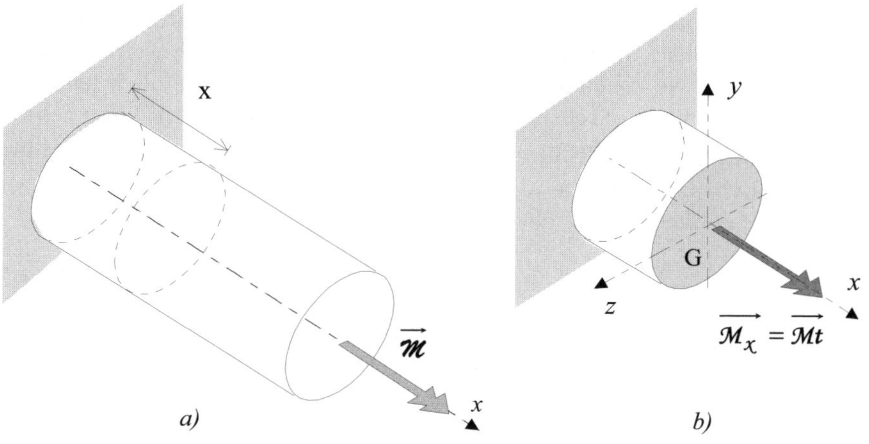

Figure 9.22. *Circular cross-section*

O Deformation of an elementary slice of beam

We consider an elementary slice of this beam with a basic volume $S \times dx$, before and after applying a longitudinal moment Mt. The generators parallel to $\vec{x}$ of the beam slice become helical as represented in Figure 9.23a.

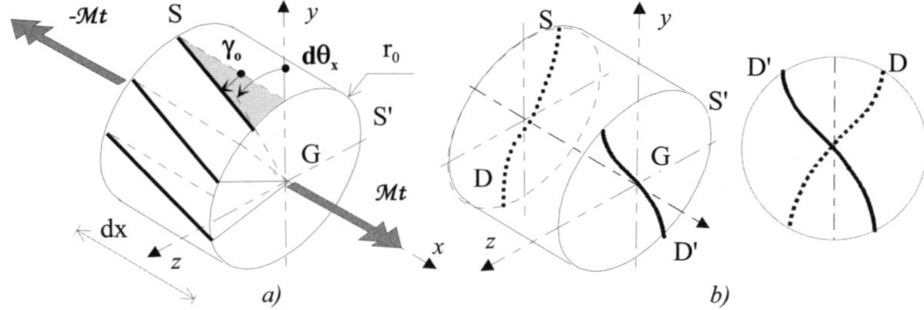

Figure 9.23. *Deformation in the case of a circular cross-section*

The same figure shows the slice dx limited by cross-sections which should deform in an identical manner. Let us suppose that a diameter D' of section S' deforms as shown in Figure 9.23b. Then if the section is looked at from the other side of the slice, it can be seen that the corresponding diameter D of the section S deforms along the dotted curve. Whereas irrespective of the section observed all the diameters should deform in the same manner, we should observe curves D and D' which can be superposed by rotation $d\theta_x(x)$. It can be concluded that the only possibility for this to happen is when the diameters remain rectilinear after deformation. By a similar argument it can be observed that the circular plane cross-sections remain plane after torsion[23]. From this, in the case of a circular section (solid or hollow), we conclude that there is no warping. On the other hand, since there exists a symmetry of repetition of any angle (or "axisymmetry"), the deformed contours remain circular. One of the consequences is that the distortion γ_0 visible in Figure 9.23a changes with the radius. If r_0 is the radius of the section, due to the smallness of this distortion we obtain:

$$\gamma_0 \cong \tan\gamma_0 = \frac{r_0 \times d\theta_x(x)}{dx}$$

Thus, for a current radius $0 \le r \le r_0$, the diameters remaining rectilinear:

$$\gamma = r \times \frac{d\theta_x(x)}{dx}$$

O   Stresses on a cross-section

The presence of distortions γ indicates the presence of shear stresses which are proportional to them. Figure 9.24 illustrates the distribution of these stresses on a radius of the circular section. For a current radius r, the shear stress is denoted as $\tau(r)$. It is perpendicular to the radial passing through the point where it is applied. In fact, it does not have any component along the radial itself because there is no angular distortion in the meridian plane defined by this radial and the $\vec{x}$ axis.

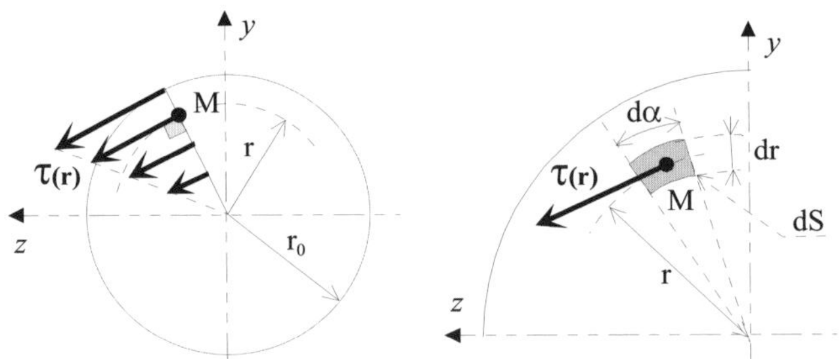

Figure 9.24. *Torsional shear stresses*

---

23 Let us suppose that S' takes the axisymmetric form of the surface S represented. Thus, S is the symmetric surface here. However, all the sections should deform in the same manner. It should then be possible to superimpose the deformed surfaces S and S'. Therefore they can only remain plane. This property attached to circular sections is a particular case. In fact, sections which do not have any geometry of revolution (also known as "non-axisymmetric") do not remain plane during the torsional deformation, they warp (see Figure 9.21). 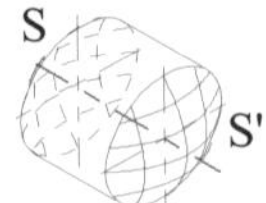

The behavior relation derived from Hooke's law for the shear is written as:

$$\tau(r) = G \times \gamma = G \times r \times \frac{d\theta_x(x)}{dx} \qquad [9.17]$$

It is verified that resultant force and moment due to the cohesion forces $\tau(r) \times dS$ at the center of the section become a pure moment along $\vec{x}$, which is the torsional moment. This moment is written as:

$$\mathcal{M}t = \int_S \tau(r) \times dS \times r$$

i.e. using the notations of Figure 9.24:

$$\mathcal{M}t = \int_{r=0}^{r=r_0} \int_{\alpha=0}^{\alpha=2\pi} \left(G \times r \times \frac{d\theta_x}{dx} \right) \times (r \times d\alpha \times dr) \times r$$

$$\mathcal{M}t = G \times \left(\pi \times \frac{r_0^4}{2} \right) \times \frac{d\theta_x}{dx} \qquad [9.18]$$

where the form of [9.16] is recognizable. Constant J can be derived as[24]:

$$J = \pi \times \frac{r_0^4}{2}$$

To conclude, the distribution of shear stresses is given by [9.17] and [9.18] in the form:

$$\tau_M(r) = \frac{\mathcal{M}t}{J} \times r = \frac{\mathcal{M}t}{\pi \times \dfrac{r_0^4}{2}} \times r \qquad [9.19]$$

9.3.2.4. *Case of a non-circular cross-sectional shape*

O Torsion center

Figures 9.21 and 9.23a illustrate rotations of two different cross-sectional shapes around the $\vec{x}$ axis under the action of a longitudinal moment $\mathcal{M}_x$. In these two

---

24 We note that for this particular case of circular section, J coincides with $I_0 = \pi \times \dfrac{r_0^4}{2}$, quadratic moment of the section with respect to its geometric center, or so-called polar quadratic moment.

special cases (rectangular and circular sections) we observe the existence of a center of symmetry merging with the geometric center G.

When the cross-sections do not have a center of symmetry, they pivot around an axis parallel to but different from $\vec{x}$. Figure 9.25b shows sections, rotating around an axis that cuts through the plane (yz) of the cross-section at a point C, called the "torsion center". Thus, during torsion all the generators of the beam become helical curves except a particular generator, remaining straight, which is the locus of the torsion centers.

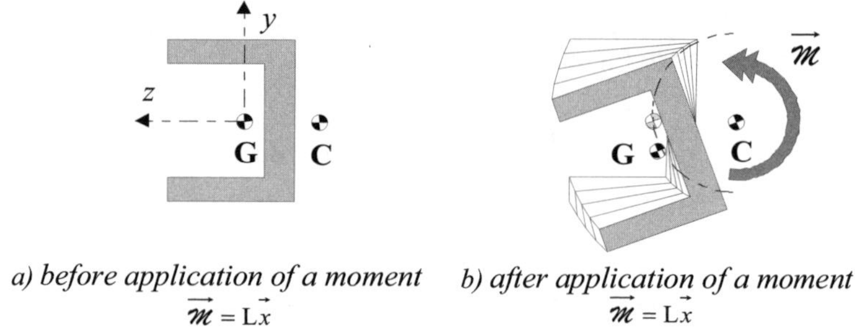

a) before application of a moment   b) after application of a moment
$$\vec{\mathcal{M}} = \text{L}\vec{x}$$ $$\vec{\mathcal{M}} = \text{L}\vec{x}$$

Figure 9.25. *Torsion of a beam with non-symmetrical cross-sectional shape*

Let us now delete the moment $\vec{\mathcal{M}} = \text{L}\vec{x}$ applied at the extremity of the beam and exert a load $\vec{F} = \text{Y}\vec{y}$ at the geometric center of this same end cross-section. The deflected shape of the beam is outlined in Figure 9.26a. The beam bends in the (xy) plane and this bending is accompanied by a torsion, i.e., the sections translate vertically along y and rotate around x (angle $\theta_x(x)$). Nevertheless, the corresponding rotation is in a direction opposite to the one represented in Figure 9.25b.

Let us suppose that it is possible to physically displace the application point of $\vec{F}$ along the $\vec{z}$ axis. Figures 9.26b and 9.26c show the displacements observed when this point of application moves towards the right. We note that when the line support of $\vec{F}$ passes through the torsion center C, the pivoting $\theta_x(x)$ of sections disappears. The torsion disappears and the beam just bends in the (xy) plane.

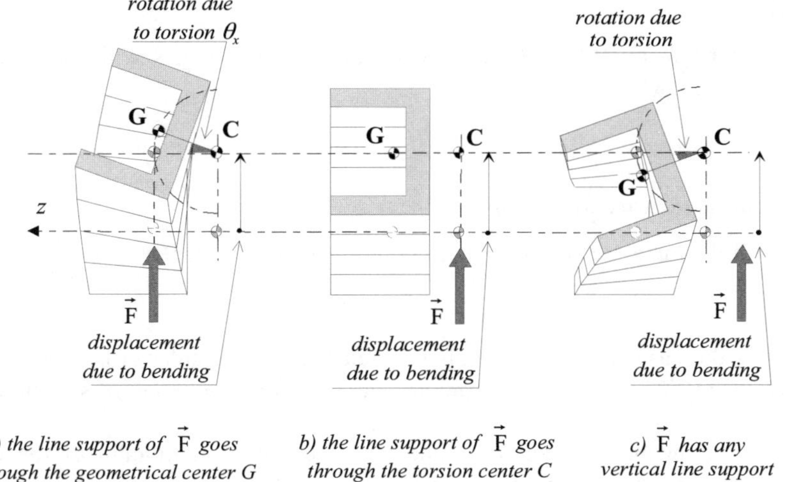

a) the line support of $\vec{F}$ goes through the geometrical center G   *b) the line support of $\vec{F}$ goes through the torsion center C*   *c) $\vec{F}$ has any vertical line support*

Figure 9.26. *Displacements of the end section under a force $\vec{F}$*

This shows that when the resultant force and moment of cohesion forces are calculated, the longitudinal moment $\mathcal{M}_x$ that appears in [9.6] should not be always considered as the torsional moment $\mathcal{M}t$. The torsional moment, i.e. the one that produces torsion, is in fact the component along $\vec{x}$ axis of the moment of cohesion forces taken *at torsion center C*.

Using [9.6] we can write:

$$\{Coh._{\mathrm{II/I}}\}_G = \left\{ \frac{\overrightarrow{\mathcal{R}_{\mathrm{II/I}}}}{\overrightarrow{\mathcal{M}_{G(\mathrm{II/I})}}} \right\}_G \Rightarrow \{Coh._{\mathrm{II/I}}\}_C = \left\{ \frac{\overrightarrow{\mathcal{R}_{\mathrm{II/I}}}}{\overrightarrow{\mathcal{M}_{C(\mathrm{II/I})}} = \overrightarrow{\mathcal{M}_{G(\mathrm{II/I})}} + \overrightarrow{CG} \wedge \overrightarrow{\mathcal{R}_{\mathrm{II/I}}}} \right\}_C$$

Let y_C and z_C be the coordinates of the torsion center in the main axes $\vec{y}$ and $\vec{z}$ of the section. Using [9.6] we have:

$$\overrightarrow{\mathcal{M}_{C(\mathrm{II}\to\mathrm{I})}} = \overrightarrow{\mathcal{M}_{G(\mathrm{II}\to\mathrm{I})}} - \left(y_C\,\vec{y} + z_C\,\vec{z} \right) \wedge \left(\mathcal{N}\,\vec{x} + \mathcal{T}_y\,\vec{y} + \mathcal{T}_z\,\vec{z} \right)$$

whose projection on the $\vec{x}$ axis is:

$$\mathcal{M}_x - y_C\mathcal{T}_z + z_C\mathcal{T}_y$$

The torsional moment can therefore be defined by the equation:

$$\mathcal{M}t = \mathcal{M}_x - y_C\mathcal{T}_z + z_C\mathcal{T}_y \tag{9.20}$$

NOTES

❑ According to [9.20], the torsional moment $\mathcal{M}t$ merges with the longitudinal moment $\mathcal{M}_x$ when:

$$-y_C \mathcal{T}_z + z_C \mathcal{T}_y = 0$$

This is the case for example [25]:

- if the torsion center is merged with the geometric center, then:

$$y_C = z_C = 0$$

as was the case with the preceding circular and rectangular cross-sections, or for sections having two axes of symmetry and therefore one center of symmetry;

- if the projections $\mathcal{T}_y$ and $\mathcal{T}_z$ (or shear resultants) of the cohesive resultant force and moment are zero:

$$\mathcal{T}_y = \mathcal{T}_z = 0$$

- if $y_C = 0$ (the $\vec{z}$ axis is the axis of symmetry for the section) and $\mathcal{T}_y = 0$;

- if $z_C = 0$ (the $\vec{y}$ axis is the axis of symmetry for the section) and $\mathcal{T}_z = 0$;

This can be summarized in the following table.

The torsional moment is merged with the longitudinal moment, $\mathcal{M}t = \mathcal{M}_x$ in any of the following cases
the geometric center G of the section is also the center of symmetry
the shear resultants are zero: $\mathcal{T}_y = \mathcal{T}_z = 0$
the $\vec{z}$ axis is an axis of symmetry for the section and $\mathcal{T}_y = 0$
the $\vec{y}$ axis is an axis of symmetry for the section and $\mathcal{T}_z = 0$

$$[9.21]$$

---

25 Here we make the exception of the very particular case that verifies this relation with non-zero values for the four quantities, i.e. yc/Ty=zc/Tz.

❏ When the section accepts an axis of symmetry, points C and G are on this axis of symmetry (for example, profiles with U or equal-flanged L section shapes; see Figure 9.27).

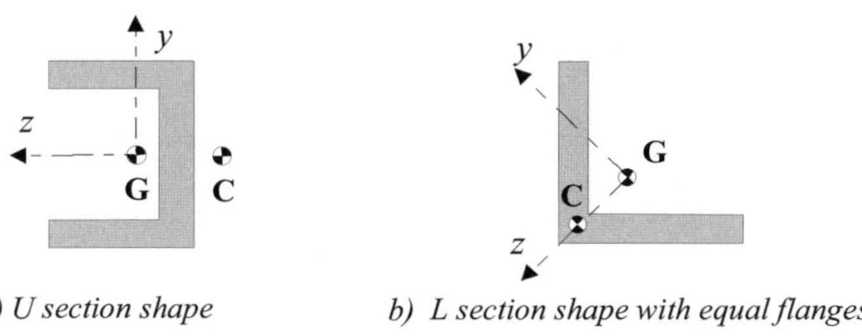

a) U section shape   *b) L section shape with equal flanges*

Figure 9.27. *Cross-sectional shapes with one axis of symmetry*

❏ Figure 9.28a shows a pictorial view of the deflected beam shape of Figure 9.25b. The deflected shape for bending without torsion, already indicated in Figure 9.26b, is outlined in Figure 9.28b. Figure 9.28c shows the same beam clamped at one end. A pure moment $\overrightarrow{\mathcal{M}}$ is exerted at the free end. Then the clamped section cannot warp (see the form of the warp in Figure 9.29b. In the clamping zone, torsion is said to be "constrained" and this phenomenon modifies deformations and stresses. "Sufficiently far" from this clamping, the torsion can be considered as uniform, i.e., similar to the one in Figure 9.28a.

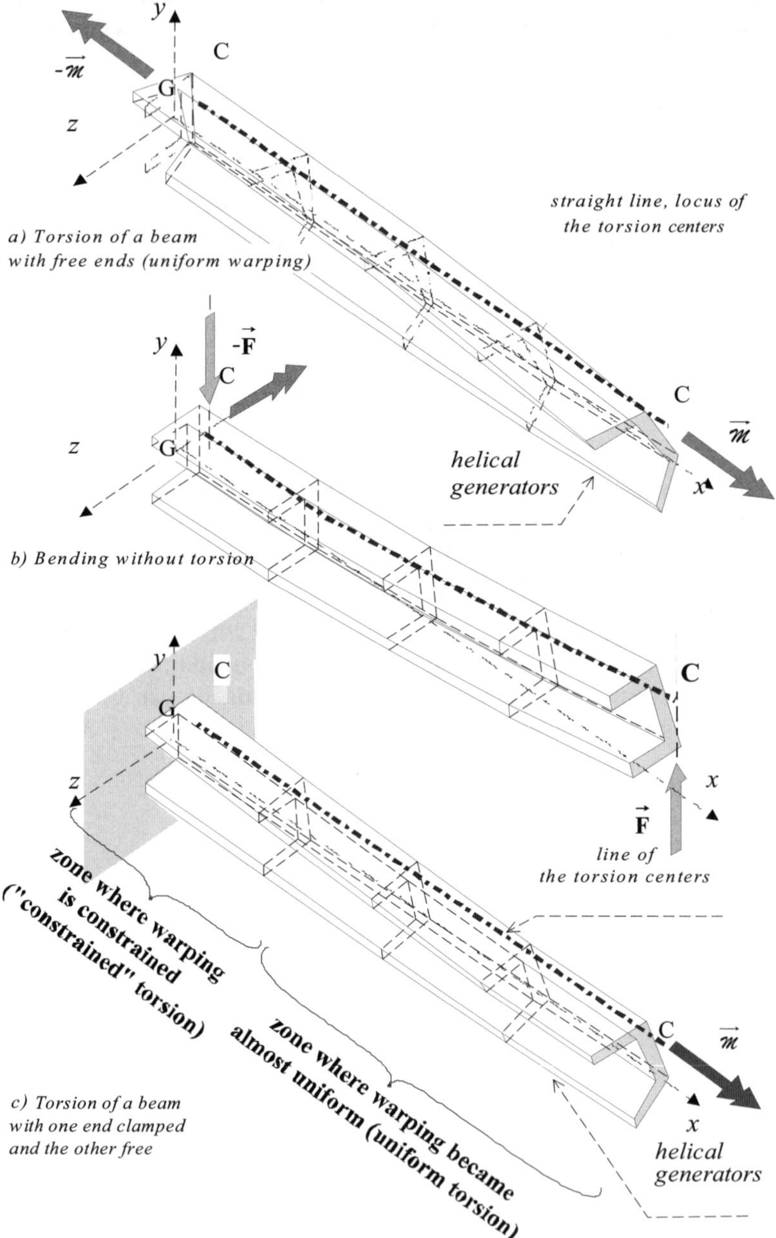

Figure 9.28. *Torsion of a U profile*

○ Displacement field

Let us take the case where $Mt = \mathcal{M}_x = C^{te} \; (\forall\, x)$ with a torsion center $C \neq G$. As per [9.21], this case involves zero shear resultants $\mathcal{T}_y = \mathcal{T}_z = 0$. Let y_C and z_C be the coordinates of the torsion center C in the main[26] quadratic axes. It is assumed that the displacement of every point M of the cross-section having coordinates y and z results not only in a rotation around the axis of the torsion centers, as has already been seen, but also in a displacement outside the initial plane of the section, directed along $\vec{x}$ axis, and called the "warping displacement"[27]. These two types of displacements are illustrated in Figure 9.29.

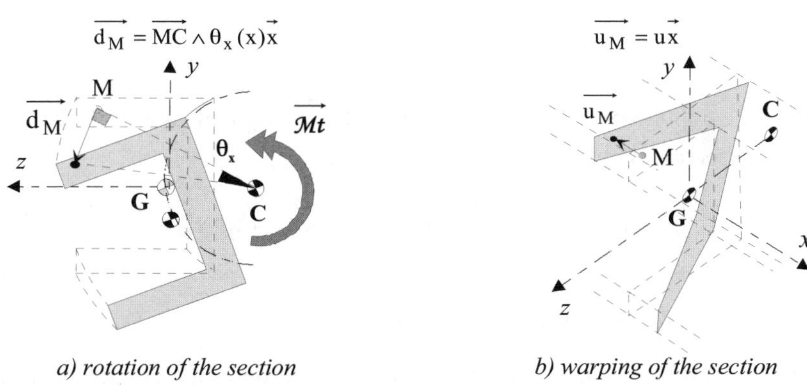

<table>
<tr><td align="center">a) rotation of the section</td><td align="center">b) warping of the section</td></tr>
</table>

Figure 9.29. *Displacement of every point M:* $\overrightarrow{d_M} + \overrightarrow{u_M}$

The displacement due to the rotation y is $\overrightarrow{d_M}$ (Figure 9.29a). We have:

$$\overrightarrow{d_M} = \overrightarrow{MC} \wedge \theta_x\,\vec{x} = (y_C - y)\vec{y} \wedge \theta_x\,\vec{x} + (z_C - z)\vec{z} \wedge \theta_x\,\vec{x}$$

i.e.:

$$\overrightarrow{d_M} = -(y_C - y)\theta_x\,\vec{z} + (z_C - z)\theta_x\,\vec{y}$$

---

26 See [9.1].

27 By examining previously the behavior of a beam with circular cross-section under torsion (section 9.3.2.1), the axisymmetric considerations led to the conclusion that such cross-sections remained plane and perpendicular to the longitudinal axis $\vec{x}$. The longitudinal displacement due to warping did not exist then. In the case of any section, the absence of axisymmetry does not allow such an affirmation. That is the reason why such a possibility of warping should be *a priori* foreseen.

The displacement due to warping of the section is $\overrightarrow{u_M}$ (Figure 9.29b). The torsion being uniform, the corresponding twisted form is the same for every cross-section. Thus, the displacement $\overrightarrow{u_M}$ should be independent of x. It varies with the position of the point M in the section. It is therefore a function of variables y and z. To simplify the calculations that follow, the warping displacement will be written in the following form:

$$\overrightarrow{u_M} = u\vec{x} = \frac{d\theta_x}{dx}\left(\Phi(y,z) - y\times z_C + z\times y_C\right)\vec{x}$$

where the constant rotation rate $\dfrac{d\theta_x}{dx}$ as well as a function $\Phi(y,z)$ called the "*warping function*" appear.

In brief, components of the displacement $\left(\overrightarrow{d_M} + \overrightarrow{u_M}\right)$ of any point M on the current cross-section with abscissa x are written as:

$$u(x,y,z) = \frac{d\theta_x}{dx}\left(\Phi(y,z) - y\times z_C + z\times y_C\right)$$

$$v(x,y,z) = -(z-z_C)\theta_x \qquad\qquad [9.22]$$

$$w(x,y,z) = (y-y_C)\theta_x$$

With this displacement field, let us calculate the deformations that appear in behavior relation [9.8]. By expressing the distortions based on [1.18] and by observing that displacements u, v, w are functions of several variables we have:

$$\varepsilon_x = \frac{\partial u}{\partial x} = 0$$

$$\gamma_{xy} = \frac{\partial v}{\partial x} + \frac{\partial u}{\partial y} = \frac{d\theta_x}{dx}\left(\frac{\partial \Phi}{\partial y} - z\right) \qquad\qquad [9.23]$$

$$\gamma_{xz} = \frac{\partial w}{\partial x} + \frac{\partial u}{\partial z} = \frac{d\theta_x}{dx}\left(\frac{\partial \Phi}{\partial z} + y\right)$$

where we recall the rotation rate $\dfrac{d\theta_x}{dx} = C^{te}$. Behavior relation [9.8] shows that only shear stresses remain (Figure 9.30):

$$\sigma_x = 0$$

$$\tau_{xy} = G\frac{d\theta_x}{dx}\left(\frac{\partial\Phi}{\partial y} - z\right)$$

$$\tau_{xz} = G\frac{d\theta_x}{dx}\left(\frac{\partial\Phi}{\partial z} + y\right)$$

[9.24]

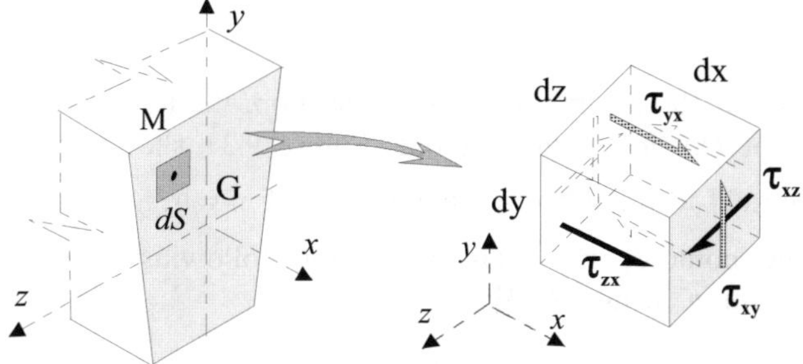

Figure 9.30. *Stresses due to torsion*

The torsional moment results from the moment of elementary cohesion forces, which are purely tangential here, i.e.:

$$\mathcal{M}t\,\vec{x} = \int_S \overrightarrow{GM} \wedge \left(\tau_{xy}\,\vec{y} + \tau_{xz}\,\vec{z}\right)dS$$

$$\mathcal{M}t = \int_S \left(y \times \tau_{xz} - z \times \tau_{xy}\right)dS = G\frac{d\theta_x}{dx}\int_S\left(\frac{\partial\Phi}{\partial z}\,y - \frac{\partial\Phi}{\partial y}\,z + y^2 + z^2\right)dS$$

where the form [9.16] is found again. The general form of the torsion constant is thus obtained as:

$$J = \int_S\left(\frac{\partial\Phi}{\partial z}\,y - \frac{\partial\Phi}{\partial y}\,z + y^2 + z^2\right)dS$$

O Determination of the function $\Phi(y, z)$

Let us use the longitudinal local equilibrium equation [9.9] recalled below:

$$\frac{\partial\sigma_x}{\partial x} + \frac{\partial\tau_{xy}}{\partial y} + \frac{\partial\tau_{xz}}{\partial z} = 0$$

with the stress expressions in [9.24], this equation reduces to:

$$\frac{\partial^2 \Phi}{\partial y^2} + \frac{\partial^2 \Phi}{\partial z^2} = 0 \qquad\qquad [9.25]$$

The "Laplacian" of the function $\Phi(y, z)$ is said to be zero in the domain of the cross-section.

NOTES

❏ On the contour (boundary) of the cross-section, the torsion shear stresses can only be zero or tangent to the contour. Figure 9.31 shows that a shear stress component normal to the contour cannot exist. If $\vec{n} = n_y \vec{y} + n_z \vec{z}$ is the outward normal to the contour, we shall obtain the following equation, called the "boundary condition" using [9.24]:

$$\vec{\tau} \cdot \vec{n} = 0 \Rightarrow \left(\frac{\partial \Phi}{\partial y} - z\right) n_y + \left(\frac{\partial \Phi}{\partial z} + y\right) n_z = 0 \qquad\qquad [9.26]$$

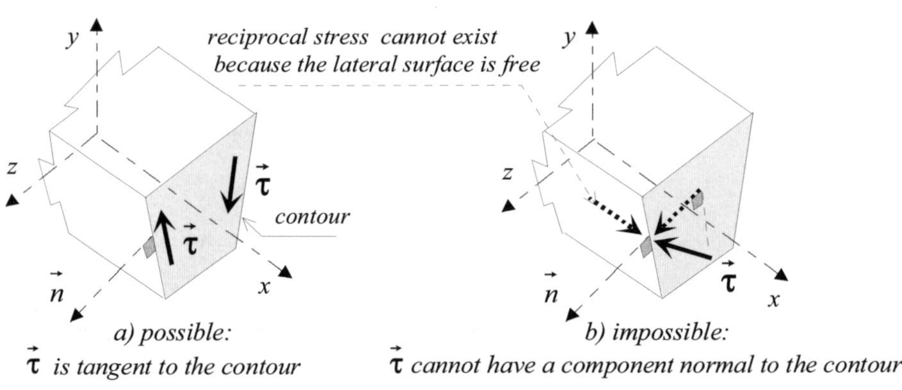

Figure 9.31. *Torsion shear stresses at boundaries*

❏ Figure 9.32 indicates the distribution pattern of the shear stresses for a rectangular section when we solve the problem of differential equation [9.25] associated with the boundary condition [9.26]. For a small number of sections with simple geometric shapes (rectangle, ellipse, triangle) the analytical solution of the problem is written with relatively heavy calculations. For any cross-sectional shapes, solution $\Phi(y, z)$ should be numerically calculated. Specific additional software, working from the input of the geometry of the section based on CAD, are available for this.

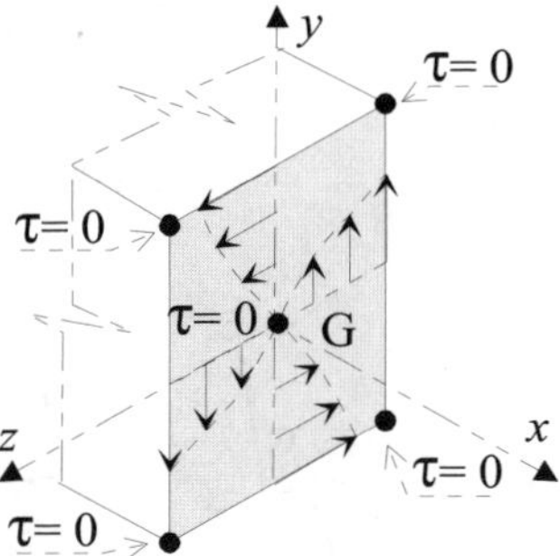

Figure 9.32. *Torsion shear stresses for a rectangular section shape*

❐ Equations [9.25] and [9.26] involve the derivatives of $\Phi(y, z)$. Thus, the latter is defined to the nearest additive constant[28].

❐ We have seen above that the general torsion constant J is written as:

$$J = \int_S \left(\frac{\partial \Phi}{\partial z} y - \frac{\partial \Phi}{\partial y} z + y^2 + z^2 \right) dS$$

In the particular case of a circular (or annular) section which does not warp[29], the displacement field [9.22] is simplified because $u(x, y, z) = 0$. This converts into $\Phi(y, z) \equiv 0$ because $y_C = z_C = 0$. Therefore, constant J becomes $J = \int_S \left(y^2 + z^2 \right) dS = I_0$, polar quadratic moment of the section that we have already seen.

O   Coordinates of the torsion center

In Figure 9.33 let us take a portion of a beam whose left end is clamped. This portion is between a current cross-section S, sufficiently distanced from the clamping, and the free end section S'.

---

28 This prevents neither the numerical calculation of torsion constant J nor that of shear stresses, because only the derivatives of $\Phi(y, z)$ appear in associated equations (see form of J above and equation [9.24]).
29 See section 9.3.2.3.

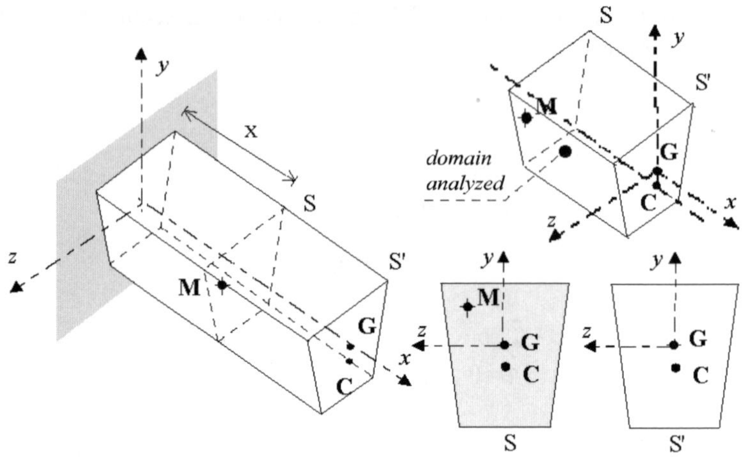

Figure 9.33. *Domain between cross-sections S and S'*

Let us apply successively on the end section S' the following loads:

♦ loading 1: a force $\vec{F} = Y\vec{y} + Z\vec{z}$ passing through the torsion center C (Figure 9.34);

♦ loading 2: a moment $\vec{\mathcal{M}} = L\vec{x}$ (Figure 9.35).

Let us describe the following works developed on the beam domain between sections S and S' (see Figure 9.33), and which correspond to increases of potential energy of deformation of this domain:

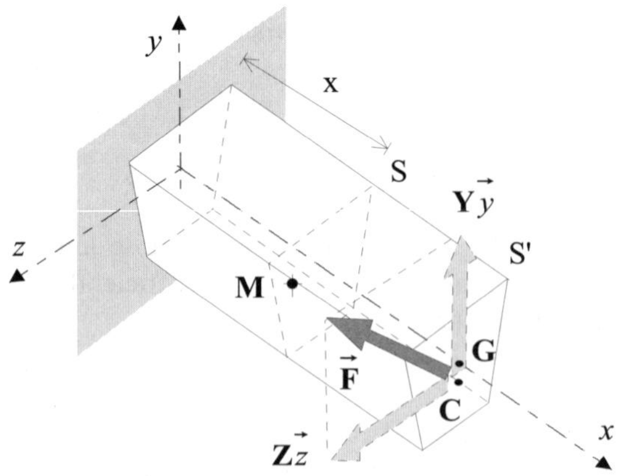

Figure 9.34. *Applying load 1*

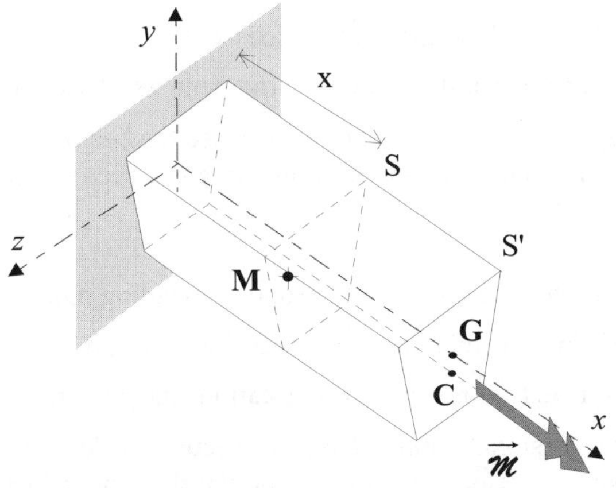

Figure 9.35. *Applying load 2*

- starting with a progressive application of $\vec{F}$, the corresponding work developed on the beam portion SS' is denoted by W_F ;

- then, $\vec{F}$ staying in place, a gradual application of $\vec{m}$ gives a work developed on the beam portion SS': $W_m + W_{(\text{effort}\,\vec{F}\times\text{displacements due to }\,\vec{m})}$.

Let us now modify the order of application of the loads:

- starting with a progressive application of $\vec{m}$, the corresponding work developed on the beam portion SS' is W_m ;

- then, $\vec{m}$ staying in place, a gradual application of $\vec{F}$ gives a work developed on the beam portion SS': $W_F + W_{(\text{moment}\,\vec{m}\times\text{displacements due to}\,\vec{F})}$

The final state being independent of the order of loading, the same increase in potential energy is observed:

$$W_F + W_m + W_{(\text{force}\,\vec{F}\times\text{displa}^{\text{ts}}\text{ due to }\,\vec{m})} = W_m + W_F + W_{(\text{moment}\,\vec{m}\times\text{displa}^{\text{ts}}\text{ due to }\,\vec{F})}$$

i.e.:

$$W_{(\text{force}\,\vec{F}\times\text{displa}^{\text{ts}}\text{ due to }\,\vec{m})} = W_{(\text{moment}\,\vec{m}\times\text{displa}^{\text{ts}}\text{ due to}\,\vec{F})}$$

These works are developed on cross-sections S and S' because there is no force acting on the lateral surface of the portion. We note that:

– on section S' force $\vec{F}$ applied at C does not produce any work because the section pivots around $C\vec{x}$. In the same way moment $\vec{\mathcal{M}}$ does not produce any work when force $\vec{F}$ is applied at torsion center C because the beam bends without rotation as indicated in the beginning of the section (no cross-section turns around an axis parallel to $\vec{x}$);

– on section S force $\vec{F}$ applied on S' produces bending moments $\mathcal{M}f_y$ and $\mathcal{M}f_z$ which cause a distribution of normal stresses[30] and a corresponding displacement field $u_{M\,bending}$ directed along $\vec{x}$ [31] (no rotation due to torsion). Then the shear stresses due to the torsional moment on the section S do not work. The second member of the above equation is then zero, and the first one reduces to:

$$W_{(\text{force}\,\vec{F}\times\text{displa}^{\text{ts}}\text{ due to }\vec{\mathcal{M}})} = \int_S - \underset{bending}{\sigma_x} \times dS \times \underset{torsion}{u_M} = 0 \quad [32]$$

i.e. with [9.22] and [9.36] [30]:

$$\int_S \left(\frac{\mathcal{M}f_y}{I_y} \times z - \frac{\mathcal{M}f_z}{I_z} \times y \right) \times dS \times \frac{d\theta_x}{dx}\left(\Phi(y,z) - y \times z_C + z \times y_C \right) = 0$$

which, keeping in mind the properties of the main axes [9.1], leads to:

$$\frac{\mathcal{M}f_y}{I_y} \int_S \left(z \times \Phi(y,z) + z^2 \times y_C \right) dS - \frac{\mathcal{M}f_z}{I_z} \int_S \left(y \times \Phi(y,z) - y^2 \times z_C \right) dS = 0$$

This equation should be verified when force $\vec{F}$, while still acting at C, varies in magnitude and in direction in the plane of the cross-section. In other words it remains valid when $\mathcal{M}f_y$ and $\mathcal{M}f_z$ vary in an independent manner. Based on this comment the coordinates of the torsion center C^{33} are derived:

---

30 The expression used here for normal stresses will be justified in the following section (see section 9.3.3).

31 See section 9.3.4.2.

32 We note on the figure below the justification of the term $\left(-\sigma_x\right)$.

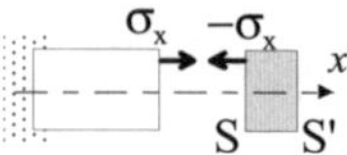

33 We can recall (see [9.1]) that $\int_S z^2 dS = I_y$ and $\int_S y^2 dS = I_z$.

$$y_C = -\frac{1}{I_y}\int_S z \times \Phi(y,z) \times dS \; ; \quad z_C = \frac{1}{I_z}\int_S y \times \Phi(y,z) \times dS$$

➤ *In brief*: uniform torsion of a beam with a non-circular cross-section is characterized by relations in the following table.

Uniform torsion of a beam with any cross-section
$\vec{x}$, $\vec{y}$, $\vec{z}$ are the main axes of a current cross-section

In every cross-section: only non-zero resultant elementary cohesion forces: $Mt = \mathcal{M}_x = C^{te}$: torsional moment along $\vec{x}$	displacement of a cross-section: $\theta_x(x)$ rotation of the cross-section about axis parallel to $\vec{x}$

behavior relation for the torsional moment: $$Mt = G \times J \times \frac{d\theta_x(x)}{dx}$$
torsional constant J: $$J = \int_S \left(\frac{\partial \Phi}{\partial z} y - \frac{\partial \Phi}{\partial y} z + y^2 + z^2 \right) dS$$
torsion shear stresses: $$\tau_{xy} = G \frac{d\theta_x}{dx}\left(\frac{\partial \Phi}{\partial y} - z \right) \; ; \; \tau_{xz} = G \frac{d\theta_x}{dx}\left(\frac{\partial \Phi}{\partial z} + y \right)$$
torsion center C: its coordinates in main axes are: $$y_C = -\frac{1}{I_y}\int_S z \times \Phi(y,z) \times dS \; ; \quad z_C = \frac{1}{I_z}\int_S y \times \Phi(y,z) \times dS$$
warping function $\Phi(y,z)$:   It is (to a nearest additive constant) the solution to the problem: $$\frac{\partial^2 \Phi}{\partial y^2} + \frac{\partial^2 \Phi}{\partial z^2} = 0 \text{ in the cross-section}$$ $\dfrac{\partial \Phi}{\partial y} \times n_y + \dfrac{\partial \Phi}{\partial z} \times n_z = z \times n_y - y \times n_z$ on the contour of the cross-section domain

$$[9.27]$$

NOTE

❐ Right from the beginning of this study[34] we have emphasized the case where $Mt = M_x$, and we have noticed, according to [9.21], that this was equivalent to considering zero shear resultants $T_y = T_z = 0$.

In other cases, the torsional moment should be calculated with the help of equation [9.20], recalled below:

$$Mt = M_x - y_C T_y + z_C T_z$$

Then, for such cases, results [9.27] given in the previous table provide only the "uniform torsion" aspect of the overall behavior of the beam. Apart from getting twisted, the latter should also:

– bend, as indicated in the example in Figure 9.26;

– be subjected to the shear resultants T_y and T_z.

These two phenomena, analyzed in sections 9.3.3 and 9.3.4 are given further on. We shall see that they result in specific displacements and stresses in addition to the results given in [9.27].

9.3.2.5. *Torsion characteristics for some particular cross-sectional shapes*

The following table indicates the calculation results for some geometrically simple sections shapes. For any other cross-sectional shapes, loading the CAD geometry in specific additional software allows calculation of the warping function $\Phi(y, z)$ and associated quantities given in [9.27].

---

34 See section 9.3.2.4.

section shape	type	torsional shear stresses	torsion constant J
	two axes of symmetry $\vec{y}$ and $\vec{z}$	$\tau = \dfrac{Mt \times r}{I_0}$	$J = I_0 = \dfrac{\pi \times r_0^{4}}{2}$
	thin open section which can be considered as a deformed thin rectangle $\ell \times e$	$\tau_{max} \cong \dfrac{Mt \times e}{J}$	$J = \beta \times \ell \times e^{3}$ with $\beta = f\left(\dfrac{\ell}{e}\right)$ $\beta = \dfrac{1}{3}$ if $\dfrac{\ell}{e} \gg 1$
	thin tube of any form and of thickness e (which can vary along the length)	$\tau = \dfrac{Mt}{2eA}$	$J = \dfrac{4 \times A^{2}}{\oint \dfrac{d\ell}{e}}$ (if $e = C^{te}$: $J = \dfrac{4e \times A^{2}}{\ell}$)

NOTE

❐ Comparison of torsional constants of two thin-walled tubes.

Let us consider two tubes with the same radius r, same thickness e, and same length ℓ, one of which is closed and the other open along a generator, as is represented in the table below. The sections are thin, i.e. $\dfrac{e}{r} \ll 1$.

type of section shape	shear stress		torsional constant J
closed tube (1)	$\tau_1 = \dfrac{Mt}{2eA} = \dfrac{1}{2\pi}\dfrac{Mt}{r^2 e}$		$J_1 = 2\pi r^3 e = I_0$
open tube (2)	$\tau_2 = \dfrac{Mt}{J}e = \dfrac{3}{2\pi}\dfrac{Mt}{re^2}$		$J_2 = \dfrac{2}{3}\pi re^3 \ll I_0$

Let us apply the same torsional moment on both the tubes. The extremities of each tube rotate by an angle θ:

– for the closed tube (1): $Mt = G\,J_1\,\dfrac{\theta_1}{\ell}$

– for the open tube (2): $Mt = G\,J_2\,\dfrac{\theta_2}{\ell}$

from which we can derive: $\theta_1 J_1 = \theta_2 J_2$.

Using the values of the previous table, the following ratios can be noted:

torsion rigidity ratio	maximum shear stress ratio
$\dfrac{\theta_2}{\theta_1} = \dfrac{J_1}{J_2} = \dfrac{3r^2}{e^2} \gg 1$	$\dfrac{\tau_2}{\tau_1} = \dfrac{3r}{e} \gg 1$
the open tube is much more flexible under torsion than the closed tube	*the tangential stresses are much higher in the open tube*

We can have an idea of the increased torsion flexibility of the split tube by analyzing the deflected shapes due to torsion in Figure 9.36.

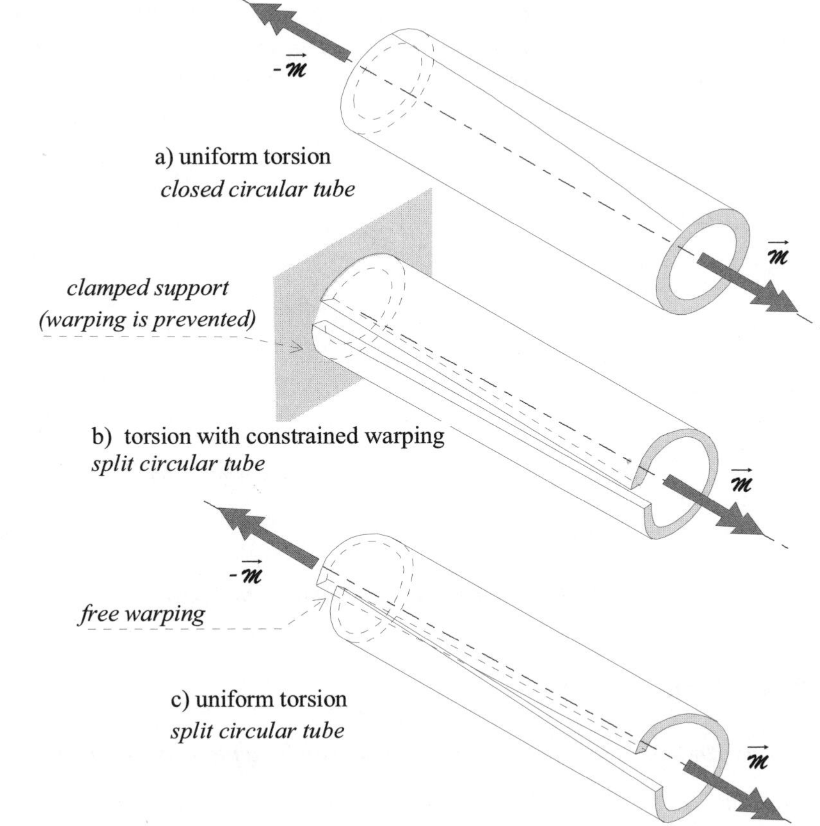

Figure 9.36. *Deflected shapes due to torsion. The torsional rigidities decrease from a) to c)*

9.3.2.6. *Torsion with constrained warping*

Figure 9.36b shows the nature of the deflected shape for the split tube whose one end is clamped. The clamped section cannot warp. It can then be easily conceived that, for an identical torsional moment, rotation due to torsion in Figure 9.36 shall be less in case b) than in case c)[35]. The torsional rigidity increases when the relative

---

35 As a general note, when we eliminate the displacement possibilities of certain zones of a linear elastic structure, it is evidently less deformable, i.e. more rigid.

displacements along the sides of the split are prevented. Concerning the closed circular tube, it is also seen that when the corresponding end is clamped (Figure 9.36a), the torsional rigidity will not be affected in any way because the circular sections do not warp. This rigidification phenomenon is accentuated in the case of thin-walled cross-section profiles called "open", i.e., there are no box profiles. A typical case is the split tube above, or the I profile of Figure 9.37.

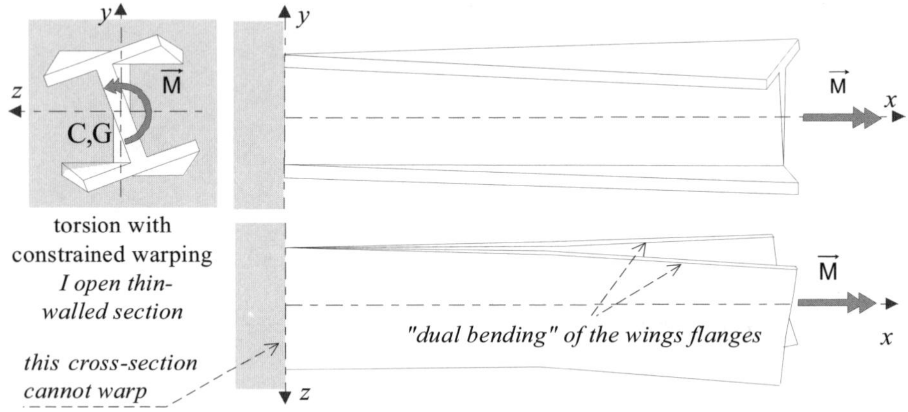

Figure 9.37. *Constrained torsion warping of an I-profile*

9.3.3. *Pure bending*

9.3.3.1. *Pure bending in the specific case of a beam with a plane of symmetry*

O Definition

In Figure 9.38a, we are looking at a beam with a plane of symmetry (plane (xy)). The $\vec{x}$ axis passes through the geometric centers of the sections marked G. The $\vec{y}$ axis is a main[36] quadratic axis. The left end of the beam is clamped and is loaded on its right end by moment $\vec{\mathcal{M}} = N\vec{z}$.

---

36 See [9.1].

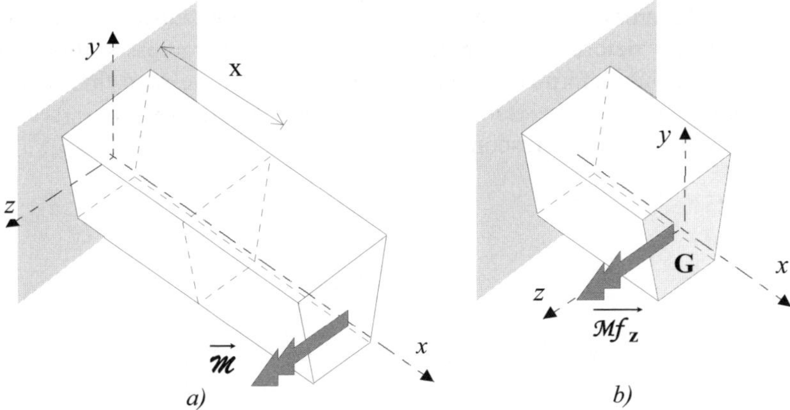

Figure 9.38. *Pure bending*

Then on the current section of abscissa x, the resultant force and moment of the cohesive forces are written at the cross-section center G as:

$$\left\{Coh_{(x^+/x^-)}\right\}_G = \left\{\begin{array}{c} \overrightarrow{\mathcal{R}_{(x^+/x^-)}} = \vec{0} \\ \hline \overrightarrow{\mathcal{M}_{G(x^+/x^-)}} = N\vec{z} \end{array}\right\}_{G,r}$$

The only non-zero projection (see section 9.1.4) is the bending moment $\mathcal{M}f_z$ which is identical on any cross-section.

$$\mathcal{M}f_z = N$$

This is the so-called *pure bending* phenomenon.

O Deformation of an elementary beam slice

Given the existence of symmetry plane (*xy*) for the beam, the deflected shape of the beam after bending also admits this plane as symmetry plane. It is said that the beam bends in its plane of symmetry.

In Figure 9.39 let us concentrate on two adjacent elementary beam slices of same length dx. They are loaded in an identical manner (same bending moment $\mathcal{M}f_z$) which means that the deflected shapes should be identical and each one should admit a plane of symmetry. Starting from such considerations Figure 9.39a represents the two deformed elements. It is obvious that with such a deformed shape it is impossible to join them, or "stick" to them. Thus these deformations are termed "incompatible". The only case of "compatible deformation" is shown in Figure

9.39b, where the cross-sections remain planes. In addition, ensuring the curvature continuity of the deformed centerline involves that the cross-sections remain perpendicular to this deformed centerline. Consequently, we note that in the pure bending case, the restrictive Bernoulli's hypothesis on deformations (see section 9.1.5.2 and Figure 9.11c) corresponds to reality.

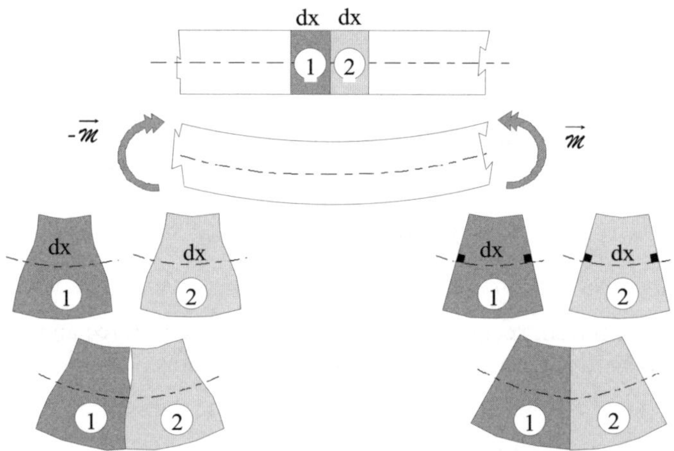

a) deformations of adjacent slices are identical and each with a plane of symmetry; nevertheless, these deformations are incompatibles

b) deformations of adjacent slices are now compatible; moreover, curvature continuity of the deformed centerline is ensured

Figure 9.39. *Deformation of elementary beam slices*

The two cross-sections limiting an elementary slice of beam pivot one with respect to the other with a rotation angle $d\theta_z$ about $\vec{z}$ axis, that we assume to be the main axis passing through center G[37] (see Figure 9.40). From this beam slice let us isolate a small prismatic domain of length dx and of base area dS around a point M of coordinates y and z as represented in Figure 9.40a. In Figure 9.40b, we highlight the contraction of this small domain as well as the elongation of another similar small domain. These domains can be considered as minuscule testing samples under uniaxial strain.

---

37 This hypothesis will be proved in the *"NOTES"* that follow.

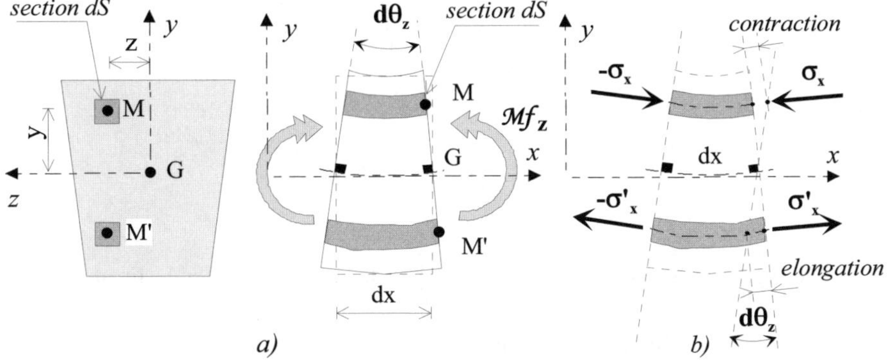

Figure 9.40. *Deformation of testing micro-samples in a slice of beam*

As the displacements are very small, the contraction value marked as du is:

$$\mathrm{d}u\vec{x} = \overrightarrow{MG} \wedge \mathrm{d}\theta_z \vec{z} = -\left(y\vec{y} + z\vec{z}\right) \wedge \mathrm{d}\theta_z \vec{z} = -\left(y\mathrm{d}\theta_z\right)\vec{x}$$

The initial length being dx, the relative elongation is written as[38]:

$$\frac{\mathrm{d}u}{\mathrm{d}x} = \varepsilon_x = -y\frac{\mathrm{d}\theta_z}{\mathrm{d}x} \quad [39]$$

Applying Hooke's law for this small element, we have:

$$\varepsilon_x = \frac{\sigma_x}{E} = -y\frac{\mathrm{d}\theta_z}{\mathrm{d}x} \Rightarrow \sigma_x = -E \times y\frac{\mathrm{d}\theta_z}{\mathrm{d}x} \qquad [9.28]$$

The basic cohesion force which acts on this "testing micro-sample" is:

$$\overrightarrow{\mathrm{d}f} = \sigma_x \vec{x} \times \mathrm{d}S$$

Let us calculate the moment with respect to the $\vec{z}$ axis passing through G (Figure 9.40a) of elementary forces of this type acting on all the micro-samples constituting the beam slice. Since they are cohesion forces, the bending moment Mf_z should be found. Therefore, we have:

---

38 See [1.6].

39 It is clear from Figure 9.40b that the concerned micro-sample *contracts*, $\mathrm{du(x)} < 0$, because y>0 and $\mathrm{d}\theta_z(x) > 0$, while the other lower micro-sample *elongates*, $\mathrm{du(x)} > 0$, because y<0 and $\mathrm{d}\theta_z(x) > 0$.

$$\mathcal{M}f_z\,\vec{z} = \int_S y\vec{y} \wedge \vec{df} = \int_S -y\sigma_x\,dS\vec{z}$$

i.e. with [9.28]:

$$\mathcal{M}f_z = \int_S E \times y^2 \frac{d\theta_z}{dx}\,dS$$

where the main quadratic moment $I_z = \int_S y^2 dS$ can be noticed. From this the behavior relation for the bending moment $\mathcal{M}f_z$ can be written as:

$$\mathcal{M}f_z = E \times I_z \frac{d\theta_z}{dx} \qquad [9.29]$$

○ Stresses on a cross-section

The following normal bending stress σ_x is derived from [9.28]:

$$\sigma_x = -\frac{\mathcal{M}f_z}{I_z} \times y \qquad [9.30]$$

Thus, if $\mathcal{M}f_z$ is known, the distribution of normal stresses is known. It is linearly distributed on the section. This is illustrated in Figure 9.41b for a rectangular cross-section.

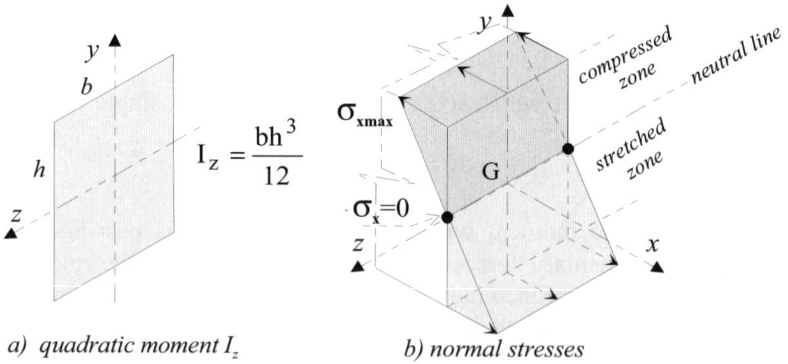

Figure 9.41. *Normal bending stresses on a rectangular cross-section*

NOTES

❑ From equation [9.29] we can observe the proportionality between the bending moment $\mathcal{M}f_z$ and what is called the bending deformation $\dfrac{d\theta_z}{dx}$. The proportionality coefficient EI_z can be designated by the term "bending stiffness". Thus, for a beam to be rigid during bending, it should have:

• a material with high Young's modulus E ("rigid material");

• a section shape to match the largest possible quadratic moment I_z , and nevertheless suitable for the proposed application. For example, if the rectangular section of Figure 9.41a should retain the same area $b \times h$, it would be better that in the expression $I_z = \dfrac{bh^3}{12}$ height h be increased to the disadvantage of the width b to obtain a quadratic moment[40] as high as possible.

❑ We can also note in equation [9.30] that a large quadratic moment can reduce the normal stresses.

❑ We assumed that the cross-sections limiting the beam slices represented in 9.36 pivoted around the main axis $\vec{z}$. This property needs to be verified. The only non-zero projection of the resultant force and moment of cohesive forces is the moment $\mathcal{M}f_z$. It is therefore observed that the normal resultant $\mathcal{N}$ [41] is zero. Besides, the latter is expressed starting from elementary cohesive forces by[42]:

$$\mathcal{N} = \int_S \sigma_x \, dS$$

With σ_x given by equation [9.28], we have:

$$0 = \mathcal{N} = \int_S -E \times y \frac{d\theta_z}{dx} \, dS = -E \frac{d\theta_z}{dx} \int_S y \, dS$$

---

40 The form of a cross-section of constant area S can be "shaped" to increase the quadratic moment I_z as is indicated qualitatively below:

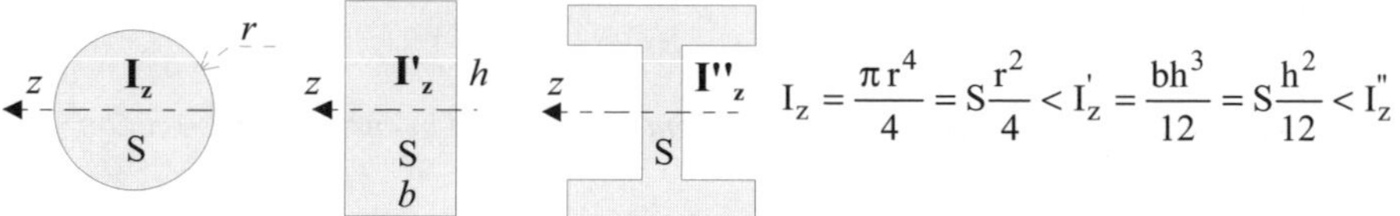

$$I_z = \frac{\pi r^4}{4} = S\frac{r^2}{4} < I'_z = \frac{bh^3}{12} = S\frac{h^2}{12} < I''_z$$

41 See [9.6].
42 See [9.13].

The zero value of $\mathcal{N}$ implies that of the integral $\int_S y\,dS$. Thus, the origin of ordinates y is at G: cross-sections are pivoting around the main axis $\vec{z}$ passing through G. In this case the $\vec{z}$ axis is also called the *neutral* axis: it is the location of the points where $\sigma_x = 0$.

❑ Another approach can be made through the displacement field.

We have seen (Figure 9.40) that in every elementary beam slice the cross-sections are subjected to a relative rotation $d\theta_z$. A current cross-section turns then by an angle $\theta_z(x)$. Let u(x), v(x), w(x) be the displacement components of a point $M(y, z)$ in this section. We have:

$$\overrightarrow{MM'} = \overrightarrow{GG'} + \overrightarrow{MG} \wedge \theta_z(x)\vec{z}$$

where M' and G' are respectively the new positions of M and G after bending (see Figure 9.42). That is with $\overrightarrow{MG} = -y\vec{y} - z\vec{z}$:

$$\overrightarrow{MM'} = u\vec{x} + v\vec{y} + w\vec{z} = \left(u(x)_G\,\vec{x} + v(x)_G\,\vec{y}\right) - y\theta_z(x)\vec{x}$$

after projection on axes $\vec{x}$, $\vec{y}$, $\vec{z}$:

$$\begin{bmatrix} u = u(x)_G - y\theta_z(x) \\ v = v(x)_G \\ w = 0 \end{bmatrix}$$

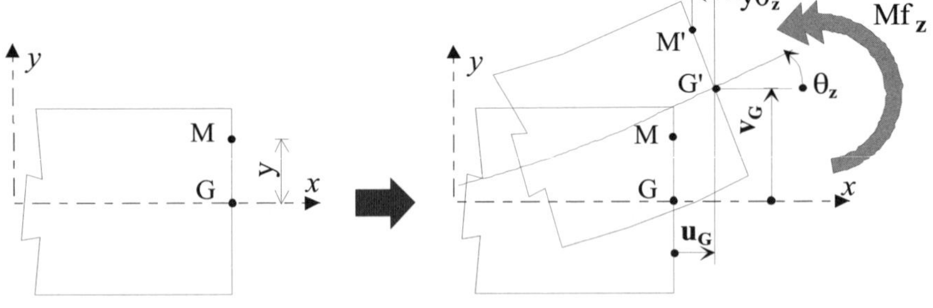

Figure 9.42. *Plane bending: displacement of a point M in current cross-section*

Let us calculate the deformation along the $\vec{x}$ axis. We have[43]:

$$\varepsilon_x = \frac{\partial u}{\partial x} = \frac{du_G}{dx} - y\frac{d\theta_z}{dx}$$

which along with Hooke's law corresponds to normal stress:

$$\sigma_x = E \times \varepsilon_x = E\frac{\partial u}{\partial x} = E\frac{du_G}{dx} - E \times y\frac{d\theta_z}{dx}$$

The only non-zero resultant force here is the bending moment $\mathcal{M}f_z$. In particular, the normal resultant $\mathcal{N}$ is zero. We therefore have[44]:

$$\mathcal{N} = \int_S \sigma_x \, dS = E\frac{du_G}{dx}S - E\frac{d\theta_z}{dx}\int_S y\,dS = 0 \ ^{45}$$

From this, $\dfrac{du_G}{dx} = 0 \Rightarrow u(x)_G = C^{te} = 0$ is derived, the zero value due to the presence of a clamped support on the left. The deformed centerline does not undergo any length variation. It is therefore also called the *"neutral line"*.

Thus, the displacement of every point M, represented in Figure 9.43, becomes:

$$\begin{cases} u(x, y, z) = -y\theta_z(x) \\ v(x, y, z) = v(x)_G \\ w(x, y, z) = 0 \end{cases} \qquad [9.31]$$

This is represented in Figure 9.43.

---

43 See [1.18].

44 See [9.13].

45 Point G is the geometric center of the section, thus $\displaystyle\int_S y\,dS = 0$.

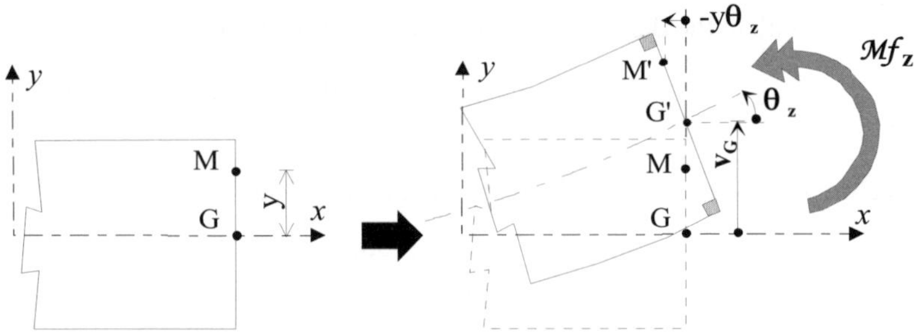

Figure 9.43. *Pure bending: displacement of a point M in the current cross-section*

Based on this displacement field let us again find the behavior relation [9.29]. For this, the moment with respect to the $\vec{Gz}$ axis of elementary cohesion forces should be calculated. By definition the bending moment Mf_z is:

$$Mf_z\,\vec{z} = \int_S y\vec{y} \wedge \sigma_x\,\vec{x}dS$$

the normal stress σ_x is given by Hooke's law [1.6], that is $\sigma_x = E \times \varepsilon_x$. With [9.31] we have:

$$\varepsilon_x = \frac{\partial u(x)}{\partial x} - y\frac{d\theta_z}{dx}\,.$$

from where:

$$\sigma_x = -E \times y\frac{d\theta_z}{dx}$$

From the previous vectorial equation, we have:

$$Mf_z = \int_S -y \times \sigma_x\,dS = E\frac{d\theta_z}{dx}\int_S y^2 dS$$

behavior relation [9.29] becomes:

$$Mf_z = E \times I_z\frac{d\theta_z}{dx}$$

❏ Given the low values of $v(x)$ and $\theta_z(x)$, it can be noticed in Figure 9.43 that:

$$\theta_z(x) \cong \tan\theta_z(x) = \frac{dv_G}{dx}$$

where $\dfrac{dv_G}{dx}$ represents the inclination of the tangent with respect to the deformed mean line.

Behavior relation [9.29] can also be written as:

$$Mf_z = E \times I_z \frac{d\theta_z}{dx} = E \times I_z \frac{d^2 v_G}{dx^2} \qquad [9.32]$$

❑ Without recalculating the above, it is possible to foresee the results of a similar study for pure bending which could take place in the main plane (xz). Instead of equation [9.32] the following similar equation can be obtained:

$$Mf_y = E \times I_y \frac{d\theta_y}{dx} = -E \times I_y \frac{d^2 w_G}{dx^2} \quad {}^{46}$$

and instead of equation [9.30]:

$$\sigma_x = \frac{Mf_y}{I_y} \times z$$

---

46 The sign "-" is justified by the sign conventions corresponding to the main plane (xz), and which are recalled in the figure below:

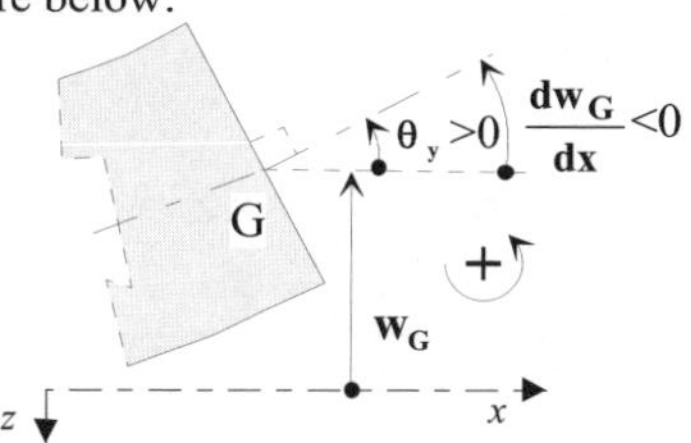

➢ *Summary:*

<table>
<tr><td colspan="2" align="center">Pure bending in the main plane (xy)</td></tr>
<tr>
<td>

resultant of the cohesion forces on a cross-section:

$\mathcal{M}f_z(x)$: bending moment along $\vec{z}$

</td>
<td>

displacements of a section:

v(x) translation along $\vec{y}$ of cross-section geometric center G

$\theta_z(x)$ rotation about $\vec{z}$ of the cross-section

</td>
</tr>
<tr>
<td colspan="2">

behavior relation for the bending moment $\mathcal{M}f_z(x)$:

$$\mathcal{M}f_z = E \times I_z \frac{d\theta_z}{dx} = E \times I_z \frac{d^2 v}{dx^2}$$

</td>
</tr>
<tr>
<td colspan="2">

normal stresses:

$$\sigma_x = -\frac{\mathcal{M}f_z}{I_z} \times y$$

</td>
</tr>
<tr><td colspan="2" align="center">Pure bending in the main plane (xz)</td></tr>
<tr>
<td>

resultant cohesion forces on a cross-section:

$\mathcal{M}f_y(x)$: bending moment along $\vec{y}$

</td>
<td>

displacements of a section:

w(x) translation along $\vec{z}$ of cross-section geometric center G

$\theta_y(x)$ rotation about $\vec{y}$ of the cross-section

</td>
</tr>
<tr>
<td colspan="2">

behavior relation for the bending moment $\mathcal{M}f_y(x)$:

$$\mathcal{M}f_y = E \times I_y \frac{d\theta_y}{dx} = -E \times I_y \frac{d^2 w}{dx^2}$$

</td>
</tr>
<tr>
<td colspan="2">

normal stresses:

$$\sigma_x = \frac{\mathcal{M}f_y}{I_y} \times z$$

</td>
</tr>
</table>

[9.33]

9.3.3.2. *The general case of pure bending*

○ Definition

A beam with any cross-section shape and without any particular symmetry is taken for study. This beam is clamped on the left side and has a coordinate system constituted by the main axes $(\vec{x}, \vec{y}, \vec{z})$ [9.1]. The $\vec{x}$ axis is the centerline passing through geometric center G of every section while $\vec{y}$ and $\vec{z}$ axes are the main axes. The displacement of the right end of the beam is free. This end is subjected to a moment within the plane (*yz*) represented in Figure 9.44.

$$\vec{\mathcal{M}} = M\vec{y} + N\vec{z}$$

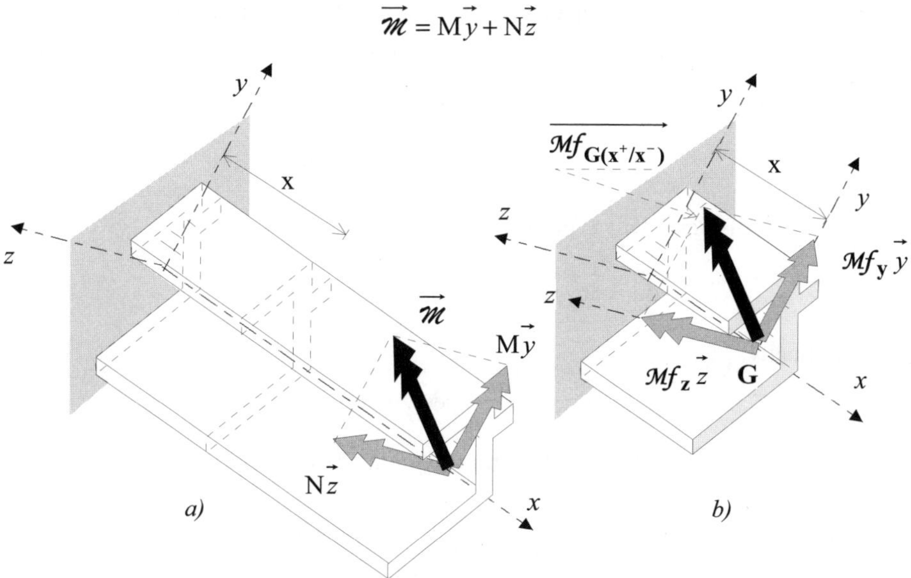

Figure 9.44. *General case of pure bending*

In the current cross-section with x abscissa, the resultant force and moment of elementary cohesion forces are written as:

$$\left\{Coh_{(x^+/x^-)}\right\}_G = \left\{\begin{array}{l} \overrightarrow{\mathcal{R}_{(x^+/x^-)}} = \vec{0} \\ \hline \overrightarrow{\mathcal{M}_{G(x^+/x^-)}} = M\vec{y} + N\vec{z} \end{array}\right\}_{G,r}$$

Bearing in mind equation [9.6], the only non-zero resultant cohesion forces has two bending moments as projections:

$$\overrightarrow{\mathcal{M}f_{G(x^+/x^-)}} = \mathcal{M}f_y\,\vec{y} + \mathcal{M}f_z\,\vec{z}$$

with:

$$\mathcal{M}f_y = M$$

$$\mathcal{M}f_z = N$$

◯ **Deformation of a beam element**

The comments of the previous section on the compatibility of deformations are valid and lead us to consider that cross-sections move while remaining plane and perpendicular to the deformed centerline. The rotation of a section is written as $\vec{\theta} = \theta_y\,\vec{y} + \theta_z\,\vec{z}$. A small elementary test sample considered in a beam slice in a similar manner to the above (see Figure 9.40) elongates by:

$$\overrightarrow{du} = \overrightarrow{du_G} + \overrightarrow{MG} \wedge \overrightarrow{d\theta}$$

where $\overrightarrow{du_G}$ is displacement at point G, i.e.:

$$\overrightarrow{du} = \overrightarrow{du_G} - \left(y\vec{y} + z\vec{z}\right) \wedge \left(d\theta_y\,\vec{y} + d\theta_z\,\vec{z}\right)$$

or algebraically:

$$du = du_G - yd\theta_z + zd\theta_y$$

Using [1.8], Hooke's law [1.6] for this element[47] is written as:

$$\frac{\sigma_x}{E} = \varepsilon_x = \frac{\partial u}{\partial x} = \frac{du_G}{dx} - y\frac{d\theta_z}{dx} + z\frac{d\theta_y}{dx}$$

➤ Point of zero displacement.

In the cross-section, let us look for a point which does not move. The normal resultant being zero, we can write on the $\vec{x}$ axis (see [9.13]):

$$\mathcal{N} = \int_S \sigma_x\,dS = E\int_S \left(\frac{du_G}{dx} - y\frac{d\theta_z}{dx} + z\frac{d\theta_y}{dx}\right) = 0$$

i.e.:

---

47 In fact we have: $du = \dfrac{\partial u}{\partial x}\,dx$.

$$\frac{du_G}{dx}S - \frac{d\theta_z}{dx}\int_S y\,dS + \frac{d\theta_y}{dx}\int_S z\,dS = 0$$

The origin of the axes being geometric center G of the section, we have (see [9.1]):

$$\int_S y\,dS = \int_S z\,dS = 0$$

from where: $\dfrac{du_G}{dx} = 0 \Rightarrow u_G(x) = C^{te} = 0$, with the zero-value due to the clamping at the left end of the beam. Here also we find the property according to which the deformed centerline (or mean line) does not undergo any length variation. It is therefore also called the "*neutral line*".

Taking into account the equation of σ_x written above, the basic force which acts on the elementary "test sample" having a section dS similar to the one defined in Figure 9.40a is:

$$\overrightarrow{df} = \sigma_x\,\vec{x}\,dS = E\left(-y\frac{d\theta_z}{dx} + z\frac{d\theta_y}{dx}\right)\vec{x}\,dS \qquad [9.34]$$

Let us calculate the moment with respect to the geometric center G of a current cross-section of all elementary cohesive forces acting on this section. The resultant moment of the cohesion forces is found as:

$$\overrightarrow{\mathcal{M}f_{G(x^+/x^-)}} = \mathcal{M}f_y\,\vec{y} + \mathcal{M}f_z\,\vec{z} = \int_S \overrightarrow{GM}\wedge\overrightarrow{df}\,dS$$

$$= \int_S \left(y\vec{y} + z\vec{z}\right)\wedge E\left(-y\frac{d\theta_z}{dx} + z\frac{d\theta_y}{dx}\right)\vec{x}\,dS$$

$$\mathcal{M}f_y\,\vec{y} + \mathcal{M}f_z\,\vec{z} = E\left(\frac{d\theta_y}{dx}\int_S z^2\,dS - \frac{d\theta_z}{dx}\int_S yz\,dS\right)\vec{y}.......$$

$$.......+ E\left(\frac{d\theta_z}{dx}\int_S y^2\,dS - \frac{d\theta_y}{dx}\int_S yz\,dS\right)\vec{z}$$

In line with the properties of the main quadratic axes [9.1], this equation is reduced to:

$$\mathcal{M}f_y\,\vec{y} + \mathcal{M}f_z\,\vec{z} = \mathrm{EI}_y\,\frac{d\theta_y}{dx}\,\vec{y} + \mathrm{EI}_z\,\frac{d\theta_z}{dx}\,\vec{z}$$

from where the behavior relations for the bending moments $\mathcal{M}f_z$ and $\mathcal{M}f_y$ on a beam cross-section are:

$$\mathcal{M}f_y = \mathrm{EI}_y\,\frac{d\theta_y}{dx} \quad ; \quad \mathcal{M}f_z = \mathrm{EI}_z\,\frac{d\theta_z}{dx} \qquad [9.35]$$

The main quadratic axes [9.1] thus prove to be useful because they bring back the study of any pure bending to that of two independent or "decoupled" pure bending in the two planes called main planes $(\vec{x}, \vec{y})$ and $(\vec{x}, \vec{z})$ (see Figure 9.45):

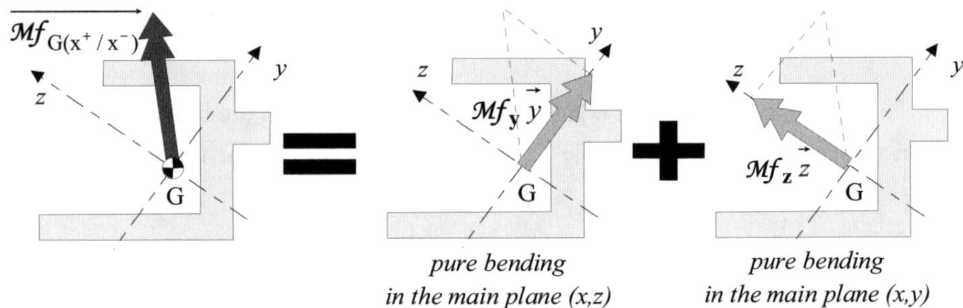

Figure 9.45. *Any pure bending*

○ Stresses on a cross-section

By reverting to equation [9.34]:

$$\sigma_x = \mathrm{E}\left(-y\,\frac{d\theta_z}{dx} + z\,\frac{d\theta_y}{dx} \right)$$

we obtain using [9.35]:

$$\sigma_x = \frac{\mathcal{M}f_y}{\mathrm{I}_y}\times z - \frac{\mathcal{M}f_z}{\mathrm{I}_z}\times y \qquad [9.36]$$

Figure 9.46 gives the nature of these normal stresses in the particular case of a rectangular section.

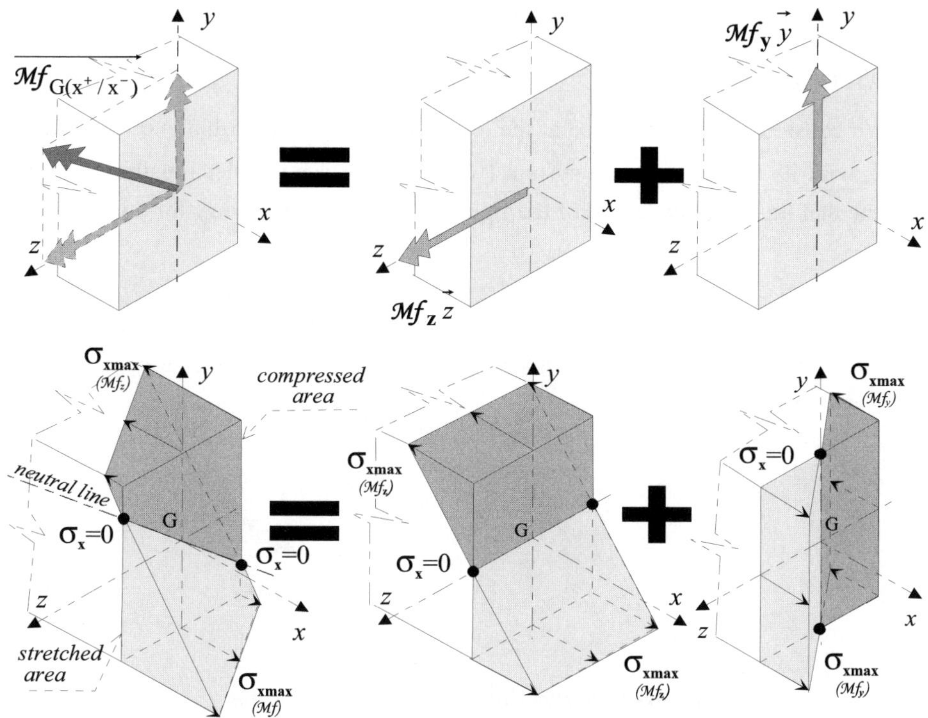

Figure 9.46. *General case of pure bending: nature of the normal stresses*

NOTE

❑ With an analogous note to [9.32], the behavior relations can also be written as in [9.33]:

$$\mathcal{M}f_z = E \times I_z \frac{d\theta_z}{dx} = E \times I_z \frac{d^2 v}{dx^2} \quad \text{and} \quad \mathcal{M}f_y = E \times I_y \frac{d\theta_y}{dx} = -E \times I_y \frac{d^2 w}{dx^2}$$

❑ As per [9.36] the straight line characterized by the equation $\frac{\mathcal{M}f_y}{I_y} \times z - \frac{\mathcal{M}f_z}{I_z} \times y = 0$ in the plane (yz) is the loci of points with zero normal stress. This is the *"neutral line"* of the section, separating the compressed area from the stretched area.

9.3.4. *Plane bending with shear resultant*

9.3.4.1. *Definition*

We consider a beam clamped at the left end and free at the right end (see Figure 9.47), with its associated main system ($\vec{x}$, $\vec{y}$, $\vec{z}$). For a simpler illustration let us assume that the main plane (*xy*) of the figure is a plane of symmetry for the beam.

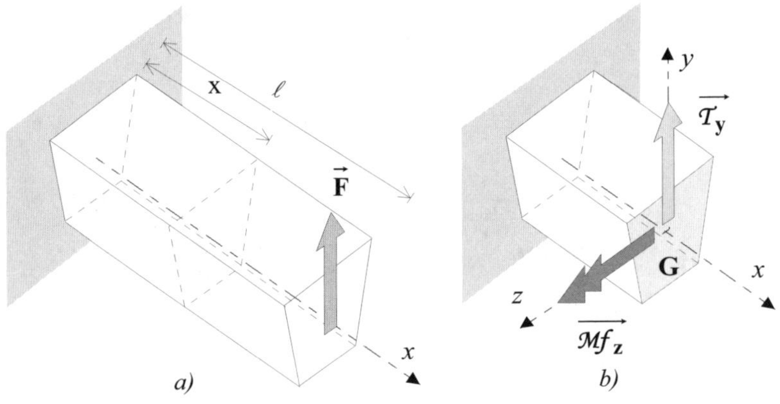

Figure 9.47. *Plane bending with shear resultant*

The beam is loaded by a force $\vec{F} = Y\vec{y}$ at its free right end. Then on the current cross-section of x abscissa, the resultant force and moment of the cohesion forces are written as:

$$\left\{Coh_{(x^+/x^-)}\right\}_G = \left\{\begin{array}{l} \overrightarrow{\mathcal{R}_{(x^+/x^-)}} = Y\vec{y} \\[2mm] \overrightarrow{\mathcal{M}_{G(x^+/x^-)}} = Y(\ell-x)\vec{z} \end{array}\right\}_{G,r}$$

The only non-zero components are (see [9.6]):

$$\left[\begin{array}{l} \mathcal{T}_y = Y \\[2mm] \mathcal{M}f_z = Y(\ell-x) \end{array}\right.$$

The effects due to the bending moment $\mathcal{M}f_z$ (deformations, stresses) are known and were dealt with in the previous section. Here a supplementary shear resultant $\mathcal{T}_y$ is observed. This is a resultant force resulting from the elementary cohesive forces applied on all facets dS of the considered cross-section. As $\mathcal{T}_y$ is located in the plane of the section, we shall assume from now on that the corresponding elementary cohesive forces are also contained in the plane of the section, i.e., they are linked to tangential – or shear – stresses. Therefore, here they involve the shear stresses τ_{xy} and τ_{xz}[48]. We shall study below these stresses and the effects they produce.

9.3.4.2. *Displacement field*

As an introduction to the analysis of the displacements, Figure 9.48 gives the nature of the deformation of a rectangular section subjected to the only shear resultant $\mathcal{T}_y$.

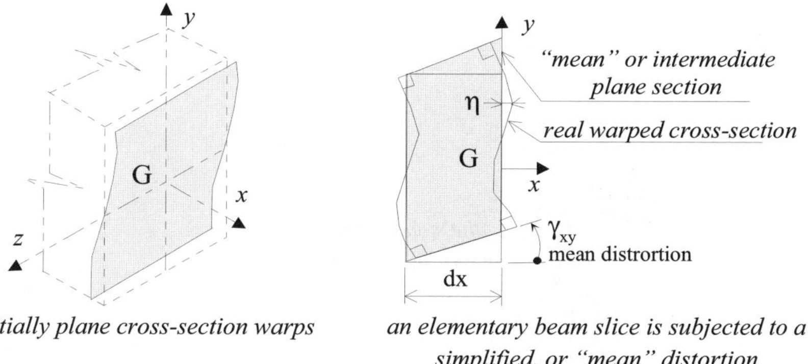

an initially plane cross-section warps   *an elementary beam slice is subjected to a simplified, or "mean" distortion*

Figure 9.48. *Deformation of a rectangular section beam under the shear resultant* $\mathcal{T}_y$

We recall that the displacement of point M(y, z) of a current cross-section under pure bending moment $\mathcal{M}f_z$ was written as (see [9.31]):

$$\begin{bmatrix} u(x, y, z) = -y\theta_z(x) \\ v(x, y, z) = v(x)_G \\ w(x, y, z) = 0 \end{bmatrix} \qquad [9.37]$$

Here the shear stresses linked to the shear resultant produce a supplementary deformation (distortion) on every small element of the beam, as can be seen in Figure 9.49b. These distortions induce a change in the shape of the section as indicated in Figures 9.48 and 9.49a. Every point of the real warped cross-section is

------

48 See Figure 9.9.

derived from a point of the "mean" plane section through a (small) displacement in the direction $\vec{x}$, noted as $\eta(x, y, z)$ in Figure 9.48 as well as in Figure 9.49a.

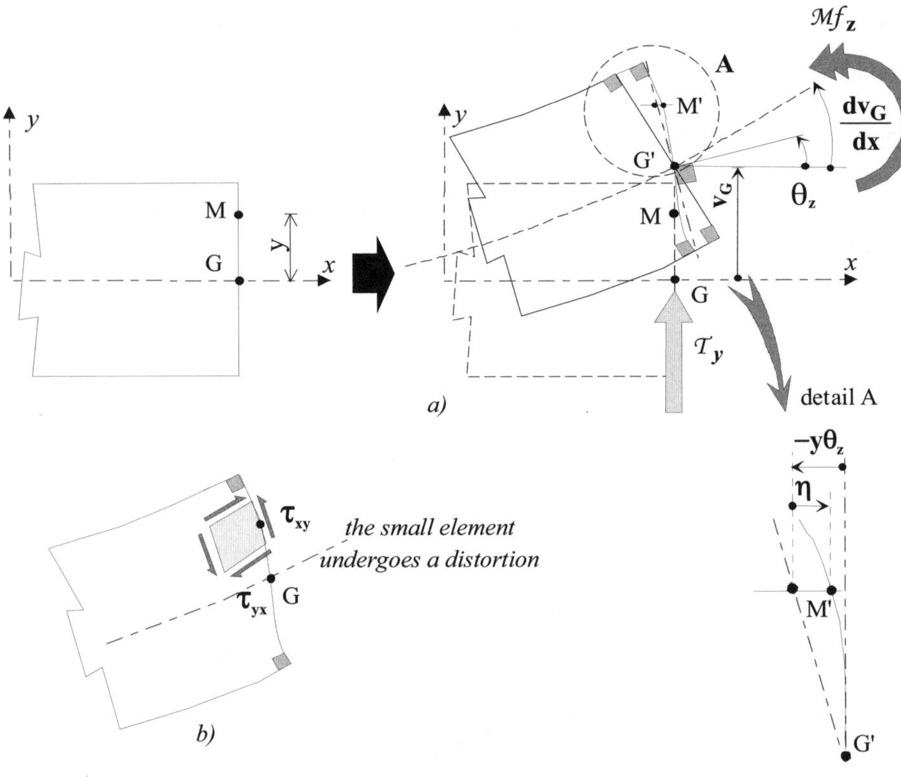

Figure 9.49. *Shear stress deformation*

We note in this figure that the normal to the "mean" plane section makes an angle θ_z with the $\vec{x}$ axis such that $\theta_z \neq \dfrac{d v_G}{d x}$. Bernoulli's hypothesis that we illustrated in Figure 9.11c is replaced here by the principle of the "almost plane" section of Figure 9.11b. Displacement field [9.37] is then modified as follows (see Figure 9.49)[49]:

---

49 The slope represented by $\dfrac{d v_G}{d x}$ is highly exaggerated. In reality it is not perceptible. This enables us to write equation [9.38].

$$\begin{bmatrix} u = -y\theta_z(x) + \eta \\ v = v_G(x) \\ w = 0 \end{bmatrix} \qquad [9.38]$$

where, in order to simplify matters, the following notations are used:

$$u \Rightarrow u(x, y, z) \; ; \; \eta \Rightarrow \eta(x, y, z).$$

We have seen in the previous section that the resultant force and moment of the cohesion forces on the cross-section had $\mathcal{T}_y$ and $\mathcal{M}f_z$ as the only non-zero projections. Mainly the normal resultant is zero. We therefore have $\mathcal{N} = \int_S \sigma_x \, dS = 0$.

The normal stress is expressed using[50]:

$$\sigma_x = E \times \varepsilon_x = E\frac{\partial u}{\partial x} = -E \times y\frac{d\theta_z}{dx} + E\frac{\partial \eta}{\partial x}$$

from where:

$$0 = \mathcal{N} = \int_S \sigma_x \, dS = E\int_S \varepsilon_x \, dS = -E\frac{d\theta_z}{dx}\int_S y \times dS + E\frac{\partial}{\partial x}\int_S \eta \times dS$$

i.e. using [9.1]:

$$\int_S \eta \times dS = 0 \qquad [9.39]$$

Let us now write that the bending moment $\mathcal{M}f_z$ is the moment with respect to $\vec{G z}$ of elementary cohesion forces:

$$\mathcal{M}f_z \vec{z} = \int_S y\vec{y} \wedge \sigma_x \vec{x} \, dS$$

$$\mathcal{M}f_z = \int_S -y \times \sigma_x \times dS = EI_z\frac{d\theta_z}{dx} - E\frac{\partial}{\partial x}\int_S y \times \eta \times dS$$

We have seen in section 9.3.3 the form of the behavior relation for the bending moment (see [9.29]). We can also note that it is possible to vary the bending moment on the cross-section while still maintaining the shear resultant constant. For example, it is possible to apply a pure supplementary moment at the extremity of the beam: the

---

50 See [1.18].

bending moment will be modified but the shear resultant will not change. The shear resultant should affect the above relation only through displacement η, being the result of the distortions represented in Figure 9.49b. Thus by considering [9.29] we have: $\int_S y \times \eta \times dS = Cte$, $\forall x$. Let us define the "mean" section (Figures 9.48 and 9.49) such that:

$$\int_S y \times \eta \times dS = 0 \qquad [9.40]$$

finally the displacement field can be written as:

$$\begin{bmatrix} u = -y\theta_z + \eta \\ v = v_G(x) \\ w = 0 \\ \text{with } \int_S \eta \times dS = 0 \text{ and } \int_S y \times \eta \times dS = 0 \end{bmatrix} \qquad [9.41]$$

9.3.4.3. *Analysis of displacement $\eta(x, y, z)$*

Considering a point M in the cross-section[51] the shear stress is written as:

$$\vec{\tau} = \tau_{xy}\,\vec{y} + \tau_{xz}\,\vec{z}$$

Component τ_{xy} induces a distortion γ_{xy} such that[52]:

$$\gamma_{xy} = \frac{\partial u}{\partial y} + \frac{\partial v}{\partial x} = \frac{\partial(-y\theta_z + \eta)}{\partial y} + \frac{\partial v}{\partial x} = -\theta_z + \frac{\partial \eta}{\partial y} + \frac{dv_G}{dx}$$

Behavior relation $\tau_{xy} = G \times \gamma_{xy}$ leads to:

$$\tau_{xy} = G \times \gamma_{xy} \qquad [9.42]$$

Component τ_{xz} induces a distortion γ_{xz} such that[52]:

$$\gamma_{xz} = \frac{\partial u}{\partial z} + \frac{\partial w}{\partial x} = \frac{\partial(-y\theta_z(x) + \eta)}{\partial z} + \frac{\partial w}{\partial x} = \frac{\partial \eta}{\partial z}$$

Behavior relation $\tau_{xz} = G \times \gamma_{xz}$ leads to:

---

51 See Figure 9.8.
52 See Figure 9.10 and equation [1.18].

$$\tau_{xz} = G\frac{\partial \eta}{\partial z} \qquad [9.43]$$

In the local equilibrium equation [9.9] let us replace the shear stresses τ_{xy} and τ_{xz} using the values found in [9.42] and [9.43]. We arrive at:

$$G\frac{\partial^2 \eta}{\partial y^2} + G\frac{\partial^2 \eta}{\partial z^2} = -\frac{\partial \sigma_x}{\partial x}$$

As per [9.30] we have $\sigma_x = -\frac{\mathcal{M}f_z}{I_z}\times y$ and as per [9.12] we have

$$\frac{d\mathcal{M}f_z}{dx} = -\mathcal{T}_y\,.$$

From where:

$$-\frac{\partial \sigma_x}{\partial x} = \frac{y}{I_z}\frac{d\mathcal{M}f_z}{dx} = -\frac{y}{I_z}\times \mathcal{T}_y$$

the local equilibrium equation becomes:

$$\frac{\partial^2 \eta}{\partial y^2} + \frac{\partial^2 \eta}{\partial z^2} = -\frac{y}{G\times I_z}\times \mathcal{T}_y \qquad [9.44]$$

The supplementary axial displacement $\eta(x,y,z)$ as well as the angular difference $\left(\frac{dv_G}{dx} - \theta_z\right)$ that appear in Figure 9.49a are directly related to the presence of the shear resultant $\mathcal{T}_y$.

The material of the beam being elastic and linear, the three quantities η, $\left(\frac{dv_G}{dx} - \theta_z\right)$ and $\mathcal{T}_y$ are proportional for the current section under study. Due to the nature of the shear stresses [9.42] and [9.43] we express η as follows[53]:

$$\eta = \frac{\mathcal{T}_y}{G}\times f(y,z) - \left(\frac{dv_G}{dx} - \theta_z\right)\times y \qquad [9.45]$$

The tangential stresses τ_{xy} and τ_{xz} are thus simplified to:

____

[53] The proportionality coefficients adopted to write η enable simplification of the subsequent calculations.

$$\tau_{xy} = G\left(\frac{dv}{dx} - \theta_z\right) + G\frac{\partial \eta}{\partial y} = \mathcal{T}_y \frac{\partial f}{\partial y}$$

$$\tau_{xz} = G\frac{\partial \eta}{\partial z} = \mathcal{T}_y \frac{\partial f}{\partial z}$$

[9.46]

In addition, when using the form [9.45] for η , [9.44] is written as:

$$\frac{\partial^2 f(y,z)}{\partial y^2} + \frac{\partial^2 f(y,z)}{\partial z^2} = -\frac{y}{I_z}$$

[9.47]

where the first member is designated by the term "Laplacian" of function f(y,z) (marked as f from now on).

The second member of equation [9.47] is not zero. This equation is said to be a "Poisson's equation". The solution to such an equation should be associated with the boundary conditions. Here, the lateral surface of the beam is supposed to be free of any force. Then the reciprocity property of shear stresses leads to a tangential stress $\vec{\tau} = \tau_{xy}\,\vec{y} + \tau_{xz}\,\vec{z}$ which can only be tangent to the contour or zero as indicated[54] earlier and is shown again in Figure 9.50.

Therefore, if the outward normal to the contour is $\vec{n} = n_y\,\vec{y} + n_z\,\vec{z}$ (Figure 9.50a) we have:

$$\vec{\tau} \cdot \vec{n} = 0$$

or with [9.46]:

$$\overrightarrow{\text{grad}\, f} \cdot \vec{n} = 0 \iff \frac{\partial f}{\partial y} n_y + \frac{\partial f}{\partial z} n_z = 0$$

[9.48]

---

54 See Figure 9.31.

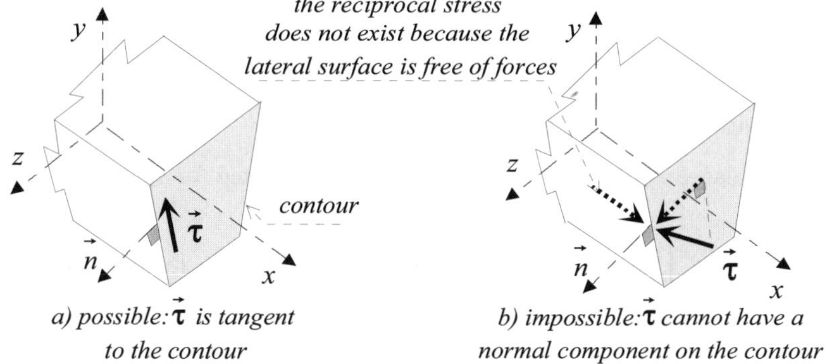

Figure 9.50. *Bending shear stresses at boundaries*

Furthermore, function f(y, z) obeys the condition given in [9.41] i.e.[55]:

$$\int_S \eta\, dS = \frac{\mathcal{T}_y}{G}\int_S f \times dS = 0 \;\Rightarrow\; \int_S f \times dS = 0$$

In brief, the function f(y,z) is known after solving the following problem:

$$\frac{\partial^2 f}{\partial y^2} + \frac{\partial^2 f}{\partial z^2} = -\frac{y}{I_z}\ \ \text{in the section}$$

$$\frac{\partial f}{\partial y}n_y + \frac{\partial f}{\partial z}n_z = 0\ \ \text{on the contour of the section}$$

$$\text{with}\ \int_S f \times dS = 0\ \ \text{(condition of uniqueness)}$$

9.3.4.4. *Shear stresses*

They are written based on [9.46]:

$$\tau_{xy} = \mathcal{T}_y\,\frac{\partial f}{\partial y}\ ;\ \ \tau_{xz} = \mathcal{T}_y\,\frac{\partial f}{\partial z}$$

---

55 In fact using form [9.45] for η gives: $\int_S \eta\, dS = \dfrac{\mathcal{T}_y}{G}\int_S f \times dS - \left(\dfrac{dv_G}{dx} - \theta_z\right)\int_S y \times dS$. The second term is zero as a result of the choice for the origin of the axes (geometric center G).

i.e.:

$$\vec{\tau} = \tau_{xy}\,\vec{y} + \tau_{xz}\,\vec{z} = \mathcal{T}_y \times \overrightarrow{\text{grad } f(y,z)} \tag{9.49}$$

9.3.4.5. *Behavior relation for the shear resultant*

Using displacement field [9.41] and form [9.45] adopted for $\eta(x, y, z)$, we have:

$$0 = \int_S y \times \eta \times dS = \int_S y\left[\frac{\mathcal{T}_y}{G} \times f(y,z) - \left(\frac{dv_G}{dx} - \theta_z\right) \times y\right] dS$$

$$\left(\frac{dv_G}{dx} - \theta_z\right) \times I_z = \frac{\mathcal{T}_y}{G} \int_S y \times f(y,z)\,dS$$

By taking:

$$\frac{1}{k_y} = \frac{S}{I_z} \int_S y \times f(y,z)\,dS \tag{9.50}$$

The behavior relation for shear resultant $\mathcal{T}_y$ is obtained:

$$\mathcal{T}_y = G \times k_y \times S \times \left(\frac{dv_G}{dx} - \theta_z\right) \tag{9.51}$$

where k_y is a dimensionless coefficient called shear coefficient. It is specific to the cross-section shape. It can be shown that its values are less than 1. In finite element software, the product $k_y \times S$ is often known as the "*shear section*" for the shear resultant $\mathcal{T}_y$. In section 9.3.4.7 the values of k_y are given for some standard section shapes.

NOTES

❒ We find again in [9.51] the term $\left(\dfrac{dv_G}{dx} - \theta_z\right)$ which was already noted in Figure 9.49 as the "mean distortion" $\gamma_{xy\,mean}$ of Figure 9.48:

$$\text{mean distortion} = \left(\frac{dv_G}{dx} - \theta_z\right) \tag{9.52}$$

This can again be observed in Figure 9.51. Equation [9.51] shows that it is proportional to shear resultant $\mathcal{T}_y$.

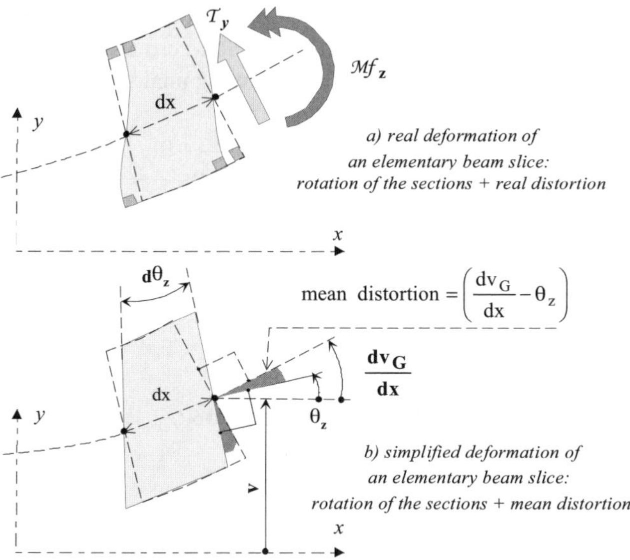

Figure 9.51. *Real and simplified bending shear deformation*

❏ Simplification: Bernoulli's hypothesis.

For the cases of slender homogenous beams and in the static domain, distortion due to the shear resultant is small and its influence is often negligible[56]. Thus we can use the standard Bernoulli's hypothesis which has already been mentioned (see section 9.1.5.2). With this hypothesis we ignore the distortion, enabling equation [9.52] to be written as:

$$\frac{dv_G}{dx} - \theta_z \cong 0 \;\Rightarrow\; \frac{dv_G}{dx} \cong \theta_z \tag{9.53}$$

The behavior relation for the bending moment [9.35] becomes:

$$Mf_z = EI_z \frac{d^2 v_G}{dx^2} \tag{9.54}$$

---

56 For homogenous beams the additional deflection caused by the shear force is of the order of some % of the deflection due to the bending moment and is therefore negligible. It is however advised to consider the distortion due to the shear force:
– for static problems: when we deal with beams which are no longer homogenous but composite. It even becomes indispensable for sandwich structures (the two deflections – due to Mf and due to T – are similar in size);
– for dynamic problems: when we need to consider a large number of eigenfrequencies and modes of the structure.

❏ Without going into the previous calculations, it should be possible to foresee the results of a similar study of the bending with non-zero shear resultant in the main plane (xz). Instead of equation [9.51] the following analogous equation is obtained:

$$\mathcal{T}_z = G \times k_z \times S \times \left(\frac{dw_G}{dx} + \theta_y \right)$$

where a mean distortion $\left(\dfrac{dw_G}{dx} + \theta_y \right)^{57}$ appears which when ignored gives:

$$\frac{dw_G}{dx} + \theta_y \cong 0 \ \Rightarrow \ \frac{dw_G}{dx} \cong -\theta_y \qquad [9.55]$$

The behavior relation for the bending moment $\mathcal{M}f_y$ [9.35] becomes:

$$\mathcal{M}f_y = -EI_y \frac{d^2 w_G}{dx^2} \qquad [9.56]$$

❏ With the already used displacement field [9.41], the normal stress was written as:

$$\sigma_x = E \times \varepsilon_x = E \frac{\partial u}{\partial x} = -E \times y \frac{d\theta_z(x)}{dx} + E \frac{\partial \eta}{\partial x}$$

By taking into account behavior relation [9.51], displacement [9.45] is written as:

$$\eta = \frac{\mathcal{T}_y}{G} \times f(y,z) - y \frac{\mathcal{T}_y}{G \times k_y \times S} = \frac{\mathcal{T}_y}{G} \left[f(y,z) - \frac{y}{k_y \times S} \right]$$

The shear resultant being the same in every current cross-section[58]:

---

57 We shall note here the presence of a sign + within the bracket instead of the sign − when bending occurred in the (xy) plane. This particularity is justified by the sign conventions recalled in the figure below.

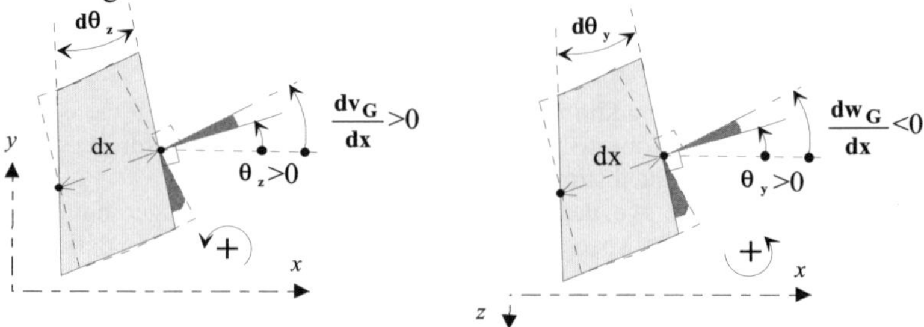

58 See section 9.3.4.1.

$$\frac{\partial \eta}{\partial x} = 0 \Rightarrow \sigma_x = -E \times y \frac{d\theta_z}{dx}$$

We thus verify that the normal stress retains the value [9.28], and remains well linked to the only bending moment $\mathcal{M}f_z$.

9.3.4.6. *Application: case of a rectangular section*

Let us examine a current rectangular cross-section $(b \times h)$, with height h along $\vec{y}$ of a beam loaded as before. Then (see Figure 9.50) let us assume that the shear stresses are directed only along $\vec{y}$, i.e. as per [9.46]:

$$\tau_{xz} = \mathcal{T}_y \frac{\partial f}{\partial z} = 0 \Rightarrow f = f(y)$$

Equation [9.47] becomes:

$$\frac{d^2 f(y)}{dy^2} = -\frac{y}{I_z} \Rightarrow f = \left(-\frac{y^3}{6} + a_1 y + a_0 \right) \times \frac{1}{I_z}$$

Condition [9.48] is written as:

$$\frac{df}{dy} = 0 \quad \text{for} \quad y = \pm \frac{h}{2} \Rightarrow a_1 = \frac{h^2}{8}$$

By taking into account the uniqueness relation:

$$\int_S f \times dS = 0 \Rightarrow a_0 = 0$$

we get:

$$f = \frac{y}{2 \times I_z} \left(\frac{h^2}{4} - \frac{y^2}{3} \right)$$

from where the value of the shear stress τ_{xy} for the ordinate y:

$$\tau_{xy} = \mathcal{T}_y \frac{df}{dy} = \frac{\mathcal{T}_y}{I_z} \times \frac{1}{2} \left(\frac{h^2}{4} - y^2 \right)$$

and with $I_z = \int_S y^2 \times dS = \frac{bh^3}{12}$:

$$\tau_{xy} = \frac{T_y}{bh} \times \frac{3}{2}\left(1 - 4\frac{y^2}{h^2}\right) \qquad [9.57]$$

The distribution of shear stresses τ_{xy} for a rectangular section appears to vary in a parabolic manner along the $\vec{y}$ axis (see Figure 9.52).

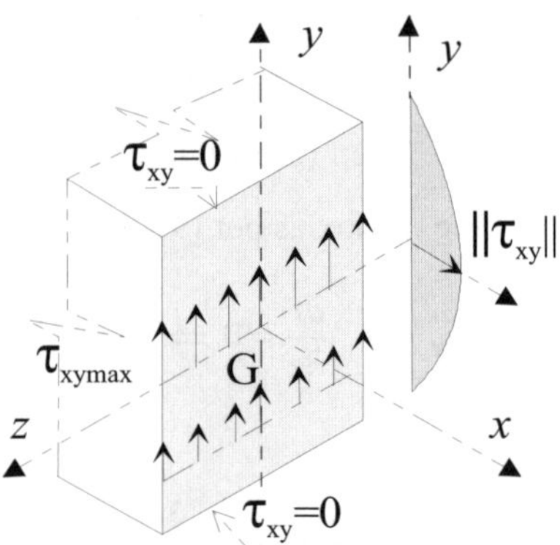

Figure 9.52. *Distribution of shear stresses due to shear resultant* T_y
on a rectangular section

The tangential stress τ_{xy} assumes its maximum value when $y = 0$, i.e.:

$$\tau_{xy_{max}} = \frac{3}{2} \times \frac{T_y}{bh}$$

NOTES

❏ We have just seen that in the case of the rectangular section we had $\tau_{xz} = 0$ (Figure 9.53a). As can be observed in Figure 9.53b the case of a circular section will be different.

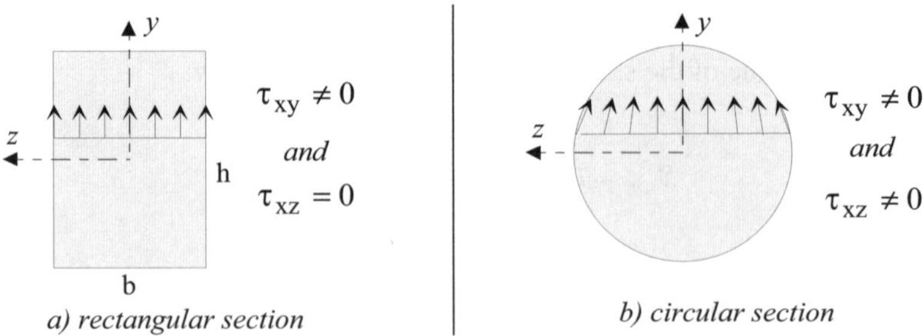

a) rectangular section   *b) circular section*

Figure 9.53. *Comparison of shear stresses due to shear resultant* T_y

In a similar manner when bending in the main plane (*xz*) with a non-zero shear resultant, the configuration shown in Figure 9.54 will be obtained:

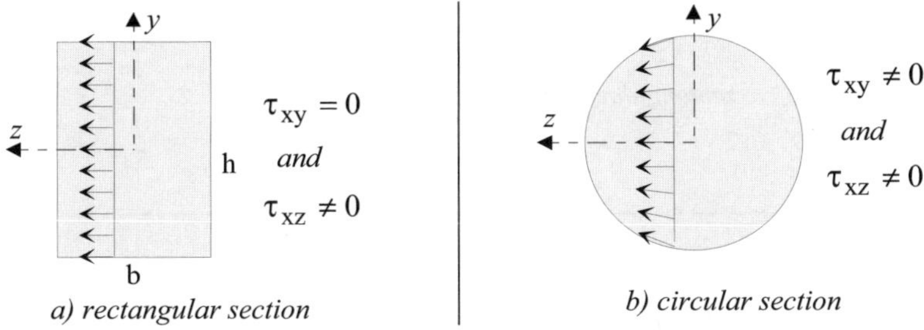

a) rectangular section   *b) circular section*

Figure 9.54. *Comparison of shear stresses due to shear resultant* T_z

❏ Function f above being known, it possible to calculate the shear coefficient in equation [9.51] using formula [9.50]. For the rectangular section we have:

$$\frac{1}{k_y} = \frac{S}{I_z} \int_S y \times f \times dS = \frac{S}{2I_z^2} \int_{-h/2}^{h/2} y^2 \left(\frac{h^2}{4} - \frac{y^2}{3} \right) \times b \times dy$$

$$\frac{1}{k_y} = \frac{S \times b}{2I_z^2} \times \frac{h^5}{60}$$

i.e., with $S = b \times h$ and $I_z = \dfrac{b \times h^3}{12}$:

$$k_y = \frac{5}{6}$$

9.3.4.7. *Values of the shear coefficient k_y and shear section S_{ry} for some section shapes*

The shear coefficient k_y of behavior relation [9.51] is calculated based on the function f(y, z) using equation [9.50] referred to here:

$$\frac{1}{k_y \times S} = \frac{1}{I_z} \int_S y \times f(y, z) dS$$

The following table gives some values of k_y and corresponding shear section $S_{ry} = k_y \times S$.

cross-section under shear resultant $\mathcal{T}_y$	maximum shear stress	shear coefficient k_y	shear section $S_{ry} = k_y \times S$
rectangular	$\tau_{max} = \dfrac{3}{2} \times \dfrac{\mathcal{T}_y}{S}$	$\dfrac{5}{6}$	$\dfrac{5}{6} \times bh$
I section		$\dfrac{5}{6}$	$\dfrac{5}{6} \times S_{real}$
I section	$\tau_{max} \cong \dfrac{\mathcal{T}_y}{S_{web}}$	$\dfrac{S_{web}}{S}$	S_{web} *(approximation)*
circular	$\tau_{max} = \dfrac{4}{3} \times \dfrac{\mathcal{T}_y}{S}$	$\dfrac{6}{7}$	$\dfrac{6}{7} \times S_{real}$
circular thin tube	$\tau_{max} = 2 \times \dfrac{\mathcal{T}_y}{S}$	$\dfrac{1}{2}$	$\dfrac{1}{2} \times S_{real}$
rectangular thin tube	$\tau_{max} \cong \dfrac{\mathcal{T}_y}{S_{real}}$	$\dfrac{S_{real}}{S}$	S_{real} *(approximation)*

$$[9.58]$$

9.3.4.8. *Summary*

For plane bending, the typical resultant force and moment on a cross-section are:

– a bending moment;

– a shear resultant.

Then, the characteristic relations can be summarized as follows.

Plane bending in the main plane (xy)	
resultant force and moment of cohesive forces on a cross-section: $\mathcal{M}f_z(x)$: bending moment along $\vec{z}$ $\mathcal{T}_y(x)$: shear resultant along $\vec{y}$	*displacements of a section:* $v(x)$ translation along $\vec{y}$ of the geometric center G of the section $\theta_z(x)$ rotation around $\vec{z}$ of the cross-section
local equilibrium equation [9.12]:	
$\dfrac{d\mathcal{T}_y}{dx} + q_y = 0$	$\dfrac{d\mathcal{M}f_z}{dx} + \mathcal{T}_y = 0$
behavior relation:	
for the bending moment: $\mathcal{M}f_z = EI_z \dfrac{d\theta_z}{dx}$	for the shear resultant: $\mathcal{T}_y = G \times k_y \times S \times \left(\dfrac{dv}{dx} - \theta_z \right)$
simplified form if we ignore the deformation due to resultant shear: $$\mathcal{M}f_z = EI_z \dfrac{d^2 v}{dx^2}$$	
shear coefficient: $$\dfrac{1}{k_y} = \dfrac{S}{I_z} \int_S y \times f(y, z)\,dS$$	

stresses in a cross-section:

normal stresses: $\sigma_x = -\dfrac{\mathcal{M}f_z}{I_z} \times y$

shear stresses: $\vec{\tau} = \tau_{xy}\,\vec{y} + \tau_{xz}\,\vec{z} = \mathcal{T}_y \times \overrightarrow{\mathrm{grad}\, f(y,z)}$

---

deformations in the plane (xy):

longitudinal deformation: $\varepsilon_x = -y\dfrac{d\theta_z}{dx}$

mean distortion $\gamma_{xy\,\mathrm{mean}} = \left(\dfrac{dv}{dx} - \theta_z\right)$

---

characteristic function for the effects of resultant shear $\mathcal{T}_y(x) : f(y,z)$

f(y,z) is the solution to the problem:

$$\frac{\partial^2 f}{\partial y^2} + \frac{\partial^2 f}{\partial z^2} = -\frac{y}{I_z} \quad \text{in section S}$$

$\overrightarrow{\mathrm{grad}\, f} \cdot \vec{n} = 0$ on the contour of the section *(boundary condition)*

with: $\displaystyle\int_S f(y,z)\,dS = 0$ condition for uniqueness of function f

$$[9.59]$$

From the previous results we can derive the characteristic form of the equations when the plane bending takes place in the main plane (xz)[59].

<table>
<tr><td colspan="2" align="center">Plane bending in the main plane (xz)</td></tr>
<tr>
<td>

resultant force and moment of cohesive forces on a cross-section:

$\mathcal{M}f_y(x)$: bending moment along $\vec{y}$

$\mathcal{T}_z(x)$: shear resultant along $\vec{z}$

</td>
<td>

displacements of a section:

$w(x)$ translation along $\vec{z}$ of geometric center G of the section

$\theta_y(x)$ rotation around $\vec{y}$ of the cross-section

</td>
</tr>
<tr>
<td colspan="2">

local equilibrium equation [9.12]:

$$\frac{d\mathcal{T}_z}{dx} + q_z = 0 \qquad\qquad \frac{d\mathcal{M}f_y}{dx} - \mathcal{T}_z = 0$$

</td>
</tr>
<tr>
<td colspan="2">

behavior relation:

for the bending moment:

$$\mathcal{M}f_y = EI_y \frac{d\theta_y}{dx}$$

for the shear resultant:

$$\mathcal{T}_z = G \times k_z \times S \times \left(\frac{dw}{dx} + \theta_y \right)$$

</td>
</tr>
<tr>
<td colspan="2">

simplified form if we ignore the deformation due to shear resultant:

$$\mathcal{M}f_y = -EI_y \frac{d^2 w}{dx^2}$$

</td>
</tr>
<tr>
<td colspan="2">

shear coefficient:

$$\frac{1}{k_z} = \frac{S}{I_y} \int_S z \times g(y, z) dS$$

</td>
</tr>
</table>

<table>
<tr>
<td>

stresses in a cross-section:

normal stresses: $\sigma_x = \dfrac{\mathcal{M}f_y}{I_y} \times z$

tangential shear stresses: $\vec{\tau} = \tau_{xy}\, \vec{y} + \tau_{xz}\, \vec{z} = \mathcal{T}_z \times \overrightarrow{\text{grad}\, g(y, z)}$

</td>
</tr>
</table>

---

59 See *NOTE* and footnotes at the end of section 9.3.4.5.

deformations in the plane (xz):

longitudinal deformation: $\varepsilon_x = z\dfrac{d\theta_y}{dx}$

mean distortion $\gamma_{xz\ mean} = \left(\dfrac{dw}{dx} + \theta_y\right)$

---

characteristic function for the effects of shear resultant $T_z(x) : g(y,z)$.

$g(y,z)$ is the solution to the problem:

$$\frac{\partial^2 g}{\partial y^2} + \frac{\partial^2 g}{\partial z^2} = -\frac{z}{I_y} \quad \text{in section S}$$

$\overrightarrow{\text{grad}\ g} \cdot \vec{n} = 0$ on the contour of the section *(boundary condition)*

with: $\displaystyle\int_S g(y,z)dS = 0$ condition for uniqueness of function g

[9.60]

NOTES

❑ It could be noted that the bending moment could exist independently of the shear resultant. It was the particular case of the pure bending (see section 9.3.3) where the bending moment existed while the shear resultant was zero.

However the reverse is not always possible. In practice, the bending moment can be canceled for discrete-values of the x abscissa but can never be zero whatever the value of x is. The presence of a shear resultant implies that there must be a bending moment[60] as could be seen in Figure 9.47.

❑ Torsion with bending shear resultants:

We have studied in section 9.3.2.4, the torsion of a beam with any cross-section shape where it was assumed that $Mt = M_x = C^{te}$. Generally, as the beam had a torsion center C different from the geometric center G of the section, this hypothesis implied that the shear resultants were zero (see [9.21]). Then additional presence of resultant shears T_y and T_z requires us:

---

60 It is nevertheless possible to imagine a load which can always maintain a zero bending moment, with a non-zero shear force (figure below) but such a load (distributed moments) appears artificial in the static domain, and far from reality.

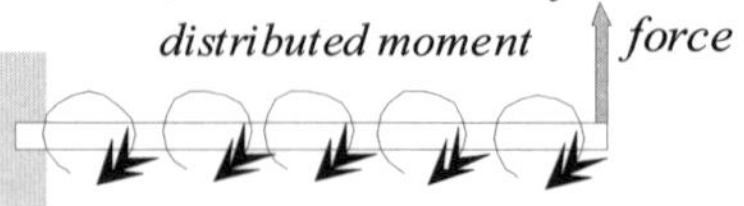

– to express the torsional moment which leads to the results [9.27] with the help of equation [9.20], i.e. $Mt = M_x - y_C T_z + z_C T_y$. This enables us mainly to calculate the torsional shear stresses;

– to then calculate the shear stresses due to resultant shears using the previous results [9.59] and [9.60].

The total shear stress distribution appears as the superposition of the previous distributions: torsion shear stresses *and* shear stresses due to shear resultants[61].

9.3.5. *Any loading*

When the six components of resultant forces and moments of cohesive forces [9.6] act simultaneously, we observe the superposition of effects described in the previous sections. Figure 9.55 represents these components in the simple case of a section beam with a center of symmetry, i.e. for which the torsion center is merged with the elastic center G (see [9.21]). Then, we recall that $Mt = M_x$. Table [9.61] summarizes the main results concerning the small displacements of a current cross-section and the cohesion stresses acting on this section.

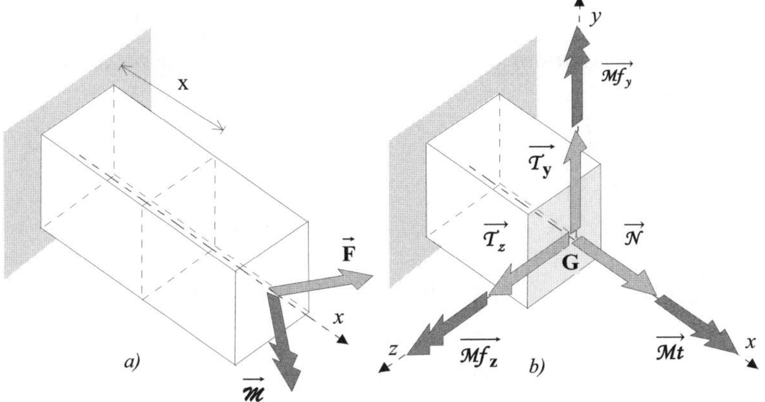

Figure 9.55. *Any loading*

---

61 Note that the approaches to studying torsion in section 9.3.2.4 and bending with shear force above are different. The first is "rigorous" as per the laws of elasticity and leads to the exact solution of a certain elasticity problem. The "exact" solution of the problem of simple bending of a clamped-free beam with a shear resultant also exists. Such a solution is not detailed here as it is considered to be too heavy and requires "extended" notions of elasticity which are not considered in this work. Instead, the original approach (see bibliography D. Gay) suggested in this section 9.3.4 requires only the basic knowledge that is available in section 9.1.

Small displacements of a current cross-section:

3 linear displacements (translations) of geometric center G: u(x), v(x), w(x)

3 angular displacements (rotation of the sections): $\theta_x(x)$, $\theta_y(x)$, $\theta_z(x)$

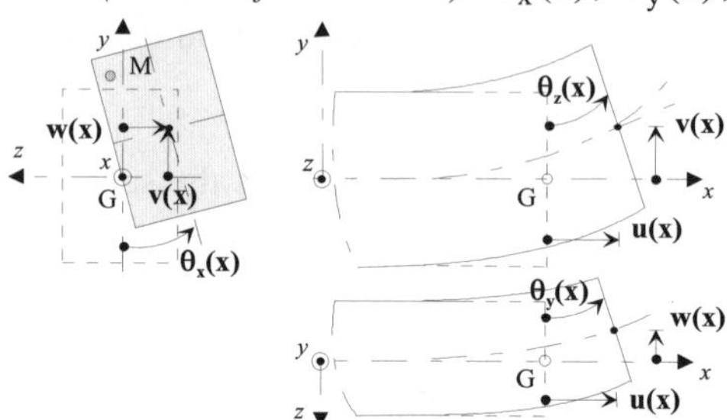

traction compression

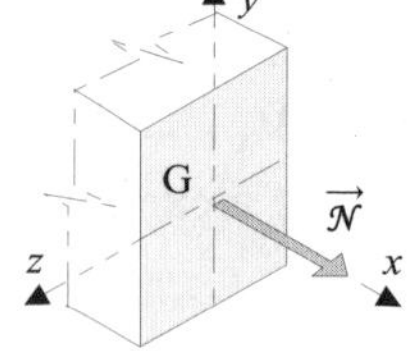

$$\mathcal{N} = ES\frac{du}{dx}$$

torsion

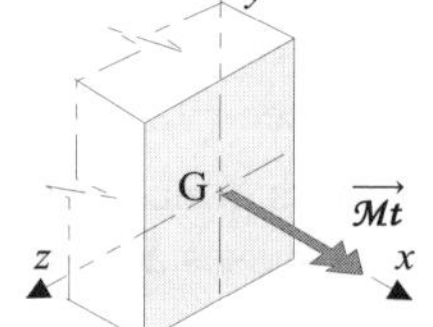

$$\mathcal{M}t = G \times J \times \frac{d\theta_x}{dx}$$

bending with shear resultant

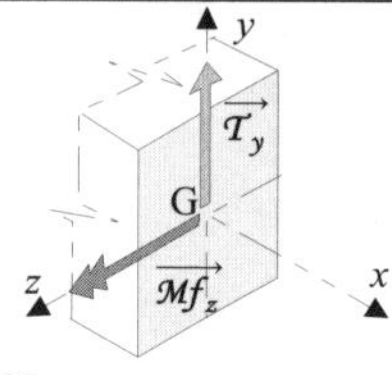

$$\mathcal{M}f_z = EI_z\frac{d\theta_z}{dx} \;;\; \mathcal{T}_y = G \times k_y \times S \times \left(\frac{dv}{dx} - \theta_z\right)$$

simplified form: $\mathcal{M}f_z = EI_z\dfrac{d^2v}{d^2x}$

bending with shear resultant

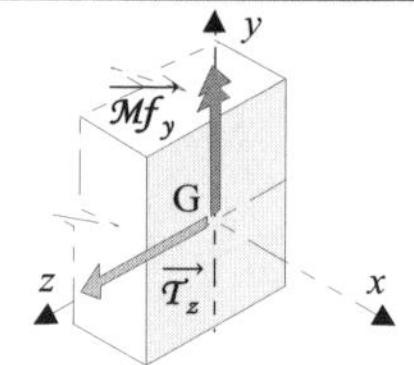

$$\mathcal{M}f_y = EI_y\frac{d\theta_y}{dx} \;;\; \mathcal{T}_z = G \times k_z \times S \times \left(\frac{dw}{dx} + \theta_y\right)$$

simplified form: $\mathcal{M}f_y = -EI_y\dfrac{d^2w}{d^2x}$

stresses on a section

normal stress at M(y,z): algebraic sum along $\vec{x}$:

$$\sigma_x = \underset{\mathcal{N}}{\sigma_x} + \underset{\mathcal{M}f_y}{\sigma_x} + \underset{\mathcal{M}f_z}{\sigma_x} = \frac{\mathcal{N}}{S} + \frac{\mathcal{M}f_y}{I_y} \times z - \frac{\mathcal{M}f_z}{I_z} \times y$$

shear stress at M(y,z): vectorial sum in the plane (yz): $\vec{\tau} = \underset{\mathcal{T}_y}{\vec{\tau}} + \underset{\mathcal{T}_z}{\vec{\tau}} + \underset{\mathcal{M}t}{\vec{\tau}}$

[9.61]

9.4. Application: example of detailed calculation of the resultant forces and moments of cohesive forces

The following example is meant to describe in detail the method of evaluating the resulting forces and moments on the current cross-sections of a structure consisting of straight rectangular beams.

We consider the case of a support arm embedded in a foundation and with its free end loaded (see Figure 9.56a). We wish to write:

– the transmittable resultant forces and moments through the clamped linking;

– the resultant forces and moments of cohesive forces on every cross-section of each one of the two beams which form the support arm.

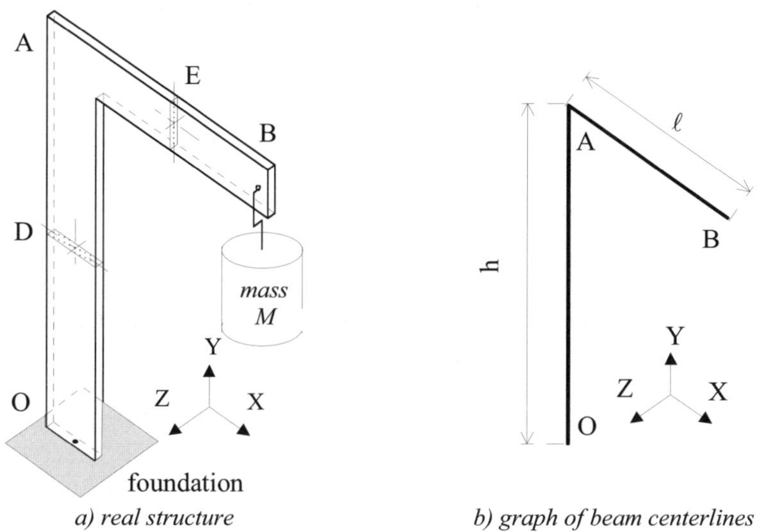

a) real structure   *b) graph of beam centerlines*

Figure 9.56. *Support arm*

9.4.1. *Preliminary static analysis*

Based on the model in Figure 9.56b we propose to calculate the resultant force and moment of the transmittable mechanical actions from the foundation on the support arm at geometric center O of the clamped cross-section. For this we take a coordinate system $R\,(\vec{X}, \vec{Y}, \vec{Z})$ valid for the entire structure, already known as the "global coordinate system" or "structural system".

a) Isolated system: the support arm

⇨ Analysis of mechanical actions:

– the resultant force and moment of mechanical actions exerted at point B is at B^{62}:

$$\{\mathcal{F}_{\text{weight/arm}}\}_B = \begin{Bmatrix} \vec{B} = -Mg\vec{Y} \\ \vec{0} \end{Bmatrix}_{B,R}$$

where $\vec{B}$ models the action on the arm of the hook belonging to mass M as in Figure 9.56a;

– the resultant force and moment of the transmittable mechanical actions through the clamped linking is written at O as:

$$\{\mathcal{F}_{\text{foundation/support arm}}\}_O = \begin{Bmatrix} \overrightarrow{\mathcal{R}_O} = X_O\vec{X} + Y_O\vec{Y} + Z_O\vec{Z} \\ \overrightarrow{\mathcal{M}_O} = L_O\vec{X} + M_O\vec{Y} + N_O\vec{Z} \end{Bmatrix}_{O,R}$$

that is a total of 6 algebraic unknowns.

b) Solution:

– on the isolated support arm of Figure 9.57, let us apply the fundamental principle of statics with $AB = \ell$ and $OA = h$.

$$\{\mathcal{F}_{\text{foundation/support arm}}\}_O + \{\mathcal{F}_{\text{weight/support arm}}\}_O \equiv \{0\}$$

– corresponding to the 2 vectorial equations:

$$X_O\vec{X} + Y_O\vec{Y} + Z_O\vec{Z} - Mg\vec{Y} = \vec{0}$$

$$L_O\vec{X} + M_O\vec{Y} + N_O\vec{Z} + \overrightarrow{OB} \wedge -Mg\vec{Y} = \vec{0}$$

---

62 Preliminary study of the mass equilibrium is necessary to determine the action of the hook on the arm at point B. This is very simple.

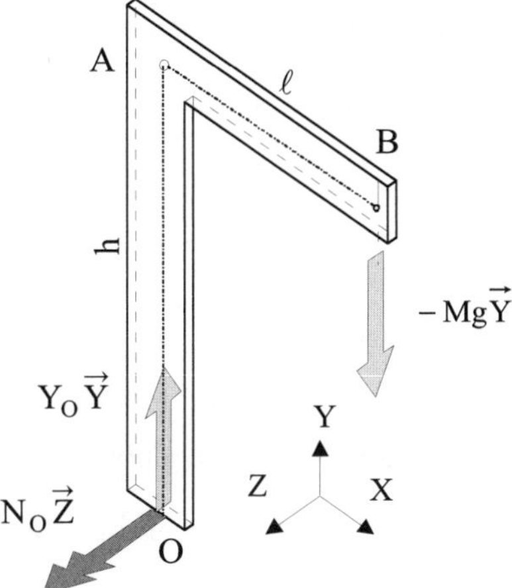

Figure 9.57. *Balanced support arm*

The components of the resultant force and moment at point O transmittable from the foundation to the support arm are obtained as:

$$\left\{ \mathcal{F}_{foundation/support\ arm} \right\}_O = \left\{ \begin{array}{l} \overrightarrow{\mathcal{R}_O} = Y_O \vec{Y} = Mg\vec{Y} \\ \overrightarrow{\mathcal{M}_O} = N_O \vec{Z} = \ell Mg\vec{Z} \end{array} \right\}_{O,R}$$

9.4.2. Resultant force and moment on every cross-section

PRELIMINARY NOTE

❏ Local and global coordinate system

In Figure 9.57 the global coordinate system $R\,(\vec{X}, \vec{Y}, \vec{Z})$ or structural system previously mentioned is unique for the structure under study (support arm). However, this structure consists of two perpendicular beams clamped at A. According to the usual method of study for beams, a coordinate system $r\,(\vec{x}, \vec{y}, \vec{z})$ is linked to each of the beams (one of the axes, $\vec{x}$ axis for example, merged with the centerline of the considered beam, the other two being the main axes of the rectangular cross-section). Thus we assign a local coordinate system to the concerned beam.

◯ Study of the horizontal beam AB of the support arm

We propose to calculate the resultant force and moment in the current cross-section E (see Figure 9.58).

Let us isolate part OAE.

a) In the global system

Let us calculate the resultant force and moment at E of the external mechanical actions applied to part EB expressed in the global coordinate system $R\,(\vec{X}, \vec{Y}, \vec{Z})$:

– with the resultant force and moment on the cross-section noted as $\{\mathcal{F}_{ext/EB}\}_E$ we have:

$$\{\mathcal{F}_{ext/EB}\}_E = \left\{ \begin{array}{l} \overrightarrow{\mathcal{R}_E} = Y_E\,\vec{Y} = \text{-}Mg\vec{Y} \\[4pt] \overrightarrow{\mathcal{M}_E} = \overrightarrow{EB} \wedge \text{-}Mg\vec{Y} \end{array} \right\}_{E,\,R}$$

using $\overrightarrow{EB} = (\ell - X)\vec{X}$ we obtain:

$$\{\mathcal{F}_{ext/EB}\}_E = \left\{ \begin{array}{l} \overrightarrow{\mathcal{R}_E} = Y_E\,\vec{Y} = \text{-}Mg\vec{Y} \\[4pt] \overrightarrow{\mathcal{M}_E} = N_E\,\vec{Z} = -(\ell - X)Mg\vec{Z} \end{array} \right\}_{E,\,R} \qquad [9.62]$$

as per [9.4] the resultant force and moment of cohesive forces is thus obtained:

$$\{Coh_{EB/OE}\}_E = \{\mathcal{F}_{ext/EB}\}_E{}^{[63]} \qquad [9.63]$$

---

[63] From [9.5], we can also write $\{Coh_{EB/OE}\}_E = -\{\mathcal{F}_{ext/OE}\}_E$.

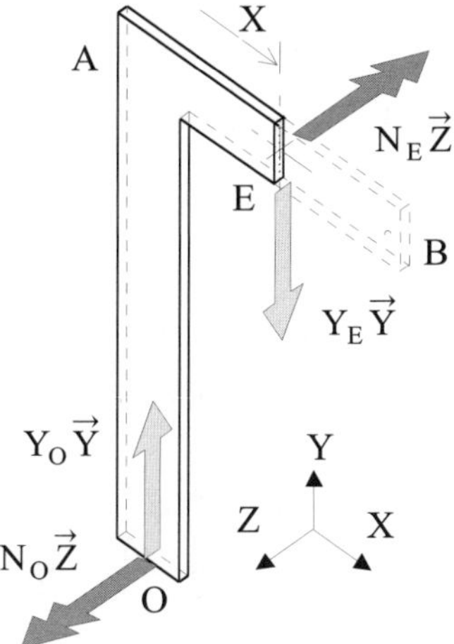

Figure 9.58. *Equilibrium of portion OAE*

b) In the local system

Let us calculate the actions exerted by part EB on portion AE expressed in the local system $r\,(\vec{x},\vec{y},\vec{z})$ linked to the remaining portion AE (see Figure 9.59).

From [9.62] the resultant force and moment $\left\{\mathcal{F}_{ext/EB}\right\}_E$ expressed in the global coordinate system $R\,(\vec{X},\vec{Y},\vec{Z})$ are known. They can be easily expressed in the local system $r\,(\vec{x},\vec{y},\vec{z})$. In fact $\vec{X}=\vec{x}$, $\vec{Y}=\vec{y}$, $\vec{Z}=\vec{z}$, or in matrix form:

$$\left\{\begin{matrix}\vec{X}\\\vec{Y}\\\vec{Z}\end{matrix}\right\}=\begin{bmatrix}1&0&0\\0&1&0\\0&0&1\end{bmatrix}\bullet\left\{\begin{matrix}\vec{x}\\\vec{y}\\\vec{z}\end{matrix}\right\}$$

from where, as per [9.63]:

$$\left\{Coh_{EB/OE}\right\}_E=\left\{\mathcal{F}_{ext/EB}\right\}_E=\left\{\begin{matrix}-Mg\vec{y}\\-(\ell-X)Mg\vec{z}\end{matrix}\right\}_{E,\,r}\qquad[9.64]$$

whereas in the local coordinate system, the resultant force and moment has the form of [9.6] recalled here:

$$\left\{Coh_{EB/OE}\right\}_E = \left\{\begin{array}{l} \overrightarrow{\mathcal{R}_{EB/OE}} = \mathcal{N}_x\,\vec{x} + \mathcal{T}_y\,\vec{y} + \mathcal{T}_z\,\vec{z} \\ \overrightarrow{\mathcal{M}_{E(EB/OE)}} = \mathcal{M}_x\,\vec{x} + \mathcal{M}f_y\,\vec{y} + \mathcal{M}f_z\,\vec{z} \end{array}\right\}_{E,r}$$

from where by comparing with [9.64]:

$$\begin{bmatrix} \mathcal{N}_x = 0 \\ \mathcal{T}_y = -\mathrm{Mg} \\ \mathcal{T}_z = 0 \end{bmatrix} \quad \text{and} \quad \begin{bmatrix} \mathcal{M}_x = 0 \\ \mathcal{M}f_y = 0 \\ \mathcal{M}f_z = -(\ell - \mathrm{X})\mathrm{Mg} \end{bmatrix}$$

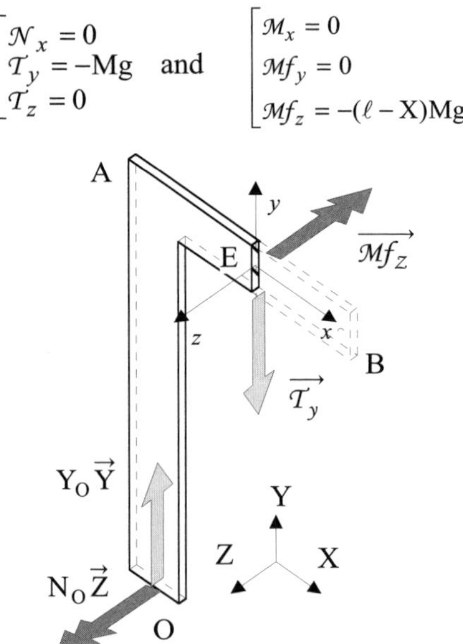

Figure 9.59. *Resultant force and moment around E on portion OAE*

The only non-zero resultant forces here are:

$\mathcal{T}_y$: shear resultant along $\vec{y}$;

$\mathcal{M}f_z$: bending moment around $\vec{z}$.

NOTE

❏ In the above, portion AE of beam AB is isolated. Then the local coordinate system $r(\vec{x}, \vec{y}, \vec{z})$ is defined based on the outward normal at E of isolated portion AE ($\vec{x}$ axis). The main axes $\vec{y}$ and $\vec{z}$ are then chosen in such a manner that the trihedral remains direct.

If we wish to isolate portion EB instead, the same steps should be taken to define the local coordinate system with an outward normal at E of portion EB, which is the $\vec{x'}$ axis of Figure 9.60, and proper $\vec{y'}$ and $\vec{z'}$ axes.

In such a case, let us calculate the actions of portion OE on portion EB in the local coordinate system $r'(\vec{x'},\vec{y'},\vec{z'})$:

$$\{Coh_{OE/EB}\}_E = \{\mathcal{F}_{ext/OE}\}_E \text{ or } \{Coh_{OE/EB}\}_E = -\{\mathcal{F}_{ext/EB}\}_E \ ;$$

$\{\mathcal{F}_{ext/EB}\}_E$ is always given by [9.62]. In addition it is observed that $\vec{X} = -\vec{x'}$, $\vec{Y} = -\vec{y'}$, $\vec{Z} = \vec{z'}$, which in the matrix form is written as:

$$\begin{Bmatrix} \vec{X} \\ \vec{Y} \\ \vec{Z} \end{Bmatrix} = \begin{bmatrix} -1 & 0 & 0 \\ 0 & -1 & 0 \\ 0 & 0 & 1 \end{bmatrix} \bullet \begin{Bmatrix} \vec{x'} \\ \vec{y'} \\ \vec{z'} \end{Bmatrix}$$

from where by using property [9.5]:

$$\{Coh_{OE/EB}\}_E = -\{\mathcal{F}_{ext/EB}\}_E = \begin{Bmatrix} -Mg\vec{y'} \\ (\ell - X)Mg\vec{z'} \end{Bmatrix}_{E,\,r'} \qquad [9.65]$$

In the local coordinate system the resultant force and moment takes the form [9.6] recalled here:

$$\{Coh_{OE/EB}\}_E = \begin{Bmatrix} \overrightarrow{\mathcal{R}_{OE/EB}} = \mathcal{N}_{x'}\,\vec{x'} + \mathcal{T}_{y'}\,\vec{y'} + \mathcal{T}_{z'}\,\vec{z'} \\ \overrightarrow{\mathcal{M}_{E(OE/EB)}} = \mathcal{M}_{x'}\,\vec{x'} + \mathcal{M}f_{y'}\,\vec{y'} + \mathcal{M}f_{z'}\,\vec{z'} \end{Bmatrix}_{E,\,r'}$$

from where by comparing with [9.65]:

$$\begin{bmatrix} \mathcal{N}_{x'} = 0 \\ \mathcal{T}_{y'} = -Mg \\ \mathcal{T}_{z'} = 0 \end{bmatrix} \text{ and } \begin{bmatrix} \mathcal{M}_{x'} = 0 \\ \mathcal{M}f_{y'} = 0 \\ \mathcal{M}f_{z'} = (\ell - X)Mg \end{bmatrix}$$

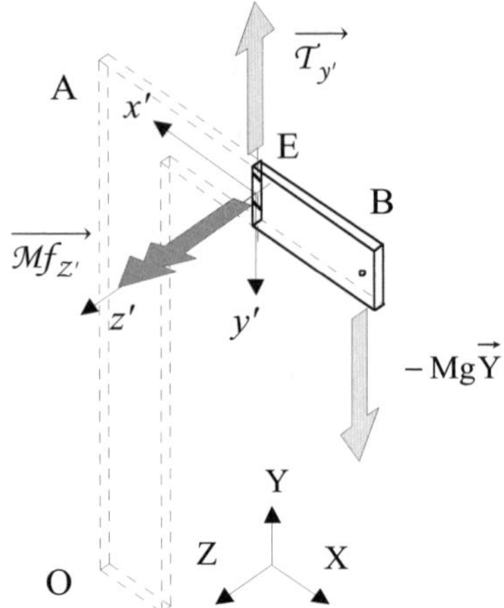

Figure 9.60. *Resultant force and moment around E on portion EB*

The non-zero resultant forces here are:

$\mathcal{T}_{y'}$: shear resultant along $\vec{y}'$;

$\mathcal{M}f_{z'}$: bending moment around $\vec{z}'$.

○   Study of vertical beam OA

We consider a current cross-section D (see Figure 9.61).

Let us isolate portion OD.

a) In the global system

Let us calculate in the global coordinate system $R\,(\vec{X}, \vec{Y}, \vec{Z})$ the resultant force and moment at D of external mechanical actions exerted on section DB:

– with the resultant force and moment on the cross-section noted as $\{\mathcal{F}_{\text{ext/DB}}\}_D$ we have:

$$\{\mathcal{F}_{\text{ext/DB}}\}_D = \left\{ \begin{array}{l} \overrightarrow{\mathcal{R}_D} = Y_D\vec{Y} = -Mg\vec{Y} \\ \overrightarrow{\mathcal{M}_D} = \overrightarrow{DB} \wedge -Mg\vec{Y} \end{array} \right\}_{D,\,R}$$

using $\overrightarrow{DB} = \ell\vec{X} + (h-\ell)\vec{Y}$ we obtain:

$$\{\mathcal{F}_{ext/DB}\}_D = \left\{ \begin{array}{l} \overrightarrow{\mathcal{R}_D} = Y_D \vec{Y} = -Mg\vec{Y} \\ \overrightarrow{\mathcal{M}_D} = N_D \vec{Z} = -\ell Mg\vec{Z} \end{array} \right\}_{D,R} \qquad [9.66]$$

as per [9.4], the resultant force and moment on cross-section D of cohesive forces can then be written as:

$$\{Coh_{DB/OD}\}_D = \{\mathcal{F}_{ext/DB}\}_D \quad ^{64} \qquad [9.67]$$

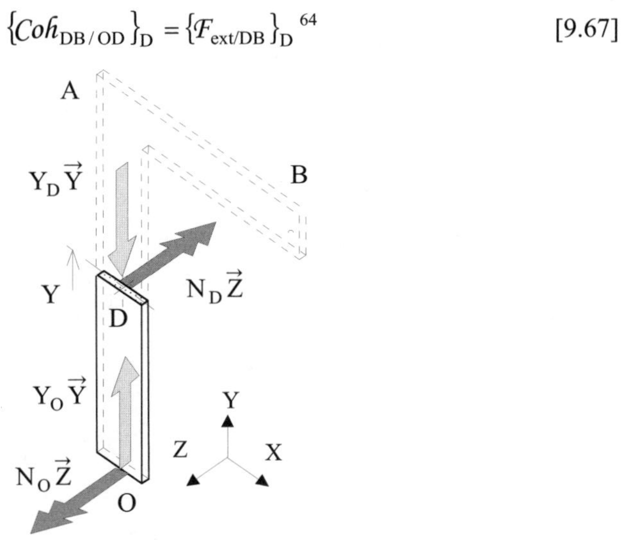

Figure 9.61. *Equilibrium of portion OD*

b) In the local system

From [9.66] the resultant force and moment expressed in the global coordinate system is known. Thus they can be expressed easily in the local coordinate system $r(\vec{x}, \vec{y}, \vec{z})$, drawn in Figure 9.62. In fact, $\vec{X} = -\vec{y}$, $\vec{Y} = \vec{x}$, $\vec{Z} = \vec{z}$, or in matrix form:

$$\left\{ \begin{array}{c} \vec{X} \\ \vec{Y} \\ \vec{Z} \end{array} \right\} = \begin{bmatrix} 0 & -1 & 0 \\ 1 & 0 & 0 \\ 0 & 0 & 1 \end{bmatrix} \bullet \left\{ \begin{array}{c} \vec{x} \\ \vec{y} \\ \vec{z} \end{array} \right\}$$

from where , as per [9.67]:

---

64 Or even as per [9.5], we can write: $\{Coh_{DB/OD}\}_D = -\{\mathcal{F}_{ext/OD}\}_D$.

$$\left\{Coh_{DB/OD}\right\}_D = \left\{\mathcal{F}_{ext/DB}\right\}_D = \left\{\begin{array}{c} -Mg\vec{x} \\ \\ -Mg\ell\vec{z} \end{array}\right\}_{D,r} \tag{9.68}$$

In the local coordinate system, the resultant force and moment of cohesive forces on cross-section D takes the form [9.6] recalled here:

$$\left\{Coh_{DB/OD}\right\}_D = \left\{\begin{array}{c} \overrightarrow{\mathcal{R}_{DB/OD}} = \mathcal{N}_x\,\vec{x} + \mathcal{T}_y\,\vec{y} + \mathcal{T}_z\,\vec{z} \\ \\ \overrightarrow{\mathcal{M}_{D(DB/OD)}} = \mathcal{M}_x\,\vec{x} + \mathcal{M}f_y\,\vec{y} + \mathcal{M}f_z\,\vec{z} \end{array}\right\}_{D,r}$$

from where by comparison with [9.68]:

$$\left[\begin{array}{l} \mathcal{N}_x = -Mg \\ \mathcal{T}_y = 0 \\ \mathcal{T}_z = 0 \end{array}\right. \quad \text{and} \quad \left[\begin{array}{l} \mathcal{M}_x = 0 \\ \mathcal{M}f_y = 0 \\ \mathcal{M}f_z = -Mg\ell \end{array}\right.$$

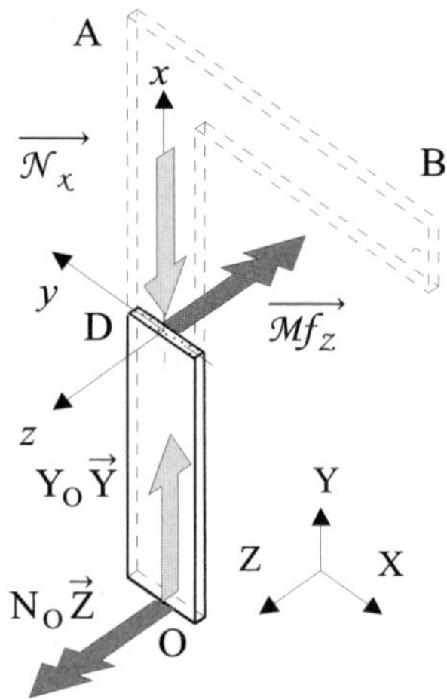

Figure 9.62. *Resultant force and moment around D on portion OD*

The only non-zero resultant forces here are:

$\mathcal{N}_x$: normal resultant along $\vec{x}$;

$\mathcal{M}f_z$: bending moment around $\vec{z}$.

NOTE

❏ In the above we have isolated portion OD of vertical beam OA . Then the local coordinate system $r(\vec{x},\vec{y},\vec{z})$ is defined based on the outward normal at D of isolated part OD; ($\vec{x}$ axis). The main axes $\vec{y}$ and $\vec{z}$ can then be chosen to complete the direct trihedral.

If we prefer to isolate part DAB instead, the same steps can be taken to define the local coordinate system with an outward normal at D of part AD, i.e. the $\vec{x'}$ axis of Figure 9.63, and proper axes $\vec{y'}$ and $\vec{z'}$.

In such a case, let us calculate the actions of part OD on part DAB expressed in the local coordinate system $r'(\vec{x'},\vec{y'},\vec{z'})$:

$$\{Coh_{OD/DB}\}_D = \{F_{ext/OD}\}_D \text{ or } \{Coh_{OD/DB}\}_D = -\{F_{ext/DB}\}_D \text{ ;}$$

$\{F_{ext/DB}\}_D$ is given by [9.68]. In addition, $\vec{X}=\vec{y'}$, $\vec{Y}=-\vec{x'}$, $\vec{Z}=\vec{z'}$, or in matrix form:

$$\begin{Bmatrix}\vec{X}\\\vec{Y}\\\vec{Z}\end{Bmatrix} = \begin{bmatrix}0 & 1 & 0\\-1 & 0 & 0\\0 & 0 & 1\end{bmatrix} \bullet \begin{Bmatrix}\vec{x'}\\\vec{y'}\\\vec{z'}\end{Bmatrix}$$

from where by using the properties in [9.5]:

$$\{Coh_{OD/DAB}\}_D = -\{F_{ext/DB}\}_D = \begin{Bmatrix}-Mg\vec{x'}\\Mg\ell\vec{z'}\end{Bmatrix}_{D,r'} \qquad [9.69]$$

In the local coordinate system, the resultant force and moment takes the form of [9.6] recalled here:

$$\{Coh_{OD/DAB}\}_D = \begin{Bmatrix}\overrightarrow{R_{OD/DAB}} = \mathcal{N}_{x'}\vec{x'}+\mathcal{T}_{y'}\vec{y'}+\mathcal{T}_{z'}\vec{z'}\\\overrightarrow{\mathcal{M}_{D(OD/DAB)}} = \mathcal{M}t_{x'}\vec{x'}+\mathcal{M}f_{y'}\vec{y'}+\mathcal{M}f_{z'}\vec{z'}\end{Bmatrix}_{D,r'}$$

after comparing with [9.69]:

$$\begin{bmatrix}\mathcal{N}_{x'} = -Mg\\\mathcal{T}_{y'} = 0\\\mathcal{T}_{z'} = 0\end{bmatrix} \quad \text{and} \quad \begin{bmatrix}\mathcal{M}_{x'} = 0\\\mathcal{M}f_{y'} = 0\\\mathcal{M}f_{z'} = Mg\ell\end{bmatrix}$$

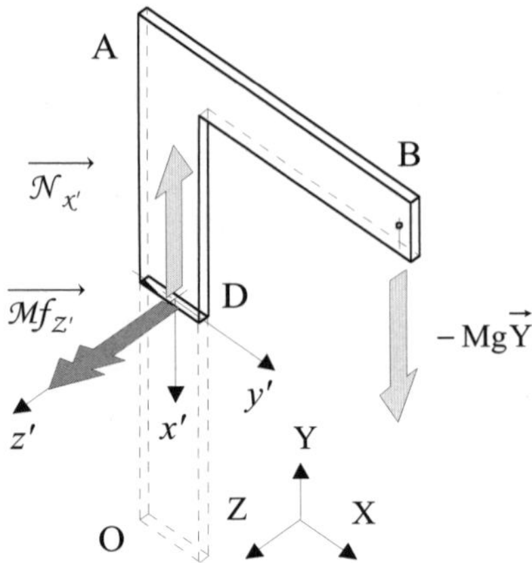

Figure 9.63. *Resultant force and moment around D on portion DAB*

The only non-zero resultant forces are:

$\mathcal{N}_{x'}$: normal resultant along $\vec{x'}$;

$\mathcal{M}f_{z'}$: bending moment around $\vec{z'}$.

Chapter 10

Additional Elements of Elasticity

In this chapter we return to the states of stresses within a loaded structure with the purpose of now establishing some properties previously mentioned and used, but without justification, particularly:

– the reciprocity property of shear stresses;

– the description of any state of stress and associated behavior relation;

– the decomposition of the deformation – or strain – energy with the aim of establishing the notion of equivalent stress.

10.1. Reverting to the plane state of stresses

10.1.1. *Influence of the coordinate system*

We analyzed in section 1.4.2 the plane state of stresses. We see again in Figure 10.1a the representation of stresses σ_x, σ_y, τ_{xy} on a small parallelepipedic balanced domain ($dx \times dy \times e$) surrounding current point M. This plane state of stresses is defined in the (xy) system. Figure 10.1b is obtained after having divided the small element into two parts. An oblique cross-section with area $dS = ds \times e$ having an outward normal $\vec{n}$ is highlighted. A cohesion force $\vec{df}$ [1] acts on this section.

---

1 It should be remembered that the "infinitely small" dy and dx are not necessarily equal; the small element taken here is a rectangle.

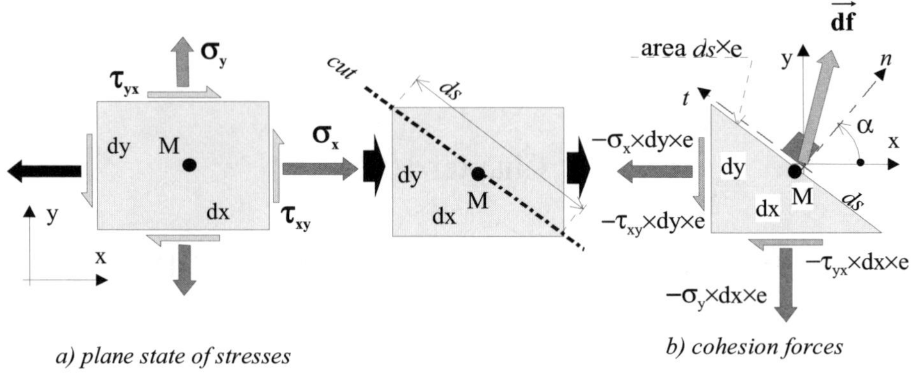

Figure 10.1. *Cohesion force on an oblique cross-section*

It can be seen that $\overrightarrow{df}$ can project along $\vec{n}$ and a direction perpendicular to $\vec{n}$, denoted as $\vec{t}$ in Figure 10.1b, which is obtained by rotating $\vec{n}$ through $(\vec{t},\vec{n}) = +\dfrac{\pi}{2}$. We therefore have (Figure 10.2a):

$$\overrightarrow{df} = \overrightarrow{df_n} + \overrightarrow{df_t}$$

or, by taking into account the definition [1.1] of stress[2]:

$$\overrightarrow{C}_{(M,\vec{n})} = \frac{\overrightarrow{df}}{dS} = \frac{\overrightarrow{df_n}}{dS} + \frac{\overrightarrow{df_t}}{dS} = \vec{\sigma} + \vec{\tau} = \sigma\vec{n} + \tau\vec{t}$$

where σ and τ are the normal and tangential (or shear) components of stress $\overrightarrow{C}_{(M,\vec{n})}$ on the facet with normal $\vec{n}$.

---

2 See section 1.2.2, Figure 1.4.

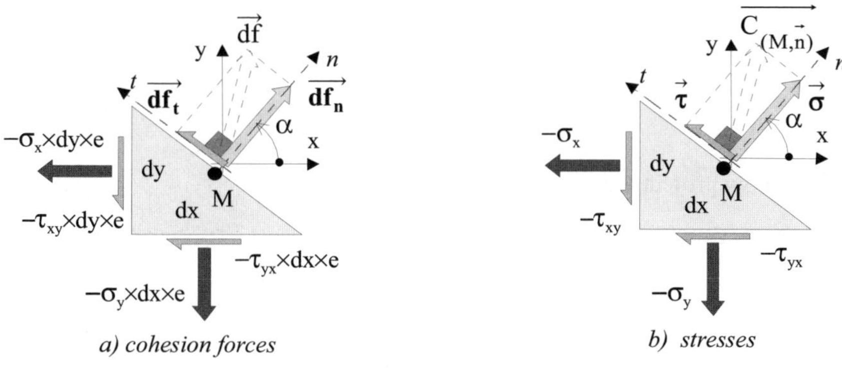

Figure 10.2. *Equilibrium of the domain; stresses on a facet defined through* $\vec{n}$

We propose to calculate σ and τ. For this, let us consider the equilibrium of the loaded domain according to Figure 10.2a.

a) Preliminary note: reciprocity of shear stresses

We propose calculating around point M the moments of all cohesion forces that appear on Figure 10.2a. We note immediately that the supports of forces passing through M give zero moment values. There only remains the moments of shear forces, giving at equilibrium:

$$\left(\tau_{xy} \times dy \times e\right) \times \frac{dx}{2} - \left(\tau_{yx} \times dx \times e\right) \times \frac{dy}{2} = 0$$

From which we deduce the reciprocity property which has often been used until now[3]:

$$\tau_{xy} = \tau_{yx}$$

b) Calculus of σ *and* τ

By projecting along $\vec{n}$, and by denoting $c = \cos\alpha$ and $s = \sin\alpha$ we obtain:

$$df_n - \sigma_x \times dy \times e \times c - \tau_{xy} \times dy \times e \times s - \tau_{yx} \times dx \times e \times c - \sigma_y \times dx \times e \times s = 0$$

recalling that $\tau_{xy} = \tau_{yx}$, and with $df_n = \sigma \times dS = \sigma \times ds \times e$:

$$\sigma = \sigma_x \times \frac{dy}{ds} \times c + \tau_{xy} \times \frac{dy}{ds} \times s + \sigma_y \times \frac{dx}{ds} \times s + \tau_{xy} \times \frac{dx}{ds} \times c$$

---

3 This property is generalized in section 10.2.2 (see [10.19]).

by noting that $\dfrac{dy}{ds} = c$ and $\dfrac{dx}{ds} = s$, we have:

$$\sigma = \sigma_x \times c^2 + \sigma_y \times s^2 + 2\tau_{xy} \times c \times s$$

Let us now project along $\vec{t}$:

$$df_t - \tau_{xy} \times dy \times e \times c - \sigma_y \times dx \times e \times c - \sigma_x \times dy \times e \times (-s) - \tau_{yx} \times dx \times e \times (-s) = 0$$

and with $df_t = \tau \times dS = \tau \times ds \times e$:

$$\tau = -\sigma_x \times \frac{dy}{ds} \times s + \sigma_y \times \frac{dx}{ds} \times c + \tau_{xy} \times \frac{dy}{ds} \times c - \tau_{xy} \times \frac{dx}{ds} \times s$$

i.e.: $\tau = -\sigma_x \times c \times s + \sigma_y \times c \times s + \tau_{xy}\left(c^2 - s^2\right)$

The above expressions for σ and τ, valid on a facet of any orientation α, can be regrouped as:

$$\begin{Bmatrix} \sigma \\ \tau \end{Bmatrix} = \begin{bmatrix} c^2 & s^2 & 2cs \\ -cs & cs & \left(c^2 - s^2\right) \end{bmatrix} \bullet \begin{Bmatrix} \sigma_x \\ \sigma_y \\ \tau_{xy} \end{Bmatrix} \qquad [10.1]$$

10.1.2. *Principal directions and stresses*

In Chapter 1^4 we indicated the existence of principal directions $\vec{1}$, $\vec{2}$ and principal associated stresses. Figure 10.3 refers to Figure 1.22, from now on with the common stress notation (σ_1, σ_2).

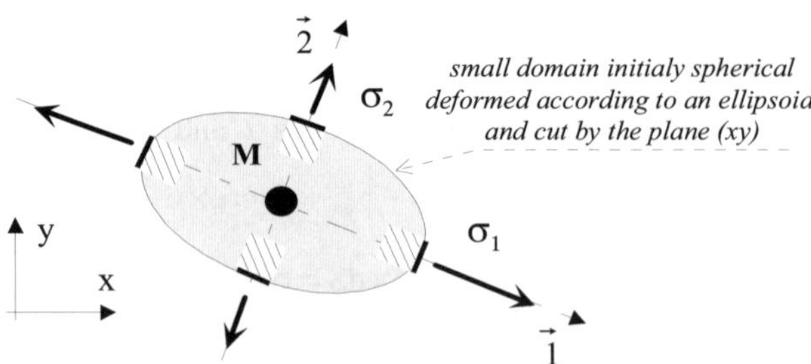

Figure 10.3. *Principal directions and stresses for a plane state of stresses*

---

4 See *NOTE* of section 1.4.2.4.

Starting from the knowledge of σ_x, σ_y and τ_{xy}, we analyze below how to determine these directions and the associated values σ_1 and σ_2.

On the particular facet having a principal direction as outward normal, the stress will be, by definition, reduced to its normal component. Thus, on this facet we shall have: $\tau = 0$.

Using [10.1]:

$$0 = -c \times s\left(\sigma_x - \sigma_y\right) + \tau_{xy}\left(c^2 - s^2\right)$$

$$\text{or} \quad \frac{c \times s}{c^2 - s^2} = \frac{\tau_{xy}}{\sigma_x - \sigma_y}$$

We obtain the equation:

$$\tan 2\alpha_0 = \frac{2\tau_{xy}}{\sigma_x - \sigma_y} \qquad\qquad [10.2]$$

in which α_0 is the particular angle such that:

$$2\frac{\cos\alpha_0 \times \sin\alpha_0}{\cos\alpha_0{}^2 - \sin\alpha_0{}^2} = \frac{\sin 2\alpha_0}{\cos 2\alpha_0} = \tan 2\alpha_0 .$$

Two values of α_0 that verify this equation are:

$$\alpha_{01} = \frac{1}{2}\arctan\frac{2\tau_{xy}}{\sigma_x - \sigma_y}, \quad \text{and} \quad \alpha_{02} = \alpha_{01} + \frac{\pi}{2}$$

These two angles which give the principal directions are represented in Figure 10.4. Here we make sure that the principal directions $\vec{1}$, $\vec{2}$, $\vec{3}$ are perpendicular with one another[5].

____

5 $\vec{z}$ is also a principal direction ($\vec{z} \equiv \vec{3}$) because the stress on a facet with normal $\vec{z}$ is zero.

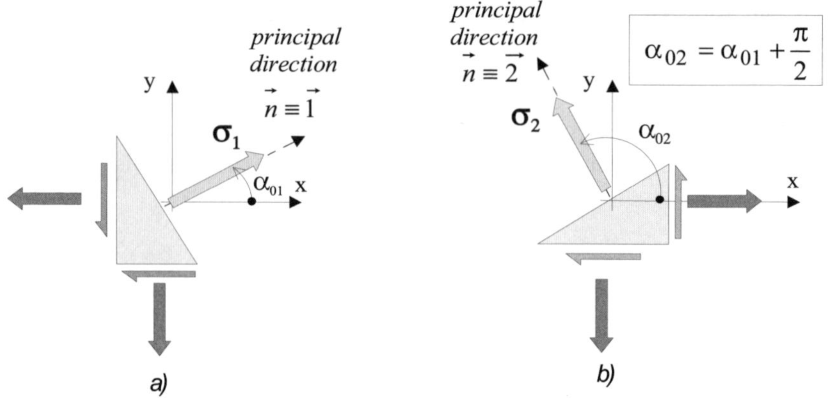

Figure 10.4. *Principal directions*

The principal stresses are obtained from [10.1] in the form:

$$\sigma_1 = c_1{}^2 \sigma_x + s_1{}^2 \sigma_y + 2 c_1 \times s_1 \tau_{xy} \qquad [10.3]$$

with $c_1 = \cos \alpha_{01}$; $s_1 = \sin \alpha_{01}$

$$\sigma_2 = c_2{}^2 \sigma_x + s_2{}^2 \sigma_y + 2 c_2 \times s_2 \tau_{xy} \quad \text{with} \quad \begin{cases} c_2 = \cos \alpha_{02} = \cos\left(\alpha_{01} + \dfrac{\pi}{2} \right) = -s_1 \\[2mm] s_2 = \sin \alpha_{02} = \sin\left(\alpha_{01} + \dfrac{\pi}{2} \right) = c_1 \end{cases}$$

i.e.:

$$\sigma_2 = s_1{}^2 \sigma_x + c_1{}^2 \sigma_y - 2 c_1 \times s_1 \tau_{xy} \qquad [10.4]$$

10.1.3. *Mohr graphical representation*

Instead of $\vec{x}$ and $\vec{y}$, let us choose the coordinate system defined by the principal directions $\vec{1}$, $\vec{2}$. Then considering the absence of shear on the principal facets ($\tau_{12} = 0$), [10.1] can be written as:

$$\begin{Bmatrix} \sigma \\ \tau \end{Bmatrix} = \begin{bmatrix} c^2 & s^2 \\ -cs & cs \end{bmatrix} \bullet \begin{Bmatrix} \sigma_1 \\ \sigma_2 \end{Bmatrix}$$

The angle $\left(\vec{n},\vec{1}\right)$ which is formed by the normal under study with the direction $\vec{1}$ is noted as α_1 in Figure 10.5 (we have $c = \cos\alpha_1$ and $s = \sin\alpha_1$).

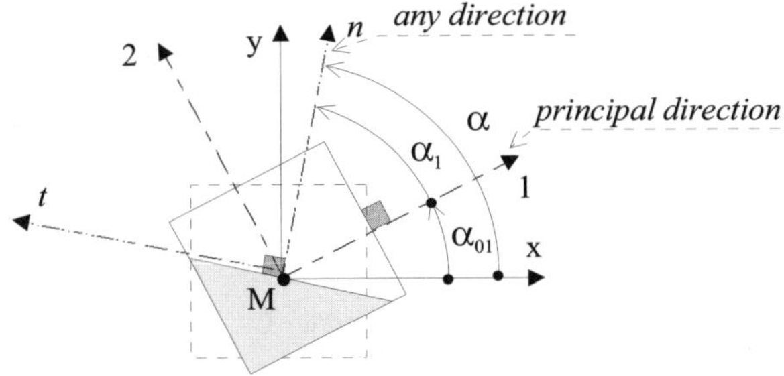

Figure 10.5. *Principal directions as intermediate references*

The previous equation can be written in detail as follows:

$$\sigma = c^2\sigma_1 + s^2\sigma_2 = \frac{1}{2}\left(c^2 + s^2\right)(\sigma_1 + \sigma_2) + \frac{1}{2}\left(c^2 - s^2\right)(\sigma_1 - \sigma_2)$$
$$\tau = -c \times s\,(\sigma_1 - \sigma_2)$$

i.e.:

$$\sigma = \frac{\sigma_1 + \sigma_2}{2} + \frac{\sigma_1 - \sigma_2}{2}\cos 2\alpha_1$$
$$\tau = -\frac{\sigma_1 - \sigma_2}{2}\sin 2\alpha_1 \qquad\qquad [10.5]$$

Equations [10.5] provide σ and τ on a facet making an angle α_1 with the principal direction $\vec{1}$. They can also be obtained through a graphic construction whose steps are given in detail in Figure 10.6, where for fixing the concept, it is assumed that $\sigma_1 > \sigma_2 > 0$.

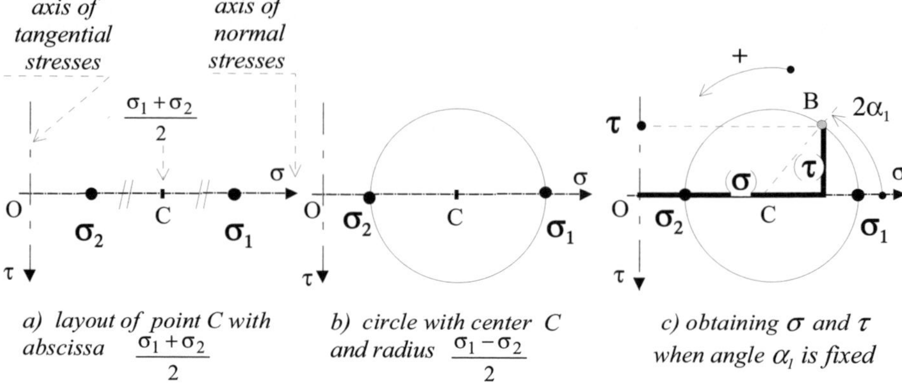

a) layout of point C with abscissa $\dfrac{\sigma_1+\sigma_2}{2}$ b) circle with center C and radius $\dfrac{\sigma_1-\sigma_2}{2}$ c) obtaining σ and τ when angle α_1 is fixed

Figure 10.6. *Graphical construction of σ and τ from the principal stresses*

This graphical construction is based on the two perpendicular axes represented in Figure 10.6, with a non-standard orientation of the vertical[6] axis. The circle thus defined is called the "Mohr circle". The relation between the Mohr representation and the real phenomenon is represented in Figure 10.7.

---

6 According to the authors, other representation conventions can be found. They relate mainly to the orientation of the vertical axis in liaison with the direction of rotation of angle $2\alpha_1$. The following representation can also be adopted:

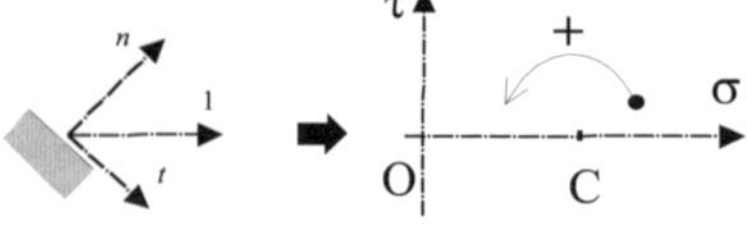

We can also find in the literature an orientation of the $\vec{t}$ axis of the facet such that $\left(\vec{t},\vec{n}\right)=-\dfrac{\pi}{2}$. When this convention is taken into account, it leads to the following graphic convention for a Mohr representation:

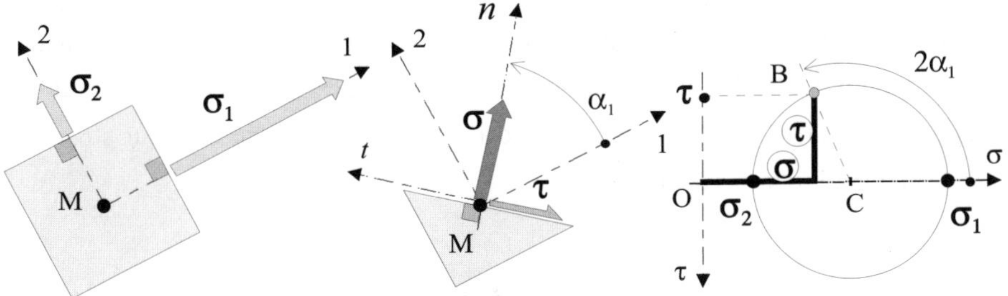

Figure 10.7. *Real phenomenom and Mohr representation*

NOTES

❏ In fact, the construction of the Mohr circle does not require any previous knowledge of the main stresses σ_1 and σ_2. σ and τ on the facet of Figure 10.7 can be obtained starting from equations [10.1], which can be rewritten as:

$$\sigma = c^2\sigma_x + s^2\sigma_y + 2cs\tau_{xy} = \frac{\sigma_x + \sigma_y}{2} + \frac{\sigma_x - \sigma_y}{2}\cos 2\alpha + \tau_{xy}\sin 2\alpha \quad [10.6]$$

$$\tau = -cs(\sigma_x - \sigma_y) + \tau_{xy}(c^2 - s^2) = \frac{\sigma_x - \sigma_y}{2}\sin 2\alpha + \tau_{xy}\cos 2\alpha \quad [10.7]$$

An illustration of σ and τ on a facet defined by the angle α is detailed in Figure 10.8 where, to fix the concept, it has been assumed that $\sigma_x > \sigma_y > 0$.

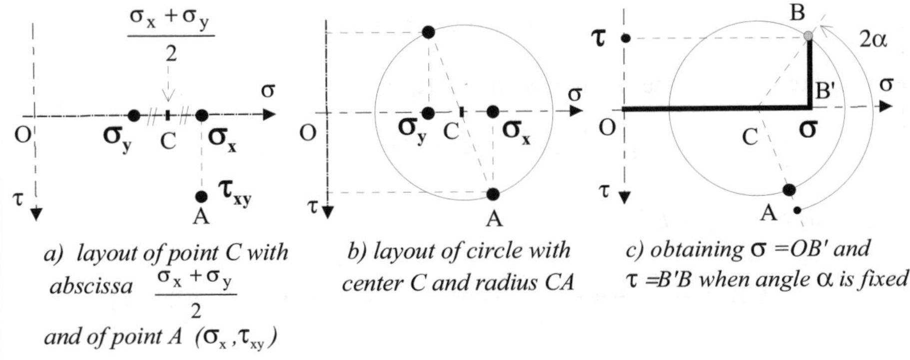

a) layout of point C with abscissa $\dfrac{\sigma_x + \sigma_y}{2}$ and of point A (σ_x, τ_{xy})

b) layout of circle with center C and radius CA

c) obtaining $\sigma = OB'$ and $\tau = B'B$ when angle α is fixed

Figure 10.8. *Illustration of σ and τ for any orientation starting from given stresses $\sigma_x, \sigma_y, \tau_{xy}$ in the x, y axes*

◆ As proof of this illustration, let us assume that we follow the procedure in Figure 10.8 to determine the particular angle α_0 corresponding to a facet such as $\tau_{xy} = 0$. Figure 10.8c allows us to obtain the angles α_{01} and α_{02} plotted in Figure 10.9a. The directions determined in this way (absence of shear stress) are the principal directions $\vec{1}$ and $\vec{2}$ [7].

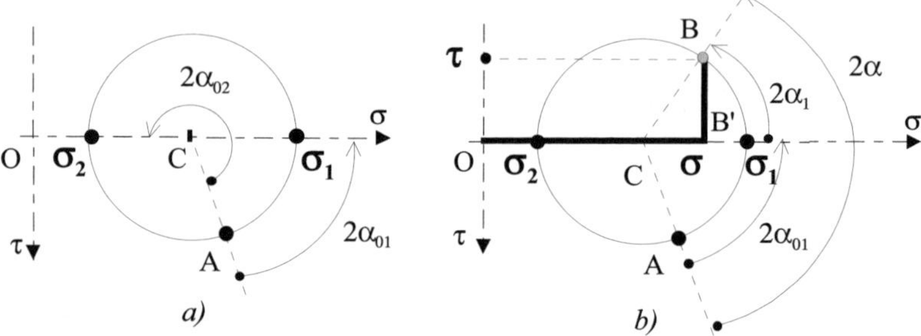

Figure 10.9. *Justification of the above graphical obtention*

Here we find the same circle as that in Figure 10.6 because, due to [10.6], it has the same center and diameter. At the same time we note that the abscissa of center C is:

$$OC = \frac{\sigma_x + \sigma_y}{2} = \frac{\sigma_1 + \sigma_2}{2}$$

which leads to the following characteristic property:

$$\sigma_x + \sigma_y = \sigma_1 + \sigma_2$$

Let us calculate the length OB' in Figure 10.9b:

$$OB' = \frac{\sigma_1 + \sigma_2}{2} + \frac{\sigma_1 - \sigma_2}{2} \cos 2(\alpha - \alpha_{01})$$

i.e. with $\alpha - \alpha_{01} = \alpha_1$ [8]:

$$OB' = \frac{\sigma_1 + \sigma_2}{2} + \frac{\sigma_1 - \sigma_2}{2} \cos 2\alpha_1 = \sigma \quad \text{as per equation [10.5]}$$

then:

---

7 See [10.2].
8 See Figures 10.5 and 10.7.

$$B'B = \frac{\sigma_1 - \sigma_2}{2}\sin(2\alpha - 2\alpha_{01}) = \frac{\sigma_1 - \sigma_2}{2}\sin 2\alpha_1 = -\tau \text{ as per equation [10.5]}$$

❏ Justification of algebraic signs for shear stresses: using the Mohr representation, let us deduce the given state of stresses $\sigma_x, \sigma_y, \tau_{xy}$ in the x, y axes from the state of principal stresses. By completing Figure 10.6b using Figure 10.8b, the following Figure 10.10a is obtained.

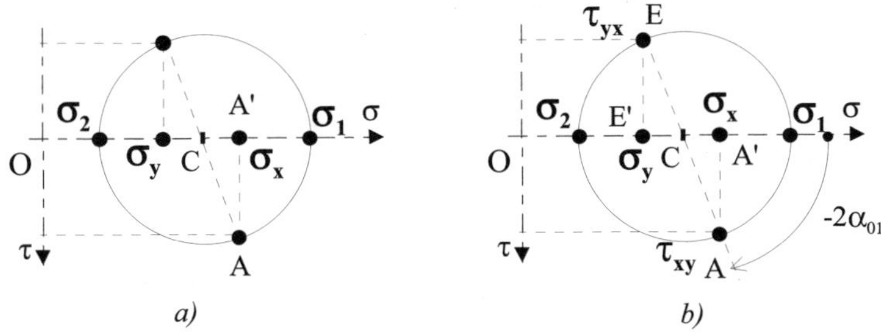

Figure 10.10. *Given state of stresses* $\sigma_x, \sigma_y, \tau_{xy}$ *in the x, y axes deduced from principal state of stresses*

Lengths OA' and A'A can then be interpreted respectively as normal and tangential stress on a facet with normal $\overrightarrow{n_x} \equiv \overrightarrow{x}$ which is derived from that of normal $\overrightarrow{1}$ through rotation $\left(\overrightarrow{x}, \overrightarrow{1}\right) = -\alpha_{01}$. Thus, in addition to $OA' = \sigma_x$ the following is also derived:

$$A'A = \tau_{xy}$$

where σ_x and τ_{xy} are the projections of the stress vector on local axes linked to the facet (Figure 10.11a). In the same way lengths OE' and E'E are respectively the normal and tangential stress on a facet with normal $\overrightarrow{n_y} \equiv \overrightarrow{y}$ which is derived from that of the normal $\overrightarrow{1}$ through rotation $\left(\overrightarrow{y}, \overrightarrow{1}\right) = -\alpha_{01} + \frac{\pi}{2}$. Thus, in addition to $OE' = \sigma_y$ the following can also be derived:

$$E'E = \tau_{yx}$$

where σ_y and τ_{yx} are the projections of the stress vector on local axes linked to the facet (Figure 10.11b)[9].

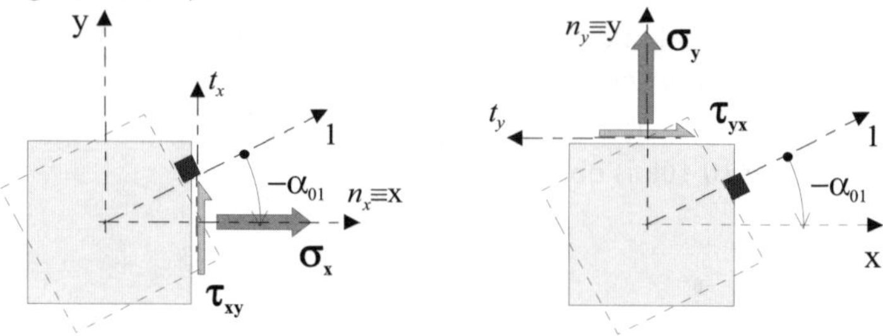

a) local axes linked to the facet
with normal $\overrightarrow{n_x} \equiv \overrightarrow{x}$

b) local axes linked to the facet
with normal $\overrightarrow{n_y} \equiv \overrightarrow{y}$

Figure 10.11. *Signs of shear stresses in local axes linked to facets*

❑ We can immediately deduce from Figure 10.10 the following results:

– radius of the Mohr circle:

$$CA = \sqrt{\left(\frac{\sigma_x - \sigma_y}{2}\right)^2 + \tau_{xy}{}^2}$$

– principal stresses:

$$\sigma_1 = \frac{\sigma_x + \sigma_y}{2} + \sqrt{\left(\frac{\sigma_x - \sigma_y}{2}\right)^2 + \tau_{xy}{}^2}$$

$$\sigma_2 = \frac{\sigma_x + \sigma_y}{2} - \sqrt{\left(\frac{\sigma_x - \sigma_y}{2}\right)^2 + \tau_{xy}{}^2}$$

[10.8]

---

9 From Figure 10.10b it is obvious that stress $E'E = \tau_{yx}$ is negative on axis $\overrightarrow{\tau}$ which corroborates its representation in Figure 10.11b. On the other hand it should be positive on axis $\overrightarrow{x}$ because, on the facet referred to (with normal $\overrightarrow{y}$), we have $\overrightarrow{t_y} = -\overrightarrow{x}$.

10.1.4. *Summary*

Stresses on a facet of any orientation

given the plane state of stresses (plane (xy)) on the elementary domain below:

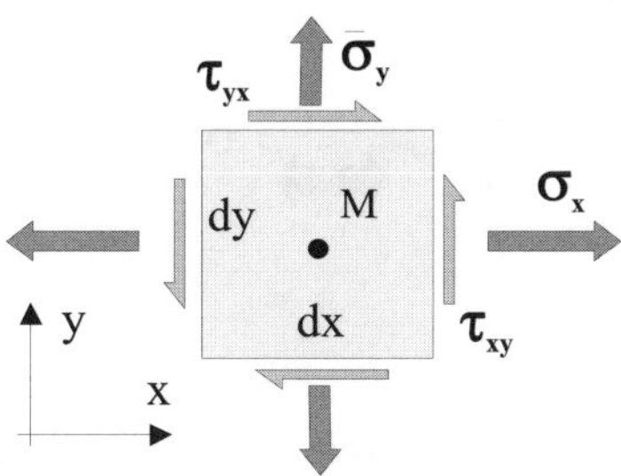

The stress acting on a facet whose outward normal $\vec{n}$ is within the plane (xy) with any orientation α, is obtained through its normal component σ along $\vec{n}$ and tangential component τ along $\vec{t}$ with $(\vec{t},\vec{n})=+\dfrac{\pi}{2}$) by using the relations:

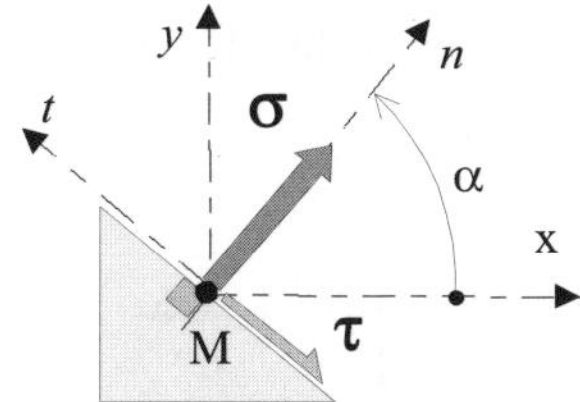

$$\begin{Bmatrix}\sigma \\ \tau\end{Bmatrix}=\begin{bmatrix}c^2 & s^2 & 2cs \\ -cs & cs & (c^2-s^2)\end{bmatrix}\bullet\begin{Bmatrix}\sigma_x \\ \sigma_y \\ \tau_{xy}\end{Bmatrix} \quad with \quad \begin{pmatrix}c=\cos\alpha \\ s=\sin\alpha\end{pmatrix}$$

σ and τ can also be obtained using the Mohr representation summarized below:

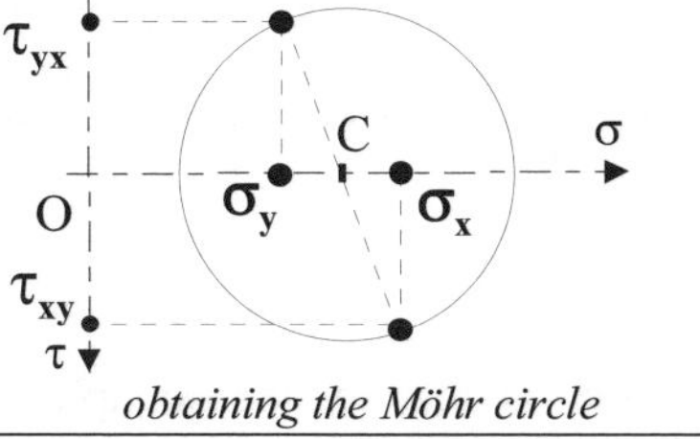

obtaining the Möhr circle

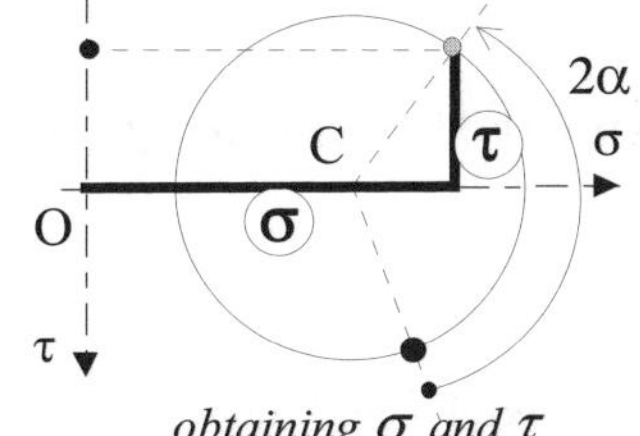

obtaining σ and τ

[10.9]

10.1.5. *Some remarkable plane states of stresses with their Mohr representation*

10.1.5.1. *Case 1*

Let us study the plane state of stresses of Figure 10.12a reduced to a single normal stress σ_x along $\vec{x}$ [10]. This state is called a "uniaxial" state of stress.

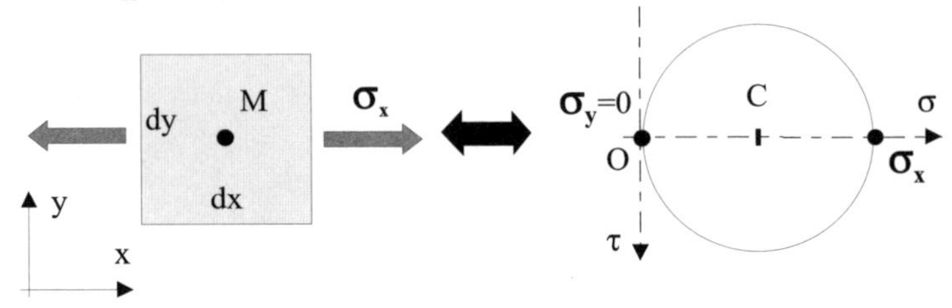

a) *particular case of plane state of stress*   b) *associated Mohr representation*

Figure 10.12. *Uniaxial state of stress*

Figure 10.12b represents the corresponding Mohr circle obtained using the method [10.9]. Stresses σ and τ on a facet whose normal $\vec{n}$ makes an angle α with $\vec{x}$ are represented in Figure 10.13.

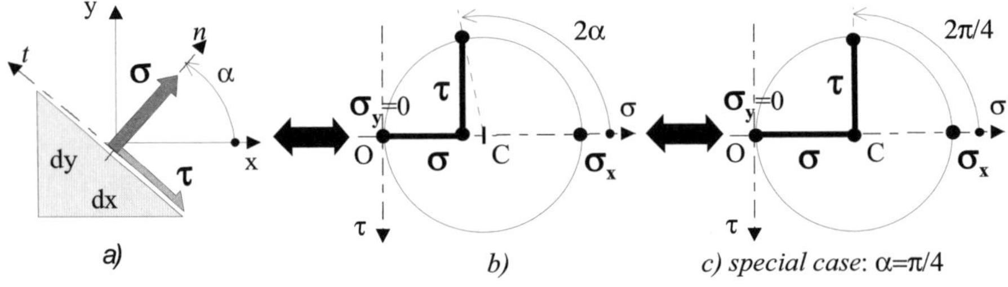

a)   b)   c) *special case*: $\alpha=\pi/4$

Figure 10.13. *Stresses on a facet whose normal $\vec{n}$ makes any angle α with $\vec{x}$*

NOTES

❏ This uniaxial – and plane – state of stresses is obtained for example when a flat test sample is loaded by means of a traction testing machine (Figure 10.14).

---

10 It should be observed that due to the absence of shear on the facets perpendicular to $\vec{x}$ and $\vec{y}$ (without forgetting $\vec{z}$), these axes become the principal directions.

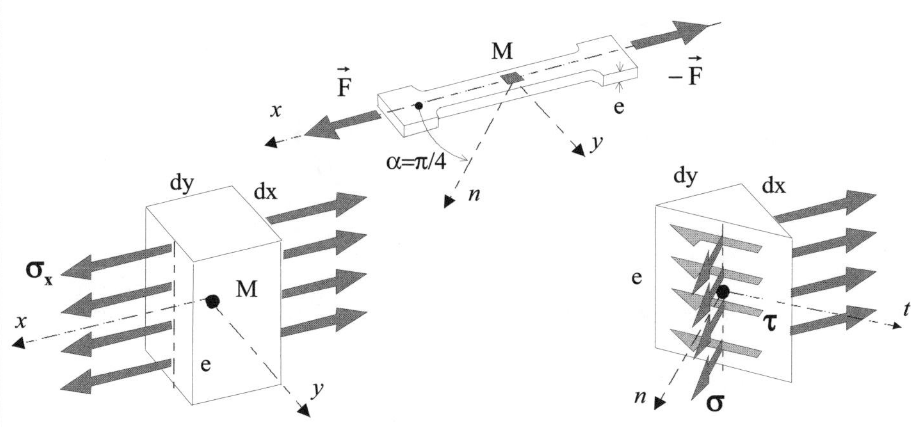

Figure 10.14. *Maximum shear value for the uniaxial state of stress*

❑ When we analyze the stresses on a facet such that $\alpha = \dfrac{\pi}{4}$, the following specific values are observed on the associated Mohr circle (Figure 10.13c):

$$\tau = -\sigma = -\frac{\sigma_x}{2} \qquad\qquad [10.10]$$

The shear stress τ then attains its maximum absolute value.

10.1.5.2. *Case 2*

Let us consider the special plane state of stresses represented in Figure 10.15a. It is characterized by the only normal stresses σ_x and σ_y[11] such that $\sigma_x > 0$ and $\sigma_y = -\sigma_x < 0$. Figure 10.15b represents the associated Mohr circle.

---

11 It should be noted that the absence of shear stresses on the facets perpendicular to $\vec{x}$ and $\vec{y}$ indicates that directions $\vec{x}$ and $\vec{y}$ are the principal directions. For example: $\vec{x} \equiv \vec{1}$; $\vec{y} \equiv \vec{2} \Rightarrow \sigma_x = \sigma_1$; $\sigma_y = \sigma_2 = -\sigma_1$.

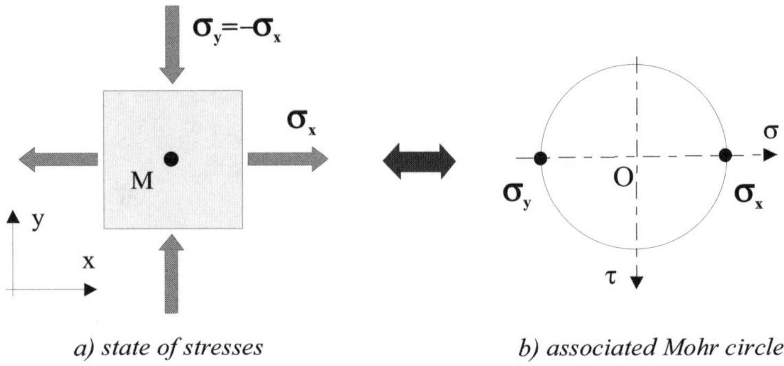

a) state of stresses   b) associated Mohr circle

Figure 10.15. *Only normal stresses:* $\sigma_x > 0$ *and* $\sigma_y = -\sigma_x < 0$

We can observe in Figure 10.16 that the stress on a facet whose normal $\vec{n}$ makes an angle $\dfrac{\pi}{4}$ with $\vec{x}$ is reduced to:

$$\sigma = 0 \; ; \; \tau = -\sigma_x < 0 \; (\text{or } \tau = \sigma_y)$$

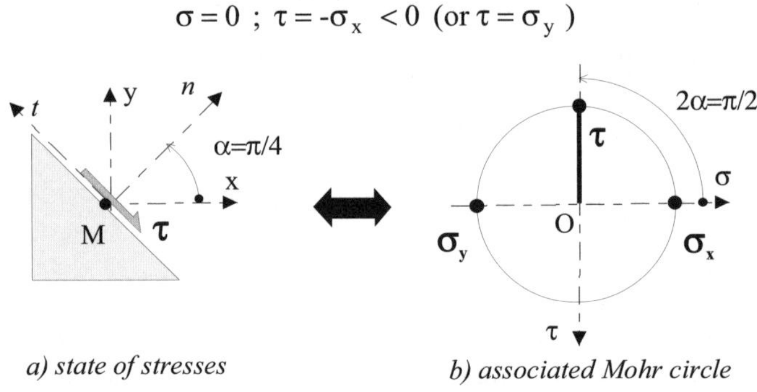

a) state of stresses   b) associated Mohr circle

Figure 10.16. *Pure shear*

This facet is therefore subjected to a pure shear. A similar condition should be ensured for the facet whose normal makes the angle $-\dfrac{\pi}{4}$ with $\vec{x}$, which is in accordance with the reciprocity property of the shear stresses[12].

---

12 See sections 10.1.1 and 10.2.2 (see [10.9]).

NOTES

❐ The above results enable us to note the equivalence between the states of stresses acting on each of the two square elementary domains, one cut along the directions $\vec{x}$ and $\vec{y}$ and the other along the directions $\vec{X}$ and $\vec{Y}$ which are derived from the earlier ones through a rotation $\alpha = \dfrac{\pi}{4}$, as shown in Figure 10.17.

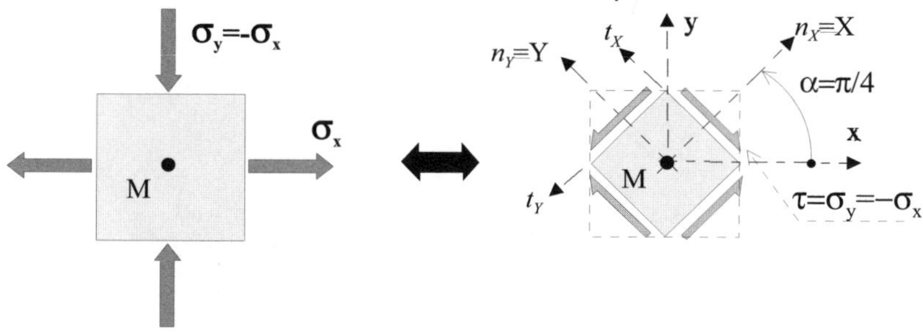

Figure 10.17. *Equivalent states of stresses*

❐ This equivalence enables the representation of small elastic displacements and thus the deflected shape of each of the small domains of Figure 10.17. This is shown in Figure 10.18.

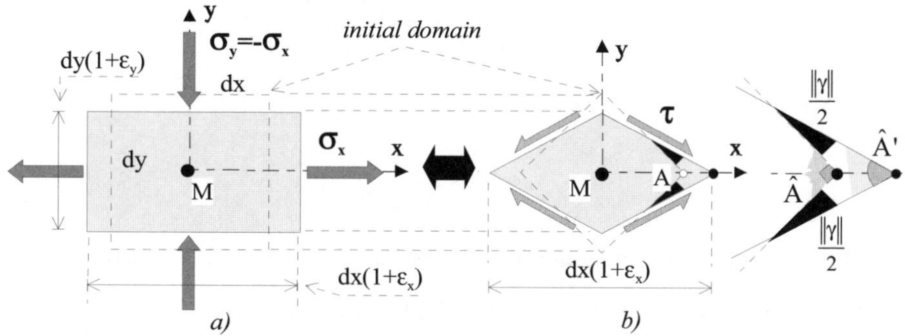

Figure 10.18. *Deflected shapes*

Figure 10.18b shows the mechanism of angular distortion γ. The length variations are written as[13]:

---

13 See [1.6].

$$dx \rightarrow dx + \varepsilon_x \times dx = dx(1+\varepsilon_x)$$
$$dy \rightarrow dy + \varepsilon_y \times dy = dy(1+\varepsilon_y)$$

Angle $\hat{A}$ initially equal to $\dfrac{\pi}{2}$ becomes $\hat{A}' < \hat{A}$ such that:

$$\hat{A}' = \frac{\pi}{2} + \gamma = 2\left(\frac{\pi}{4} + \frac{\gamma}{2}\right) \text{ with } \gamma < 0$$

whereas taking into account the smallness of distortion γ, Figure 10.18b should be read as:

$$\tan\frac{\hat{A}'}{2} = \tan\left(\frac{\pi}{4} + \frac{\gamma}{2}\right) \cong \frac{1+\dfrac{\gamma}{2}}{1-\dfrac{\gamma}{2}} = \frac{\dfrac{dy}{2}(1+\varepsilon_y)}{\dfrac{dx}{2}(1+\varepsilon_x)}$$

from where with $dx = dy$ and by comparing the two fractions given above:

$$\frac{\gamma}{2} = \varepsilon_y = -\varepsilon_x .$$

Deformation ε_x can be expressed with behavior relation [1.12] as follows:

$$\varepsilon_x = \frac{\sigma_x}{E} - \frac{\nu}{E}\sigma_y = \sigma_x\left(\frac{1+\nu}{E}\right) = -\frac{\gamma}{2}$$

and by recalling $\sigma_x = -\tau$ we have:

$$\gamma = \tau \times 2\left(\frac{1+\nu}{E}\right)$$

Let us compare this equation with behavior relation [1.12]. It is written here in the $\vec{X}$ and $\vec{Y}$ axes (Figure 10.17) and should be read as: $\gamma_{XY} = \dfrac{\tau_{XY}}{G}$ [14]. The comparison leads to the identification of:

---

14 It should be noted that in the $\vec{X}$ and $\vec{Y}$ axes the shear stress τ_{XY} is negative, as represented below, where we also note the reciprocity property $\tau_{XY} = \tau_{YX}$.

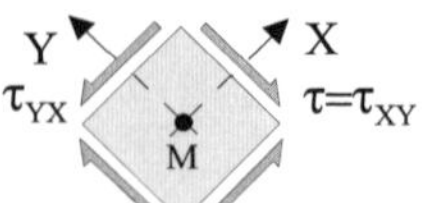

$$G = \frac{E}{2(1+\nu)}$$ [10.11]

The shear modulus appears as an elastic quantity which depends on the longitudinal modulus E and Poisson's ratio ν.

10.1.5.3. *Case 3*

Let us consider the particular plane state of stresses represented in Figure 10.19a) characterized by the only normal stresses σ_x and σ_y such that, for example, $\sigma_x = \sigma_y > 0$ [15]. Then, and as in the earlier case, directions $\vec{x}$ and $\vec{y}$ (and $\vec{z}$) are the principal directions. Figure 10.19b shows that the associated Mohr circle is characterized by a zero diameter: $(\sigma_x - \sigma_y) = 0$. The circle is therefore reduced to its center C.

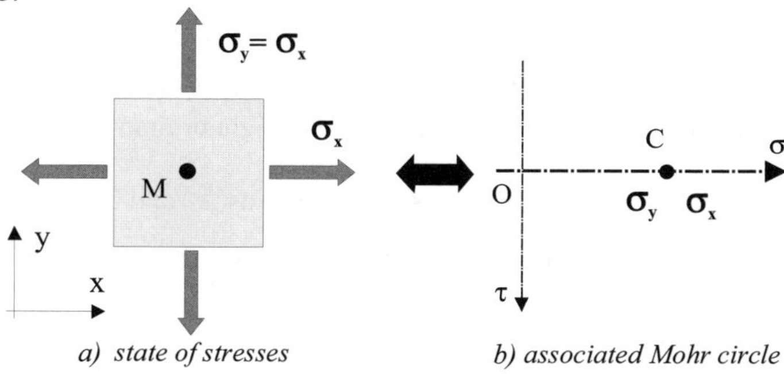

a) state of stresses   b) associated Mohr circle

Figure 10.19. *Particular plane state of stresses* $\sigma_x = \sigma_y > 0$

This implies that the state of stresses on a facet whose normal $\vec{n}$ makes any angle α with $\vec{x}$ (Figure 10.20) is reduced to: $\begin{bmatrix} \sigma = \sigma_x \\ \tau = 0 \end{bmatrix}$.

In other words, in this particular case, any direction $\vec{n}$ of the plane (xy) is a principal direction, or it can even be said that there is no shear on any facet whose normal $\vec{n}$ is in the plane (xy). From there it can be inferred that there is no distortion during the deformation of any prismatic element of $\vec{z}$ axis. This element undergoes

---

15 Similar properties can be obtained by taking $\sigma_x = \sigma_y < 0$.

only dilatations (case where $\sigma_x = \sigma_y > 0$) or contractions (case where $\sigma_x = \sigma_y < 0$) along the normals to its facets.

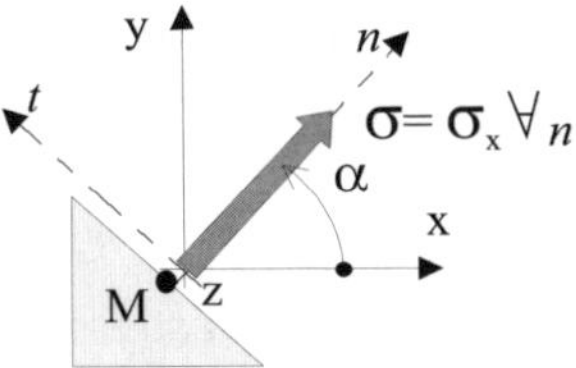

Figure 10.20. *Any direction in the plane x,y is a principal direction*

10.1.5.4. *Case 4: cylindrical vessel under pressure*

Figure 10.21 represents a tubular vessel under internal pressure $\|p_0\|$ [16] denoted hereafter as p_0. Note on this figure zones (1) and (2) as marked on the reservoir. The vessel is assumed to "float" in space, without weight or linking forces.

The cylindrical zone of the vessel has a mean radius "r" and a thickness "e". The envelope thus formed is supposed to be thin, i.e. $\dfrac{e}{r} \ll 1$.

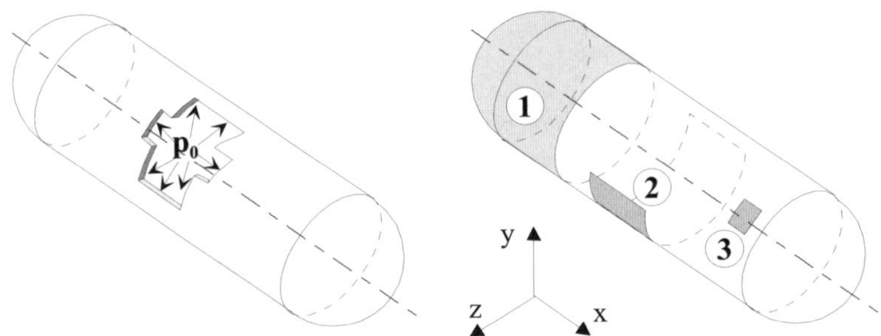

Figure 10.21. *Cylindrical vessel under pressure*

• The isolated zone (1) in Figure 10.22a is balanced along direction $\vec{x}$.

---

16 The pressure vector considered as a stress exerted by a liquid on a wall with outward normal $\vec{n}$ is expressed as $\vec{p} = -\|p_0\|\vec{n}$.

– On the annular section of area $2\pi re$ with a normal parallel to $\vec{x}$, we can observe normal cohesion forces parallel to $\vec{x}$ [17] characterized by a stress distribution σ_x which is uniform, due to the axisymmetry. The sum along $\vec{x}$ of these cohesion forces is written as: $2\pi \times r \times e \times \sigma_x$.

– Forces act on the bottom of the reservoir due to pressure. The "projected" [18] area of the bottom is: $\pi \times r^2$.

The result of these pressure forces is $-p_0 \times \pi \times r^2$. The balancing of zone (1) (Figure 10.22a) along $\vec{x}$ gives:

$$2\pi \times r \times e \times \sigma_x - p_0 \times \pi \times r^2 = 0$$

leading to:

$$\sigma_x = p_0 \frac{r}{2e} \qquad\qquad [10.12]$$

---

17 In fact, these forces cannot have any tangential component because their distribution is axisymmetric.

18 On a facet dS whose normal $\vec{n}$ is at an angle α with $\vec{x}$, the elementary pressure force is $-p_0 \times dS \times \vec{n}$. When projecting this along $\vec{x}$, we obtain:

$-p_0 \times dS \times \vec{n} \cdot \vec{x} = -p_0 \times dS \times \cos\alpha$. The facet of area $dS \times \cos\alpha$ appears to be the projection of dS parallel to $\vec{x}$ on a plane perpendicular to $\vec{x}$. This is known as the projected surface. The set of facets dS which forms the bottom of the reservoir has a projected area: $\int_{base} dS \times \cos\alpha = \pi \times r^2$

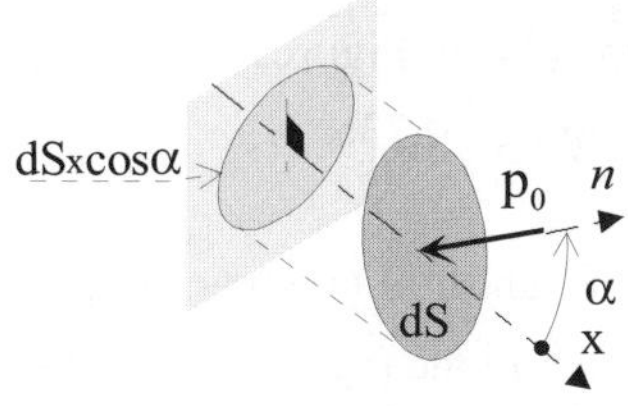

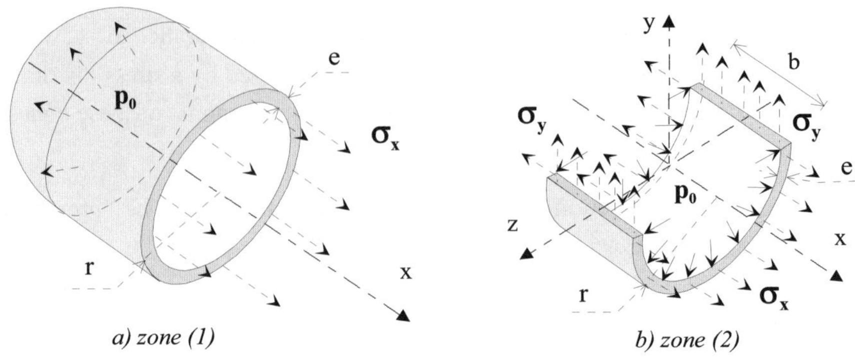

Figure 10.22. *Balancing of zones (1) and (2)*

♦ Zone (2) in Figure 10.21 is isolated in Figure 10.22b.

– On the two sections represented with normals $\vec{y}$, the cohesion forces are characterized by a similar distribution of stresses σ_y due to axisymmetry. The summation along the $\vec{y}$ axis for the two sections is written as:

$$\sigma_y \times e \times b \times 2 .$$

– Forces act on the internal wall of the reservoir due to pressure. The "projected" area along $\vec{y}$ of the internal wall being $2r \times b$, the resultant of these pressure forces is: $-p_0 \times 2r \times b$. The equilibrium along $\vec{y}$ can thus be written as:

$$\sigma_y \times e \times b \times 2 - p_0 \times 2r \times b = 0$$

leading to:

$$\sigma_y = p_0 \frac{r}{e}$$

and with [10.12]:

$$\sigma_y = 2\sigma_x = p_0 \frac{r}{e} \qquad [10.13]$$

♦ The small wall element (3) of Figure 10.21 is subjected to stresses σ_x and $\sigma_y = 2\sigma_x$ [19] as is shown in Figure 10.23a.

---

19 The small wall element is also subjected to a pressure p_0 corresponding to a normal stress p_0 along direction $\vec{z}$ perpendicular to the (xy) plane. Therefore, this case does not logically

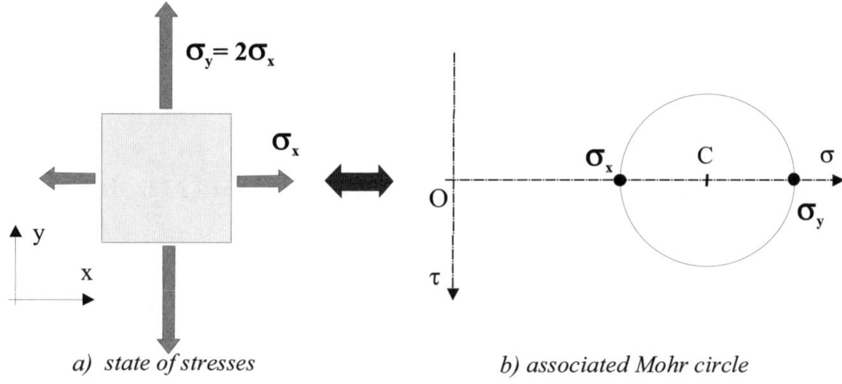

a) state of stresses

b) associated Mohr circle

Figure 10.23. *Wall element of the vessel*

It can be seen that directions $\vec{x}$ and $\vec{y}$ are still the principal directions. The associated Mohr circle is shown in Figure 10.23b.

Figure 10.24 represents the state of stresses derived from the Mohr representation on a section of the vessel whose normal is at 45° with the $\vec{x}$ axis.

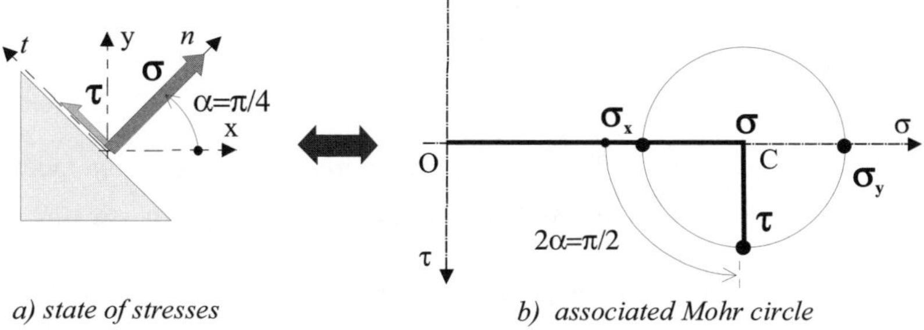

a) state of stresses

b) associated Mohr circle

Figure 10.24. *Stresses on a particular facet*

The following values are noted:

---

constitute a plane state of stresses. However, we have assumed a thin wall for this reservoir, characterized by $\dfrac{r}{e} \gg 1$. Thus, [10.13] shows that σ_x and $\sigma_y \gg p_0$. We can therefore conveniently ignore p_0 as against σ_x and σ_y and find the status of the plane state of stresses.

$$\sigma = \frac{\sigma_x + \sigma_y}{2} = 1.5\,\sigma_x \qquad ; \qquad \tau = \frac{\sigma_y - \sigma_x}{2} = 0.5\,\sigma_x$$

10.1.5.5. *Numerical example*

We consider the plane state of stresses shown in Figure 10.25 in the coordinate system (xy) with the values (in MPa).

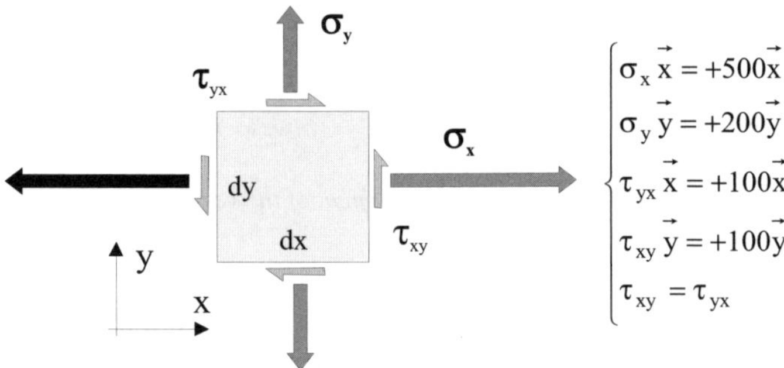

Figure 10.25.

We expect to determine the principal directions and stresses, as well as the stress on a facet of normal $\vec{n}$ such that $\left(\vec{n}, \vec{x}\right) = \alpha$.

NOTE

❑ Facet local axes: from Figure 10.25 we know the stresses on facets with normals $\vec{n_x} \equiv \vec{x}$ and $\vec{n_y} \equiv \vec{y}$. With the Mohr representation of stresses on any facet using the "local" axes $\vec{n}$ and $\vec{t}$ $\left(\vec{t}, \vec{n}\right) = \dfrac{\pi}{2}$, we obtain for the facets in question the local axes shown in Figure 10.26.

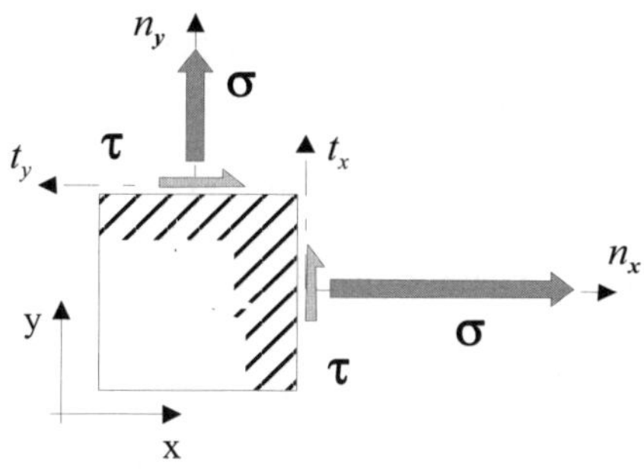

Figure 10.26. *Facet local axes*

From there by simple comparison with Figure 10.25, we can derive:

– on the facet with normal $\overrightarrow{n_x} \equiv \overrightarrow{x}$ (local system $\overrightarrow{n_x}, \overrightarrow{t_x}$):

$$\begin{cases} \sigma = \sigma_x \\ \tau = \tau_{xy} \end{cases}$$

– on the facet with normal $\overrightarrow{n_y} \equiv \overrightarrow{y}$ (local system $\overrightarrow{n_y}, \overrightarrow{t_y}$):

$$\begin{cases} \sigma = \sigma_y \\ \tau = -\tau_{yx} \end{cases}$$

Determining the principal stresses and directions: let us follow the steps indicated in [10.9] for the Mohr representation. The circle in Figure 10.27a is obtained.

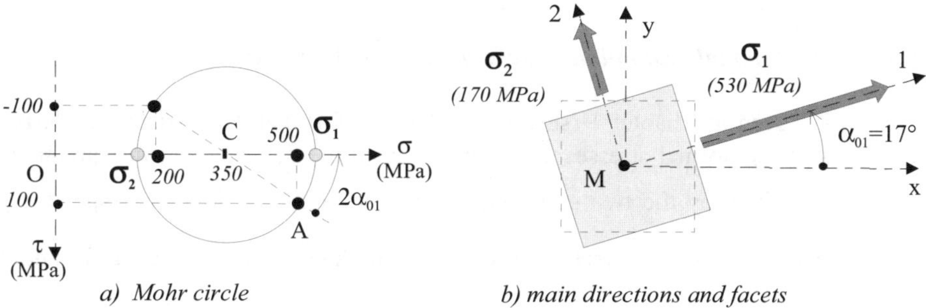

a) *Mohr circle*   b) *main directions and facets*

Figure 10.27. *Mohr representation*

The radius of the circle is derived: $CA = \sqrt{150^2 + 100^2} \cong 180\,\text{MPa}$, which leads to the values of the main stresses:

$$\sigma_1 = 350 + 180 = 530\,\text{MPa} \quad ; \quad \sigma_2 = 350 - 180 = 170\,\text{MPa}$$

The main direction is obtained by rotating by an angle α_{01} from $\vec{x}$, as (see Figure 10.27): $\tan 2\alpha_{01} = \dfrac{100}{150} \Rightarrow \alpha_{01} = 17^\circ$. We then obtain the diagram given in Figure 10.27b for the principal[20] directions and associated facets.

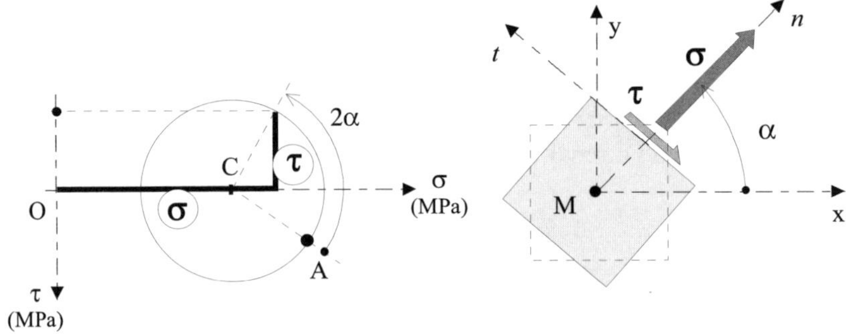

Figure 10.28. *Stresses on a facet such that* $\left(\vec{n}, \vec{x}\right) = \alpha$

On every facet of normal $\vec{n}$ such that $\left(\vec{n}, \vec{x}\right) = \alpha$, the stress can be obtained through its components σ and τ in the local axes of the facet with the construction as in Figure 10.28.

10.1.6. *Experimental evaluation of deformations to define stresses*

We have seen in Chapter 1 (section 1.4.2.4/*NOTES*) that it is possible to derive the values of the normal stresses σ_x and σ_y from the deformation measurements ε_x and ε_y given by the two strain gauges whose grids are oriented respectively along directions $\vec{x}$ and $\vec{y}$ (see Figure 1.23). However, since the gauges cannot

---

20 The same values of α_{01}, σ_1 and σ_2 could be obtained from equations [10.2], [10.3] and [10.4].

directly measure the distortion, we cannot access the value γ_{xy} which enables calculating the shear stress τ_{xy} using equation [1.14]:

$$\tau_{xy} = G \times \gamma_{xy} \tag{10.14}$$

Let us consider again the area around point M (Figure 1.12 or 10.1) and a domain of small dimensions $\Delta x \times \Delta y \times e$ around this point. On the top side of this domain the strain gauges J_X and J_Y are bonded respectively along $\vec{X}$ and $\vec{Y}$ which are obtained from $\vec{x}$ and $\vec{y}$ through a rotation of angle $\dfrac{\pi}{4}$ about the $\vec{z}$ axis. The arrangement is shown in Figure 10.29.

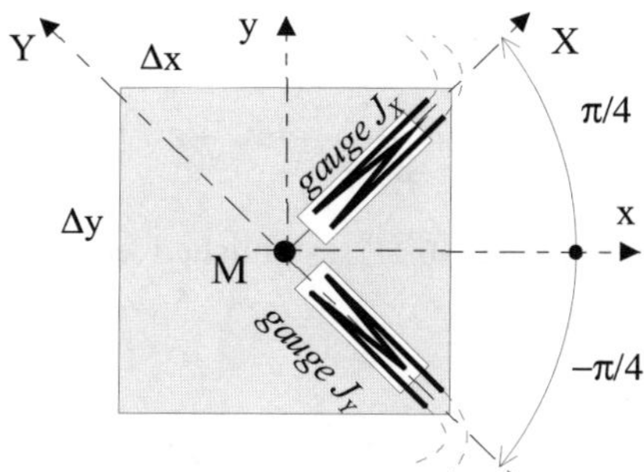

Figure 10.29. *Strain gauges to access* τ_{xy}

Thus, we can measure:

$$\varepsilon_X \text{ with } J_X$$

$$\varepsilon_Y \text{ with } J_Y$$

σ_X and σ_Y in the $\vec{X}$ and $\vec{Y}$ axes can be obtained using equations [1.18] which are recalled here:

$$\begin{Bmatrix} \sigma_X \\ \sigma_Y \end{Bmatrix} = \frac{E}{1-\nu^2} \begin{bmatrix} 1 & \nu \\ \nu & 1 \end{bmatrix} \bullet \begin{Bmatrix} \varepsilon_X \\ \varepsilon_Y \end{Bmatrix} \tag{10.15}$$

Using [10.6], we can write

$$\sigma_X = \frac{\sigma_x + \sigma_y}{2} + \tau_{xy} \qquad (\text{as } \alpha = \frac{\pi}{4} \Rightarrow c = s = \frac{\sqrt{2}}{2})$$

$$\sigma_Y = \frac{\sigma_x + \sigma_y}{2} - \tau_{xy} \qquad \left(\text{as } \alpha = \frac{3\pi}{4} \Rightarrow c = -s = -\frac{\sqrt{2}}{2}\right)$$

From where by calculating the difference:

$$\sigma_X - \sigma_Y = 2\tau_{xy}$$

$$\sigma_X - \sigma_Y = 2\tau_{xy}$$

The same calculation using [10.15] gives:

$$\sigma_X - \sigma_Y = \frac{E}{1+\nu}\left(\varepsilon_X - \varepsilon_Y\right)$$

and by comparing the above two equations:

$$\tau_{xy} = \frac{E}{2(1+\nu)}\left(\varepsilon_X - \varepsilon_Y\right)$$

where we can recognize equation [10.11] established earlier for the shear modulus, i.e. $G = \frac{E}{2(1+\nu)}$. Therefore:

$$\tau_{xy} = G\left(\varepsilon_X - \varepsilon_Y\right)$$

Comparison with behavior relation [10.14] leads to distortion γ_{xy}, i.e.:

$$\gamma_{xy} = \varepsilon_X - \varepsilon_Y \qquad\qquad [10.16]$$

NOTES

❑ The above shows that the values of distortions can be obtained indirectly from the axial deformation of gauges. In addition, taking into account the notes in section 1.4.2.4 concerning the evaluation of normal stresses, we would need four strain gauges placed as shown in Figure 10.30a to evaluate the stresses σ_x, σ_y and τ_{xy}.

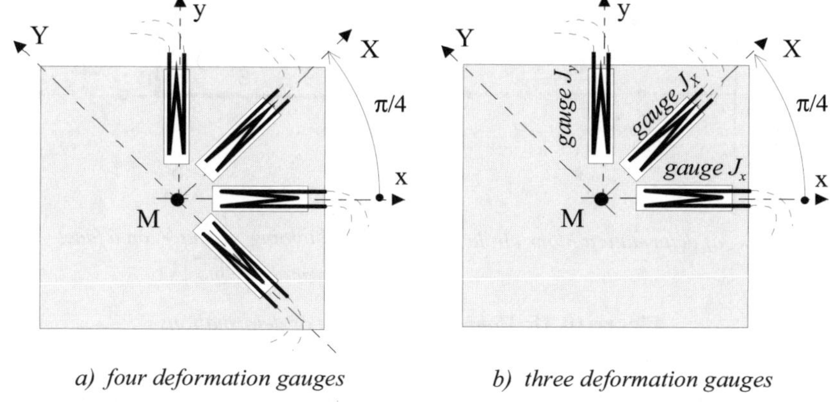

a) four deformation gauges   *b) three deformation gauges*

Figure 10.30. *Evaluation of stresses* σ_x, σ_y *and* τ_{xy}

❏ It can be shown[21] that the three gauges in Figure 10.30b are sufficient because the deformations in axes other than $\vec{x}$ and $\vec{y}$ can be derived from deformations ε_x, ε_y, γ_{xy} using an equation of the same type as equation [10.1] for the stresses, which is written here as:

$$\left\{\begin{array}{c} \varepsilon \\ \dfrac{\gamma}{2} \end{array}\right\} = \begin{bmatrix} c^2 & s^2 & 2cs \\ -cs & cs & (c^2 - s^2) \end{bmatrix} \bullet \left\{\begin{array}{c} \varepsilon_x \\ \varepsilon_y \\ \dfrac{\gamma_{xy}}{2} \end{array}\right\} \quad \text{with} \quad \left(\begin{array}{c} c = \cos\alpha \\ s = \sin\alpha \end{array}\right) \quad [10.17]$$

where ε (dilatation or contraction) represents the axial deformation along the normal $\vec{n}$ at an angle $(\vec{n},\vec{x}) = \alpha$ to the $\vec{x}$ axis and γ is the distortion in the $\vec{n}$ and $\vec{t}$ axes.

Using the same principle as for the stresses, a "deformation" Mohr circle can be drawn as represented in Figure 10.31.

---

21 The proof of this property is not dealt with in this book.

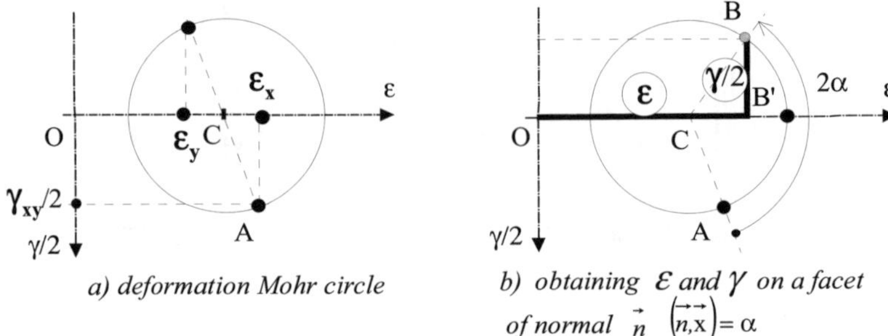

a) deformation Mohr circle   *b) obtaining $\mathcal{E}$ and γ on a facet*
of normal $\vec{n}$ $\left(\vec{n},\vec{x}\right)=\alpha$

Figure 10.31. *Mohr representation of deformations*

In these conditions, let us suppose that we have all the information given by the three strain gauges of Figure 10.30b), i.e.:

$$J_x \to \varepsilon_x \; ; J_y \to \varepsilon_y \; ; J_X \to \varepsilon_X$$

Equation [10.17] helps in writing the following, using $\alpha = \dfrac{\pi}{4}$:

$$\varepsilon_X = \frac{\varepsilon_x + \varepsilon_y}{2} + \frac{\gamma_{xy}}{2}$$

i.e.:

$$\gamma_{xy} = 2\varepsilon_X - \left(\varepsilon_x + \varepsilon_y\right)$$

Thus, the values ε_x, ε_y and γ_{xy} are obtained based on the information from the three gauges at 45° and the corresponding graph plotted using these measurements is shown in Figure 10.32a.

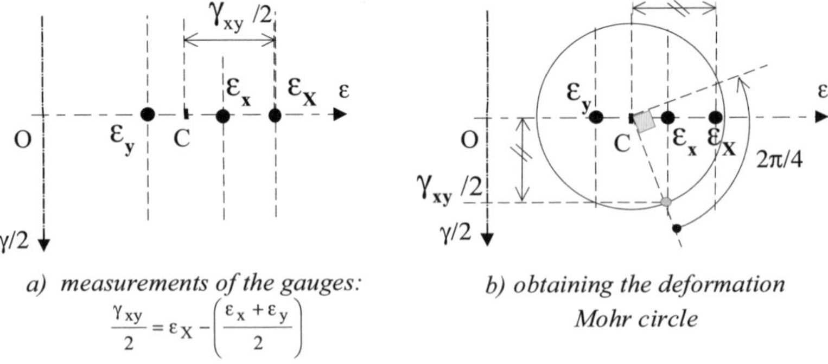

a) measurements of the gauges:
$$\frac{\gamma_{xy}}{2} = \varepsilon_X - \left(\frac{\varepsilon_x + \varepsilon_y}{2}\right)$$

b) obtaining the deformation Mohr circle

Figure 10.32. *Mohr representation for deformations deduced from ε_x, ε_y and γ_{xy}*

10.1.7. *Deformation energy in principal axes*

If we refer to expressions [2.44] which give the strain potential energy of an elementary volume dV for a plane state of stresses, i.e.:

$$dE_{pot.} = \frac{1}{2}\left(\frac{\sigma_x^{\,2}}{E} + \frac{\sigma_y^{\,2}}{E} - 2\frac{\nu}{E}\sigma_x\sigma_y + \frac{\tau_{xy}^{\,2}}{G} \right) \times dV$$

We can simplify by writing the stresses in the principal axes:

$$\sigma_x \rightarrow \sigma_1 \; ; \; \sigma_y \rightarrow \sigma_2 \; ; \; \tau_{xy} = 0$$

leading to:

$$dE_{pot.} = \frac{1}{2E}\left(\sigma_1^{\,2} + \sigma_2^{\,2} - 2\nu\sigma_1\sigma_2 \right) \times dV \qquad\qquad [10.18]$$

Again from equation [2.44], the following can also be obtained:

$$dE_{pot.} = \frac{1}{2}\left(\sigma_1 \times \varepsilon_1 + \sigma_2 \times \varepsilon_2 \right) \times dV$$

where ε_1 and ε_2 are the deformations (see section 1.4.3) along $\vec{1}$ and $\vec{2}$, which can be named as principal deformations (dilatations or contractions).

10.2. Complete state of stresses

10.2.1. *Principal directions and stresses*

Let us recall the general case of any loaded[22] structure: a very small sphere of isolated matter around point M before loading deformed into a small ellipsoid after loading. We had indicated[23] that the three axes of this ellipsoid corresponded to the three principal directions $\vec{1}$, $\vec{2}$ and $\vec{3}$ for the stresses. On the facets normal to these directions, the principal stresses had the special characteristic of being purely normals. Figure 10.33a reverts to Figure 1.6, using the common notations $\vec{\sigma_1}$, $\vec{\sigma_2}$ and $\vec{\sigma_3}$ to refer to the stresses. Figure 10.33b represents a small parallelepipedic domain dV oriented properly so that its faces are parallel to the principal facets of Figure 10.33a. Then, under the loading, the facets of this parallelepiped either stretch or come closer as represented.

---

22 See section 1.1 and Figure 1.1.
23 See Figure 1.6.

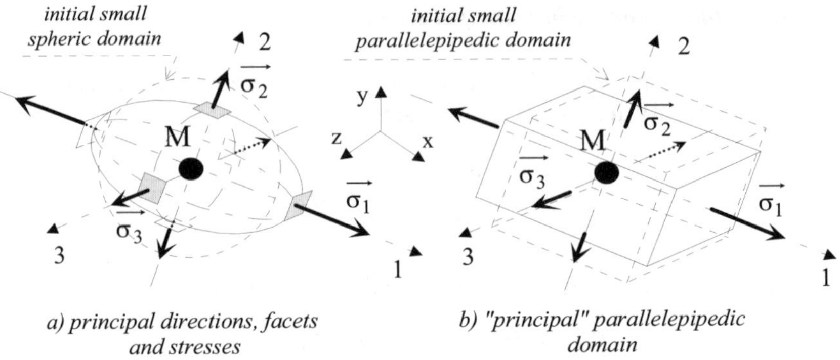

Figure 10.33. *Two elementary domains in principal axes*

10.2.2. *Stresses in any $\vec{x}$, $\vec{y}$, $\vec{z}$ axes*

In Figure 10.34a the normals $\vec{x}$, $\vec{y}$, $\vec{z}$ to the facets of an elementary parallelepiped volume around point M are distinct from the main directions $\vec{1}$, $\vec{2}$, $\vec{3}$. Then a stress vector acts on each one of the facets, characterized by a normal and a tangential component. Figure 10.34b shows stress $\overrightarrow{C}_{(M,\vec{x})}$ on the facet of normal $\vec{x}$ when projected on the three axes, i.e.:

– normal stress along $\vec{x}$: σ_x

– tangential stress along $\vec{y}$: τ_{xy}

– tangential stress along $\vec{z}$: τ_{xz}

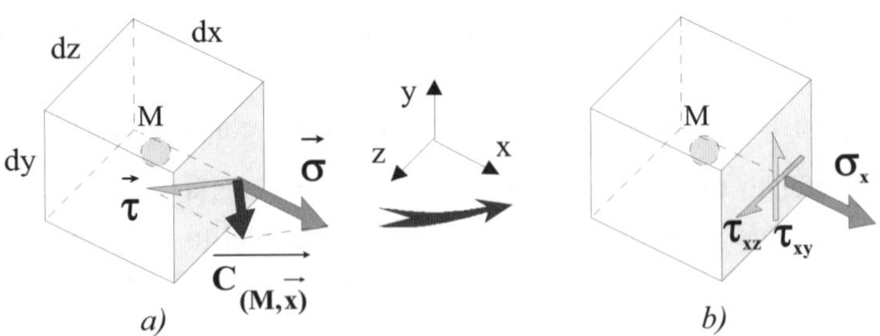

Figure 10.34. *Projections of the stress* $\overrightarrow{C}_{(M,\vec{x})}$

By proceeding in the same manner for the other facets of the domain, we obtain the components marked in Figure 10.35.

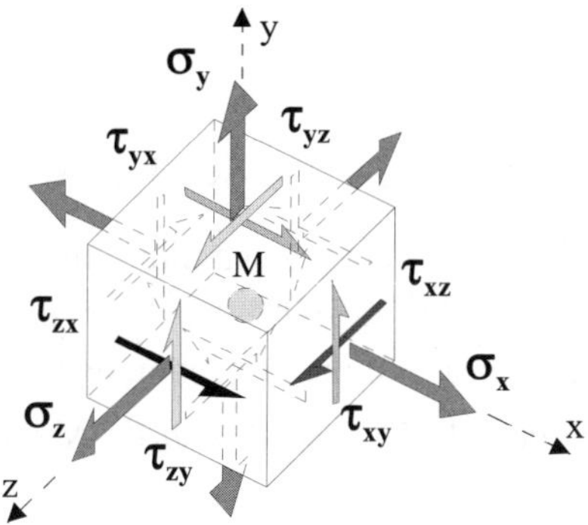

Figure 10.35. *The (six) components of the complete state of stresses*

Let us simplify by assuming that these components remain constant in a certain finite domain of the structure. They are therefore the same for all the points near point M in question[24]. For example Figure 10.36 shows two domains around points M and M' that are so close to each other that they will be subjected to identical stress components.

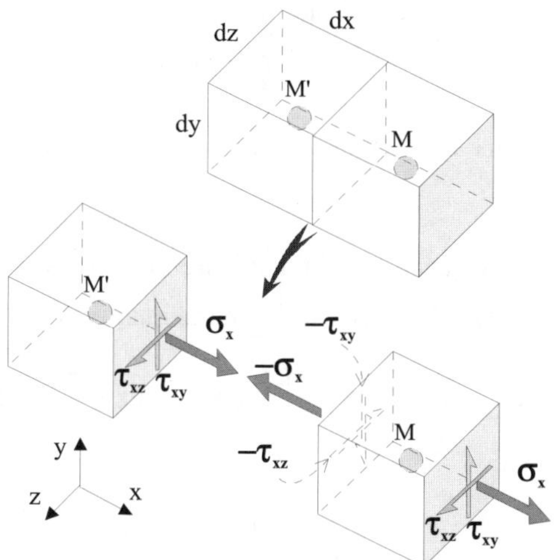

Figure 10.36. *Two contiguous elementary domains*

---

24 This hypothesis is not indispensable. This is done mainly to simplify the calculations that follow.

In Figure 10.36, we have represented, as an example, the stresses on the facets perpendicular to $\vec{x}$. With such a representation, we obtain the following for the domain in Figure 10.35:

– three normal stresses: σ_x, σ_y, σ_z

– six tangential stresses: τ_{xy}, τ_{xz}, τ_{yx}, τ_{yz}, τ_{zx}, τ_{zy}

The domain in Figure 10.35 being in equilibrium, the sum of the moments around center M of the corresponding cohesion forces, projected on the three axes, should be zero. Let us take the case of the projection on the $\vec{Mz}$ axis. For this, let us represent the domain in Figure 10.37 below, with its $\vec{Mz}$ axis perpendicular to the plane of the figure.

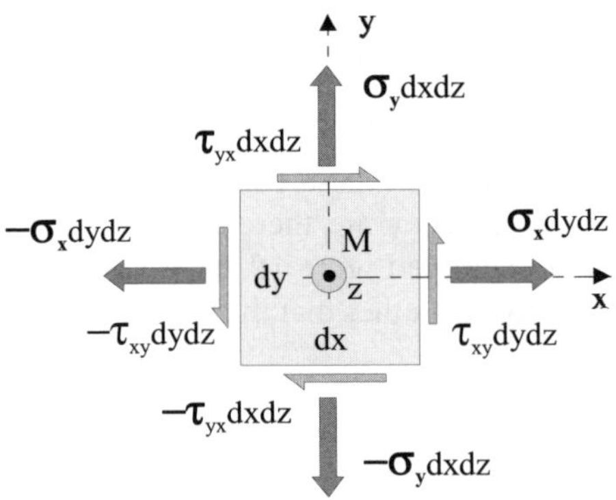

Figure 10.37. *Moment equilibrium about z axis*

The cohesion forces on facets z^+ and z^- are not represented. They create zero moment projections on $\vec{Mz}$. The sum of the moments gives:

$$\left(\tau_{xy} \times dy \times dz \times \frac{dx}{2} \right) \times 2 - \left(\tau_{yx} \times dx \times dz \times \frac{dy}{2} \right) \times 2 = 0$$

i.e.:

$$\tau_{xy} = \tau_{yx}$$

if we examine now the sum of the moments around M projected on the $\vec{My}$ axis, we have:

$$\left(\tau_{zx} \times dx \times dy \times \frac{dz}{2}\right) \times 2 - \left(\tau_{xz} \times dy \times dz \times \frac{dx}{2}\right) \times 2 = 0$$

i.e.:

$$\tau_{zx} = \tau_{xz}$$

Finally, the moments around M projected on the $\overrightarrow{Mx}$ axis leads to:

$$\left(\tau_{yz} \times dx \times dz \times \frac{dy}{2}\right) \times 2 - \left(\tau_{zy} \times dx \times dy \times \frac{dz}{2}\right) \times 2 = 0$$

i.e.:

$$\tau_{yz} = \tau_{zy}$$

Thus, we can verify the generality of the reciprocity property of shear stresses, that was partially justified in the above (see section 10.1.1).

As a result, it is now possible to define the cohesion forces on the six facets of the parallelepiped using the *six* stress components represented in Figure 10.35, i.e.:

- three normal stress components: $\sigma_x, \sigma_y, \sigma_z$

- three shear stress components: $\tau_{xy} = \tau_{yx}$, $\tau_{yz} = \tau_{zy}$, $\tau_{zx} = \tau_{xz}$

$$[10.19]$$

10.2.3. *Deformations*

Let us examine the stress components on the two facets perpendicular to $\overrightarrow{x}$ axis marked as $\sigma_x, \tau_{xy}, \tau_{xz}$ in Figure 10.38. We can observe that, with the aim of balancing the elementary domain, it is necessary to apply the reciprocal stresses $\tau_{yx} = \tau_{xy}$ and $\tau_{zx} = \tau_{xz}$.

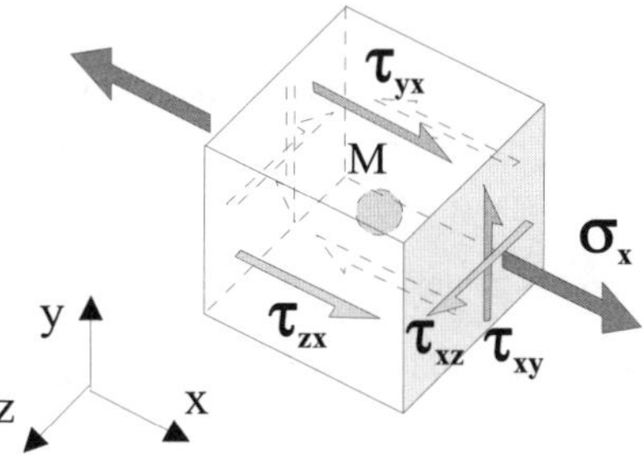

Figure 10.38. *Balancing of the elementary domain along x axis*

The deformed shapes of the small element corresponding to each of the above stresses σ_x, τ_{xy}, τ_{xz} [25] are illustrated in Figure 10.39.

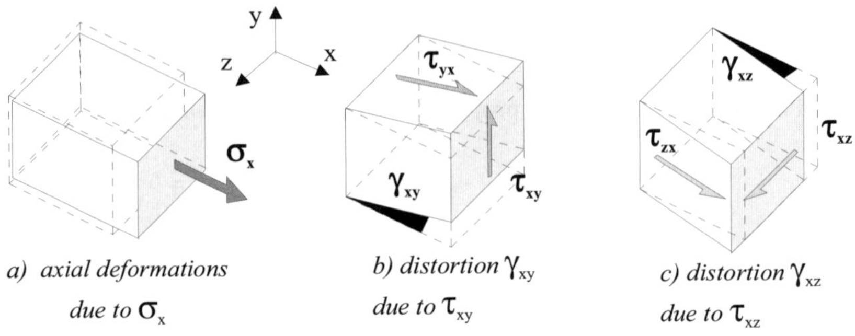

a) axial deformations due to σ_x

b) distortion γ_{xy} due to τ_{xy}

c) distortion γ_{xz} due to τ_{xz}

Figure 10.39. *Deformations of the small element under the stresses* σ_x, τ_{xy}, τ_{xz}

Figure 10.39 shows the types of deformation already described in section 1.4.2, that is:

a) due to σ_x : a main linear deformation ε_x along $\vec{x}$, and two secondary linear deformations along $\vec{y}$ and $\vec{z}$, respectively $\varepsilon_y = -\nu\varepsilon_x$ and $\varepsilon_z = -\nu\varepsilon_x$, where ν is the already seen Poisson's coefficient[26];

b) due to the shear stress τ_{xy} : an angular distortion γ_{xy} ;

c) due to the shear stress τ_{xy} : an angular distortion γ_{xz} .

---

25 The balancing of the small domain thus represented in (Figure 10.38) can be easily checked. In fact:

• the resultant of forces is zero on the three axes $\vec{x}$, $\vec{y}$, $\vec{z}$, for example, along $\vec{x}$:
$\sigma_x\,dydz - \sigma_x\,dydz + \tau_{yx}\,dxdz - \tau_{yx}\,dxdz + \tau_{zx}\,dxdy - \tau_{zx}\,dxdy = 0$. The same is applicable along $\vec{y}$ and $\vec{z}$;

• the resultant of moments around center M of the domain is zero when projected on the three axes, for example, about $\vec{Mz}$: $\left(\tau_{xy}dydz\dfrac{dx}{2}\right)\times 2 - \left(\tau_{yx}dxdz\dfrac{dy}{2}\right)\times 2 = 0$. It is the same about $\vec{My}$ and $\vec{Mx}$.

26 See section 1.4.2.1, equations [1.7] and [1.8] and NOTE at the end of section 1.4.2.4.

Thus:

$$\left\{\varepsilon_x \; ; \; \varepsilon_y = -\nu\varepsilon_x \; ; \; \varepsilon_z = -\nu\varepsilon_x\right\} \leftrightarrow \sigma_x \; ; \; \gamma_{xy} \leftrightarrow \tau_{xy}; \; \gamma_{xz} \leftrightarrow \tau_{xz}$$

that can now be completed as follows when taking into account the other stress components indicated in Figure 10.35:

$$\left\{\varepsilon_y \; ; \; \varepsilon_x = -\nu\varepsilon_y \; ; \; \varepsilon_z = -\nu\varepsilon_y\right\} \leftrightarrow \sigma_y \; ; \; \gamma_{yx} \leftrightarrow \tau_{yx} \; ; \; \gamma_{yz} \leftrightarrow \tau_{yz}$$

$$\left\{\varepsilon_z \; ; \; \varepsilon_x = -\nu\varepsilon_z \; ; \; \varepsilon_y = -\nu\varepsilon_z\right\} \leftrightarrow \sigma_z \; ; \; \gamma_{yx} \leftrightarrow \tau_{yx} \; ; \; \gamma_{yz} \leftrightarrow \tau_{yz}$$

$$[10.20]$$

NOTES

❏ Let us recall the deformation expressions [1.18] for the plane state of stresses. The small displacements of any point M(x,y) were denoted as u(x,y) and v(x,y). Then the deformations were written:

$$\varepsilon_x = \frac{\partial u}{\partial x} \; ; \quad \varepsilon_y = \frac{\partial v}{\partial y} \quad ; \quad \gamma_{xy} = \frac{\partial u}{\partial y} + \frac{\partial v}{\partial x}$$

These expressions can be generalized as follows: the small displacements of any point M(x,y,z) are now denoted as:

$$u(x,y,z); \quad v(x,y,z); \quad w(x,y,z)$$

Thus the state of deformation around this point can be written as:

$$\varepsilon_x = \frac{\partial u}{\partial x} \; ; \; \varepsilon_y = \frac{\partial v}{\partial y} \; ; \quad \varepsilon_z = \frac{\partial w}{\partial z}$$

$$\gamma_{xy} = \frac{\partial u}{\partial y} + \frac{\partial v}{\partial x} \; ; \; \gamma_{yz} = \frac{\partial v}{\partial z} + \frac{\partial w}{\partial y} \; ; \; \gamma_{zx} = \frac{\partial u}{\partial z} + \frac{\partial w}{\partial x}$$

10.2.4. *Behavior relations*

In section 1.4.2 we saw for a plane state of stresses how to define the deformations observed for every type of stress using a law of proportionality. This law brings into play the modulus of elasticity E and Poisson's ratio ν during the application of normal stress, and shear modulus G during application of the tangential or shear stress.

When the state of stresses is complete, with the six components [10.19], it is possible to correlate in a similar manner the six corresponding simple states of stresses 1 to 6 to the six types of deformations [10.20] that they produce.

For that, the method mentioned in section 1.4.2 is applied to obtain the results indicated below[27]:

a) deformations due to σ_x (load case 1):

$$\varepsilon_{x1} = \frac{1}{E}\sigma_x \; ; \quad \varepsilon_{y1} = -\nu\times\varepsilon_{x1} = -\frac{\nu}{E}\sigma_x \; ; \quad \varepsilon_{z1} = -\nu\times\varepsilon_{x1} = -\frac{\nu}{E}\sigma_x$$

b) then for σ_y (load case 2)[28]:

$$\varepsilon_{x2} = -\nu\times\varepsilon_{y2} = -\frac{\nu}{E}\sigma_y \; ; \quad \varepsilon_{y2} = \frac{1}{E}\sigma_y \; ; \quad \varepsilon_{z2} = -\nu\times\varepsilon_{y2} = -\frac{\nu}{E}\sigma_y$$

c) now applying stress σ_z (load case 3) will give an analogous result for deformations:

$$\varepsilon_{x3} = -\nu\times\varepsilon_{z3} = -\frac{\nu}{E}\sigma_z \; ; \quad \varepsilon_{y3} = -\nu\times\varepsilon_{z3} = -\frac{\nu}{E}\sigma_z \; ; \quad \varepsilon_{z3} = \frac{1}{E}\sigma_z$$

For the shear stresses, the method mentioned in section 1.4.2.3 for τ_{xy} can be used to arrive at[29]:

$$\gamma_{xy} = \frac{1}{G}\tau_{xy}$$

which is completed, using a similar procedure when applying τ_{xz} alone, by a distortion (see Figure 10.39):

$$\gamma_{xz} = \frac{1}{G}\tau_{xz}$$

Then when applying τ_{yz} alone:

$$\gamma_{yz} = \frac{1}{G}\tau_{yz}$$

The total deformations result in the superposition of the partial deformations detailed below for every simple state of stresses.

The following table summarizes these simple states of stresses and shows the deformations resulting from their superposition.

---

27 See section 1.4.2.1, equations [1.7] and [1.8] and NOTE the end of section 1.4.2.4.
28 See section 1.4.2.2, equations [1.9] and [1.10].
29 See section 1.4.2.3, equation [1.12].

loading case	ε_x	ε_y	ε_z
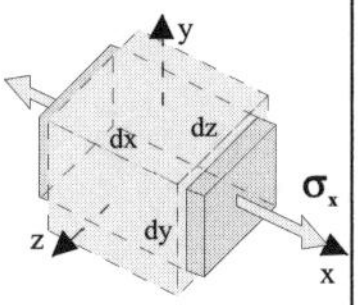	$\varepsilon_{x1} = \dfrac{1}{E}\sigma_x$	$\varepsilon_{y1} = -\nu \times \varepsilon_{x1} = -\dfrac{\nu}{E}\sigma_x$	$\varepsilon_{z1} = -\nu \times \varepsilon_{x1} = -\dfrac{\nu}{E}\sigma_x$
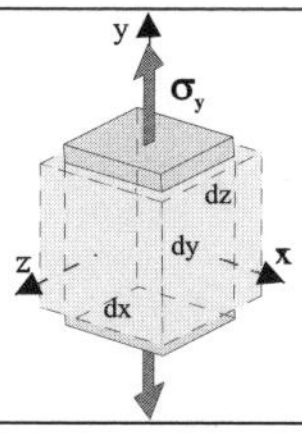	$\varepsilon_{x2} = -\nu \times \varepsilon_{y2} = -\dfrac{\nu}{E}\sigma_y$	$\varepsilon_{y2} = \dfrac{1}{E}\sigma_y$	$\varepsilon_{z2} = -\nu \times \varepsilon_{y2} = -\dfrac{\nu}{E}\sigma_y$
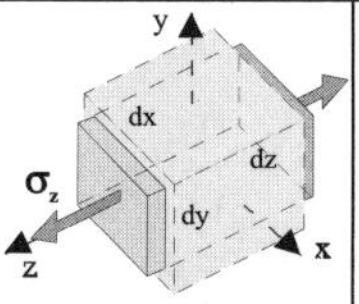	$\varepsilon_{x3} = -\nu \times \varepsilon_{z3} = -\dfrac{\nu}{E}\sigma_z$	$\varepsilon_{y3} = -\nu \times \varepsilon_{z3} = -\dfrac{\nu}{E}\sigma_z$	$\varepsilon_{z3} = \dfrac{1}{E}\sigma_z$
superposition of the three previous loadings	$\begin{aligned}\varepsilon_x &= \varepsilon_{x1} + \varepsilon_{x2} + \varepsilon_{x3}\\ &= \dfrac{1}{E}\sigma_x - \dfrac{\nu}{E}\sigma_y - \dfrac{\nu}{E}\sigma_z\end{aligned}$	$\begin{aligned}\varepsilon_y &= \varepsilon_{y1} + \varepsilon_{y2} + \varepsilon_{y3}\\ &= -\dfrac{\nu}{E}\sigma_x + \dfrac{1}{E}\sigma_y - \dfrac{\nu}{E}\sigma_z\end{aligned}$	$\begin{aligned}\varepsilon_z &= \varepsilon_{z1} + \varepsilon_{z2} + \varepsilon_{z3}\\ &= -\dfrac{\nu}{E}\sigma_x - \dfrac{\nu}{E}\sigma_y + \dfrac{1}{E}\sigma_z\end{aligned}$

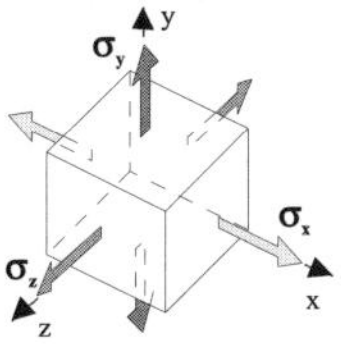

$$\begin{Bmatrix} \varepsilon_x \\ \varepsilon_y \\ \varepsilon_z \end{Bmatrix} = \begin{bmatrix} \dfrac{1}{E} & -\dfrac{\nu}{E} & -\dfrac{\nu}{E} \\[2mm] -\dfrac{\nu}{E} & \dfrac{1}{E} & -\dfrac{\nu}{E} \\[2mm] -\dfrac{\nu}{E} & -\dfrac{\nu}{E} & \dfrac{1}{E} \end{bmatrix} \bullet \begin{Bmatrix} \sigma_x \\ \sigma_y \\ \sigma_z \end{Bmatrix}$$

loading case	γ_{xy}	γ_{yz}	γ_{zx}
	$\gamma_{xy} = \dfrac{1}{G}\tau_{xy}$	0	0
	0	$\gamma_{yz} = \dfrac{1}{G}\tau_{yz}$	0
	0	0	$\gamma_{zx} = \dfrac{1}{G}\tau_{zx}$
superposition of the three previous loadings	$\gamma_{xy} = \dfrac{1}{G}\tau_{xy}$	$\gamma_{yz} = \dfrac{1}{G}\tau_{yz}$	$\gamma_{zx} = \dfrac{1}{G}\tau_{zx}$
	$\begin{Bmatrix} \gamma_{xy} \\ \gamma_{yz} \\ \gamma_{xz} \end{Bmatrix} = \begin{bmatrix} \dfrac{1}{G} & 0 & 0 \\ 0 & \dfrac{1}{G} & 0 \\ 0 & 0 & \dfrac{1}{G} \end{bmatrix} \bullet \begin{Bmatrix} \tau_{xy} \\ \tau_{yz} \\ \tau_{xz} \end{Bmatrix}$		

The results given in these tables show that it is possible to regroup the relationships between deformations ε, γ, and stresses, in a single equation, the *"behavior relation"*, written in the matrix form.

complete state of stresses	"deformations-stresses" behavior relation
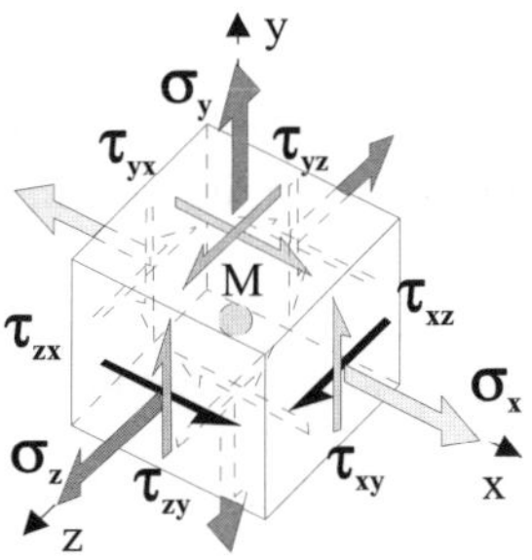 y, σ_y, τ_{yx}, τ_{yz}, τ_{zx}, M, τ_{xz}, σ_x, σ_z, τ_{zy}, τ_{xy}, x, z	$$\begin{Bmatrix} \varepsilon_x \\ \varepsilon_y \\ \varepsilon_z \\ \gamma_{xy} \\ \gamma_{yz} \\ \gamma_{zx} \end{Bmatrix} = \begin{bmatrix} \dfrac{1}{E} & -\dfrac{\nu}{E} & -\dfrac{\nu}{E} & 0 & 0 & 0 \\ -\dfrac{\nu}{E} & \dfrac{1}{E} & -\dfrac{\nu}{E} & 0 & 0 & 0 \\ -\dfrac{\nu}{E} & -\dfrac{\nu}{E} & \dfrac{1}{E} & 0 & 0 & 0 \\ 0 & 0 & 0 & \dfrac{1}{G} & 0 & 0 \\ 0 & 0 & 0 & 0 & \dfrac{1}{G} & 0 \\ 0 & 0 & 0 & 0 & 0 & \dfrac{1}{G} \end{bmatrix} \bullet \begin{Bmatrix} \sigma_x \\ \sigma_y \\ \sigma_z \\ \tau_{xy} \\ \tau_{yz} \\ \tau_{zx} \end{Bmatrix}$$

$$[10.21]$$

NOTES

❒ The behavior relation above is written based on a square and symmetric matrix of elastic coefficients.

❒ This matrix consists of only two distinct elastic coefficients: E and ν. In fact, it was observed in [10.11] that modulus G could be written as $G = \dfrac{E}{2(1+\nu)}$.

❒ Let us cite the universally known simple case of a state of stress caused by the action of a uniform pressure when a structure is totally immersed in a fluid at rest (air, water). Irrespective of the orientation of the system (xyz) chosen for a small domain of the structure and by taking the intensity of this pressure as $\|p_0\| = p_0$, we can verify that[30]:

$$\sigma_x = \sigma_y = \sigma_z = -p_0$$

---

30 Such a verification is not detailed here.

In such a case there are no tangential stresses and, therefore, any spatial direction can be considered as a principal direction. Such a state of stress is considered to be "isotropic" or in this case "hydrostatic". Behavior relation [10.21] indicates in this case:

$$\varepsilon_x = \varepsilon_y = \varepsilon_z = -p_0 \left(\frac{1-2v}{E} \right)$$

$$\gamma_{xy} = \gamma_{yz} = \gamma_{zx} = 0$$

[10.22]

The small domain around the point under study contracts "uniformly" without undergoing any angular distortion. An elementary sphere around a point becomes a smaller sphere and an elementary cube becomes a smaller cube.

This happens with almost all the materials. The only exception being the materials characterized by Poisson's ratio such that:

$$1-2v = 0 \quad \Rightarrow \quad v = 0.5$$

Thus, the previous equation [10.22] indicates that the contraction of the small domain is zero and its volume does not change under pressure. Thus, this domain is of a material which is said to be "incompressible" or "hyper-elastic". This is mainly the case with the elastomers (synthetic rubber).

❑ Under any general state of stresses [10.19], a small domain of the structure stores potential energy of deformation, or "strain potential energy". It is this aspect that will be studied further in this book[31].

10.2.5. *Strain potential energy*

The method used in section 2.3.2 to write the strain potential energy in the case of plane stresses can be reviewed here for successive cases of similar loads:

• First let us take the loading case of Figure 2.17, or loading 1, corresponding to the application of a basic force dF_x along $\vec{x}$. It is represented again in Figure 10.40, and results in a strain potential energy:

$$dE_{pot.1} = \frac{1}{2} dF_x \times du_1 = \frac{1}{2} dF_x \times \varepsilon_{x1} \times dx = \frac{1}{2} (\sigma_x \times dy \times dz) \times \varepsilon_{x1} \times dx = \frac{1}{2} \sigma_x \times \varepsilon_{x1} \times dV$$

---

31 In this book we shall not examine in detail the evolution of any general state of stresses when the local coordinate system varies (particularly the principal directions and Mohr tri-circles).

where $dV = dx \times dy \times dz$ is the elementary volume of the domain. With behavior relation [10.21]:

$$dE_{pot.1} = \frac{1}{2} \frac{{\sigma_x}^2}{E} \times dV$$

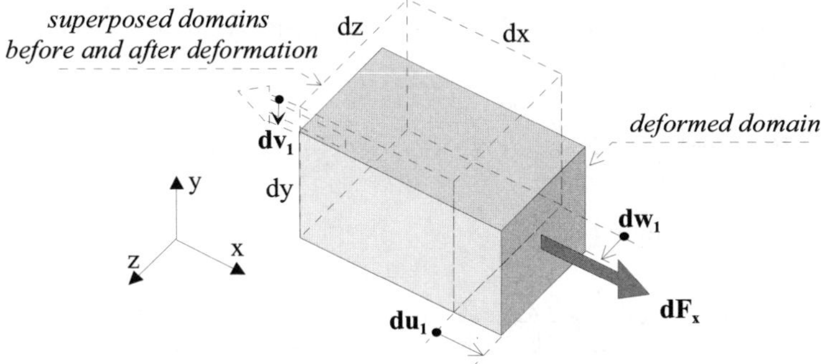

Figure 10.40. *Displacements when applying dF$_x$*

$\bullet$ If the loading case 2 corresponding to the application of force dF_y along $\vec{y}$ is superposed, the force dF_x staying in place, a supplementary energy is noted:

$$dE_{pot.2} = \frac{1}{2} dF_y \times dv_2 + dF_x \times du_2$$

where du_2 represents the displacement along $\vec{x}$ due to Poisson's effect when a load is exerted along $\vec{y}$. i.e.[32]:

$$dE_{pot.2} = \left(\frac{1}{2} \times \sigma_y \times \varepsilon_{y2} + \sigma_x \times \left(-\nu \times \varepsilon_{y2} \right) \right) \times dV$$

and with the behavior relation [10.21]:

$$dE_{pot.2} = \left(\frac{1}{2} \times \frac{{\sigma_y}^2}{E} - \frac{\nu}{E} \times \sigma_x \times \sigma_y \right) \times dV$$

---

32 See section 2.3.2.

◆ Finally if we consider the loading case 3 corresponding to the application of dF_z along $\vec{z}$, forces dF_x and dF_y remaining in place, a supplementary energy is noticed:

$$dE_{pot.3} = \frac{1}{2} dF_z \times dw_3 + dF_y \times dv_3 + dF_x \times du_3$$

where du_3 and dv_3 are the displacements respectively along $\vec{x}$ and $\vec{y}$ due to Poisson's effect, i.e.:

$$dE_{pot.3} = \left(\frac{1}{2} \times \sigma_z \times \varepsilon_{z3} + \sigma_y \times \left(-\nu \times \varepsilon_{z3}\right) + \sigma_x \times \left(-\nu \times \varepsilon_{z3}\right) \right) \times dV$$

and with behavior relation [10.21]:

$$dE_{pot.3} = \left(\frac{1}{2} \times \frac{\sigma_z{}^2}{E} - \frac{\nu}{E} \times \sigma_x \times \sigma_z - \frac{\nu}{E} \times \sigma_y \times \sigma_z \right) \times dV$$

That is, when simultaneously all the forces dF_x, dF_y and dF_z are applied:

$$dE_{pot.} = dE_{pot.1} + dE_{pot.2} + dE_{pot.3}$$

$$dE_{pot.} = \frac{1}{2} \times \frac{1}{E} \left[\sigma_x{}^2 + \sigma_y{}^2 + \sigma_z{}^2 - 2\nu\left(\sigma_x \times \sigma_y + \sigma_y \times \sigma_z + \sigma_z \times \sigma_x \right) \right] \times dV$$

The application of tangential forces is carried out as in section 2.3.2.3 (see also Figure 10.35). Thus we have:

◆ case 4: application of dF_{xy}: $dE_{pot.4} = \dfrac{1}{2} \times \dfrac{\tau_{xy}{}^2}{G} \times dV$

◆ case 5: application of dF_{xz}: $dE_{pot.5} = \dfrac{1}{2} \times \dfrac{\tau_{xz}{}^2}{G} \times dV$

◆ case 6: application of dF_{yz}: $dE_{pot.6} = \dfrac{1}{2} \times \dfrac{\tau_{yz}{}^2}{G} \times dV$

i.e. with six load cases in all:

$$dE_{pot.} = \frac{1}{2} \left\{ \frac{1}{E} \left[\sigma_x{}^2 + \sigma_y{}^2 + \sigma_z{}^2 - 2\nu\left(\sigma_x \times \sigma_y + \sigma_y \times \sigma_z + \sigma_z \times \sigma_x \right) \right] \dots \right.$$
$$\left. \dots + \frac{1}{G} \left(\tau_{xy}{}^2 + \tau_{yz}{}^2 + \tau_{zx}{}^2 \right) \right\} \times dV \qquad [10.23]$$

NOTES

❐ Different ways of writing the strain potential energy.

It can be observed that [10.23] is a second degree equation of the stresses. It is a "quadratic form" and can be rewritten as[33]:

$$dE_{pot.} = \frac{1}{2}\begin{bmatrix}\sigma_x & \sigma_y & \sigma_z & \tau_{xy} & \tau_{yz} & \tau_{zx}\end{bmatrix} \bullet \begin{Bmatrix} \dfrac{1}{E}\sigma_x - \dfrac{\nu}{E}\sigma_y - \dfrac{\nu}{E}\sigma_z \\[2mm] -\dfrac{\nu}{E}\sigma_x + \dfrac{1}{E}\sigma_y - \dfrac{\nu}{E}\sigma_z \\[2mm] -\dfrac{\nu}{E}\sigma_x - \dfrac{\nu}{E}\sigma_y + \dfrac{1}{E}\sigma_z \\[2mm] \dfrac{1}{G}\tau_{xy} \\[2mm] \dfrac{1}{G}\tau_{yz} \\[2mm] \dfrac{1}{G}\tau_{zx} \end{Bmatrix} \times dV$$

i.e.:

$$dE_{pot.} = \frac{1}{2}\begin{Bmatrix}\sigma_x \\ \sigma_y \\ \sigma_z \\ \tau_{xy} \\ \tau_{yz} \\ \tau_{zx}\end{Bmatrix}^T \bullet \begin{bmatrix} \dfrac{1}{E} & -\dfrac{\nu}{E} & -\dfrac{\nu}{E} & 0 & 0 & 0 \\[2mm] -\dfrac{\nu}{E} & \dfrac{1}{E} & -\dfrac{\nu}{E} & 0 & 0 & 0 \\[2mm] -\dfrac{\nu}{E} & -\dfrac{\nu}{E} & \dfrac{1}{E} & 0 & 0 & 0 \\[2mm] 0 & 0 & 0 & \dfrac{1}{G} & 0 & 0 \\[2mm] 0 & 0 & 0 & 0 & \dfrac{1}{G} & 0 \\[2mm] 0 & 0 & 0 & 0 & 0 & \dfrac{1}{G} \end{bmatrix} \bullet \begin{Bmatrix}\sigma_x \\ \sigma_y \\ \sigma_z \\ \tau_{xy} \\ \tau_{yz} \\ \tau_{zx}\end{Bmatrix} \times dV \quad [10.24]$$

where we recognize the square and symmetric matrix of elastic coefficients of behavior relation [10.21].

On comparing with relation [10.21], we can observe that the previous equation can also be written as:

---

[33] See section 12.3.

$$dE_{pot.} = \frac{1}{2} \begin{Bmatrix} \sigma_x \\ \sigma_y \\ \sigma_z \\ \tau_{xy} \\ \tau_{yz} \\ \tau_{zx} \end{Bmatrix}^T \bullet \begin{Bmatrix} \varepsilon_x \\ \varepsilon_y \\ \varepsilon_z \\ \gamma_{xy} \\ \gamma_{yz} \\ \gamma_{zx} \end{Bmatrix} \times dV$$

that is:

$$dE_{pot.} = \frac{1}{2}\left(\sigma_x \times \varepsilon_x + \sigma_y \times \varepsilon_y + \sigma_z \times \varepsilon_z + \tau_{xy} \times \gamma_{xy} + \tau_{yz} \times \gamma_{yz} + \tau_{zx} \times \gamma_{zx}\right) \times dV$$

$$[10.25]$$

❏ When the $\vec{x}$, $\vec{y}$, $\vec{z}$, axes of the elementary volume dV coincide with the principal directions $\vec{1}$, $\vec{2}$, $\vec{3}$ we have[34]:

$$\begin{bmatrix} \sigma_x \rightarrow \sigma_1 \\ \sigma_y \rightarrow \sigma_2 \quad \tau_{xy} = \tau_{yz} = \tau_{zx} = 0 \\ \sigma_z \rightarrow \sigma_3 \end{bmatrix}$$

and equation [10.23] becomes:

$$dE_{pot.} = \frac{1}{2E}\left[\sigma_1^2 + \sigma_2^2 + \sigma_3^2 - 2v\left(\sigma_1 \times \sigma_2 + \sigma_2 \times \sigma_3 + \sigma_3 \times \sigma_1\right)\right] \times dV \quad [10.26]$$

---

34 See Figure 10.33.

10.2.6. *Summary*

For any general state of stresses (known as a "three-dimensional" or "triaxial" state of stresses), the behavior of an isotropic, elastic and linear material is summarized below.

Any complete state of stresses
the mechanical loading on the structure creates
a field of small elastic displacements: the displacements of every point M(x,y,z) are marked: $u(x,y,z)$ $v(x,y,z)$ $w(x,y,z)$
a state of stresses at every point M defined by the: ◆ *normal stresses:* $$\sigma_x(x,y,z) \; ; \; \sigma_y(x,y,z) \; ; \; \sigma_z(x,y,z)$$ ◆ *tangential or shear stresses:* $$\tau_{xy}(x,y,z) \; ; \; \tau_{yz}(x,y,z) \; ; \; \tau_{zx}(x,y,z)$$
these stresses create
a state of deformation at every point M defined by the: ◆ *axial deformations along* $\vec{x}, \vec{y}, \vec{z}$ *respectively:* $$\varepsilon_x = \frac{\partial u}{\partial x} \; ; \; \varepsilon_y = \frac{\partial v}{\partial y} \; ; \; \varepsilon_z = \frac{\partial w}{\partial z}$$ ◆ *angular distortions:* *in the plane (xy):* $\gamma_{xy} = \dfrac{\partial u}{\partial y} + \dfrac{\partial v}{\partial x}$ *in the plane (yz):* $\gamma_{yz} = \dfrac{\partial v}{\partial z} + \dfrac{\partial w}{\partial y}$ *in the plane (zx):* $\gamma_{zx} = \dfrac{\partial u}{\partial z} + \dfrac{\partial w}{\partial x}$

Stresses and deformations are linked by a behavior relation

here is the "deformations-stresses" behavior relation:

$$
\begin{Bmatrix} \varepsilon_x \\ \varepsilon_y \\ \varepsilon_z \\ \gamma_{xy} \\ \gamma_{yz} \\ \gamma_{zx} \end{Bmatrix} =
\begin{bmatrix}
\dfrac{1}{E} & -\dfrac{\nu}{E} & -\dfrac{\nu}{E} & 0 & 0 & 0 \\
-\dfrac{\nu}{E} & \dfrac{1}{E} & -\dfrac{\nu}{E} & 0 & 0 & 0 \\
-\dfrac{\nu}{E} & -\dfrac{\nu}{E} & \dfrac{1}{E} & 0 & 0 & 0 \\
0 & 0 & 0 & \dfrac{1}{G} & 0 & 0 \\
0 & 0 & 0 & 0 & \dfrac{1}{G} & 0 \\
0 & 0 & 0 & 0 & 0 & \dfrac{1}{G}
\end{bmatrix}
\bullet
\begin{Bmatrix} \sigma_x \\ \sigma_y \\ \sigma_z \\ \tau_{xy} \\ \tau_{yz} \\ \tau_{zx} \end{Bmatrix}
$$

that can be inversed to obtain the "stresses-deformations" behavior relation:

$$
\begin{Bmatrix} \sigma_x \\ \sigma_y \\ \sigma_z \\ \tau_{xy} \\ \tau_{yz} \\ \tau_{zx} \end{Bmatrix} =
\dfrac{E}{(1+\nu)(1-2\nu)}
\begin{bmatrix}
1-\nu & \nu & \nu & 0 & 0 & 0 \\
\nu & 1-\nu & \nu & 0 & 0 & 0 \\
\nu & \nu & 1-\nu & 0 & 0 & 0 \\
0 & 0 & 0 & \dfrac{1-2\nu}{2} & 0 & 0 \\
0 & 0 & 0 & 0 & \dfrac{1-2\nu}{2} & 0 \\
0 & 0 & 0 & 0 & 0 & \dfrac{1-2\nu}{2}
\end{bmatrix}
\bullet
\begin{Bmatrix} \varepsilon_x \\ \varepsilon_y \\ \varepsilon_z \\ \gamma_{xy} \\ \gamma_{yz} \\ \gamma_{zx} \end{Bmatrix}
$$

note: it should be remembered that $\qquad G = \dfrac{E}{2(1+\nu)}$

The potential energy of deformation or strain potential energy of an element of volume dV can take the following forms

$dE_{pot.} = ...$

$$\frac{1}{2}\left[\frac{\sigma_x^2}{E} + \frac{\sigma_y^2}{E} + \frac{\sigma_z^2}{E} - 2\frac{\nu}{E}\left(\sigma_x\sigma_y + \sigma_y\sigma_z + \sigma_z\sigma_x\right) + \frac{\tau_{xy}^2}{G} + \frac{\tau_{yz}^2}{G} + \frac{\tau_{zx}^2}{G}\right]dV$$

or can be expressed in matrix form:

$$dE_{pot.} = \frac{1}{2}\begin{Bmatrix}\sigma_x \\ \sigma_y \\ \sigma_z \\ \tau_{xy} \\ \tau_{yz} \\ \tau_{zx}\end{Bmatrix}^T \bullet \begin{bmatrix} \frac{1}{E} & -\frac{\nu}{E} & -\frac{\nu}{E} & 0 & 0 & 0 \\ -\frac{\nu}{E} & \frac{1}{E} & -\frac{\nu}{E} & 0 & 0 & 0 \\ -\frac{\nu}{E} & -\frac{\nu}{E} & \frac{1}{E} & 0 & 0 & 0 \\ 0 & 0 & 0 & \frac{1}{G} & 0 & 0 \\ 0 & 0 & 0 & 0 & \frac{1}{G} & 0 \\ 0 & 0 & 0 & 0 & 0 & \frac{1}{G} \end{bmatrix} \bullet \begin{Bmatrix}\sigma_x \\ \sigma_y \\ \sigma_z \\ \tau_{xy} \\ \tau_{yz} \\ \tau_{zx}\end{Bmatrix} \times dV$$

$$dE_{pot.} = \frac{1}{2}\left(\sigma_x \times \varepsilon_x + \sigma_y \times \varepsilon_y + \sigma_z \times \varepsilon_z + \tau_{xy} \times \gamma_{xy} + \tau_{yz} \times \gamma_{yz} + \tau_{zx} \times \gamma_{zx}\right) \times dV$$

$dE_{pot.} = ...$

$$\frac{1}{2}\frac{E}{(1+\nu)(1-2\nu)}\left[(1-\nu)\left(\varepsilon_x^2 + \varepsilon_y^2 + \varepsilon_z^2\right) + 2\nu\left(\varepsilon_x\varepsilon_y + \varepsilon_y\varepsilon_z + \varepsilon_z\varepsilon_x\right) + (1-2\nu)\left(\gamma_{xy}^2 + \gamma_{yz}^2 + \gamma_{zx}^2\right)\right]dV$$

or in matrix form:

$$dE_{pot.} = \frac{1}{2}\begin{Bmatrix}\varepsilon_x \\ \varepsilon_y \\ \varepsilon_z \\ \gamma_{xy} \\ \gamma_{yz} \\ \gamma_{zx}\end{Bmatrix}^T \bullet \frac{E}{(1+\nu)(1-2\nu)} \begin{bmatrix} 1-\nu & \nu & \nu & 0 & 0 & 0 \\ \nu & 1-\nu & \nu & 0 & 0 & 0 \\ \nu & \nu & 1-\nu & 0 & 0 & 0 \\ 0 & 0 & 0 & \frac{1-2\nu}{2} & 0 & 0 \\ 0 & 0 & 0 & 0 & \frac{1-2\nu}{2} & 0 \\ 0 & 0 & 0 & 0 & 0 & \frac{1-2\nu}{2} \end{bmatrix} \bullet \begin{Bmatrix}\varepsilon_x \\ \varepsilon_y \\ \varepsilon_z \\ \gamma_{xy} \\ \gamma_{yz} \\ \gamma_{zx}\end{Bmatrix} \times dV$$

[10.27]

10.2.7. *Components of the strain potential energy*

In the following pages, we propose differentiating two distinct composition mechanisms of the total strain potential energy given in the previous section.

10.2.7.1. *Strain energy without distortion*

O Returning to distortion

Let us again take the case of the deformed small domain of Figure 1.6 (Chapter 1) along its principal directions $\vec{1}$, $\vec{2}$, $\vec{3}$. It is again shown in Figure 10.41a in plane representation (plane $\vec{1}$, $\vec{2}$).

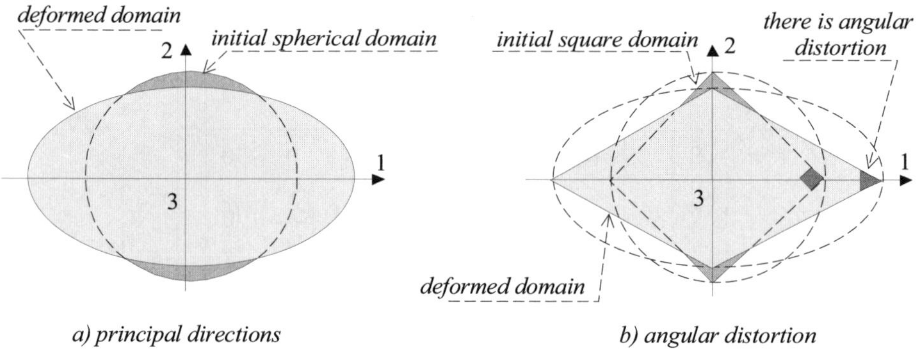

a) principal directions   b) angular distortion

Figure 10.41. *Deformation with distortion of a small element*

This figure shows the angular distortion of a cross-section which was originally a square. The deformation of the domain which was initially a sphere (Figure 10.41a) implies internal distortions (Figure 10.41b). Whereas, if we examine the particular state of the principal stresses such that $\sigma_1 = \sigma_2 = \sigma_3$ (Figure 10.42a), it can be easily imagined that the deflected shape of the initially spherical domain will also be a sphere[35] (Figure 10.42b). This transformation (homothetic) preserves the angles of the parallelepipedic domain of Figure 10.42c without causing any internal distortions. However, there are contractions (or dilatations).

---

35 This is because the lengths of the axes of the ellipsoid are identical, see also section 10.2.4/*NOTE*.

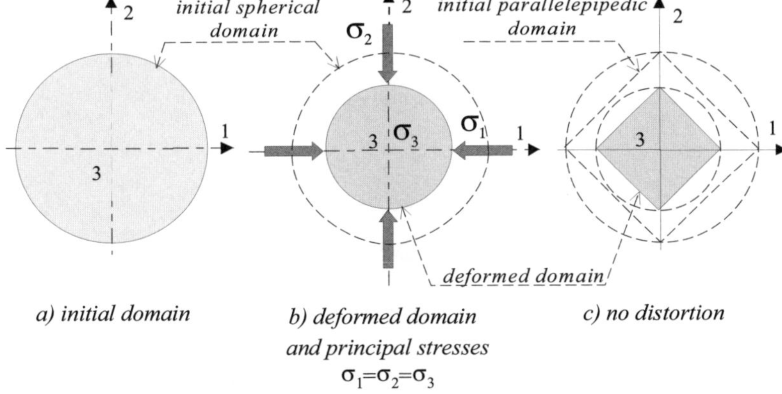

a) initial domain

b) deformed domain
and principal stresses
$\sigma_1 = \sigma_2 = \sigma_3$

c) no distortion

Figure 10.42. *Deformation without distortion of a small element:*
"hydrostatic" state of stresses

As already mentioned[36], this isotropic or "hydrostatic" state of stresses is possible in reality because it is created, for example, by immersing a body in a fluid. The intensity of the fluid pressure $\|p_0\|$, marked as p_0, is uniform, and we have:

$$\sigma_1 = \sigma_2 = \sigma_3 = -p_0$$

In this case the deformations take the unique value of the dilatation [10.22] recalled here:

$$\varepsilon_1 = \varepsilon_2 = \varepsilon_3 = -p_0\left(\frac{1-2v}{E}\right)$$

O   Expression of the strain energy without distortion

As we have just seen, the absence of distortion corresponds to the absence of angular variation of an elementary domain. The edge lengths of this domain vary while conserving right angles. The domain evolves by dilating or contracting. We indicated in section 10.1.5.3 that the particular plane state of stresses with identical principal stresses $\sigma_1 = \sigma_2$ in plane (xy) was characterized by the absence of shear, therefore, of angular distortion. In the same way we have noticed that the "hydrostatic" state of stresses showed the same characteristic concerning absence of shear. Thus, let us take the case of three principal identical stresses. Figure 10.43 illustrates the deformation of a small domain due to the unique value: $\sigma_1 = \sigma_2 = \sigma_3 = \sigma > 0$.

---

36 See section 10.2.4/*NOTE.*

For such an isotropic state of stresses we have observed that the behavior relation reduced to [10.22], i.e.:

$$\varepsilon_1 = \varepsilon_2 = \varepsilon_3 = \sigma\left(\frac{1-2\nu}{E}\right)$$

[10.28]

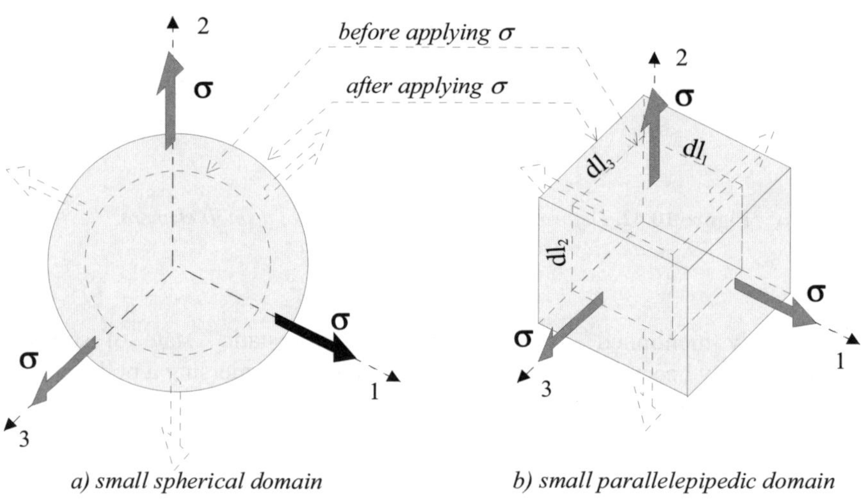

Figure 10.43. *Isotropic state of stresses*

The edges of the domain in Figure 10.43b whose initial values were $d\ell_1$, $d\ell_2$, $d\ell_3$ become $d\ell_1(1+\varepsilon_1)$, $d\ell_2(1+\varepsilon_2)$, $d\ell_3(1+\varepsilon_3)$[37].

The initial volume of the small domain:

$$dV = d\ell_1 \times d\ell_2 \times d\ell_3$$

becomes:

$$dV' = d\ell_1 \times d\ell_2 \times d\ell_3 (1+\varepsilon_1) \times (1+\varepsilon_2) \times (1+\varepsilon_3)$$

i.e.:

$$dV' = dV(1+\varepsilon_1 + \varepsilon_2 + \varepsilon_3)\ ^{38}$$

____

37 See [1.6].

We can observe a relative increase in the volume dV, known as "cubical dilatation":

$$\Delta = \varepsilon_1 + \varepsilon_2 + \varepsilon_3 \qquad [10.29]$$

which is written using [10.28]:

$$\varepsilon_1 + \varepsilon_2 + \varepsilon_3 = 3\sigma\left(\frac{1-2\nu}{E}\right) \qquad [10.30]$$

Let us write the corresponding strain potential energy after noting that the phenomenon is without distortion.

We obtain using [10.27] and by taking $\sigma_1 = \sigma_2 = \sigma_3 = \sigma$:

$$dE_{\substack{\text{pot.}\\ \textit{without distortion}}} = \frac{1}{2E}\left(3\sigma^2 - 2\nu \times 3\sigma^2\right) \times dV = \frac{3}{2}\sigma^2\left(\frac{1-2\nu}{E}\right) \times dV \qquad [10.31]$$

NOTES

❏ For any state of stresses, the principal stresses being marked as σ_1, σ_2, σ_3 behavior relation [10.21] is written as:

$$\varepsilon_1 = \frac{\sigma_1}{E} - \nu\frac{\sigma_2}{E} - \nu\frac{\sigma_3}{E} \; ; \; \varepsilon_2 = -\nu\frac{\sigma_1}{E} + \frac{\sigma_2}{E} - \nu\frac{\sigma_3}{E} \; ; \; \varepsilon_3 = -\nu\frac{\sigma_1}{E} - \nu\frac{\sigma_2}{E} + \frac{\sigma_3}{E}$$

and the cubic dilatation [10.29] in this case assumes the value:

$$\Delta = \varepsilon_1 + \varepsilon_2 + \varepsilon_3 = (\sigma_1 + \sigma_2 + \sigma_3)\left(\frac{1-2\nu}{E}\right)$$

By comparing with [10.30], we observe that the value of the "equivalent isotropic stress" σ which creates the same dilatation is:

$$\sigma = \frac{(\sigma_1 + \sigma_2 + \sigma_3)}{3}$$

Using [10.31] the strain potential energy without distortion shall then be written as:

$$dE_{\substack{\text{pot.}\\ \textit{without distortion}}} = \frac{(\sigma_1 + \sigma_2 + \sigma_3)^2}{6}\left(\frac{1-2\nu}{E}\right) \times dV \qquad [10.32]$$

---

38 The deformations ε being very small with respect to one, the infinitesimal terms with a degree greater than one are ignored in this limited development.

10.2.7.2. *Distortion strain energy*

Considering the above, we can note that:

$$dE_{\substack{pot.\\ distortion}} = dE_{\substack{pot.\\ total}} - dE_{\substack{pot.\\ without\ distortion}}$$

i.e., with principal stresses as σ_1, σ_2, σ_3, and with [10.26] and [10.32]

$$dE_{\substack{pot.\\ distortion}} = \ldots\ldots$$

$$\frac{1}{2E}\left[\sigma_1{}^2 + \sigma_2{}^2 + \sigma_3{}^2 - 2v(\sigma_1\sigma_2 + \sigma_2\sigma_3 + \sigma_3\sigma_1) - \frac{(\sigma_1 + \sigma_2 + \sigma_3)^2}{3}(1 - 2v)\right] \times dV$$

$$dE_{\substack{pot.\\ distortion}} = \ldots\ldots$$

$$\frac{1}{6E}\left[3(\sigma_1{}^2 + \sigma_2{}^2 + \sigma_3{}^2) - 6v(\sigma_1\sigma_2 + \sigma_2\sigma_3 + \sigma_3\sigma_1) - (\sigma_1 + \sigma_2 + \sigma_3)^2(1 - 2v)\right] \times dV$$

which after calculation becomes:

$$dE_{\substack{pot.\\ distortion}} = \frac{1+v}{6E}\left[(\sigma_1 - \sigma_2)^2 + (\sigma_2 - \sigma_3)^2 + (\sigma_3 - \sigma_1)^2\right] \times dV \qquad [10.33]$$

NOTES

❑ For a plane state of stresses in plane (xy), we have seen that direction $\vec{z}$ is a principal direction[39]. The principal stresses being σ_1 and σ_2 ($\sigma_3 = 0$), equation [10.33] becomes:

$$dE_{\substack{pot.\\ distortion}} = \frac{1+v}{3E}\left(\sigma_1{}^2 + \sigma_2{}^2 - \sigma_1\sigma_2\right) \times dV \qquad [10.34]$$

❑ Any axes:

for a plane state of stresses, we can[40] express σ_1 and σ_2 in relation to stresses σ_x, σ_y, τ_{xy} when local axes $\vec{x}$, $\vec{y}$ are coplanar but different from $\vec{1}$, $\vec{2}$. By carrying expressions [10.8] into [10.34] we obtain:

$$dE_{\substack{pot.\\ distortion}} = \frac{1+v}{3E}\left(\sigma_x{}^2 + \sigma_y{}^2 - \sigma_x\sigma_y + 3\tau_{xy}{}^2\right) \times dV \qquad [10.35]$$

---

39 In fact we obtain $\sigma_z = \tau_{zx} = \tau_{zy} = 0$.

40 See [10.8].

which can also be written as:

$$dE_{pot.\ distortion} = \frac{1+\nu}{6E}\left[\left(\sigma_x - \sigma_y\right)^2 + \sigma_y^2 + \sigma_x^2 + 6\tau_{xy}^2\right] \times dV$$

When the state of stresses is no longer plane, but is three-dimensional, and is expressed in a local coordinate system (xyz) distinct from the principal directions, it can be shown[41] that equation [10.33] is modified as follows:

$$dE_{pot.\ distortion} = \frac{1+\nu}{6E}\left[\left(\sigma_x - \sigma_y\right)^2 + \left(\sigma_y - \sigma_z\right)^2 + \left(\sigma_z - \sigma_x\right)^2 + 6\left(\tau_{xy}^2 + \tau_{yz}^2 + \tau_{zx}^2\right)\right] \times dV \quad [42]$$

$$[10.36]$$

10.2.7.3. *Summary*

The potential strain energy of an elementary volume dV of elastic material can be divided into:

$$dE_{pot.\ total} = dE_{pot.\ without\ distortion} + dE_{pot.\ distortion} \qquad [10.37]$$

$$\begin{array}{ccc} [10.26] & [10.32] & [10.33] \\ \text{or } [10.27] & & \\ \text{or } [10.36] & & \end{array}$$

---

41 The justification is not dealt with in this book.

42 It can thus be observed that during the uniaxial traction test (see section [10.8]) where the only non-zero component of stress is reduced to σ_x, the distortion strain energy is not zero.

Chapter 11

Structural Joints

As we understand it in this book, structures generally consist of basic parts of different shapes (beams, plates, massive parts), or of the same nature but with large dimensions so that they cannot be fabricated with only one holding portion. This is typically the case of aircraft sections that are made up of multiple assembled components, or of metallic frameworks assembled on-site.

These assemblies are made with localized joints of "clamped" type that may be disassembled or not, which we call "structural joints". They provide structural "continuity". In particular, they must be:

– sufficiently rigid, not allowing any localized flexibility in the global structure;

– sufficiently resistant so as to allow the transmission of loads. These loads are generally forces and moments transmitted through the localized interfaces of such structural joints. The preliminary knowledge of these transmittable forces and moments is provided through the calculation results based on the finite element model of the structure (section 2.2.2/*NOTES*).

Here we plan to examine three traditional solutions to realize "clamped joints" with plane interfaces between the two structural parts to be joined. These three solutions hold the parts together and sometimes can ensure that elements stay in position. They are, in order[1]:

– bolted joints (can be disassembled);

– riveted joints;

____

1 Joining by adhesion (bonding) of different parts is not studied in this book.

– welded joints.

These three methods of assembly are characterized by very localized "material components" between assembled parts, which means components of small size compared to the dimensions of the junction zone. Therefore, we must expect the joints themselves and their immediate vicinity to experience:

– high stresses, because it is at these spots that forces are transmitted;

– and consequently, resulting localized deformations that are greater than those of the assembled parts in the same area.

The first two methods of joining involve fasteners characterized by cylindrical bodies whose lateral surfaces are partially or entirely smooth.

11.1. General information on connections by means of cylindrical fasteners

In this section we shall restrict ourselves to a description of behavior characteristics specific to such types of joining. The verification of the strength capabilities of each of these types will be detailed in following sections.

11.1.1. *Contact pressure*

In practice, the joining of parts based on cylindrical bodies involves:

– the assembly of two or more structural parts with the help of "rivets", so-called riveted joint or riveted assembly;

– the assembly of two or more structural parts by means of screws, bolts or studs, so-called assembly with "threaded fasteners" , or bolted joints.

Generally speaking, these fasteners are characterized by a cylindrical geometry, with a lateral surface that may be completely or partially smooth: the "shank". For the case of bolts, "threads" are machined to allow "screwing".

The working of these elements may differ notably according to:

– the fastener type: rivet or screw shank;

– the initial assembly conditions, or the accidental variations of these initial conditions during time.

The cylindrical fasteners fit through bored housings. Let us consider the cylindrical surface of a rod which is pressed on a housing under the action of a radial force $\vec{F}$ as shown in Figure 11.1a. We assume that the diameter of this rod is slightly smaller than that of its housing (weak play). The contact between the two elements occupies a finite area, for, as we know, the constituting elastic materials are deformable. The rod subjected to $\vec{F}$ is balanced by elementary contact forces labeled $\vec{df}$ in Figure 11.1a, acting upon every facet dS of this contact area. It is thus possible to define an interface stress at every point of that contact area, as: $\dfrac{\vec{df}}{dS}$.

Normally, we assume that force $\vec{df}$ is perpendicular to the contact facet. The interface stress is thus normal to the contact facet. This is called contact pressure.

$$\overrightarrow{p_{contact}} = \frac{\vec{df}}{dS}$$

The following figure shows the appearance of the contact pressure distribution[2].

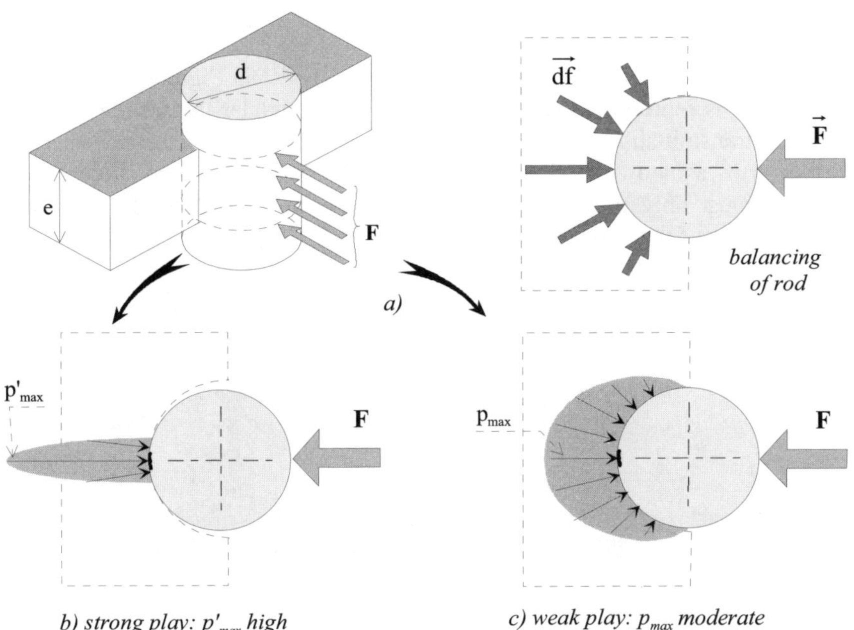

Figure 11.1. *Contact pressure between rod and housing*

2 We must remember that the forces $\vec{df} = \vec{p} \times dS$, and not the interface stresses, contribute to maintain equilibrium. In Figure 11.1, elementary contact forces are not shown; the authors prefer to show the contact stress (or pressure distribution), which they feel is more explicit.

Under the pressing force $\vec{F}$, the contact pressure between the rod and its housing seems greater if the play, i.e., the difference between the housing diameter and the rod diameter, is significant. When force $\vec{F}$ increases, this contact pressure becomes locally too high for one of the two elements in contact (or even for both). Localized plastification occurs (or permanent local deformation). This is due to the nature of the material, the surface roughness and the surface treatment of the component in question. This is especially important when the housing is machined in a metal sheet of little thickness or "thin sheet". Then there is mushrooming or "bearing" of the housing. As this phenomenon is repeated during the life of the structure, the hole deformation increases. Then the initially weak play between rod and housing is increasing. This local deterioration of the structure may eventually initiate a larger deterioration, and even a global structure failure.

11.1.2. *General information on riveting*

11.1.2.1. *Transmission of mechanical loads in riveted joints*

Take two parts assembled by riveting, labeled 1 and 2 in Figure 11.2a. The plane joint interface is represented in Figure 11.2b. Then consider the resultant force and moment of transmittable mechanical forces around point M of the interface. They have the forms as shown in the figure. In fact, rivets must not be subject to axial forces, but must only prevent the relative movement parallel to the interface, or relative slipping of the parts, as described below.

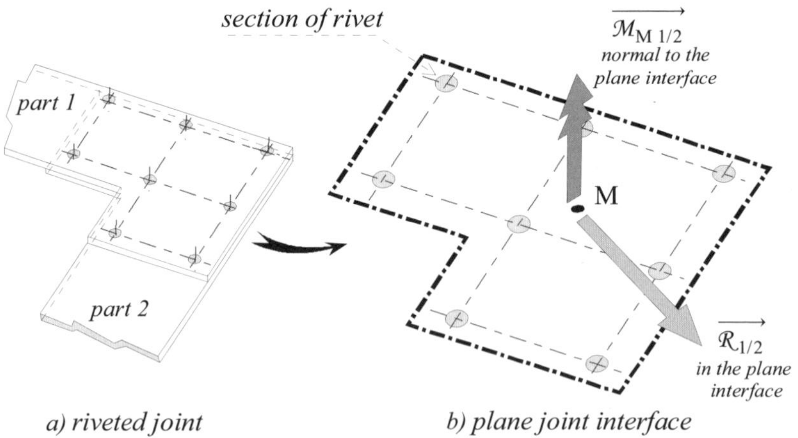

Figure 11.2. *Riveted joint: resultant force and moment of transmittable forces, around a point M of the interface*

11.1.2.2. *Functioning of a rivet*

In contemporary structural riveting (as in aeronautics, for example), the rivets fit through tight-fitting holes, without any play. During their installation, a light tension is produced in the rivet, which "tightens" the two parts to be joined one against the other. This tightening is weak and we shall ignore it[3]. The essential role of modern rivets is to prevent any relative slipping of the parts or components that they join.

The most loaded zone of a rivet is in the vicinity of its cross-section by the joining interface between the two assembled parts. This section tends to be sheared, as represented in Figure 11.3.

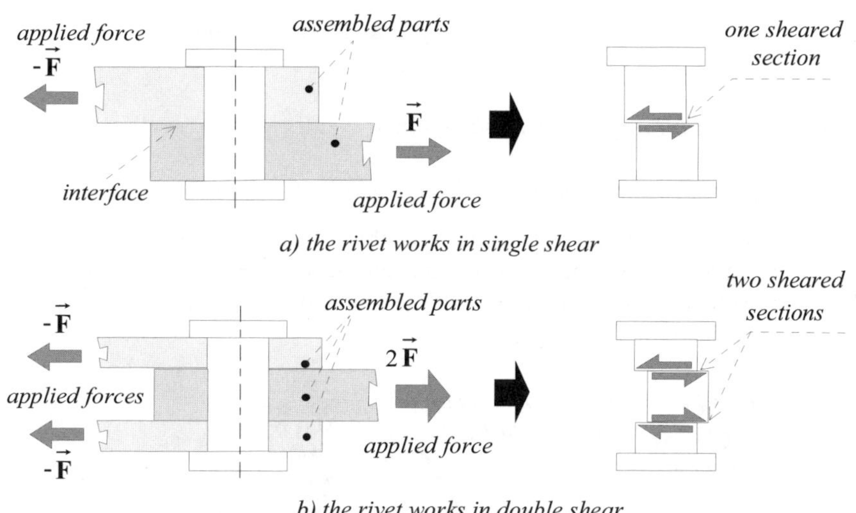

Figure 11.3. *Functioning of a rivet*

NOTES

❑ The presence of normal contact pressure on the cylindrical shank of the rivet, associated with the friction coefficient, results in allowing the installation of shear stresses on the cylindrical shank as shown in Figure 11.4. The reciprocity of shear stresses – at least locally – is thus assured, especially in zone (A) in Figure 11.4b. The phenomenon is thus more complex here than it would be for the "bending

---

3 This was not the case in the past, when the forged steel rivets with brazier heads were fitted when hot (the Eiffel tower, works of art such as viaducts, etc.). The contraction that occurred during cooling "tightened" (compressed) the parts thus assembled. Certain rivets these days have heads that are reinforced, allowing them to also work in traction. Such cases are not dealt with here.

with shear" of a circular beam. Nonetheless, it results in "uniformization" of shear stresses on the sheared rivet section.

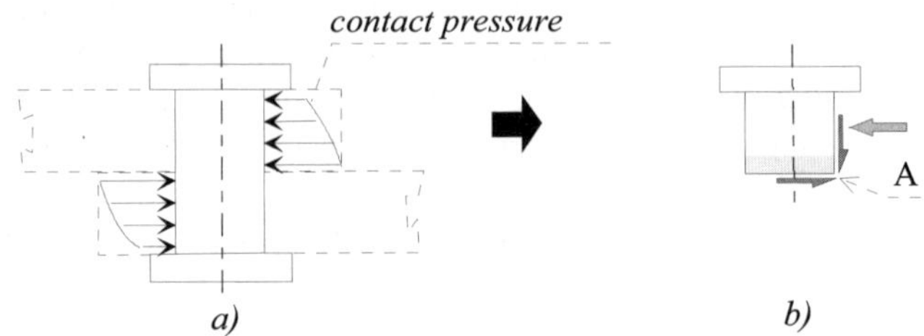

Figure 11.4. *Reciprocity of shear stresses*

11.1.3. *General information on bolted joints*

11.1.3.1. *Transmission of mechanical loads in a bolted joint*

Consider two parts of a structure assembled with threaded fasteners, labeled 1 and 2 in Figure 11.5a. The plane joining interface is represented in Figure 11.5b. Thus the resultant force and moment of the transmittable forces around a point M of the interface have the form represented.

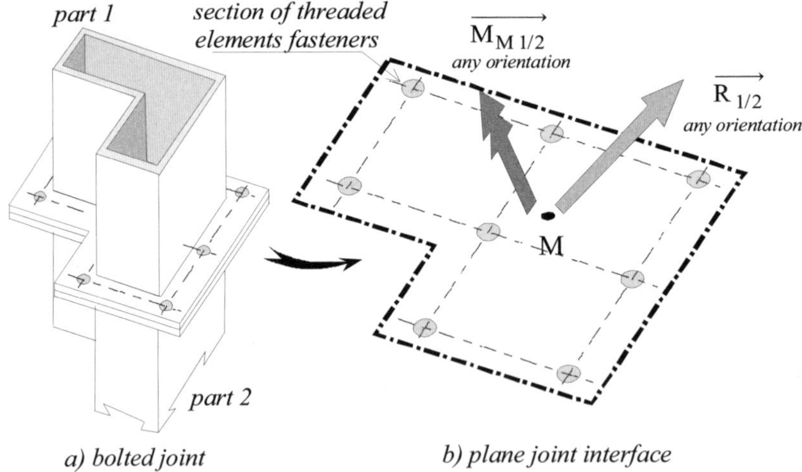

Figure 11.5. *Bolted joint: resultant force and moment of transmittable forces, around a point M of the interface*

11.1.3.2. *Functioning of threaded fasteners*

This is a universal fastening mean, with multiple practical configurations which lead to different behaviors. Figure 11.6 shows a few of these configurations. Their common property is the length of the working screw shank which is equal to the total thickness of the assembled parts. To every screw there corresponds a rotating nut. The screw and nut set is called a "bolt". Thus, these are called "bolted joints".

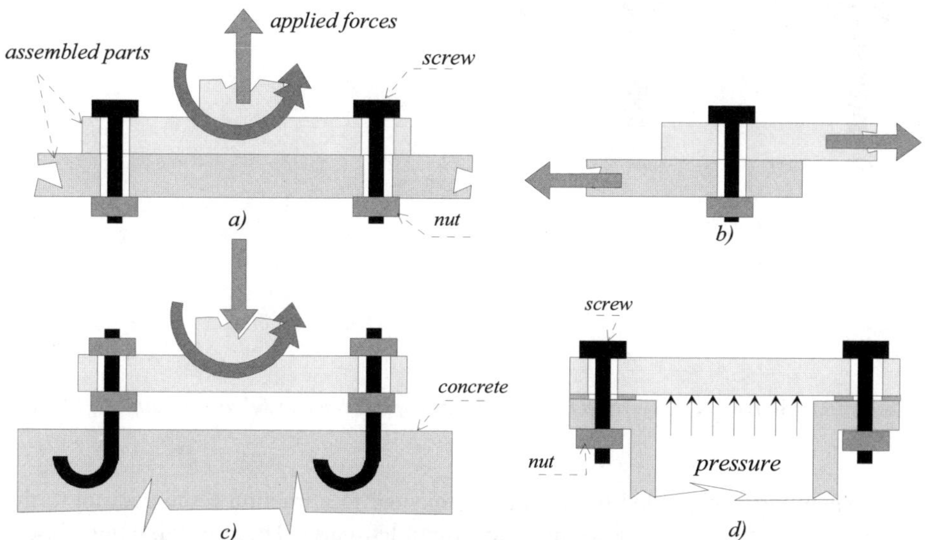

Figure 11.6. *Different types of bolted assemblies*

The essential characteristic of such assemblies is the installation (except in particular cases, as in Figure 11.6c, for example) of a tensile preload, or "pre-tightening" on the screw shank.

This pre-tightening has several consequences:

– it creates contact pressure at the interface of assembled parts (Figure 11.7b). When these parts are also subject to a force trying to "slide" a part relative to the other (Figure 11.7c), the existence of a coefficient of interface friction between the materials brings about a resistance to the relative sliding of assembled parts, which prevents this kind of movement;

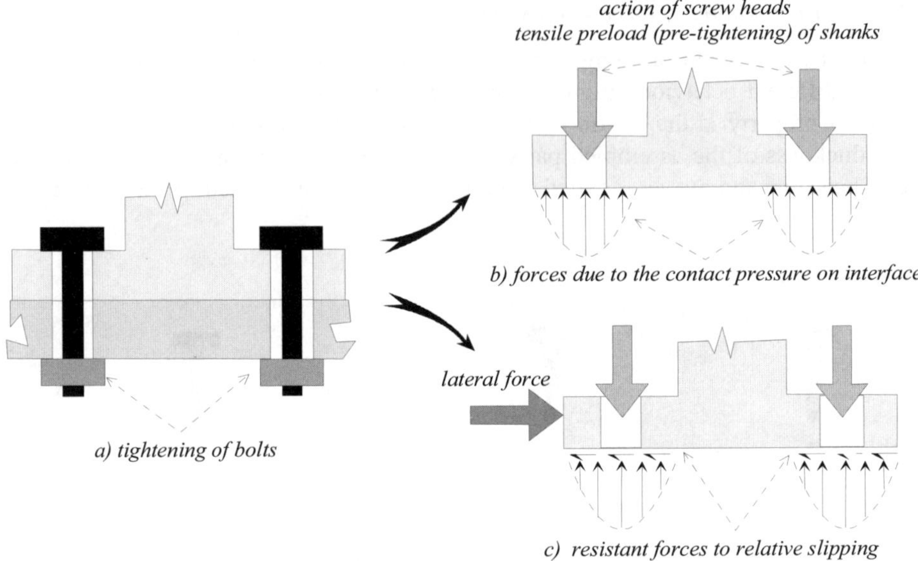

Figure 11.7. *The pre-tightening leads to resistance to relative sliding*

– when parts assembled by bolts are subjected to a traction force parallel to the screw shanks, it tends to "separate" the assembled parts. The pre-tightening should be sufficient to prevent the assembled parts coming apart;

– in practice, when the joint is subject to such a "separation", this force is often a "fluctuating" force. It oscillates around an average value, with a certain fluctuation amplitude, showing a "waved" form[4]. Pre-tightening attenuates the fluctuation magnitude of tension in the screw shank. In fact, the fluctuation magnitude of the tensile force which acts upon the screw shank is lower than the actual fluctuation magnitude upon the joint[5]. This, of course, is beneficial to the fatigue resistance of the screw shank.

11.1.4. Deterioration of riveted and bolted joints

11.1.4.1. Rupture of fasteners

The first mode of deterioration corresponds naturally to the rupture of fasteners:

---

4 See Chapter 7, Figure 7.21.
5 See section 11.2.2.2 below.

– shanks of bolts or rivets that break due to shearing according to Figure 11.3;

– shanks of bolts that break due to high tensile force, or as a result of fatigue phenomenon[6].

Nonetheless, there are also other causes for deterioration that we shall briefly cite later.

11.1.4.2. *Bearing*

In section 11.1.1, we pointed out that the contact pressure between a cylindrical rod and its housing or bored hole after the application of a radial force $\vec{F}$ could become inadmissible because they bring about the phenomenon of bearing. Access to real contact pressures is a very difficult problem that is not considered here. In practice, we determine from tests the equivalent average pressure that brings about this kind of deterioration.

♦ Equivalent average pressure: the simplest pressure distribution that we can represent in a model for the problem of contact in Figure 11.1 consists of a uniform pressure p_0 as shown in Figure 11.8. This is an "idealized" state that does not correspond to reality, but which allows a simple and rapid pre-dimensioning.

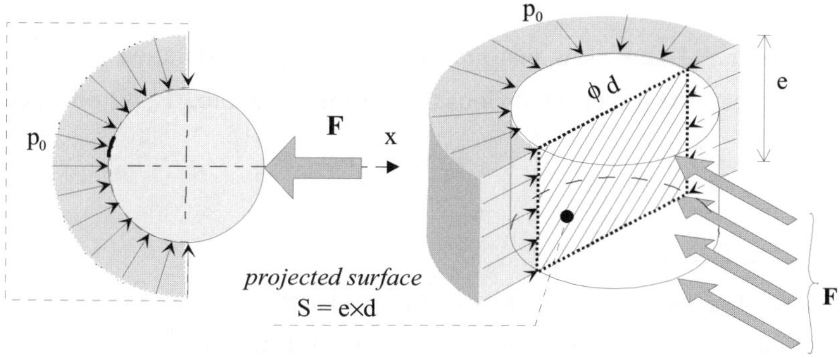

Figure 11.8. *Idealized state of uniform pressure p_0*

---

6 See Chapter 7, section 7.3.

We thus obtain the resulting pressure force[7] projected on axis $\vec{x}$:

$$p_0 \times e \times d$$

where $e \times d$ is the "projected surface" obtained from the semi-cylindrical surface upon which p_0 acts. The equilibrium of the cylindrical rod is written as projected on $\vec{x}$:

$$p_0 \times e \times d = F$$

i.e.:

$$p_0 = \frac{F}{e \times d}$$

p_0 is described as the "average pressure for bearing". In practice, we shall avoid the bearing deterioration phenomenon by observing a condition of the type:

$$p_0 < p_{\text{bearing strength}}$$

In which the limit value $p_{\text{bearing strength}}$ is obtained from tests. It is therefore necessary to know test results obtained with the same materials and the same play between the rod and its housing.

NOTE

❏ In Figure 11.1, we saw that the risk of bearing deterioration was greater in b than in c, because it seemed obvious that for the same force $\vec{F}$, we obtained:

$$p'_{\text{max}} > p_{\text{max}}$$

---

7 On elementary facet dS whose normal $\vec{n}$ makes an angle α with $\vec{x}$, the basic pressure force has the value $-p_0 \times dS \times \vec{n}$ (where we have noted $p_0 = \|\mathbf{p_0}\|$). If we project on the direction $\vec{x}$, we obtain:

$-p_0 \times dS \times \vec{n} \cdot \vec{x} = -p_0 \times dS \times \cos\alpha$ .The area $dS \times \cos(\pi - \alpha) = -dS \times \cos\alpha$ is the projection of dS parallel to $\vec{x}$ on a plane perpendicular to $\vec{x}$. This is called the projected surface. The sum of facets dS that constitute the semi-cylindrical contact above has a projected surface:

$$-\int_S dS \times \cos\alpha = e \times d$$

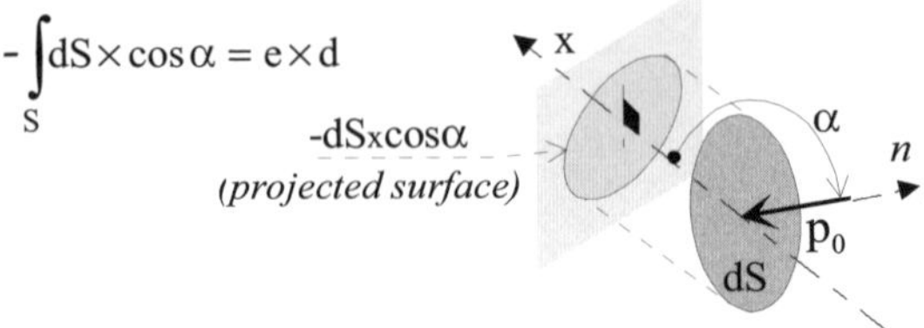

We can therefore foresee that the empirical value $p_{\text{bearing strength}}$ varies inversely to the play between the rod and housing. It reduces when the play increases.

From the above, we shall retain the following.

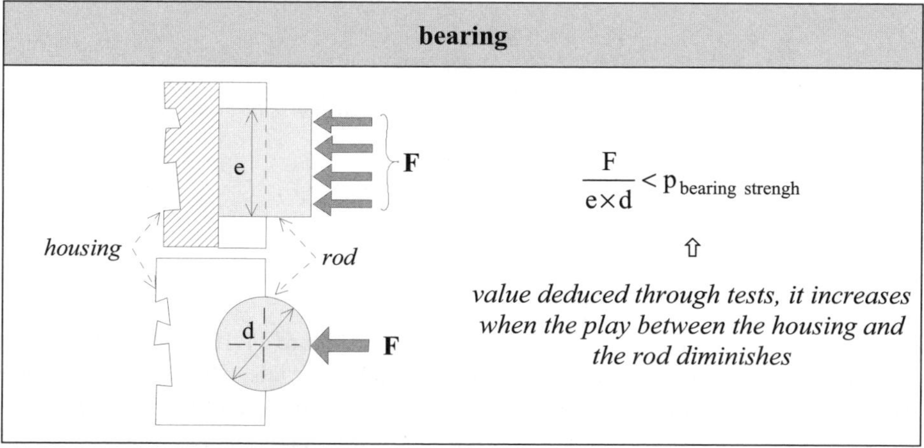

11.1.4.3. *Spacing of fasteners*

The following concerns mainly the practical aspects of assembling thin sheets. Two failure modes which may occur are represented below.

○ Punching of the sheet

In Figure 11.9, we illustrated two sheets assembled by a rivet. The distance from the center of the hole to the edge of the sheet is called the "edge distance". We see that beyond a certain value, there is a risk of shear tearing failure, which results in a "punching" of the sheet in its own plane. With the values indicated, the non-punching criterion shall be:

$$\frac{F}{2 \times e \times \text{edgedistance}} < R_{rg\ \text{sheet}}$$

where we have denoted R_{rg} the rupture shear stress of the sheet.

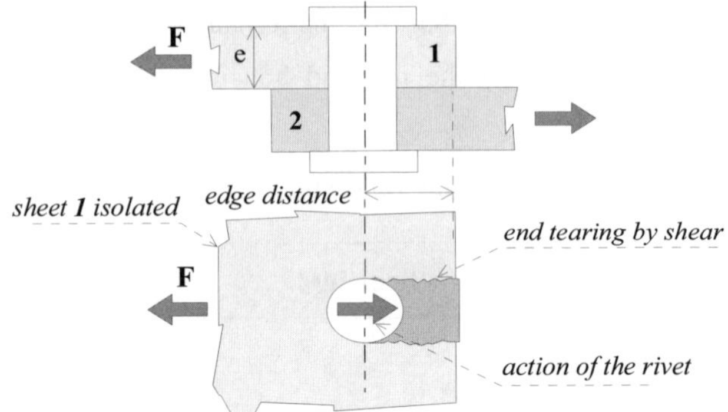

Figure 11.9. *"Punching" of the sheet*

○ Tensile rupture of the sheet

Large sheets are assembled with long overlaps (Figure 11.10a). Then numerous fasteners are aligned in rows, and placed at regular intervals. The interval between two fasteners is called the "fastener's pitch" (Figure 11.10b). Every fastener transmits a force $\vec{F}$. The contact pressure between the sheets is neglected.

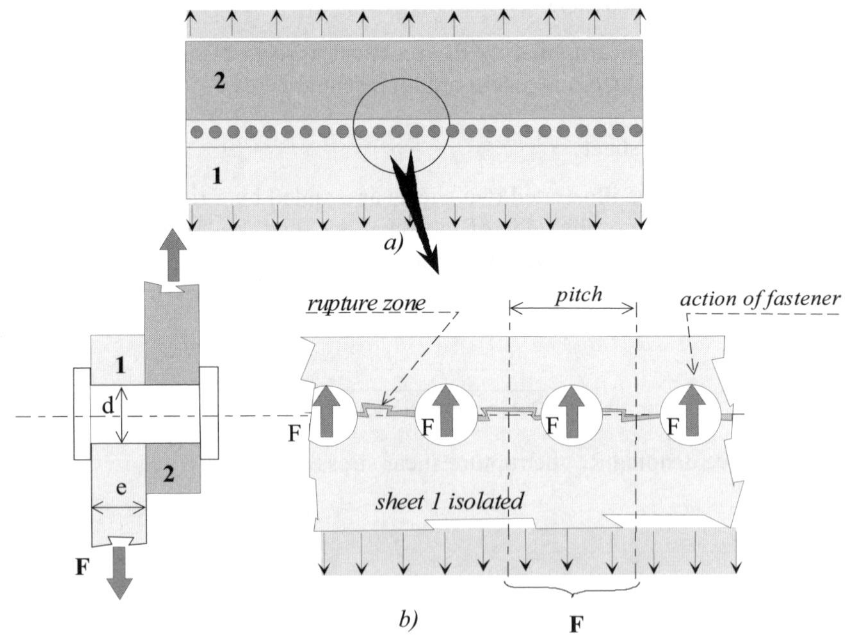

Figure 11.10. *Tensile rupture of the sheet*

Figure 11.10b shows a vulnerable zone where rupture may occur[8]. In this rupture zone, we observe a normal stress that may be, approximately, written as[9]:

$$\sigma_{sheet} = \frac{F}{e(pitch - d)}$$

We see that the stress may become very great when, for a given diameter d, the pitch is too small. This may result in tensile rupture. The criterion for non-rupture can be written as:

$$\frac{F}{e(pitch - d)} < R_{r\ steet}$$

where we have noted R_r the rupture tensile stress of the sheet.

11.2. Bolted joint

In the following, we indicate the procedure that makes possible the verification of whether a bolted joint is capable of tolerating the forces transmitted by the joint.

11.2.1. *Simplified case where the tightening is neglected*

11.2.1.1. *Hypotheses*

Here, we shall assume that:

♦ Shanks of the threaded elements are no longer subject to pre-tightening (this is, for example, the case, after the loosening of the screws).

♦ The shanks are smooth (plain-shank) and are carefully tight-fitted (close sliding fit) through bored holes in the interface zone.

It must be noted that these hypotheses define a particular case of jointing attachment. Thus, the absence of prior tension pre-tightening corresponds to an

---

8 This rupture gives the edge strip what is known as the "postage stamp" appearance.
9 In reality, the stress in the rupture zone is not uniform. It is higher near the holes (phenomenon of stress concentration):

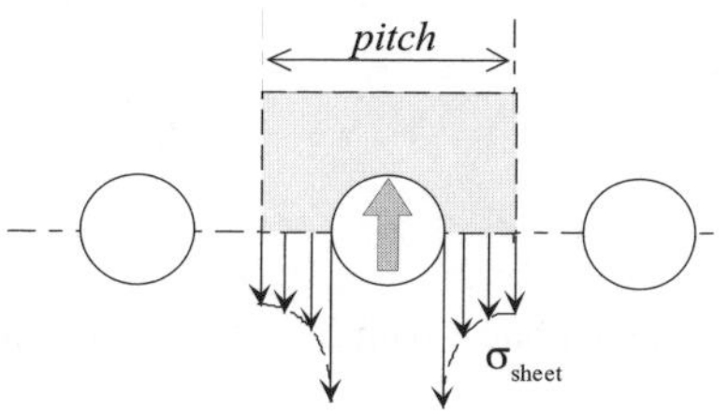

accidental loosening of the threaded elements. The plain-shanks are then subject to shearing in the interface zone of the assembled parts. However, few bolts are really tight-fitted in the two parts assembled[10]. In ordinary assemblies, the plain-shanks fit with a play in the housings of at least one of the two assembled parts. The presence of this play brings about additional inequalities in the distribution of shear forces on every shank.

♦ Finally we shall assume that the shanks may be strained by shear and also by tension and/or compression. This last hypothesis is not realistic because the shanks cannot work in compression. It makes it possible, however, to be free from dimensions of the base plate, and generally corresponds to a pessimistic or unfavorable evaluation, which is beneficial for the dimensioning. This can be seen in Figure 11.11 where we make the parallel between the real behavior in case of loosening, and the simplified behavior obtained from this hypothesis.

---

10 Generally, this concerns assemblies with accurate positioning (precision assemblies, aeronautical attachments).

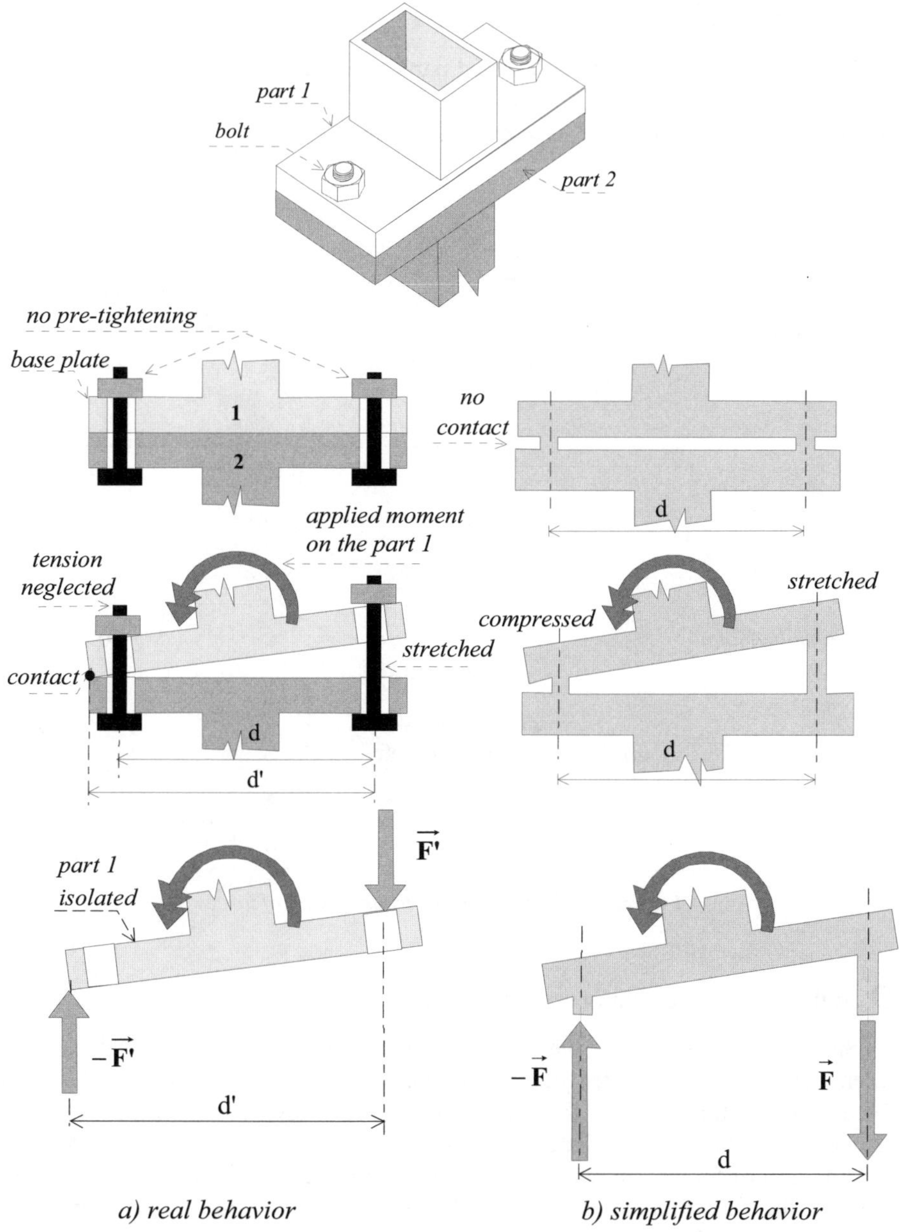

Figure 11.11. *Hypothesis of tension-compression: the simplification creates a pessimistic condition: $d<d' \Rightarrow F>F'$*

11.2.1.2. *Model of joining interface*

The interface between the assembled parts is reduced to n circular sections each working in the area "s_i", which are relatively small when compared to the dimensions of the joint area as represented in Figure 11.12.

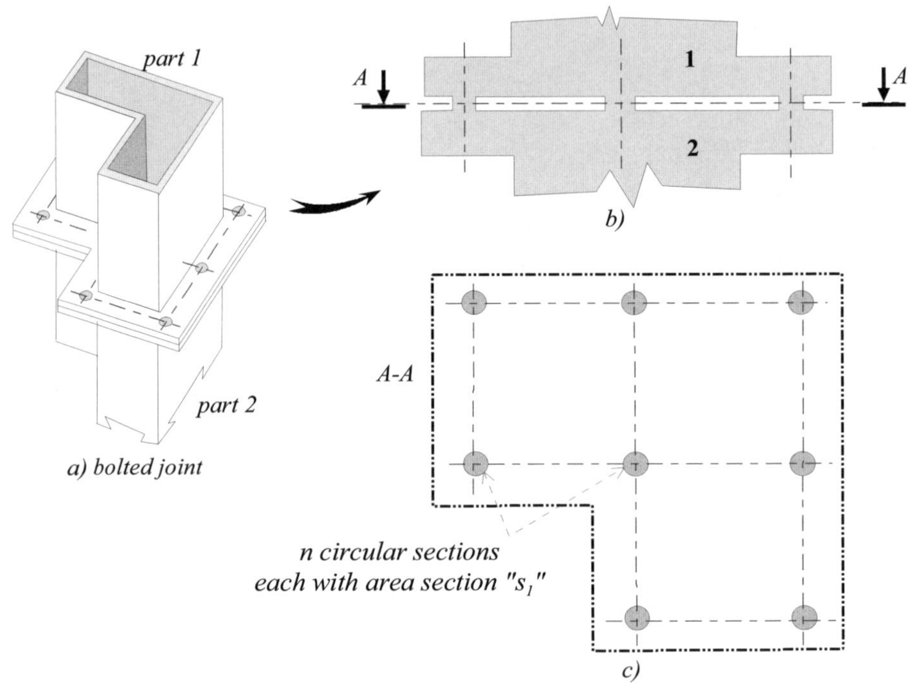

Figure 11.12. *Interface for transmission of forces*

○ Relative displacement of assembled parts

Since the dimensions of the fasteners are far smaller compared to those of the assembled parts, we consider that they are far more flexible (deformable). Under the forces to be transmitted, the assembled parts show (very) little rigid-body displacement due to the fasteners deformation. As a result, the interface for transmission of forces, shown shaded in Figure 11.12c (set of "n" circular sections), behaves, for the normal stresses in shanks, like a beam would behave with a cross-section constituted of the "n" disjointed circular sections.

This analogy facilitates the research of normal forces transmitted by each of the "n" sections "s_i".

Very simple example: let us consider in Figure 11.13 a row of three identical untightened bolts. On the upper part, supposed to be undeformable, a force $\vec{F}$ is applied as shown. The shanks of the bolts being modeled by identical springs, we can see that the deformations, and consequently the forces in the three springs, are identical. By considering the equilibrium of the upper part, we have $F = 3k \times \Delta\ell$, where k is the stiffness of the springs and $\Delta\ell$ their common extension[11].

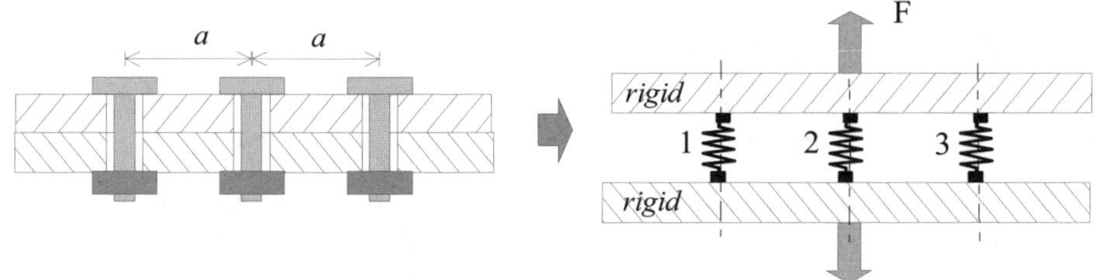

Figure 11.13. *Very simple example*

O   Characteristics of the modeled joining interface

Let us reconsider the analogy above: the interface is constituted by an equivalent beam whose cross-section is made of "n" shank sections. As for a classic beam[12], we link a set of principal axes (see Figure 11.14) to the equivalent section. This means (see Figure 11.14):

– an axis $\vec{x}$ perpendicular to the equivalent section and directed as an outward normal coming from the geometric center G, barycenter of the sections "s_i";

– principal quadratic axes $\vec{y}$ and $\vec{z}$ in the plane of the equivalent section.

11 See [2.5].
12 See [9.1].

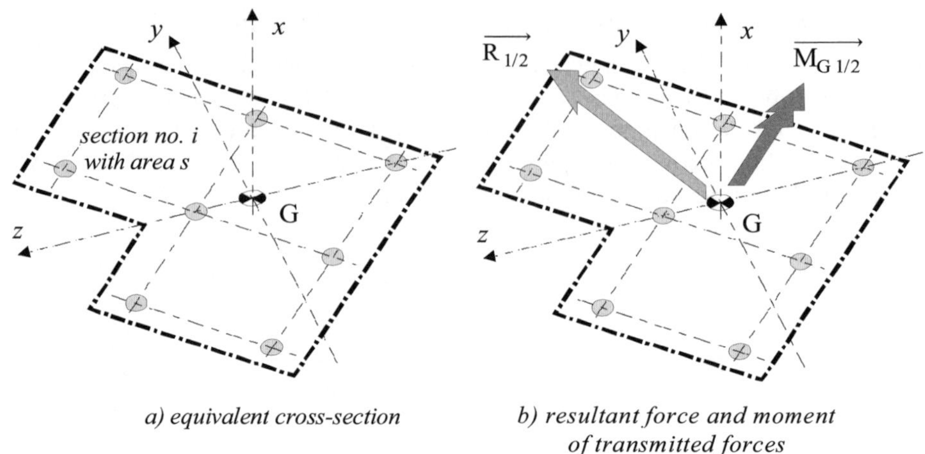

a) equivalent cross-section   b) resultant force and moment
                                        of transmitted forces

Figure 11.14. *Interface of a bolted joint*

With the notations in Figure 11.14a where points i = 1,…, n are the centers of the "n" fasteners, we thus have:

– a barycenter or geometric center G such that:

$$\sum_{i=1}^{n} y_i \times s_i = \sum_{i=1}^{n} z_i \times s_i = 0$$

i.e., for the case of identical sections $s_i = s$ [13]:

$$\sum_{i=1}^{n} y_i = \sum_{i=1}^{n} z_i = 0 \qquad [11.1]$$

– principal quadratic moments:

$$I_y = \sum_{i=1}^{n} z_i^2 \times s = s \times \sum_{i=1}^{n} z_i^2$$

$$[11.2]$$

$$I_z = \sum_{i=1}^{n} y_i^2 \times s = s \times \sum_{i=1}^{n} y_i^2$$

---

13 These definitions are applicable in most general cases of "n" attachment characterized by different sections: $s_1 \neq s_2 \neq \dots s_n$. Nevertheless, for the large majority of applications this general case is not realistic.

with, in addition: $\displaystyle\sum_{i=1}^{n} y_i \times z_i = 0$

11.2.1.3. *Forces on each fastener*

Figure 11.14b represents the resultant force and moment of transmittable forces acting on the equivalent section. As already announced at the beginning of the chapter[14], these are deduced from the loads acting upon the structure. Thus, from the resultant force and moment, already known projections are obtained on axes $\vec{x}$, $\vec{y}$, $\vec{z}$, which are recalled below[15]:

$\mathcal{N}_x$ or $\mathcal{N}$: normal resultant	$\mathcal{M}_x$: longitudinal moment
$\mathcal{T}_y$: shear resultant	$\mathcal{M}f_y$: bending moment
$\mathcal{T}_z$: shear resultant	$\mathcal{M}f_z$: bending moment

The force on fastener cross-section no. i is noted $\vec{F_i} = X_i\,\vec{x} + Y_i\,\vec{y} + Z_i\,\vec{z}$. It is deduced from the preceding projections as follows:

➢ normal resultant $\mathcal{N}$:

Normal stresses act on every fastener "i" of section "s" (Figure 11.15a), as represented in Figure 11.15b. These are written as:

$$\sigma_{\mathcal{N}} = \frac{\mathcal{N}}{n \times s}$$

They correspond to a normal "cohesive" force acting on each fastener shank (represented in Figure 11.15c):

$$X_{\mathcal{N} i} = \sigma_{\mathcal{N}} \times s = \frac{\mathcal{N}}{n} \ \forall\, i$$

---

14 Calculation results from finite element model of the structure give transmittable forces (see section 5.2.2/*NOTES*).

15 See [9.6].

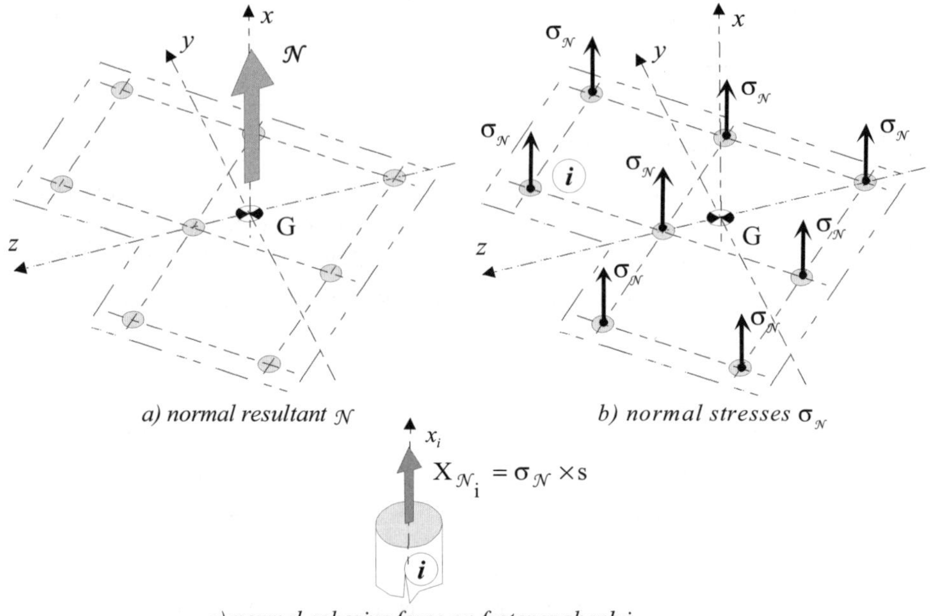

a) normal resultant $\mathcal{N}$ b) normal stresses $\sigma_{\mathcal{N}}$

c) normal cohesive force on fastener shank i

Figure 11.15. *Stresses on fastener shanks due to* $\mathcal{N}$

NOTE

❏ Let us remind ourselves once more that such a distribution of normal forces on the "n" bolts assumes that the pieces to be assembled are more rigid than the bolt shanks, as is generally the case for structural assemblies. The example of Figure 11.13 gives a very simple interpretation of such a joint with three bolts modeled by "springs".

➢ Shear resultant $\mathcal{T}_y$

From the hypothesis of carefully tight-fitted shanks, deformation due to the shear resultant is different from that of simple standard bending[16]. We then assume that this resultant shear brings to each fastener "i" of section "s" (see Figure 11.16) a uniform shear stress distribution such that:

$$\tau_{xy\,\mathcal{T}_y} = \frac{\mathcal{T}_y}{n \times s}$$

---

16 Deformation due to the resultant shear for simple bending is represented in Chapter 9, Figure 9.44. It assumes that the lateral surface of the shank is free of forces, which is not the case here, as pointed out in section 11.1.2.2 (see Figure 11.4).

This shear stress corresponds on each fastener shank to a tangential cohesive force along $\vec{y}$, which is represented in Figure 11.16c:

$$Y_{\mathcal{T}_y i} = \tau_{xy_{\mathcal{T}_y}} \times s = \frac{\mathcal{T}_y}{n} \quad \forall\, i$$

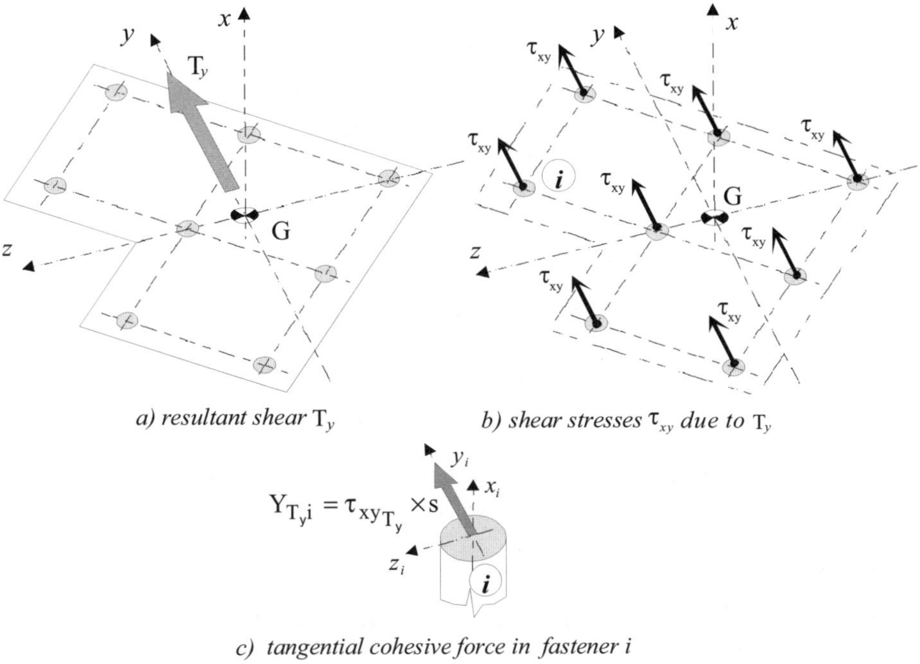

a) resultant shear T_y *b) shear stresses* τ_{xy} *due to* T_y

$$Y_{T_y i} = \tau_{xy_{T_y}} \times s$$

c) tangential cohesive force in fastener i

Figure 11.16. *Stresses on fasteners due to* $\mathcal{T}_y$

NOTE

❑ Let us remember that we have hypothesized that the shanks are carefully tight fitted in their housings. For example, let us consider the three identical, untightened, aligned bolts of Figure 11.17 below, under force $\vec{F}$. Thus, this hypothesis enables us to observe the behavior given by springs of rigidity k, that is: $F = 3k \times \Delta\ell$, where k is the stiffness of the springs and $\Delta\ell$ their common extension.

On the contrary, this model does not appear to be appropriate any longer if the shanks of the bolts are fitted with play as in Figure 11.17b, which is, however, a frequent occurrence. Thus, our hypothesis on the absence of pre-tightening leads to contacts between shanks and housings that are mostly unpredictable. We will thus not examine this case.

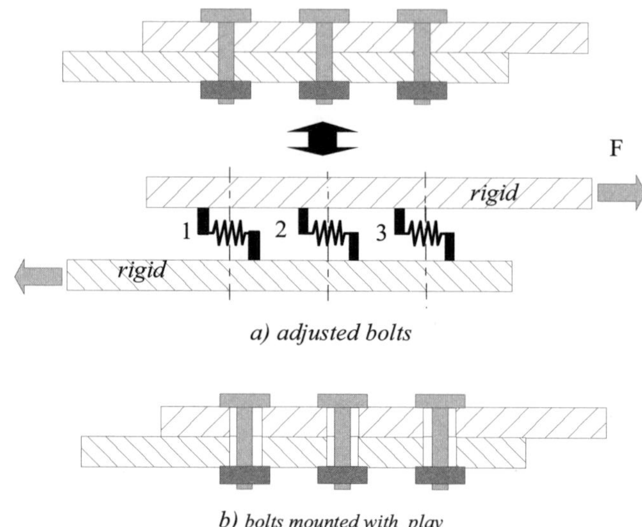

a) adjusted bolts

b) bolts mounted with play

Figure 11.17. *The model with springs (case a) is not valid when bolts are not tight fitted (case b))*

➢ Shear resultant T_z

In a similar manner, this shear resultant will give shear stresses on each fastener (see Figure 11.18):

$$\tau_{xz\,T_z} = \frac{T_z}{n \times s}$$

This shear stress corresponds, on every shank, to a tangential cohesive force following $\vec{z}$ and represented in Figure 11.18c:

$$Z_{T_z i} = \tau_{xz\,T_y} \times s = \frac{T_z}{n} \quad \forall\, i$$

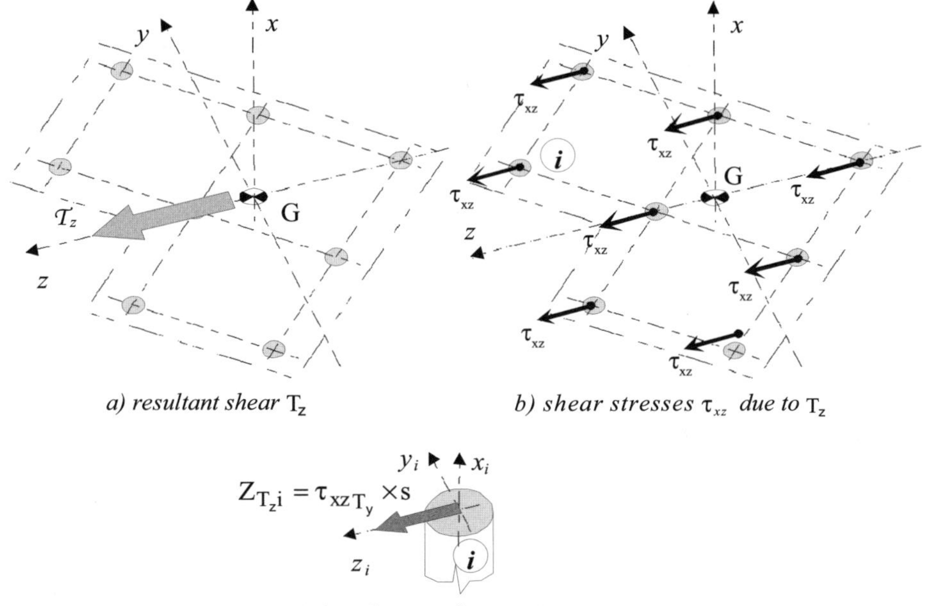

a) resultant shear T_z

b) shear stresses τ_{xz} *due to* T_z

$$Z_{T_z}i = \tau_{xzT_y} \times s$$

c) tangential on force on fastener i

Figure 11.18. *Stresses on fasteners due to* T_z

> Torsion moment $\mathcal{M}_x$

We can make the same comment here as that made for the shear resultant: the lateral surface of the shank is no longer free (contact with housing) and therefore the boundary conditions of classical torsion of beams are no longer valid[17]. The distribution of the shear stresses is modified. Considering the equivalent global section made of "n" shank sections s_i, we shall adopt a stress distribution that is identical to that of the torsion of a circular section beam[18], as indicated in Figure 11.19[19].

---

17 For standard torsion see Chapter 9, section 9.3.2.4 and [9.27].

18 See Chapter 9, section 9.3.2.3.

19 During torsion of a circular section beam, the sections do not warp, they remain plane contrary to non-circular sections; see section 9.3.2.4. Here, we consider that the equivalent section composed of "n" circular shank sections does not warp. This reinforces the analogy that we adopted.

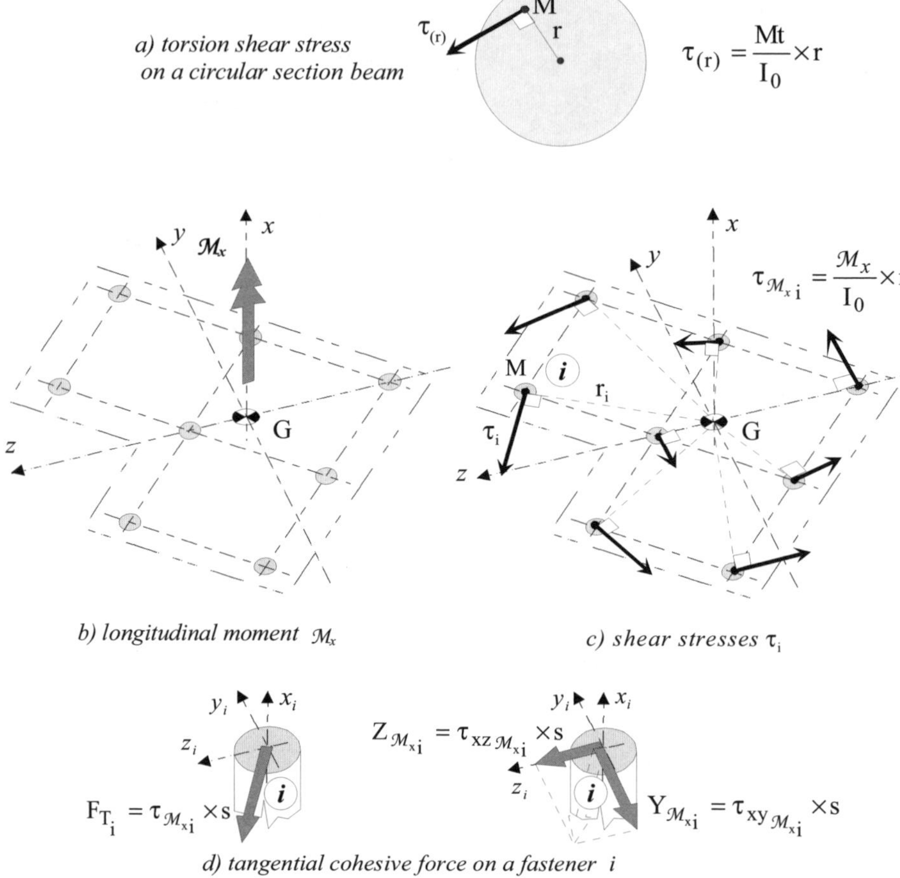

Figure 11.19. *Stresses on fasteners due to* $\mathcal{M}_x$

Stress $\tau_{\mathcal{M}_x i}$ at point (i) of Figure 11.19c can also be written in vector form:

$$\overrightarrow{\tau_{\mathcal{M}_x i}} = \left(\tau_{xy_i}\,\vec{y} + \tau_{xz_i}\,\vec{z}\right) = \overrightarrow{M_i G} \wedge \frac{\mathcal{M}_x}{I_0}\,\vec{x} = -\left(y_i\,\vec{y} + z_i\,\vec{z}\right) \wedge \frac{\mathcal{M}_x}{I_0}\,\vec{x}$$

i.e.:

$$\tau_{xy\,\mathcal{M}_x i} = -\frac{\mathcal{M}_x}{I_0} \times z_i\,; \qquad \tau_{xz\,\mathcal{M}_x i} = \frac{\mathcal{M}_x}{I_0} \times y_i\,; \qquad [11.3]$$

Here, the quadratic polar moment is written as:

$$I_0 = \sum_{i=1}^{n} r_i^2 \times s = \sum_{i=1}^{n} \left(y_i^2 + z_i^2 \right) \times s$$

or, according to [11.2], as: $I_0 = I_y + I_z$.

The corresponding tangential cohesive forces, shown in Figure 11.19d, are noted for each attachment:

$$Y_{\mathcal{M}_x i} = \tau_{xy\,\mathcal{M}_x i} \times s = -\mathcal{M}_x \times \frac{z_i}{\sum\limits_{i=1}^{n} \left(y_i^2 + z_i^2 \right)}$$

$$Z_{\mathcal{M}_{xi}} = \tau_{xz\,\mathcal{M}_x i} \times s = \mathcal{M}_x \times \frac{y_i}{\sum\limits_{i=1}^{n} \left(y_i^2 + z_i^2 \right)} \qquad [11.4]$$

NOTES

❏ Let us verify that the sum of these tangential forces thus defined is zero, as they should reduce in G in a longitudinal moment $\mathcal{M}_x$, that is, a pure moment. With the earlier expressions and the properties [11.1] of the geometric center G, we obtain:

$$\sum_{i=1}^{n} Y_{\mathcal{M}_x i} = \sum_{i=1}^{n} \left(-\mathcal{M}_x \times \frac{z_i}{\sum\limits_{i=1}^{n} \left(y_i^2 + z_i^2 \right)} \right) = -\frac{\mathcal{M}_x \times s}{I_0} \sum_{i=1}^{n} z_i = 0$$

$$\sum_{i=1}^{n} Z_{\mathcal{M}_x i} = \sum_{i=1}^{n} \left(\mathcal{M}_x \times \frac{y_i}{\sum\limits_{i=1}^{n} \left(y_i^2 + z_i^2 \right)} \right) = \frac{\mathcal{M}_x \times s}{I_0} \sum_{i=1}^{n} y_i = 0$$

❏ The hypothesis relating to flexible fastener shanks is in line with the analogy of classic torsion used here. For example, when we consider the torsional behavior in Figure 11.20 around axis $\vec{x}$ of the simplified joint, the deformation of the springs (identical) modeling the shanks corresponds to forces on these springs that are proportional to radius r.

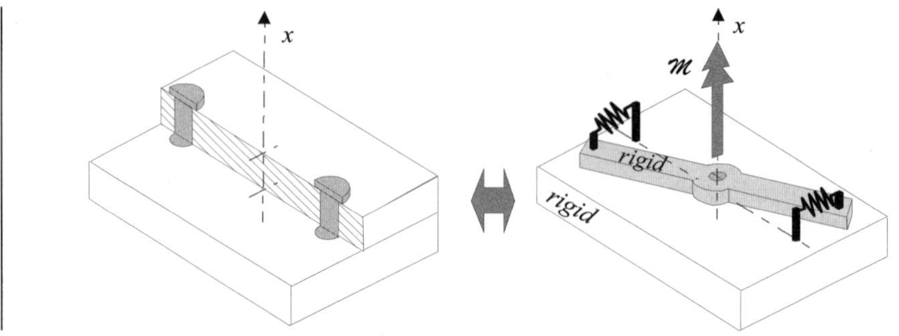

Figure 11.20. *Simplified bolted joint under torsion*

➢ Bending moment $\mathcal{M}f_y$

This is illustrated in Figure 11.21, and reveals the existence of normal stresses at centers (i) of each fastener given by[20]:

$$\sigma_{\mathcal{M}f_y\,i} = \frac{\mathcal{M}f_y}{I_y} \times z_i$$

i.e. with [11.2]:

$$\sigma_{\mathcal{M}f_y\,i} = \frac{\mathcal{M}f_y}{s} \times \frac{z_i}{\sum\limits_{i=1}^{n} z_i^{\,2}}$$

which corresponds to a normal cohesive force on fastener (i) represented in Figure 11.21c:

$$X_{\mathcal{M}f_y\,i} = \sigma_{\mathcal{M}f_y\,i} \times s = \mathcal{M}f_y \times \frac{z_i}{\sum\limits_{i=1}^{n} z_i^{\,2}}$$

---

20 See [9.36].

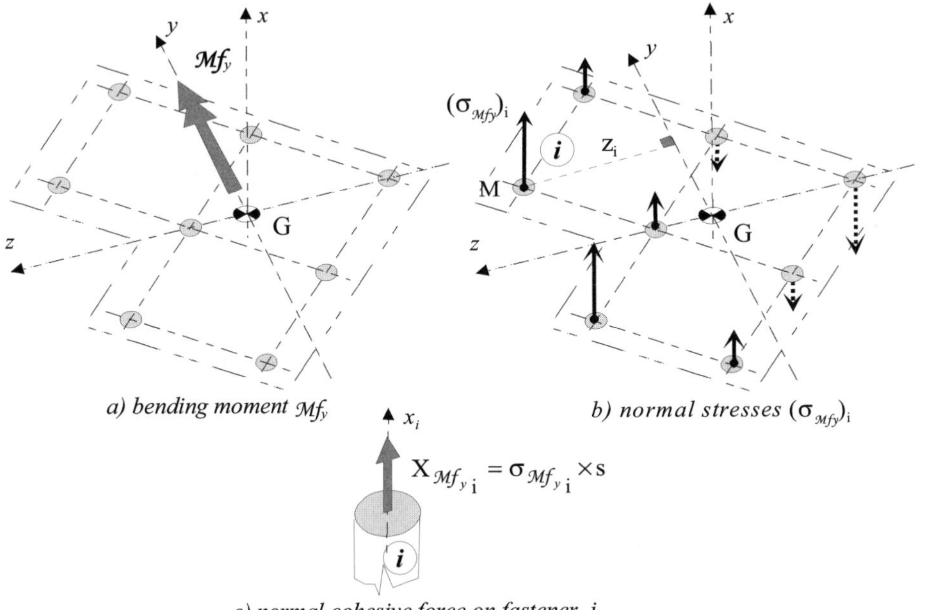

a) bending moment $\mathcal{M}f_y$

b) normal stresses $(\sigma_{\mathcal{M}fy})_i$

c) normal cohesive force on fastener i

Figure 11.21. *Stresses on fasteners due to $\mathcal{M}f_y$*

➤ Bending moment $\mathcal{M}f_z$

This is illustrated in Figure 11.22 and reveals the existence of normal stresses at centers (i) of each fastener given by[21]:

$$\sigma_{\mathcal{M}f_z\,i} = -\frac{\mathcal{M}f_z}{I_z}\times y_i$$

i.e. with [11.2]:

$$\sigma_{\mathcal{M}f_z\,i} = -\frac{\mathcal{M}f_z}{s}\times\frac{y_i}{\displaystyle\sum_{i=1}^{n} y_i^{\,2}}$$

This corresponds to a normal cohesive force on fastener (i) represented in Figure 11.22c:

____

21 See [9.30] or [9.36].

$$X_{\mathcal{M}f_{z\,i}} = \sigma_{\mathcal{M}f_{z\,i}} \times s = -\mathcal{M}f_z \times \frac{y_i}{\displaystyle\sum_{i=1}^{n} y_i^{\,2}}$$

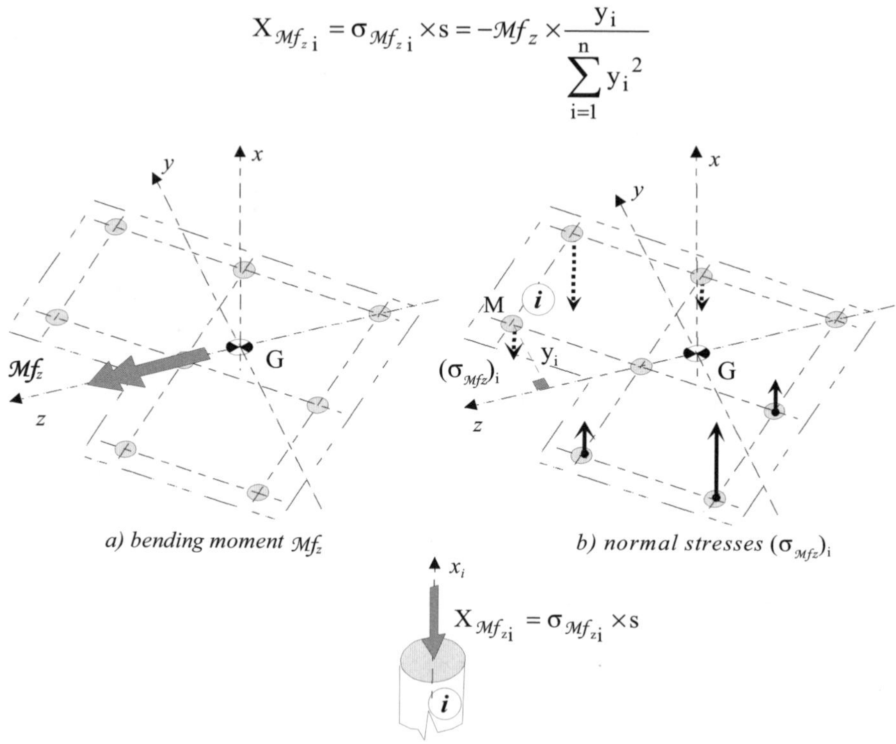

a) bending moment $\mathcal{M}f_z$ *b) normal stresses $(\sigma_{\mathcal{M}fz})_i$*

$$X_{\mathcal{M}f_{z i}} = \sigma_{\mathcal{M}f_{z i}} \times s$$

c) normal cohesive force on fastener shank i

Figure 11.22. *Stresses on fastener shanks due to $\mathcal{M}f_z$*

➤ Global stresses

These occur at center (i) of each fastener as the result of the superposition of stresses noted above for each of the projections of the resultant force and moment of the transmittable forces.

We will thus obtain:

– total normal stresses on $\vec{x}$ axis:

$$\sigma_i = \sigma_{\mathcal{N}} + \sigma_{\mathcal{M}f_{y\,i}} + \sigma_{\mathcal{M}f_{z\,i}}$$

– total shear stresses:

on $\vec{y}$ axis: $\tau_{xy_i} = \tau_{xy_{\mathcal{T}_{y\,i}}} + \tau_{xy_{\mathcal{M}_{x\,i}}}$;   on $\vec{z}$ axis: $\tau_{xz_i} = \tau_{xz_{\mathcal{T}_{z\,i}}} + \tau_{xz_{\mathcal{M}_{x\,i}}}$

➤ Total forces on fasteners

As we have already mentioned, we obtain these by multiplying the stresses mentioned above by the area "s" of the section of fastener, i.e.:

– total normal force on $\vec{x}$ axis:

$$X_i = X_{\mathcal{N}} + X_{\mathcal{M}f_y{}_i} + X_{\mathcal{M}f_z{}_i}$$

– total tangential force that has components:

on $\vec{y}$ axis: $Y_i = Y_{\mathcal{T}_y} + Y_{\mathcal{M}_x{}_i}$; on $\vec{z}$ axis: $Z_i = Z_{\mathcal{T}_z} + Z_{\mathcal{M}_x{}_i}$

We obtain for shank no. i the components X_i , Y_i , Z_i . They allow the definition of:

– an "average" normal stress that reaches its maximum in the cross-section corresponding to the minor diameter of the thread or minor core section of the threaded shank, written as s_0 (Figure 11.23):

$$\sigma_i = \frac{X_i}{s_0}$$

– a shear stress $\tau_i = \sqrt{\tau_{xy_i}{}^2 + \tau_{xz_i}{}^2}$. This is calculated in the shank cross-section found at the interface between the assembled parts. We shall see in Figure 11.23 that this section may be that of the plain shank "s", or that of the thread root section of the shank "s_0":

$$\tau_{xy_i} = \frac{Y_i}{s\,(\text{or } s_0)} ; \qquad \tau_{xz_i} = \frac{Z_i}{s\,(\text{or } s_0)}$$

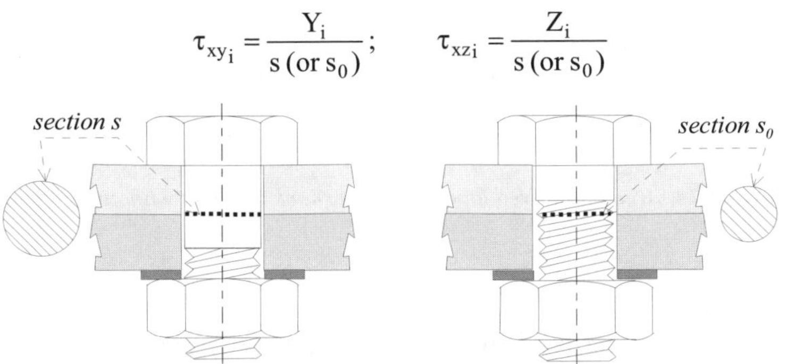

a) section of the smooth plain shank   *b) thread root cross-section of the shank*

Figure 11.23. *Shear of a screw shank of a bolt*

11.2.1.4. *Resistance criteria*

The stresses thus calculated in the fastener shanks are subjected to limits prescribed by the standards and rules. We do not calculate equivalent stress, as Von Mises[22] equivalent stress since the values τ and σ are estimated in *a priori* distinct shank areas (plain shank and thread root section of threaded shank).

For example, standards supply limit values that are not to be exceeded:

– for tensile stress and shear stress when one or the other of the corresponding stresses is present:

$$\sigma \leq 0.6 \times R_r \quad ; \quad \tau \leq 0.8 R_{rg} \qquad [11.5]$$

– when these two stresses exist simultaneously, we must verify a relation of type:

$$\frac{\sigma}{0.84 \times R_r} + \frac{\tau}{0.8 \times R_{rg}} \leq 1$$

where the coefficients are prescribed by the rules[23] obtained from experimental results. The relations described here above are illustrated graphically by the shaded area in Figure 11.24.

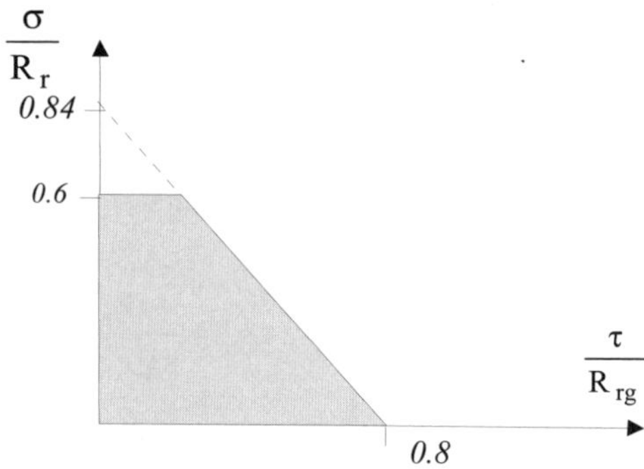

Figure 11.24. *Area of allowed loading on the shank of a bolt without tightening*

NOTE

❐ In addition to the earlier verification, we must also examine the other risks of deterioration mentioned in section 11.1.4.2, and especially verify the bearing condition.

---

22 Von Mises criteria; see Chapter 7, section 7.2.2.3.
23 See bibliography: J. Morel. The standard rules are presented here in a simplified manner. For more detailed information, see the rules of Eurocode 3.

11.2.1.5. *Summary*

Bolted joint of two parts 1 and 2

estimation of forces on "n" fasteners with center (i) and section "s"

data characteristic of the modeled joining interface

geometric center G of the "n" sections: $\displaystyle\sum_{i=1}^{n} y_i = \sum_{i=1}^{n} z_i = 0$;

$\vec{y}$ *and* $\vec{z}$ *principal quadratic axes:* $\displaystyle\sum_{i=1}^{n} y_i \times z_i = 0$

$$I_y = s \times \sum_{i=1}^{n} z_i{}^2 \; ; \;\; I_z = s \times \sum_{i=1}^{n} y_i{}^2 \; ; \;\; I_0 = I_y + I_z$$

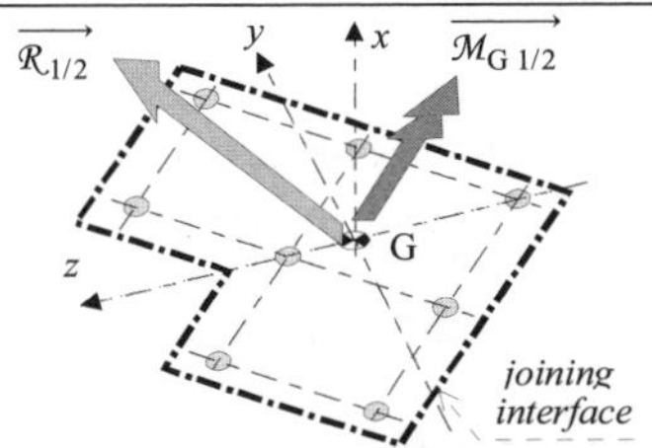

resultant force and moment of the transmittable forces (known because calculated beforehand on a finite element model)

$$\left\{\begin{array}{l} \overrightarrow{\mathcal{R}_{1/2}} = \mathcal{N}\,\vec{x} + \mathcal{T}_y\,\vec{y} + \mathcal{T}_z\,\vec{z} \\[2mm] \overrightarrow{\mathcal{M}_{G1/2}} = \mathcal{M}_x\,\vec{x} + \mathcal{M}f_y\,\vec{y} + \mathcal{M}f_z\,\vec{z} \end{array}\right\}_{G}$$

stresses at center (i) of a fastener section

♦ *normal stress:* $\quad \sigma_{x_i} = \dfrac{\mathcal{N}}{n \times s} + \dfrac{\mathcal{M}f_y}{I_y} \times z_i - \dfrac{\mathcal{M}f_z}{I_z} \times y_i$

♦ *shear stresses:* $\tau_{xy_i} = \dfrac{\mathcal{T}_y}{n \times s} - \dfrac{\mathcal{M}_x}{I_0} \times z_i$; $\tau_{xz_i} = \dfrac{\mathcal{T}_z}{n \times s} + \dfrac{\mathcal{M}_x}{I_0} \times y_i$

corresponding force on the fastener section i

♦ *normal force :*

$$X_i = \dfrac{\mathcal{N}}{n} + \mathcal{M}f_y \times \dfrac{z_i}{\displaystyle\sum_{i=1}^{n} z_i{}^2} - \mathcal{M}f_z \times \dfrac{y_i}{\displaystyle\sum_{i=1}^{n} y_i{}^2}$$

♦ *shear force components:*

$$Y_i = \dfrac{\mathcal{T}_y}{n} - \mathcal{M}_x \times \dfrac{z_i}{\displaystyle\sum_{i=1}^{n}\left(y_i{}^2 + z_i{}^2\right)} \; ; \;\; Z_i = \dfrac{\mathcal{T}_z}{n} + \mathcal{M}_x \times \dfrac{y_i}{\displaystyle\sum_{i=1}^{n}\left(y_i{}^2 + z_i{}^2\right)}$$

♦ *total shear force:* $\quad F_{T_i} = \sqrt{Y_i{}^2 + Z_i{}^2}$

[11.6]

<table>
<tr><td align="center">Case of negligible tightening</td></tr>
<tr><td align="center">stresses in fastener "i", calculated in accordance with the rules</td></tr>
</table>

normal stress:

$$\sigma_i = \frac{X_i}{s_0}$$

(s_0: minor thread root section of the threaded shank)

tangential stress:

$$\tau_i = \frac{F_{Ti}}{s\ (or\ s_0)}$$

(s or s_0: section of shank at the interface of the assembled parts)

<table>
<tr><td align="center">resistance criteria</td></tr>
</table>

$$\text{if } \sigma_i > 0 \text{ (tension)} \qquad \frac{\sigma_i}{R_r} \leq 0.6$$

$$\frac{\sigma_i}{0.84 \times R_r} + \frac{\tau_i}{0.8 \times R_{rg}} \leq 1$$

(σ_i, τ_i) must be inside the so-defined area:

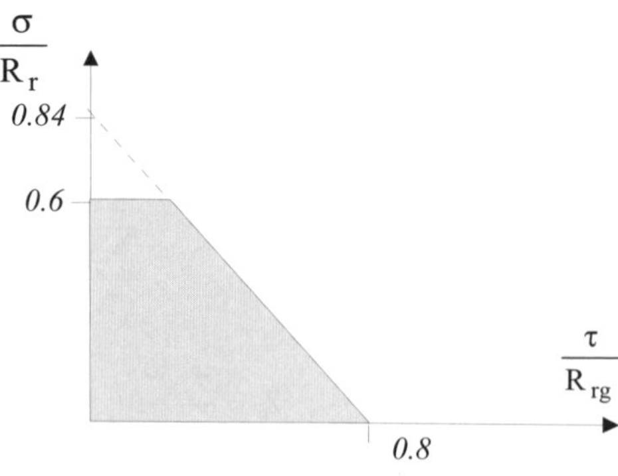

[11.7]

11.2.1.6. *Example*

We consider the bolted joint in Figure 11.25, composed of a bracket clamped on a frame and ensuring its relative position by means of four bolts with tight-fitted plain shanks, with a denoted section "s" in the joining interface.

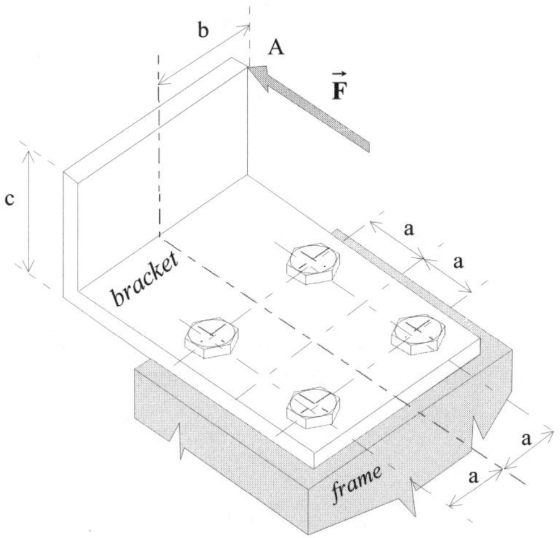

Figure 11.25. *Bolted joint with four bolts*

We neglect the pre-tightening of bolts. A force $\vec{F}$ is exerted at point A, and we will attempt to estimate the forces acting on bolt shanks by following procedure [11.6]. We obtain the modeled joining interface in Figure 11.26, characterized by its principal axes $\vec{y}$ and $\vec{z}$. The transmittable resultant force and moment act on this interface:

$$\left\{ \begin{array}{l} \overrightarrow{\mathcal{R}} = F\vec{y} \\ \overrightarrow{\mathcal{M}_G} = (F\times b)\vec{x} + (F\times c)\vec{z} \end{array} \right\}_G$$

corresponding to: $\mathcal{T}_y = F$; $\mathcal{M}_x = F\times b$; $\mathcal{M}f_z = F\times c$

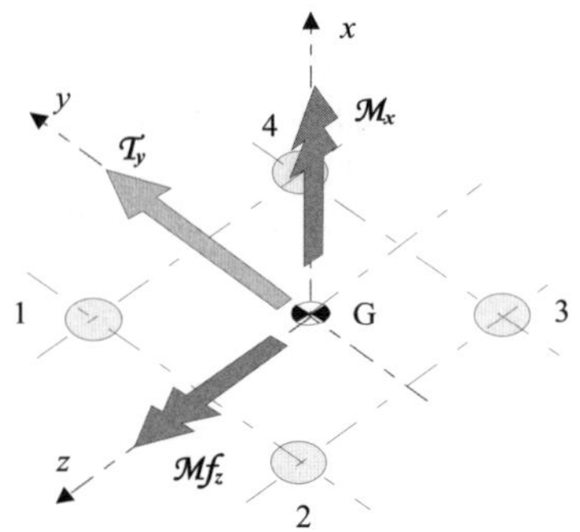

Figure 11.26. *Modeled joining interface*

Figure 11.27 shows the forces exerted on each section, deduced from the three components of the resultant force and moment around the geometric center G, defined earlier.

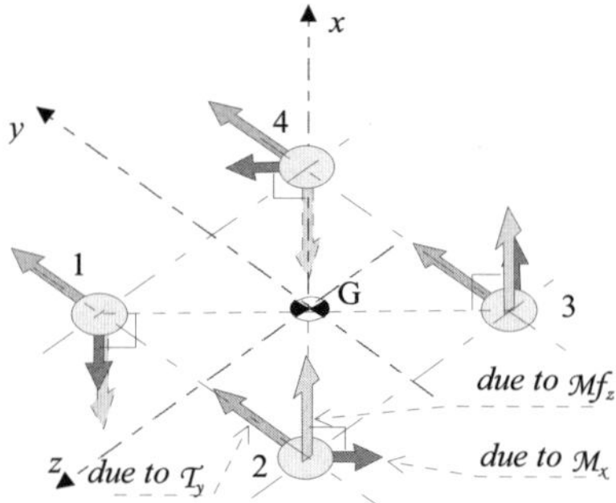

Figure 11.27. *Forces on each shank section*

The characteristic quadratic moments of the modeled joining interface are:
$$I_y = I_z = 4a^2 s \; ; \; I_0 = 8a^2 s \,.$$

Components X_i, Y_i, Z_i of the forces on shanks i, calculated following [11.6], are indicated in the following table.

shank 1	$$X_1 = -F\frac{c \times a}{4a^2} = -\frac{F \times c}{4a}$$ $$Y_1 = \frac{F}{4} - F\frac{b \times a}{8a^2} = \frac{F}{4} - \frac{F \times b}{8a}; \quad Z_1 = \frac{F \times b}{8a}$$
shank 2	$$X_2 = -F\frac{c \times (-a)}{4a^2} = \frac{F \times c}{4a}$$ $$Y_2 = \frac{F}{4} - F\frac{b \times a}{8a^2} = \frac{F}{4} - \frac{F \times b}{8a}; \quad Z_2 = -\frac{F \times b}{8a}$$
shank 3	$$X_3 = -F\frac{c \times (-a)}{4a^2} = \frac{F \times c}{4a}$$ $$Y_3 = \frac{F}{4} - F\frac{b \times (-a)}{8a^2} = \frac{F}{4} + \frac{F \times b}{8a}; \quad Z_3 = -\frac{F \times b}{8a}$$
shank 4	$$X_4 = -F\frac{c \times a}{4a^2} = -\frac{F \times c}{4a}$$ $$Y_4 = \frac{F}{4} - F\frac{b \times (-a)}{8a^2} = \frac{F}{4} + \frac{F \times b}{8a}; \quad Z_4 = \frac{F \times b}{8a}$$

We note a maximum "tensile + shear" force on shank 3, with the components:

$$X_3 = \frac{F \times c}{4a}; \quad F_{T3} = \sqrt{Y_3{}^2 + Z_3{}^2} = \frac{F}{4}\sqrt{1 + \frac{b}{a} + \frac{b^2}{2a^2}} \qquad [11.8]$$

Figure 11.28 shows a graphic representation of the forces applied on each bolt with components X_i, Y_i, Z_i.

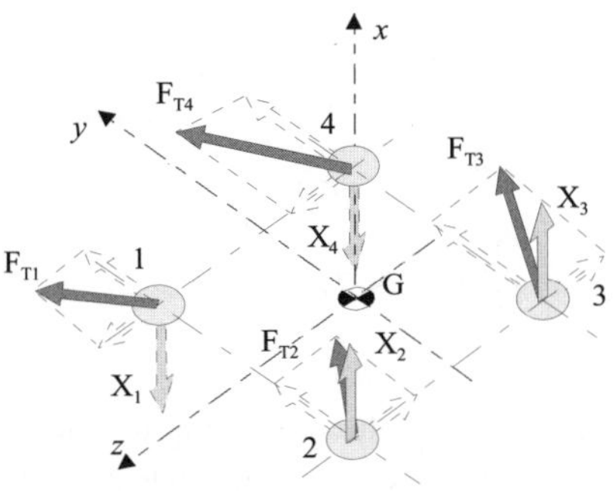

Figure 11.28. *Forces acting on each bolt*

Next we must verify the resistance criteria [11.7] with the maximum force obtained.

11.2.2. *Case of pre-tightening*

The advantage of pre-tightening has been briefly underlined in section 11.1.3. We will now describe some characteristics of pre-tightening with the sole objective of the simple verification of dimensioning. The detailed study of the various categories of applications involving bolts with pre-tightening is long and complex. This type of study can be found in specialized books[24].

11.2.2.1. *Tightening torque*

When we tighten a bolt by applying, on the rotating nut, a torque of $\overrightarrow{\mathcal{M}_0} = L_0 \vec{x}$ with a wrench, a part of this moment (roughly 50%) is picked up by the threads of the screw shank (creation of an axial force and friction on the threads) and the other part is picked up by friction under the bearing face of the rotating nut that is thus tightened (see Figure 11.29). Without going into the details of relations that can be written[25], we can in this way, generate an axial force $\overrightarrow{F_0} = X_0 \vec{x}$ in the screw shank of diameter "d" that can roughly be estimated[26].

$$X_0 \cong \frac{L_0}{0.2 \times d}$$

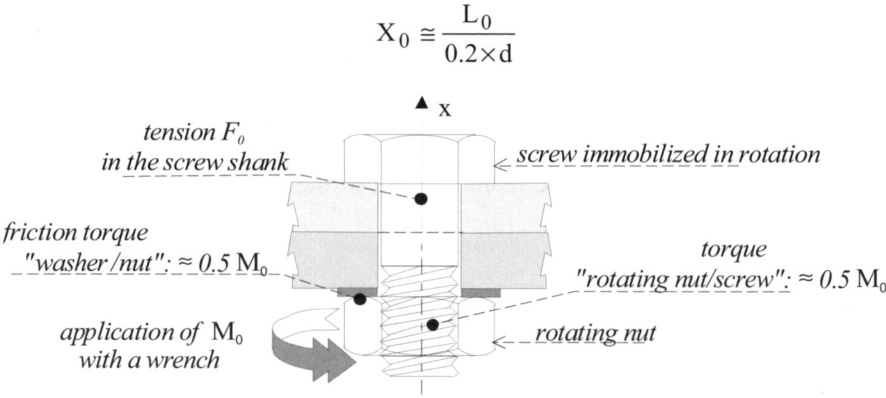

Figure 11.29. *Tightening of the nut*

The part (approximately 50%) of the tightening torque L_0 that is picked up by the threads of the screw shank constitutes a torque that generates torsion shear stresses in the shank. During the tightening of the rotating nut, the following will appear in the threaded area of the shank that is the most vulnerable (the minor thread root

---

24 See bibliography: J. Guillot.
25 These relations integrate the geometry of the screw, of the bearing surface of the rotating nut as well as the friction coefficients of the "rotating nut/screw" and "nut/washer".
26 See bibliography: M. Drouin.

cross-section is written as: $s_0 = \pi \times r_0{}^2 < s$ where s is the section of the plain shank):

– a normal stress due to the tightening that we must, conforming to rules and standards, limit to $0.7 \times R_r$[27]:

$$\sigma = \frac{X_0}{s_0} \leq 0.7 \times R_r \qquad [11.9]$$

- a maximum shear stress[28]:

$$\tau_0 = \frac{L_0}{2} \times \frac{r_0}{i_0} \qquad \text{with } i_0 = \frac{\pi r_0{}^4}{2}$$

During tightening, we must therefore consider the equivalent Von Mises[29] stress. Nevertheless, we can observe during the life-time of the assembly that the shear stress almost disappears due to an attenuation phenomenon (this causes a loss of tensile stress estimated at 20% of the initial stress).

11.2.2.2. *Behavior of a bolted joint with pre-tightening*

The axial force of pre-tightening written as X_0 should be "sufficient" so as to ensure the transmission through the joining of forces called "working forces".

As we have already mentioned in section 11.1.3, the main objective of pre-tightening is to create contact pressure between parts that have been assembled, so as to:

– prevent the separation of the assembled parts;

– change the alternating fatigue strain on the joint in a "waving" strain in screw shanks, which is better for the fatigue behavior[30];

– due to the friction coefficient, create a resistance to slipping between the assembled parts that prevents the screw shank from being sheared, as is illustrated in Figure 11.30.

---

27 See bibliography: J. Morel.
28 Distribution of shear stresses on a circular section: see Chapter 9, section 9.3.2.3 and Figures 9.21 and 11.19a.
29 See section 7.2.2.3.
30 We will not go into the details of this aspect of the behavior of a bolted joint that is nevertheless very important.

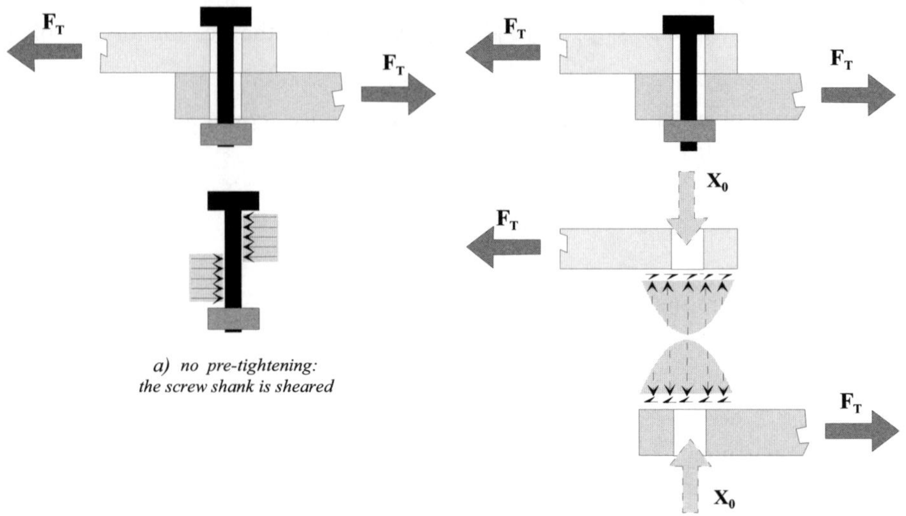

a) no pre-tightening:
the screw shank is sheared

b) pre-tightening: tangential forces (friction) must be sufficient
so as to balance external shearing force FT.
Then the screw shank is not sheared any more

Figure 11.30. *Pre-tightening of a bolt*

○   "Tangential" working force (perpendicular to the axis of the bolt)

The working tangential force, transmittable by the bolt, being noted as F_T, we must obtain the following in order to respect the rules[31]:

$$X_0 \geq 1.25 \frac{F_T}{f}$$

where f is the coefficient of mutual friction of the surfaces of the two parts in contact.

○   Tensile working force (following the axis of the bolt)

If the bolt is strained by an external tension force, written as X in Figure 11.31, the contact pressures generated by pre-tightening decreases. In order to conserve the contact between the assembled parts, and avoid the canceling of contact pressures, we should respect[32].

---

31 See bibliography: J. Morel.

32 The value 0.8X constitutes a minimum value. Particularly, we should anticipate an increase in this value so as to take into account the attenuation phenomenon mentioned in section 11.2.2.1 (certain books recommend an additional stress going up to 100 MPa in the root section s_0).

$$X_0 \geq 0.8X$$

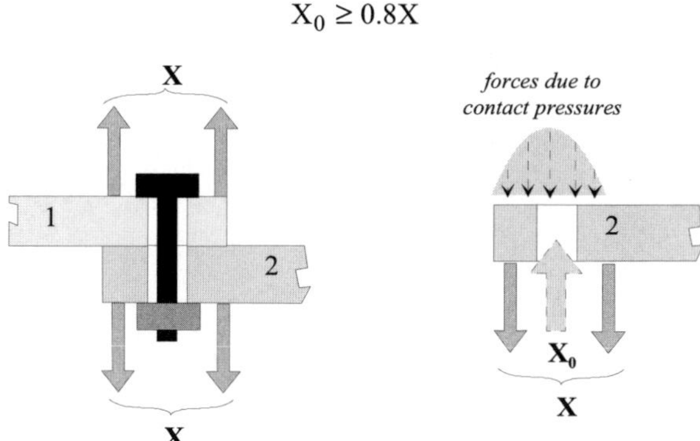

Figure 11.31. *The external tensile force X makes decrease contact pressures*

NOTE

❏ When the assembly is simultaneously subjected to a working tensile force X and a working tangential force F_T, the tangential contact forces due to the friction coefficient must be sufficient to resist the relative slipping of the plates. Thus the guiding values give[33]:

$$X_0 \geq 1.25\frac{F_T}{f} + 0.8X \qquad [11.10]$$

○ Working bending moment on the bolted joint

Example: let us consider in Figure 11.32, part 1, fixed on part 2 by two bolts with pre-tightening.

During tightening, an internal state of equilibrium in the assembly is achieved with the creation of a contact pressure shown in Figure 11.32 by a uniform value p_0[34] such that:

$$p_0 = \frac{X_0}{b \times \ell}$$

where X_0 is the tension due to pre-tightening of one bolt shank.

---

33 See bibliography: J. Morel. It would be better to anticipate an increase so as to take into account the phenomenon of attenuation already quoted.

34 This is only an approximative distribution. We can imagine indeed that with a base that is not very thick, and therefore more flexible while bending, the contact pressures will be exerted mainly in the vicinity of the bolts.

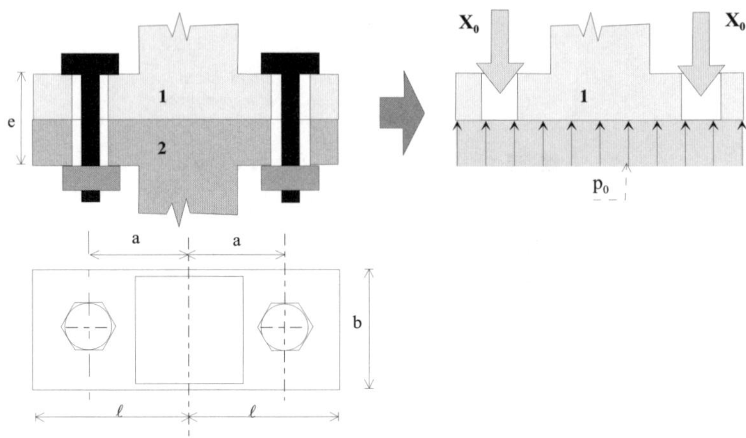

Figure 11.32. *Bolted joint with two pre-tightened bolts*

We can now subject the bolted joint to a moment $M_{ext}\ \vec{y}$ acting on the upper part, written as 1 (Figure 11.33). We assume that the application of this moment will not result in any separation in the parts involved in the joining:

– no break in contact between the two parts at the joining interface;

– no break in contact of screw heads or nuts with the assembled parts;

Nevertheless, the moment $M_{ext}\ \vec{y}$ modifies the contact pressures between the two parts as well as the tension of the screw shanks. In Figure 11.33 we have only shown the *variations* in forces ΔX_0 and in pressure Δp_0 created by the application of this moment, variations that must be added to the initial values (X_0, p_0) of Figure 11.32, not shown here.

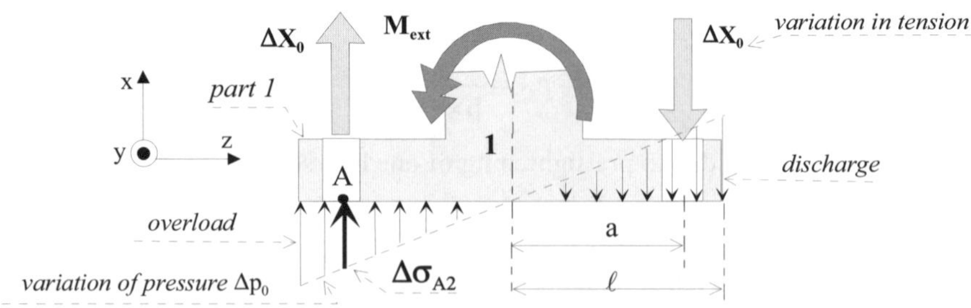

Figure 11.33. *Application of a moment M_{ext}*

The variation of force ΔX_0 causes a moment noted as:

$$\overrightarrow{\mathcal{M}_{bolt/1}} = M_{bolt}\,\vec{y}\,.$$

The variation of pressure Δp_0 causes a moment denoted as: $\overrightarrow{\mathcal{M}_{2/1}} = M_2\,\vec{y}\,.$

The equation of equilibrium of the moments projected on $\vec{y}$ axis for part 1 is noted as:

$$M_{ext} + M_2 + M_{bolt} = 0 \qquad\qquad [11.11]$$

Let us write that at point A, we have the same strain ε_x in the screw-shank and in the assembled parts that have practically the same dimension "e" in Figure 11.32[35]:

$$\varepsilon_x = \frac{\Delta\sigma_{A2}}{E_2} \cong \frac{\Delta\sigma_{bolt}}{E_{bolt}}$$

where $\Delta\sigma_{A2}$ represents the variation of stress in A in part 2 and $\Delta\sigma_{bolt}$ the variation in normal stress in the shank, both negative. If we assume that the same material has been used in part 2 and the bolt, we have at point A:

$$\Delta\sigma_{A2} = \Delta\sigma_{bolt} \qquad\qquad [11.12]$$

In addition, the linear pressure distribution as in Figure 11.33 appears to be identical to the distribution of normal bending stresses on a beam cross-section under the effect of a bending moment. We know that[36] in this case, the calculation leads to:

$$\Delta\sigma_{A2} = \frac{M_2}{\mathcal{J}_y}\times a$$

where $\mathcal{J}_y$ is the quadratic moment of the rectangular contact surface pierced with two holes of surface "s" (Figures 11.32 and 11.34):

$$\mathcal{J}_y = \frac{b\times(2\ell)^3}{12} - 2a^2\times s$$

---

35 We observe that at point A, strain ε_x corresponds to a compression. We thus have $\varepsilon_x < 0$.

36 See Chapter 9, [9.36].

In addition, we have (Figure 11.33):

$$M_{bolts} = -\Delta X_0 \times 2a = \Delta\sigma_{bolt} \times s \times 2a \quad \Rightarrow \quad \Delta\sigma_{bolt} = \frac{M_{bolts}}{2a \times s} \qquad [11.13]$$

and [11.12] is written as:

$$\frac{M_2}{J_y} \times a = \frac{M_{bolts}}{2a \times s} \quad \Rightarrow \quad \frac{M_2}{M_{bolts}} = \frac{J_y}{2a^2 \times s} \qquad [11.14]$$

We thus obtain with [11.11]:

$$M_{ext} = -M_{bolts}\left(1 + \frac{J_y}{2a^2 \times s}\right) \quad \Rightarrow \quad \frac{M_{bolts}}{M_{ext}} = \frac{-1}{1 + \dfrac{J_y}{2a^2 \times s}} \qquad [11.15]$$

or with [11.13]:

$$M_{ext} = \Delta X_0 \times 2a\left(1 + \frac{J_y}{2a^2 \times s}\right) \quad \Rightarrow \quad \Delta X_0 = \frac{M_{ext}}{2a\left(1 + \dfrac{J_y}{2a^2 \times s}\right)} \qquad [11.16]$$

The total force in the shank of the tightest bolt (the one to the right in Figure 11.33) is thus written as:

$$X_0' = X_0 + \Delta X_0 = X_0 + \frac{M_{ext}}{2a\left(1 + \dfrac{J_y}{2a^2 \times s}\right)} \qquad [11.17]$$

and the criteria of resistance [11.9] is written as:

$$X_0' = X_0 + \frac{M_{ext}}{2a\left(1 + \dfrac{J_y}{2a^2 \times s}\right)} \leq 0.7 \times R_r \times s \qquad [11.18]$$

NOTES

❐ The ratio $\dfrac{J_y}{2a^2 \times s}{}_4$ that appears in [11.15] or [11.16] has as numerator, the quadratic moment of the surface of contact and as denominator, the quadratic moment of two circular sections, as shown in Figure 11.34 where we can observe the difference in the magnitude of these two quadratic moments. We will thus obtain:

$$1 + \frac{\mathcal{J}_y}{2a^2 \times s} = 1 + \frac{\left(\dfrac{b \times (2\ell)^3}{12} - 2a^2 \times s\right)}{2a^2 \times s} = \frac{1}{3}\frac{b \times \ell^3}{a^2 \times s} \gg 1 \quad [37]$$

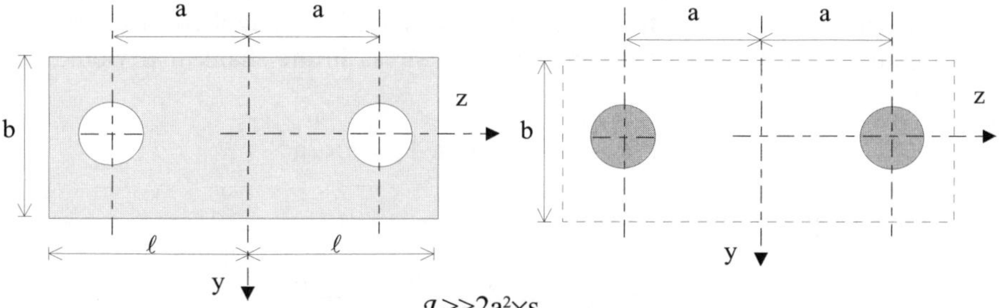

a) surface of contact   b) surfaces of bored holes

Figure 11.34. *Joining interface*

Expression [11.15] shows that the bolts take up only a very small part of moment M_{ext}, as:

$$-\frac{M_{bolts}}{M_{ext}} = \delta \ll 1 \quad [38]$$

This is also observed in [11.17] where the total force on the most loaded bolt is written as:

$$X_0' = X_0 + \delta \times \frac{M_{ext}}{2a} \qquad [11.19]$$

Thus, if we implement a pre-tightening such that:

---

37 For example, with a bolt shank of diameter 6 mm, and b = 18 mm, ℓ = 30 mm, a = 15mm,

we get: $1 + \dfrac{\mathcal{J}_y}{2a^2 \times s} = 25$.

38 With the previous example: $\delta = \dfrac{1}{25} = 0.04$.

$$X_0 = \frac{M_{ext}}{2a} \qquad [11.20]$$

we will observe, by applying the external moment M_{ext}, that the initial tension in the shank of the bolt varies little, since relation [11.19] then indicates:

$$X_0' = X_0(1+\delta) \cong X_0$$

❐ Associated with this example, we can examine the force that the model with negligible tightening, as in section 11.2.1, would induce in the same shank for the corresponding applied moment M_{ext}. With method [11.6] applied to the interface of Figure 11.34, we would obtain a normal stress in the shank that would be as follows:

$$\sigma_x = \frac{M_{ext}}{I_y} \times a \ \text{ with } I_y = s \times 2a^2$$

corresponding to a force:

$$X = \sigma_x \times s = M_{ext}\frac{a}{2a^2} = \frac{M_{ext}}{2a}$$

Let us suppose that a working force of the same value acts on the present pre-tightened joint. We have seen at the beginning of this section that under the influence of such an external tensile force, the preload in the shank should correspond to:

$$X_0 \geq 0.8X$$

In this case it should correspond to the minimum:

$$X_0 = 0.8X = 0.8\frac{M_{ext}}{2a} \qquad [11.21]$$

even for this minimum value, relation [11.19] also indicates the following:

$$X_0' = X_0(1+1.25\times\delta) \cong X_0$$

We can thus retain as a preliminary value of tightening, the value deduced from the procedure [11.6][39].

❐ We can also compare in the same example, the resulting force in the shanks that would be observed in the case of a real tilting of part 1, and without pre-tightening. This tilting was illustrated in Figure 11.11a. We are repeating it in Figure 11.35 hereafter with the notes of our example.

---

39 With an increase for the attenuation phenomenon (see the end of section 11.2.2.1).

We observe a force in the shanks:

$$X' = \frac{M_{ext}}{a + \ell} < \frac{M_{ext}}{2a} \;\Rightarrow\; X' < X_0$$

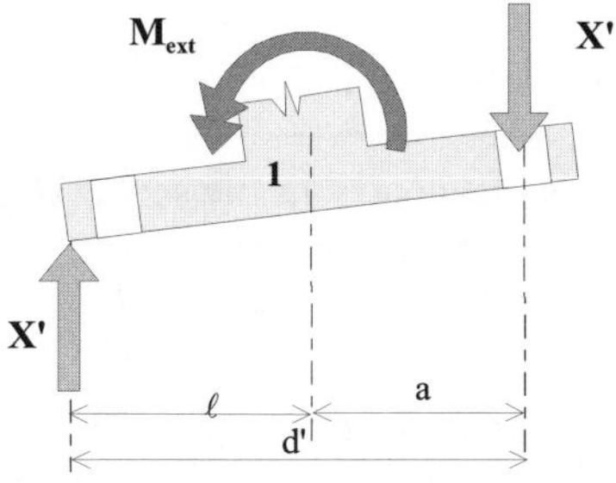

Figure 11.35

Using procedure [11.6] to estimate the forces on a fastener thus provides interesting pessimistic values to fix the initial tension values in the shanks.

11.2.2.3. *Summary*

We can illustrate as follows a procedure for dimensioning a pre-tightened bolt. Following a chronological order:

♦ we determine on the screw shank, the working forces, normal X and tangential (Y, Z, F_T), by using for example procedure [11.6];

♦ we deduce from these values, with the help of standards, the initial tensile force in the shank to transmit the working forces;

♦ with the help of standards, we verify the strength criteria for the shank thus stretched.

<table>
<tr><td align="center">Dimensioning of a pre-tightened bolt</td></tr>
</table>

estimation of the working forces on a fastener: using procedure[11.6] leads to:

    – a working tensile force: X

    – a working tangential force: $F_T = \sqrt{Y^2 + Z^2}$

<table>
<tr><td align="center">initial tensile force necessary in the screw shank (see [11.10])</td></tr>
</table>

$$X_0 \geq 0.8\,X + 1.25\frac{F_T}{f}$$

(f: coefficient of mutual friction of the surfaces in contact)

<table>
<tr><td align="center">tightening torque necessary to create this tension (approximate value)</td></tr>
</table>

$$L_0 \cong 0.2 \times X_0 \times d \qquad \text{(d nominal diameter of the shank)}$$

<table>
<tr><td align="center">initial stresses in the screw shank when tightening</td></tr>
</table>

♦*normal stress:*

$$\sigma = \frac{X_0}{s_0} \quad (s_0 = \pi\, r_0^{\,2} \text{ minor thread root cross-section)}$$

♦*maximum shear stress (approximate value):*

$$\tau = \frac{L_0}{2} \times \frac{r_0}{i_0} \quad (i_0 = \frac{\pi\, r_0^{\,4}}{2})$$

♦*equivalent normal stress of Von Mises:*

$$\sigma_{eq.\text{V.Mises}} = \sqrt{\sigma^2 + 3\tau^2}$$

<table>
<tr><td align="center">criteria of resistance</td></tr>
</table>

we must verify:

$$\Rightarrow \sigma \leq 0.7\, R_r$$

$$\Rightarrow \sigma_{eq.\text{V.Mises}} \leq R_e$$

[11.22]

11.2.2.4. *Example*

In Figure 11.36, the bolted joint of example 11.2.1.6 is again considered, but from now on with pre-tightening.

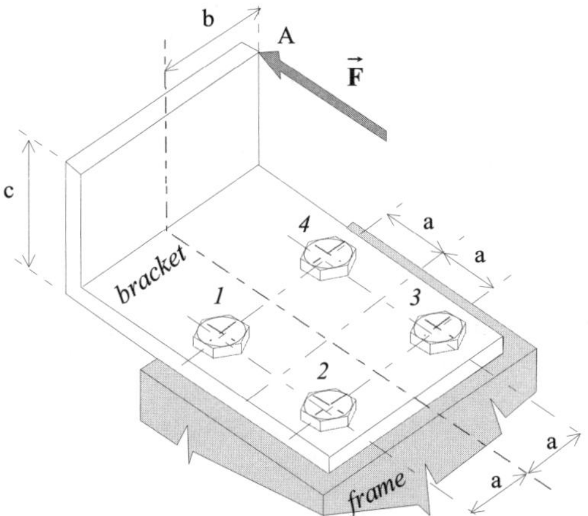

Figure 11.36. *Bolted joint with pre-tightened bolts*

We must first estimate the initial tension in the shank necessary to support force $\vec{F}$. We have already calculated in section 11.2.1.6 the forces on the four shanks of the bolts in the case of no tightening. We have observed that shank 3 was the most strained with:

– a working tensile force: $X_3 = \dfrac{F \times c}{4a}$

– a working tangential force: $F_{T3} = \dfrac{F}{4}\sqrt{1 + \dfrac{b}{a} + \dfrac{b^2}{2a^2}}$

In the present case of pre-tightening, the minimum initial tensile force necessary in the shank of this bolt according to [11.22] is:

$$X_0 = \frac{F}{4}\left(0.8\,\frac{c}{a} + \frac{1.25}{f}\sqrt{1 + \frac{b}{a} + \frac{b^2}{2a^2}}\right)$$

where f is the coefficient of interface friction between the two assembled parts. This pre-tightening induces the following stresses in the shank:

– a normal stress: $\sigma = \dfrac{X_0}{s_0}$ where s_0 is the thread root cross-section of the screw;

– a maximum shear stress: $\tau = \dfrac{L_0}{2} \times \dfrac{r_0}{i_0} \cong 0.4 \dfrac{X_0}{s_0}$.

The corresponding equivalent Von Mises stress is written as:

$$\sigma_{eq.V.Mises} = \sqrt{\sigma^2 + 3\tau^2} = 1.22 \frac{X_0}{s_0}$$

and we must verify:

$$\sigma = \frac{X_0}{s_0} \leq 0.7\, R_r\,; \qquad \sigma_{eq.V.Mises} = 1.22 \frac{X_0}{s_0} \leq R_e$$

11.3. Riveted joint

11.3.1. *Hypotheses*

As we said at the beginning of this chapter[40], modern rivets are used mainly to prevent the relative slipping of the assembled parts (sheets for example). Contrary to bolts, they are not meant to absorb strong forces perpendicular to the contact interface, called "out of plane" forces, that tend to separate the parts. The hypothesis used for bolted joints in the earlier section[41] can be summarized as follows:

– the shanks are smooth and tight fitted in the vicinity of the interface;

– the shanks work in shear.

11.3.2. *Characteristics of the modeled joining interface*

The joint interface between assembled parts (for example, two sheets denoted as 1 and 2 in Figure 11.37a) is reduced to "n" identical circular working sections "s", as in the case of a bolted assembly. The two parts are assumed to be more rigid than the rivets. We define an equivalent interface working section in the same manner. It is shown in Figure 11.37b with its barycenter G described by relations [11.1].

---

40 See section 11.1.2 and Figure 11.3.
41 See section 11.2.1.1.

11.3.3. *Forces on each attachment*

The rivets do not support forces outside the plane of the jointing interface. Thus, as already seen in section 11.1.2.1 and Figure 11.2, the resultant force and moment of forces transmittable by the riveting are characterized by a resultant $\overrightarrow{\mathcal{R}_{1/2}}$ contained entirely in the plane of the interface, and a moment $\overrightarrow{\mathcal{M}_{G1/2}}$, perpendicular to this same interface plane. This is again shown in Figure 11.37. This means that the normal force and the bending moments, which created normal stresses in the interface plane of a bolted joint, disappear. It is therefore no longer necessary to define the principal axes and quadratic moments. The result is that the orthogonal axes $\vec{y}$ and $\vec{z}$ in the interface plane could have any direction[42].

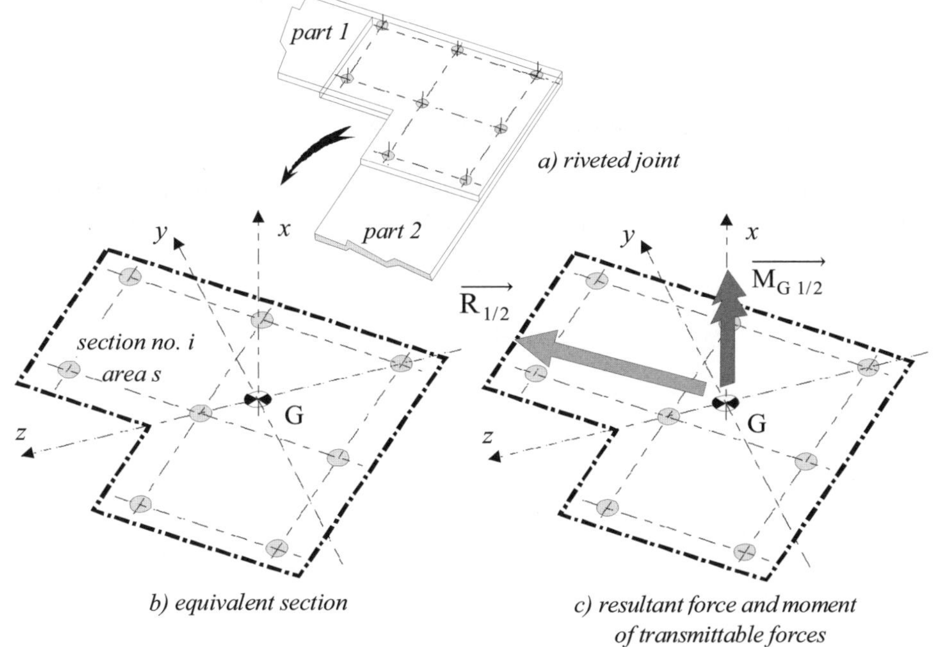

Figure 11.37. *Interface of a riveted joint*

---

42 Nevertheless, we have left the principal axes of the earlier section in the figures so as to retain homogenity in the presentation (we therefore base ourselves on the standard projections of resultant forces and moments of the transmittable forces).

The sections "s" therefore are sheared because of:

– the shear resultant force T_y as in Figure 11.16 with the same result, i.e.:

$$\tau_{xy_{T_y}} = \frac{T_y}{n \times s}$$

– the shear resultant T_z as in Figure 11.18 with the same result, i.e.:

$$\tau_{xz_{T_z}} = \frac{T_z}{n \times s}$$

– the longitudinal moment $\mathcal{M}_x$ as in Figure 11.19 with the same results, i.e.:

$$\tau_{xy_{\mathcal{M}_x i}} = -\frac{\mathcal{M}_x}{I_0} \times z_i \; ; \qquad\qquad \tau_{xz_{\mathcal{M}_x i}} = \frac{\mathcal{M}_x}{I_0} \times y_i$$

11.3.4. *Graphic representation of the shear stresses*

We will thus find shear stresses that have already been described in the earlier section and that are summarized in Figure 11.38.

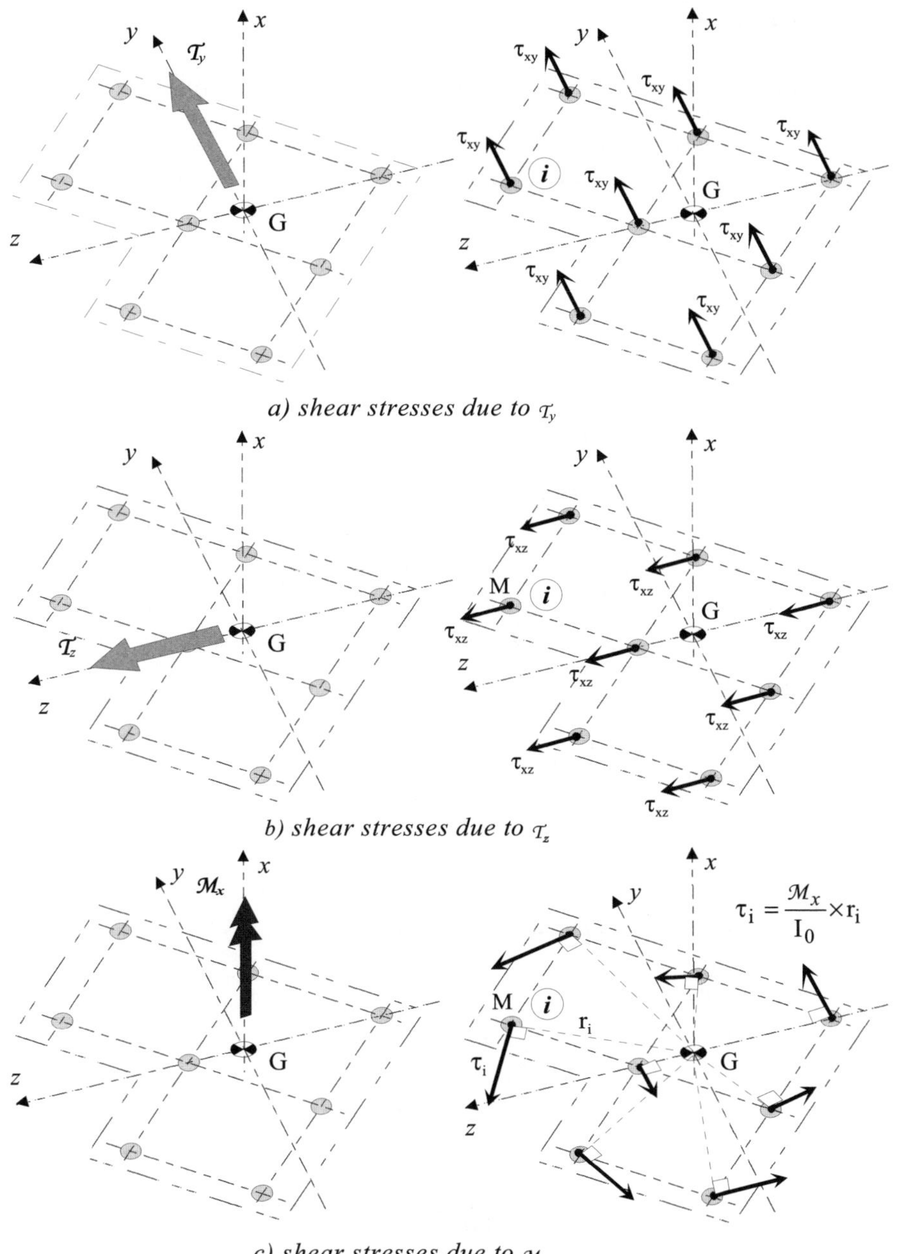

Figure 11.38. *Shear stresses on rivets*

11.3.5. *Summary*

The behavior of the riveted assembly may be summarized as follows.

<table>
<tr><td colspan="1" align="center">Riveted joint fastening together parts 1 and 2</td></tr>
</table>

Riveted joint fastening together parts 1 and 2

forces on "n" fasteners of center (i), each one with shank section "s"

G geometric center of "n" sections:
$$\sum_{i=1}^{n} y_i = \sum_{i=1}^{n} z_i = 0 \; ;$$

polar quadratic moment:
$$I_0 = s \times \sum_{i=1}^{n} (y_i + z_i)^2$$

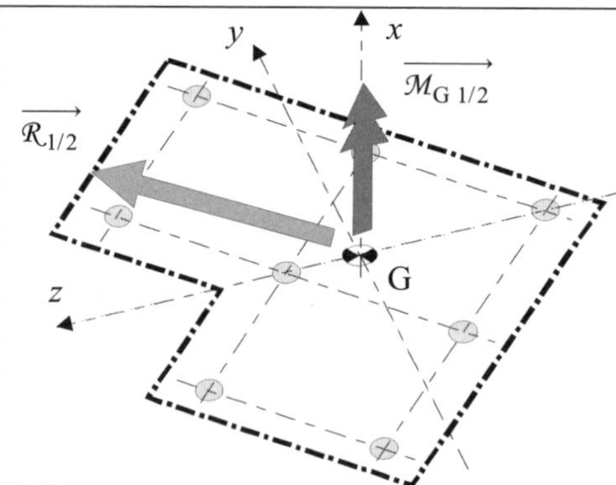

resultant force and moment of transmittable forces (already known; this is data)

$$\left\{ \begin{aligned} \overrightarrow{\mathcal{R}_{1/2}} &= \mathcal{T}_y \,\vec{y} + \mathcal{T}_z \,\vec{z} \\ \overrightarrow{\mathcal{M}_{G1/2}} &= \mathcal{M}_x \,\vec{x} \end{aligned} \right\}_G$$

shear stress on a section of center i:

components:
$$\tau_{xy_i} = \frac{\mathcal{T}_y}{n \times s} - \frac{\mathcal{M}_x}{I_0} \times z_i \; ; \quad \tau_{xz_i} = \frac{\mathcal{T}_z}{n \times s} + \frac{\mathcal{M}_x}{I_0} \times y_i$$

modulus:
$$\tau_i = \sqrt{\tau_{xy_i}{}^2 + \tau_{xz_i}{}^2}$$

corresponding shear force on section i:

components:
$$Y_i = \frac{\mathcal{T}_y}{n} - \mathcal{M}_x \times \frac{z_i}{\sum_{i=1}^{n}\left(y_i{}^2 + z_i{}^2\right)} \; ; \quad Z_i = \frac{\mathcal{T}_z}{n} + \mathcal{M}_x \times \frac{y_i}{\sum_{i=1}^{n}\left(y_i{}^2 + z_i{}^2\right)}$$

modulus:
$$F_{T_i} = \sqrt{Y_i{}^2 + Z_i{}^2}$$

criterion of resistance

$$\tau_i \leq \frac{R_{rg}}{C_s} \qquad C_s: \text{ safety factor}$$

[11.23]

11.4. *Welded joints*

11.4.1. *Preliminary observations and hypotheses*

As in the case of bolted and riveted joints, we shall not go into a detailed study of welding. We shall only provide indications regarding the dimensioning. For the earlier assemblies, a prior calculation on the modeled structure (finite element analysis) provided the resultant force and moment of transmittable forces through the joint. The same will hold true for a welded joint. This joint is assumed to already be geometrically defined and consists of a certain number of weld beads.

11.4.1.1. *State of stresses in a weld bead*

The reality of the state of stresses in the weld beads is obviously very complex, not only because of its geometry but also because of what has happened during welding (localized overheating, structural modifications of the welded alloys). We "reduce" these states of stresses in the cross-section of the bead height[43] as indicated in Figure 11.39.

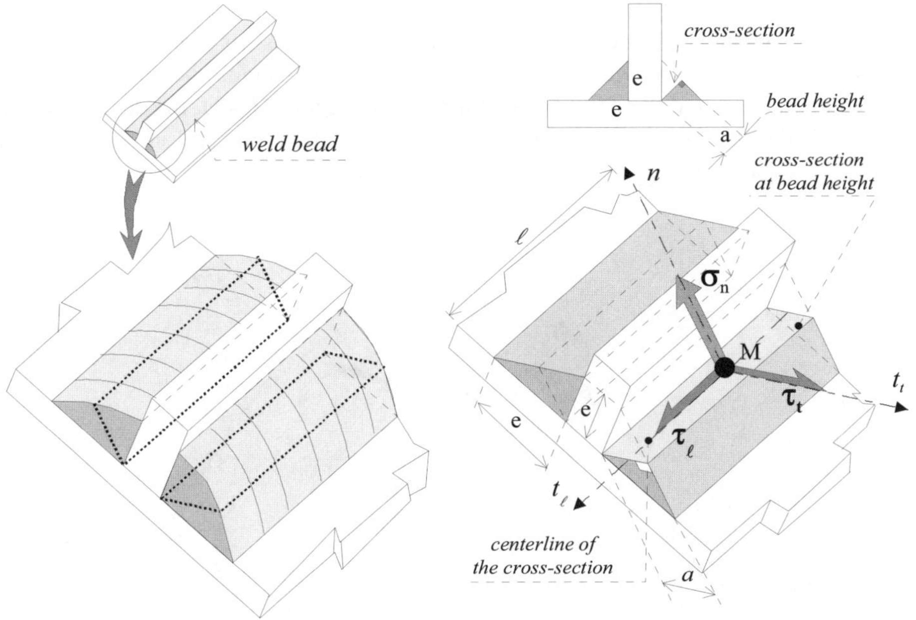

Figure 11.39. *Reduction of the state of stress in a weld bead*

43 Experiments show that when the weld bead is of the same nature as the assembled parts, the failure occurs in the bead height cross-section.

The bead height is defined by:

$$a = e\frac{\sqrt{2}}{2}$$

[11.24]

We also define the "centerline" of the cross-section of the weld bead at bead height "a".

This section is equipped with a local coordinate system denoted as $\vec{n}$, $\vec{t_\ell}$, $\vec{t_t}$. In this cross-section, we assume that the state of stresses everywhere is identical to the state of stresses on the centerline. On any point M of this centerline, we therefore note:

– a normal stress σ_n ;

– a tangential or shear stress characterized by its components τ_ℓ (longitudinal component) and τ_t (transverse component).

11.4.1.2. *Dimensioning criterion*

We immediately observe that such a state of stress is very approximate. The stresses can only be average. In fact with this model, we should be able to observe on a small area around the point M, as illustrated in Figure 11.40a, the distribution in Figure 11.40b. However, in reality, we can see that the reciprocal stresses τ_ℓ and τ_t do not exist on the vertical free faces of the bead.

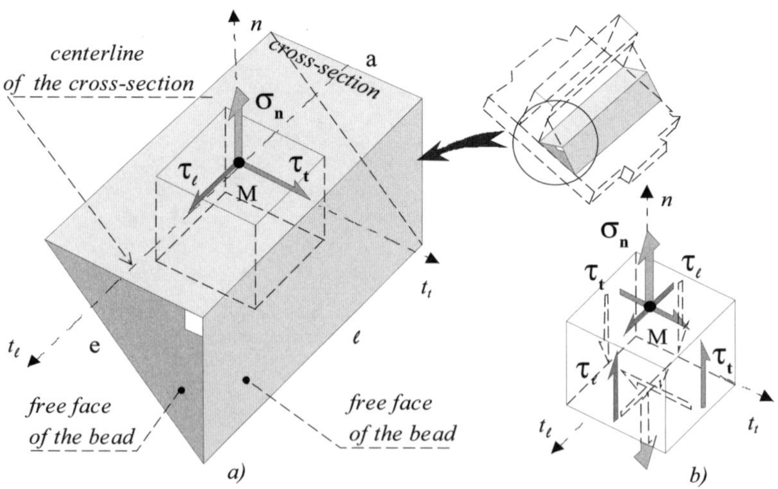

Figure 11.40. *The state of stresses in the bead can only be approximate*

Nevertheless, we define this state of average stress σ_n, τ_ℓ and τ_t and we use it in a regulation context to estimate an equivalent stress of the Von Mises type, here[44]:

$$\sigma_{eq.V.Mises} = \sqrt{\sigma_n{}^2 + 3\left(\tau_\ell{}^2 + \tau_t{}^2\right)}$$

The verification of the weld bead will consist of respecting a criterion of the form[45]:

$$\sigma_{eq.V.Mises} \leq \frac{R_{r\ bead}}{C_s}$$

where C_s is the safety factor prescribed by the regulations[46]:

$$\sqrt{\sigma_n{}^2 + 3\left(\tau_\ell{}^2 + \tau_t{}^2\right)} \leq \frac{R_{r\ bead}}{C_s} \qquad [11.25]$$

11.4.2. *Determination of the stresses in the weld bead cross-section*

11.4.2.1. *Statutory aspect*

There are several approaches that may be adopted, all approximate, differing for example from one coast to the other of the Atlantic. The European standards (Eurocode 3) specify that stresses σ_n, τ_ℓ and τ_t must be deducted, in a very simple manner, from force $\vec{F}$ acting in the middle of cross-section $a \times \ell$ of the weld bead in question. It is therefore sufficient to write the ratio $\dfrac{\vec{F}}{a \times \ell}$ that is an average stress, then to project this stress on the local axes of the bead cross-section $\vec{n}$, $\vec{t_t}$, $\vec{t_\ell}$ (Figure 11.41).

---

44 See Chapter 7, section 7.2.2.3.

45 Here we observe that the resistance criterion is not written in the usual manner, where $\sigma_{eq.V.Mises}$ marks the passage to the elasticity limit. In fact, the considerable simplification (approximation) of the state of stresses results in the use of the criterion by extension until the rupture limit, as though the bead material were fragile (see Figure 7.3a).

46 Thus, with Eurocode 3 we have for a weld bead made of steel and following the steel grades: $1 \leq C_s \leq 1.2$, ($C_s = 1$ for mild steel).

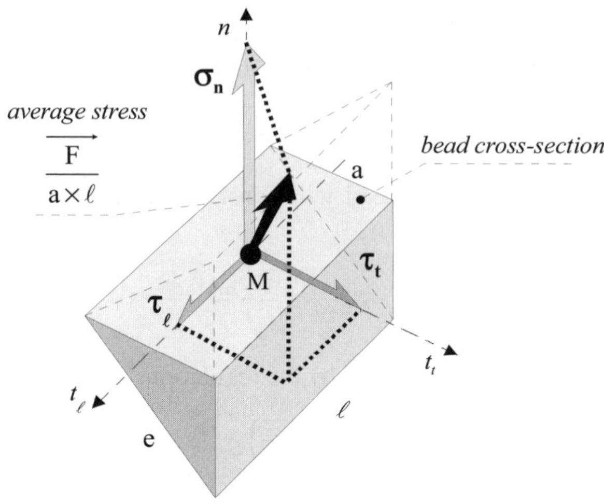

Figure 11.41. *Calculation of* σ_n, τ_ℓ *and* τ_t *from* $\vec{F}$

This projection leading to the estimation of σ_n, τ_ℓ and τ_t is not difficult when the average stress is known. The problem instead consists of the estimation of the average stress acting in the middle of the cross-section of each weld bead. When the welded joint, as well as the loading, are simple, such a stress can be easily determined, as shown in the following examples.

♦ *Example 1*: Figure 11.42 represents an angle weld bead loaded with force $\vec{F} = X\vec{x}$.

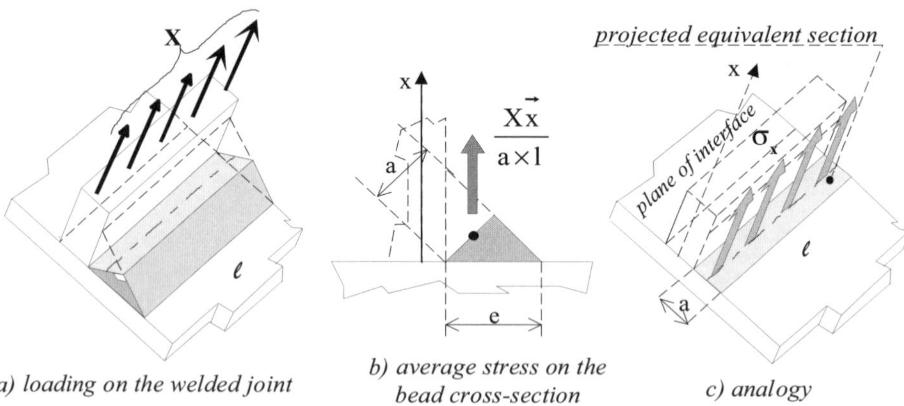

a) loading on the welded joint

b) average stress on the bead cross-section

c) analogy

Figure 11.42. *Definition of a projected section for the estimation of an average normal stress*

On the centerline of the bead cross-section of area $a \times \ell$, we can observe an average stress[47]:

$$\frac{\overrightarrow{Xx}}{a \times \ell}$$

We can note that we would obtain the same result from Figure 11.42c , where we have "flattened" the bead cross-section against the plane of interface, on which we could obtain a normal average stress due to $\overrightarrow{F} = \overrightarrow{Xx}$: $\sigma_x = \dfrac{X}{a \times \ell}$.

We must next project σ_x on the local axes $\overrightarrow{n}$, $\overrightarrow{t_t}$, $\overrightarrow{t_\ell}$ of the bead cross-section, which does not pose a problem. We can see the following in Figure 11.43.

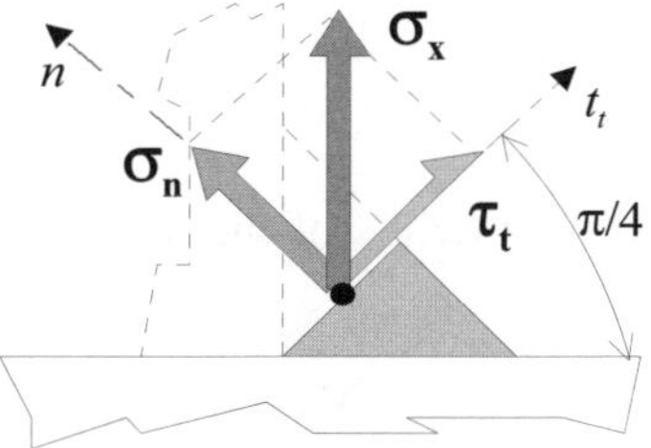

Figure 11.43. *Average stresses in local axes of bead cross-section*

$$\sigma_n = \tau_t = \frac{X}{a \times \ell}\frac{\sqrt{2}}{2} ; \quad \tau_\ell = 0 .$$

The criterion for non-rupture [11.25] is written as:

$$\sqrt{\sigma_n{}^2 + 3\left(\tau_\ell{}^2 + \tau_t{}^2\right)} = \sqrt{4\sigma_n{}^2} = \frac{X}{a\ell}\sqrt{2} < \frac{R_{r\ bead}}{C_s}$$

♦ *Example 2*: in Figure 11.44, a weld bead must transmit force $\overrightarrow{F} = \overrightarrow{Yy}$.

---

47 Here we ignore the influence of "moment arm" δ assumed to be negligible.

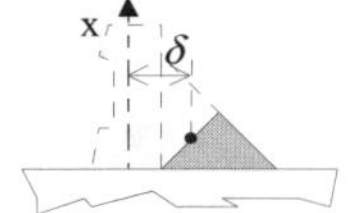

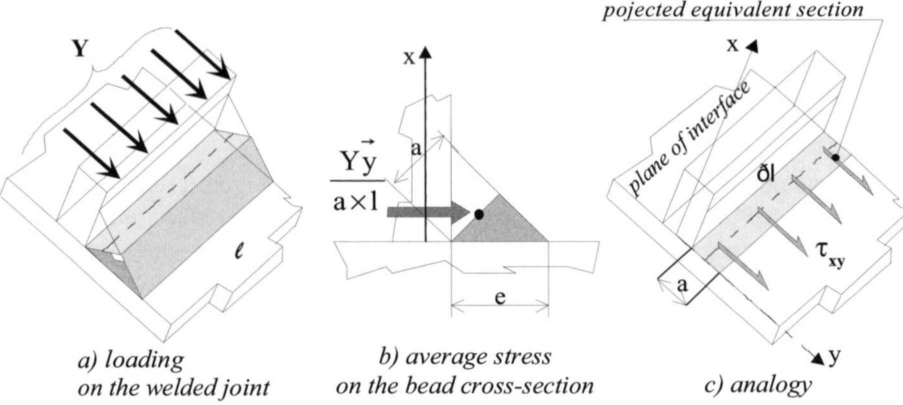

Figure 11.44. *Definition of a projected section for the estimation of an average shear stress*

The average stress on the centerline of the bead cross-section of area $a \times \ell$, is equal to[48]:

$$\frac{\overrightarrow{Y\,y}}{a \times \ell}$$

We can note that we would obtain the same result by examining the diagram in Figure 11.44c where we have "flattened" the bead cross-section against the plane of interface, on which we can observe an average shear stress due to $\overrightarrow{F} = Y\overrightarrow{y}$:

$$\tau_{xy} = \frac{Y}{a \times \ell}$$

We must next project τ_{xy} on the local axes $\overrightarrow{n}$, $\overrightarrow{t_t}$, $\overrightarrow{t_\ell}$ of the groove section , that does not pose a problem. We can read the following in Figure 11.45:

---

48 Here we ignore the influence of the "moment arm" δ assumed to be negligible.

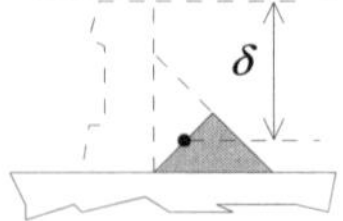

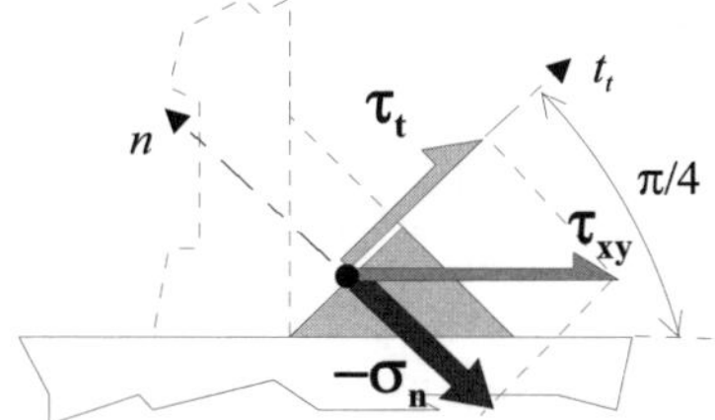

Figure 11.45. *Average stresses in local axes of bead cross-section*

$$-\sigma_n = \tau_t = \frac{Y}{a \times \ell} \frac{\sqrt{2}}{2} \, ; \quad \tau_\ell = 0 \, .$$

The criterion for non-rupture [11.25] is written as:

$$\sqrt{\sigma_n^{\,2} + 3\!\left(\tau_\ell^{\,2} + \tau_t^{\,2}\right)} = \sqrt{4\sigma_n^{\,2}} = \frac{Y}{a\ell}\sqrt{2} < \frac{R_{r\ bead}}{C_s}$$

◆ *Example 3*: in Figure 11.46 we have shown the same welding bead that must now transmit force $\vec{F} = Z\vec{z}$ that has been indicated.

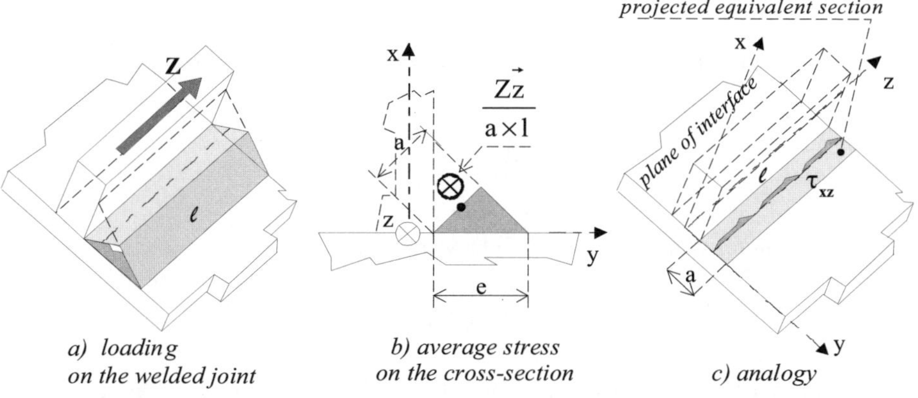

a) loading
on the welded joint

b) average stress
on the cross-section

c) analogy

Figure 11.46. *Definition of a projected section for the estimation of an average shear stress*

The average stress on the centerline of the bead cross-section of area $a \times \ell$ is equal to[49]:

$$\frac{Z\vec{z}}{a \times \ell}$$

We would obtain the same result by examining the diagram in Figure 11.46c where we have "flattened" the bead cross-section against the plane of interface, from which we can obtain an average shear stress due to $\vec{F} = Z\vec{z}$:

$$\tau_{xz} = \frac{Z}{a \times \ell}$$

We must next project τ_{xz} on the local axes $\vec{n}$, $\vec{t_t}$, $\vec{t_\ell}$ of the bead cross-section. We can read the following in Figure 11.47:

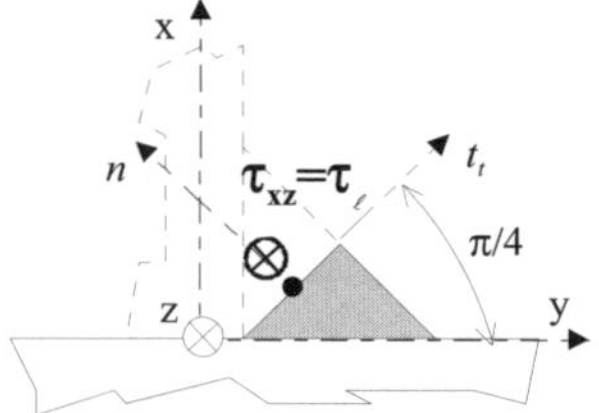

Figure 11.47. *Average stresses in local axes of bead cross-section*

$$\tau_\ell = \frac{Z}{a \times \ell} \; ; \quad \sigma_n = \tau_t = 0 \, .$$

The criteria for non-rupture [11.25] is written as:

$$\sqrt{\sigma_n^2 + 3\left(\tau_\ell^2 + \tau_t^2\right)} = \sqrt{3\tau_\ell^2} = \frac{Z}{a\ell}\sqrt{3} < \frac{R_{r\;bead}}{C_s}$$

The special "basic" cases, dealt with above, illustrate a common approach that allows the stresses in the bead cross-section to be determined. Nevertheless, in the more general case of any resultant force and moment of transmittable force, with several weld beads of different directions, this estimation cannot be made instantly.

---

49 Here we ignore the influence of "moment arm" δ assumed to be negligible.

We have been able to see in each of the examples mentioned above, the advantage of an analogy that consists of "flattening" the bead cross-section against the plane of interface to calculate the average stresses on the centerline. We now propose to use such an approach in a more general context. It could be used to estimate the stresses that are the most detrimental in weld beads.

11.4.2.2. *Definition of a model for the dimensioning of a weld interface*

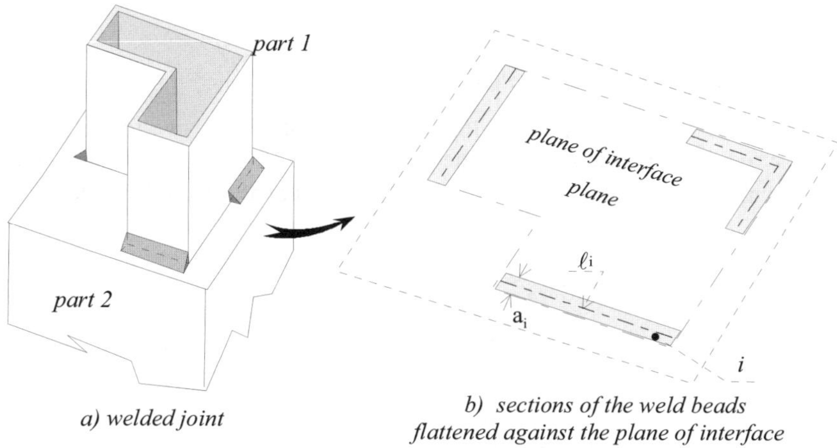

a) *welded joint*

b) *sections of the weld beads*
flattened against the plane of interface

Figure 11.48

We shall presuppose the interface reduced to the equivalent colored areas in Figure 11.48b composed of all the cross-sections of the weld beads flattened against the plane of interface, each one having its centerline ℓ_i. We shall consider it as an elastic "equivalent" interface, made of "equivalent" beads, having a global section S, connecting rigid parts 1 and 2. The approach will be similar to that used for the bolted interface in section 11.2.1.2 (see Figure 11.14). We are therefore led to define the following elements:

– G, geometric center of the equivalent area, such that:

$$\int_S y\,dS = 0 \, ; \qquad \int_S z\,dS = 0$$

– $\vec{y}$ and $\vec{z}$, main quadratic axes in the plane of interface of the welded joint:

$$\int_S yz\,dS = 0$$

– main quadratic moments:

$$I_y = \int_S z^2 dS \quad ; \quad I_z = \int_S y^2 dS \qquad [11.26]$$

– quadratic polar moment:

$$I_0 = I_y + I_z = \int_S \left(y^2 + z^2\right) dS \qquad [11.27]$$

11.4.2.3. *Stresses on each "equivalent" bead*

Figure 11.49 shows the resultant force and moment of cohesive forces, acting on the earlier "equivalent" section, with the projections $\mathcal{N}_x$, $\mathcal{T}_y$, $\mathcal{T}_z$, $\mathcal{M}_x$, $\mathcal{M}f_y$, $\mathcal{M}f_z$, assumed to be known, by means of a preliminary finite element analysis on the structure.

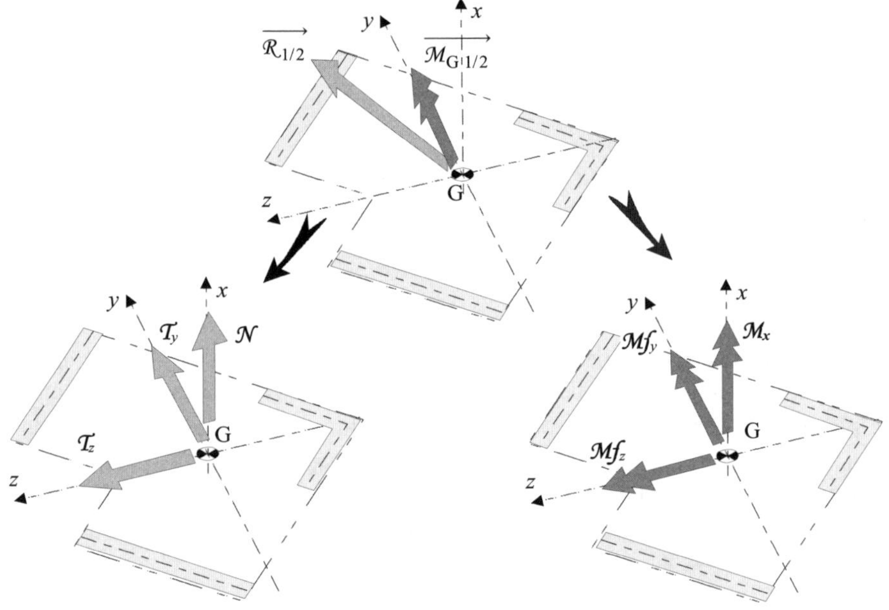

Figure 11.49. *Resultant force and moment on the "equivalent" weld section*

We must estimate on each centerline ℓ_i, the stresses corresponding to each of the projections of the forces of cohesion and then obtain the sum of the stresses thus found.

Let us successively examine each of the previous projections:

➢ normal force $\mathcal{N}$

This represents a distribution of normal uniform stresses on the "equivalent" weld section:

$$\sigma_{\mathcal{N}} = \frac{\mathcal{N}}{S}$$

➢ shear resultant $\mathcal{T}_y$

Taking into account the approximate nature of these estimations, we assume a uniform distribution of shear stress on the "equivalent" weld section, i.e.:

$$\tau_{xy} = \frac{\mathcal{T}_y}{S}$$

➢ shear resultant $\mathcal{T}_z$

Similarly, we assume a uniform distribution of shear stress on the "equivalent" weld section, i.e.:

$$\tau_{xz} = \frac{\mathcal{T}_z}{S}$$

➢ longitudinal moment $\mathcal{M}_x$

We adopt a stress distribution on the equivalent joint that is similar to that found for the torsion of a circular section beam[50], as shown in Figure 11.50 in a point M of a centerline of an "equivalent" weld section:

$$\tau_{\mathcal{M}_x} = \frac{\mathcal{M}_x}{I_o} \times r$$

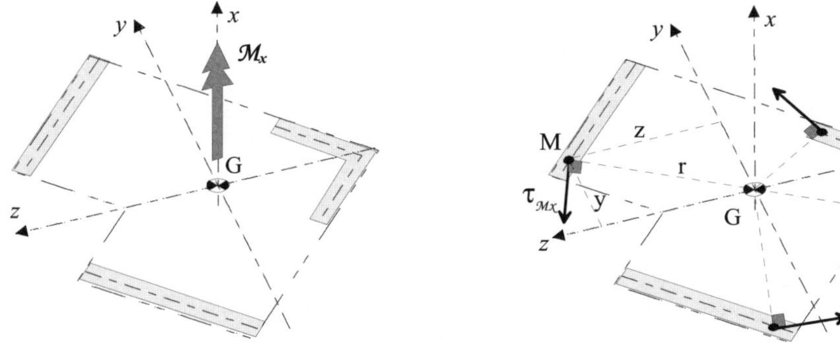

Figure 11.50. *Shear stresses due to $\mathcal{M}_x$*

---

50 See section 9.3.2.3.

with components[51]:

$$\tau_{xy\,\mathcal{M}_x} = -\frac{\mathcal{M}_x}{I_0} \times z \quad ; \quad \tau_{xz\,\mathcal{M}_x} = \frac{\mathcal{M}_x}{I_0} \times y$$

where I_0 is the quadratic polar moment [11.27].

➢ bending moment $\mathcal{M}f_y$[52]

This indicates the existence of normal stresses on each "equivalent" weld section given by:

$$\sigma_{\mathcal{M}f_y} = \frac{\mathcal{M}f_y}{I_y} \times z$$

where I_y is given by [11.26];

➢ bending moment $\mathcal{M}f_z$

In a similar manner, this indicates the existence of normal stresses on each "equivalent" weld section:

$$\sigma_{\mathcal{M}f_z} = -\frac{\mathcal{M}f_z}{I_z} \times y$$

where I_z is given by [11.26];

➢ resulting stresses at a point M(y,z) of an "equivalent" weld section

These are obtained by superposition of the earlier stresses. We shall thus obtain:

$$- \text{ in projection on axis } \vec{x} \; : \; \sigma_x = \frac{\mathcal{N}}{S} + \frac{\mathcal{M}f_y}{I_y} \times z - \frac{\mathcal{M}f_z}{I_z} \times y$$

$$- \text{ in projection on axis } \vec{y} \; : \; \tau_{xy} = \frac{\mathcal{T}_y}{S} - \frac{\mathcal{M}_x}{I_0} \times z$$

$$- \text{ in projection on axis } \vec{z} \; : \; \tau_{xz} = \frac{\mathcal{T}_z}{S} + \frac{\mathcal{M}_x}{I_0} \times y$$

---

51 See [11.3]. We can verify that the basic cohesion forces corresponding to these shear stresses have a resultant, on the whole section, equal to zero, by using a procedure identical to that in section 11.2.1.3.
52 See [9.36].

These are the components of the stress vector $\overrightarrow{C_M}$ at point M (see Figure 11.51) in the local axes (*xyz*). We can then localize the areas having the greatest stresses on the centerlines of the "equivalent" weld section.

11.4.2.4. *Stresses* σ_n, τ_ℓ, τ_t *in the bead cross-sections*

As we have been able to see in the examples in section 11.4.2.1, once we have obtained components σ_x, τ_{xy}, τ_{xz} we need to project them in the local coordinate system of the real bead cross-section in question. Let us remember that we had in fact, with the purpose of making the calculation easier, flattened the bead cross-section against the plane of interface in order to create an "equivalent" weld section. In reality, the bead cross-sections are inclined at 45° on the plane of interface. With the markings in Figure 11.51, we will obtain:

$$\left\{\begin{array}{c}\overrightarrow{n}\\\overrightarrow{t_\ell}\\\overrightarrow{t_t}\end{array}\right\} = \frac{1}{\sqrt{2}}\begin{bmatrix}1 & s & -c\\0 & c\sqrt{2} & s\sqrt{2}\\1 & -s & c\end{bmatrix}\bullet\left\{\begin{array}{c}\overrightarrow{x}\\\overrightarrow{y}\\\overrightarrow{z}\end{array}\right\} = [P]\bullet\left\{\begin{array}{c}\overrightarrow{x}\\\overrightarrow{y}\\\overrightarrow{z}\end{array}\right\}$$

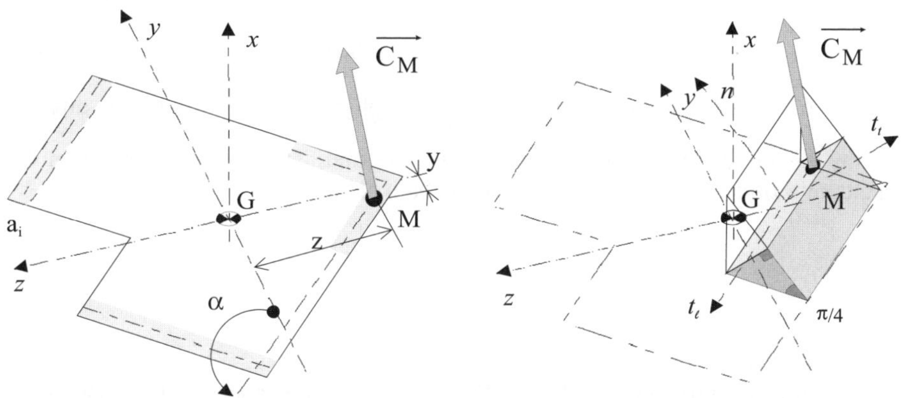

a) result of the "equivalent"   *b) setting of the stress on the bead cross-section*
interface analogy   *in the proper orientation*

Figure 11.51. *Stress* $\overrightarrow{C_M}$

a relation in which we have written $c = \cos\alpha$ and $s = \sin\alpha$ $\alpha = \left(\vec{t}_\ell, \vec{y}\right)$. We deduce the "regulation" stresses at point M in the axes of the bead cross-section in question[53]:

$$\begin{Bmatrix} \sigma_n \\ \tau_\ell \\ \tau_t \end{Bmatrix} = [P] \bullet \begin{Bmatrix} \sigma_x \\ \tau_{xy} \\ \tau_{xz} \end{Bmatrix}$$

stresses with which we must verify criterion [11.25].

NOTES

❏ The real state of stresses in the weld beads is too complex for the earlier procedure to be considered accurate and always reliable. In addition, this state of stress undergoes a rapid change when we move from one welded part to the other.

❏ In addition, this change depends on the geometry of the part. Let us consider the examples of welded joints in Figure 11.52. The estimation that we have just created is rather well adapted to the joints in Figure 11.52a. In fact, the joint in Figure 11.52b ensures the continuity of two tubes. We can thus consider the bead as a very short joining tube, with thickness "a", equal to that of the bead cross-section and in general, different to the one noted as a' of the tubes to be assembled. The state of stresses in the bead cross-section is thus similar to the one that exists in the cross-section of the assembled tubes, weighted by $\dfrac{a'}{a}$. We can thus directly use the values of the normal stresses in the tubes that have been obtained from the results of a finite element calculation with beam elements[54].

---

53 See Chapter 12, relations [12.7] and [12.9]. Note that this case is not the traditional problem of expressing a given state of stresses on two facets, deduced one from the other by a rotation of 45°. The calculation done beforehand, leading to σ_x, τ_{xy}, τ_{xz} is a "standard statutory" estimation that is convenient, as it is based on an "equivalent section" of a weld bead equal to the bead cross-section defined above, which has been artificially flattened against the plane of interface (see section 11.4.2.1).

54 As we have already mentioned (see section 5.2.2/*NOTES*), the post processors of finite element software do not provide the shear stresses due to either shear resultant or torsional moment. Knowledge of these shear stresses is nevertheless necessary for such an evaluation (see [9.27] and [9.59]). In the case of Figure 11.52b, the shear stress is the one that exists in a tubular beam. Its estimation is more difficult, even by the simple method suggested above for a joint of the type that is shown in Figure 11.52a.

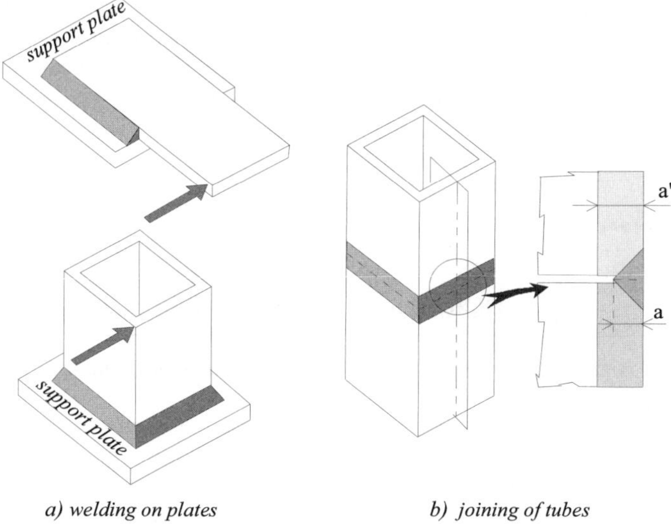

a) welding on plates   b) joining of tubes

Figure 11.52. *Examples of welded joints*

☐ When in a welded joint as shown in Figure 11.53, a part of the "equivalent" section is subjected to compression (as is frequent), the two assembled parts, being in contact, will also bear the compression. Consequently, to resist the compression, we note a surface higher than the bead cross-section, as we can see in Figure 11.53b. This adds to the approximate nature of the above procedure for estimation of stresses. Nevertheless, such a procedure leads, in this particular case, to pessimistic stress estimations, and therefore tends to be safer.

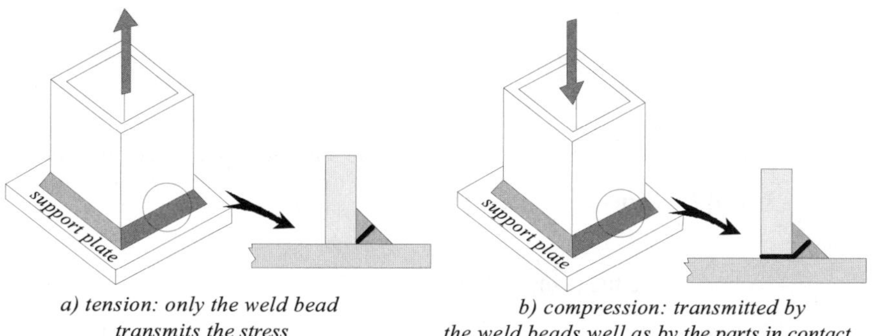

a) tension: only the weld bead   b) compression: transmitted by
transmits the stress   the weld beads well as by the parts in contact

Figure 11.53. *Pessimistic aspect of the calculation model*

❐ For all practical purposes, we can rapidly characterize an "equivalent" weld section by means of specific additional software working from input of the geometry of the section, and giving the characteristics of a beam cross-section. Such software is often integrated as a specific function in a finite element software.

11.4.3. *Summary*

Dimensioning of a welded joint

a weld bead is reduced to its bead cross-section that is flattened against the plane of interface:

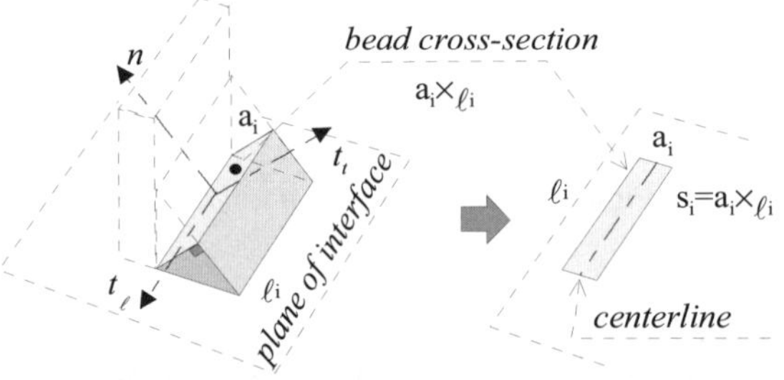

**a welded joint (between parts 1 and 2) is reduced
to an "equivalent" weld section**

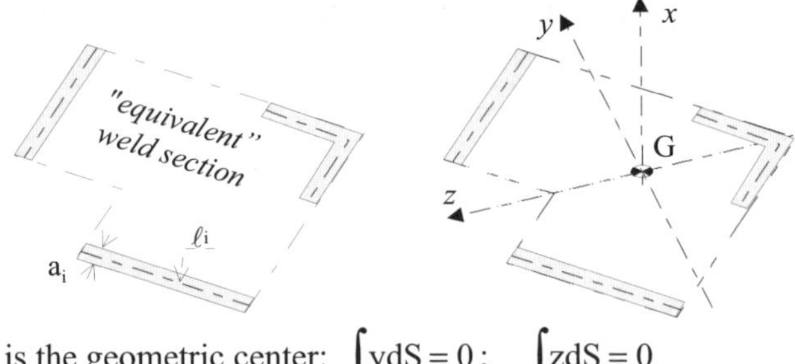

- G is the geometric center: $\displaystyle\int_S y\,dS = 0$; $\displaystyle\int_S z\,dS = 0$

- $\vec{y}$ and $\vec{z}$ are the principal quadratic axes:

$$\int_S yz\,dS = 0 \quad ; \quad I_y = \int_S z^2\,dS \quad ; \quad I_z = \int_S y^2\,dS \;;\; I_0 = I_y + I_z$$

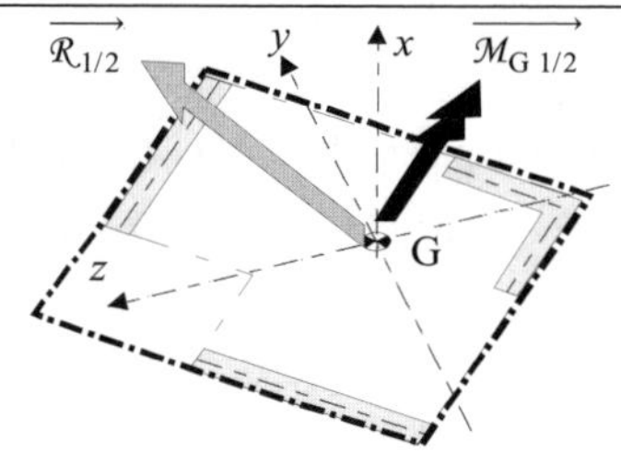

resultant force and moment of the transmittable forces (already known as data)

$$\left\{ \begin{array}{l} \overrightarrow{\mathcal{R}_{1/2}} = \mathcal{N}\ \vec{x} + \mathcal{T}_y\ \vec{y} + \mathcal{T}_z\ \vec{z} \\[2mm] \overrightarrow{\mathcal{M}_{G1/2}} = \mathcal{M}_x\ \vec{x} + \mathcal{M}f_y\ \vec{y} + \mathcal{M}f_z\ \vec{z} \end{array} \right\}_G$$

stresses on the centerlines of the "equivalent" weld section

$$\sigma_x = \frac{\mathcal{N}}{S} + \frac{\mathcal{M}f_y}{I_y} \times z - \frac{\mathcal{M}f_z}{I_z} \times y$$

$$\tau_{xy} = \frac{\mathcal{T}_y}{S} - \frac{\mathcal{M}_x}{I_0} \times z$$

$$\tau_{xz} = \frac{\mathcal{T}_z}{S} + \frac{\mathcal{M}_x}{I_0} \times y$$

regulation stresses in a bead

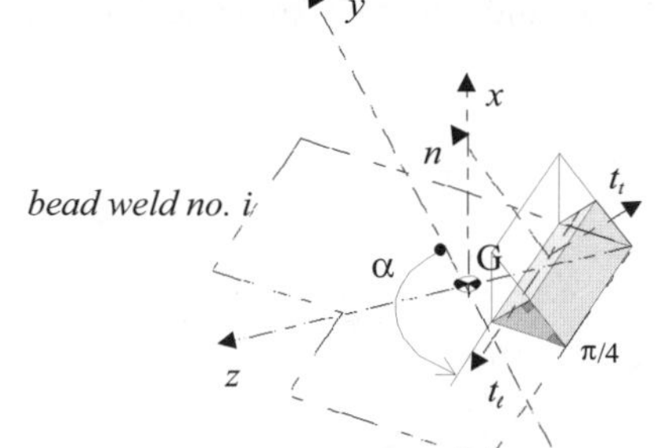

$$\left\{ \begin{array}{c} \sigma_n \\ \tau_\ell \\ \tau_t \end{array} \right\} = \frac{1}{\sqrt{2}} \begin{bmatrix} 1 & s & -c \\ 0 & c\sqrt{2} & s\sqrt{2} \\ 1 & -s & c \end{bmatrix} \bullet \left\{ \begin{array}{c} \sigma_x \\ \tau_{xy} \\ \tau_{xz} \end{array} \right\}$$

with $\alpha = \left(\vec{t_\ell}, \vec{y}\right)$, $s = \sin\alpha$; $c = \cos\alpha$

resistance criterion for the bead

$$\sqrt{\sigma_n{}^2 + 3\left(\tau_\ell{}^2 + \tau_t{}^2\right)} \leq \frac{R_{r\ bead}}{C_s} \quad \text{with } 1 \leq C_s \leq 1.2$$

[11.28]

11.4.4. *Example*

We shall consider the welded joint between two parts shown as (1) and (2) in Figure 11.54 with a bead of thickness "e". The upper part (1) is subjected to force $\vec{F}$ as shown.

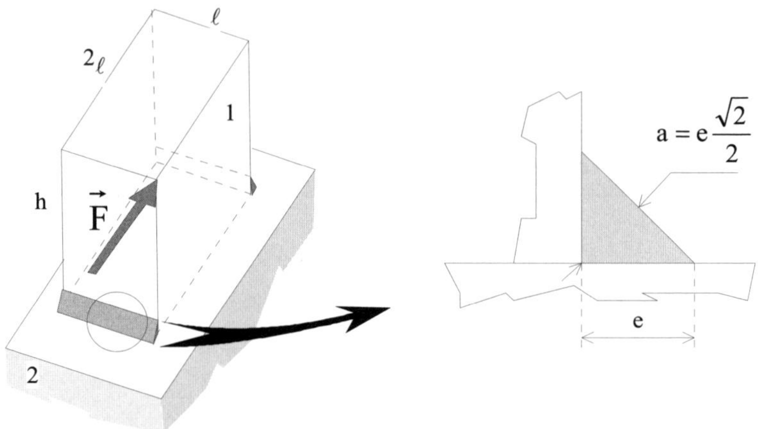

Figure 11.54. *Example*

We obtain the "equivalent" weld section of Figure 11.55, where we have written the non-zero projections of the resultant force and moment of cohesion forces between (1) and (2):

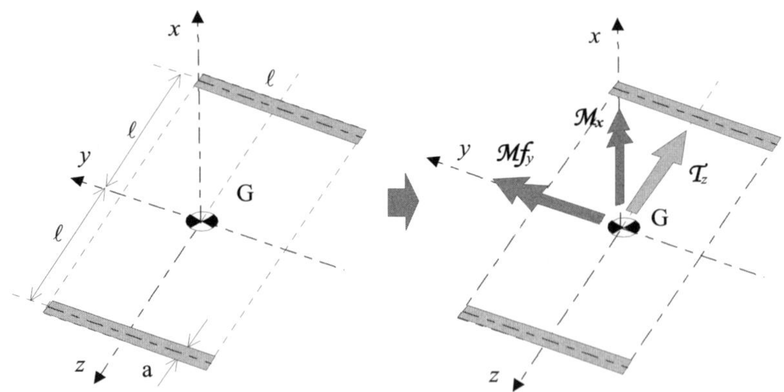

Figure 11.55. *Resultant force and moment on "equivalent" weld section*

$$\mathcal{T}_z = -F; \quad \mathcal{M}_x = F\frac{\ell}{2}; \quad \mathcal{M}f_y = F \times h$$

The characteristics of the "equivalent" weld section are written as:

$$S = 2a \times \ell$$

$$I_y \cong 2a \times \ell^3 \ ; \ I_z = \frac{a \times \ell^3}{6} \ ; \ I_0 \cong \frac{13}{6} a \times \ell^3$$

By following the results summarized in [11.28]:

• the maximum tensile stress is ($z = \ell$):

$$\sigma_x = \frac{F \times h}{I_y} \times \ell = \frac{F \times h}{2a \times \ell^2}$$

• the maximum shear stress on the earlier bead is picked up at point M_0 of Figure 11.56 shown here after ($y_{M_0} = -\frac{\ell}{2}$; $z_{M_0} = \ell$).

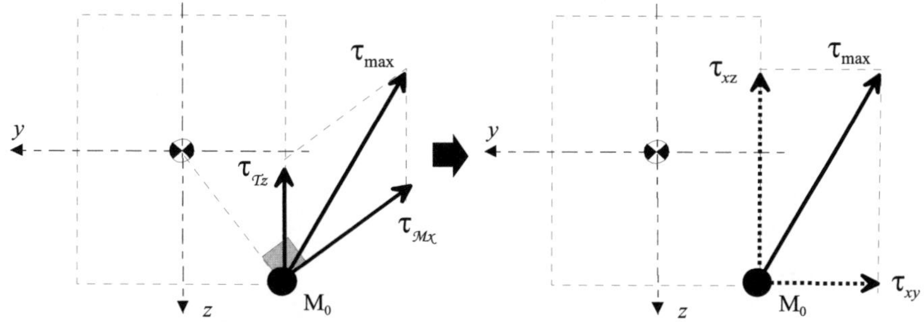

Figure 11.56. *Maximum shear stress*

At this point, we have:

$$\tau_{xy} = -\frac{F \times \ell}{2I_0} \times \ell = -\frac{3}{13}\frac{F}{a \times \ell}$$

$$\tau_{xz} = -\frac{F}{2a \times \ell} + \frac{F \times \ell}{2I_0} \times -\frac{\ell}{2} = -\frac{8}{13}\frac{F}{a \times \ell}$$

The standard stresses at the point M_0 ($y = -\dfrac{\ell}{2}$; $z = \ell$) are calculated here with $\alpha = 0$ [55]:

$$\sigma_n = \frac{F}{a \times \ell \sqrt{2}}\left[\frac{h}{2\ell} + \frac{8}{13}\right]; \quad \tau_\ell = -\frac{3}{13}\frac{F}{a \times \ell}; \quad \tau_t = \frac{F}{a \times \ell \sqrt{2}}\left[\frac{h}{2\ell} - \frac{8}{13}\right]$$

that enables us to write the criterion:

$$\sqrt{\sigma_n{}^2 + 3\left(\tau_\ell{}^2 + \tau_t{}^2\right)} \leq \frac{R_{r\ \text{cordon}}}{C_s}$$

---

55 Pay attention to the definition of the local orthonormal trihedral $\vec{n}$, $\vec{t}_\ell$, $\vec{t}_t$; see [11.28]

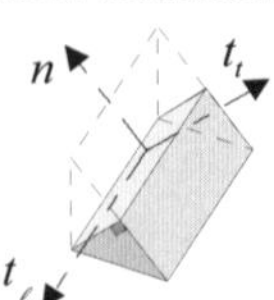

Chapter 12

Mathematical Prerequisites

The following outlines the principal (and traditional) rules and properties of matrix calculus used in this book.

12.1. Matrix calculus

12.1.1. *General information*

12.1.1.1. *Definition of a matrix*

A matrix is a table constituted of n rows and m columns of algebraic numbers.

It is convenient to locate a term "a" of a matrix by its position at the intersection of the row number i and column number j. The position is designated by the notation a_{ij}.

$$matrix\ (n \times m) : [a] = \begin{bmatrix} a_{11} & a_{12} & .. & a_{1j} & .. & a_{1m} \\ .. & .. & .. & .. & .. & .. \\ a_{i1} & .. & .. & a_{ij} & .. & a_{im} \\ .. & .. & .. & .. & .. & .. \\ a_{n1} & .. & .. & a_{nj} & .. & a_{nm} \end{bmatrix} \right\} n\ rows$$

$$m\ columns$$

When n = m, the matrix is called a "square" matrix (n × n).
When m = 1, the matrix is called a "column" matrix (n × 1), or "vector" with "n" coordinates (or components).

$$\text{Example:} \quad \begin{Bmatrix} a_{11} \\ a_{21} \\ a_{31} \end{Bmatrix} \quad \text{that we can write simply as} \quad \begin{Bmatrix} A_1 \\ A_2 \\ A_3 \end{Bmatrix}$$

12.1.1.2. *Symmetric matrix*

This is a square matrix such that $a_{ij} = a_{ji}$

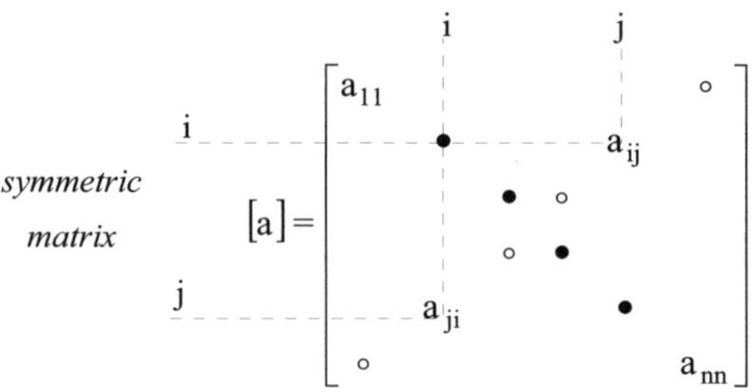

12.1.1.3. *Transposition of a matrix [a]*

The transposed matrix of a matrix [a] is obtained by permutation of the rows and the columns of [a]. The transposed matrix is denoted $[a]^T$. For example:

$$[a] = \begin{bmatrix} a_{11} & a_{12} \\ a_{21} & a_{22} \\ a_{31} & a_{32} \\ a_{41} & a_{42} \end{bmatrix} \quad \Rightarrow \quad [a]^T = \begin{bmatrix} a_{11} & a_{21} & a_{31} & a_{41} \\ a_{12} & a_{22} & a_{32} & a_{42} \end{bmatrix}$$

NOTE

❏ The transposed matrix of a square symmetric matrix is the initial matrix itself.

12.1.2. *Matrix operations*

12.1.2.1. *Addition of two matrices*

For addition to occur, it is necessary that the two matrices to be added have *the same number of rows (m) and the same number of columns (n)*. The sum of two matrices [a] and [b] is matrix [c]. Matrix additions are commutative: $[a] + [b] = [b] + [a]$.

$$\begin{bmatrix} a_{11} & a_{12} \\ a_{21} & a_{22} \end{bmatrix} + \begin{bmatrix} b_{11} & b_{12} \\ b_{21} & b_{22} \end{bmatrix} = \begin{bmatrix} (a_{11}+b_{11}) & (a_{12}+b_{12}) \\ (a_{21}+b_{21}) & (a_{22}+b_{22}) \end{bmatrix}$$

NOTE

❑ Matrix [a] can be decomposed into sub-matrices.

For example:

$$\underset{6\times6}{[a]} = \left[\begin{array}{cc|cccc} a_{11} & a_{12} & a_{13} & a_{14} & a_{15} & a_{16} \\ a_{21} & a_{22} & a_{23} & a_{24} & a_{25} & a_{26} \\ \hline a_{31} & a_{32} & a_{33} & a_{34} & a_{35} & a_{36} \\ a_{41} & a_{42} & a_{43} & a_{44} & a_{45} & a_{46} \\ a_{51} & a_{52} & a_{53} & a_{54} & a_{55} & a_{56} \\ a_{61} & a_{62} & a_{63} & a_{64} & a_{65} & a_{66} \end{array}\right]$$

can be written as:

$$\underset{6\times6}{[a]} = \left[\begin{array}{c|c} \underset{2\times2}{[b]} & \underset{2\times4}{[c]} \\ \hline \underset{4\times2}{[d]} & \underset{4\times4}{[e]} \end{array}\right]$$

Matrix [a] is not the sum of the four sub-matrices defined above, but is the sum of four 6×6 matrices built starting from the four sub-matrices. Actually, if we write:

$$\underset{6\times6}{[B]} = \left[\begin{array}{cc|cccc} a_{11} & a_{12} & 0 & 0 & 0 & 0 \\ a_{21} & a_{22} & 0 & 0 & 0 & 0 \\ \hline 0 & 0 & 0 & 0 & 0 & 0 \\ 0 & 0 & 0 & 0 & 0 & 0 \\ 0 & 0 & 0 & 0 & 0 & 0 \\ 0 & 0 & 0 & 0 & 0 & 0 \end{array}\right]$$

etc., then we can write:

$$\underset{6\times6}{[a]} = \underset{6\times6}{[B]} + \underset{6\times6}{[C]} + \underset{6\times6}{[D]} + \underset{6\times6}{[E]} \qquad [12.1]$$

by noting in capital letters the 6×6 matrices thus obtained.

12.1.2.2. *Product of a matrix by a scalar*

The product of a matrix $[a]_{(n \times m)}$ by a scalar λ is a matrix obtained by multiplying each element by λ:

$$\lambda \bullet \begin{bmatrix} a_{11} & a_{12} \\ a_{21} & a_{22} \end{bmatrix} = \begin{bmatrix} \lambda a_{11} & \lambda a_{12} \\ \lambda a_{21} & \lambda a_{22} \end{bmatrix}$$

12.1.2.3. *Product of two matrices*

The product of two matrices $[a]_{(n \times m)}$ and $[b]_{(m \times p)}$ is matrix $[c]_{(n \times p)}$.

For multiplication to be possible, the number of columns of the first matrix must be equal to the number of rows of the second matrix. For example:

$$[a]_{(5 \times 3)} \bullet [b]_{(3 \times 2)} = [c]_{(5 \times 2)}$$

the first terms of matrix $[c]_{(5 \times 2)}$ being written as:

$$c_{11} = a_{11} \times b_{11} + a_{12} \times b_{21} + a_{13} \times b_{31}$$
$$c_{12} = a_{11} \times b_{12} + a_{12} \times b_{22} + a_{13} \times b_{32}$$

and the general term c_{ij} of matrix $[c]_{(5 \times 2)}$ is written as:

$$c_{ij} = a_{i1} \times b_{1j} + a_{i2} \times b_{2j} + a_{i3} \times b_{3j}$$

Particular cases

 • The product of a square matrix and a column matrix is a column matrix.

Example: the product of the square matrix (2×2) and a column matrix (2×1) is column matrix (2×1):

$$\begin{bmatrix} a_{11} & a_{12} \\ a_{21} & a_{22} \end{bmatrix} \bullet \begin{Bmatrix} b_{11} \\ b_{21} \end{Bmatrix} = \begin{Bmatrix} (a_{11} \times b_{11}) + (a_{12} \times b_{21}) \\ (a_{21} \times b_{11}) + (a_{22} \times b_{21}) \end{Bmatrix}$$

 • The product of a square matrix and another square matrix is a square matrix

Example: the product of a matrix (2×2) and a square matrix (2×2) is a square matrix (2×2).

$$\begin{bmatrix} a_{11} & a_{12} \\ a_{21} & a_{22} \end{bmatrix} \bullet \begin{bmatrix} b_{11} & b_{12} \\ b_{21} & b_{22} \end{bmatrix} = \begin{bmatrix} \{(a_{11} \times b_{11}) + (a_{12} \times b_{21})\} & \{(a_{11} \times b_{12}) + (a_{12} \times b_{22})\} \\ \{(a_{21} \times b_{11}) + (a_{22} \times b_{21})\} & \{(a_{21} \times b_{12}) + (a_{22} \times b_{22})\} \end{bmatrix}$$

NOTES

❏ The matrix product is, in general:

- non-commutative:

$$[a] \bullet [b] \neq [b] \bullet [a].$$

- associative:

$$[a] \bullet ([b] \bullet [c]) = ([a] \bullet [b]) \bullet [c]$$

❏ The transposed matrix of the product of the two matrices is written as:

$$\left[[a] \bullet [c] \right]^T = [c]^T \bullet [a]^T$$

12.1.2.4. *Inverse of a matrix*

O Two square matrices [a] and [c] are inverse to each other if [a]•[c] = [I] where [I] is the unit matrix.

$$[I] = \begin{bmatrix} 1 & & & & \\ & 1 & & 0 & \\ & & \bullet & & \\ 0 & & & \bullet & \\ & & & & 1 \end{bmatrix} \quad \longleftarrow \text{zero terms}$$

The inverse matrix of matrix [a] is written as $[a]^{-1}$. There are methods and algorithms for inversion of matrices but their use for manual inversion is limited to very small sizes (2×2) or (3×3) [1]. To inverse larger matrices, we use:

---

1 Knowledge of such algorithms is not necessary for small sizes: for example, to find the inverse of the matrix $[a] = \begin{bmatrix} a_{11} & a_{12} \\ a_{21} & a_{22} \end{bmatrix}$, it is enough to solve the system of equations:

$$\begin{bmatrix} a_{11}x_1 + a_{12}x_2 = b_1 \\ a_{21}x_1 + a_{22}x_2 = b_2 \end{bmatrix} \Leftrightarrow [a] \bullet \begin{Bmatrix} x_1 \\ x_2 \end{Bmatrix} = \begin{Bmatrix} b_1 \\ b_2 \end{Bmatrix}$$

We find, using standard and simple methods:

$$\begin{bmatrix} x_1 = \alpha_{11}b_1 + \alpha_{12}b_2 \\ x_2 = \alpha_{21}b_1 + \alpha_{22}b_2 \end{bmatrix} \Leftrightarrow [\alpha] \bullet \begin{Bmatrix} b_1 \\ b_2 \end{Bmatrix} = \begin{Bmatrix} x_1 \\ x_2 \end{Bmatrix}$$

We have calculated $[\alpha] = [a]^{-1}$ because $[a] \bullet [\alpha] \bullet \begin{Bmatrix} b_1 \\ b_2 \end{Bmatrix} = [a] \bullet \begin{Bmatrix} x_1 \\ x_2 \end{Bmatrix} \Rightarrow [a] \bullet [\alpha] = [I]$

– calculators or specific software utilities of formal calculus when we wish to work on literal expressions;

– digital algorithms integrated in finite element software to invert large-sized matrices (several thousands to tens of thousands of rows and columns).

O A square matrix with zero determinant has no inverse. It is then called "singular".

O A square matrix [a] is called "orthogonal" when its transposed matrix is equal to its inverse[2] $[a]^{-1}$:

$$[a]^{T} = [a]^{-1}$$

Its determinant is then equal to ± 1.

O When it exists, the inverse of a symmetric matrix is a symmetric matrix.

12.1.3. *Quadratic form*

An expression of the following type (second degree of two variables x_1 and x_2) is called a quadratic form, thus:

$$W = a_1 x_1^2 + a_2 x_2^2 + a_3 x_1 x_2$$

Let us consider such an expression rewritten as follows:

$$W = \frac{1}{2}(a_{11} x_1^2 + a_{22} x_2^2 + 2a_{12} x_1 x_2)$$

We note that it may also appear as a product matrix:

$$W = \frac{1}{2}\underbrace{[x_1 \quad x_2]}_{\substack{\text{matrix line} \\ (1\times 2)}} \bullet \underbrace{\left\{ \begin{matrix} (a_{11}x_1 + a_{12}x_2) \\ (a_{12}x_1 + a_{22}x_2) \end{matrix} \right\}}_{\substack{\text{matrix column} \\ (2\times 1)}}$$

or even

$$W = \frac{1}{2}\left\{ \begin{matrix} x_1 \\ x_2 \end{matrix} \right\}^{T} \bullet \begin{bmatrix} a_{11} & a_{12} \\ a_{21} & a_{22} \end{bmatrix} \bullet \left\{ \begin{matrix} x_1 \\ x_2 \end{matrix} \right\} \qquad [12.2]$$

---

2 See also section 12.2.

where $a_{21} = a_{12}$.

We shall say that the preceding particular quadratic form is built on a square and symmetric matrix constituted with coefficients a_{ij}

12.1.4. *Eigenvalues and eigenvectors of a matrix*

12.1.4.1. *Eigenvalues*

Take for example a square matrix:

$$\underset{3\times3}{[A]} = \begin{bmatrix} a_{11} & a_{12} & a_{13} \\ a_{21} & a_{22} & a_{23} \\ a_{31} & a_{32} & a_{33} \end{bmatrix}$$

We define the characteristic polynomial $P(\lambda)$ of this matrix as:

$$P(\lambda) = \det([A] - \lambda[I]) = \underset{determinant}{\det} \begin{bmatrix} a_{11} - \lambda & a_{12} & a_{13} \\ a_{21} & a_{22} - \lambda & a_{23} \\ a_{31} & a_{32} & a_{33} - \lambda \end{bmatrix} = 0$$

By developing this determinant:

$$P(\lambda) = (a_{11} - \lambda)\det\begin{bmatrix} a_{22} - \lambda & a_{23} \\ a_{32} & a_{33} - \lambda \end{bmatrix} - a_{12}\det\begin{bmatrix} a_{21} & a_{23} \\ a_{31} & a_{33} - \lambda \end{bmatrix} + a_{13}\det\begin{bmatrix} a_{21} & a_{22} - \lambda \\ a_{31} & a_{32} \end{bmatrix} = 0$$

we obtain a polynomial of degree 3, which admits 3 roots λ_1, λ_2, λ_3. Let us call them "eigenvalues" of the matrix.

12.1.4.2. *Eigenvectors*

λ being an eigenvalue of $\underset{3\times3}{[A]}$, we shall say that a vector represented in matrix form by $\vec{V} = \begin{Bmatrix} X \\ Y \\ Z \end{Bmatrix}$ is the eigenvector associated with λ if:

$$[A] \bullet \{V\} = \lambda\{V\}$$

i.e., if:

$$([A] - \lambda[I]) \bullet \{V\} = \{0\} \quad \Leftrightarrow \quad \begin{cases} (a_{11} - \lambda)X + a_{12}Y + a_{13}Z = 0 \\ a_{21}X + (a_{22} - \lambda)Y + a_{23}Z = 0 \\ a_{31}X + a_{32}Y + (a_{33} - \lambda)Z = 0 \end{cases}$$

For the eigenvalue λ_1, the corresponding eigenvector is written as $\overrightarrow{V_1} = \begin{Bmatrix} X_1 \\ Y_1 \\ Z_1 \end{Bmatrix}$,

the components of this vector verify the linear and homogenous system:

$$\begin{cases} (a_{11} - \lambda_1)X_1 + a_{12}Y_1 + a_{13}Z_1 = 0 \\ a_{21}X_1 + (a_{22} - \lambda_1)Y_1 + a_{23}Z_1 = 0 \\ a_{31}X_1 + a_{32}Y_1 + (a_{33} - \lambda_1)Z_1 = 0 \end{cases}$$

Similarly, the eigenvalues λ_2, λ_3, make the calculation of associated eigenvectors $\overrightarrow{V_2}$ and $\overrightarrow{V_3}$ possible.

12.2. Change in orthonormal coordinate system

12.2.1. *Case of coplanar coordinate systems*

We wish to link components V_x and V_y of a vector labeled $\overrightarrow{V} = V_x \vec{x} + V_y \vec{y}$ in a direct orthonormal coordinate system "L" (or local) and components V_X and V_Y of the same vector expressed in another direct orthonormal coordinate system "G" (or global), i.e. $\overrightarrow{V} = V_X \overrightarrow{X} + V_Y \overrightarrow{Y}$.

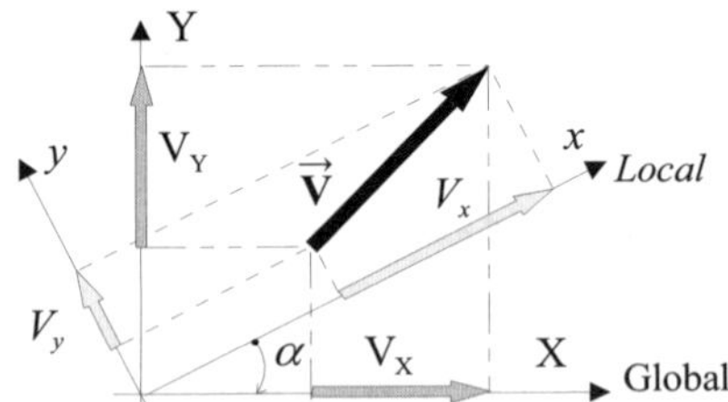

Figure 12.1. *Representation of the same vector in two coplanar coordinate systems*

Figure 12.1 allows us to write: $\overrightarrow{X} = \cos\alpha \vec{x} - \sin\alpha \vec{y}$ and $\overrightarrow{Y} = \sin\alpha \vec{x} + \cos\alpha \vec{y}$, from which:

$$\vec{V} = V_X\left(\cos\alpha\,\vec{x} - \sin\alpha\,\vec{y}\right) + V_Y\left(\sin\alpha\,\vec{x} + \cos\alpha\,\vec{y}\right)$$

$$\vec{V} = \left(V_X\cos\alpha + V_Y\sin\alpha\right)\vec{x} + V_Y\left(-V_X\sin\alpha + \cos\alpha\right)\vec{y}$$

and by identification:

$$V_x = V_X\cos\alpha + V_Y\sin\alpha$$
$$V_y = -V_X\sin\alpha + V_Y\cos\alpha$$

which is written in the matrix form:

$$\underset{Local}{\begin{Bmatrix} V_x \\ V_y \end{Bmatrix}} = \underset{\underset{2\times2}{[P]}}{\begin{bmatrix} \cos\alpha & \sin\alpha \\ -\sin\alpha & \cos\alpha \end{bmatrix}} \bullet \underset{Global}{\begin{Bmatrix} V_X \\ V_Y \end{Bmatrix}} \qquad [12.3]$$

$\underset{2\times2}{[P]}$ is called the transfer matrix from the Global to the *Local* coordinate system.

NOTE

❏ In the preceding expression [12.3], we can interchange the Global and *Local* on the condition that the angle α changes to $-\alpha$. Thus:

$$\underset{Local}{\begin{Bmatrix} V_x \\ V_y \end{Bmatrix}} = \underset{\underset{2\times2}{[P]}}{\begin{bmatrix} \cos\alpha & \sin\alpha \\ -\sin\alpha & \cos\alpha \end{bmatrix}} \bullet \underset{Global}{\begin{Bmatrix} V_X \\ V_Y \end{Bmatrix}} \qquad [12.4]$$

The new transfer matrix appearing in relation [12.4] is the inverse of $\underset{2\times2}{[P]}$. We can see that it is, at the same time, equal to the transposed matrix of $\underset{2\times2}{[P]}$ (see [12.3]). $\underset{2\times2}{[P]}$ is an *orthogonal* matrix as indicated in section 12.1.2.4 (we can verify particularly that its determinant is equal to 1):

$$\underset{2\times2}{[P]}^{-1} = \underset{2\times2}{[P]}^{T} \qquad [12.5]$$

12.2.2. *Cases of any general coordinate systems*

Here the Local and Global coordinate systems are direct orthonormal bases distinct in space.

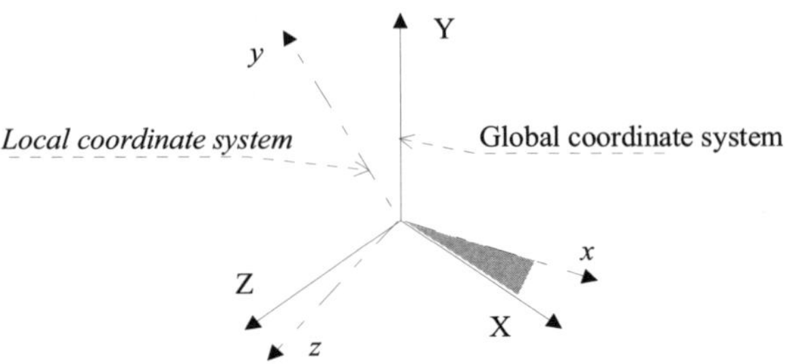

Figure 12.2. *Any general coordinate systems*

Vectorial relations between unit vectors of the two coordinate systems are written as[3]:

$$\vec{x} = \cos(\vec{x}, \vec{X})\,\vec{X} + \cos(\vec{x}, \vec{Y})\,\vec{Y} + \cos(\vec{x}, \vec{Z})\vec{Z}$$
$$\vec{y} = \cos(\vec{y}, \vec{X})\,\vec{X} + \cos(\vec{y}, \vec{Y})\,\vec{Y} + \cos(\vec{y}, \vec{Z})\vec{Z} \qquad [12.6]$$
$$\vec{z} = \cos(\vec{z}, \vec{X})\,\vec{X} + \cos(\vec{z}, \vec{Y})\,\vec{Y} + \cos(\vec{z}, \vec{Z})\vec{Z}$$

Relation [12.6] can thus be written in matrix form:

$$\begin{Bmatrix} \vec{x} \\ \vec{y} \\ \vec{z} \end{Bmatrix} = [P] \bullet \begin{Bmatrix} \vec{X} \\ \vec{Y} \\ \vec{Z} \end{Bmatrix} \qquad [12.7]$$

Transfer matrix $[P]$ is made up of director cosines, allowing the passage from Global reference (XYZ) to Local reference (*xyz*).

When we consider the reverse passage, from the Local coordinate system to the Global system, we see that [12.6] becomes:

---

3 We have followed the writing convention: $\left(\vec{x}, \vec{X}\right) = \theta$.

$$\vec{X} = \cos(\vec{X}, \vec{x})\,\vec{x} + \cos(\vec{X}, \vec{y})\,\vec{y} + \cos(\vec{X}, \vec{z})\vec{z}$$

$$\vec{Y} = \cos(\vec{Y}, \vec{x})\,\vec{x} + \cos(\vec{Y}, \vec{y})\,\vec{y} + \cos(\vec{Z}, \vec{z})\vec{z}$$

$$\vec{Z} = \cos(\vec{Z}, \vec{x})\,\vec{x} + \cos(\vec{Z}, \vec{y})\,\vec{y} + \cos(\vec{Z}, \vec{z})\vec{z}$$

or, in the matrix form, taking into account that $\cos(\vec{i}, \vec{j}) = \cos(\vec{j}, \vec{i})$:

$$\left\{\begin{array}{c}\vec{X}\\\vec{Y}\\\vec{Z}\end{array}\right\} = [\mathrm{P}]^{\mathrm{T}} \bullet \left\{\begin{array}{c}\vec{x}\\\vec{y}\\\vec{z}\end{array}\right\} \qquad [12.8]$$

we see that we have written relation [12.7] in the reverse sense, and we therefore verify the property:

$$[\mathrm{P}]^{-1} = [\mathrm{P}]^{\mathrm{T}}$$

Let us then express the same vector $\vec{V}$ in each of these coordinate systems. We have:

$$\vec{V} = V_X\,\vec{X} + V_Y\,\vec{Y} + V_Z\,\vec{Z} = \left\{\begin{array}{c}V_X\\V_Y\\V_Z\end{array}\right\}_{Global}^{\mathrm{T}} \bullet \left\{\begin{array}{c}\vec{X}\\\vec{Y}\\\vec{Z}\end{array}\right\}$$

or with [12.8]:

$$\vec{V} = \left\{\begin{array}{c}V_X\\V_Y\\V_Z\end{array}\right\}_{Global}^{\mathrm{T}} \bullet [\mathrm{P}]^{\mathrm{T}} \bullet \left\{\begin{array}{c}\vec{x}\\\vec{y}\\\vec{z}\end{array}\right\}$$

while we can also write:

$$\vec{V} = V_x\,\vec{x} + V_y\,\vec{y} + V_z\,\vec{z} = \left\{\begin{array}{c}V_x\\V_y\\V_z\end{array}\right\}_{Local}^{\mathrm{T}} \bullet \left\{\begin{array}{c}\vec{x}\\\vec{y}\\\vec{z}\end{array}\right\}$$

Identifying the two expressions of $\vec{V}$ leads to:

$$\left\{\begin{array}{c}V_x\\V_y\\V_z\end{array}\right\}_{Local}^{\mathrm{T}} = \left\{\begin{array}{c}V_X\\V_Y\\V_Z\end{array}\right\}_{Global}^{\mathrm{T}} \bullet [\mathrm{P}]^{\mathrm{T}}$$

which is also written as (see section 12.1.2.3):

$$\left\{ \begin{matrix} V_x \\ V_y \\ V_z \end{matrix} \right\}_{Local}^{T} = \left[[P] \bullet \left\{ \begin{matrix} V_X \\ V_Y \\ V_Z \end{matrix} \right\}_{Global} \right]^{T}$$

i.e.:

$$\left\{ \begin{matrix} V_x \\ V_y \\ V_z \end{matrix} \right\}_{Local} = [P] \bullet \left\{ \begin{matrix} V_X \\ V_Y \\ V_Z \end{matrix} \right\}_{Global} \qquad [12.9]$$

Appendix A

Modeling of Common Mechanical Joints

Useful characteristics of standardized joints are presented below, without going into their detailed mechanical study.

A.1. Definition

A.1.1. *Monolithic unit*

When a completely clamped and permanent joint is created between two or more parts of a structure through a traditional mechanical process[1], by definition, there is no possibility of relative motion between these parts. We say that a clamped joint has been created. We also say that these parts, joined in such a manner, constitute a kinematically "monolithic" unit.

A.1.2. *Joints*

As completely opposed to the preceding case, we can study two parts (1 and 2) without any contact. In such a case, one of the parts can be considered as a reference (part 1 for instance). Then, we know that the displacement of part 2 relative to part 1 is characterized by six traditional components in a coordinate system associated with 1:

- three translations following the coordinate axes;
- three rotations around the coordinate axes.

---

1 See Chapter 8.

We still say that there is no connection between parts 1 and 2. Between the two extremes mentioned, the clamped joint, and the absence of joint, "intermediate" linking can be carried out, which allows relative displacements of parts 1 and 2 to be more or less "controlled". We shall limit ourselves to the description of simplified intermediate linkings, which are, furthermore, standardized. Such joints make it possible to consider a mechanical assembly, or a structure, as a whole unit, with the aim of a complete modeling.

A.1.3. *Perfect joints*

A joint between two parts is characterized by contact surfaces which are more or less extended between these parts. It is referred to as "perfect" if it has the following characteristics:

- the contact surfaces are geometrically perfect and assumed to be undeformable;

- the play between the surfaces is zero;

- the contacts are without friction.

A.2. Common standardized mechanical joints (ISO 3952)

NOTES

❑ The following tables contain an inventory of standardized joints between parts 1 and 2 allowing the modeling of mechanical assemblies.

We have given a technological example associated with each type of joint.

- The coordinate system linked to part 1 has a particular point of origin, called the "center of the joint".

- The displacement of part 2 relative to part 1 is characterized by the allowable relative displacements mentioned in these tables. The parts thus joined are subjected to mechanical loadings. Without examining this aspect in detail, we will simply say that each of these joints is necessarily capable of transmitting these forces. We define the resultant force and moment of these transmittable forces at the center of the joint.

❑ From a practical point of view, we can note in these tables the complementarity of:

- components of allowable displacements;

- components of the resultant force and moment of transmittable forces.

In fact, zero relative displacement corresponds to a non-zero transmittable force, non-zero relative displacement corresponds to zero transmittable force.

clamping or **clamped support** *(standardized name)*		
technological example	*technological solutions*	*symbolic representation*
	welding joint, bolted joint, riveted joint, bonded joint, binding.	
allowable relative displacement: *(translations are denoted: T_x, T_y, T_z)* *(rotations are denoted: R_x, R_y, R_z)*		*resultant force and moment of transmittable forces:* *(forces are denoted: X, Y, Z)* *(moments are denoted: L, M, N)*
$\begin{Bmatrix} 0 \\ 0 \\ 0 \\ 0 \\ 0 \\ 0 \end{Bmatrix}$		$\begin{Bmatrix} X \\ Y \\ Z \\ L \\ M \\ N \end{Bmatrix}$

revolute joint *(standardized name)*		
technological example	*technological solutions*	*symbolic representation*
	plain bearing with thrust bearing, shaft bearing	
allowable relative displacement: *(translations are denoted: T_x, T_y, T_z)* *(rotations are denoted: R_x, R_y, R_z)*		*resultant force and moment of transmittable forces:* *(forces are denoted: X, Y, Z)* *(moments are denoted: L, M, N)*
$\begin{Bmatrix} 0 \\ 0 \\ 0 \\ R_x \\ 0 \\ 0 \end{Bmatrix}$		$\begin{Bmatrix} X \\ Y \\ Z \\ 0 \\ M \\ N \end{Bmatrix}$

prismatic joint, slider *(standardized name)*		
technological example	*technological solutions*	*symbolic representation*
	slider in track, rack gear, two parallel recirculating ball-type gears	
allowable relative displacement: (translations are denoted: T_x, T_y, T_z) (rotations are denoted: R_x, R_y, R_z)		*resultant force and moment of transmittable forces:* (forces are denoted: X, Y, Z) (moments are denoted: L, M, N)
$\begin{Bmatrix} T_x \\ 0 \\ 0 \\ 0 \\ 0 \\ 0 \end{Bmatrix}$		$\begin{Bmatrix} 0 \\ Y \\ Z \\ L \\ M \\ N \end{Bmatrix}$

screw joint, helical joint *(standardized name)*		
technological example	*technological solutions*	*symbolic representation*
	screw-nut system, ball-screw assembly *(translation and rotation combined)*	
allowable relative displacement: (translations are denoted: T_x, T_y, T_z) (rotations are denoted: R_x, R_y, R_z)		*resultant force and moment of transmittable forces:* (forces are denoted: X, Y, Z) (moments are denoted: L, M, N)
$\begin{Bmatrix} T_x \\ 0 \\ 0 \\ R_x \\ 0 \\ 0 \end{Bmatrix}$		$\begin{Bmatrix} 0 \\ Y \\ Z \\ 0 \\ M \\ N \end{Bmatrix}$

cylinder joint, cylindrical joint *(standardized name)*		
technological example	*technological solutions*	*symbolic representation*
	plain bearing, recirculating ball-type gears	
allowable relative displacement: *(translations are denoted: T_x, T_y, T_z)* *(rotations are denoted: R_x, R_y, R_z)*		*resultant force and moment of transmittable forces:* *(forces are denoted: X, Y, Z)* *(moments are denoted: L, M, N)*
$\begin{Bmatrix} T_x \\ 0 \\ 0 \\ R_x \\ 0 \\ 0 \end{Bmatrix}$		$\begin{Bmatrix} 0 \\ Y \\ Z \\ 0 \\ M \\ N \end{Bmatrix}$

slotted spherical joint, spherical-in-slot joint *(standardized name)*		
technological example	*technological solutions*	*symbolic representation*
	Cardan joint, gimbal joint, universal joint *(here slot is in plane $O\vec{x}\vec{z}$)*	
allowable relative displacement: *(translations are denoted: T_x, T_y, T_z)* *(rotations are denoted: R_x, R_y, R_z)*		*resultant force and moment of transmittable forces:* *(forces are denoted: X, Y, Z)* *(moments are denoted: L, M, N)*
$\begin{Bmatrix} 0 \\ 0 \\ 0 \\ 0 \\ R_y \\ R_z \end{Bmatrix}$		$\begin{Bmatrix} X \\ Y \\ Z \\ L \\ 0 \\ 0 \end{Bmatrix}$

<table>
<tr><td colspan="3" align="center">spherical pair, spherical pair joint, spherical joint (standardized name)</td></tr>
<tr><td align="center">technological example</td><td align="center">technological solutions</td><td align="center">symbolic representation</td></tr>
<tr><td></td><td align="center">ball and socket, swivel bearing, spherical bearing</td><td></td></tr>
<tr><td colspan="2" align="center">allowable relative displacement:
(translations are denoted: T_x, T_y, T_z)
(rotations are denoted: R_x, R_y, R_z)</td><td align="center">resultant force and moment
of transmittable forces:
(forces are denoted: X, Y, Z)
(moments are denoted: L, M, N)</td></tr>
<tr><td colspan="2" align="center">$$\begin{Bmatrix} 0 \\ 0 \\ 0 \\ R_x \\ R_y \\ R_z \end{Bmatrix}$$</td><td align="center">$$\begin{Bmatrix} X \\ Y \\ Z \\ 0 \\ 0 \\ 0 \end{Bmatrix}$$</td></tr>
</table>

<table>
<tr><td colspan="3" align="center">plane pair joint, planar joint (standardized name)</td></tr>
<tr><td align="center">technological example</td><td align="center">technological solutions</td><td align="center">symbolic representation</td></tr>
<tr><td></td><td align="center">lubricated contact between plane surfaces</td><td></td></tr>
<tr><td colspan="2" align="center">allowable relative displacement:
(translations are denoted: T_x, T_y, T_z)
(rotations are denoted: R_x, R_y, R_z)</td><td align="center">resultant force and moment
of transmittable forces:
(forces are denoted: X, Y, Z)
(moments are denoted: L, M, N)</td></tr>
<tr><td colspan="2" align="center">$$\begin{Bmatrix} T_x \\ T_y \\ 0 \\ 0 \\ 0 \\ R_z \end{Bmatrix}$$</td><td align="center">$$\begin{Bmatrix} 0 \\ 0 \\ Z \\ L \\ M \\ 0 \end{Bmatrix}$$</td></tr>
</table>

<table>
<tr><td colspan="3" align="center">cylinder plane pair joint (standardized name)</td></tr>
<tr><td align="center">technological example</td><td align="center">technological solutions</td><td align="center">symbolic representation</td></tr>
<tr><td></td><td align="center">knife-edge of scale, cylinder on plane surface, locking cam</td><td></td></tr>
<tr><td colspan="2" align="center">allowable relative displacement:
(translations are denoted: T_x, T_y, T_z)
(rotations are denoted: R_x, R_y, R_z)</td><td align="center">resultant force and moment of transmittable forces:
(forces are denoted: X, Y, Z)
(moments are denoted: L, M, N)</td></tr>
<tr><td colspan="2" align="center">$$\begin{Bmatrix} T_x \\ T_y \\ 0 \\ R_x \\ 0 \\ R_z \end{Bmatrix}$$</td><td align="center">$$\begin{Bmatrix} 0 \\ 0 \\ Z \\ 0 \\ M \\ 0 \end{Bmatrix}$$</td></tr>
</table>

<table>
<tr><td colspan="3" align="center">sphere groove joint, spherical and sliding joint (standardized name)</td></tr>
<tr><td align="center">technological example</td><td align="center">technological solutions</td><td align="center">symbolic representation</td></tr>
<tr><td></td><td align="center">sphere fitted in cylindrical housing,
spherical bearing with outer race slide-fitted</td><td></td></tr>
<tr><td colspan="2" align="center">allowable relative displacement:
(translations are denoted: T_x, T_y, T_z)
(rotations are denoted: R_x, R_y, R_z)</td><td align="center">resultant force and moment of transmittable forces:
(forces are denoted: X, Y, Z)
(moments are denoted: L, M, N)</td></tr>
<tr><td colspan="2" align="center">$$\begin{Bmatrix} T_x \\ 0 \\ 0 \\ R_x \\ R_y \\ R_z \end{Bmatrix}$$</td><td align="center">$$\begin{Bmatrix} 0 \\ Y \\ Z \\ 0 \\ 0 \\ 0 \end{Bmatrix}$$</td></tr>
</table>

simple support, pointwise support, sphere plane joint *(standardized name)*		
technological example	*technological solutions*	*symbolic representation*
	sphere on plane	
allowable relative displacement: *(translations are denoted: T_x, T_y, T_z)* *(rotations are denoted: R_x, R_y, R_z)*		*resultant force and moment* *of transmittable forces:* *(forces are denoted: X, Y, Z)* *(moments are denoted: L, M, N)*
$\left\{\begin{array}{c} 0 \\ T_y \\ T_z \\ R_x \\ R_y \\ R_z \end{array}\right\}$		$\left\{\begin{array}{c} X \\ 0 \\ 0 \\ 0 \\ 0 \\ 0 \end{array}\right\}$

Appendix B

Mechanical Properties of Materials

B.1. Mechanical properties of some materials used for structures

B.1.1. *Steels and casting*

Standardized nuances EN (Europe)	Standardized nuances ASTM (USA)	Elasticity modulus E (MPa)	Poisson's ratio v	Volumic mass ρ (kg/m^3)	Tensile breaking strength at traction R_r (MPa)	Elastic limit at traction R_e (MPa)
General purpose structural steels						
S 235 JR	A 283 C	205,000	0.3	7,800	340	235
S 355 JR	A 572 Gr.50	205,000	0.3	7,800	490	355
E 295		205,000	0.3	7,800	470	295
Low alloy steel (no added element exceeds 5% in mass)						
34 Cr Mo 4	AISI 4135	205,000	0.3	7,800	700-1,100	450-750
36 Ni Cr Mo 16 UNI 7845		205,000	0.3	7,800	1,000-1,750	800-1,250
High alloy steel (stainless steel)						
X 2 Cr Ni 19-11	S30403 (304L)	205,000	0.3	7,800	440-640	185
Spheroidal graphite cast iron						
EN-GJS-400-15	Gr 65-45-12	165,000	0.3	7,200	400	250

B.1.2. *Non-ferrous metals*

Standardized nuances EN (Europe)	Standardized nuances ASTM (USA)	Elasticity modulus E (MPa)	Poisson's ratio v	Volumic mass ρ (kg/m^3)	Tensile strength R_r (MPa)	Elastic limit at traction R_e (MPa)
Aluminum alloys						
EN-AW-2017	2017 A	70,000	0.3	2,800	470	295
4047	4047	70,000	0.3	2,700	160	80
EN-AC-Al Mg5 EN-AC-51300		70,000	0.3	2,800	180	100
Copper alloy						
CuSn8	C52100	110,000	0.3	8,800	480	170
Titanium alloy						
Ti-6Al-4V	B265/B348 Gr.5	105,000	0.3	4,400	1,050	950
Magnesium alloy						
EN-MC Mg Al Zn		44,000	0.3	1,800	160	80

Appendix C

List of Summaries

Part I

Chapter 1. The Basics of Linear Elastic Behavior

Materials homogenous, isotropic, elastic, linear: Hooke's law for uniaxial traction or compression (along the $\vec{x}$ axis) [1.6]
Materials homogenous, isotropic, elastic, linear: plane state of stresses in the plane (xy) [1.18]
Normal resultant $\mathcal{N}_x$ and its consequences [1.30]
Shear resultant $\mathcal{T}_y$ and its consequences [1.31]
Shear resultant $\mathcal{T}_z$ and its consequences [1.32]
Torsion moment $\mathcal{M}t$ and its consequences [1.33]
Bending moment $\mathcal{M}f_y$ and its consequences [1.34]
Bending moment $\mathcal{M}f_z$ and its consequences [1.35]

Chapter 2. Mechanical Behavior of Structures: An Energy Approach

Elementary potential energies in the domain ($S \times dx$) of a straight beam [2.34]
Different expressions for potential energy under plane stress [2.44]
Loading and degrees of freedom of a structure [2.91]
Same structure, same nodes (1) and (2), same loads F_1 and F_2, different linking conditions [2.39]
"Flexibility" approach of a structure; "stiffness" approach of a structure [2.122]

Chapter 3. Discretization of a Structure into Finite Elements

Case of coplanar local and global systems of coordinates behavior equation of the element Figure 3.9
Behavior of the truss element under traction-compression, in the local and global coordinate systems Figure 3.8
Behavior of the beam element under torsion, in the local and global coplanar coordinate systems Figure 3.10
Behavior of the beam element bending in the plane (xy) in the local and global coplanar coordinate systems Figure 3.14
Behavior of the triangular element working as membrane, in the local and global coplanar coordinate systems Figure 3.23
Topology of the main types of finite elements Figure 3.36

Part II

Chapter 5. Other Types of Finite Elements

Behavior relation of the element: case of any local and global coordinate systems [5.1]
Behavior of the beam element: in the local and in the global coordinate system Figure 5.8
Behavior of the triangular element for the plane state of stress: in the local and in the global coordinate system Figure 5.11
Behavior of the quadrilateral element in plane state of stress: in the local and in the global coordinate system Figure 5.15
Behavior of the complete plate elements (plane stress + bending): in the local and in the global coordinate system Figure 5.24
Tetrahedric and hexahedric solid elements: in the local and in the global coordinate system Figure 5.31

Chapter 6. Introduction to Finite Elements for Structural Dynamics

Dynamic behavior of a structure (free vibrations; without damping; structure properly linked) [6.29]

Chapter 7. Criteria for Dimensioning

Dimensioning of a structure (static loading case, linear elastic domain) Figure 7.2
Approximative curve of the fatigue test [7.21]

Chapter 8. Practical Aspects of Finite Element Modeling

Beam element Figure 8.1
Complete plate elements (membrane + bending) Figure 8.2
Solid 3D elements Figure 8.3
The same structure modeled by each of the three element types Figure 8.4
Use of the software; section 8.5.3

Part III

Chapter 9. Behavior of Straight Beams

Coordinate system linked to a current cross-section [9.1]
Traction-compression [9.15]
The torsional moment is merged with the longitudinal moment, $Mt = M_x$ in any of
the following cases [9.21]
Uniform torsion of a beam with any cross-section [9.27]
Pure bending in the main plane (xy) [9.33]
Plane bending in the main plane (xy) [9.59]
Plane bending in the main plane (xz) [9.60]
Small displacements of a current cross-section [9.61]

Chapter 10. Additional Elements of Elasticity

Stresses on a facet of any orientation [10.9]
Complete state of stresses; "deformations-stresses" behavior relation [10.21]
Any complete state of stresses [10.27]

Chapter 11. Structural Joints

Bolted joint of two parts 1 and 2; estimation of forces on "n" fasteners with center
(i) and section "s" [11.6]
Dimensioning of a pre-tightened bolt [11.22]
Dimensioning of a riveted joint [11.23]
Dimensioning of a welded joint [11.28]

Appendices

A: Modeling of Common Mechanical Joints
B: Mechanical Properties of Materials

Bibliography

BAZERGUI, André, *Résistance des Matériaux*, Ecole Polytechnique de Montréal, 1993, Canada.

BLEUZEN, Claude, *Comportement en Fatigue des Matériaux Métalliques*, Cours du Diplôme d'Etudes Approfondies de Génie Mécanique; Paul Sabatier University, Toulouse III, 1994.

COURBON, Jean, *Résistance des Matériaux*, Dunod, Paris, 1964.

DROUIN, Gilbert, *Eléments de Machines*, Editions de l'Ecole Polytechnique de Montréal, 1986.

FAURIE, Jean-Pierre, *Guide du Dessinateur: les Concentrations de Contraintes*, Editions du CETIM, 1977.

GAY, Daniel, *Matériaux Composites*, 5th Edition Hermes Science – Lavoisier, Paris, 2005.

GUILLOT, Jean, *Assemblages par Eléments Filetés; Modélisation et Calcul*, Techniques de l'Ingénieur; BM5 563, 10-1997.

IMBERT, Jean-François, *Analyse des Structures par Eléments Finis*, ENSAE, Cépadues Editions, Toulouse, 1984.

MOREL, Jean, *Calcul des Structures Métalliques selon l'Eurocode 3*, Editions Eyrolles, Paris, 1994.

YOUNG, WARREN C., *Roark's Formulas for Stress and Strain*, 6th edition, McGraw-Hill International Editions, 1989.